Boston Studies in the Philosophy and History of Science

Volume 336

The series *Boston Studies in the Philosophy and History of Science* was conceived in the broadest framework of interdisciplinary and international concerns. Natural scientists, mathematicians, social scientists and philosophers have contributed to the series, as have historians and sociologists of science, linguists, psychologists, physicians, and literary critics.

The series has been able to include works by authors from many other countries around the world.

The editors believe that the history and philosophy of science should itself be scientific, self-consciously critical, humane as well as rational, sceptical and undogmatic while also receptive to discussion of first principles. One of the aims of Boston Studies, therefore, is to develop collaboration among scientists, historians and philosophers.

Boston Studies in the Philosophy and History of Science looks into and reflects on interactions between epistemological and historical dimensions in an effort to understand the scientific enterprise from every viewpoint.

More information about this series at http://www.springer.com/series/5710

Jordi Cat • Adam Tamas Tuboly
Editors

Neurath Reconsidered

New Sources and Perspectives

Springer

Editors
Jordi Cat
Department of History and Philosophy of Science and Medicine
Indiana University Bloomington
Bloomington, IN, USA

Adam Tamas Tuboly
Institute of Philosophy
Hungarian Academy of Sciences
Budapest, Hungary

ISSN 0068-0346 ISSN 2214-7942 (electronic)
Boston Studies in the Philosophy and History of Science
ISBN 978-3-030-02127-6 ISBN 978-3-030-02128-3 (eBook)
https://doi.org/10.1007/978-3-030-02128-3

Library of Congress Control Number: 2018964942

This Springer imprint is published by the registered company Springer Nature Switzerland AG.
The registered company address is: Gewerbestrasse 11, 6330 Cham, Switzerland

Contents

1 **Introduction** 1
Jordi Cat and Adam Tamas Tuboly

Part I Otto Neurath: Life and Work

2 **A Viennese Library in Exile: Otto Neurath and the Heritage of Central European Culture in the Anglo-Saxon World** 23
Friedrich Stadler

3 **Otto Neurath: The Philosopher in the Cave** 45
Don Howard

4 **Science and Socialism: Otto Neurath as a Political Writer (1919–1932)** 67
Günther Sandner

5 **United by Action: Neurath in England** 89
Adam Tamas Tuboly

Part II Science, Society, and Method

6 **Visualizing Relations in Society and Economics: Otto Neurath's Isotype-Method Against the Background of his Economic Thought** 117
Elisabeth Nemeth

7 **Traveling Exhibitions in the Field: Settlements, War-Economy, and the Collaborative Practice of Seeing, 1919–1925** 141
Sophie Hochhäusl

8 **Generating Cognitive Tools: Neurath's Educational Ideal and the Concept of ISOTYPE** 177
Angélique Groß

9 Rationality and Pseudo-Rationality in Political Economy: Neurath, Mises, Weber ... 197
Thomas Uebel

10 Neurath's Theory of Theory Classification: History, Optics & Epistemology ... 217
Gábor Á. Zemplén

Part III Logic and Semantics

11 Neurath and the Legacy of Algebraic Logic ... 241
Jordi Cat

12 Neurath and Carnap on Semantics ... 339
A. W. Carus

13 Rejecting Semantic Truth: On the Significance of Neurath's Syntacticism ... 363
Derek Anderson

Part IV Context and Influence

14 What a Difference a Decade Makes: The Planning Debates and the Fate of the Unity of Science Movement ... 385
George A. Reisch

15 Of Tennis Courts and Fireplaces: Neurath's Internment on the Isle of Man and his Politics of Design ... 413
Michelle Henning

16 Does Understanding Mean Forgiveness? Otto Neurath and Plato's "Republic" in 1944–45 ... 435
Antonia Soulez

17 Thinking About the "Common Reader:" Otto Neurath, L. Susan Stebbing and the (Modern) Picture-Text Style ... 451
Silke Körber

18 Logical Empiricism and Art: The Correspondence Otto Neurath/Meyer Schapiro ... 471
Hans-Joachim Dahms

Appendices ... 489

References ... 687

Index ... 695

About the Editors and Contributors

Editors

Jordi Cat Department of History and Philosophy of Science and Medicine, Indiana University Bloomington, Bloomington, IN, USA

Adam Tamas Tuboly Institute of Philosophy, Hungarian Academy of Sciences, Budapest, Hungary

Contributors

Derek Anderson Philosophy Department, Boston University, Boston, MA, USA

A. W. Carus Munich Center for Mathematical Philosophy, Ludwig-Maximilians Universität, Munich, Germany

Jordi Cat Department of History and Philosophy of Science and Medicine, Indiana University Bloomington, Bloomington, IN, USA

Hans-Joachim Dahms Institute Vienna Circle, Vienna, Austria

Angélique Groß Adult Education Centre, Lebach, Germany

Michelle Henning London School of Film, Media and Design, University of West London, London, UK

Sophie Hochhäusl University of Pennsylvania, Philadelphia, PA, USA

Radcliffe Institute of Advanced Studies, Harvard University, Cambridge, MA, USA

Don Howard Department of Philosophy, University of Notre Dame, Notre Dame, IN, USA

Silke Körber Institute of Philosophy, Doctoral Candidate, Humboldt-Universität zu Berlin, Berlin, Germany

Elisabeth Nemeth Department of Philosophy, University of Vienna, Vienna, Austria

George A. Reisch Chicago, IL, USA

Günther Sandner Institute Vienna Circle, University of Vienna, Vienna, Austria

Antonia Soulez Université de Paris VIII, Paris, France

Friedrich Stadler Vienna Circle Institute, University of Vienna, Vienna, Austria

Adam Tamas Tuboly Institute of Philosophy, Hungarian Academy of Sciences, Budapest, Hungary

Thomas Uebel Philosophy, School of Social Sciences, University of Manchester, Manchester, UK

Gábor Á. Zemplén Department of Argumentation and Marketing Research, Institute of Business Economics, Eötvös University (ELTE), Budapest, Hungary

Morals and Science Research Group (MTA Lendület), Hungarian Academy of Sciences, Budapest, Hungary

Department of Philosophy and History of Science, Budapest University of Technology and Economics (BME), Budapest, Hungary

List of Figures

Fig. 6.1 Methods of Visualization 128
Fig. 6.2 Unemployed in Berlin 129
Fig. 6.3 Latin America's main products of export 132
Fig. 6.4 Automobile industry 133
Fig. 6.5 Symbols 134
Fig. 6.6 Public signs (the one with the cars) 135
Fig. 6.7 Infant mortality 135

Fig. 7.1 Otto Neurath sitting on the balcony of a communal building at the Hoffingergasse settlement during a speech by Adolf Müller, Vienna, 1921. (Source: Archives of the Cooperative Altmannsdorf-Hetzendorf) 142
Fig. 7.2 Construction at Rosenhügel settlement designed by Hugo Mayer, Vienna, 1921. (Source: Archives of the Cooperative Altmannsdorf-Hetzendorf) 146
Fig. 7.3 Organized allotment gardens on the outskirts, Vienna, 1919. (Source: Austrian Horticultural Society) 149
Fig. 7.4 Construction at Siedlung Hoffingergasse designed by Josef Frank, Vienna, 1921. (Source: Archives of the Cooperative Altmannsdorf-Hetzendorf) 151
Fig. 7.5 Construction at Siedlung at Lainz designed by Adolf Loos and Margarete Lihotzky, Vienna, 1921. (Source: Archives of the Horst Zecha, Siedlung Friedenstadt) 152
Fig. 7.6 Construction with Pax bricks at Siedlung Hoffingergasse designed by Josef Frank, Vienna, 1921. (Source: Archives of the Cooperative Altmannsdorf-Hetzendorf) 153
Fig. 7.7 Construction of row houses at Siedlung Hoffingergasse with allotments in the foreground and background, designed by Josef Frank, Vienna, 1921. (Source: Archives of the Cooperative Altmannsdorf-Hetzendorf) 154

Fig. 7.8 Visitors at the allotment garden, settlement, and housing exposition in front of core houses with fully functioning allotment gardens. The crowd reaches back to the *Ringstraße*, Vienna, 1923. (Source: Austrian League of Allotment Gardeners) ... 156
Fig. 7.9 Members of the exhibition committee in front of the flower exhibit at the allotment garden, settlement and housing exhibition. Neurath third on the left, back, Adolf Müller, second on the right, back, Kampffmeyer fourth on the right, front, Vienna, 1923. (Source: Austrian League of Allotment Gardeners) ... 162
Fig. 7.10 Displays by GESIBA (on the left) with core-house models in City Hall, Vienna, 1923. (Source: Austrian League of Allotment Gardeners) ... 164
Fig. 7.11 Chart of allotment gardeners' material achievements compiled by the municipality's garden section in the midst of beekeeping display, Vienna, 1923. (Source: Austrian League of Allotment Gardeners) ... 165
Fig. 7.12 Grouping of various vegetables presented by allotment gardeners of the Rosental cooperative in the courtyard of City Hall, Vienna, 1923. (Source: Austrian League of Allotment Gardeners) ... 166
Fig. 7.13 Otto Neurath presenting the *Atlas for Society and Economy* to Viennese mayor Karl Seitz, Vienna, 1930. (Source: Otto and Marie Neurath Isotype Collection, University of Reading N-Files, N 958b) ... 167
Fig. 7.14 Worker making shoes, Vienna, 1925–1932. (Source: Otto and Marie Neurath Isotype Collection, University of Reading N-Files, N 175) ... 170
Fig. 7.15 Production of socks, Vienna, 1925–1932. (Source: Otto and Marie Neurath Isotype Collection, University of Reading N-Files, N 176) ... 171
Fig. 7.16 Foundation stone setting ceremony, Siedlung Hoffingergasse, Müller, Neurath, and Kampffmeyer in the crowd, Vienna, 1921. (Source: Archives of the Cooperative Altmannsdorf-Hetzendorf) ... 173

Fig. 8.1 Wissenschaftliche Weltauffassung ... 178
Fig. 8.2 Structure of the concept of enlightenment ... 180
Fig. 8.3 Specification of Wissenschaftliche Weltauffassung ... 183
Fig. 8.4 Polizeiliche Einschreitungen in Wien im Febr. 1925. (© Otto and Marie Neurath Isotype Collection, University of Reading) ... 187
Fig. 8.5 Völkergruppen der Erde. (© Otto and Marie Neurath Isotype Collection, University of Reading) ... 187

Fig. 8.6 Kraftwagenbestand der Erde. (© Österreichisches Gesellschafts- und Wirtschaftsmuseum, Wien) 188
Fig. 8.7 Rationalization and reduction in the number of workers. (© Otto and Marie Neurath Isotype Collection, University of Reading) 189
Fig. 8.8 Death-rate/Suicide-rate. (© Otto and Marie Neurath Isotype Collection, University of Reading) 189
Fig. 8.9 Transformation 190
Fig. 8.10 Operation of ISOTYPE 191

Fig. 10.1 Neurath's (1916/1983, 24) diagram of and commentary on blurred margins 225
Fig. 10.2 Constellations of theories 228

Fig. 11.1 Dual statements in *Vorlesungen* (1890a, 196), vol. 1 258
Fig. 11.2 Example of O. Hahn and Neurath's notation 288
Fig. 11.3 Dual presentation of contrasted statements about rent, wages and interest in Henry George's *Progress and Poverty* (1882, 197), reprinted in Neurath and Schapire's *Lesebuch* (1910), vol. 2, p. 275 291
Fig. 11.4 Jevons' logical abecedarium, Jevons 1874, vol. 1, p. 109 301
Fig. 11.5 Jevons' logical machine, Jevons 1874, vol.1, frontispice 302
Fig. 11.6 Symbolic representation of the eight combinations of three classes (properties) and their negations, Jevons 1874, p. 367 302
Fig. 11.7 Fragment of table of symbolic representations of logical combinations of four classes, Schröder (1890a, b, 563) 303
Fig. 11.8 Fragment of table of symbolic representations of logical combinations of four elements, Schröder (1890a, b, 564) 303
Fig. 11.9 Neurath's symbolic presentation of the eight combinations of three theoretical elements and their negations forming a classification of systems of hypotheses (original publication in Neurath 1915, 376) 303
Fig. 11.10 Neurath's table of optical theories, or systems of hypotheses, Neurath (1915, 385) 304
Fig. 11.11 Neurath's handwritten "List of Dualities" on a copy of C.G. Field's article "Plato's 'Republic' and Its Use in Education," *The Journal of Education*, 77 (909), April 1945, pp. 161–162. (I'm grateful to Adam Tamas Tuboly for drawing my attention to this file) 309

Fig. 14.1 One of Kaempffert's more prominent features about Neurath and the *International Encyclopedia of Unified Science*, in the New York Times Book Review, Aug. 7, 1938 389
Fig. 14.2 Kaempffert's coverage of Neurath's ISOTYPE project, Jan 22, 1933 390

Fig. 14.3 U.S. Government posters utilizing and adapting Neurath's Vienna Method of pictorial statistics in the 1930s 392
Fig. 14.4 Advertisement in the *Boston Globe* for Thomas Kuhn's Lowell Lectures (clipping from Kuhn's papers, TSK-MIT, box 3, folder 10) 405

Fig. 17.1 *The New Democracy: Education. The New Horizon* 460
Fig. 17.2 *The New Democracy: Woman and a New Society* 461
Fig. 17.3 *The Soviets and Ourselves: How do you do Tovarish?* 461
Fig. 17.4 Double page spread from the *The New Democracy series: Women and a New Society* (1946) 463
Fig. 17.5 Double page spread from the first volume of the *Future Book series* (1946) 464

Fig. A1 First page of Olga Hahn and Otto Neurath's "On Duality in Logic" (1909) 492
Fig. A2 First page of Neurath's letter to Carnap 512

List of Tables

Table 10.1 Neurath's elementary notions discussed in KH and PO................. 231

Chapter 1
Introduction

Jordi Cat and Adam Tamas Tuboly

The Viennese-born polymath Otto Neurath died on 22 December 1945 in Oxford, a few months after the end of World War Two. A social engineer and sociologist of happiness, Neurath was not only a socially sensitive educator, advocating for any institute and organization that was concerned with the well-being of people; he was also a trained scientist and philosopher. Studying mathematics, economics, history, philosophy, and physics in Vienna and then in Berlin during the early years of the long twentieth century, Neurath became involved in many of the disputes among social and natural scientists that shaped the course of the fields.

Through sociological and historical inquiries about how actual science historically developed, Neurath unswervingly propagated in various forums his ideal of the socially relevant "unified science," his fears about the dangers of over-systematization, his misgivings about the concept of neutral "data" and "facts" (also of a neutral "observation language"), and about putting too much weight on theoretical argumentations and logical calculation. There were no exceptions: Neurath debated and poked his friends, fellow logical empiricists and supposed or real enemies with the same vehemence.

His character, style of writing and the priority he gave to social action over intellectual contemplation resulted, however, in a fragile legacy, with no easy classification or safe place in collective memory, whether in the media or in academia, sciences or humanities. This volume – with its essays, translations, and the edited 1940–1945 scientific and personal correspondence of Neurath with Carnap – aims

J. Cat (✉)
Department of History and Philosophy of Science and Medicine, Indiana University Bloomington, Bloomington, IN, USA
e-mail: jcat@indiana.edu

A. T. Tuboly
Institute of Philosophy, Hungarian Academy of Sciences, Budapest, Hungary
e-mail: tuboly.adam@btk.mta.hu

J. Cat, A. T. Tuboly (eds.), *Neurath Reconsidered*,
Boston Studies in the Philosophy and History of Science 336,
https://doi.org/10.1007/978-3-030-02128-3_1

at furthering our understanding of Neurath's life and works in order to see how scientific and philosophical ideas and social projects change over time and across places and what relevance they could still have in our time. The essays collected here also contribute a new range of perspectives and objects of study, from attention to physical objects and collections to attention to logic and its image.

1.1 Preliminary Remarks

How did Neurath's life and writings regain interest, relevance and visibility? And, to whom? In 1947, Philipp Frank, the physicist-turned-philosopher and old friend of Neurath became the director of the *Institute for the Unity of Science* in Cambridge, MA that was after all a continuation of Neurath's earlier Institute in the Netherlands. After Frank died on 21 July 1966 in distressing mental and physical conditions in a nursing home, the American pragmatist Charles Morris approached the Trustees and Committee members to "seriously consider the future of the Institute." Morris also suggested that the Institute, "if this proves necessary, [should aid somehow] the publication of the Neurath volume which Mary [sic] Neurath and Robert Cohen have edited. The book has not yet found a publisher. We might contribute toward the publication or agree to buy and distribute as gifts a number of copies of the book."[1]

Robert S. Cohen was the unrelenting driving force behind Otto Neurath's restoration to a former academic presence in the English-speaking world. He had announced already in 1963 – in the volume *The Philosophy of Rudolf Carnap* (Open Court's Library of Living Philosophers) – that the publication of the *Selected Works of Otto Neurath* was forthcoming (Cohen 1963, 151, n.113). And the volume, prepared by Neurath's widow, Marie Neurath, and Cohen himself, was promptly completed in 1966.[2]

Carnap had been "definitely in favor" of the publication of the Neurath volume, and after a meeting of the Institute in January 1967, Morris reported that "[t]here was unanimous agreement that the Institute should consider a sum of from $500 to $1000 as subsidization of the collection of papers of Otto Neurath [...]. The manuscript is now being submitted to publishers. Carl Hempel alone expressed some doubts about the publication of such a volume."[3] As it is well known, the book did

[1] Morris to Professors and Trustees, Institute for the Unity of Science, 25 August 1966 (RC 102-48-18)

[2] Given that most of the papers for the Carnap volume were delivered during the 1950s, the Neurath volume was presumably under construction already in the mid-1950s. See Morris to Carnap, 20 April 1966 (RC 102-48-26).

[3] Morris to Professors and Trustees, Institute for the Unity of Science, 25 January 1967 (RC 102-48-12). On Hempel's case see his letters to Ernest Nagel (2 and 7 September 1966, CH 28–02) and Nagel to Hempel, 3 September 1966 (CH 28–02). Hempel and Nagel were actually against another idea of Morris: to publish a selection of essays (invited and essay-prize style) on the history of the Unity of Science Movement. Both Hempel and Nagel claimed (in the aforementioned letters) that given the interest and commitment to publish the lectures of "John Hopkins Seminars in Philosophy" devoted to logical empiricism it "militates further against the idea of a volume of

not appear until 1973, under the title of *Empiricism and Sociology* as the first volume of the *Vienna Circle Collection*.

The first volume of Neurath's essays was followed by a second (1983); but for a few reviews, the English-speaking academic world did not respond favorably to Neurath's work or the attached memory of his political activities. Two things might be mentioned here. First, even though Neurath had published several monographs and brochures in English, many of them were simply unavailable to the average reader: his books on and using ISOTYPE (1936, 1937, 1939) were not republished for almost three decades. Though the monographs of the *International Encyclopedia of Unified Science* (with Neurath's monograph on the social sciences) appeared as an oversize joint volume in 1971, the reputation of Unified Science was already in ruins.[4] So, even though Neurath's *Einheitswissenschaft* series was published in the *Vienna Circle Collection* under the editorship of Brian McGuinness in 1987, it did not help much towards the recognition and understanding of the movement.

Second, regarding the literature on Neurath, the Anglo-American scholarship was considerably thinner than the German-speaking world's. While there were a few articles by Carl Hempel, Willard Quine, Donald Davidson and others, there wasn't any monograph-length account of his work (see Uebel 1991, 10–14). By contrast, by 1990 a number of books had been published on Carnap's various phases and ideas.[5]

1.2 Phases of the English Neurath Reception

We will focus on the growing English-speaking reception of Neurath's works over the last few decades.[6] We distinguish five periods, each of them documenting a difference in the philosophical interest and the broader context of the Neurath reception.

The first phase opened with the publication of the English translation of Neurath's selected sociological and political essays in the above-mentioned *Empiricism and Sociology*. The volume contained not only 80 pages of important memoirs, but classical papers, pamphlets and mini-monographs such as *Anti-Spengler*, *Personal Life and Class Struggle*, *Empirical Sociology*, the famous manifesto ("The Scientific Conception of the World: The Vienna Circle"), and many more items on pictorial

essays on the Unity of Science." The collection appeared in 1969 as Achinstein and Barker (1969), and without actually mentioning the unity of science.

[4] Not just the whole project decayed slowly, but Neurath's monograph was considered as the "black eye" of the series already when it was published. See Reisch (2003, 208).

[5] See e.g. Kazemier and Vuyjse (1962), Schilpp (1963), Hausman and Wilson (1967), Butrick (1970), Hintikka (1975), Norton (1977), Runggaldier (1984), Proust (1989).

[6] The German-speaking literature also has many interesting items about semantics, protocol-sentences, pictorial education and about the socio-political context of Neurath's philosophical worldview. Nonetheless, since many of these works did not have a major and lasting influence on the secondary English literature, we will not discuss them here.

education, war economy, socialization and planning. This colorful mixture was tied together by the idea that Neurath's scientific and scholarly work had not just a special social and political context, but broader relevance as well. The content of the papers reflected the social embeddedness of his thought in an often elusive tangle of philosophy, science and politics that posed interpretative challenges for the general readers.

A certain A.S.C. (1974, 132) wrote a quite neutral descriptive review of the book, emphasizing that the "anthology displays a many sided character concerned with both conceptual clarity and human happiness." The fact, however, that nothing of philosophical relevance surfaced in the review may have documented an interesting pattern of how scholars read the collection. Others were keener on pointing out the peculiar non-philosophical character of the volume. In another review, Alan Ryan (1976, 194) complained that "the selection of philosophical and sociological work is a bit strange; or, rather, it would be strange, if the audience for whom it is intended were an audience of professional philosophers." The most radical was J. W. N. Watkins on the pages of *The British Journal for the Philosophy of Science*, who complained pointedly about the editors' selection principles: why they included certain items and why they did not include others. Watkins (1974, 343) missed especially the *philosophical* papers of Neurath, ("[f]or the opening volume in a 'Vienna Circle Collection', there is surprisingly little of Neurath's Vienna Circle work"); while on the other hand, he regarded the included items, or at least bigger portions of them, as "Marxist propaganda," "near-banality," or loose and sloppy revelations often prefixed as "to a certain degree," "under certain conditions," indicating some reservations as signs of scientific seriousness but lacking actual explanations.

The most important feature of Watkins's (1974, 345) review was his unwitting "discovery" of a "persistent and striking parallelism, or rather linkage, between Neurath's philosophical ideas and his political ideas." He suggested four main sets of parallels: (1) physicalism and Marxism, (2) physicalism, socialism and economic planning, (3) unified physicalist language and proletarian socialism, (4) international socialism and pictorial education/language. The general parallelism and correspondence, and these particular strands have surfaced many times since in the secondary literature and are treated in detail by the contributors to this volume as well.

But the point was made: when the first portion of Neurath's long awaited collection of essays finally appeared in 1973, the direct reception was more unfavorable than it might have been expected. The above-mentioned complaints document, however, an interesting trend: scholars repudiated Neurath as a serious philosopher due to the lack of formally or rigurously formulated theses and arguments and a politically neutral philosophical look on the world; a few years had still to pass until the English-speaking audience discovered that Neurath's strength might be found exactly in those weaknesses that others did not hesitate to point out repeatedly.[7]

---

[7] It should be noted as well, though, that Neurath's reception might have been hindered by the lack of conscientious editorial work on his essay-collections. The reviewers of the volume complained that there were no editorial introductions: not a general one for the volume, or shorter ones for the

The second period was distinguished by a well-informed shift towards the *philosophical* relevance of Neurath's ideas, initiated by the publication of his second selected writings, *Otto Neurath: Philosophical Papers 1913–1946*, edited again by Robert S. Cohen and Marie Neurath in 1983. The volume contained essays mainly on theory building, physicalism and the idea of unified science. The detailed elaboration and commentary on these often short, ambiguous, puzzling, almost mysterious revelations would follow in the forthcoming years. The explanatory secondary literature was never meant to provide merely historical contextualizations; it consistently aimed to emphasize the contemporary philosophical relevance of the material. This came also in stages.

The very first monograph on Neurath's work in English was Danilo Zolo's short book, *Reflexive Epistemology: The Philosophical Legacy of Otto Neurath*, written originally in Italian.[8] Zolo (1989, 167) aimed at showing one version of what he called – in Neurathian fashion – "pluri-Neurath": it was a picture of an anti-positivist positivist who could be accepted also by most post-positivists. Utilizing Neurath's famous nautical metaphor, Zolo presented Neurath as a thinker who merited being in the focus of research as much as Schlick or Carnap were, since, like them, he defended important and exciting theses on science, language and method. Though a great deal of the book contributed valuable scholarship, the volume did not receive a good reception overall. Zolo preserved Neurath's ambiguous and somewhat cryptic language thus he did not succeed in replanting Neurath's ideas into the contemporary settings. While almost all the reviews welcomed the *effort* to rehabilitate the unforgotten Neurath, the first line of Richard Creath's (1993, 359) (quite devastating) review summarized the general feeling well: "This is not a very good book, but it has its virtues."

Two years later came a defining breakthrough in the Neurath literature: Thomas Uebel (1991) *translated, edited* and *introduced* a volume of papers about Neurath (*Rediscovering the Forgotten Vienna Circle: Austrian Studies on Otto Neurath and the Vienna Circle*), written by leading Austrian scholars in 1991. The papers by Rudolf Haller, Friedrich Stadler, Elisabeth Nemeth, Heiner Rutte, Eckehart Köhler,

individual essays; no editorial notes except a few general ones; and there was no systematic statement of purpose on their behalf regarding the selection principles and the general philosophical-historical aim of the edition.

[8] One interesting exception should be mentioned though: in 1966, the logician David L. Székely published a 66 pages long booklet in his monograph series ("Association for Unification and Automation in Science") under the title "Otto Neurath and the Unity of Science Movement." It contained some pages from Charles Morris and a reprint of his 1962 article, "On the History of the International Encyclopedia of Unified Science," along with some papers and remarks of Székely on Neurath and on his own ideas of unification. See Székely and Morris (1966). In 1967, Morris sent out a circular letter to the Institute for the Unity of Science, because Székely wanted to republish it with the help of the Institute to distribute it in America as well (Morris to the Institute, 20 February 1967, RC 102-48-09). In return, Carnap answered that he was happy to have some of Morris's papers in one volume, though Székely's papers were quite weak, thus he did not support the republication of the volume with the help of the Institute (Carnap to Morris, 3 March 1967, RC 102-48-08).

Lola Fleck, Karl Müller, Johann Dvorak, and finally Neurath's son, Paul Neurath, were outstanding scholarly papers about the context, development, and significance of Neurath's life and works.

A year later, Uebel published his own book-length philosophical appreciation of Neurath's significance. The groundbreaking *Overcoming Logical Positivism From Within* (1992) was a major step in the philosophical articulation of Neurath's thought: while the main line of bearing consists of the famous protocol sentence debate, Uebel's effective contribution was far-reaching. Reconstructing the often implicit arguments in sophisticated epistemological terms and filling in the gaps of the various interrelated topics raised new questions, or at least raised them in new terms, about the nature of philosophy, science and their relation. Neurath surfaced as an exciting scholar who had to say something relevant also for our contemporary debates and discussions. *Overcoming* was as historical and philosophical: what Zolo aimed at was achieved by Uebel by relying on the same archival sources but using them as reconstructed items in painstakingly nuanced arguments.[9]

Uebel's reconstruction and the essay collection set in motion a self-propelled machine. First off, several articles appeared about the positions taken by Carnap, Schlick, and especially Neurath in the protocol sentence debate. One approach deserves special attention: it was Thomas Oberdan's (1993) *Protocols, Truth and Convention* that appeared in the same series of *Studien Zur Österreichischen Philosophie* in which Uebel's was published a year before, though there is no reference to Uebel's work in the book. Oberdan (1993, 2) argued that actually much of the "philosophical substance" in and around the protocol sentence debate "derives from the underlying conceptions of language." Though Oberdan's monograph could be read more profitably from the angle of Carnap's development and provides a refined picture of Schlick (it is still underappreciated from *these* angles), *Protocols, Truth and Convention* presented another station towards the contemporary rereading of Neurath and his context.[10]

While both Uebel's and Oberdan's projects started from the protocol sentence debate, they had a bigger aim and a broader perspective as well: namely placing Neurath on the map of twentieth-century intellectual history and putting his ideas back on the philosophical table. These motivations were picked up and developed by various authors in 1996, initiating thus the third phase of the reception. The new phase, however, had a quite peculiar twist in itself: they tried to combine both the socio-political *and* the philosophical angles of Neurath's life and work.

[9] Fifteen years later, building on an enormous secondary literature on logical empiricism, Uebel (2007) published an extended, updated and revised version of his book as *Empiricism at Crossroads*. For a longer review essay see Cat (2012).

[10] In his review of Oberdan's book, Thomas Uebel (1995, 312) called the reader's attention to the consideration that "admirers of Neurath are bound to find their man less charitably treated then his ideas deserve." A few years later Uebel (1996, 1999) and Oberdan (1998, 1999) had a debate (one of a few among the logical-empiricism scholars) over anti-foundationalism in the Vienna Circle: though that discussion centered mainly on the interpretation of Schlick, it has also relevance for the contextualization of Neurath as well.

In 1996, two important volumes appeared, approaching somewhat similar and connected themes, thus marking the beginnings of the third period which connected Neurath's political and philosophical ideas through the conception of unified science. *Otto Neurath: Philosophy Between Science and Politics* was written jointly by Nancy Cartwright, Jordi Cat, Lola Fleck, and Thomas Uebel (1996). Besides providing the first detailed English biography of Neurath, and a detailed treatment of the context and philosophical significance of Neurath's various famous ship-metaphors, the authors presented a complex and nuanced version of Neurath's unity of science ideal. They argued that Neurath held a peculiar position regarding the nature and relevance of science: even though he was a pluralist with some relativistic underpinnings, having a broad historical and sociological approach towards the contingency of scientific theories, he still firmly believed in the social and theoretic role of science, and aimed towards the unification and integration of the human and natural sciences accordingly. In the book, they emphasized distinctive features of Neurath's project of unity, method and language and grounded them on a corresponding emphasis on his political context and the inseparable commitments to social goals and the social sciences. In their narrative, this is a quite important and novel account, especially after the science wars and the age of post-modernism of the 1980s and 1990s.

The other book, *Encyclopedia and Utopia: The Life and Work of Otto Neurath (1882–1945)* was a collection of essays, edited by Elisabeth Nemeth and Friedrich Stadler (1996) as the 4th volume of the *Vienna Circle Institute Yearbook* series. It featured 22 essays from Neurath's short biography, through his philosophical ideas to pictorial education and economy. Approaching Neurath's work, which is "linked with each other on several levels" (p. xi.), the collection might be read as providing a more general context and application of the Cartwright-Cat-Fleck-Uebel volume.

During the same period, when Neurath's philosophical papers were published in English, a new tendency appeared in philosophy of science, or at least got a bigger and more seductive voice: numerous scholars argued for the metaphysical and/or methodological disunity of science; this tendency culminated in John Dupré's "The Disunity of Science" (1993) and Peter Galison's and David Stumpf's *The Disunity of Science: Boundaries, Context, and Power* (1996). Against this background, others began investigating what Neurath had meant by such expressions as "unity of science" and "unified science." As it turned out quickly and became evident in the 1996 volumes, it both had a philosophical dimension (but since Neurath would have not used the word "philosophy," it might be more apt to say that "scientific" or "technical" aspect) and a social dimension. The Unity of Science Movement – led by Neurath personally on the continent and through Rudolf Carnap, Charles Morris, and Philipp Frank in the United States – promised a pluralist and permissive alternative to the absolutism and reductionism of other unified and disunified conceptions.[11]

[11] From the early 1990s on, it was George Reisch who published several articles about the Unity of Science Movement, the *International Encyclopedia of Unified Science*, and Neurath's own "planning-account" of unified science. See especially Reisch (1994). Recently John Symons, Olga Pombo, and Juan Manuel Torres (2011) edited a volume on Neurath and the unity of science,

The early and mid-2000s saw another turn in the literature on Neurath. The English bibliography of Neurath's work at the end of his 1983 selected writings promised a volume of Neurath's translated economical works, but it did not appear until 2004 under Thomas Uebel's and Robert S. Cohen's (2004) editorship (*Otto Neurath: Economic Writings. Selections 1904–1945*). The selection featured 19 items from the time of Neurath's doctoral dissertation until his 1946 posthumous writings, relating his most important works on economy, society and social theory. It was supplemented in 2007 by a volume of eight essays discussing *Otto Neurath's Economics in Context* in the Yearbook series, providing some more explanatory remarks on the subject.[12]

The editors (Elisabeth Nemeth, Stefan W. Schmitz and Thomas Uebel (2007)) explained the renewed interest in Neurath's economical thoughts by the fact that Neurath's role in the socialist calculation debate with Ludwig von Mises and F. A. Hayek might have some interesting arguments and refreshing new insights into the current political-economic debates. His notions of moneyless economy and calculation-in-kind proved to be especially important for recent debates and struggles over social and economic discussions. Finally, Neurath's ideal of life-conditions, the pursuit of happiness and ecological economics are also promising fields of research both within the context of logical empiricism and outside.[13]

The fifth and so far final major shift in the reception history came in the late 2000s. It was widely known that Neurath was not a regular philosopher after the linguistic turn, but he was keenly interested in pictures, pictorial education and he established (and also directed) various museums across the globe (among others from Vienna and Berlin through Moscow and The Hague, to Chicago and Mexico). Nonetheless, for a very long time, Neurath's ideas on how one could utilize pictures and museums – in order to disseminate empirical information and emancipate those in need of help – were considered remarkable but cognitively irrelevant anecdotes of his life. Among many analytic philosophers, *education* was perhaps just a practical task beyond first-order theoretical investigation; at least the long neglect of the subject indicates something like that. In these shifts in perspective, new work on Neurath kept tracking broader changes in academic interests.

Such a "neglect" of the pictorial had been justified neither from a bibliographical nor from a philosophical point of view, but it could be explained by the alternative priorities. The surprising aspect is that Neurath's international projection had long been based on his work on visual education. In fact, in the course of his lecture tour of England already in November 1933 (just one year after Schlick's London lectures on "Form and Content" and before Carnap's arrival), Neurath gave various talks on pictorial education and on the importance of his own Viennese method at Bristol's

---

though Thomas Uebel (2013) suggested that most of the volume should be treated with a grain of salt. On the relations between science and politics of Neurath see Cartwright, Cat, and Chang (1991) and (1996).

[12] Though it should be mentioned that Thomas Uebel's more than one hundred pages long editorial introduction to the selection of Neurath's economical writings is still a landmark in the field.

[13] On Neurath and the theoretical and practical idea of happiness see Whyte (2007), Stuchlik (2011) and Sandner (2019).

Education Department of the Bristol Co-operative Society; at Fircroft College, Birmingham University and presumably other places as well, including London.[14] Later in 1937, the Dundee-based *The Courier and Advertiser* newspaper published a short summary of Neurath's method, now under the name "ISOTYPE" as "You Draw This New Language." Besides having a few pictures and a regular description of Neurath's approach, the article ends with the followings: "Give it to a man in a ship's fo'c'sle or an Einstein in his study and in a few minutes it will teach more than millions of words in an ordinary encyclopedia."[15]

Neurath's first visits to England were connected to the interest in his ideal of pictorial education and its practical context, and not in the theoretical philosophy and ideal of analysis of logical empiricism. Then, in the 2000s, the academic presence and influence of mass and visual media became even more widespread, interest in Neurath's visual education got more and more intense. Neurath had named his new method of representing mainly statistical data as *Vienna Method of Pictorial Statistics* (*Wiener Methode der Bildstatistik*) after its geographical origins. Later, in the 1930s, when the whole business became internationally famous, he renamed his brainchild after his later third wife's suggestion (Marie Reidemeister/Neurath) as ISOTYPE, that is, *International System of TYpographic Picture Education.* Representing statistical data visually became so easy and comprehensible, that Neurath believed that his ISOTYPE method could be applied also in the post-war reconstruction era since it does not require any specific linguistic capacities so it could reach a wider and broader fragment of society.

Fortunately, a number of illuminating volumes and monographs have appeared on Neurath's ideal of how pictures, statistical data and museums could be related and integrated in order to yield the most available and standardized method for the dissemination of knowledge to emancipate the masses for the future. The specific method of creating ISOTYPE charts and transforming pure statistical data into pictures was called "transformation" and Robin Kinross published a book on the process (Kinross and Neurath 2009) that contains materials also from Marie Neurath and thus makes it an essential first-person reconstruction of the world of ISOTYPE. Another such recollection is Neurath's (2010) long waited visual autobiography that was published recently under the editorship of Christopher Burke and Matthew Eve. Certain extracts (Neurath 1946) were published already after Neurath's death: they were edited by Marie Neurath and the famous British filmmaker Paul Rotha who also collaborated with the Neuraths during World War Two.[16] An important and unique collection of essays on the history of ISOTYPE (from the 1920s to the 1970s when Marie Neurath retired from the Isotype Institute in

[14] Neurath's lectures were advertised every day before the lectures at the local newspapers: *Western Daily Press* and the *Bristol Mirror* (2, 3, 6 November 1933). Neurath's Birmingham lecture was interconnected with the 1933 Workers Educational Association's National Conference.

[15] In 1933 (17 November, *Daily Herald*), as it was claimed in a journal article, Neurath called his method as "The Mickey Mouse of Social Facts."

[16] On the significance of Rotha's films in Britain and his relation to Neurath, see recently Boon (2016).

England) appeared recently by Christopher Burke, Eric Kindel, and Sue Walker (2013).

Finally there are two monographs that especially aimed at the reconstruction and discussion of Neurath's ideas of how knowledge-dissemination, democracy, pictures and education are related: Hadwig Kraeutler (2008) published a book (actually the only one on the subject) on Neurath's museum work and his contemporary museological relevance; the other (highly pictorial) treatise was published by Nader Vossoughian (2008) depicting Neurath's ideal of the global polis. Both volumes depicted a range of Neurath's interests long regarded as interesting side-events in his life without major theoretical or practical importance, worth careful and detail studies. Things keep changing.

1.3 Chapters of the Volume

In his editorial introduction to *Rediscovering the Forgotten Vienna Circle,* Thomas Uebel (1991, 4) pointed out that "[t]he case of Neurath [...] presents both a puzzle and a promise." We think that Neurath's "case" (that is, his role in logical empiricism, in the unified science movement, in ISOTYPE, and still-neglected remarks) still shows promise: presumably now more than ever. In the age of heavily questioning not just science or scientific procedure, but even the bare fact of "facts" and "truth," one might raise the question with increasing scruples: how could science have any social and theoretical relevance? As it was emphasized by Nancy Cartwright, Jordi Cat, Lola Fleck and Thomas Uebel (1996, 3), even if some remarks of Neurath are "familiar in contemporary post-modern discourse [...] Neurath had one striking difference from most post-moderns. Although he was a pluralist regarding knowledge-systems and took seriously their historical and cultural roots, he trusted firmly in the power of science."

Therefore the volume serves two aims. Firstly, by pointing out new perspectives on Neurath's life and work, the various chapters will demonstrate how one can think and act in a society as a scholar without believing in "the science" or "the system of science," or in "facts" and "truth." The argument is also historical insofar as it features an actual, thus possible, model, embedded in evolving sets of particular circumstances. Secondly, the essays collected here also point to where the fruitful *puzzles* in the life and work of Neurath were and still are, and, in passing, what makes them so.

Part I of the volume includes four essays and aims at a general introduction to Neurath's life and work, emphasizing those elements that shed light on Neurath's oeuvre from both familiar and new perspectives. In Chap. 2, the volume opens with Friedrich Stadler's introduction to Neurath's life of intellectual, literary and other sorts of risks, dangers and adventures. Stadler calls attention to Neurath's libraries and collections. Already in his childhood, Neurath was fascinated by his father's enormous library, where he was especially drawn to visual illustrations some dating back to Ancient times and the French *Encyclopédie*, and inspiring lifelong dealings with pictorial communication. Cases in point are the founding of his Social and Economic Museum of Vienna and the invention of the so-called Vienna Method of Pictorial Language, later on renamed ISOTYPE. During the flourishing "Red Vienna" period he added to his father's library numerous books in several languages that he used as aids in his intense new life as a social reformer, teacher and museologist, philosopher, sociologist and historian of science. With his forced migrations from Vienna and The Hague, this most valuable collection of books and brochures was partly destroyed, dispersed, and also *aryanized* when Austro-Fascism came into power in Austria and Hitler's troops invaded into the Netherlands. Fortunately, Neurath himself and a few family members and collaborators survived and barely managed to save parts of this unique collection of books that offers a different glimpse into the education and intellectual background of the polymath Neurath in Dutch and British exile. Also while in exile he bought and published books, amassing some 3000 volumes, all in the process of continuing his activities for the causes of Logical Empiricism, the Unity of Science and Isotype. Stadler presents three typical case studies from the Neurath library provided spanning his interests economics, philosophy and literature: Neurath's annotations in books of Friedrich A. von Hayek, Karl Popper, and the writer Stefan Zweig.

Don Howard considers in Chap. 3 the question of philosophy's relevance to extra-academic concerns that is still much with us today. Plato told us that once the philosopher has seen the truth in the full light of the sun, she must return to the cave, there to put knowledge to work in making a better world, even though, being temporarily unaccustomed to the dark, she risks ridicule from those still in thrall to illusion. Howard reflects upon the life and career of Neurath as a modern exemplification of this ideal of philosophical engagement. In spite of, or, perhaps, because of having never held an academic appointment beyond his early post at the New Vienna Trade School, Neurath made a considerable difference to the education and living conditions of many. The key components of what Howard terms Neurath's "philosophy of science in action" are explicated in order to understand how that could be. Foregrounded are Neurath's socialism, his own version of the thesis of the empirical underdetermination of theory by evidence, his anti-metaphysical stance, and his commitment to physicalism and the unity of science. The paper concludes with a discussion of the contemporary relevance of Neurath's model of engaged philosophy of science.

In Neurath's work, there was always a close relationship between science and politics; it is worth noting, however, that he also intervened in the politics of the day. In Chap. 4 Günther Sandner focuses on Neurath's time in Revolutionary Bavaria and Red Vienna and analyzes his articles in two newspapers in the interwar period: the German periodical *Economy and Order of Life* (a supplement of the *Art Guardian*) and the Austrian socialist daily *Worker's Newspaper*. The examination starts when Neurath became a socialist and member of the Social Democratic Party (SPD) after the war in Germany and ends shortly before his forced migration from Austria in 1934. The two series of articles differed in several respects. While in *Economy and Order of Life* Neurath focused only on socialization, he addressed a number of different topics in the *Worker's Newspaper*. The question of his role and self-image was a common theme in many of these articles. Beyond that, however, there were intimations to closely related themes that he developed more precisely only later in his life: the question of experts, citizens, and democracy. All of these provide new perspectives on Neurath's life and works.

Finally, the general aim in Chap. 5 of Adam Tamas Tuboly's essay is to give a biographical, historical, and philosophical reconstruction of Neurath's final years in England. Besides reconstructing Neurath's arrival in England, he argues that since the 1930s, Neurath was eager to promote further the brand and fortunes of logical empiricism. His conception of the brand was based not on attention to theoretical commitments, but on practical considerations and decisions. Using a detailed case study of Neurath's relation to the Hungarian sociologist of knowledge Karl Mannheim, Tuboly shows that the development of their connections documents how Neurath gave increasing priority to practical aims. The concluding section points to some further considerations on Neurath's legacy.

Part II contains essays that are bound together by the idea that Neurath's method and ideal of science are unthinkable without the social context and its social goals. In a paper, updated and extensively revised for this volume, in Chap. 6, Elisabeth Nemeth shows how two domains of Neurath's broad and multifaceted work are intimately related: the concepts and methods he wanted to implement in political economics, on the one hand, and the methods of visualization that he and his interdisciplinary team developed at the Social and Economic Museum of Vienna, on the other. Some of Neurath's suggestions in both domains, as Nemeth argues in detail, seem surprisingly modern even today.

Neurath made unique contributions to the fields of museology and curation, which culminated in the founding of Social and Economic Museum of Vienna and its mobile exhibitions in the 1920s. But until today, Neurath's involvement in the organization of portable "field exhibitions" that predated those at the Social and Economic Museum by at least half a decade remains understudied. In Chap. 7, Sophie Hochhäusl argues that field exhibitions, which were informed by Neurath's theories on war economy, are instructive in analyzing his overall curatorial ideas. Staged on the outskirts of the city in collaboration with allotment garden and settlement cooperatives, these exhibitions utilized plans, diagrams, and pictorial statistics to convey social and political statements of facts but also everyday objects. By pairing abstract graphic information with everyday objects, they invited inhabitants into a conversation about the material world as well the future drawing on personal experience. As such, field exhibitions created a communal environment for viewing and debating information and championed what she calls "a collaborative practice of seeing."

Angélique Groß considers Neurath's famous ISOTYPE and practical concerns related to pictures in Chap. 8. In relation to education, she argues, Neurath did not articulate a philosophy or a theory; he just practiced education and arranged it for employees institutionally, personally, didactically and methodically. In a similar way, his methodical procedure of transformation, which is manifested in the pictorial presentation system of ISOTYPE, was not based on an explicit theory of depiction. Instead, according to Groß, Neurath developed a concept whose realization evolved through a successful practice of transmission. However, in the practice of educational depiction, Neurath resorted to constants that are describable analytically as theoretical implications. In this respect, Groß's contribution points out the context of the educational and descriptive practice and its evolution out of Neurath's ideals and concrete standards.

In Chap. 9, Thomas Uebel draws attention to elements of Neurath's economic works that make them relevant even to contemporary discussions. Uebel confronts Neurath's controversial contributions to the socialist calculation debate with the criticism of two well-known opponents, Ludwig von Mises and Max Weber. Uebel examines each side's arguments at a certain level of abstraction so as to allow what are lasting points of significance in Neurath's proposals to shine through more clearly. While these points are closely interwoven with Neurath's schemes for marketless socialism, they are conceptually independent. Suitably so, they have proven influential in just this independent capacity.

Neurath's early work on the classification of systems of hypotheses in optics provided some of the key insights of Neurath's later philosophy of science. In Chap. 10, Gábor Á. Zemplén investigates how Neurath developed his theory of theory-classification in response to inconsistencies he stumbled upon while studying the historical theories. Neurath's empiricism and thoroughgoing fallibilism, according to Zemplén, informed his mapping of the group of theories, locating "elementary notions" of theories and taking into account the "blurred margins" of theories. To replace false dichotomies the project provided a finer-grained analysis of theories and could be utilized to locate unconceived alternatives. Zemplén discusses the close links between Neurath's optical essays, his notion of an "auxiliary motive," and his attack on pseudorationalism in the "Lost Wanderers of Descartes" paper. He also provides a comparison of the two essays, with an extended table of the elementary notions Neurath listed, and discusses Neurath's two-tier methodology for historical reconstruction.

In the three essays of Part III, hitherto almost unknown or unfortunately neglected aspects and fields of Neurath's works are taken up from new perspectives. In Chap. 11, Jordi Cat introduces a broader intellectual context and sketches an integrated account with the purpose of examining the significance of Neurath's attention to logic in early works and subsequent positions. In the process he draws attention to the significance of work by Olga Hahn and Susan Stebbing. The specific attention to algebraic logic is important in integrating Neurath's own interest in mathematics and combining, since Leibniz, the ideals of a universal language and of a calculus of reasoning. The interest in universal languages constitutes a much broader, so-called tradition of pasigraphy that extended beyond philosophical projects. Cat argues (1) that Neurath's works can be embedded in a richer intellectual landscape that includes developments in logic and their local reception in Vienna, and that his attention to logic developed a sustained symbolic standpoint – with semiotic and typographic expressions; (2) that specific aspects of the work in algebraic logic became a standard and a resource in subsequent work often thought independent, while its value was steadily challenged by the separate goal of empirical theorizing and practical application in social domains – including in the areas of economics, history and visual communication; (3) that, in particular, the presentation of systems of algebraic logic by Neurath's sources such as Stanley Jevons and Schröder was not isolated from discussions of political economy; and finally (4) that some of Neurath's positions in matters of language, unity and epistemology in the articulation of logical empiricism and its debates are better understood in terms of shared but diversified acquaintance with pasigraphy, formal standards and logical projects.

Carnap is still often portrayed as a "representationalist." While the genealogy of this prejudice may not actually go back to Neurath's response to Carnap's embrace of Tarskian semantics, there is a continuity of motivation and rhetoric. However, based on a reading of the later Neurath-Carnap correspondence reproduced in this volume, A. W. Carus argues in Chap. 12 that the apparent dispute between them over semantics really was largely terminological, with certain differences of emphasis amplified by personality differences and the long interruption of personal contact due to the war. Carus claims that their conceptions of a language of science can be reconciled. Carnap was neither a representationalist nor an anti-representationalist nor an inferentialist (though it may appear that he can legitimately be portrayed as any of these), since ultimately to embrace one of these positions is to endorse an "order of explanation" or ontological primacy, and, argues Carus, Carnap rejected ontology.

In Chap. 13 Derek Anderson argues for the centrality of syntacticism to Neurath's physicalism, encyclopedism, and the unity of science program. He defends the intelligibility of Neurath's opposition to semantics and underlines the role syntacticism plays in his anti-metaphysical empiricism. As Anderson argues, Neurath was correct to fear that the semantic turn would spell the end of logical empiricism. Many of the most influential metaphysical arguments of the twentieth century turn on premises that would be undermined by a Neurathian rejection of semantics.

The last part of essays are devoted to the context and influence of Neurath's ideas, emphasizing again topics and issues that were either rarely discussed in the secondary literature so far or, if they are well-known, they are represented under new lights also with regards Neurath's legacy.

In Chap. 14, George Reisch examines selected writings of the American science writer Waldemar Kaempffert, Science Editor for the *New York Times*, in public support of Otto Neurath, his Isotype projects, and his Unity of Science Movement. His attention focuses first on Kaempffert's writings in the 1930s, when some intellectuals, the American public, and their elected leaders were relatively sympathetic with Neurath's quest to unify the sciences in ways that would advance and direct scientific research toward practical goals. Then he turns to the 1940s to examine the debate over the nature, scope, and limits of wartime research and Vannevar Bush's call for a national institution to support research. Against Bush, James Bryant Conant, and others, Kaempffert argued vigorously for a foundation that would adopt values and methods of the Unity of Science Movement, but he lost that argument as the National Science Foundation finally took shape. To suggest that this public debate influenced not only the decline of Neurath's Unity of Science Movement but the scholarly development of history and philosophy of science after the war, the paper considers early writings and events in the life of Conant's protégé

Thomas Kuhn, whose *Structure of Scientific Revolutions* helped shape that development.

Neurath's version of functionalism is one that begins with people "as we find them," a proposition first set out in his 1917 essay "The Converse Taylor System." Any attempt to redesign the existing furnishings of everyday life must take into account "functions" that go beyond the obvious purpose of objects: functions that are to do with sociability, happiness, familiarity, the love of "coziness," and that address the diversity and contradictoriness of people. In Chap. 15, Michelle Henning considers how Neurath applied and made use of these ideas about design in 1940s Britain, during and after his internment on the Isle of Man between 1940 and 1941 and in talks, papers and correspondence from this period. Henning does not focus on the Isotype Institute, which would usually be considered his principal intervention in design, but on his commentary on everyday objects and practices. In particular, Henning focus on four objects (if they can be called that) – tennis courts, fireplaces, chairs and shoes – and through these elaborate some of the connections between Neurath's ideas about the design of everyday life, and the significance of everyday practices, and his logical empiricism.

In Chap. 16, Antonia Soulez considers Otto Neurath's late discussion of the political and social context of Plato's *Republic*, especially how Neurath conceived them in the 1940s. Neurath's argumentation is contrasted with the ideas of Karl Popper, both with regard to the latter's reading of Plato and to his general methodology. The distinction between Neurath's treatments of epistemology and politics is also discussed, by highlighting how these two were interwoven in the discussion, and how they differentiated Neurath's articles from Popper's considerations in the *Open Society*.

Silke Körber (Chap. 17) picks up a rarely discussed episode in Neurath's life, namely his relation to L. Susan Stebbing and their joint efforts to educate the "common reader." Stebbing supported Neurath's pragmatic ideas on the "humanization" of knowledge, and both of them were looking for ways to apply modern logic and linguistic analysis, not only to the transfer of information in science and teaching but above all in publication projects for the "common reader." In 1941, Stebbing became the first president of the Isotype Institute in Oxford, which Neurath directed until 1945. Soon after ISOTYPE was founded, long-term relations began between it and the book-packaging company Adprint, managed by German-speaking emigrés in London, as well as its successors and British clients (publishers). A technically and organizationally sophisticated process for the production of illustrated non-fiction books was gradually established. Körber argues that the "picture-text style" developed by Neurath and epitomized in *Modern Man in the Making* (1939) was applied to non-fiction books and series with "integrated layouts," then professionalized and successively transformed into a production model for illustrated books which enabled scientific information to be prepared for the mass market and the "common reader" – in the service of a modern, democratic (post-war) society.

Logical positivists had a very lively interest in the revolutionary science of their time, but also in modern art and especially in 'international style' architecture. Surprisingly they never published a representative volume or longer statement on art and architecture. In Chap. 18, Hans-Joachim Dahms points out that it is not well known that Otto Neurath, the leading organizer and spokesman of logical positivsm, invited the (later on) eminent art historian and critic Meyer Shapiro (professor at Columbia University NY) to contribute a volume on art to the *International Encyclopedia of Unified Science*. Shapiro failed to deliver the promised book. But from the extended correspondence and some material in the Shapiro papers, Dahms describes the general direction the project might have taken. The correspondence deserves attention also because it also covers questions of the endangered peace and the approaching war, the academic scene in Europe and in the USA, and surprisingly: Martin Heidegger. Neurath died in 1945, but, as Dahms points out, Shapiro came back to the Heidegger theme in 1968 when he wrote his famous harsh criticism of Heidegger's programmatic long paper "Das Kunstwerk" (the work of art) where he interprets one of Van Gogh's shoe-paintings.

The volume ends with two appendices. Appendix 1 contains the English translation of Neurath's logic papers: there are three papers with Neurath as the sole author, one co-authored with his wife-to-be Olga Hahn – the sister of his logical empiricists colleague and mathematician Hans Hahn –, and one whose sole author was Olga Hahn. The translations are completed with Jordi Cat's editorial comments that aim to clarify and contextualize these important papers, which appear here in English for the very first time.

Appendix 2 contains the 1940–1945 Neurath-Carnap English correspondence. Though there are many (side) issues recurring throughout the letters, seven themes stand out as the core issues of Neurath and Carnap's late correspondence.[17] These are, in (sometimes overlapping) temporal order, the following: (a) *emigration* (b) *the relation between Platonism in the history of philosophy and the oppressive German climate that lead to Nazism*, (c) *semantics*, (d) *Popper's philosophy*, (e) *Russell's philosophy*, (f) *Neurath's* Encyclopedia *monograph*, and (g) *Neurath's own place in the movement*.

After Neurath fled The Hague towards England in May 1940 he was interned on the Isle of Man as a so-called enemy alien. This "prison-experience" became fixed in Neurath's experience of emigration more broadly, shaping the way he, for instance, evaluated comparatively British and Continental standards and experiences of social life. Some of these events are reconstructed and discussed by Michele Henning (Chap. 15.) and Adam Tamas Tuboly (Chap. 5.). Also the fact that the letters that make up the correspondence were written in English constitutes a telling expression of Neurath and Carnap's new life in exile.

One of the last major preoccupations Neurath addressed in print and in private was the relation of Platonism and Kantianism to the so-called German climate, stemming from his reading and personal experiences before and during World War Two. Even though he published some papers under a pseudonym and under his own name with J. A. Lauwerys, his correspondence constitutes an even more valuable source of insight: it functioned as a reading diary as well, and Neurath commented

---

[17] The 1923–1940 Neurath-Carnap correspondence is being now transcribed and edited for publication by Johannes Friedl, Ulf Höfer, Christian Damböck, and Adam Tamas Tuboly.

on many important events of his life. He described at various places in his letters how the irrational inclination towards the idealism and perfectionism extolled by Platonism and the relentless pressure from duties and imperatives of the Kantian doctrines contributed to a very special cultural climate that played an important role nourishing the broad adoption of oppressive ideologies in the 1930s. These questions are discussed in Antonia Soulez's paper (Chap. 16.).

The absolutist elements in the history of philosophy are highlighted not just in the classical works of the field, but in Neurath's contemporaries as well. He is unsparing towards Carnap's new field of semantic investigations: while he addressed more technical epistemological and logical arguments against semantics from a pragmatist ground, Neurath's main concern was more political. He was afraid that talking about "the truth" and "the meaning" of statements might easily lead to absolutist conceptions in philosophy, and these conceptions could become new tendencies that might be used (at the end of the slippery slope) for intolerant social and political views. Neurath's arguments against semantics are discussed by André Carus (Chap. 12.) and Derek Anderson (Chap. 13.), while Jordi Cat (Chap. 11.) also considers the background of Neurath's logical works that led finally to his mature views on formalism, truth, systematization, and semantics.[18]

Besides Carnap, Karl Popper and Bertrand Russell weren't treated with any patience or sympathy either. Popper was the methodological boogeyman for Neurath since the early 1930s when the former came up with the idea of the *Logik der Forschung* book and Neurath reviewed it critically. He detected many pseudo-scientific and absolutist elements in Popper's ideal of falsification, deductive-schemes and ambition towards systematization and lawfulness.[19] On the other hand, though Russell was considered to be the grandfather of logical empiricism, in the 1940s, he was viewed with much more skepticism: as he embraced the notions of "truth," "meaning," "semantics," and "facts," Neurath distanced himself even more from him. Though in his published articles we find only a few remarks and hints (Neurath 1941/1983), his correspondence with Carnap was centered often on the discussion of Russell's new *An Inquiry into Meaning and Truth*. The movement was facing a crisis that pitted disagreement and change against the desire for a stable, visible and inspiring identity, or brand.

The *International Encyclopedia of Unified Science* was a grandiose enterprise that started in 1938 under the general editorship of Neurath, and as inevitably happens, its scope, standards and reception kept changing over time. Neurath's own 1944 monograph was perceived to be knotty by many: Carnap withdrew his name from the cover, indicating that he has nothing to do with this special volume. While the reason was that Carnap saw the page-proofs at too late a point and had too many objections to the contents, Neurath feared that darker motives hid behind Carnap's attack and the ensuing (yet another) confrontation would not help either Neurath's

---

[18] The political dimension of Neurath's debate with respect to semantical analysis is taken up by George Reisch (2005, Chap. 15.) in his earlier book.

[19] Neurath's relation to Popper is discussed in details by Cartwright, Cat, Fleck, Uebel (1996, Part 3) and Cat (1995).

leadership, friendship or the intellectual movement. The correspondence documents many such latter-day tensions between Carnap and Neurath. Their debates were not just personal, that is, not just a difference in attitudes is manifest in these letters: there were much more at stake, namely different conceptions of philosophy in general, not unrelated to the difference in attitudes and politics, and the very future and fate of logical empiricism.

With regard to the nature and dissemination of the movement, Neurath also tried to clarify his own leadership role. The orthodox picture of Neurath presents him as a "dynamo": who utilized half-baked arguments and showed much more agility in organizing, and thus become the "big locomotive" of logical empiricism, as once Carnap called him. The Neurath-Carnap correspondence, however, points to different motives: Neurath considered himself as a theoretician as well, having in store important ideas on science and its methodology. In recent scholarship thus Neurath emerged as an original thinker, often ahead of his time: our difficulties in understanding him stems from the fact that much of Neurath's writings were polemical, and thus needs much contextualization. These questions are taken up by Don Howard (Chap. 3.), Günther Sandner (Chap. 4.) and Adam Tamas Tuboly (Chap. 5.), who argue for a middle-way picture that finds places the sought-after theoretical significance in Neurath's practical efforts.

Otto Neurath is one of those few figures from the twentieth century who achieved various theoretical results and defended important intellectual and social positions that have been later taken up by others in numerous fields. Starting with history, economics, sociology, education, theory of pictures, housing and museums, along with logic, semantics, and general classifications of theories, Neurath could be seen as a special non-philosophical scholar. However, as the following chapters will make clear, there was always something deeply philosophical in how Neurath tried to extend his points of views outside the classical territory of philosophy. Still, even when assessing just Neurath's intellectual significance, the philosophical investigation shouldn't become the sole task and measure.

Acknowledgements During the preparation and edition of this volume, Adam Tamas Tuboly was supported by the *MTA BTK Lendület Morals and Science* Research Group; by the MTA Premium Postdoctoral, the *János Bolyai Research Scholarship* and by the "Empiricism and atomism in the twentieth-century Anglo-Saxon philosophy" NKFI project (124970). We are indebted to the Carnap Archive (Rudolf Carnap Papers, 1905–1970, ASP.1974.01, Special Collections Department, University of Pittsburgh) and to the Hempel Archive (Carl Gustav Hempel Papers, 1903–1997, ASP.1999.01, Special Collections Department, University of Pittsburgh) for the permission to quote their materials.

References

A.S.C. 1974. Review of Neurath's Empiricism and Sociology. *The Review of Metaphysics* 28 (1): 131–132.

Achinstein, Peter, and Stephen F. Barker, eds. 1969. *The Legacy of Logical Positivism*. Baltimore: The John Hopkins Press.

Boon, Timothy. 2016. 'To formulate a plan for better living': Visual communication and scientific planning in Paul Rotha's documentary films. In *Pursuing the Unity of Science: Ideology and Scientific Practice from the Great War to the Cold War*, ed. Harmke Kamminga and Geert Somsen, 156–181. London: Routledge.
Burke, Christopher, Eric Kindel, and Sue Walker, eds. 2013. *Isotype: Design and Contexts, 1925–1971*. London: Hyphen Press.
Butrick, Richard. 1970. *Carnap on Meaning and Analyticity*. The Hague: Mouton.
Cartwright, Nancy, Jordi Cat, and Hasok Chang. 1991. Otto Neurath: Unification as the Way to Socialism. In *Einheit der Wissenschaften*, ed. J. Mittelstrass, 91–110. Berlin: de Gruyter.
———. 1996. Otto Neurath: Politics and the Unity of Science. In Galison and Stump (1996): 347–369.
Cartwright, Nancy, Jordi Cat, Lola Fleck, and Thomas Uebel. 1996. *Otto Neurath: Philosophy Between Science and Politics*. Cambridge: Cambridge University Press.
Cat, Jordi. 1995. The Popper-Neurath Debate and Neurath's Attack on Scientific Method. *Studies in History and Philosophy of Science Part A* 26 (2): 219–250.
———. 2012. Review of Uebel's Empiricism at the Crossroads. *HOPOS* 2 (2): 354–360.
Cohen, Robert S. 1963. Dialectical Materialism and Carnap's Logical Empiricism. In *The Philosophy of Rudolf Carnap*, ed. Paul A. Schilpp, 99–158. La Salle: Open Court.
Creath, Richard. 1993. Review of Zolo's Reflexive Epistemology. *Philosophy of Science* 60 (2): 359–360.
Dupré, John. 1993. The Disunity of Science. *Mind* 92: 321–346.
Galison, Peter, and David Stump, eds. 1996. *The Disunity of Science. Boundaries, Contexts and Power*. Stanford: Stanford University Press.
Hausman, Alan, and Fred Wilson. 1967. *Carnap and Goodman - Two Formalists*. Iowa: University of Iowa.
Hintikka, Jaakko, ed. 1975. *Rudolf Carnap, Logical Empiricist: Materials and Perspectives*. Dordrecht: D. Reidel.
Kazemier, B.H., and D. Vuyjse, eds. 1962. *Logic and Language: Studies Dedicated to Professor Rudolf Carnap on the Occasion of His Seventieth Birthday*. Dordrecht: D. Reidel.
Kinross, Robin, and Marie Neurath. 2009. *The Transformer: Principles of Making Isotype Charts*. London: Hyphen Press.
Kraeutler, Hadwig. 2008. *Otto Neurath. Museum and Exhibition Work: Spaces (Designed) for Communication*. Frankfurt am Main: Peter. Lang.
Nemeth, Elisabeth, and Friedrich Stadler, eds. 1996. *Encyclopedia and Utopia: The Life and Work of Otto Neurath (1882-1945)*. Dordrecht: Kluwer.
Nemeth, Elisabeth, Stefan W. Schmitz, and Thomas Uebel, eds. 2007. *Otto Neurath's Economics in Context*. Dordrecht: Springer.
Neurath, Otto. 1936. *International picture language. The first rules of Isotype*. London: Kegan Paul, Trench, Trubner.
———. 1937. *Basic by Isotype*. London: Kegan Paul, Trench, Trubner & Co..
———. 1939. *Modern Man in the Making*. New York: Alfred A. Knopf.
———. 1941/1983. Universal Jargon and Terminology. In *Empiricism and Sociology*, ed. Robert S. Cohen and Marie Neurath, 213–229. Dordrecht: D. Reidel.
———. 1946. From Hieroglyphics to Isotype: Extracts from the Manuscript (edited by Marie Neurath and Paul Rotha). *Future Books* (3): 92–100.
———. 2010. *From Hieroglyphics to Isotype: A Visual Autobiography*, ed. Christopher Burke and Matthew Eve. London: Hyphen Press.
Norton, Bryan G. 1977. *Linguistic Frameworks and Ontology. A Re-Examination of Carnap's Metaphilosophy*. The Hague: Mouton.
Oberdan, Thomas. 1993. *Protocols, Truth and Convention*. Amsterdam: Rodopi.
———. 1998. Discussion: The Vienna Circle's 'Anti-Foundationalism'. *British Journal for the Philosophy of Science* 49: 297–308.

———. 1999. Deconstructing Protocols: Reply to Uebel. *British Journal for the Philosophy of Science* 50 (2): 301–304.
Proust, Joëlle. 1989. *Questions of Form: Logic and Analytic Proposition from Kant to Carnap*. Minneapolis: University of Minnesota Press.
Reisch, George. 1994. Planning Science: Otto Neurath and the International Encyclopedia of Unified Science. *British Journal for the History of Science* 27: 153–175.
———. 2003. Disunity in the International Encyclopedia of Unified Science. In *Logical Empiricism in North America*, ed. Gary L. Hardcastle and Alan W. Richardson, 197–215. Minneapolis: University of Minnesota Press.
———. 2005. *How the Cold War Transformed Philosophy of Science. To the Icy Slopes of Logic*. New York: Cambridge University Press.
Runggaldier, Edmund. 1984. *Carnap's Early Conventionalism: an Inquiry into the historical background of the Vienna Circle*. Amsterdam: Rodopi.
Ryan, Alan. 1976. Positivism and Sociology [review]. *The Philosophical Quarterly* 26 (103): 194–195.
Sandner, Günther. 2019. Bringing Happiness. Otto Neurath and the Debates on War Economy, Socialization and Social Economy. In *Philosophy of Social Science*, ed. Friedrich Stadler, Sami Philström, and Niels Weidtmann. Cham: Springer. Forthcoming.
Schilpp, Paul, ed. 1963. *The Philosophy of Rudolf Carnap*. La Salle: Open Court.
Stuchlik, Joshua. 2011. Felicitology: Neurath's Naturalization of Ethics. *HOPOS* 1 (2): 183–208.
Symons, John, Olga Pombo, and Juan Manuel Torres, eds. 2011. *Otto Neurath and the Unity of Science*. Dordrecht: Springer.
Székely, D.L., and Charles Morris. 1966. *Otto Neurath and the Unity of Science Movement*. Jerusalem: Mimeographed monograph of the Association for Unification and Automation in Science.
Uebel, Thomas, ed. 1991. *Rediscovering the Forgotten Vienna Circle: Austrian Studies on Otto Neurath and the Vienna Circle*. Dordrecht: Kluwer.
———. 1992. *Overcoming Logical Positivism from Within – The Emergence of Neurath's Naturalism in the Vienna Circle's Protocol Sentence Debate*. Amsterdam: Rodopi.
———. 1995. Review of Oberdan's *Protocols, Truth and Convention*. In *The Foundational Debate: Complexity and Constructivity in Mathematics and Physics*, ed. W. Depauli-Schimanovicz, Eckehart Köhler, and Friedrich Stadler, 310–312. Dordrecht: Kluwer.
———. 1996. Anti-Foundationalism and the Vienna Circle's Revolution in Philosophy. *British Journal for the Philosophy of Science* 47: 415–440.
———. 1999. Protocols, Affirmations, and Foundations: Reply to Oberdan. *British Journal for the Philosophy of Science* 50: 297–300.
———. 2007. *Empiricism at the Crossroads. The Vienna Circle's Protocol-Sentence Debate*. Chicago: Open Court.
———. 2013. Two Halves of Unity. *Metascience* 22: 481–484.
Uebel, Thomas, and Robert S. Cohen, eds. 2004. *Otto Neurath: Economic Writings. Selections 1904-1945*. Dordrecht: Springer.
Vossoughian, Nader. 2008. *The Language of the Global Polis*. Rotterdam: NAi Publishers.
Watkins, J.W.N. 1974. Review of Neurath's Empiricism and Sociology. *The British Journal for the Philosophy of Science* 25 (4): 343–352.
Whyte, I.B. 2007. Otto Neurath and the sociology of happiness. In *Man-Made Future: Planning, Education and Design in Mid-Twentieth- Century Britain*, ed. I.B. Whyte, 16–38. London: Routledge.
Zolo, Danilo. 1989. *Reflexive Epistemology: The Philosophical Legacy of Otto Neurath*. Dordrecht/ Boston: Kluwer.

Part I
Otto Neurath: Life and Work

Chapter 2
A Viennese Library in Exile: Otto Neurath and the Heritage of Central European Culture in the Anglo-Saxon World

Friedrich Stadler

Abstract Otto Neurath experienced an adventurous as well as dangerous life. Already in his childhood, he was fascinated by his father's huge library. He was especially impressed by images and illustrations since Ancient times and the French *Encyclopédie*, which inspired his lifelong dealing with picture language. This became manifest with the founding of his "Social and Economic Museum of Vienna" and the invention of his "Vienna Method of Pictorial Language," later on renamed ISOTYPE. In the flourishing period of "Red Vienna" he acquired a lot of books in several languages covering his research fields and practical activities as a social reformer, teacher and museologist, philosopher, sociologist and historian of science. With his forced migrations from Vienna and The Hague, this most valuable collection of books and brochures was partly destroyed, dispersed, and *aryanized*, when Austro-Fascism came into power in Austria and Hitler's troops invaded into the Netherlands. Luckily, Otto Neurath and some members of his family and collaborators could survive and save parts of this unique collection of books which mirrors the education and intellectual background of the polymath Neurath in Dutch and British exile. Here, he again bought and published books continuing his activities for the Vienna Circle of Logical Empiricism as well as the Unity of Science and the Isotype movements covering some 3000 books in 1945. Three typical case studies from the Neurath library are provided covering economics, philosophy, and literature: Neurath's annotations in books of Friedrich A. von Hayek, Karl Popper, and the writer Stefan Zweig.

F. Stadler (✉)
Vienna Circle Institute, University of Vienna, Vienna, Austria
e-mail: friedrich.stadler@univie.ac.at

J. Cat, A. T. Tuboly (eds.), *Neurath Reconsidered*,
Boston Studies in the Philosophy and History of Science 336,
https://doi.org/10.1007/978-3-030-02128-3_2

2.1 A Brief Overview of Neurath's Life[1]

Who was Otto Neurath? William Johnston (1972, 192) had introduced him already in 1972 as "one of the neglected geniuses of the twentieth century." Without doubt, he was a most productive and innovative intellectual with a wide range of interests and gifts, which he inherited from his social-liberal assimilated Jewish family in the Habsburg monarchy. Emancipation and Enlightenment were the pillars of this significant education till the expulsion and destruction of Jewish life after the "Anschluss." This was accompanied by an increasing "Cultural Exodus" with the culmination after 1938 between forced migration and the Shoah.

Otto was the first of two sons of Gertrud Kaempffert (1847–1914), a Protestant woman, and the Viennese social reformer and political economist Wilhelm Neurath, (1840–1901), who exerted an enduring influence on the young student in Vienna and Berlin. Wilhelm Neurath, together with his friends the social reformer Josef Popper-Lynkeus and the philosopher-scientist Ernst Mach, served as role model for Otto's lifework.[2] They all represented Viennese progressive liberalism and socialism with philosophy and science as an anti-metaphysical "scientific world-conception." Especially, his father's library with some 13.000 volumes turned out to become an intellectual cosmos for his son Otto, to be dealt with later on.

Otto Neurath attended the school in Vienna, studied mathematics, political economy, and history at the universities in Vienna and Berlin (at the recommendation of economist Ferdinand Tönnies). He completed his Ph.D. *summa cum laude* with the dissertation *Zur Anschauung der Antike über Handel, Gewerbe und Landwirtschaft* (*On Commerce, Trade, and Agriculture in Ancient Times*) in 1906 under Eduard Meyer, which was published subsequently. In parallel, he submitted a second study as dissertation entitled *Antike Wirtschaftsgeschichte* (*Economic History of Antiquity*), which appeared in 1909. Already in his student years, Neurath became a member of the renowned "Goethe Society" for his edition and introduction of *Faust* by F. Marlow (= Ludwig Hermann Wolfram). In 1906 he completed his military service and from 1907 to 1914 he taught political economy in the tradition of his father at the *Neue Wiener Handelsakademie* (New Vienna Trade School).

In 1907 Neurath married the social scientist, author and feminist Anna Schapire (1877–1911), who died from complications caused by the birth of their son Paul in 1911. Paul Neurath became a successful sociologist (demography) after he escaped the concentration camps and could emigrate to the US, where he was advised by Paul Lazarsfeld (the founder of empirical social research) at Columbia University in N.Y.C. Only recently, we enjoy a double biography on Anna Schapire and her sister, the art historian Rosa Schapire (1874–1954), who escaped to the UK in 1939 (see Dogromaci and Sandner 2017).

---

[1]This article is based on two lectures which I delivered at the Bard Graduate Center, N.Y.C.: Library Lecture April 10 and Brown Bag Lunch, April 11, 2018. I am grateful to Dean Peter N. Miller for this invitation and the related discussion. Special thanks go to Adam Tuboly for his redaction and copy editing.

[2]On Wilhelm and Otto Neurath, see Uebel (1995).

In 1912 Otto Neurath married the blind mathematician Olga Hahn (1882–1937), the sister of the famous mathematician Hans Hahn, who became an important member of the later Vienna Circle. They jointly published mathematical and logical articles before World War One.[3] This period also encompassed the early philosophical and scientific activities in the so-called "First Vienna Circle" of Logical Empiricists with Hahn and the physicist and Einstein biographer Philipp Frank, with both of whom he kept private and scholarly contacts until the inter-war years.

Prior to 1914 Otto Neurath also published sociological studies, *inter alia*, on the so-called "theory of war economy" and on the economic situation in the Balkans, to which he took a number of study trips with the support of a grant awarded by the Carnegie Foundation for International Peace. Following the outbreak of World War One, Neurath served on the east front and was afterward appointed director Museum on War Economy in Leipzig, at the same time working in the war economy section of the war ministry in Vienna. In this period he developed his theory and practice of graphic representation of socio-economic relations and his models on economies in kind or *in natura* calculation (*Naturalwirtschaft*). He also completed his *Habilitation* and became private lecturer (Privatdozent) in political economy at the University of Heidelberg in 1917 – a position which, due to his war service and later his involvement in the Munich socialist republic (*Räterepublik*) he was never to actively exercise.

In the revolutionary phase of the post-war period, Neurath became president of the Central Economic Planning Office in Bavaria, where as a "social engineer" he had unsuccessfully tried to carry out his vision of full socialization on the basis of his planned economy in 1919. After the movement was defeated by the Right military troops, Neurath was sentenced to one and a half years imprisonment for "aiding high treason." After 6 weeks of detention, he was able to travel to Austria on Otto Bauer's intervention (who was then Secretary of State), but he was no longer permitted to enter Germany (until 1926) and thus also lost his lectureship in Heidelberg.

After his return to Vienna, Neurath devoted his efforts to the housing and settlement movement in Viennese communal politics, taking over the job of a secretary general of the organization for settlement movement. In 1923 he founded the "Museum for Settlement and City Planning," which was the basis for the "Social and Economic Museum of Vienna" (Gesellschafts- und Wirtschaftsmuseum in Wien, or GWM, 1925–1934), of which he was the founder and director till its dissolution in the Civil War of 1934.[4] This new kind of institution, which was to be an "educational museum of the present day" for the knowledge and understanding of socio-economic correlations, ran regular exhibitions in the new city hall, as well as in the first and twelfth districts of Vienna. Up to 1933 thirty-six national and international exhibitions were organized there or supplied with material. The topics presented there convey an impressive picture of systematic encyclopaedism, the work

[3] For a discussion of the logic-papers, see Jordi Cat's chapter in the present volume; for a translation of these papers, see Appendix 1. in the present volume as well.

[4] On Neurath's involvement in the Settlement and Housing Movement, see Sophie Hochhäusl's chapter in the present volume.

towards popular education that had been carried out in the spirit of social reform: health, women and children, social politics, housing and urban planning, peace education, schooling, the workers' movement, art, social security, architecture, etc. Along with this, there were independent touring exhibitions at home and abroad, as well as separate publications and articles in various periodicals and books.

In 1927, Philipp Frank's brother, Josef Frank was acquired for the museum as an architect, and a year later Neurath employed the artist and designer Gerd Arntz, who drew the characteristic symbols and systematized the production techniques. The scientific department under Aloys Fischer, the "transformation" department under Marie Reidemeister (later Marie Neurath) and the technical collaborators completed the team.[5] For several years the innovative Social and Economic Museum [SEM] cultivated a working partnership with Otto Glöckel's social-democrat school-reform. This enabled the SEM to contribute to the cultural life of Vienna by making an impact on visual education. While the Vienna Method of Pictorial Statistics was relatively well known in its theoretical and practical application in school contexts, a parallel initiative in adult education received less attention: from 1931 to 1933 the SEM in Vienna regularly published the *Fernunterricht* (correspondence course/distant learning) as distance education booklets (retitled *Bildstatistik* from 1932) as subscription magazines, each dedicated to an individual topic. From today's perspective, one can treat these brochures with their changing topics as elements in the field of social history. As it emerges from the preface, there was particular emphasis on user-friendly orientation and the invitation to provide feedback, which helps in creating an interest-oriented mode of communication that puts the participants on an equal footing. In these brochures, Neurath provides an account of the interconnectedness between science and popular education, which, according to his thinking, could be brought about through the empirical view of science.

Additionally, in Holland in 1931 the "Mundaneum" was founded with the aim of intensifying international collaboration. Branches were set up in Berlin, Amsterdam, Prague, New York, London, and Moscow. Following this phase of productive work at home and abroad came the end of the SEM after the political events of 12th February 1934. Several functionaries were arrested, and valuable fittings were confiscated. The renamed "Austrian Institute for Pictorial Statistics" came under the control of the Austro-fascist corporative state until the National Socialists seized this institution for their own propaganda purposes. Despite these confiscations, Neurath was able to transfer a large number of valuable holdings to Holland.

The fundamental objective of the "Vienna method of pictorial statistics" was to represent socio-economic facts and correlations, particularly with regard to their historical development, in a simple, easily graspable system of symbolic figures. A range of real things and complex facts was to be represented by means of a fixed range of signs and symbols, in which the same sign would always be used for the same object. This method of visualization, therefore, arose from the coinciding of content and size as well as from a mapping of sets, such that a larger set of objects

[5] On the process of transformation, see Neurath and Kinross (2009) and Angélique Groß's chapter in the present volume.

was represented by a larger set of signs without perspective. In this way – said Neurath – the facts about society could be reflected quantitatively. The method, developed and improved following his emigration to Holland, was, with the change in circumstances, renamed the "International System of Typographic Picture Education," with the acronym "Isotype" (which was also the Greek for "always the same sign").[6]

The programmatic titles of Neurath's specific writings alone convey a sense of the intentions and socio-political orientation of this conception, which became an international one from 1930. The pictorial representation of social facts with the help of "statistical hieroglyphics" would concisely and informatively explain society in all its aspects with a view to improving "living conditions." The scope of topics, therefore, extends from home, the reality of the world of work, housing and urban planning, to economic considerations on social welfare. From a modern-day perspective the monographs *Bildstatistik nach Wiener Methode in der Schule* (1933), *International Picture Language* (1936), and finally his most mature work *Modern Man in the Making* (1939) are of special interest. In these numerous publications, pictograms are used to treat the theory and application of picture language along with written language. In addition, with his essay "Museums of the Future" (1933/1973) Neurath laid out a program for a modern sociological and economic museum that remains impressive to this day, and which can be interpreted as an alternative to postmodern museums, the collections of devotional objects, and the contrived "total works of art" (*Gesamtkunstwerke*). As Neurath wrote in a paper originally written for the American *Survey Graphic* magazine:

> From Comenius' *Orbis Pictus* an uninterrupted movement leads to modern visual education. A picture made according to the Vienna method shows at the first glance the most important aspect of the subject; obvious differences must be at once distinguishable. At the second glance, it should be possible to see the more important details; and at the third glance, whatever details there may be. A picture that has still further information to give at the fourth and fifth glance is, from the point of view of the Vienna school, to be rejected as pedagogically unsuitable.
>
> Thus a new clarity and purposefulness is developing in communication that may be regarded as preparation for more incisive social planning. Teachers and other groups of people concerned in social education, directors of museums, and editors of periodicals are confronted with the responsibility of placing their energies at the service of this common international task. (Neurath 1933/1973, 223.)

Still in Vienna, Neurath attempted, in view of the emerging fascism and NS, to promote the internationalization of pictorial statistics by founding branches and institutions abroad.[7] After he had to emigrate in the wake of the events on February 12, 1934 to the Netherlands, the "Foundation for Visual Education" in The Hague was to become the platform for the further activities of the encyclopedia movement and pictorial education. There, under the most difficult conditions, he continued his

---

[6] On pictorial education, the Viennese Method and on ISOTYPE see Angélique Groß's, Elisabeth Nemeth's, and Silke Körber's chapters in the present volume.

[7] In parallel, he entertained contacts with the German Bauhaus in Dessau and the CIAM movement of architects between the wars; see Dahms (2004).

work on the project known as *Isotype*. His second wife Olga Hahn died in 1937 in Holland from complications after a kidney operation (Sandner forthcoming).

At the beginning of the 1920s, still during his Vienna years, Neurath resumed his philosophical work from before World War One and was actively involved in the Vienna Circle around Moritz Schlick and further circles in Neurath's private apartment together with Olga and in his museum. Important signs of these activities was the publication of the manifesto *Wissenschaftliche Weltauffassung: Der Wiener Kreis* (Scientific World-Conception: The Vienna Circle) in 1929, which he wrote together with Hans Hahn and Rudolf Carnap (1929/1973).[8] In addition, he founded the "Verein Ernst Mach" (1928–1934), the society for the popularization of scientific world-conception in the context of Vienna's reform movements. Neurath was also active as a teacher at Vienna's Workers' University, in the labor union movement, and at diverse adult education institutes. In philosophy of science, he was the main representative of the so-called "left wing" of the Vienna Circle (Uebel 2007), and he served as a leading figure for public relations and – increasingly after 1934 – for the internationalization of Logical Empiricism (1935–1941: e.g., with six International Congresses for the Unity of Science in Paris, Copenhagen, Cambridge, Harvard, and Chicago). In addition, in his Dutch exile, the untiring man founded the Institute for the Unity of Science in The Hague in 1937.

In 1940, when German troops invaded the Netherlands, Neurath and his long-term collaborator and partner Marie Reidemeister, had to flee again in a boat dramatically to England, where they were interned on the Isle of Man as "enemy alien." After his release, he subsequently established another Isotype Institute together with Marie, whom he married in 1941.[9] From 1941 to 1945 he served as a lecturer at the University of Oxford at the same time – a position, which he could have never achieved in Europe. And Neurath returned also back to his Viennese roots when acting as a consultant of the city of Bilston with a bottom-up town renewal project (Nikolow 2004). Till the unexpected end of his life in England (he never thought of returning to Austria again) he wrote books and articles on Logical Empiricism and inspired by the six congresses he edited (together with Rudolf Carnap and Charles Morris) the book series *International Encyclopedia of Unified Science* from 1938 on. His last monograph within this series was the contested book *Foundations of the Social Sciences*, which, by the way, led to the dramatic break with his good old friend Rudolf Carnap.[10]

In the age of 63 years, Otto Neurath died of heart failure on December 22, 1945, at his desk in Oxford, when he looked for a passage in Goethe's "Iphigenie in Tauris," a symbolic return to his literary interests some 45 years after his edition of Marlow's "Faust."

---

[8] The manifesto was reprinted recently with its various translations and commentaries in Stadler and Uebel (2012).

[9] On Neurath's way to and work in England, see Adam Tamas Tuboly's, Silke Körber's, and Michelle Henning's chapters in the present volume.

[10] On Neurath's break up with Carnap and the reasons involved, see André Carus's chapter in the present volume.

2.2 A Brief Overview of Neurath's Work

The research on one of the most brilliant all-round intellectuals of Viennese modernity amounts to a substantial sum total, which can hardly be given in a brief account. Neurath's life and work between "encyclopaedia and utopia" points to the continuities and breaks in social development since the turn of the century (during the struggle between revolution and reform) and the project of creating a scientific picture of the world in the spirit of the Enlightenment and in context of modern "civil society." In dealing with this exemplary life-story of Austrian intellectual migration, there emerge here relevant and essential elements of his future work in education and understanding of science, which can be outlined only by a few key words: a non-hierarchical picture of science (for instance with the ship metaphor popularized by Quine, see Cartwright, Cat, Fleck and Uebel 1996) together with a relativistic and non-reductionist epistemology, an empiricism (naturalism) that regarded the difference between everyday and scientific knowledge as one of degree, an overarching perspective on knowledge in words *and* pictures, and, especially, research and education as the subject and object of a cooperative scientific praxis with a constant demand for societal change. Accordingly, this unfulfilled and fragmentary ambitious project of modernity is thematized, in all its facets, in the present day: in the philosophy of science and science studies, in the ecological perspective in political economy, in architecture and the social housing movement, in modern commercial art and typography, and finally in the present-day vision of museums as social "museums of the future."

It is no accident that the most recent literature on Neurath's life and work is international and inter-disciplinary and, similarly, lies at the points of tension between modern and postmodern narratives. These disparate perspectives, which demonstrate the absurdity of the long-prevailing quarrel about "positivism," are gradually providing the building blocks for the conception of his life and research. It can be understood as a dynamic undertaking in his contemporary environment such that the "producers" of science were themselves made part of the epistemological process (the "Republic of Letters"). This was accompanied by abandoning an absolute meta-perspective. From this vantage point alone it is easy to imagine how this conception would have interfered in past debates on the "science wars" of the 1990s. Neurath's approach means renouncing every secure system of science and knowledge, connected with thinking in alternative utopias of science/society, which certainly corresponds with a conception of possibility (*Möglichkeitssinn*) according to Robert Musil's novel *The Man without Qualities*. This is the background to Neurath's postulating a plan for the establishment of freedom, happiness, and prosperity – written against Spengler's cultural pessimism. It seems that this also today in an age of globalization and planning corporations again, remains a highly topical demand, appealing both as an alternative to laissez-faire capitalism, and to a challenged New Economy following Hayek and the Chicago School.

For Neurath knowledge was always a collective enterprise embedded in a social context. Throughout his life, he never abandoned his enlightened view of a social and cognitive totality.[11] For him, the discovery and explication of seemingly disparate connections between phenomena, both in terms of text and image and the argumentation based thereon was an essential precondition for all intellectual work. To view science as an end in itself was totally alien to him. He believed that knowledge – even as part of a hardly predictable historic process – should serve life in all the areas between the poles of everyday life and specialist research activity. This concept of reflective science will be surprising only to those who have been influenced by the ideologized so-called *positivism debate* of the 1960s. Both these distortions generalized single elements of Logical Empiricism and used them to form a hostile portrait by a dialectic magic. On the other hand, traditional academic metaphysics, obsessed with the subtleties of meaning and oblivious to the consequences of material beings feels equally challenged by the radical and down-to-earth "scientific world conception" of the neo-Enlightenment, which Neurath promoted so strongly.

It is not easy to approach the complex and highly diverse life work of Otto Neurath without getting lost in details, on the one hand, or without resorting to inadequate generalizing descriptions such as "positivist," "physicalist," "Austro-Marxist," etc., on the other. How can one obtain an overview of a man, both of theory and practice, who covered the entire spectrum of knowledge in about 300 publications (including around 30 monographs and books), and who worked in many countries (mainly in Austria, Germany, The Netherlands, England) as a philosopher, social scientist and teacher, a man who pursued a demanding profession (as museum director), yet also figured centrally in the Vienna Circle?

It seems that the answer to this question lies in Neurath's principle of thinking and acting: to proceed from incomplete and unstable elements and its related practice, and to employ, under the constraints of uncertainty and mere probability, the criteria of empiricism without ever losing sight of the fragmentary starting point or the contemplation of a historical process. This approach in itself constitutes a criticism of the idea of a fundamental and hierarchical *system* and at the same time an appeal to think in ways differing from those which prevail in society and the sciences, an appeal to risk a concrete *utopia* and to contribute to the promotion of the common good by means of cooperative *planning* for freedom and happiness.

After World War One, Neurath envisaged an international "republic of scholars," much in the tradition of the French Encyclopedists (Dahms 1996). Yet despite his optimism, he realized that the growth of Fascism and National Socialism threatened to end all these young democracies and he was not naïve enough to believe that the proper proportion of rationalism and empiricism alone reverse the "demise of reason." In his back then unpublished response to Max Horkheimer's "The Latest Attack on Metaphysics" (1937/1972) Neurath, (anticipating the main arguments against the dialectic of Enlightenment), referred to the limitations of social criticism which result from the fragility of knowledge including sociology and its history:

[11] On this idea in Neurath's work, see Don Howards's chapter in the present volume.

> That which has gone completely unnoticed at one time becomes conspicuous in another, what has been noticed but judged as unimportant, may become the center of important considerations in another time. [...] One has to take into account here that "constructions" and "raw materials" are difficult to separate. Some of our observation sentences and perceptions [Anschauungen] turn out to be very stable, but – principally speaking – nothing is certain, everything flows. (Neurath 2011, 16.)

Moreover, claimed Neurath (2011, 21), that "we do not recognize a tribunal beyond science, with [sic] sits in judgment of science and investigates its foundations." This means, there also is not available some form of transcendental reason. A central affinity to the standpoint of Wittgenstein's middle and late periods is apparent: philosophy is viewed as a language game, a game from which we cannot escape but about which we can certainly speak intelligently.

From the start, Neurath himself had underlined the unity of his thoughts and of actions. The methodological holism in philosophy of science (Duhem-Neurath-Quine principle) and naturalism extends to the "orchestration of the science by the encyclopedism of Logical Empiricism" (Neurath 1946b/1983). His interest in visual education and the Viennese method of picture statistics led up to *Isotype*. Loose but deliberate analogies can be found in Neurath's original writings, as can attempts to apply his theoretical concepts to education, social reform and the politics of knowledge. The following areas must be seen from this point of view: economy in kind, war economics, planning theory as well as his involvement in the Bavarian Revolution, the Viennese housing movement and town planning projects in England, not to mention his visual education projects and the founding of the "Social and Economic Museum of Vienna" as well as of the "International Foundation for Visual Education" in Holland (The Hague).

Thus, Neurath also emphasized several times, with a reference to Leibniz and Comenius, that the future encyclopedia did not aim merely at a standardization of the *language of science* but also at a standardization of the *visual representation* (he projected 260 volumes of text and a 10 volume "Visual Thesaurus"). This combination of text and pictures had already been realized in part in his publications *Fernunterricht* (1931–1933) and his successful book *Modern Man in the Making* (1939). Towards the end of his life, Neurath referred to the empiricist "mosaic of the sciences." In the spirit of this formulation, we can arrive at an understanding of his work by means of a kind of *collage,* employing the regulative idea of the unity of science and society. Neurath's encyclopedism also contains (long before C. P. Snow's well-known critique) an implied criticism of separated scientific cultures.

The biography of Otto Neurath is marked by the historical caesuras of 1918/1919, 1934 and 1938/1940, which brought major interruptions to his work. The fact that he nevertheless carried on between destruction and reconstruction under the most trying circumstances can offer encouragement and indicate that, even in a hostile environment the humanization of knowledge can make a modest contribution to the democratization of society.

First, there is Neurath's vision of a joint enterprise of science, which, poetically formulated as boat metaphor, reappears in different variations throughout his work: "We are like sailors who on the open sea must reconstruct their ship but are never able to start afresh from the bottom. Where a beam is taken away a new one must at once be put there, and for this the rest of the ship is used as support. In this way, by using the old beams and driftwood the ship can be shaped entirely anew, but only by gradual reconstruction" (Neurath 1921/1973, 199).

Then there is the ethos with which Neurath pursued this reconstruction. In the middle of the revolutionary post-war phase, Neurath (1919/1981, 137f) delivered an analysis of the *Zeitgeist* in his short article on "Utopien" which, particularly in our present times of change, gives us reason to reflect on utopias. And finally, there is the consistency and continuity of his thought. In his intellectual testament, Neurath states:

> I always promoted monism as a means of empiricist communication and I promoted pluralism as an attitude in making hypotheses. I have contended with thinkers of all kinds who tried to declare one system as marked out before others and therefore I tried to convey the insight that we need a kind of "decision" wherever we have to make a "choice", even when we are trying a scientific theory [...]. One may evolve more than one theory of light starting from the same basis, as one may plan more than one holiday tour from the same starting point. (Neurath 1946a, 526–527.)

2.3 Otto Neurath: Visual Autobiography and His Library

Neurath's posthumously published visual autobiography was written in the last 2 years of his life in Oxford in parallel to his manuscript "Visual Education: Humanization versus Popularization" (1996). It was conceived of as a sort of pictorial atlas following his previous brochure *International Picture Language* (1936). The manuscript was extended and further developed as a special form of autobiography, which is nicely commented on by Neurath himself reproduced in the volume of 2010: "The visual material presented in this book is to support the case for consistent visual information by means of language-like technique. Since matters such as these involve the whole way of life in one way or another, it is hardly possible to conduct experiments" (reproduced by Burke 2010, xxii). And in the synopsis for the projected publisher Adprint, Neurath writes on the purpose: "The book will show different sources from which Isotype has evolved. It does not deal with visual education as a whole. Even persons not interested in the question as such should be attracted by the combination of pictures and their explanations" (quoted from Burke 2010, xxvi).

The reason why he wrote a visual autobiography reads nicely with a high topicality: "Today an ever-changing visual stream flows before our eyes. [...] How rightly can our period called the century of the eye" (Neurath 2010, 3). In order to evaluate Neurath's writings on visual education and communication, we have to add, that at the same time he published the monograph *Foundations of the Social Sciences* (1944a) and worked on a manuscript entitled "Tolerance and Persecution."

Let me now draw attention only to some passages of his visual autobiography, in which he refers to his father's library – as a typical feature of Neurath's oeuvre:

> I was brought up in a scholarly home in Vienna. Even as a child I was already looking at shelves and cupboards full of books, and the impression they made has accompanied me throughout my life. In the entrance hall of our apartment stood extraordinarily large glass cases crammed with books and pamphlets; bookshelves up to the ceiling covered all the walls of our drawing room, even the space between the windows. [...]
>
> I think I very soon started to count the number of books in the library. I believe I made at this time my first mathematical estimates [...] I counted the number of shelves and the average number of books in each shelf – mostly arranged two rows deep. I reached a peak number of about 13,000 books altogether. They were all at my disposal, since I was allowed to rummage in the library to my heart's content. (Neurath 2010, 23.)

In a footnote to this report (that is, to a part of it which was published in *Empiricism and Sociology*, Marie Neurath comments remarkably: "Otto had read most of Kant and other philosophers when he still enjoyed playing with tin soldiers" (page 80, note 2).[12]

It is no coincidence that it was in England, where the enlightenment fostered egalitarian concepts of education, that Neurath's ambitions struck fertile soil. Thus, he was able to present the importance of visual education vis-à-vis popular education in several articles in the periodicals of adult education organizations. E.g., the article "Visual aids in adult education," Neurath (1944b) describes the range from Comenius to modern picture language and its role in the struggle against superficial knowledge, illiteracy and the confinement to simplistic reading material; his argument was supported by a web of various "visual arguments," and ended with a realistic, but still unrealized mode of visual communication promoting an "atmosphere of argumentative meditation and of some peacefulness."

Current international research into the field of visual communication and general semiotics confirms to an impressive degree the viability of developing Neurath's approach further. Following the subsequent development made in the Anglo-American world, slowly but surely in German-speaking areas, too, attention is turning to this innovative tradition of the "Vienna Method," as demonstrated by the discovery anew, and the rediscovery, of Neurath's life and work. And the question arises on significance in the digital age of Internet and Social Media.

When Wilhelm Neurath died in 1901, his rich library (located at Am Heumarkt 4, later Marxergasse 34) with 13.000 volumes was still at the disposal of the young Otto. Nevertheless, we do not know exactly, whether he could save this most valuable collection given his financial problems as a student till his appointment as a teacher in 1907. E.g., we know that some 1000 volumes of this library were bought by the Marx-Engels Institute Moscow in 1926:

> After Wilhelm Neurath's death in 1901, the library was at his son's disposal and he had to sell parts of it in 1904 to finance his meagre existence as a doctoral student in Berlin. He still possessed some of his father's books during his years in the Netherlands and made inquiries about recovering them after the Second World War. But it seems that they did not survive the German occupation. (Burke 2010, xi.)

---

[12] The all too early death of Neurath prevented the completion of his visual autobiography, which was published posthumously only in 2010 with Hyphen Press, London. Before that, only some minor parts were published at different places. See Neurath (1945/1973) and (1946b).

After his return from Munich to Vienna Neurath had begun to establish his own library, additionally. He regularly bought books from a wide range of cultural, intellectual and philosophical areas, mostly in German and English. During his Vienna Circle years up to 1934 his own books and those on the international movement of Logical Empiricism and later of the *Encyclopedia* project complemented this intellectual collection. To date, we are not sure how many of the books from his Vienna time were transferred to The Hague after the Civil War in February 1934. Obviously, the later arrival of Olga and Marie enabled a restricted transfer, maybe facilitated by Otto's son Paul, who could stay in Vienna till his imprisonment and deportation to the concentration camp in 1938. Paul himself reports vaguely on his father that "he finally organized from Prague the transfer of his Institute and his own belongings and family to The Hague" (P. Neurath 1973, 31). His memories on the first apartment of the Neuraths in the Schlossgasse (fourth district Margareten) refer also to the library: "The place was enormous. It consisted essentially of five big rooms which my father transformed into six and a corridor, with the help of book shelves and partitions. He needed practically all of it, because of his large library, a good deal of which was inherited from his father" (P. Neurath 1973, 32). And he remembers that the "whole room looked like Faust's studio" (1973, 34).

It is evident from Marie's unpublished autobiography that Neurath, given his untiring nature, again continued to buy books mostly in the many antiquarian bookstores in the Netherlands. Given the adventurous flight from The Hague to England in 1940 all private belongings, the Isotype collection and the library had to be left. But his Dutch friends succeeded to store most parts of the valuable owning and Isotype collection even if other materials were confiscated by German troops ("Amt Rosenberg") and later on re-appeared in Moscow archives. Paul Neurath (1973, 41) remembers that "three pictures and one or two smaller pieces were bought by friends of my parents at The Hague when the contents of the house were sold at an auction by the Nazis." Luckily, Paul donated the old Chinese Gobelin to the Institute Vienna Circle. Parts of the Neurath estate came back from Moscow to the Archive of the Republic of Austria without books, while the Austrian National Library bought parts of the Marie Neurath archives. So, the puzzle of the fate of Neurath's *Nachlass* is being solved gradually.

Again luckily, what remained in The Hague after 1940 was sent by friends to the Neuraths after the war before Marie moved from Oxford to London in 1948 (Mulder 1985, 384). I had the pleasure to meet Marie (1981/1982) once in a nice typical British house in the beautiful neighborhood of Hampstead in midst of émigrés from Austria and Germany – where also Anna Freud lived. The continuation of the Isotype movement and the donation of the Isotype collection was arranged by Marie Neurath with Michael Twyman from the University of Reading, Department of Typography and Graphic Communication, where the whole Otto and Marie Isotype collection is still located and researched in called "Isotype revisited": https://isotyperevisited.org/.

From there the scientific books went back again to the Netherlands, this time my old friend, the Dutch scholar Henk Mulder in Amsterdam, who in parallel had contacted Marie and established the Vienna Circle Archives with the estates of Moritz

Schlick (from Barbara van de Velde-Schlick) and Otto Neurath. In Amsterdam Henk Mulder founded privately the Vienna Circle Archives (Wiener Kreis Archiv), legally owned by the Vienna Circle Foundation, which transferred the whole archives via the Dutch Academy of Sciences to the Rijksarchief Noord in Haarlem (NL). From here, the Institute Vienna Circle received a collection of ca. 1350 books in 1995, now being available to the scientific community as the Neurath exile library ("Otto Neurath Arbeitsbibliothek"). This was a late and symbolic coming home of one of the most brilliant and impressive intellectuals of the twentieth century, a real humanist and polymath in very hard times.

To reconstruct Neurath's English years, his personal library and books could be also useful. To see that, I will present three short case studies.

In the few years in England between 1940-45 good old contacts with a remarkable intellectual and practical manifestation were re-established – and one is inclined to ask in the sense of counterfactual history: What would have been, if Neurath had survived the World War Two period? First of all, in Oxford, he initiated Central European disputes on plan vs. market or socialism vs. liberalism (with F. A. Hayek) and philosophical relativism vs. absolutism (with Karl Popper) again in the new context of the envisioned liberated postwar society. Therefore, their relationships significantly emerged already in the Viennese years, can be described as more or less conflict-ridden communication between family resemblance and distance. Let me allude to Neurath's relation to Popper and Hayek before coming to his related library copies.

Besides sharing a rejection of Platonic social philosophy (*Republic*), seen as a legitimation for authoritarian and totalitarian ideas, including the *Führerkult* (which according to Hayek's Plato-interpretation was also part of a specific English controversy) there was the controversial encounter of both personalities that began in the early twenties. Neurath immediately criticized Popper after the publication of his *Logik der Forschung*, accusing him of being an advocate of an absolutist "pseudorationalism "(Neurath 1935/1983) – as he, incidentally, also rejected the verification endorsed by the "Wittgenstein camp." The option of "unity of science" or "unity of method" (which, by the way, was also refuted by Hayek) appeared as the main alternative. But there are also uncontested familiarities: it is just Popper's appeal to planning for institutions in his exile publications *The Open Society* (1945) and *The Poverty of Historicism* (1957; originally published in 1944/1945), which highlight the differences with Hayek – marginalized by Popper facing Hayek's total opposition towards any form of planning theory and practice.

Neurath bought the first edition of Popper's *The Open Society and its Enemies* in 1945 and wrote some annotations in his copy of the two volumes. Despite a lot of remarkable agreements, like on Plato, Hegel, Fichte, and Wittgenstein, he commented critically the Marx interpretation of the author with reference to determinism and predictions, regarding the sociology of knowledge: e.g., on Popper's conclusion "a social technology is needed whose results can be tested by social engineering" (p. 210), he notes: "*how making tests? One history!!*"

And Neurath agrees with a thick "*ON,*" where Popper writes in the chapter "The Revolt Against Reason": "The rational and imaginative analysis of the consequences of a moral theory has a certain analogy in scientific method. For in science, too, we do not accept an abstract theory because it is convincing itself; we rather decide to accept or reject it after we have investigated those concrete and practical consequences which can be more directly by experiment" (p. 220).

In the concluding chapter "Has History Any Meaning?" again Neurath confirms the statement "that all scientific descriptions of facts are highly selective, that they always depend on theories" (p. 247) and Popper's claim that "the history of power politics is nothing but the history of international crime and mass murder" (p. 257). But in contrary, he also sees a totalitarian ideal in Popper's reference to Christianity (p. 264) and misses arguments instead of names in Popper's criticism of conventionalists' view of experiments.

Regarding the term "social engineering" Neurath mentions his own concept of "Gesellschaftstechnik" not mentioned by Popper (1945, 185) and he continues with the question why one should not employ "utopian engineering" as a viable method of social technology (p. 242). And it is not surprising that Neurath at the end of volume 1 underlines Popper's footnote: "*Money* is one of the symbols as well as one of the difficulties of the open society. There is no doubt that we have not yet mastered the rational control of its use" (p. 266; original emphasis).

To sum up, Neurath's comments and criticisms seem rather mild (with the exception of utopianism and some Marx interpretation), given the preceding characterization of Popper as an oppositional philosopher of the Vienna Circle in the 1930s, which is acknowledged by the latter in his short memories of Neurath, where we can read:

> Neurath and I had disagreed deeply on many and important matters, historical, political, and philosophical; in fact on almost all matters which interested us both except one – the view that the theory of knowledge was important for an understanding of history and of political problems. Yet though we had disagreed so deeply about so many and so important matters, I shall always feel that he was one of the strongest personalities I ever met; a real original thinker, and an undaunted fighter who dreamt of a better and more humane world. (Popper 1973, 56).

Mainstream historiography obscures the difference of these "Ambivalent Brothers in Mind": first, Popper's insistence on the unity of method for natural and social sciences, second, his preference for a limited planning for institutions, third, his adherence to a socially oriented welfare economy in the tradition of Austrian social reform. Probably because of his personal indebtedness to, and acquaintance with Hayek, who essentially enabled Popper's position at LSE, Popper himself played down his differences with Hayek's social philosophy in New Zealand. This conclusion can be drawn on the basis of the published comments to Hayek's "Scientism." And although both directly/indirectly argue against Neurath under the (Cold War) labels of "objectivism," "collectivism" and "historicism," there can be no doubt about the shortcomings of the equation "Scientism = Historicism." In this field we lack further studies on the renewed interaction and relation of the Austrian School and Vienna Circle after their emigration, taking into account the controversial issues in "Red Vienna" between the wars.

The life and work of Otto Neurath and Marie Reidemeister (-Neurath) continued in their second exile systematically with personal and scientific relations during the 1930s. The memories of Marie illustrate their permanent efforts in the promotion of the Encyclopedic movement and adult education via the *Isotype*-movement, and also his initiatives continuing the housing and settlement movement in his Vienna days. Neurath renewed all contacts (with Friedrich Waismann, Rose Rand, Friedrich Hayek, and most importantly with Susan Stebbing) and for a short time pursued his academic ambitions with significant publications and research projects that remained uncompleted.[13] A closer look at his exile library signals the continuity to develop the interwar plans including his experience with the fascist decade: his book projects on "International Planning for Freedom," "Visual Education" and on "Persecution and Toleration" were again on the agenda. And he focused on Hitler-Germany education with reference to Plato and Kant – a topic which continued to be discussed in exile (see Sandner 2011 and Antonia Soulez's chapter in the present volume). We realize Neurath's cooperation in a newly founded Fabian Society, production of Isotype-films with Paul Rotha (see Boon 2016) for the anti-Nazi education and *Isotype* as a contribution to the fight against totalitarianism.

Karl Popper was aware of Neurath's life and work after World War One, which becomes obvious from his autobiographical remarks (1973). He remembered Neurath's involvement in the Bavarian revolution (1919/1920) in the connection with a planned economy based on full socialization and with reference to the semi-socialization program of Josef Popper-Lynkeus. Popper was inclined to sympathize with the latter's (utopian) project, who was a distant relative. Apart from differences in personality and mentality, on the one hand, the Marxist dissenter and politically oriented encyclopedist, and the critical rationalist philosopher on the other, Popper accused Neurath of having succumbed to utopianism, historicism and scientism as represented by the Vienna Circle and the *Ernst Mach Society*.

Popper was very flattered that Neurath published the criticism of his *Logik der Forschung* as "Pseudorationalism of Falsification" (1935/1983) and was not unpleased ("nicht unzufrieden") with this honorable attack. Surprisingly, he never replied in a systematic way. Maybe because Neurath (1935/1983, 131) criticized in light of his methodological holism: "the absolutism of falsification [...] is in many ways a counterpart against the *absolutism of verification* which Popper attacks" (original emphasis). Popper's attempt to characterize Neurath's *Empirische Soziologie* (1931) in the context of historical prophecy fails to assess the author's foundation of social science. From the outset Neurath remained very skeptical of explanations on the basis of *one* method and *one* image of science without pragmatically relativizing the field of "Prediction and Induction" (1946a/1983): "Unity of Science" as represented in the ambitious project of the *International Encyclopedia of Unified Science* or "Unity of Method" as explicated from Popper's *Logik der Forschung* to the *Open Society* and the *Poverty of Historicism* seemed to be an alternative approach in the history and philosophy of science.

---

[13] On Neurath's relation to Susan Stebbing, see Silke Körber's chapter in the present volume.

The direct confrontation of both opponents in their still unpublished correspondence shows a discussion in philosophy of science at a high level. (Stadler 2001/2015, 243–276). At the same time, one might wonder whether Popper didn't exaggerate the real differences between him and the so-called "positivists" – a designation which Neurath so strongly opposed as a cliché – and underestimate any form of scientific cooperation between the new "encyclopedists" (Cat 1995; Cartwright, Cat, Fleck and Uebel 1996). In this connection it is, indeed, surprising that also Critical Rationalism can be a suitable tool for a planning methodology (Faludi 1986). And it is this aspect – namely Popper's appeal to planning for freedom or institutions (in his *Open Society* and *Poverty of Historicism*), which highlights the differences which were overstated by Popper due to Hayek's total opposition towards each form of planning theory and practice.

The convinced "Scottish liberal" Hayek was in earlier times fond of Mach's epistemology and Schlick's *General Theory of Knowledge*, before he then began to oppose the Vienna Circle because he thought its (philosophy of) social science was dominated by Otto Neurath. Hayek felt himself from the beginning of the 1920s on to be Neurath's opponent regarding economics: inspired by the Carl Menger's "conception of the spontaneous generation of institutions," by Ludwig von Mises' *Gemeinwirtschaft* and by Popper's anti-inductivist *Logik der Forschung*, he lobbied against the so-called "positivist economics" – with Neurath as the most appropriate target. The intellectual divorce centered on planning theory, *in natura* calculation, and generally speaking on the concept of value, which Hayek – erroneously – missed in Neurath's social science.

All these indirect oppositions culminated in the short dispute in the 1940s in English exile (although Hayek had emigrated to London already in 1931). This conflict sheds more light on the alternative conceptions of social science and its methodology. Hayek's articles on "Scientism and the Study of Society" (1942–1944) in *Economica* and the subsequent publication of his *Road to Serfdom* (1944) was the starting point for a renewed *Methodenstreit*: two cultures (natural science vs. social sciences and the humanities) were the main options in the scientific enterprise. Neurath took the initiative, as of 1945, in communicating via correspondence: "Enclosing I am sending you a review of your book. I tried to discover, what we have in common – unfortunately you are rather 'absolute' in your EITHER-OR attitude. On Plato you may find some remarks in the article enclosed."[14]

Neurath (1945/2004) published in his last year a remarkably moderate review of *Road to Serfdom*, by showing that his Logical Empiricism is providing a "through and through" pluralist view towards "International Planning for Freedom" (Neurath 1942/1973). In the short book review in the *London Quarterly of World Affairs* we read:

> We agree with Hayek that there are people who like planning as a means of obtaining totalitarian leadership and there are others who unwittingly support fascism by promoting certain principles of planning. But we cannot go all the way with Hayek in his relegation of all planning to this category. (Neurath 1945/2004, 546.)

---

[14] Neurath to F. A. Hayek, 11 January 1945; Otto Neurath Nachlass.

And he concludes rhetorically:

> What would Professor Hayek answer if the tables were turned on him? Is it so unlikely that some people, seeing only Hayek's [...] alternative of totalitarianism with full employment to a free market with the usual booms and slumps, will choose the former with tears in their eyes? Is it so unlikely that people who think of planning for freedom will be in a better position, when stating their case against the painful market society of the past and against dictatorial planning, based on totalitarian fascism? (Neurath 1945/2004, 548.)

But Neurath's personal annotations in his own copy of *Road to Serfdom* are much more critical and bitter: "*His technique: Overstate a case, create car(r)ricature of it, then fight it and then kill it is either German or immoral etc.*" He also wrote in his personal copy the following:

> There is some danger that planning as a fashion/may/be used by totalitarian groups for weakening the democratic behavior, which implies – muddle. Democracy – muddle – and victory. But that is **not** the muddle of slums, distressed wars, depressions etc. but multiplicity of decisions, freedom of societies, local authorities. The fascists try to discredit muddle and to praise order, unification, subordination as such, otherwise they cannot run the show!!
>
> Therefore we need an analysis of planning with reckoning in kind **plus** muddle!
>
> *That lacks – therefore* ***danger***. (Original underlining.)

And he concludes at the last pages with the confession:

> ***Not*** *acceptable the hypothesis: that 'exploitation' as such creates business cycle (Marx). I can imagine a 'crisis-free' society with highest production and consumption* ***plus*** *'exploitation'....*
>
> *Reckoning in kind: a) democratic society Possible b) totalitarian society Possible.*

As I indicated, there emerged also a controversy over the concept of "Scientism" between Hayek and Neurath, which later included Popper. There, Hayek dealt extensively with the applicability of the methods of natural science and "social engineering" to the problems of man and society (as directed against Karl Mannheim, Neurath and maybe also in some sense against Popper).

In his "Scientism and the Study of Society" (1942–1944/1979) – re-published in *The Counter-revolution in Science* – Hayek condemned the appraisal of natural science methodology as the only "scientific method." According to Hayek, this does not hold, because "facts" in the natural and social sciences are totally different: on the one hand causally explicable, on the other, they are mere unobservable "opinions" of the actors producing their "objects." Common sense via analogy is the central key for understanding in social science. These essays distanced Hayek from all forms of "objectivism," "behaviorism," directly referring to Neurath's "physicalism," accusing him of supporting *in natura* calculation (instead of calculation in terms of price and value) and taking "naïvely for granted that what appears alike to us will also appear alike to other people" (Hayek 1942–1944/1979, 79). What surprises us, at first sight, is the lack of a critique of language and the merging of the theoretical and meta-theoretical levels of speaking about the external world. Despite of all these misunderstandings regarding "methods of science," Neurath seems ultimately willing to agree with Hayek's conclusion, who quoted Morris R. Cohen that "the great lesson of humility which science teaches us, that we can never be omnipotent or omniscient, is the same as that of all great religions: man is not and never will be the god before whom he must bow down" (Hayek 1942–1944/1979, 182).

Although Neurath tried to start a discussion in which he referred to theoretical contributions on social science, Hayek refused to enter into a detailed exchange: from Cambridge (where the London School of Economics had its wartime address), to Oxford, Hayek wrote that he is "by no means so much opposed to 'Logical Positivism' as you appear to think and with some members of your former group, particularly with Karl Popper, I find myself in complete agreement" and – alluding to physicalism and *in natura* calculation – he continued to articulate his skeptical position towards Neurath, at the same time agreeing "entirely with what you say on Plato. He certainly was the arch-totalitarian."[15] Hayek was busy with lecturing abroad and moved back to London only later. He was convinced that he had already dealt exhaustively with the issue of "Scientism." Two weeks after Hayek's last hesitant letter, Neurath died unexpectedly of a heart attack at December 22, 1945: the started dialogue between the adherents of plan and market shimmered through in the ensuing Hayek-Popper communication.

For a better understanding of the lasting *Methodenstreit* in the 1930s and 1940s represented by the triangle Hayek-Popper-Neurath we first have to reconstruct the discussions in their socio-historical context, second to examine unpublished sources, third to distance ourselves from clichés about schools of thought and, finally to confront these results with today's research. In doing so we could fully appreciate the historical background together with internal theory dynamics without producing myths of partisanship. This would provide a rational option, namely a pluralist way for positioning this unsolved debate in an evolutionary context of theoretical fields. And Neurath's library provides a valuable additional source for these Central European disputes in British exile.

Finally, a short note on Stefan Zweig (1881–1942), one of the most renowned authors and intellectuals of the twentieth century. He was a proponent of a typical Viennese and Central European culture covering biographical, fictional and historical fields up to his last remembrance book *Die Welt von Gestern. Erinnerungen eines Europäers* (Stockholm: Bermann-Fischer 1942)/*The World of Yesterday* (London: Cassell and New York: Viking Press 1943) – a sort of autobiography. This book is the story of the rise, glory, and fall of the European, mostly Jewish Culture in the period of Fascism and National Socialism. An impressive narrative of the destruction of human life and the decline of a civil society in a world of nationalism, racism and anti-Semitism which ended up in the Shoah. His last book was published before he consciously committed suicide with his wife Lotte in Brazilian exile (February 23, 1942) caused by his personal fate and the vanishing hope for a better world.

Neurath in Oxford was immediately puzzled by this book as a story on the world of his youth and his life in Vienna. He bought the English translation of this literary obituary on a lost culture and read it with increasing interest page by page with

[15] See F. A. Hayek to Neurath, 2 February 1945. (ONN).

continuous handwritten annotations from the beginning to the end of this most influential publication, to be translated into many languages. Neurath was fascinated but also provoked as a historian and sociologist by the content of that autobiographical document which was, and still is appreciated also a historical study of the late Habsburg Empire and the First Austrian Republic till the rise Austro-Fascism and NS. Given his intellectual background, this was not really surprising because Neurath had also written a critical monograph against Oswald Spengler's most influential book *Decline of the West* already in 1921, entitled *Anti-Spengler*, mainly directed to Spengler's cultural pessimism. I will call attention only to some selected pages as typical manifestations of the indirect review of this bestseller of Zweig, whom he most likely and surprisingly never had met in person: Zweig's *The World of Yesterday* (1943) is a memory book (*Erinnerungen eines Europäers*) on the decline of the epoch of *fin de siècle* Vienna from Habsburg to Hitler. Till to the present it is praised as a literary document of a vanished epoch and as a historical study of this epoch. Already on the first page Neurath notes:

> *Zweig usually speaks of writers, etc, etc no knowledge of social correlations; no reading of sociology. A human and kind pacifist, who, as a well-to-do, has no contacts with the masses knows writers, composers etc.*
>
> *Unpleasant neighborhood Emil Ludwig, St. George, Elis. Förster Nietzsche, Rich. Strauß, Werfel, G. Hauptmann, Haushofer. Heine not mentioned! Not mentioned: Lagarde, Gobineau, Rembrandt als Erzieher Chamberlain but Spengler*
>
> *Romain Rolland, Shaw, Wells, Duhamel the freedom-West.*

And he continues on the back side harshly: *"Always wrong in looking at men and world. P.60, 146, 148, 154, 156, 112, 214, 234, 313."*

In the Preface we read: *"what kind of 'flower' 1840, 1850, 1860? Persecution!"* And still in the last page: *"always are purgatories and hells, but one does not think of them, Albigenses, Torquemada, Bartholomew, Armenia, China, India, Lynching, unemployment."*

Here it is worth mentioning that Neurath also purchased another book of Zweig, namely *The Right to Heresy. Castellio against Calvin* (1936), which was also a biographical story on 2 opponents, written as a coded fiction and analogy to Hitler's rise and tyranny in Nazi-Germany.

Again, Neurath is commenting it from the historical and sociological point of view, e.g.: *"what about the persecuted Albigenses etc.?"* (p. 22.) or references to witch burning. But there is also the confirmation of Max Weber's thesis on the Protestant origin of capitalism.

The background of his interest was certainly his own projected monograph entitled "Tolerance and Persecution," which he could not finish. But it seems, that – despite the many differences regarding the historiography of the last century, he was the complementing intellectual twin of Zweig. Both were right, the former as an author with his semi-fiction, the latter as a scholar and social historian.

At the end, I will note the fact that Neurath also read Bertrand Russell's books from the beginning of the Vienna Circle days and he met him in connection with the unity of science congresses. As a social Epicurean Neurath was interested in

Russell's *Mysticism and Logic* (1910/1932) and *Conquest of Happiness* (1930). Herein we find a paragraph in the chapter on competition as follows.

> I think it should be admitted that an element of genuine though irrational fear as to the consequences of ruin frequently enters into a business man's anxieties. Arnold Bennett's Clayhanger, however rich he became, continued to be afraid of dying in the workhouse. I have no doubt that those who have suffered greatly through poverty in their childhood are haunted by terrors lest their children should suffer similarly, and feel that it is hardly possible to build up enough millions as a bulwark against this disaster. (Russell 1930, 49–50.)

Here Neurath wrote next to this paragraph simply "St. Zweig"!

References

Boon, Timothy. 2016. 'To Formulate a Plan for Better Living': Visual Communication and Scientific Planning in Paul Rotha's Documentary Films, 1935–45. In *Pursuing the Unity of Science: Ideology and Scientific Practice from the Great War to the Cold War*, ed. Harmke Kamminga and Geert Somsen, 156–181. London/New York: Routledge.

Burke, Christopher. 2010. Introduction. In Neurath (2010): vii–xviii.

Cartwright, Nancy, Jordi Cat, Lola Fleck, and Thomas Uebel. 1996. *Otto Neurath: Philosophy Between Science and Politics*. Cambridge: Cambridge University Press.

Cat, Jordi. 1995. The Popper-Neurath Debate and Neurath's Attack on Scientific Method. *Studies in History and Philosophy of Science Part A* 26 (2): 219–250.

Dahms, Hans-Joachim. 1996. Vienna Circle and French Enlightenment. A Comparison of Diderots Encylopédia with Neuraths International Encyclopedia of Unified Science. In *Encyclopedia and Utopia. The Life and Work of Otto Neurath (1882–1945)*, ed. Elisabeth Nemeth and Friedrich Stadler, 53–61. Dordrecht: Kluwer.

———. 2004. Neue Sachlichkeit in the Architecture and Philosophy of the 1920s. In *Carnap Brought Home. The View from Jena*, ed. Steve Awodey and Carsten Klein, 357–375. Chicago: Open Court.

Dogromaci, Burcu, and Günther Sandner, eds. 2017. *Rosa und Anna Schapire. Sozialwissenschaft, Kunstgeschichte und Feminismus um 1900*. Berlin: AvivA.

Faludi, Andreas. 1986. *Critical Rationalism and Planning Methodology*. London: Pion.

Hayek, F.A. 1942–1944/1979. Scientism and the Study of Society. In *The Counter-Revolution of Science: Studies on the Abuse of Reason*. Second ed, 17–182. Indianapolis: Liberty Press.

———. 1944. *Road to Serfdom*. London: Routledge.

Horkheimer, Max. 1937/1972. The Latest Attack on Metaphysics. In *Critical Theory. Selected Essays*, ed. M. O'Connell, 132–187. New York: Continuum.

Johnston, William M. 1972. *The Austrian Mind. An Intellectual and Social History, 1848–1938*. Berkeley: University of California Press.

Mulder, Henk. 1985. The 'Vienna Circle Archive' and the Literary Remains of Moritz Schlick and Otto Neurath. *Synthese* 64 (3): 375–387.

Neurath, Otto. 1919/1981. Utopien. In *Gesammelte philosophische and metodologische Schriften*, ed. Rudolf Haller and Heiner Rutte, 137–139. Wien: Hölder-Pichler-Tempsky.

———. 1921/1973. Anti-Spengler. In *Empiricism and Sociology*, ed. Marie Neurath and Robert S. Cohen, 158–213. Dordrecht: D. Reidel.

———. 1933. *Bildstatistik nach Wiener Methode in der Schule*. Wien und Leipzig: Deutsche Verlag für Jugend und Volk.

———. 1933/1973. Museums of the Future. In *Empiricism and Sociology*, ed. Marie Neurath and Robert S. Cohen, 218–223. Dordrecht: D. Reidel.

———. 1935/1983. Pseudorationalism of Falsification. In *Otto Neurath: Philosophical Papers 1913–1946*, ed. Robert S. Cohen and Marie Neurath, 132–138. Dordrecht: D. Reidel.
———. 1936. *International Picture Language*. London: Kegan Paul, Trench, Trubner & Co..
———. 1939. *Modern Man in the Making*. New York: Alfred A. Knopf.
———. 1942/1973. International Planning for Freedom. In *Empiricism and Sociology*, ed. Marie Neurath and Robert S. Cohen, 422–440. Dordrecht: D. Reidel.
———. 1944a. *Foundations of the Social Sciences*. Chicago: University of Chicago Press.
———. 1944b. Visual Aids in Adult Education. *Highway* 35: 88–90.
———. 1945/1973. Otto Neurath's Childhood from Autobiographical Notes. In *Empiricism and Sociology*, ed. Marie Neurath and Robert S. Cohen, 4–6. Dordrecht: D. Reidel.
———. 1945/2004. Alternatives to Market Competition. In *Economic Writings. Selections 1904–1945*, ed. Thomas E. Uebel and Robert S. Cohen, 546–548. Dordrecht: D. Reidel.
———. 1946a/1983. Prediction and Induction. In *Otto Neurath: Philosophical Papers 1913–1946*, ed. Marie Neurath and Robert S. Cohen, 243–246. Dordrecht: D. Reidel.
———. 1946b/1983. The Orchestration of the Sciences by the Encyclopedism of Logical Empiricism. In *Otto Neurath: Philosophical Papers 1913–1946*, ed. Marie Neurath and Robert S. Cohen, 230–242. Dordrecht: D. Reidel.
———. 1946a. For the Discussion: Just Annotations, Not a Reply. *Philosophy and Phenomenological Research* 6 (4): 526–528.
———. 1946b. From Hieroglyphics to Isotype. Extracts from the Manuscript; With an Introduction by Paul Rotha. *Future Books* 3: 92–100.
———. 1996. Visual Education: Humanisation versus Popularisation. In *Encyclopedia and Utopia. The Life and Work of Otto Neurath (1882–1945)*, ed. Elisabeth Nemeth and Friedrich Stadler, 245–335. Dordrecht: Kluwer.
———. 2010. *From Hieroglyphics to Isotype: A Visual Autobiography*, ed. Matthew Eve and Christopher Burke. London: Hyphen Press.
———. 2011. Unity of Science and Logical Empiricism: A Reply. In *Otto Neurath and the Unity of Science*, ed. John Symons, Olga Pombo, and Juan Manuel Torres, 15–30. Dordrecht: Springer.
Neurath, Paul. 1973. From Otto Neurath's Son, the Sociologist Paul Neurath. In *Empiricism and Sociology*, ed. Marie Neurath and Robert S. Cohen, 29–41. Dordrecht: D. Reidel.
Neurath, Marie, and Robin Kinross. 2009. *The Transformer: Principles of Making Isotype Charts*. London: Hyphen Press.
Nikolow, Sybilla. 2004. Planning, Democratization and Popularization with Isotype ca.1945: A Study of Otto Neurath's Pictorial Statistics with the Example of Bilston, England. In *Induction and Deduction in the Sciences*, ed. Friedrich Stadler, 299–329. Dordrecht: Kluwer.
Popper, Karl. 1945. *The Open Society and its Enemies*. London: Routledge.
———. 1957. *The Poverty of Historicism*. London: Routledge.
———. 1973. On Otto Neurath. In *Empiricism and Sociology*, ed. Marie Neurath and Robert S. Cohen, 51–56. Dordrecht: D. Reidel.
Russell, Bertrand. 1910/1932. *Mysticism and Logic and Other Essays*. London: George Allen & Unwin Ltd..
———. 1930. *Conquest of Happiness*. London: George Allen & Unwin Ltd.
Sandner, Günther. 2011. The German Climate and Its Opposite: Otto Neurath in England, 1940–45. In *Political Exile and Exile Politics in Britain after 1933*, ed. A. Grenville and A. Reiter, 67–85. Amsterdam/New York: Rodopi.
———. forthcoming. A Political Recess: Otto Neurath in the Netherlands (1934–40). In *Significs and the Vienna Circle: Intersections*, ed. Gerard Alberts, Luc Bergmans, and Fred Muller. Dordrecht: Springer.
Stadler, Friedrich. 2001/2015. *The Vienna Circle. Studies in the Origins, Development, and Influence of Logical Empiricism*. 2nd ed. Dordrecht: Springer.
Stadler, Friedrich, and Thomas Uebel, eds. 2012. *Wissenschaftliche Weltauffassung. Der Wiener Kreis. Hrsg. vom Verein Ernst Mach (1929)*. Vienna/New York: Springer.

Uebel, Thomas. 1995. Otto Neurath's Idealist Inheritence: The Social and Economic Thought of Wilhelm Neurath. *Synthese* 103 (1): 87–121.

———. 2007. *Empiricism at the Crossroads. The Vienna Circle's Protocol-Sentence Debate*. Chicago: Open Court.

Zweig, Stefan. 1936. *The Right to Heresy. Castellio against Calvin*. New York: The Viking Press.

———. 1943. *The World of Yesterday*. London/New York: Cassell/Viking Press.

Chapter 3
Otto Neurath: The Philosopher in the Cave

Don Howard

Abstract The question of philosophy's relevance to extra-academic concerns is much with us today. Plato tells us that, once the philosopher has seen the truth in the full light of the sun, she must return to the cave, there to put knowledge to work in making a better world, even though, being temporarily unaccustomed to the dark, she risks ridicule from those still in thrall to illusion. This paper reflects upon the life and career of Otto Neurath as a modern exemplification of this ideal of philosophical engagement. In spite of, or, perhaps, because of his never having held an academic appointment, Neurath made a difference for the good in human affairs. The key components of what I term Neurath's "philosophy of science in action" are explicated in order to understand how that could be. Foregrounded are Neurath's socialism, his own version of the thesis of the empirical underdetermination of theory by evidence, his anti-metaphysical stance, and his commitment to physicalism and the unity of science. The paper concludes with a discussion of the contemporary relevance of Neurath's model of engaged philosophy of science.

3.1 Introduction: The Philosopher's Return to the Cave

Plato's allegory of the cave, presented at the beginning of Book VII in *The Republic*, is widely remembered as an evocative story of enlightenment. A person long held prisoner in a deep, dark cave, bound to see only shadows of silhouettes cast upon the wall of the cave, and knowing nothing more, takes those shadows to be reality. Then, one day, she frees herself from her chains, turns, is temporarily blinded by the light of the fire producing the shadowy images, regains her vision and has the illusion revealed to her. She struggles in the semi-darkness to climb out of the cave, only to be blinded again by the light of the sun. When, finally, her eyes adjust to the daylight, ultimate truth and reality stand before her in plain view. This is the

D. Howard (✉)
Department of Philosophy, University of Notre Dame, Notre Dame, IN, USA
e-mail: dhoward1@nd.edu

J. Cat, A. T. Tuboly (eds.), *Neurath Reconsidered*,
Boston Studies in the Philosophy and History of Science 336,
https://doi.org/10.1007/978-3-030-02128-3_3

aspiring philosopher's path to the highest kind of knowledge, which, in Plato's version, is found in the heaven of the forms.

But the allegory does not end there, with the philosopher in happy, enlightened repose. Less well remembered is that, after seeing the truth in the full light of day, Plato's true philosopher, the philosopher-king, descends back into the cave to enlighten the lives of the other prisoners with some small part of the truth that she has won. This turns out to be a difficult task.

> And when he remembered his old habitation, and the wisdom of the den and his fellow-prisoners, do you not suppose that he would felicitate himself on the change, and pity them?
>
> Certainly, he would.
>
> And if they were in the habit of conferring honours among themselves on those who were quickest to observe the passing shadows and to remark which of them went before, and which followed after, and which were together; and who were therefore best able to draw conclusions as to the future, do you think that he would care for such honours and glories, or envy the possessors of them? Would he not say with Homer,
>
> 'Better to be the poor servant of a poor master,'
>
> and to endure anything, rather than think as they do and live after their manner?
>
> Yes, he said, I think that he would rather suffer anything than entertain these false notions and live in this miserable manner.
>
> Imagine once more, I said, such an one coming suddenly out of the sun to be replaced in his old situation; would he not be certain to have his eyes full of darkness?
>
> To be sure, he said.
>
> And if there were a contest, and he had to compete in measuring the shadows with the prisoners who had never moved out of the den, while his sight was still weak, and before his eyes had become steady (and the time which they would be needed to acquire this new habit of sight might be very considerable), would he not be ridiculous? Men would say of him that up he went and down he came without his eyes; and that it was better not even to think of ascending; and if any one tried to loose another and lead him up to the light, let them only catch the offender, and they would put him to death.
>
> No question, he said. (Plato 1888, 216–217.)

Plato's point is that the true philosopher recognizes an obligation to go back into the cave, however much she risks ridicule or even death. This is the role of the philosopher-king in the ideal state. One seeks the truth not merely for its own sake, but also for the purpose of an enlightened leadership of the *polis*. Would that the philosopher's obligation to immerse herself in public affairs were more widely recognized and rewarded.

Otto Neurath was hardly a philosopher-king. He deliberately avoided political office and conventional party politics. His life and career nonetheless stand as models of socially and politically engaged philosophical practice. Moreover, unlike most of the professional philosophers who have held public office, figures like Sarvepalli Radhakrishnan, Tomáš Masaryk, José Ortega y Gasset, or Julián Besteiro,[1] Neurath's philosophy made specific and substantial contributions to his

[1] Radakrishnan served as Vice President and President of India after independence, 1952–1967. Masaryk was the first President of Czechoslovakia after World War One, 1918–1937. Besteiro was twice member of the Spanish Congress of Deputies, 1918–1923 and 1931–1937, and he served as President of the Congress from 1931 to 1933. Ortega was, likewise, a member of the Congress of Deputies, 1931–1933.

political work and was, in turn, constructed in part with an eye toward understanding the place of politics in philosophy and science. The only other example known to me of a prominent philosopher whose philosophical and political work is so deeply entangled with one another is Neurath's contemporary, György Lukács.[2]

How are we to understand Neurath's compelling example of socially and politically engaged philosophy? This paper reflects upon Neurath's life and career as a modern exemplification of this ideal of philosophical engagement. In spite of, or, perhaps, because of his never having held an academic appointment, Neurath made a difference for the good in human affairs. How did these contextual circumstances shape the content of his philosophical project? Did it make a difference to his philosophy of science that his day job in the 1920s was that of the director of the Social and Economic Museum of Vienna, where he developed the "Isotype" system for the visual communication of information? And how did that philosophical program inform his social and political projects, from worker education to urban design? How did his broadly Duhemian, holist, underdeterminationist epistemology of science shape his understanding of how science, especially social science, affects social reform? And does Neurath's example have continuing relevance to contemporary discussions of possible roles for the socially engaged philosopher of science?[3]

3.2 Neurath in the Agora

Neurath's entire life was devoted to public service. Early in World War One, he directed the Department of War Economy for the Austro-Hungarian government, after which he was appointed director of the German Museum of War Economy in Leipzig. In March of 1919, he was appointed as president of the Central Economic Administration of the short-lived Bavarian Soviet Republic, which was overthrown in May, after which Neurath was tried, convicted, and sentenced to prison for the crime of assisting high treason. He was released from prison thanks to the intercession of the Austrian Foreign Minister. Exiled from Germany, Neurath returned to "Red" Vienna, where, under the socialist municipal government that ruled for more than a decade, he occupied several prominent positions, starting with his serving as secretary of the Austrian Association for Settlements and Small Gardens, which played a role in developing new kinds of housing for the poor and the working class in Vienna.[4]

[2] Lukács was the Minister of Culture for the post-World War One Hungarian Soviet Republic, from March until August 1919.

[3] For the record, Neurath was not, himself, sympathetic with Plato's conception of the ideal state in *The Republic*. Writing toward the end of World War Two, in February 1945, Neurath and Joseph Lauwerys (1945) emphasize the extent to which the totalitarian aspects of Plato's state resemble Nazi Germany (Neurath and Lauwerys 1945). In a manner strikingly similar to Popper's (1945) critique in his *The Open Society and Its Enemies*, Neurath and Lauwerys deplore Plato's subordination of the individual to the collective.

[4] For Neurath's involvement and work in the Association see Sophie Hochhäusl's chapter in the present volume.

In 1923, Neurath established the Social and Economic Museum of Vienna, which he directed until his forced exile from Austria in 1934. This was his most significant exercise in public engagement and the one with, perhaps, the most enduring legacy. The museum functioned mainly as an institution for public education, intended to serve primarily, again, the poor and working class. It was in this setting that Neurath developed his later, world famous "Isotype" system for the visual communication of social and economic information.[5] In 1928, Neurath and Moritz Schlick started the Ernst Mach Society in Vienna. Designed to disseminate new scientific ideas and the logical empiricist philosophy of science of the Vienna Circle to a broader public, including the working classes, the society also served as a forum for socialist intellectuals.

When Austrofascist dictator, Engelbert Dollfuss, began his violent suppression of the socialists in February 1934, the Ernst Mach Society was forced to close, and Neurath, who, by a happy accident was in Moscow, fled to the Netherlands, where he had already established a branch of the Social and Economic Museum, known as the Mundaneum Institute. Forced to flee again when the Germans invaded The Netherlands in May of 1940, Neurath sought refuge in England, where he founded the Isotype Institute in Oxford and did consulting work on a number of public projects. He died in December of 1945.[6]

Neurath's commitment to public engagement is seen in his writings, as well. Much of his work was, of course, directed to an audience of professional philosophers, economists, social theorists, historians, and educators. But some of his most impactful books were written for a public audience, such as his 1921 critique of the then-popular "decline of the West" analysis of Europe's post-war malaise, *Anti-Spengler* (Neurath 1921/1973) or his 1939 book *Modern Man in the Making* (Neurath 1939). Prominent among his hundreds of articles are the many essays that he wrote for political journals, especially *Der Kampf*, which was the official journal of the Austrian Social-Democratic party. Some of his most interesting work directly explored the connections between politics and science, his 1928 *Lebensgestaltung und Klassenkampf* [*Personal Life and Class Struggle*] (Neurath 1928/1973), being among the best.[7]

Lastly, Neurath was constantly in motion and on the move, lecturing, consulting, organizing, and collaborating. He enjoyed and excelled at building new institutions, establishing new journals and book series, and bringing people together, as with his crucial role in establishing and maintaining the Vienna Circle, itself. He seemed to be possessed of inexhaustible reserves of energy, enthusiasm, and force of will. All

---

[5] For more on "Isotype," see Neurath (2010) and Vossoughian (2008). See Angélique Groß's and Sophie Hochhäusl's chapter in the present volume as well.

[6] The definitive biography of Neurath is Sandner (2014). Cartwright, Cat, Fleck, Uebel (1996) is another excellent source. Very interesting are the biographical memoirs collected in Neurath and Cohen (1973). Stadler (2001/2015) is also helpful. On Neurath's rather neglected English-period see the chapters of Michelle Henning, Antonia Soulez and Adam Tamas Tuboly in the present volume.

[7] Neurath's early political writings are analyzed by Günther Sandner in the present volume.

of these attributes were on display in his leading the effort to launch yet another of his enduring legacies, *The International Encyclopedia of Unified Science* (Neurath, Morris, and Carnap 1936–1969 – which, while never completed as planned, had a significant impact on the philosophy of science and other disciplines, and this not merely because it was the place where Thomas Kuhn's *Structure of Scientific Revolutions* was first published (1962).[8]

3.3 Science in Action: Neurath as a Theorist of Socially and Politically Engaged Science and Philosophy

Neurath's career as a public intellectual, deeply and consistently engaging large and important questions of social and economic justice, war and peace, the rise of Fascism and the challenges thereby posed to democracy and freedom, was grounded in a theoretical analysis of the place of science and philosophy in society and politics. That theoretical understanding was shaped by Neurath's experiences as an activist intellectual, and his activism was shaped, in turn, by the way he theorized science as a means to the betterment of humankind.

Many thinkers, from Francis Bacon to the leading minds of the French Enlightenment, from Auguste Comte to American progressives of the late-nineteenth and early twentieth centuries have looked to science as a force for good in human affairs. But Neurath was not at all an Enlightenment intellectual. Both his philosophy of science and his socialist politics left him deeply skeptical of naive notions of scientific truth and progress, especially for their neglect of the ineluctable social, economic, political, and historical embedding of science. But he was, at the same time, a critic of the "scientific materialism," the Marxist materialism of Friedrich Engels and Lenin, that denied a leading role for ideas in social change. Neurath's social-political philosophy of science is something *sui generis* and stands to this day as an inspiring alternative to neo-liberal views of science in the political arena, such as those of Philip Kitcher (2001, 2011), and to Marxist-inspired social studies of science deriving from the so-called "Strong Programme" of David Bloor and Barry Barnes (Bloor 1976; Barnes et al. 1996). What were the key features of Neurath's distinctive theory of science, and the philosophy of science, in action?[9]

[8] On the history of the *International Encyclopedia of Unified Science*, see Morris (1962), Nemeth and Stadler (1996), and Reisch (1994).

[9] The unflattering contrast with Latour (1987) is intentional.

3.3.1 Socialism

Neurath was a socialist, though not a Marxist of the Leninist variety. He was a product of the distinctive political-intellectual tradition of "Austro-Marxism," which was strongly associated with the Austrian Social-Democratic Party, the co-founder of which, Victor Adler, was one of the leading theorists of Austro-Marxism, a legacy carried on by his son, Friedrich Adler. Among the distiguishing features of Austro-Marxism is its critique of the naive notions of science and materialism that defined the scientific materialism of Engels and Lenin. Austro-Marxism came in two flavors, a Kantian one and a Machian one (see M. Adler 1925 and F. Adler 1907, 1918), but both versions of Austro-Marxism agreed that socialist theory required a more sophisticated philosophy of science than Engels had provided, and they agreed that socialist theory had to allow for ideas or theory to play a leading role in social change, as opposed to mere material circumstances, as had been argued by orthodox Marxist-Leninists. According ideas or theory, including scientific theory, a leading role in social change was necessary in order to legitimate social-democratic parties that sought change via electoral politics and parliamentary debate, by contrast with revolutionary Marxist-Leninists who sought change at the barrel of a gun. It is for this reason that Lenin (1909/1947) viciously attacked the Austro-Marxists, especially those of the Machian variety, in his book, *Materialism and Empirio-Criticism*, singling out by name Neurath's friend and fellow member of the left-wing of the Vienna Circle, the physicist, Philipp Frank.[10]

While a product of the Austro-Marxist milieu in Vienna, Neurath came to his own commitment to socialism slowly and by a somewhat surprising route. It was really his work in war planning during World War One that led him to prize the advantages of a planned economy, not only for purposes of war but also for achieving social justice. And he put his training in economics to work in pursuing that goal in various settings in the 1920s and beyond. He was, of course, interested in experiments in economic planning in the Soviet Union, but was not a communist, remaining a Marxist social democrat to the end of his life.[11]

For our purposes, it is especially the distinctive Austro-Marxist theme of giving scientific theory – economic theory, social theory, psychological theory, even theory in the natural sciences – a leading role in social change that is important for understanding Neurath's conception of science in action. Of course, even the scientific materialists prized the role of science in social change, but Neurath, like his fellow Austro-Marxists, theorized that role for science very differently than did Lenin. For Neurath, science is not simply driven by material circumstance and the class struggle. No, scientific research and scientific theory play their own, independent role. Consider this remark from Neurath's *Lebensgestaltung und Klassenkampf* [*Personal Life and Class Struggle*]:

[10] Blum (1985) is a good source for more on Austro-Marxism.

[11] On Neurath's economics see Thomas Uebel's chapter in the present volume.

> Nothing would be further amiss than to think that a Marxist-minded representative of the proletarian class struggle would respect only such scientific work which relates directly to the strategy of the class struggle. It is precisely Marxism that uncovers indirect relations and detours, and thus might ascertain that cultivating pure logic and the most general problems of mathematics and physics is especially favorable to revolutionary thinking. The Marxist will tend to regard it not as a mere accident that among the representatives of just these abstract disciplines ordinarily thought to be impractical, there are so many socialists as well as bourgeois in opposition, as for instance, the English logician and mathematician Bertrand Russell or the German physicist Albert Einstein. A cultivation of this kind of scientific thought seems almost a form of dissolution of metaphysical and half-theological thought, which under many disguises and masks is more alive today among the bourgeoisie than two generations ago. This is quite understandable, for bourgeois groups are closing ranks against the proletariat which has no traditions, and they must make their peace with the powers of yesterday, above all with clerical groups. [...] The cultivation of scientific, unmetaphysical thought, its application above all to social occurrences, is quite Marxist. (Neurath 1928/1973, 295.)

While according science a leading role in social change, Neurath also appreciated the social, political, and historical embedding of science. Bourgeois science was, for Neurath, importantly different from science in the service of the working classes. But even bourgeois science could, under some circumstances, advance the cause of progressive social change.

> If then Marxists wish to maintain that Marxism is more scientific than bourgeois science, they might present this historically as follows: the better the proletariat grasps the social engineering relations of our order and surveys its own chances, the more successfully it can fight. One might think that the same is true for the bourgeois front. If it is correct that the changes of history will bring the decline of today's order and the ruling classes, then the doctrine which maintains this truth is indeed suitable to provide the winners with courage and adherents, as long as they can grasp that they belong to the rising class; but it will weaken the front which contains groups that abandon this front because of the doctrine of the class struggle and its effects. (Neurath 1928/1973, 296.)

Among the reasons why, according to Neurath, Marxism is more scientific than bourgeois science is what, today, we would term a version of standpoint theory:

> The workers who lack a rich bourgeois education, can become superior to the bourgeois precisely in the field of social life in that they have a greater understanding for social connections and can apply even a smaller amount of knowledge more significantly. Marxism shows the proletarians who are engaged in the class struggle what is especially important to know; and it preserves adherents from the often disorganized educational endeavour of bourgeois enlightenment, which from the outset sees in merely increasing knowledge something worth striving for as such. (Neurath 1928/1973, 292–293.)

It is precisely the oppressed status of the working classes that affords them a privileged epistemic status, more clearly grasping social relations and seeing the lie in rationalizations of bourgeois privilege, rationalizations the falsity of which bourgeois thinkers cannot see as lies because their class status places them in an epistemically disadvantaged state. They cannot see through those lies because their doing so would undermine the power and prerogatives of their own class.

Neurath's philosophy of science in action thus paints a picture of politically engaged, indeed revolutionary science in service to the achievement of justice. But how, exactly, does Neurath think that science can function in such a setting? If we put science in service to a political agenda, do we not risk fatal compromise to the objectivity and epistemic integrity of science?

3.3.2 Underdetermination and Pseudorationalism

Central to a more detailed understanding of Neurath's philosophy of science in action is his version of the theory holism, semantic holism, and the associated doctrine of the empirical underdetermination of theory choice that is better known in the form that it was given by Pierre Duhem (1906/1954). Neurath (1913/1983) first elaborated these ideas in a remarkable 1913 essay, "Die Verirrten des Cartesius und das Auxiliarmotiv" ["The Lost Wanderers of Descartes and the Auxiliary Motive"].[12]

Neurath begins by recalling a passage from Descartes' *Discourse on Method* where, in order to contrast theoretical and practical reason, Descartes considers the example of people lost in a dense wood. No evidence or reasons point to a way out. Instead, they have to make an empirically and theoretically ungrounded choice of a direction, any direction, and then keep moving in that direction with firm resolve. Otherwise they could wander about forever and die. The demands of life, in other words, require action. For Descartes (1637/1903, 26), this is the essence of practical reason. But for Neurath, theory and praxis are the same in this respect. It is the universal lot of reason in all of its forms, including in the sciences, that logic and evidence do not suffice for theory choice. This is because, as thinkers and researchers, we do not and cannot occupy a view from nowhere. We are embedded in history, culture, and tradition:

> There are fundamental objections to the Cartesian view. Whoever wants to create a world-view or a scientific system must operate with doubtful premises. Each attempt to create a world-picture by starting from a *tabula rasa* and making a series of statements which are recognized as definitively true, is necessarily full of trickeries. The phenomena that we encounter are so much interconnected that they cannot be described by a one-dimensional chain of statements. The correctness of each statement is related to that of all the others. It is absolutely impossible to formulate a single statement about the world without making tacit use at the same time of countless others. Also we cannot express any statement without applying all of our preceding concept formation. On the one hand, we must state the connection of each statement dealing with the world with all the other statements that deal with it, and on the other hand we must state the connection of each train of thought with all our earlier trains of thought. We can vary the world of concepts present in us, but we cannot discard it. Each attempt to renew it from the bottom up is by its very nature a child of the concepts at hand. (Neurath 1913/1983, 3.)

It follows that even theoretical inquiry can go in various directions consistent with available facts:

[12] Neurath's paper is analyzed also in Gábor Zemplén's chapter with regard Neurath's ideas on optics.

> In order to make progress one very often finds oneself in the position of having to choose one of several hypotheses of equal probability. The necessity of provisional rules in the field of thinking is usually less clearly understood; this may be related to the fact that one can, so to speak, lead several theoretical lives simultaneously. [...] Starting from the same initial point one can always develop different theories of light, just as one can undertake different excursions. But one should not overlook the fact that it is certainly of consequence which trains of thought one has once had before a certain investigation. (Neurath 1913/1983, 3.)

Choose we must among the alternatives available to us. But how do we choose? Sometimes, especially in exigent circumstances, we make a random choice. More commonly, however, we choose on the basis of what Neurath terms "auxiliary motives." Sometimes it is custom or tradition, sometimes instinct, sometimes superstition, sometimes the authority of an esteemed figure, sometimes it is the view of the majority. Especially in the sciences, it is sometimes simplicity that decides. And, of course, it is frequently a political agenda that fills the void left by logic and evidence.

It is noteworthy that Neurath terms these factors auxiliary *motives*, not reasons. He means deliberately to make this an issue about the psychology of judgment and not pure reason alone. Neurath's epistemology of science is a kind of naturalistic epistemology. What he gives us here are supposed to be psychological and, thus, scientific facts about how reason operates, not apriori norms. Still, our recognizing the role of auxiliary motives has normative implications because of the widespread failure to discern or admit the work that such motives do.

Neurath invents the term, "pseudorationalism," to deride the view that unaided reason can lead us to the truth. The pseudorationalists are those who disguise and deny, consciously or unconsciously, the role of auxiliary motives. He would later level this charge not only against demagogues and apologists for greed and class interest but also against philosophers of science who promoted a naively decontextualized view of the logic of science, as with his devastating review of Karl Popper's *Logic of Scientific Discovery* (Popper 1934/1959; Neurath 1935/1983).

All of these themes – empirical underdetermination of theory choice, auxiliary motives, and pseudorationalism – were taken up again by Neurath in his 1921 book, *Anti-Spengler*:

> The wish to found action on perfect insight means to nip it in the bud. Politics are actions, always built on an inadequate survey. But a world-view, too, is action; embracing the manifold universe is an anticipation of unpredictable efforts. In the end, all our thinking depends on such inadequacies. We must advance, even without certainty. The only question is whether we are aware of it or not.
>
> Our pseudo-rationalists dare not face this fact. Frivolity! they cry when it is found that even with the most developed insight more than one way remains open for important decisions and that casting lots can thus become meaningful. They will not admit, precisely when some great task is to be undertaken, that insight becomes awareness of its own limits. (Neurath 1921/1973, 158–159.)

It is intellectual humility and openness about underdetermination and the ineluctable role of auxiliary motives that most advantages reason in action. In "The Lost Wanderers of Descartes," Neurath (1913/1983, 11) described the auxiliary motive as "the culmination of rationalism." There is sweet irony in the idea that the cause of scientific objectivity is furthered by the frank admission of our biases and agendas. At the very least, when we are honest with one another about the work of the auxiliary motives, we can subject them to public, critical scrutiny.

In *Anti-Spengler* Neurath also deploys his famous "boat" metaphor, which nicely embodies all of the core ideas of his philosophy of science in action, and he here acknowledges Duhem as a progenitor:

> Duhem has shown with special emphasis that every statement about any happening is saturated with hypotheses of all sorts and that these in the end are derived from our whole world-view. We are like sailors who on the open sea must reconstruct their ship but are never able to start afresh from the bottom. Where a beam is taken away a new one must at once be put there, and for this the rest of the ship is used as support. In this way, by using the old beams and driftwood, the ship can be shaped entirely anew, but only by gradual reconstruction. (Neurath 1921/1973, 199.)

The implied contrast with Descartes' often-used metaphor of rebuilding the edifice of knowledge on solid foundations is, of course, intentional.

For the purposes of understanding Neurath's philosophy of science in action, what is most important is his argument about the role of the auxiliary motive, for this is what provides legitimation for Neurath's politically engaged science. That politics can and in many instances should play a crucial role in science in no way demeans the intellectual integrity of science. On the contrary, since, in the end, we must choose on the basis of non-empirical factors, we enhance the intellectual integrity of science by frankly asserting the agendas that motivate science in action. Political agendas might have less bearing on theory choice in physics, but they very commonly can and must play a role in fields like economics and social theory, the fields in which Neurath was trained and the fields in which he acted.[13]

3.3.3 The Critique of Metaphysics

If there was any one doctrine that united all of the members of the Vienna Circle and logical empiricism, more generally, it was their repudiation of metaphysics. The best known argument was the one advanced by Rudolf Carnap (1931/1959) in his classic, 1931 essay, "The Elimination of Metaphysics through Logical Analysis of Language." It was here that Carnap premiered his famous verifiability criterion of meaningfulness, according to which a proposition possesses "cognitive meaning" only if it is either analytically true or false or capable of empirical confirmation or falsification. The propositions of traditional metaphysics either fail to be expressible in a proper logical form in the first place, as with Martin Heidegger's infamous assertion that "the nothing nothings," or, if they can be so analyzed and expressed, they fail to satisfy the stated criterion. Ethical judgments likewise lack cognitive

[13] The question of political agendas and the concept formation, theory acceptance and rejection in physics was discussed in detail by Philipp Frank (1957).

meaning, but they might still possess emotive meaning and be valuable as expressions of emotional attitudes, as might some metaphysical claims. Carnap also notes that some metaphysical abstractions, such as "das Volk" ["the people"] have been put to use in the service of regressive or pernicious political agendas. Still, his main point is a logical one about the semantics of metaphysical discourse.

It was Neurath, however, who was best known within the Vienna Circle for his antipathy toward metaphysics. He famously proposed a list of prohibited metaphysical words, the *Index Verborum Prohibitorum* (see Uebel 1992, 76), and it is said that Schlick grew so weary of Neurath's interrupting meetings of the Vienna Circle to complain about their lapsing into metaphysics that Hans Hahn proposed that Neurath should just say, "M," to which Neurath responded that it would save everyone a lot of time if, instead, he said "non-M" on those rare occasions when the others were not indulging in sin (Neurath and Cohen 1973, 82–83). But Neurath was, at the same time, highly skeptical of Carnap's formalistic approach to eliminating metaphysics, holding that the ideal of perfect formal languages was a will-o'-the-wisp. What, then, were Neurath's reasons for excoriating metaphysics?

Part of the answer is that, though this fact is not well remembered today, the elimination of metaphysics because of the baleful political work that it had long done was a major theme in Marxist theory. This is one of the most important issues pressed by Friedrich Engels in his *Anti-Dühring* (1878/1939) and it is echoed in Lenin's *Materialism and Empirio-Criticism* (1909/1947). Engels and Lenin held that their own scientific materialism should not be seen as a damnable form of metaphysics, precisely because it was scientific and dialectical, by contrast with the simply metaphysical materialism – Engels calls it "mechanical materialism" – of Julien Offray de la Mettrie, Ludwig Büchner, Karl Vogt, Jacob Moleschott, and Ludwig Feuerbach.[14] The Austro-Marxists, especially the Machian Austro-Marxists disagreed. They argued that even the materialism of Lenin was suspect, and, following the lead of Ernst Mach's antimetaphysical philosophy of science, they urged a more metaphysically austere form of Marxism. The clearest voice on this point was Neurath's Viennese contemporary, Friedrich Adler.

Particularly helpful for understanding the political context of the anti-metaphysics movement in Viennese socialist circles is Adler's 1907 essay, "Friedrich Engels und die Naturwissenschaft" (1907/1925), which was reprinted in 1925 in a collection of papers, *Marxismus und Naturwissenschaft*, that was published to honor the thirtieth anniversary of Engels's death (Jenssen 1925). The second section of Adler's essay is entitled, simply, "Anti-Metaphysik" (1907/1925, 153–159). Adler here recalls the long history of Marxist critiques of metaphysics starting with Engels's *Anti-Dühring* (1878/1939) and his "Ludwig Feuerbach und die Ausgang der klassischen deutschen Philosophie" (1886). The central point had always been that classical metaphysics, through its positing of static entities, is insufficiently dialectical. But Adler then traces the development of anti-metaphysical philosophy of science since Engels's day to make the point that, at the beginning of the twentieth century, we had achieved

[14] For more on the history of mechanical materialism from a Marxist point of view, see Wittich (1971).

still greater clarity about the nature and dangerous political tendency of metaphysics in all of its guises. Two important socialist philosophers of science play the leading role in Adler's story, the positivists, Richard Avenarius and Ernst Mach. Most significant for Adler are Mach's historical-critical works on mechanics and the theory of heat (Mach 1883/1919, 1896/1986) and his late book, *Kultur und Mechanik* (Mach 1915). Adler argues that the historical-critical approach of Mach represents the culmination of the dialectical approach in a philosophy of science that eschews all metaphysics.

Adler was not alone among socialist, Viennese intellectuals in seeing the situation in this way. In a 1936 essay on "Logisierender Empirismus in der Philosophie der U.S.S.R.," Neurath's old friend, fellow student, and fellow member of the left-wing of the Vienna Circle, the physicist, Philipp Frank (1936/1949, 202), faulted dialectical materialism for "bear[ing] within itself the germ of idealism." If dialectical materialism were really sincere about "waging war" against both idealism and mechanistic materialism,

> it would have to avoid the description of matter as something existing objectively – which is also, in the last analysis, an idealistic conception – and instead would speak of intersubjective propositions. Then it would approach more and more closely the conception represented by logical empiricism, especially by the Vienna Circle. For these groups carry on the same two-front war, against the idealistic school philosophy and against the belief that Newtonian mechanics in its original form is a basis of all science. (Frank 1936/1949, 203.)

Given this context for the anti-metaphysical emphasis in Neurath's thinking, it is instructive to read this early passage from the Vienna Circle's 1929 manifesto, co-written by Neurath, Hahn, and Carnap, *Wissenschaftliche Weltauffassung: Der Wiener Kreis*:

> That Vienna was specially suitable ground for this development is historically understandable. In the second half of the nineteenth century, liberalism was long the dominant political current. Its world of ideas stems from the enlightenment, from empiricism, utilitarianism and the free trade movement of England. [...]
>
> Thanks to this spirit of enlightenment, Vienna has been leading in a scientifically oriented people's education. With the collaboration of Victor Adler and Friedrich Jodl, the society for popular education was founded and carried forth; "popular university courses" and the "people's college" were set up by the well-known historian Ludo Hartmann whose anti-metaphysical attitude and materialist conception of history expressed itself in all his actions. [...]
>
> In this liberal atmosphere lived Ernst Mach (born 1838) who was in Vienna as student and as privatdozent (1861–64). He returned to Vienna only at an advanced age when a special chair of the philosophy of the inductive sciences was created for him (1895). He was especially intent on cleansing empirical science, and in the first place, physics, of metaphysical notions. We recall his critique of absolute space which made him a forerunner of Einstein, his struggle against the metaphysics of the thing-in-itself and of the concept of substance, and his investigations of the construction of scientific concepts from ultimate elements, namely sense data. [...]
>
> Roughly at the same time as Mach, his contemporary and friend Josef Popper-Lynkeus worked in Vienna. Beside his physical and technical achievements we mention his large-scale, if unsystematic philosophical reflections (1899) and his rational economic plan (*A General Peacetime Labour Draft*, 1878). He consciously served the spirit of enlightenment, as is also evident from his book on Voltaire. His rejection of metaphysics was shared by

> many other Viennese sociologists, for example Rudolf Goldscheid. [...] Marxist theory likewise was cultivated and extended with special emphasis in Vienna (Otto Bauer, Rudolf Hilferding, Max Adler and others).
>
> These influences from various sides had the result, especially since 1900, that there was in Vienna a sizeable number of people who frequently and assiduously discussed more general problems in close connection with empirical sciences. Above all these were epistemological and methodological problems of physics, for instance Poincaré's conventionalism, Duhem's conception of the aim and structure of physical theories (his translator was the Viennese Friedrich Adler, a follower of Mach, at that time privatdozent in Zurich); also questions about the foundations of mathematics, problems of axiomatics, logistic and the like. (Neurath et al. 1929/1973, 301–303.)

These paragraphs seem clearly the work of Neurath, who first drafted the manifesto. Later paragraphs, reflecting Carnap's editorial influence, stress the logical arguments against metaphysics. But here Neurath focuses precisely on the manner in which the anti-metaphysical movement grew in the "liberal" political atmosphere of Vienna, and he highlights, by name, several of the prominent, Austro-Marxist intellectuals who were among the leaders of the movement. As Neurath (1928/1973, 295) wrote one year earlier in his *Lebensgestaltung und Klassenkampf*, "[t]he cultivation of scientific, unmetaphysical thought, its application above all to social occurrences, is quite Marxist."

Thus, for Neurath, as for a number of other members of the Vienna Circle, the assault on metaphysics was not just an exercise in logic and the philosophy of language. It was, instead, one of the key political commitments of logical empiricism.

3.3.4 Physicalism

Many first-time readers of Neurath and many who know about logical empiricism from secondary sources are puzzled by the fact that the great opponent of metaphysics was, at the same time, a staunch supporter of physicalism. Why is that not an outright contradiction? The answer is that, while physicalism is a metaphysical thesis for many philosophers, the claim that everything in the universe is ultimately physical in nature, for Neurath, physicalism is a very different kind of claim, a claim not about substance, but about language.

Physicalism, for Neurath, is the proscriptive claim that we should adopt a physicalist observation language, or protocol language, for all of the sciences, meaning a language whose observational primitive terms putatively refer to medium-size physical objects. The contrast is with phenomenalism, which, like physicalism, is not a metaphysical thesis, but the proposal that we adopt a phenomenalist observation language, one whose primitive observational terms putatively refer to the subjective contents of immediate sense experience. The Vienna Circle's famous "protocol sentence debate" of the early 1930s was an argument between the proponents of these two views, with Neurath (1932/1983, 1934/1983) leading the physicalist side and Schlick (1934/1979) leading the phenomenalist side. Carnap (1931/1934, 1931/1987) stood in the middle, having already argued in his *Der logische Aufbau der Welt*

(1928/2003) that the choice between a physicalist or a phenomenalist basis for the construction of higher scientific concepts was a matter of convention, though he opted for a phenomenalist basis in the detailed execution of the project.

The seeming advantage of a phenomenalist protocol language was veridicality in our epistemic access to the subjective contents of immediate experience. Such a grounding in certainty had been a leitmotif of empiricism going all the way back to David Hume who, while denying certainty to all "matters of fact" held that all of our ideas must ultimately be grounded in simple ideas that are copies of the "impressions" whose immediacy and vivacity guarantee their truth (Hume 1739–1740, 1748). But Neurath argued that, even if we chose a phenomenalist protocol language, the moment we try to form sentences out of phenomenal primitive terms those phenomenalist protocol sentences will already be deeply entangled in the web of belief and, thus, will lose whatever foundational status we might have hoped they would have possessed. As we saw above, as early as 1913 Neurath (1913/1983, 3) had been arguing that "the phenomena that we encounter are so much interconnected that they cannot be described by a one-dimensional chain of statements. The correctness of each statement is related to that of all the others." In the context of the early-1930s protocol sentence debate, he elaborates on that same theme:

> There is no way to establish fully secured, neat protocol statements as starting points of the sciences. There is no *tabula rasa*. We are like sailors who have to rebuild their ship on the open sea, without ever being able to dismantle it in dry-dock and reconstruct it from the best components. Only metaphysics can disappear without trace. Imprecise "verbal clusters" ["Ballungen"] are somehow always part of the ship. If imprecision is diminished at one place, it may well appear at another place to a stronger degree. (Neurath 1932/1983, 92)

He concludes: "The fiction of an ideal language composed of neat atomic statements is as metaphysical as the fiction of Laplace's spirit" (Neurath 1932/1983, 91). The fact that a protocol language is a language of statements means that it can never be grounded in phenomenal primitive experience. There might be such primitive experience, but it cannot be expressed in its primitive form.

If a phenomenalist protocol language is a metaphysical fiction, what is the alternative? Better, according to Neurath, to take as the universal, basic language of science one built out of terms referring to ordinary, medium-sized physical objects. At the very least, such a language has the advantage of speaking about publicly accessible objects rather than only privately accessible, subjective experience. Moreover, such a protocol language reflects the reality of genuinely human discourse. We can clean up ordinary language, purge the metaphysical elements, and strive for such precision and clarity as we can achieve. But complete clarity and precision is another Cartesian illusion, something in principle impossible to achieve.

Neurath's emphasis on the public nature of physicalist protocols is a crucial part of his argument and another key element of his philosophy of science in action. This is because Neurath views science not as an activity pursued by individual scientists, but as a collective enterprise. In *Lebensgestaltung und Klassenkampf* he wrote: "It is not a single individual who can really think new notions through to the end, but only whole groups or generations. Thinking, too, is a collective occurrence" (Neurath 1928/1973, 293). This emphasis on the collective nature of intellectual labor had been prominent in the work of Marxist theorists from the earliest years of the movement. It was a point central to the work of Engels (1878/1939, 96), who wrote: "What is human thought? Is it the thought of the individual human being? No. But it exists only as the individual thought of many billions of past, present and future men." The idea of the collective nature of intellectual labor was also, at the time, well known on the left through the work of the now little remembered, German-American, socialist philosopher, Joseph Dietzgen (1869), whose book, *Das Wesen der menschlichen Kopfarbeit* [*The Essence of Human Mental Labor*] was widely read and cited. For Neurath then, believing that science serves the interests of the class struggle, the language of science must facilitate scientific action that is collective and collaborative as is all striving for progress. Solving the problem of food production and distribution, for example, requires the efforts of scientists and engineers from many different disciplines. How can they do that work if they do not speak a common language? How can the architect, the urban designer, and the economist solve the problem of public housing if they cannot communicate.

Neurath's concern for a universal language that would facilitate inclusive, collective class struggle extended beyond science as a field of mere research to science in application, and this not merely application by technical experts but application involving directly those whose lives would be changed. This was the primary motivation behind Neurath's development of "Isotype," the universal symbolic language for efficiently communicating social and economic information to the working classes and people more generally. The basic symbolic forms – silhouettes of factories, workers, tractors, and sheaves of wheat – are literally physicalist protocols. Providing crucial information to workers in this form enables them to participate democratically in the great scientific projects of eradicating poverty, educating everyone, and bringing peace and justice to the world (see Neurath 1927).

3.3.5 Unity of Science

Neurath's advocacy of a physicalist protocol language is closely connected with a final prominent feature of his socially and politically engaged philosophy of science, namely, his championing the ideal of the unity of science. As with his defense of physicalism, Neurath's stress on the importance of the unity of science also occasions serious confusion. There are many different unity of science theses (Cat 2017). One might seek a vertical integration, all higher-order sciences, from sociology and psychology, down through biology and chemistry, being said to be reducible, ultimately, to physics. But Neurath dismissed this mode of unification as another case of metaphysics. One might seek a methodological unification of all of the sciences. This, too, Neurath repudiated as vestigal metaphysics. One might seek a lateral unification among different branches of science, say by subsuming them all under some one or a few fundamental laws, as when, in the nineteenth century, mechanics and thermodynamics were unified under the umbrella of the law of the conservation of energy and the entropy law. There is nothing objectionable about such unification, per se, unless, as with the dream of total, vertical unification, one projects this as a goal for all of the sciences, in which case it becomes, again, metaphysics.

Unity for Neurath meant something strikingly different. It meant, first, the linguistic unification afforded by the choice of a physicalist protocol language, one that could be spoken by everyone, from the particle physicist to the social psychologist. It meant, second, an encyclopedic unification, making the essential contents of all of the sciences universally available in the form of a comprehensive collection of accessible monographs authored by the best people in every field. This is the vision that stood behind the last great, unfinished project of Neurath's life, the *International Encyclopedia of Unified Science* (Neurath et al. 1938–1969; see also Neurath 1936/1983, 1937a/1983, 1937b/1983).

Neurath always contrasted the ideal of encyclopedic unity with that of the systematic unity of science envisioned many years earlier by Auguste Comte. Like Neurath, Comte wanted to put science to use in the betterment of the human condition, a commitment that he shared with most of the utopian socialists of the first half of the nineteenth century. He envisioned as the culmination of scientific development a "social physics," that would establish the fundamental laws of social phenomena as physics had done for mechanics. This social physics would stand atop the great systematic structure of all of the sciences and would be the key to the right ordering of society and the achievement of ultimate human well being. The striving for systematic unification was to become a hallmark of positivist thinking for many decades after Comte. But in this notion of systematization, Neurath saw yet another vestige of metaphysical thinking.

Neurath's concept of encyclopedic unification is to be contrasted, as well, with the Soviet model of the centrally planned organization of science. Neurath understood that science thrived where diversity was promoted. This is a direct implication of the empirical underdetermination of theory choice. There is never one and only one right way forward in science. We must allow the exploration of different pathways. This is Neurath's reason for believing that science requires democracy and freedom to flourish. In one of his very last papers, published a few months after his death, Neurath replied to a lengthy, critical discussion of his notion of the unity of science by Horace Kallen (1940, 1946a) In the face of Kallen's worries about the possible "totalitarian" implications of the "unity of science," Neurath celebrated the "pluralism" of his own view, and invoked the musical metaphor of "orchestration" as a better way of expressing the implications of encyclopedic unification. While there is an overall harmony in the scientific enterprise, different groups of thinkers must be allowed their own voices. Even occasional dissonance can be for the better (Neurath 1946).

In every discussion of the encyclopedia project, Neurath stressed the value of encyclopedic unification for fostering more collaborative scientific work. Thus, in a 1937 article introducing the *International Encyclopedia of Unified Science* in the journal, *Philosophy of Science*, Neurath (1937a/1983, 181) wrote: "Such an encyclopedia will show that scientists, though working in different fields and different countries, may nevertheless cooperate as successfully within this wide field as when they normally cooperate within such special fields as physics or mathematics." We see here the same emphasis on the collective nature of scientific investigation that was an important premise in the argument for a physicalist protocol language. For Neurath, then, the quest for encyclopedic unity in science was yet another essential feature of a philosophy of science designed to theorize, legitimate, and actuate science in service of progressive political change.

3.4 Conclusion: Neurath Today

Neurath affords us an uncommon example of the philosophical life lived not in the clouds but in the rough and tumble of social and political action. In today's jargon, we would call him an activist philosopher of science, someone whose paramount goal was to effect progressive, social and political change. Neurath thought that science, properly understood, would play a crucial role in striving for that goal. I have reviewed, here, the central features of his philosophy of science in action so as to understand how, in Neurath's view, science can play a transformative role in the quest for a just social order without compromising the objectivity and the intellectual integrity of science.

Neurath's vision was not that of a Soviet, scientific, managerial elite that too often simply subordinated science to a political agenda, as with the Soviet Union's promotion of Trofim Lysenko to a leading role in the biosciences because the leadership wrongly thought that Lysenko's resurrection of Lamarckianism – belief in the inheritance of acquired characteristics – would facilitate their aim of producing the "new Soviet man," which venture set back Russian bioscience by a generation (see Jarovsky (1970) and Sofer (1994)). Neurath's vision was that of a Viennese democratic socialist in the broad tradition of Austro-Marxism. It was a pluralist and democratic vision, not a totalitarian one.

Neurath's vision was also not that of the early- and mid-twentieth century critical theorists, such as Max Horkheimer and Theodor Adorno. They were socialists and shared with Neurath and the Austro-Marxists a dissent from Marxist-Leninist orthodoxy, along with the belief that ideas, in the form of critique, can play a leading role in social change. But, unlike Neurath, they indicted all of science as theorized in the empiricist tradition as implicated in the regressive work of merely "traditional theory," by contrast with properly progressive "critical theory" (see Horkheimer 1937/1999 and Dahms 1994), thereby totally misunderstanding the new, politically-engaged theory of science that had developed on Neurath's left wing of the Vienna Circle. Neurath and Max Horkheimer were in dialogue with one another, but Horkheimer's blinkered view of empiricism as part of the legacy of the Enlightenment made it hard for him to grasp the new form of empiricism that Neurath was championing.

Lastly, as mentioned above, Neurath's vision was neither the neo-liberal view of science in democracy promoted by Kitcher nor the neo-Marxist way of regarding science developed by strong programme sociologists of scientific knowledge. Neurath's theory of science in action stands apart, as something radically novel in its own day and still radically novel today. Is its significance now merely historical, or is it or should it be an integral part of the contemporary conversation about the social and political embedding of science and the role of values in science?

Elsewhere I have answered that last question with a resounding, "Yes" (Howard 2009). Let me here briefly summarize the argument. It is gratifying to see the recent upsurge in literature on such topics as the role of values in science. Clearly philosophers of science are once again feeling the need to make the discipline relevant to the social and political challenges of the day. Much of this work is first class, Heather Douglas (2009) and Kevin Elliott (2017) being two of the top thinkers in this space. But no one, in the contemporary literature, has produced the kind of highly-articulated, comprehensive, theoretical framework, spanning everything from the history of science to science communication, and embedded in both a sophisticated political theory and an equally sophisticated, fundamental philosophy of science, that Neurath produced. Nor has anyone, today, deployed such a theoretical framework, as did Neurath, in a wide array of scientific, technical, and policy arenas ranging from physics, economics, sociology, and psychology to architecture, housing, urban planning, and worker education. Moreover, no one thinker today has woven together as did Neurath specific conceptual resources of fundamental importance, such as the notions of the auxiliary motive, underdetermination of theory choice by evidence, pseudorationalism, physicalism, the unity of science, and the collective nature of scientific knowledge making. No, Neurath's philosophy of science in action still stands as an exceptional philosophical achievement.

Why did Neurath's impact wane so rapidly in the post-World War Two era? Partly it was the accident of his untimely death in December 1945. The force Neurath's personality was as important to his project as were any of the specific theoretical components. Philosophy for Neurath was not simply a body of theory, it was a way of life. More important, however, was the dramatically changed political climate in the post-war period, especially the tensions of the Cold War and the rise of McCarthying in the United States and strident anti-communist movements more generally. Neurath's old friend, Philipp Frank (1957), did his best to try to carry on the legacy of Neurath, but the intellectual and political forces at work in the 1950s made this impossible.[15]

Others have argued for the continuing relevance of Neurath's philosophical project (see, for example, Cartwright et al. 1996, Nemeth and Stadler 1996, Uebel 2000, and Potochnik 2011). No one did more to revive Neurath's legacy than the editor of his collected papers, Rudolf Haller (1981–1998). A few contemporary scholars have respected the enduring import of Neurath's project sufficiently so as to have subjected it to serious critique, especially the claim that the core methodological notions of underdetermination and the auxiliary motive provide the optimal framework for theorizing the place of values in science (see Brown 2013). Still, while Neurath's work on Isotype is once again very much in vogue (see Neurath 2010, Vossoughian 2008), his exemplary philosophy of science in action still does not draw the attention that it deserves. More's the pity.

[15] On the Cold War context see Howard (2003) and Reisch (2005). On Frank's marginalization see Reisch (2005, Chap. 15) and Tuboly (2017).

References

Adler, Friedrich. 1907/1925. Friedrich Engels und die Naturwissenschaft. In Jenssen (1925): 146–177.

———. 1918. *Ernst Machs Ueberwindung des mechanischen Materialismus*. Wien: Verlag der Wiener Volksbuchhandlung Ignaz Brand & Co.

Adler, Max. 1925. *Kant und der Marxismus. Gesammelte Aufsätze zur Erkenntniskritik und Theorie des Sozialen*. Berlin: E. Laub.

Barnes, Barry, David Bloor, and John Henry. 1996. *Scientific Knowledge: A Sociological Analysis*. Chicago: University of Chicago Press.

Bloor, David. 1976. *Knowledge and Social Imagery*. Chicago: University of Chicago Press.

Blum, Mark E. 1985. *The Austro-Marxists, 1890–1918: A Psychobiographical Study*. Lexington, KY: University Press of Kentucky.

Brown, Matthew J. 2013. Values in Science beyond Underdetermination and Inductive Risk. *Philosophy of Science* 80: 829–839.

Carnap, Rudolf. 1928/2003 *The Logical Structure of the World.* Trans. R. George. Chicago: Open Court.

———. 1931/1934. *The Unity of Science*. Trans. Max Black. London: K. Paul, Trench, Trubner & Co.

———. 1931/1959. The Elimination of Metaphysics Through Logical Analysis of Language. In *Logical Positivism*, ed. A. J. Ayer, 60–81. New York: The Free Press.

———. 1931/1987. On Protocol Sentences. Trans. Richard Creath and Richard Nollan. *Noûs* 21(4):457–470.

Cartwright, Nancy, Jordi Cat, Lola Fleck, and Thomas Uebel. 1996. *Otto Neurath: Philosophy Between Science and Politics*. Cambridge: Cambridge University Press.

Cat, Jordi. 2017. The Unity of Science. In *The Stanford Encyclopedia of Philosophy*, ed. Edward N. Zalta, Fall 2017 Edition. https://plato.stanford.edu/entries/scientific-unity/.

Dahms, Hans-Joachim. 1994. *Positivismusstreit. Die Auseinandersetzung der Frankfurter Schuler mit dem logischen Positivismus, dem amerikanischen Pragmatismus und dem kritischen Rationalismus*. Frankfurt am Main: Suhrkamp.

Descartes, René. 1637/1903. *Discourse on the Method of Rightly Conducting the Reason, and Seeking Truth in the Sciences*. Trans. John Veitch. Chicago: Open Court; London: Kegan Paul, Trench, Trübner & Co.

Dietzgen, Joseph. 1869. *Das Wesen der menschlichen Kopfarbeit. Dargestellt von einem Handarbeiter. Eine abermalige Kritik der reinen und praktischen Vernunft*. Hamburg: Otto Weißner.

Douglas, Heather. 2009. *Science, Policy, and the Value-Free Ideal*. Pittsburgh: University of Pittsburgh Press.

Duhem, Pierre. 1906/1954. *The Aim and Structure of Physical Theory*. Trans. P. P. Wiener. Princeton: Princeton University Press.

Elliott, Kevin. 2017. *A Tapestry of Values: An Introduction to Values in Science*. New York: Oxford University Press.

Engels, Friedrich. 1878/1939. *Herr Eugen Dühring's Revolution in Science (Anti-Dühring)*, Marxist Library: Works of Marxism-Leninism. Vol. 43. New York: International Publishers.

———. 1886/1888. *Ludwig Feuerbach und die Ausgang der klassischen deutschen Philosophie*. Stuttgart: J.H.W. Dietz.

Frank, Philipp. 1936/1949. Logical Empiricism and the Philosophy of the Soviet Union. In *Modern Science and Its Philosophy*, 198–206. Cambridge, MA: Harvard University Press.

———. 1957. *Philosophy of Science: The Link Between Science and Philosophy*. Englewood Cliffs: Prentice-Hall.

Haller, Rudolf, ed. 1981–1998. *Otto Neurath. Gesammelte Schriften*. Vol. 4. Wien: Hölder-Pichler-Tempsky.

Horkheimer, Max. 1937/1999. Traditional and Critical Theory. In *Critical Theory: Selected Essays*, ed. M. O'Connell, 188–243. New York: Continuum Press.
Howard, Don. 2003. Two Left Turns Make a Right: On the Curious Political Career of North American Philosophy of Science at Mid-century. In *Logical Empiricism in North America*, ed. Alan Richardson and Gary Hardcastle, 25–93. Minneapolis: University of Minnesota Press.
———. 2009. Better Red than Dead – Putting an End to the Social Irrelevance of Postwar Philosophy of Science. *Science and Education* 18: 199–220.
Hume, David. 1739–1740. *A Treatise of Human Nature: Being an Attempt to Introduce the Experimental Method of Reasoning into Moral Subjects*. Vol. 3. London: John Noon.
———. 1748. *Philosophical Essays Concerning Human Understanding*. London: A. Millar.
Jenssen, Otto, ed. 1925. *Marxismus und Naturwissenschaft. Gedenkschrift zum 30. Todestage des Naturwissenschaftlers Friedrich Engels*. Berlin: E. Laub.
Joravsky, David. 1970. *The Lysenko Affair*. Cambridge, MA: Harvard University Press.
Kallen, Horace. 1940. The Meaning of 'Unity' among the Sciences. *Educational Administration and Supervision* 26 (2): 81–97.
———. 1946. The Meanings of 'Unity' among the Sciences, Once More. *Philosophy and Phenomenological Research* 6: 493–496.
Kitcher, Philip. 2001. *Science, Truth, and Democracy*. New York: Oxford University Press.
———. 2011. *Science in a Democratic Society*. Amherst: Prometheus Books.
Kuhn, Thomas. 1962. *The Structure of Scientific Revolutions*. International Encyclopedia of Unified Science, vol. 2, no. 2. Chicago: University of Chicago Press.
Latour, Bruno. 1987. *Science in Action: How to Follow Scientists and Engineers through Society*. Cambridge, MA: Harvard University Press.
Lenin, V.I. 1909/1947. *Materialism and Empirio-Criticism: Critical Comments on a Reactionary Philosophy*. Moscow: Foreign Languages Publishing House.
Mach, Ernst. 1883/1919. *The Science of Mechanics. A Critical and Historical Account of ist Development*. 4th ed. Chicago and London: Open Court.
———. 1896/1986. *Principles of the Theory of Heat. Historically and Critically Elucidated*. Dordrecht: D. Reidel.
———. 1915. *Kultur und Mechanik*. Stuttgart: W. Spemann.
Morris, Charles W. 1962. On the History of the International Encyclopedia of Unified Science. In *Logic and language: Studies Dedicated to Professor Rudolf Carnap on the Occasion of His Seventieth Birthday*, ed. B.H. Kazemier and D. Vuysje, 242–246. Dordrecht: D. Reidel.
Nemeth, Elisabeth, and Friedrich Stadler, eds. 1996. *Encyclopedia and Utopia: The Life and Work of Otto Neurath (1882–1945)*. Dordrecht: Kluwer.
Neurath, Otto. 1913/1983. The Lost Wanderers of Descartes and the Auxiliary Motive (On the Psychology of Decision). In *Otto Neurath: Philosophical Papers 1913–1946*, ed. Robert S. Cohen and Marie Neurath, 1–12. Dordrecht: D. Reidel.
———. 1921/1973. Anti-Spengler. In Neurath and Cohen (1973): 158–213.
———. 1927. Statistik und Proletariat. *Kulturwille, Statistik und Klassenkampf* 5 (9): 186.
———. 1928/1973. Personal Life and Class Struggle. In Neurath and Cohen (1973): 249–298.
———. 1932/1983 Protocol Sentences. In *Otto Neurath: Philosophical Papers 1913–1946*, ed. Robert S. Cohen and Marie Neurath, 91–99. Dordrecht: D. Reidel.
———. 1934/1983 Radical Physicalism and the 'Real World'. In *Otto Neurath: Philosophical Papers 1913–1946*, ed. Robert S. Cohen and Marie Neurath, 100–114. Dordrecht: D. Reidel.
———. 1935/1983 Pseudorationalism of Falsification. In *Otto Neurath: Philosophical Papers 1913–1946*, ed. Robert S. Cohen and Marie Neurath, 132–138. Dordrecht: D. Reidel.
———. 1936/1983 Encyclopedia as 'Model'. In *Otto Neurath: Philosophical Papers 1913–1946*, ed. Robert S. Cohen and Marie Neurath, 145–158. Dordrecht: D. Reidel.
———. 1937a/1983. Unified Science and Its Encyclopedia. In *Otto Neurath: Philosophical Papers 1913–1946*, ed. Robert S. Cohen and Marie Neurath, 172–182. Dordrecht: D. Reidel.

———. 1937b/1983 The New Encyclopedia of Scientific Empiricism. In *Otto Neurath: Philosophical Papers 1913–1946*, ed. Robert S. Cohen and Marie Neurath, 189–199. Dordrecht: D. Reidel.

———. 1939. *Modern Man in the Making*. New York: Alfred A. Knopf; London: Secker and Warburg.

———. 1946 The Orchestration of the Sciences by the Encyclopedia of Logical Empiricism. In *Otto Neurath: Philosophical Papers 1913–1946*, ed. Robert S. Cohen and Marie Neurath, 230–242. Dordrecht: D. Reidel.

———. 2010. *From Hieroglyphics to Isotype: A Visual Autobiography*. Eds. Matthew Eve and Christopher Burke. London: Hyphen Press.

Neurath, Marie, and Robert S. Cohen, eds. 1973. *Empiricism and Sociology*. Dordrecht: D. Reidel.

Neurath, Otto, and Joseph A. Lauwerys. 1945. Plato's Republic and German Education I. *The Journal of Education* 77 (907): 57–59.

Neurath, Otto, Hans Hahn, and Rudolf Carnap. 1929/1973. The Scientific Conception of the World: The Vienna Circle. In Neurath and Cohen (1973): 299–318.

Neurath, Otto, Charles Morris, and Rudolf Carnap, eds. 1938–1969. *International Encyclopedia of Unified Science*. Chicago: University of Chicago Press.

Plato. 1888. *The Republic of Plato*. Trans. B. Jowett. Third edition. Oxford: Clarendon.

Popper, Karl. 1934/1959. *The Logic of Scientific Discovery*. New York, Basic Books; London: Hutchinson & Co.

———. 1945. *The Open Society and Its Enemies*. London: Routledge.

Potochnik, Angela. 2011. A Neurathian conception of the unity of science. *Erkenntnis* 74: 305–319.

Reisch, George. 1994. Planning science: Otto Neurath and the International Encyclopedia of Unified Science. *The British Journal for the History of Science* 27: 153–175.

———. 2005. *How the Cold War Transformed Philosophy of Science: To the Icy Slopes of Logic*. New York: Cambridge University Press.

Sandner, Günther. 2014. *Otto Neurath: Eine politische Biographie*. Wien: Paul Zsolnay.

Schlick, Moritz. 1934/1979. On the Foundation of Knowledge. In *Moritz Schlick: Philosophical Papers. Vol. II*, ed. H.L. Mulder and B.F. van de Velde-Schlick, 370–387. Dordrecht: D. Reidel.

Soyfer, Valery N. 1994. *Lysenko and the Tragedy of Soviet Science*. New Brunswick: Rutgers University Press.

Stadler, Friedrich. 2001/2015. *The Vienna Circle: Studies of the Origin, Development, and Influence of Logical Empiricism*. 2nd ed. Dordrecht: Springer.

Tuboly, Adam Tamas. 2017. Philipp Frank's decline and the crisis of logical empiricism. *Studies in East European Thought* 69 (3): 257–276.

Uebel, Thomas. 1992. *Overcoming Logical Empiricism from Within: The Emergence of Neurath's Naturalism in the Vienna Circle's Protocol Sentence Debate*. Amsterdam and Atlanta: Rodopi.

———. 2000. *Vernunftkritik und Wissenschaft. Otto Neurath und der erste Wiener Kreis*. Wien/New York: Springer.

Vossoughian, Nader. 2008. *Otto Neurath: The Language of the Global Polis*. New York: NAI/D.A.P. Distributed Art Publishers.

Wittich, Dieter, ed. 1971. *Vogt, Moleschott, Büchner. Schriften zum kleinbürgerlichen Materialismus in Deutschland*. Vol. 2. Berlin: Akademie-Verlag.

Chapter 4
Science and Socialism: Otto Neurath as a Political Writer (1919–1932)

Günther Sandner

Abstract In the case of Otto Neurath, there was always a close relationship between science and politics. It is worth noting, however, that he also intervened in politics of the day. The essay focuses on Neurath's time in Revolutionary Bavaria and Red Vienna and analyzes his articles in two newspapers in the interwar period: the German periodical *Economy and Order of Life* (a supplement of the *Art Guardian*) and the Austrian socialist daily *Workers' Newspaper*. The examination starts when Neurath definitely became a socialist and member of the Social Democratic Party (SPD) after the war in Germany and ends shortly before his forced migration from Austria in 1934. The two series of articles differed in some respects. While in *Economy and Order of Life* Neurath focused only socialization, he addressed a number of different topics in the *Arbeiter-Zeitung* (*AZ – Workers' Newspaper*) ranging from guild socialism over housing, architecture and settlement to education and the scientific world-conception. The question of his role and his self-image was a common theme in many of these articles. Beyond that, however, there were even approaches to closely related themes that he developed more precisely only in his later life: the question of experts, citizens, and democracy.

4.1 Intellectual Interventions into Politics

As most readers of this book would probably agree, Otto Neurath was the most effective political character among the Viennese philosophers of Logical Empiricism. Although there was a politicized left wing within the Vienna Circle (Carnap 1963; Uebel 2005), which included Rudolf Carnap, Philipp Frank, Hans Hahn and – at its periphery – Edgar Zilsel, even among these, nobody ever played a role in politics comparable to that of Otto Neurath. His intellectual biography demonstrates that he had developed a strong profile in scientific and pedagogical fields such as the

---

G. Sandner (✉)
Institute Vienna Circle, University of Vienna, Vienna, Austria
e-mail: guenther.sandner@univie.ac.at

J. Cat, A. T. Tuboly (eds.), *Neurath Reconsidered*,
Boston Studies in the Philosophy and History of Science 336,
https://doi.org/10.1007/978-3-030-02128-3_4

history of economy and political economy, sociology, the philosophy of science and visual education, and even more (cf. Sandner 2014a). None of his scientific and/or educational projects, however, can be strictly separated from politics. In Neurath's case, there was always a close relationship between science and politics. Politics, however, existed not only subcutaneously in his well-known intellectual projects such as unity of science or international picture language. Moreover, he actually intervened in politics of the day. He did so not only on a theoretical level but as an active part of political projects such as those in revolutionary Bavaria after the Great War or in Red Vienna of the interwar years. This is an obvious difference from both the other left-wingers in the Vienna Circle and many political philosophers and/or philosophers of science.[1]

This essay focuses on a particular category of Otto Neurath's political publications: his newspaper articles from the interwar period. The examination starts when Neurath definitely became a socialist and member of the Social Democratic Party (SPD) after the war in Germany (Neurath 1920/1973, 21; Sandner 2014a, 127) and ends shortly before his forced migration from "New Vienna" or "Red Vienna" in 1934. In Vienna, he belonged to the Austro-Marxist intellectual group and played different roles in culture, education and politics (Sandner 2006).

Otto Neurath often and continuously published in newspapers and the daily press, starting as a young scholar and continuing more or less until the end of his life. A number of publications on the political Neurath have already focused on this issue (e.g. Sandner 2014a, b; Cartwright et al. 1996). This essay, however, concentrates on political writings in a narrower sense – on his role as a public disputer, and as someone who wanted to convince larger audiences as both an expert and a political activist. These publications are little known, and they are not part of any edition of collected writings. In addition, they were never translated into English.[2]

The publications in question represent two different series of articles published between 1919 and 1932: Firstly, his contributions for a supplement of the German journal *Deutscher Wille des Kunstwart* (*German Will of the Art Guardian*) called *Wirtschaft und Lebensordnung* (*Economy and Order of Life*) in 1919. In this period of less than a year, Otto Neurath explained and defended his ideas of a future socialized economy. Most of these articles were published very shortly before Neurath became president of the Central Economic Office in Munich. Although he developed similar thoughts and ideas on the issue of socialization in books, booklets and scientific essays, there are some peculiarities of these articles due to this particular medium and its expected readership. Secondly, there are his contributions in the socialist daily *Arbeiter-Zeitung* (*AZ – Workers' Newspaper*) in Vienna. The timeframe is considerably longer, between 1919 and 1932, and Neurath addressed a number of *different* topics, such as guild socialism, settlement, education, and the

[1] On the political level of Neurath's philosophical thought, see Don Howard's chapter in the present volume.

[2] The most important collections of Otto Neurath's writings are the following: Neurath and Cohen (1973); Hegselmann (1979); Neurath and Cohen (1983); Haller and Rutte (1981); Haller and Kinross (1991); Haller and Höfer (1998); Uebel and Cohen (2004).

scientific world-conception. Although his forced migration from Vienna in 1934 did not mean the end of his political writings, it effected a deep impact, even in his political life.[3]

On the one hand, the present essay shows the range of topics that Neurath addressed as a political commentator and analyst. In most cases, it is obvious how these articles relate to his scientific and philosophical writings. On the other hand, Neurath's writings from these years reflect a theme that was starting to gain ground: the question of experts and citizens in a democracy. Of course, this was also a self-reflection on his role as an economist, philosopher of science and visual educator. Neurath's contemporary writings on the subject, however, did not make things easier. He often, and in many different times and contexts, insisted on his role as an apolitical social engineer who always aimed to work on and realize his utopian ideas independent of political power. Was he exclusively an expert or a partisan political commentator? Or did he assume both roles? What exactly was his idea and definition of politics? What did he want to achieve, and how were his social and political aims related to his scientific program? Did he just translate the scientific and philosophical program into political statements and positions, or was there an interdependency between his fields?

Why these two particular periodicals? A few words on the criteria for their selection may be added: Otto Neurath published in many different periodicals, and even in many different *socialist* ones. His essays and articles in Red Vienna, for instance, appeared in a number of party political or syndicalist-related newspapers and journals, such as *Arbeit und Wirtschaft* (*Labor and Economy*), *Bildungsarbeit* (*Educational Work*) or *Der Kampf* (*The Campaign*). Although the intellectual level of the *Arbeiter-Zeitung* was remarkably high, its readership was much larger than that of *Der Kampf*, for instance, the programmatic and theoretical journal of the *Social Democratic Workers' Party* (SDAP). The *Workers' Newspaper* was the official daily of the Austrian socialists. Founded in the nineteenth century, the newspaper became the party's most important medium, despite being only one of many socialist periodicals in interwar Austria. In the mid-1920s, about 100,000 copies were published daily (Stadler 1990, 105; Pelinka and Scheuch 1989, 75–76). This implied a considerable readership, though not very large compared with the more than 700,000 party members at the end of the 1920s (Maderthaner 1995, 180). Anyway, it is fascinating to see in which ways Neurath developed his thoughts in his journalistic contributions and tried to impart knowledge in this political context.

The *Kunstwart* (*The Art Guardian*) had been founded in 1887 in Dresden and was edited by poet and publicist Ferdinand Avenarius (brother of philosopher Richard Avenarius). The frequency of its publication was mostly biweekly, sometimes only monthly. Although its circulation differed, there were never many more than 20,000 copies published (Kratzsch 1969). Its focus was on culture (and especially the arts), not on politics. It was definitely not a declared socialist or leftwing journal; it was targeted to the educated middle class and covered a wide range of

[3]The author has examined the "political Neurath" in both periods of emigration: Sandner (2011) and (2019a).

political orientations, including German nationalist ones. Nevertheless, Otto Neurath used its supplement *Economy and Order of Life* and even exploited it with the help of his friend Wolfgang Schumann. During the first half of the year 1919, they used the medium to influence a politically interested and educated bourgeois readership. They decided to back their ideas for a wide-ranging socialization of the economy with the help of this periodical. Both Otto Neurath and Wolfgang Schumann had already been active with the traditional *Der Kunstwart* before.

4.2 Socialization, Social Economy and Socialism

4.2.1 The War Economist and Socialization Theoretician

Before the Great War, Otto Neurath was primarily known in the intellectual world as an economist. Although he had also addressed subjects in other academic fields, such as the philosophy of science or, as a young man, even German literature, it was his theory of war economy that drew broad attention to his intellectual work. In this context, his focus was twofold: He both examined war economy historically (his doctoral thesis was on the history of economy) and studied contemporary wars and their economic consequences, especially the Balkan Wars of 1912–1913 (Sandner 2017). Based upon this research, he developed his theory on war economy. In Neurath's view, political economy had previously paid insufficient attention to war economy. Aside from obvious and undisputable negative effects, he was convinced that war could also stimulate an economy. To evaluate and systematize these different effects of war on the economy, a special discipline of war economy was needed (as a sub-discipline of political economy). Although he was not the first to address the topic, as he himself admitted, he presented himself as the founder of this new approach as a special scientific discipline (Neurath 1913). Well-established scholars such as Max Weber respected and recognized him as an expert in this field, and economist Franz Eulenburg (1916/17, 1918) discussed his approach in the well-established journal *Archive for Social Science and Social Policy* (*Archiv für Sozialwissenschaft und Sozialpolitik*).[4] Franz Eulenburg, certainly, fundamentally rejected Neurath's thesis. For him, war was nothing but a rare case in economics that neither delivered useful advice nor served as any role model for a well-functioning peace-time economy.

Neurath, in contrast, was convinced that the instruments of war economy such as planning and scientific expertise, calculation in kind, and the strong focus on efficiency and productivity, were perfectly useful even and especially in peace. He was convinced that the future economy was an administrative and moneyless economy beyond the market. He insisted that there could always be a much better utilization of energies in the economy – and in his view, it was the war economy that

[4] On Neurath's economic ideas, see Jordi Cat's, Elisabeth Nemeth's, and Thomas Uebel's chapters in the present volume.

demonstrated how this could be done. A peacetime economy that used these experiences was the key for a better future society. This new economic order could achieve its most important aim: A higher quality of life for the people. Thus, paradoxically, it was the war that paved the way for human happiness.

During the Great War, Otto Neurath's ideas of war economy were both exploited and institutionalized. Although he further insisted on his role of an expert, he sided politically, implicitly at least, with the Central Powers Austria-Hungary and Germany. Parts of the military were extremely interested in his research on war economy and tried to make use of it. After his deployment as an officer of the Austrian army on the Russian front, he returned to Vienna and became leader of a group in the *Scientific Committee of War Economy* (*Wissenschaftliches Komitee für Kriegswirtschaft*) in April 1916. Almost simultaneously, he got involved in the emerging Museum of War Economy in Leipzig, whose director he became in 1918 (Sandner 2014a, 85–99 and 2014c, 389–394).

After the war, there was a widespread intellectual consensus that there must be changes in both the political and the economic order. In this time "war socialism" – the idea that war had prepared the conditions for socialism – became popular (Krüger 1997; Sandner 2014a, 109–111). In some respects, at least, this concept was similar to Neurath's approach. There was a widespread mood for change. In this context, the question of socialization was widely discussed, not only by representatives of the radical left. Walter Rathenau, for instance, was among the references Otto Neurath often mentioned in his writings, as well as Rudolf Wissell, Wichard von Moellendorf and others (cf. Rathenau 1918, Moellendorff 1916).

Neurath systematically adapted his concept for socialization from his theory of war economy. Again, he tried to implement his economic ideas practically and to intervene directly into politics. The addressees, however, were no longer the military powers of Germany and Austria. Now, he applied to the post-war revolutionary governments. Together with Wolfgang Schumann and Hermann Kranold, he designed the so-called Kranold-Neurath-Schumann plan in February 1919 (Neurath 1919). They developed their idea for a total socialization[5] and indicated concrete measures for its implementation. The booklet ended in 21 legislative proposals. The three men were convinced that socialist governed Saxony could be an ideal place to start with the necessary transformation of the economic order. Much to their regret, however, the left-wing government refused their proposals (Sandner 2014a, 111–114).

Neurath's ideas for socialization were anything but unclear. "The aim of socialization is to produce and distribute the final product socialistically," he put it (1920/2004, 377). The year before, he had explained: "Of a complete realization of socialism, however, one can generally speak only when both the socialist distribution and the planned administration of production takes places through society" (1919/1973, 137). With thoughts like this, however, he tried to convince the readership of a well-established German periodical.

[5]As Neurath (1920a, 7) pointed out, the term total socialization (*Vollsozialisierung*) was not his own but was introduced by Wolfgang Schumann.

4.2.2 *The Articles for the Supplement* Economy and Order of Life

Beginning in 1908 and continuing through the prewar period, Otto Neurath wrote a number of articles for the *Kunstwart*, even addressing the subject of war economy in them (cf. Wilhelm 1911). Moreover, he became a friend of Wolfgang Schumann, the stepson of editor Ferdinand Avenarius. It was the beginning of a collaboration between Neurath and Schumann that continued for many years. Wolfgang Schumann not only contributed continuously to the journal, later acting as its editor, but was also a fellow-campaigner in another venture of Avenarius – the *Dürerbund* (*Dürer League*), an organization that was active in cultural policy (Kratzsch 1969).

Between October 1915 and March 1919 *Der Kunstwart* was published under a new name as *Deutscher Wille des Kunstwarts* (*German Will of the Art Guardian*). This name as well as the idea behind it was obviously related to the war. Addressing a "German Will" in times of war implied both ideological militarization and politicization; it denoted the will to win the war, the German will to victory. According to the editorial concept of the *Kunstwart*, however, it was not only and even not necessarily a war on material things (such as territories, population, industries etc.) but a war between national or ethnic communities that differed in character, spiritually as well as morally (Kratzsch 1969, 364–398).

After the war, the journal appeared in a new format, and with a new supplement called *Wirtschaft und Lebensordnung* (*Economy and Order of Life*). In January 1919, it was presented as a *joint venture* between the traditional *Kunstwart* and the newly established *Deutsches Wirtschaftsmuseum* (*German Economic Museum*), the follower of the *War Museum* in Leipzig. Otto Neurath was still its director, and his museum was responsible for the supplement. The two parts of the journal, the *German Will* and its supplement, differed even in design: Old German lettering in one (*Deutscher Wille des Kunstwart*), alphabetic characters in Latin in the other (*Wirtschaft und Lebensordnung).* Even in its short history of nine months, the supplement's subtitle changed repeatedly. It was usually *Papers of the German Economic Museum* (*Blätter des Deutschen Wirtschaftsmuseums*) with or without the appendix *Biweekly of economic understanding* (*Halbmonatsschrift für Wirtschaftsverständnis*). Only in a few issues it was explicitly indicated that the periodical was directed by Otto Neurath and Wolfgang Schumann.

In an introductory essay, four men explained the character of the joint venture: Editor Ferdinand Avenarius and publisher Georg D.W. Callwey spoke for the traditional *Kunstwart* and, for the *German Economic Museum*, its director Otto Neurath and its general secretary Wolfgang Schumann chimed in (Avenarius et al. 1919).

On the whole, the supplement appeared in the journal's issues 1–14 of 1919 (published in the period between January and September). Due to the role of Otto Neurath, there was an obvious trend: The later the date of the publication of the supplement's issue, the fewer his own contributions. He was extremely active in issues 1–6, but in the later ones, he published only a single article or nothing at all.

The journal as a whole was renamed *Kunst- und Kulturwart* (*Guardian of Art and Culture*) in April 1919, and the supplement continued for only a few months.

In the introduction to the series, Otto Neurath and Wolfgang Schumann stressed their idea of economic enlightenment without any party political bias. The supplement's idea, as they put it, was to transfer social and economic knowledge to the people, not to proclaim political doctrines (Neurath and Schumann 1919). This position corresponded with their view on socialization: Both men were convinced that the advantages of a socialized economy can be demonstrated and even be proven by science. Therefore, socialization needed to be discussed objectively and reasonably. It would not be political polemic but scientific evidence that would help to transform the economy.

In fact, the supplement had only a very few authors. Wolfgang Schumann and especially Otto Neurath were by far the most frequent contributors. In Neurath's case, however, this was not obvious *prima facie*. It was the use of pseudonyms that concealed the obvious staffing shortage. Otto Neurath wrote not only with his real name but also under two other identities, as "Fonsow" and as "Karl Wilhelm." While the former name was not that easy to identify, the latter – a pseudonym he had used before and would use afterward – was obvious: The author's full name was Otto *Karl Wilhelm* Neurath. Most of the supplement's issues consisted of only eight pages and included, besides the texts of Neurath, Schumann and a few occasional authors, historical texts such as those of John Stuart Mill, Karl Marx or Karl von Vogelsang.

The supplement's main theme was, naturally, socialization. Its key text was the four-part essay *Zur Sozialisierung der Wirtschaft* (*Towards the socialization of the economy*), in which Otto Neurath (as Fonsow) developed his idea of socialization and even discussed some problems that could face a future non-market economy. In the first part (Fonsow 1919a), he contrasts two different models of the economic order: The free market economy (*Verkehrswirtschaft*) on the one hand and the future socialist economy as an administrative economy followed by a plan on the other. In a socialized economy it is community, and not the individual entrepreneur as master, who makes production decisions. In a free market economy, those enterprises were undertaken that yielded the highest money profits. In the administrative economy, in contrast, the plan replaces net profits in regulating the economy. In the end, the economy yields the same advantages to all and increased people's happiness. Neurath eventually addresses some possible problems: If there is no profit, how to motivate individuals to collaborate in a "socialized society" (Fonsow 1919a, 4)? How can the danger of a dramatic decrease in production be prevented? In answer to this question, Neurath presents systems of premiums (i.e. gratifications for hard and efficient workers) in the second part of the essay. Premiums, as Neurath understood them, would be related to the requirements of the community. That means that premiums even may be paid if noisy works are not undertaken or dust formation is avoided (Fonsow 1919b, 11).

In the third part, he explains the systems of premiums in more detail and then points out that a future economy needs to find out systematically exactly which humans are suited for which activity. What were the best activities to assign each individual? How can planners optimize the use of capabilities? How can planners bring about a combination of people and jobs such that a maximum of happiness is achieved? The socialization's aim was a socialized order of life, based upon improved efficiencies, but without any schematization of personal life. Probably, as Neurath prognosticated, there would even be greater diversification of ways of life in socialism (Fonsow 1919c, 22).

In the last part, Neurath portrayed the Central Economic Office as the decisive authority, shortly *before* he actually became president of exactly this institution in Munich, and he focused on the need for universal statistics and an economic plan as a basic principle. For Neurath, the ideal time for socialization was not sometime in the future, it was now. Socialization, he continued, could start even in a region smaller than a nation-state, and there was simply no need to wait for neighboring countries. In his view, socialization was the best system for the postwar economy because it could handle problems such as shortage and poverty with its strong focus on economic efficiency. In this context, he referred to the example of 3000 different pocketknives that seemingly existed at that time (Fonsow 1919d, 36). The setting of industrial norms and types implied the disappearance of many types that, in fact, nobody really needed. This could happen, however, without any decline in quality of life. Socialization, he continued, must be done without compromises. It always meant total socialization. To socialize or not was only a political question, he put it, but if a society decided on socialization, it needed to be done immediately, quickly and entirely (Fonsow 1919d, 37).

Beyond this core idea, he addressed a few other subjects. He explained, for instance, his concept of utopia (Neurath 1919) and stressed the importance of education for a future society (Wilhelm 1919a).

Otto Neurath played different roles in the issues of the journal. In sum, he published about 20 articles. As "Fonsow" he embodied the political economist, the established economic expert, who designed a future society, and discussed its perspectives and chances, as well as its possible problems. His argumentation was based on scientific evidence though *not* always beyond any party political bias. As "Karl Wilhelm" he examined some selected and concrete problems in greater detail, such as economy in kind and its effects on interstate relations, among others (Wilhelm 1919b, c, d). In comparison to Fonsow, who debated on principles, Karl Wilhelm was the empiricist who did little case studies. As Otto Neurath, he spoke *officially*, as editor of the journal and director of the museum. In fact, however, there was no clear separation between the authors, and, in the end, it was of course always Otto Neurath under different names.

Wolfgang Schumann, who published many articles in the journal, portrayed the *German Economic Museum*, for instance, that succeeded the Museum of War Economy in two parts (Schumann 1919a, b). Among the few other contributions, some of which are published anonymously, is one Neurath's companion Herman Kranold wrote about economic activities with and without a plan. No surprise, the former approach is by far the better one, according to the author (Kranold 1919).

Politically, two things are remarkable. Firstly, Neurath criticized the Spartacus League (the forerunner of the Communist Party) for its idealized idea of man (Fonsow 1919b, 9), and he also repeatedly rejected the politics of the Bolsheviks.

He not only explained his approach as an economic expert but also envisaged and reflected on a future socialist society. In addition, he stressed the fact that socialization, a socialist order of life, was a question of political will. However, he obviously did *not* address a particular *German Will*, as the title of the journal suggests. It was the will of the proletarian masses to realize socialism. For Neurath, there was no better way of living than in a socialized economy and socialist society. And if one thinks there is a better way, he continued, it needs to be demonstrated through scientific evidence (Fonsow 1919d, 37).

Although his essays did not represent a detailed analysis of his socialization theory (as he presented in some other publications) he delivered a summarized version of it and even discussed some practical problems as contributions to public debate. The aim was to persuade an educated readership, including multipliers such as teachers and publicists, that a new economic order was unavoidable.

In the supplement's last issue in September 1919, the revolutionary spirit had trickled away. Otto Neurath was no longer present, and Wolfgang Schumann (1919c) asserted the end of any effort to realize a planned economy, effective socialization and socialism. Interestingly, the dramatic events in Munich in spring and summer of 1919 played almost no role in *Economy and Order of Life*. Not in the supplement but in the journal *Der Kunst- und Kulturwart*, however, Wolfgang Schumann (1919d) touched on the subject. His article was, as the subtitle suggests ("personal experiences with the press") about his experiences with media reports. He complained about many false reports in the post-revolutionary Munich press, including those which reported his own assassination (!), and also responded to the defamations of his friend Otto Neurath. Moreover, both Schumann (1919e, f) and editor Avenarius (1919) looked back to Munich in three articles in which they tried to adjust and even relativize Schumann's role in revolutionary Bavaria.

Otto Neurath published only at rare intervals in the later issues of the journal's supplement, obviously because of his activities in economic planning in Munich. He became president of the Central Economic Office on 27 March 1919 under a socialist-bourgeois government and remained in office under two short-term council republics. After only a few weeks, he was dismissed in May 1919 and accused of high treason. After the trial and his condemnation to one and a half years of prison, he was instead extradited to Vienna in February 1920 thanks to the continuous interventions of the Austrian government, including and especially two social democrats, Secretary of Foreign Affairs Otto Bauer and Chancellor Karl Renner (Sandner 2014a, 132–143). In early 1920, his new career in Red Vienna started. In his booklet on *Experiences of Socialization in Bavaria* (*Bayerische Sozialisierungserfahrungen*) he recapitulated the past months as follows:

> I think we need not give up hope that all those in opposition to capitalism could join forces to replace a capitalist order with a socialist planned economy, in a peaceful manner, without submitting to compromise. I still believe that we could introduce socialism in the near future if we definitely want it. Those who say this are being sharply attacked, even hated by many; this has to be suffered, but we should not react with counter-hatred and emotional distortion. Only by remaining dispassionate ourselves can we demand objectivity from our opponents, only then can we hope that the future will led in without bitter fight. Socialism will come anyhow; may it come, not with hatred, but with love. (Neurath 1920/1973, 28.)

4.3 The Austro-Marxist

4.3.1 Back in Vienna

Back in his birthplace, Vienna, Otto Neurath began working feverishly. He organized schools and training sessions for members of the newly founded workers' council in early 1920 (Sandner 2014a, 158) and became general secretary of a *Research Institute of Social Economy* (*Foschungsinstitut für Gemeinwirtschaft*), that consulted the Austrian commission for socialization and carried out economic studies (Sandner 2014a, 162–165). Despite the existence of workers' councils and ongoing research on social economy, in fact, the Austrian efforts toward socialization ended, at the latest, in October 1920 when the Social Democratic Party lost the general elections and went into opposition. Only a very small sector of the Austrian economy was successfully socialized i.e. transformed into a social economy: the former corporations of the military apparatus in Vienna (Weissel 1976; Gerlich 1980). For Neurath (1922e), however, this was not socialization but only "social capitalism" because these enterprises were still embedded in a market economy.

Nevertheless, he did not lose his ambitions and further developed his theoretical approach: He explicitly linked his socialization theory with guild socialism, and discussed the writings of British left-winger G.D.H. Cole. Although he rejected some elements of Cole's theory, he also combined its organizational idea with total socialization (see below). By 1921, Otto Neurath had become one of the settlement movement's most important representatives and the mastermind of its unification and centralization. The settlement movement was based upon common property and thus represented social economy. Its organization was similar to the idea of guilds. Thus, Neurath became secretary of the social housing guild (*Baugilde*) and general secretary of the *Austrian Association for Settlement and Allotment* (*Österreichischer Verband für Siedlungs- und Kleingartenwesen*). The latter organized an exhibition on settlement in Vienna in autumn 1923 which resulted in the foundation of a museum for settlement (Sandner 2014a, 165–176).[6] These activities paved the way for his most important institution, the *Social and Economic Museum* (*Gesellschafts- und Wirtschaftsmuseum*), the director of which he became in 1925. In addition, he was extremely active in adult education and was, among many other activities, one of the teachers in the socialist Worker's University (*Arbeiterhochschule*), founded in 1926. From 1924 onwards, the meetings of the Schlick-Circle (resp. Vienna Circle) took place. Thus, Otto Neurath was active in many Viennese intellectual and educational fields. His numerous lectures and speeches demonstrate the astonishing variety of his themes (regularly announced in the periodical *Arbeiter-Zeitung*).

Otto Neurath was a member of the Austrian Social Democratic Workers' Party (SDAP) and collaborated with the left-wing government of the city of Vienna. Although he was never a party official he published many articles for socialist

[6] For more on Neurath's involvement in the Settlement movement, see Sophie Hochhäusl's chapter in the present volume.

periodicals. It is worth noting that some of these socialist papers, including the *AZ*, frequently used picture statistics according to his Vienna Method (though they also used non-Neurathian picture statistics). In the election campaign of 1927, the *Vienna Method of Picture Statistics* was extensively used to promote socialist slogans such as "*Einheitsschule statt Einheitsliste*" (comprehensive school instead of unified list of candidates).[7]

4.3.2 The Articles in the **Workers' Newspaper**

Between 1919 and 1932 Otto Neurath wrote 25 articles for the daily *Workers' Newspaper* (*AZ*). Thematically, these articles can be divided into four different groups (which are, nevertheless, related to each other). The first group is on socialization and guild socialism: In fact, these show a smooth transition with respect to his publications in *Economy and Order of Life*. On March 27, 1919, the day he became president of the *Central Economic Office* in Munich, Neurath published an article on the question of total socialization vs partial nationalization (Neurath 1919). Unsurprisingly, he clearly favored the first option and rejected the second. It included also a message for the Austrian socialists: Neurath developed the idea of a joint venture in socialization between Saxony, Bavaria and Austria that would, in his view, exert political pressure on the German Reich to finally implement and realize a serious socialization policy.

In February 1921, Neurath reviewed G.D.H. Cole's book on guild socialism, introduced by Wolfgang Schumann and translated by his wife Eva Schumann (Neurath 1921a). Guild socialism was an important movement in the early twentieth century, especially in Great Britain, Germany and Austria. It implied (among other things) that *different* groups of employed persons (blue collar workers, white collar workers etc.) were to be organized as guilds in their particular economic sectors. Guild organization contrasts with the organizational model of the trade unions, which focused on the same groups of employed persons from different economic firms and economic sectors. In Neurath's view, it was important to combine this organizational approach with an economic plan (and especially in this respect, he adapted or complemented Cole). A few months later, he explained in another article that, in fact, there were two different political formations of guilds: guild liberalism and guild socialism (Neurath 1921b), and he clearly favored the latter. In December 1921, he reported on his activities in the settlement movement and the making of the housing guild (*Baugilde*) (Neurath 1921c).

[7] "Einheitsschule"was the name for the socialist conception of school organization. It meant the same school for all children, independent of family background and social class. The "Einheitsliste" was a coalition of rightwing parties that ran successfully for the general election of 1927 against the feared socialist electoral victory. The *Einheitsliste* won more than 48% while the SDAP took only 42% (cf. Dachs 1995, 150). For the election campaign see the article "Wählt Einheitsschule statt Einheitsliste" (Vote for comprehensive school – not for the unified list of candidates), in: *Arbeiter-Zeitung*, 16 April 1927, 9.

The development of the settlement movement in Vienna dates back to the time of around 1918, when the city was suffering from economic hardship and severe food shortages. In postwar Austria, the movement was a self-help initiative that became a kind of small-scale realization of a non-capitalistic, social economy. Nevertheless, the Austrian economy remained far from total socialization, and even in the social democratic movement, most intellectual debaters rejected Neurath's ideas (Sandner 2014a, 147–150). In July 1922, Neurath wrote an article on class struggle and guild socialism, in which he described an existing and growing polarization between the proletarian and the bourgeois "front" in Austria (Neurath 1922a). In his view, the guild was a decisive instrument for the former because it helped to integrate social groups beyond industrial workers, including manual workers, clerks, bookkeepers, physicians etc. Guilds included working people of an entire economic sector, and therefore they broadened the proletarian front. The housing guild, for instance, represented among others construction workers, tenants and settlers. Altogether the guild included about 250,000 members (Sandner 2014a, 169–170).

With respect to the heavy Marxist class rhetoric he used, it is interesting to note that Otto Neurath had his own problem with respect to the strong polarization between the bourgeois and the proletarian front. As a representative of the settlement movement he published an article for the bourgeois liberal newspaper *Neue Freie Presse* (Neurath 1921d). The fact that a socialist author wrote for a non-socialist periodical was hereupon heavily criticized by an anonymous author in the *AZ* (Anonymous 1921), and Neurath had a hard time explaining why he had done so (Neurath 1921e; Sandner 2014a, 171–173).

Consequently, in August 1922, Neurath claimed that farmworkers and forest workers should be organized in the same guild and join this venture with the settlement movement (Neurath 1922b). A few months later, he announced an international meeting of the *Baugilden* in Vienna and developed some ideas on the unification or internationalization of national guild organizations (Neurath 1922c).

The second group of articles addressed the subjects of housing, architecture and settlement. In August 1922, Neurath intervened into debates on social policy and proposed a new measure against unemployment: Public funding for workers in particular economic sectors, for instance in cooperative housing estates. The idea was that communities, corporations etc. get the out-of-work benefits to create new jobs (Neurath 1922d). In the following year, he discussed the so-called one-kitchen house (*Einküchenhaus*). Its idea was that collectively organized meals for all tenants in a house reduced women's housework and enabled them to do paid work. Neurath discussed the idea critically, because many working class families refused to live in such houses, he argued. In fact, it were mainly bourgeois intellectuals who favored this way of living. Nevertheless, he presented it as *one* possible way of proletarian living – as was living in the settlement (Neurath 1923a). In a later article, he argued for an architectural program for the rapidly changing city of Vienna (Neurath 1923b). Consequently, he favored the idea of a general plan of architecture (*Generalarchitekturplan*) (Neurath 1923c). Especially in these articles, he demonstrated that he had remained a planning enthusiast: Only an architectural plan could bring the necessary architectural unity. Whether the city of Vienna followed this plan or not was a crucial question for the future standard of living of its people.

In January 1926, Otto Neurath reviewed two small books written by architects on proletarian architecture and proletarian houses of culture (Neurath 1926). He mostly agreed with the ideas of the authors and asserted affirmatively that a proletarian architecture is on its way forward. Class struggle, Neurath agreed, is also a cultural conflict, and the issue of architecture plays an important role in it. Otto Neurath was also active in the Viennese *Werkbund exhibition* of 1932. He published two articles on this topic in the *AZ*. In the first one (Neurath 1932a), he portrayed the *Werkbund* settlement as an adequate example for happy future living. The settlement of 70 houses designed by different architects in an outward district of the city of Vienna was presented as a model for modern living. In the second article (Neurath 1932b), he formulated some closing words on the ending exhibition and came to an ambivalent conclusion. Not everything planned was achieved. The project's initial idea to make an exhibition of a communal housing estate could not be realized. Nevertheless, it represented an exhibition of many different model houses and put up the housing problem for discussion.

The third group of articles focused on questions of education. In this context, Neurath reflected on the school of the future (Neurath 1923d). The organization of the educational system was a crucial issue for the working class. Future public schools, he pointed out, have to be accessible regardless of social class. The existing schools, however, were based on a strict separation between bourgeois and proletarian kids. He additionally stressed that the worldview or ideology of the existing teachers would also be decisive for future political development. Therefore, the question needed to be raised of how to exert appropriate political (socialist, he meant) influence on them.

Questions of school and education were among the strongest ideological conflicts in interwar Austria, pitting conservative-clericals (and/or German nationalists) on one side against liberals and Marxists on the other. In an article from 1923, Neurath reported on the case of educator Gustav Wyneken, who was accused of pederasty (Neurath 1923e). For Neurath, Wyneken was a bourgeois reformer, not a socialist, and therefore not an ally of the working class. He pointed out that Wyneken's educational reform focused on the bourgeois youth, not the proletarian one.[8] Nevertheless, he remained distanced not only with Wyneken but also with his critics, and did not join in the storm of protest against the educator. Moreover, he did *not* condemn pederasty at all. This lack of moral condemnation, however, was the reason for reactionary protests by rightwing periodicals such as *Reichspost*, which polemicized heavily against Neurath (Anonymous 1923). There was even an aftermath of this article in a libel case of a Catholic educational organization against a journalist of the *AZ*. In this case, Otto Glöckel, the former state secretary of education and head of the Vienna school board, surveyed as a witness in the trial, was faced with Neurath's article. What did he say to this article – did it represent the socialist perception he was asked? Glöckel replied only cagily (Anonymous 1924).

[8] On Otto Neurath and the Youth Movement, see Sandner (2019b).

Neurath also discussed the problem of career advice in a class society and stressed the fact that education and choice of career always depended on class, which was more important than individual capability and competence (Neurath 1923f). He also focused on the differences between proletarian and bourgeois-liberal education reforms (the title, which denoted a "proletarian-bourgeois reform" was obviously an editorial mistake), and stressed the fact that this differentiation was a crucial issue for socialist policy (Neurath 1923g).

The fourth group of articles dealt with socialism and the scientific world-conception. Neurath reviewed a collection of essays of Ernst Mach, edited by his son Ludwig Mach. Ernst Mach was, as Neurath pointed out, not only an extraordinary physician but also an important forerunner of a scientific world-conception (Neurath 1921d). Later he reviewed a book of the Jesuit Heinrich Pesch (Neurath 1924). It was a book against socialism, including any Christian version of socialist thought. Neurath critically compared the author's views with those of Karl Vogelsang and his writings in the nineteenth century. In contrast to Pesch, Vogelsang had demonstrated, in Neurath's eyes at least, strong sympathies with Christian socialism. The idea of his article was to demonstrate that contemporary Christian thought had moved away from their former social and socialist-friendly tendencies.

In the following year, he presented Epicurus, especially his lessons on happiness, as a forerunner of socialism and as a counterpart to Christian morality (Neurath 1925a). For Neurath, Epicurus had paved the way to an ethics beyond theology. As in his later book, *Personal Life and Class Struggle* (Neurath 1928/1973), he subsequently interpreted Marxism as Social-Epicureanism.

A very remarkable example of Neurath's contributions to the *AZ* is an article on the scientific world-conception (Neurath 1929). In fact, it is about scientific socialism, its development, its rise and its relevance. He claimed for an approach to workers' education based upon modern science. Neurath explained that historically, the scientific world-conception was a child of the eighteenth-century enlightenment. While enlightenment and materialism were supported by the educated bourgeoisie, in the twentieth century it was only the working class who was an ally of enlightenment and scientific thinking. The bourgeoisie, in contrast, defended reactionary thoughts of theology and metaphysics. This article was printed shortly after the publication of the famous manifesto *Scientific World Conception. The Vienna Circle*, and only a few days after philosopher Moritz Schlick had received the booklet which was dedicated to him (cf. Mulder 1968/2012). Although the title was the same as that of the manifesto, its content differed considerably.

Beyond these four groups of articles, there was only one of Otto Neurath's contributions for the *AZ* that addressed a different subject: In a rather surprising article, he wrote on the Indian reformer and pacifist Mahatma Gandhi (Neurath 1925b). Obviously, India and ideas from the Orient were very popular in these years. Neurath's article, however, was not uncritical. He criticized, for instance, Gandhi's "unworldly" proposal of stopping to beget Indian children under colonial rule; however, the article was obviously written with strong sympathy with his fight against economic exploitation against the backdrop of the colonial history of India.

Interestingly, and in contrast to many other periodicals, including socialist ones, Neurath never wrote about visual education and his *Vienna Method of Picture Statistics* in the *AZ*.

4.4 Intellectual Interventions into Politics

Between the two journals in question, *Economy and Order of Life* and *Workers' Newspaper*, there was obviously a political difference. The supplement of the *Art Guardian* resp. *German Will of the Art Guardian* (*Deutscher Wille des Kunstwart*), was a joint venture of Neurath and Schumann with a heterogeneous and even bizarre ideological formation. It is questionable in many respects to what extent the political opinions and ideas of both sides were matched at all. Both men, indeed, had worked for the *Kunstwart* before. In the midst of Neurath's controversial article of 1917 on "the reversed Taylor System," there was an announcement of an Anti-British call to sign war bonds (Neurath 1917, 21). Moreover, the journal included a number of German nationalistic articles. How to remain objective, neutral and non-partisan? Although Neurath was neither a warmonger nor a nationalist, his idea of himself as an economic expert and a social engineer seemed to help him to bridge the resulting political gap.

The situation in Vienna was different. The *AZ* was a social-democratic, leftwing periodical. Neurath was only occasionally an author there and had no editorial role, but his articles were mostly in line with the editorial policy, even though the party did never follow his ideas on socialization. In fact, he was not only a political but also a party political writer.

Therefore, the two series of articles differed in some respects. While in *Economy and Order of Life* he focused only on socialization, he addressed a number of different topics in the *AZ*. While in *Economy and Order of Life* he used pseudonyms, he published for the social democratic daily only with his real name. And while *Economy and Order of Life* was – at least in its first issues – also *his* periodical (he directed it together with Wolfgang Schumann), he only was one of many contributors to the *AZ*. Nevertheless, there were also common features and themes.

One of these subjects was the question of his role and his self-image. First and foremost, he was a social engineer, as he consistently revealed. But even as a commentator and debater, as someone who intervened into politics, he insisted on his role of an expert. Obviously, he never saw his role in terms of politics. He was convinced that science and politics could be separated clearly – and he was always on the side of science, even as a writer of articles for newspapers.

Beyond that, however, there were even approaches to closely related themes that he developed more precisely only in his later life: the question of experts, citizens, and democracy.

In his articles on socialization, for instance, he developed the idea that – although the economic plans were designed only by experts and socialization needed, in general, a large number of experts – it is, in the end, the people who decide about different options. Ordinary men, the proletariat, should choose the plan that will be put into practice. A similar idea appears in some *AZ* articles. Although the experts (architects, for instance) work out their concepts for a future way of life, it will be the working-class, the people who have to decide how they will live. The debates on the *Werkbund* settlement or the one-kitchen house are striking examples. If working-class families did not want to live in those ways, other options must be chosen.

It was only in his later life that he linked this approach systematically with visual education as an instrument that enables necessary collaboration between experts and ordinary people, presenting Isotype, the International System of Typographic Picture Education, as a way to bridge the epistemic gap (Neurath 1996).[9]

4.5 A List of Otto Neurath's Articles

4.5.1 *Otto Neurath's Articles in* Wirtschaft und Lebensordnung

Neurath, Otto, and Wolfgang Schumann. 1919. Zur Einführung. *Wirtschaft und Lebensordnung* 1 (1): 1–2.

Fonsow (= Otto Neurath). 1919. Zur Sozialisierung der Wirtschaft. *Wirtschaft und Lebensordnung* 1 (1): 2–4.

Wilhelm, Karl (= Otto Neurath). 1919. Zwischenstaatliche Naturalwirtschaft. *Wirtschaft und Lebensordnung* 1 (1): 4–5.

Nth. (= Otto Neurath). 1919. Utopien. *Wirtschaft und Lebensordnung* 1 (1): 5–6.

Fonsow (= Otto Neurath). 1919. Zur Sozialisierung der Wirtschaft. *Wirtschaft und Lebensordnung* 1 (2): 9–11.

K.W. (= Otto Neurath). 1919. Volkswirtschaftliche Lehrerkurse. *Wirtschaft und Lebensordnung* 1 (2): 13–14.

Nth. (= Otto Neurath). 1919. Der achtstündige Arbeitstag. *Wirtschaft und Lebensordnung* 1 (2): 14–15.

Wilhelm, Karl (= Otto Neurath). 1919. Papiergeldhamstern und kein Ende! *Wirtschaft und Lebensordnung* 1 (2): 15–16.

Wilhelm, Karl (= Otto Neurath). 1919. Vom "guten Gelde." *Wirtschaft und Lebensordnung* 1 (3): 20–21.

Fonsow (= Otto Neurath). Zur Sozialisierung der Wirtschaft. *Wirtschaft und Lebensordnung* 1 (3): 21–22.

Neurath, Otto. 1919. Kriegswirtschaft–Übergangswirtschaft–Verwaltungswirtschaft. *Wirtschaft und Lebensordnung* 1 (3): 23.

Wilhelm, Karl (= Otto Neurath). 1919. Steuernot, Hamsternot, Zahlungsnot. Über uneinlösliches Girogeld. *Wirtschaft und Lebensordnung* 1 (5): 33–34.

[9] On ISOTYPE and pictorial education, see Elisabeth Nemeth's, Angélique Groß's, and Silke Körber's chapters in the present volume.

Fonsow (= Otto Neurath). 1919. Zur Sozialisierung der Wirtschaft. *Wirtschaft und Lebensordnung* 1 (5): 34–37.
Wilhelm, Karl (= Otto Neurath). 1919. Waren statt Geld. *Wirtschaft und Lebensordnung* 1 (5): 37–38.
Wilhelm, Karl (= Otto Neurath). Arbeitskunst und Arbeitskunde. *Wirtschaft und Lebensordnung* 1 (6): 41–42.
Fonsow (= Otto Neurath). 1919. Staatsschulden und Volksvermögen. *Wirtschaft und Lebensordnung* 1 (6): 44–45.
Anonymous (= Otto Neurath). 1919. Das Taylorsystem in der Zukunftswirtschaft. *Wirtschaft und Lebensordnung* 1 (7): 49–50.
Neurath, Otto. 1919. Landwirtschaftlicher Kredit. *Wirtschaft und Lebensordnung* 1 (7): 53–54.
Anonymous (= Otto Neurath). 1919. Fiskus und Sozialisierung. *Wirtschaft und Lebensordnung* 1 (7): 54–56.
Neurath, Otto. 1919. Bauerngeist. *Wirtschaft und Lebensordnung* 1 (8): 62–63.
Fonsow (= Otto Neurath). 1919. Sozialisierung und Landwirtschaft. *Wirtschaft und Lebensordnung* 1 (11): 87–88.

4.5.2 Otto Neurath's Articles in Arbeiter-Zeitung

Neurath, Otto. 1919. Vollsozialisierung oder Teilverstaatlichungen. *Arbeiter-Zeitung,* 27 March 1919: 2.
Neurath, Otto. 1921. Der Gildensozialismus und unsere Zukunft. *Arbeiter-Zeitung,* 18 February 1921: 2.
Neurath, Otto. 1921. Ernst Machs Vermächtnis. *Arbeiter-Zeitung,* 27 July 1921: 5.
Neurath, Otto. 1921. Gildenliberalismus und Gildensozialismus. *Arbeiter-Zeitung,* 4. September 1921: 9.
Neurath, Otto. 1921. *Kleingärtner und Siedler.* Neue Wirtschaft, neues Leben. *Arbeiter-Zeitung,* 20 November 1921: 7.
Neurath, Otto. 1921. Kleingärtner und Siedler. Auf dem Wege zur Baugilde. *Arbeiter-Zeitung*, 18 December 1921: 12.
Neurath, Otto. 1922. Walter Rathenaus Gesellschafts- und Lebensanschauung. *Arbeiter-Zeitung,* 9 July 1922: 7–8.
Neurath, Otto. 1922. Klassenkampf und Gildensozialismus. *Arbeiter-Zeitung,* 17 July 1922: 2.
Neurath, Otto. 1922. Produktive Arbeitslosenfürsorge. *Arbeiter-Zeitung,* 9 August 1922: 5–6.
Neurath, Otto. 1922. Der Gildengedanke in der Land- und Forstarbeiterbewegung. *Arbeiter-Zeitung,* 14 August 1922: 3–4.
Neurath, Otto. 1922. Zusammenschluß der Baugilden aller Länder. *Arbeiter-Zeitung,* 3 October 1922: 3.
Neurath, Otto. 1923. Einküchenhaus. *Arbeiter-Zeitung,* 2 June 1923: 6–7.

Neurath, Otto. 1923. Zukunftschule und Arbeiterbewegung. *Arbeiter-Zeitung*, 10 July 1923: 7–8.
Neurath, Otto. 1923. Proletarisch-bürgerliche und liberale Erziehungsreform. *Arbeiter-Zeitung,* 24 July 1923: 6–7.
Neurath, Otto. 1923. Erziehung und Eros. *Arbeiter-Zeitung,* 28 July 1923: 7–8.
Neurath, Otto. 1923. Berufsberatung und Klassenkampf. *Arbeiter-Zeitung,* 9 August 1923: 7.
Neurath, Otto. 1923. Wie sollen die neuen 25.000 Wohnungen gebaut werden? *Arbeiter-Zeitung,* 24 October 1923: 6–7.
Neurath, Otto. 1923. Neu-Wiens Gesamtarchitektur – eine Aufgabe des Proletariats. *Arbeiter-Zeitung,* 28 October 1923: 5–6.
Neurath, Otto. 1924. Von Vogelsang bis Pesch. *Arbeiter-Zeitung*, 13 August 1924: 3–4.
Neurath, Otto. 1925. Im Kampfe für das Spinnrad. *Arbeiter-Zeitung*, 31 January 1925: 10.
Neurath, Otto. 1925. Auf Epikurs Wegen. *Arbeiter-Zeitung*, 22 June 1925: 7.
O.N. (= Otto Neurath). 1926. Proletarische Architektur. *Arbeiter-Zeitung.* 31 January 1926: 20.
Neurath, Otto. 1929. Wissenschaftliche Weltauffassung. *Arbeiter-Zeitung*, 13 October 1929: 17–18.
Neurath, Otto. 1932. Glückliches Wohnen. Die Bedeutung der Werkbundsiedlung für die Zukunft. *Arbeiter-Zeitung,* 19 June 1932: 10.
Neurath, Otto. 1932. Ein Schlußwort zur Werkbundsiedlung. *Arbeiter-Zeitung,* 6 August 1932: 6.

References

Anonymous. 1921. Sozialdemokraten bei der 'N.Fr.Pr.' *Arbeiter-Zeitung*, 20 April 1921: 4.
———. 1923. Seelenfang mit Steuergeldern. Wer ist Otto Neurath? *Reichspost,* 27 November 1923: 2.
———. 1924. Das sensationelle Ende eines Schulprozesses. *Reichspost* 4: 7.
Avenarius, Ferdinand. 1919. Nach den Münchner Wochen. *Der Kunst- und Kulturwart* 32. Jg. (April to June 1919): 127–128.
Avenarius, Ferdinand, Georg Callwey, Otto Neurath, and Wolfgang Schumann. 1919. Eine Erweiterung des Kunstwarts. *Deutscher Wille des Kunstwart* 32 (1): 1–2.
Carnap, Rudolf. 1963. Intellectual Autobiography. In *The Philosophy of Rudolf Carnap*, ed. Paul A. Schilpp, 3–84. La Salle: Open Court.
Cartwright, Nancy, Jordi Cat, Lola Fleck, and Thomas Uebel. 1996. *Otto Neurath. Philosophy between Science and Politics*. Cambridge: Cambridge University Press.
Dachs, Herbert. 1995. Das Parteiensystem. In *Handbuch des politischen Systems Österreichs. Erste Republik 1918–1933*, ed. Emmerich Tálos et al., 143–159. Wien: Manz.
Eulenburg, Franz. 1916/17. Zur Theorie der Kriegswirtschaft. *Archiv fur Sozialwissenschaft und Sozialpolitik* 43: 349–358.
———. 1918. Die wissenschaftliche Behandlung der Kriegswirtschaft. *Archiv fur Sozialwissenschaft und Sozialpolitik* 44: 775–785.

Fonsow (= Otto Neurath). 1919a. Zur Sozialisierung der Wirtschaft. *Wirtschaft und Lebensordnung* 1 (1): 2–4.
———. 1919b. Zur Sozialisierung der Wirtschaft. *Wirtschaft und Lebensordnung* 1 (2): 9–11.
———. 1919c. Zur Sozialisierung der Wirtschaft. *Wirtschaft und Lebensordnung* 1 (3): 21–22.
———. 1919d. Zur Sozialisierung der Wirtschaft. *Wirtschaft und Lebensordnung* 1 (5): 34–37.
Gerlich, Rudolf. 1980. *Die gescheiterte Alternative. Sozialisierung in Österreich nach dem Ersten Weltkrieg*. Wien.
Haller, Rudolf, and Ulf Höfer, eds. 1998. *Otto Neurath. Gesammelte ökonomische, soziologische und sozialpolitische Schriften*. Teil 1 and Teil 2. Wien: Hölder, Pichler, Tempsky.
Haller, Rudolf, and Robin Kinross, eds. 1991. *Otto Neurath. Gesammelte bildpädagogische Schriften*. Wien: Hölder/Pichler/Tempsky.
Haller, Rudolf, and Heiner Rutte, eds. 1981. *Otto Neurath. Gesammelte philosophische und methodologische Schriften*. Teil 1 and Teil 2. Wien: Hölder/Pichler/Tempsky.
Hegselmann, Rainer. 1979. Otto Neurath: Empiristischer Aufklärer und Sozialreformer. In *Otto Neurath: Wissenschaftliche Weltauffassung, Sozialismus und Logischer Empirismus*, 7–78. Frankfurt am Main: Suhrkamp.
Kranold, Hermann. 1919. Planloses und planmäßiges Wirtschaften. *Wirtschaft und Lebensordnung* 1 (6): 42–44.
Kratzsch, Gerhard. 1969. *Kunstwart und Dürerbund. Ein Beitrag zur Geschichte der Gebildeten im Zeitalter des Imperialismus*. Göttingen: Vandenhoeck & Ruprecht.
Krüger, Dieter. 1997. Kriegssozialismus. Die Auseinandersetzungen der Nationalökonomen mit der Kriegswirtschaft. In *Der Erste Weltkrieg. Wirkung - Wahnehmung - Analyse*, ed. Michalka, Wolfgang, 506–529. Weyarn: Seehamer Verlag.
Maderthaner, Wolfgang. 1995. Die Sozialdemokratie. In *Handbuch des politischen Systems Österreichs. Erste Republik 1918–1933*, eds. Emmerich Tálos et al., 177–194. Wien: Manz.
Moellendorff, Wichard. 1916. *Deutsche Gemeinwirtschaft*. Berlin: Siegismund.
Mulder, Henk L. 1968/2012. The Scientific World-Conception. The Vienna Circle. In *Wissenschaftliche Weltauffassung. Der Wiener Kreis*, ed. Friedrich Stadler and Thomas Uebel, 259–264. Wien and New York: Springer.
Neurath, Otto. 1913. Die Kriegswirtschaftslehre als Sonderdisziplin. *Weltwirtschaftliches Archiv* Band 1: 342–348.
———. 1917. Das umgekehrte Taylorsystem. Auch etwas zur Auslese der Tüchtigen. *Deutscher Wille des Kunstwarts* 31 (1917): 19–25.
———. 1919. Vollsozialisierung oder Teilverstaatlichungen. *Arbeiter-Zeitung*, 27 March 1919: 2–3.
———. 1919/1973. Character and Course of Socialization. In Neurath and Cohen (1973): 135–150.
———. 1920. *Bayerische Sozialisierungserfahrungen*. Wien: Neue Erde.
———. 1920/1973. Experiences of Socialization in Bavaria. In Neurath and Cohen (1973): 18–28.
———. 1920/2004. Total Socialization. Of the two Stages of the Future to Come. In Uebel and Cohen (2004): 371–404.
———. 1921a. Der Gildensozialismus und unsere Zukunft. *Arbeiter-Zeitung*, 18 February 1921: 2.
———. 1921b. Gildenliberalismus und Gildensozialismus. *Arbeiter-Zeitung*, 4 September 1921: 9.
———. 1921c. Kleingärtner und Siedler. Auf dem Wege zur Baugilde. *Arbeiter-Zeitung*, 18 December 1921: 12.
———. 1921d. Die österreichische Siedlungsbewegung. *Neue Freie Presse*, 19 April 1921.
———. 1921e. Ernst Machs Vermächtnis. *Arbeiter-Zeitung*, 27 July 1921: 5.
———. 1922a. Klassenkampf und Gildensozialismus. *Arbeiter-Zeitung*, 17 July 1922: 2.
———. 1922b. Der Gildengedanke in der Land- und Forstarbeiterbewegung. *Arbeiter-Zeitung*, 14 August 1922: 3–4.
———. 1922c. Zusammenschluß der Baugilden aller Länder. *Arbeiter-Zeitung*, 3 October 1922: 3.

———. 1922d. Produktive Arbeitslosenfürsorge. *Arbeiter-Zeitung*, 9 August 1922: 5–6.
———. 1922e. Vollsozialisierung und gemeinwirtschaftliche Anstalten. *Der Kampf* 15 (1922): 54–60.
———. 1923a. Einküchenhaus. *Arbeiter-Zeitung*, 2 June 1923: 6–7.
———. 1923b. Wie sollen die neuen 25.000 Wohnungen gebaut werden? *Arbeiter-Zeitung*, 24 October 1923: 6–7.
———. 1923c. Neu-Wiens Gesamtarchitektur – eine Aufgabe des Proletariats. *Arbeiter-Zeitung*, 28 October 1923: 5–6.
———. 1923d. Zukunftschule [sic!] und Arbeiterbewegung. *Arbeiter-Zeitung*, 10 July 1923: 7–8.
———. 1923e. Erziehung und Eros. *Arbeiter-Zeitung*, 28 July 1923: 7–8.
———. 1923f. Berufsberatung und Klassenkampf. *Arbeiter-Zeitung*, 9 August 1923: 7.
———. 1923g. Proletarisch-bürgerliche und liberale Erziehungsreform. *Arbeiter-Zeitung*, 24 July 1923: 6–7.
———. 1924. Von Vogelsang bis Pesch. *Arbeiter-Zeitung*, 13 August 1924: 3–4.
———. 1925a. Auf Epikurs Wegen. *Arbeiter-Zeitung*, 22 June 1925: 7.
———. 1925b. Im Kampfe für das Spinnrad. *Arbeiter-Zeitung*, 31 January 1925: 10.
———. 1928/1973. Personal Life and Class Struggle. In Neurath and Cohen (1973): 249–298.
———. 1929. Wissenschaftliche Weltauffassung. *Arbeiter-Zeitung*, 13 October 1929: 17–18.
———. 1932a. Glückliches Wohnen. Die Bedeutung der Werkbundsiedlung für die Zukunft. *Arbeiter-Zeitung*, 19 June 1932: 10.
———. 1932b. Ein Schlußwort zur Werkbundsiedlung. *Arbeiter-Zeitung*, 6 August 1932: 6.
———. 1996. Visual Education: Humanisation versus Popularisation. In *Encyclopedia and Utopia. The Life and Work of Otto Neurath (1882–1945)*, ed. Elisabeth Nemeth and Friedrich Stadler, 245–335. Dordrecht: Kluwer.
Neurath, Marie, and Robert S. Cohen, eds. 1973. *Empiricism and Sociology*. Dordrecht: D. Reidel.
———, eds. 1983. *Otto Neurath: Philosophical Papers 1913–1946*. Dordrecht: D. Reidel.
Neurath, Otto, and Wolfgang Schumann. 1919. Zur Einführung. *Wirtschaft und Lebensordnung* 1 (1): 1–2.
O.N. (= Otto Neurath) 1926. Proletarische Architektur. *Arbeiter-Zeitung*, 31 January 1926: 20.
Pelinka, Peter, and Manfred Scheuch. 1989. *100 Jahre AZ. Die Geschichte der Arbeiter-Zeitung*. Wien und Zürich: Europa Verlag.
Rathenau, Walter. 1918. *Die neue Wirtschaft*. Berlin: Fischer.
Sandner, Günther. 2006. *Engagierte Wissenschaft. Austromarxistische Kulturstudien und die Anfänge der britischen Cultural Studies*. Wien: Berlin: Lit.
———. 2011. The German Climate and Its Opposite: Otto Neurath in England, 1940–45. In *Political exile and exile politics in Britain after 1933*, ed. A. Grenville and A. Reiter, 67–85. Amsterdam/New York: Rodopi.
———. 2014a. *Otto Neurath: Eine politische Biographie*. Wien: Zsolnay.
———. 2014b. Political Polyphony. Otto Neurath and Politics Reconsidered. In *European Philosophy of Science – Philosophy of Science in Europe and the Viennese Heritage*, ed. Maria Carla Galavotti, Elisabeth Nemeth, and Friedrich Stadler, 211–222. Dordrecht: Springer.
———. 2014c. Was Menschenkraft zu leisten vermag: Otto Neurath und die Kriegswirtschaftslehre. In *Frontwechsel. Österreich-Ungarns "Großer Krieg" im Vergleich*, ed. Wolfram Dornik, Julia Walleczek-Fritz, and Stefan Wedrac, 377–397. Wien/Weimar: Köln/Böhlau.
———. 2017. Deviationist Perceptions of the Balkan Wars: Leon Trotsky and Otto Neurath. In *The Balkan Wars from Contemporary Perception to Historic Memory*, ed. Katrin Boeckh and Sabine Rutar, 197–215. Basingstoke: Palgrave Macmillan.
———. 2019a. A Political Recess: Otto Neurath in the Netherlands (1934–40). In *Significs and the Vienna Circle: Intersections*, ed. Gerard Alberts, Luc Bergmans, and Fred Muller. Cham: Springer. (Forthcoming).
———. 2019b. On the road to logical empiricism. Philosophers of science and young Germany. In *Logical empiricism, life reform, and the German youth movement*, ed. Christian Damböck, Günther Sandner, and Meike Werner. Cham: Springer. (Forthcoming.)

Schumann, Wolfgang. 1919a. Das deutsche Wirtschaftsmuseum. *Wirtschaft und Lebensordnung* 1 (1): 6–7.

———. 1919b. Das deutsche Wirtschaftsmuseum. *Wirtschaft und Lebensordnung* 1 (6): 45–47.

———. (= Sch.) 1919c. Zum Ende der 'Planwirtschaft'. *Wirtschaft und Lebensordnung* 1 (14): 102–103.

———. 1919d. Totgemeldet usw. Eigene Erfahrungen mit der Presse. *Der Kunst- und Kulturwart* 32 (3): 166–167.

———. (= Anonymus) 1919e. Münchner Beobachtungen und Erfahrungen 2. *Der Kunst- und Kulturwart*, 32. Jg. (June to September 1919): 77–78.

———. (= W. Sch.) 1919f. Münchner Beobachtungen und Erfahrungen 3. *Der Kunst- und Kulturwart*, 32. Jg. (June to September 1919): 128–129.

Stadler, Matthias Michael. 1990. *Die Situation der Presse in der Ersten Republik mit besonderer Berücksichtigung der Entwicklung der "Arbeiter-Zeitung" und der "Reichspost" und der politischen Auseinandersetzung zwischen Sozialdemokraten und Christlichsozialen von 1919–1934 in diesen beiden Zeitungen*. Diplomarbeit. Universität Wien.

Uebel, Thomas. 2005. Political Philosophy of science in logical empiricism: The left Vienna Circle. *Studies in History and Philosophy of Science* 36: 754–773.

Uebel, Thomas, and Robert S. Cohen, eds. 2004. *Otto Neurath: Economic Writings. Selections 1904–1945*. Dordrecht: Kluwer.

Weissel, Erwin. 1976. *Die Ohnmacht des Sieges. Arbeiterschaft und Sozialisierung nach dem Ersten Weltkrieg in Österreich*. Wien: Europaverlag.

Wilhelm, Karl. (= Otto Neurath) 1911. Kriegswesen und Volkswohlfahrt. *Der Kunstwart* 24 (3): 345–349.

———. (= Otto Neurath) (K.W.) 1919a. Volkswirtschaftliche Lehrerkurse. *Wirtschaft und Lebensordnung* 1 (2): 13–14.

———. (= Otto Neurath) 1919b. Zwischenstaatliche Naturalwirtschaft. *Wirtschaft und Lebensordnung* 1 (1): 4–5.

———. (= Otto Neurath) 1919c. Vom guten Gelde. *Wirtschaft und Lebensordnung* 1 (3): 20–21.

———. (= Otto Neurath) 1919d. Steuernot, Hamsternot, Zahlungsnot. Über uneinlösliches Girogeld. *Wirtschaft und Lebensordnung* 1 (5): 33–34.

Chapter 5
United by Action: Neurath in England

Adam Tamas Tuboly

Abstract The aim of this paper is to give a biographical, historical, and philosophical reconstruction of Neurath's final years in England. Besides reconstructing Neurath's arrival to England, in the context of his life and philosophical introduction at Oxford, I will argue that since the 1930s, Neurath was eager to develop a brand for logical empiricism. This brand was based not on theoretical commitments, but on practical considerations and decisions. Using a detailed case study on Neurath's relation to the Hungarian sociologist of knowledge Karl Mannheim, I show that the development of their connections documents how Neurath gave more and more priority to practical aims during his English years. Finally, the concluding section points to some further considerations on Neurath's legacy.

5.1 Introduction

When Otto Neurath settled down in Oxford after his 8 months long imprisonment in 1941, he arrived at a well-prepared field for logical empiricism.[1] After he attended the *Seventh International Congress of Philosophy* (1930), Schlick was invited to deliver three talks at King's College, London (1932). Just 2 years later, the British philosopher and logician Susan Stebbing invited Carnap to talk about "Philosophy

[1] This paper was supported by the MTA BTK Lendület *Morals and Science* Research Group; by the János Bolyai Research Scholarship of the Hungarian Academy of Sciences and by the "Empiricism and atomism in the twentieth-century Anglo-Saxon philosophy" NKFI project (124970), and by the MTA Premium Postdoctoral Scholarship. I am indebted to Amy Wuest for correcting many linguistic errors in the paper. I am also indebted to the Carnap Archive at Pittsburgh (Rudolf Carnap Papers, 1905–1970, ASP.1974.01, Special Collections Department, University of Pittsburgh) and to the Otto Neurath Nachlass (ONN, Wiener Kreis Archiv, Rijksarchief in Noord-Holland, Haarlem, The Netherlands) for the permission to quote the archive materials. All the translations from the archive files are mine.

A. T. Tuboly (✉)
Institute of Philosophy, Hungarian Academy of Sciences, Budapest, Hungary
e-mail: tuboly.adam@btk.mta.hu

J. Cat, A. T. Tuboly (eds.), *Neurath Reconsidered*,
Boston Studies in the Philosophy and History of Science 336,
https://doi.org/10.1007/978-3-030-02128-3_5

and Logical Syntax." Four years later, Cambridge University hosted the *Congress for the Unity of Science*, organized by Stebbing and Neurath.

In addition to these developments, individual scholars reached out to Viennese philosophers as well. A. J. Ayer attended some meetings of the Vienna Circle in 1932–1933 and published his famous text that determined the future course of logical empiricism in the English-speaking world for decades, *Language, Truth and Logic* in 1936. Max Black, who gave a talk at the *Fourth Congress for the Unity of Science*, wrote to Neurath in 1935 that he wanted to edit a collection of papers on "Logical Positivism [...] partly by members of the Viennese Circle and partly by English and American writers of empirical sympathies." Black named Ayer, Richard Braithwaite, Carl Hempel, Charles Morris, Hans Reichenbach, Schlick, Carnap, and Stebbing. Neurath was asked to write about physicalism, which he happily accepted; for unknown reasons though, nothing resulted from that project.[2]

Neurath started his new career in Oxford in the context of these communications. Nevertheless, he worked not with philosophers, but mainly with anthropologists, sociologist, and educators. Even though he published some academic papers and his monograph in the *International Encyclopedia of Unified Science*, he praised England mainly for its "muddle," i.e. for its practice-oriented, pluralist, and tolerant worldview. This is also reflected by his legacy; while he was remembered by many as the developer of ISOTYPE works, a public intellectual in debates about Germany, his philosophical legacy was overshadowed by such oversimplified accounts of logical empiricism such as Ayer's, which achieved a Bible-like status after the war in Oxford.

The aim of this paper is to reconstruct some cornerstones of this highly and unfortunately neglected period of Neurath's life and works. Section 5.2 describes Neurath's arrival to England, in the context of his life and philosophical introduction at Oxford. Section 5.3. argues for the thesis that since the 1930s, Neurath was eager to develop a brand for logical empiricism, which was based not on theoretical commitments, but on practical considerations and decisions. Section 5.3.2. gives a detailed case study on Neurath's relation to the Hungarian sociologist of knowledge, Karl Mannheim, who worked during the 1940s in London; the development of their connections documents how Neurath gave priority to practical aims during his English years. Finally, Sect. 5.4. points to some further considerations on Neurath's legacy.

5.2 Arriving and Settling Down in England

Neurath was a traveler – sometimes by his own decisions, but occasionally he was forced to move throughout Europe. These events occasionally caused permanent troubles and resulted in significant personal difficulties for him as can be seen from his letter to Rudolf Carnap that was written with some emotional overtones in 1933:

[2] Max Black to Neurath, 16 November 1935 and Neurath to Max Black, 4 December 1935 (ONN).

"I hope to see all of you before the possible longer separation begins. After the German revolution [1918/19] I did not see my friends for six years… That is a long time. It passes. […] Everything is awful and it is very difficult to keep up the hope which belief in the overall trend of history gives us."[3]

In the letter, Neurath presumably had in mind those dramatic events that took place between fascist and socialist groups in Vienna during 1933 and 1934. After all, his nightmares came true when fascist forces took power after the short-lived civil war in February 1934. At that time, Neurath was in the Soviet Union and when he heard the news from Marie Reidemeister (later his third wife, then a collaborator on pictorial statistics), he was advised not to return to Vienna but instead to visit his friends – Philipp Frank and Rudolf Carnap – in Prague. Neurath was never able to return to Austria, and his forced exile just began: traveling through Poland and Denmark, he fled to The Hague where he settled with his second wife Olga Hahn, and Marie.

During his time in The Hague, Neurath did not rest, not even for a second. Rather, he immediately started to organize the *International Congresses for the Unity of Science*, arranged the publication of the *International Encyclopedia of Unified Science*, and published popularizing accounts of his ISOTYPE-work for the English-speaking world (Neurath 1936, 1937, 1939). As the word spread about his work internationally, Neurath was invited to lecture about ISOTYPE and to establish further ISOTYPE related museums and institutions around the world: he traveled to Mexico, the Soviet Union and to the United States. Nevertheless, his work in the Netherlands was interrupted – as he used to say – by the spreading "Nazi-plague."[4]

5.2.1 From The Hague to the Isle of Man

On May 10, 1940, the Third Reich parachuted into the Netherlands. Neurath and Marie hid in their house where "[f]ood was brought to the door; many visitors even came to see [them]. In between [they] played chess and listened to the radio" (M. Neurath 1973, 68). Within 4 days, the little country, caught between Germany and Belgium, surrendered; when Neurath heard the news from the radio he knew he had to make a move. His fears were well-founded: he was half-Jewish and frequently accused of communism, so he had to flee from The Hague as soon as possible.

Since the Netherlands was officially a neutral country, many Jews sought refuge there prior 1940. When the major cities fall on the eve of May 14th, a fishing harbor (and a district) of The Hague, called Scheveningen, was filled with desperate people

[3] See Neurath to Carnap, 13 March 1933 (RC 029-11-20). Quoted and translated by Thomas Uebel (2007, 300).

[4] For more on ISOTYPE, see the chapters of Angélique Groß, Sophie Hochhäusl and Silke Körber in the present volume. On Neurath's Holland-period see Cartwright, Cat, Fleck, Uebel (1996, 82–85), Sandner (2014, 234–263 and forthcoming), and Marie Neurath's memoirs (1973, 62–64).

looking for any way to escape the advancing German troops. Neurath and Marie also reached the harbors that evening, but when they arrived no evacuation ship was left. According to Marie (1973, 69), Neurath said that "if we do not find a boat I am going on a piece of wood." "After long searching," Neurath wrote to John and Jet Pront, his friends from The Hague, that "a last corner [of the harbor] remained uninspected," and they discovered one of the last boats, called "Seaman's Hope" ("Zeemanshoop"), which was hijacked earlier that evening by a crew of four students, led by their captain, Harry Hack.

With some help from the harbor and a bit of luck the crew was able to work the engines. Shortly after, people started to occupy the little boat, which was suitable for only 20–25 people. However, 44 people ultimately boarded the vessel. When Neurath and Marie discovered familiar faces, they "jumped into the boat from the wall, because it left just the shores."[5] None of the other accounts of the event describes a couple jumping to the boot after its departure, so it appears that Neurath may have exaggerated this detail. Nonetheless, after Neurath and Marie found their places at the back of the boat, a soldier fired a warning shot over the heads of those waiting to jump from the quayside, while others fell in the water.

The heavily overcrowded lifeboat pulled out from the harbor, into the dark, cold evening, heading directly to England and watching Rotterdam burn. With the usual exhilaration and trust in the future, Marie described the night as follows:

> For us this was a great adventure indeed. We were immersed in the stories of the Huguenots and their flights in fog and night over the frontiers and the sea – so there we are, now we know how it feels. We were leaving behind us the burning Rotterdam in an occupied country about to suffer foreseeable agonies. Trusting our luck to this tiny nutshell of a boat, we felt extremely happy to have found it and to be alive together, shoulder to shoulder. (M. Neurath 1973, 70.)

With an uncertain engine and a useless compass, the tiny ship followed the stars. The boat carried entire families, including children and older people, and it even navigated through minefields. By the following afternoon, they completed two-thirds of the trip. At that time, a British Royal Navy destroyer called "Venomous" stopped them and picked up all the passengers to take them to Dover.

As many other refugees, Neurath and Marie had German passports, therefore they were treated as "enemy aliens," representing the country that British soldiers were fighting against (Kochan 1983). Since he entered the country as on official enemy, Neurath was arrested immediately and brought to Pentonville Prison in London, and later to Kempton Park. He was kept – with nearly a hundred other men – in a racecourse, sleeping on mattresses on stone floors. After a few weeks, he was shipped to the Isle of Man. Marie was first kept at the Fulham Institute, where she was able to contact Stebbing. Later, she was placed in Holloway Prison before being taken to the Isle of Man.

On the Isle of Man, there were various internment towns, nonetheless men and women lived separately on different sides of the island. Many accounts exist about daily life, general conditions, and private feelings. Neurath and Marie were doing

[5] Neurath to John and Jet Pront, 30 June 1945 (ONN).

their best to keep up their spirits. As Marie (1973, 71) later recalled, "Otto was most interested in experiencing prison life." The interment, at least for the first few months, was like an international congress for Neurath: they lived in houses (Marie and Neurath were house captains in their homes), they met many scholars, made new friendships, and Neurath even gave at least one lecture to the fellow internees.[6] They went to the cinema where they could watch films "under military control," and in the camp for men they organized plays in Neurath's attic. He wrote to the Pronts that there was in the "attic no artificial light, therefore I used Christmas candles in masses, it was a great show. A dozen of them burning – and smelling."[7]

The Neuraths were in a quite exceptional position; since they had a typewriter and other related materials, they "worked out Isotypes for the USA and got even money from there,"[8] thus they were able to buy things and bring them to each other at the regular meetings. Men and women internees were allowed to meet at first monthly. Later, more regular meetings were permitted at the women's camp. Neurath was in Onchan Camp, at the mid-east part, next to the island's capital, Douglas, while Marie was interned in Port Erin, at the southwest corner. Neurath described one of these "love meetings either on a lawn or in a big dancing palace" as follows:

> We crossed the island marching to a station, from where we had to reach our ladies. That was a day, this first meeting in this women town, imagine many thousands of women, nicely dressed assembled in one town. The railroad station overcrowded with waiting girls and women, partly in bathing suits …. I have never seen before such a mass of females together. Then we marching into the town in military order, getting ice cream from nice girls, female police in bobby uniform, and then gradually splitting up of people, who met their wives or fiancées, and then sitting down on a meadow with a view on the sea, couple shoulder on shoulder the next couple and the military authorities with dame Joana on the balcony of a big hotel looking down on this love garden ….. what a story. (Neurath to John and Jet Pront, 28 July 1945. ONN.)

Altogether Neurath was pleased; when he arrived at the camp he only had slippers for shoes. Over time, he got new shoes, clothes, and books. He recalled other small luxuries, "cakes came and even Worcester Sauce, wonderful pyjamas […] complete suits and coats from the USA from one of the great department store owners" (ibid.).

Even though Neurath and Marie developed many new friendships that turned out to be quite fruitful for Neurath in England, they got bored after a while. "In the beginning," he wrote to the Pronts, "we looked at all the imprisonment as at a kind of holidays or higher joke after the Dutch tension and the lifeboat flight…." (ibid.). As Neurath summarized the whole situation, "not tormented, not persecuted, just 'interned'" (ibid.).

The news about Neurath's internment quickly spread, and after Marie notified Stebbing about their situation, the latter asked a Leonard Marsden, a lawyer, to help them get out as quickly as possible and arrange their marriage. (Though Neurath

[6] On Neurath's lecture see Michelle Henning's chapter in the present volume.

[7] Neurath to John and Jet Pront, 28 July 1945 (ONN).

[8] Neurath to John and Jet Pront, 28 July 1945 (ONN).

and Marie were fortunate to be treated as a married couple on the Isle, they got married only after their release). Marsden did everything he could: wrote applications, forwarded other's messages and kept Neurath informed through the process. Despite the support, their release was not that easy. Neurath and Marie's case was discussed in the Society for the Protection of Science and Learning, in the British Academy Tribunal; their recommendation for release was forwarded to the Home Office on September 26, 1940. As Marsden wrote to Neurath, "[t]his seems a decided step forward and it ought not to be long now before you are both at liberty."[9]

Nevertheless, the Home Office forwarded their case to another Advisory Committee and they only came to a decision about the release in early December. Then attention turned to the fight to expedite their actual release. After everything, their release only happened on February 7th (Marie) and 8th (Neurath) in 1941, after 8 months spent in various British camps. It is well known that Neurath's application for release was supported by Stebbing and Albert Einstein, but Marsden notified Neurath (on his request) about a much longer list of supporters: the evolutionary biologist Julian Huxley, Ernest Jäckh (the international director of the *New Commonwealth Society*, later a professor at Columbia), Felix Kaufmann, John Dewey, Ernest Nagel, Charles Morris, Egon Brunswik, Rudolf Carnap, Philipps Bradley, Edgar J. Kaufmann (a German-American businessman), G. J. Laing from the National Tuberculosis Association, the publisher Alfred A. Knopf, Neurath's cousin Waldemar Kaempffert, and F. E. Compton & Co.[10]

After the Neuraths were released, they went to Oxford enjoying the hospitality of the socialist political theorist (and detective story writer) George D. H. Cole, who wrote the *History of Socialist Thought* in 7 volumes. At first, they also worked in Cole's private library that was related to the *International Institute of Social History* in its exile,[11] but then they rented an entire house overlooking the hills of Oxford. Later they had to move because the landlord came back, but their friends bought a similar house a few blocks away, which the Neuraths might have rented. They loved this new life: they grew vegetables and fruits, enjoyed visiting neighborhood cats and dogs. Their home allowed them to enjoy a view of the town and cows grazing on the surrounding meadows. In addition, their home featured nice furniture and a bathtub "which is really of my size," Neurath happily reported: "usually I could not get my full pleasure, because I could not cover me … this is now possible."[12] They also rebuilt their library with the help of friends sending articles, clippings, reprints, duplicates etc. For someone living through the terrible consequences of World War One, the hunger of the years after the war and the Great Depression, this new

[9] Leonard Marsden to Neurath, 7 October 1940 (ONN).

[10] See Leonard Marsden to Neurath, 13 December 1940 (ONN). On Kaempffert's relation and its significance to Neurath, see George Reisch's chapter in the present volume.

[11] See Neurath to Carnap, 21 September 1941 and 29 July 1942 (RC 102-55-18 and RC 115-07-58). See letters 7 and 13 in this volume. On Cole and Neurath see Günther Sandner's chapter in the present volume.

[12] See Neurath to John and Jet Pront, 25 July 1945 (ONN).

lifestyle meant a radical change (the Neuraths lived in poor conditions in The Hague as well):

> In England [there] is plenty of food, if you are not so interested in particular types of food, bread is free, as you know, there are many vegetables and a real glut in milk, we enjoy it very much, dried eggs now from the US come etc we enjoy it, too. Fine fruit etc fish free (if not canned). Not the slightest similarity with the Central European situation in the last war. Not much meat, but you can get Vitamin[s] in various forms, e.g. as pills if you are very interested in these things. (Neurath to Carnap, 17 July 1942. RC 102-56-04. See letter 11 in the volume.)

What is to be noted here is the congruence of what Neurath emphasized frequently in the letters to his friends and his theoretical-economic ideas on the standard of living: "I want they [Neurath's German friends] should drop the interest in order as such, in stressing 'rights' more than peace and happiness, in stressing 'ideals' (ecstasy, enthusiasm, arts, etc science etc impartiality, etc) more than the *daily happiness of the neighbours, the own happiness, the happiness of people more far away*" (emphasis added).[13]

Neurath loved being in England: he quickly made a name for himself among many new circles, became as active as ever and enjoyed the walks, the community, as well as supporting the fight against Nazism. As he wrote to an old friend from Vienna, they "gradually become a kind of British Furniture" (quoted by Sandner 2011, 73).

5.2.2 *Oxford Lectures, Terminology, and Life in England*

Neurath was absorbed in various projects after his release. Though he was able to reestablish the longstanding Isotype Institute in 1942 with the help of Stebbing, from a philosophical point of view the most important event was his lectures at the All Souls College, Oxford. The lectures provided both difficulties and opportunities for Neurath:

> Then I had to lecture at Oxford University […], the Professor invited me and attended kindly the lectures is an anthropologist and therefore I spoke more of anthropology as I did usually. I liked it very much, but it needs some time to see how "functionalism" goes on, and how the discussion stands just at the moment. Of course I know the main lines be heart. I had to look through the history of anthropology as you had to look another day through the history of philosophy. (Neurath to Carnap, 1 April 1944. RC 102-55-05. See letter 26 in this volume.)[14]

Neurath was invited to give a course on "Logical Empiricism and the Social Sciences" by the social anthropologist Alfred Reginald Radcliff-Brown, who was

---

[13] See Neurath to Ina Carnap, 24 September 1945 (RC 102-55-13). See letter 35 in this volume.

[14] The Oxford course was the only official academic appointment of Neurath: though he habilitated at Heidelberg in 1917, he never taught a course and his permission was suspended two years later. See Sandner (2014, 103–108).

known for his structural-functionalist theory of social relations and actions. Though Neurath has dealt (sometimes critically) with the ideas of anthropologists (especially of Lucien Lévy-Bruhl and James Frazer) on magic, religion, technique and primitive mentality during the early 1930s (1930/1983, 1931, 1931/1973, 319–330), the Oxford atmosphere was different: Radcliff-Brown (1939) just published his "Frazer Lecture" on taboos, and his vocabulary on folklore might have been new and promising to Neurath's ears.

In order to facilitate the discussion, Neurath had to dig into the newest achievements of anthropology. This task was somewhat challenging even though the discussions on anthropology and the social sciences were beneficial. Neurath adapted the terms and ideas and made use of them in his *Foundations of Social Sciences* (1944/1970) and in his last manuscript, "Visual Education: Humanisation versus Popularisation" (1996) written for Karl Mannheim.[15]

According to a letter to Cambridge University Press, he was already teaching on the 22nd of February (Hilary Term), and he continued with a discussion group through the Trinity Term between April and June. In the same letter, he asked for copies of the *International Encyclopedia of Unified Science* sending them to the following scholars: the above mentioned Julian Huxley (London), the Nobel Prize winner physiologist Archibald Vivien Hill (London), Radcliff-Brown (Oxford), G. D. H. Cole (Oxford), H. H. Price (Oxford), and Edward Stuart Cartwright (who was the organizing secretary of the new tutorial classes committee until 1945 in Oxford).[16] Though it is quite unlikely that all of these scholars attended Neurath's lectures, it is still an impressive list of interested colleagues and acquaintances. Furthermore, it is known that the Neuraths wanted to participate at Price's lectures, who gave them permission to attend.[17]

Through the lectures Neurath tried to reach out to possible collaborators, old and new friends, and sympathizers as much as he could. He even invited many of them to participate at the 1941 Chicago Congress for the Unity of Science. Due to the war, however, many European scholars were simply unable to attend a conference in the United States. For them, Neurath organized the "Terminology" workshop (often mentioned by him as "our Unity of Science meeting") with Stebbing and the pedagogue Joseph A. Lauwerys (who later played an important role in developing UNESCO), on October 3–5 of 1941 in Oxford. Almost nothing is known about this conference. Yet there is evidence that demonstrates that Neurath invited the sociologist Karl Mannheim, the philosopher C. E. M. Joad, the aforementioned H. H. Price, Leonard Russell (professor of philosophy in Birmingham), and presumably many others.[18]

[15] Neurath has sent a copy of the *Foundations* to Radcliff-Brown, thanking him all the discussions and remarks. See Neurath to Radcliff-Brown, 13 January 1945 (ONN).

[16] Neurath to Cambridge University Press, 22 February 1941 (ONN).

[17] H. H. Price to Radcliff-Brown, copy to Neurath, 13 February 1941 (ONN). During the 1941 Trinity Term Price lectured on Hume's Theory of Knowledge after publishing a book on the subject (1940); though it is unknown in which semester did Neurath attend the lectures.

[18] Neurath to Price, 13 September 1941 (ONN) and Neurath to Joad, 13 September 1941 (ONN). According to the hotel-room reservation plans, Stefan Vajda (mathematician), Martin Strauss (philosopher), Patrick Meredith (lecturer in visual education), C. H. Waddington (biologist), Charles Henry Whiteley (philosopher), Leonard Russell, and a certain A. Pinsent were attending the event. See Neurath to Talbot, 25 and 29 September 1941 (ONN).

The only indication of what might have happened at the terminology workshop comes from two articles that appeared shortly after the conference: "The Danger of Careless Terminology" (1941) and "Universal Jargon and Terminology" (1941/1983), though the same account was published later in the *Foundations of the Social Sciences* (1944/1970, 2–4). Terminology is the discipline that deals with terms, expressions, phrases and linguistic items. It might be regarded as Neurath's own version of the linguistic technique that dominated English-speaking philosophy. In these works he argued that there are various forms of terminologies, e.g. empiricist and unempiricist, pluralist and anti-pluralist, scientific and metaphysical, and that the task of logical empiricism is to analyze these expressions, and, for example, to weed out non-scientific terms. This task is a never-ending job since terminology "will always be in the making" (1944/1970, 2), thus even Neurath's (1941, 147) *index verborum prohibitorum* will undergo constant revising (cf. Reisch 1996).

This process of linguistic analysis is not meant only to identify metaphysical terms in scientific papers. Neurath also wanted to call attention to the force of words by indicating how terminology might obscure or uncover ideologies and philosophical commitments:

> [W]e propose to avoid praising and blaming words in articles and books on history: one man 'kills' another man, he does not 'murder' or 'assassinate' him. An additional statements such as 'this killing, was not in concordance with the rules of the group to which the killer belonged and was regarded as murder' would be a historical one; and we would say the same of an additional statement such as 'I – the writer of the book – and my group disapprove this actions as murder'. (Neurath 1941, 147.)

While in the case of normative terms, such as those given in this example, might seem to present a somewhat manageable task, the danger associated with normative language in purely scientific texts is harder to reveal. From Neurath's perspectives, many of his fellow logical empiricists – like Carnap, Herbert Feigl, and Carl Hempel – who worked on purely technical and formal logical issues, just tried to get rid of the practical side of science by building ideal languages with abstract rigor. Neurath explained that "sharp definition and pedantic speaking" will not help in understanding how science works, "certain vagueness is unavoidable, and a pseudo-clearness is even particularly dangerous" (1941, 146); or as he said in the other paper, presented to the Aristotelian Society, "formalization is no magic sieve" (1941/1983, 213).

This may be Neurath's attempt to get closer to various British traditions: he even mentions Jeremy Bentham, Charles Ogden, I. A. Richards, Stebbing and the Cambridge School of Analysis, and Bertrand Russell (even though Neurath was highly critical of Russell's (1940) latest book). Neurath was likely well aware of the distinct British philosophy (relying on common sense, linguistic analysis and Moore's ideas) that was becoming institutionally and professionally entrenched during the 1940s in Oxford and Cambridge. Thus "Terminology" bridged his Viennese work while also allowing him to reach out to his new colleagues in England.[19]

---

[19] Neurath gave a lecture also in Cambridge (19 November 1941) under the title "Logical Empiricism and everyday problems" (ONN, K.50).

5.3 Building a Brand for Logical Empiricism: From Reichenbach to Mannheim

Considering the wide range of published material of logical empiricists, such as brochures, pamphlets, monographs, articles, reviews and reports, one might come to understand the logical empiricists' tradition as a *unified* and *harmonized movement*. There were, of course, internal debates, divergences and infightings. Many of these debates might have seemed to be signs of a *progressive research program*, rather than of a collapsing collaboration and a fragile, thin slide of ice, heated continuously by hatred, social contempt, and disdainful mockery.

We should sit back for a moment, however, and think about the history of the Vienna Circle. Since Friedrich Stadler's (2001/2015) reconstruction, the "(non-) public phase" terminology has been rigidified in the academic literature. He locates the non-public, preparatory phase between the years 1924 and 1929, and the public phase after 1929 when Carnap, Hahn and Neurath's joint manifesto appeared. Nonetheless, it might be interesting to note that Carnap taught in Vienna from 1926 until 1931, then moved to Prague, where Philipp Frank was already a professor at the German University of Prague since 1912; as a result of his work there, he only participated in the meetings of the Vienna Circle occasionally. It is not surprising thus that many scholars of the Circle described him as just a really close outsider, who attended the meetings sometimes when he came home. After 1930, Neurath was continuously on the move, spending much of his time in the Netherlands and in the Soviet Union: in 1934 he fled to The Hague as I mentioned above. Feigl emigrated to the United States in 1930, while Menger first went to Amsterdam (1925–1927) and later to America (1930–1931). Likewise Schlick traveled to North America: he went to Stanford (1929) and to Berkeley, California (1931–1932).

All of this resulted in the overturn of power relations within the Circle. Neurath wrote to Carnap in 1930 that if neither him, nor his wife, nor Menger, Feigl or Carnap is in the Circle (which was already quite usual), then Schlick and Waismann "announce to the true people, in the wreath of their two in-the-house-phenomenologists [Felix] Kaufmann and [Robert] Neumann. Hans [Hahn], the last member of 'our group' will not hold out long."[20] Gustav Bergmann (1993, 195) wrote to Neurath in 1938 (thinking about the Wittgensteinians), that "[t]he Boltzmanngasse group [i.e. the Circle], [...] had, as I see it, already reached its highpoint in 1927/28, maintained momentum for several years and by 1931/32 already showed clear signs of splintering and, as a consequence, declining."

The Circle showed the signs of decline from Schlick's point of view as well. He wrote to the English translator of his *Fragen der Ethik* in 1933: "I am not going to have any meetings of the 'Wiener Kreis' this winter. Some of our old members have grown too dogmatic and might discredit the whole movement; so I am now trying to form a new circle out of younger men who are still free from principles."[21] The

[20] Neurath to Carnap, 15 July 1930 (RC 029-14-14).

[21] Schlick to David Rynin, 4 November 1933 (MSN).

same ideas were suggested by Bergmann (1993, 199–200) as well: "the later Circle was of a completely different sort. [...] From the original radical, rationalistic outlook there was nothing remaining [...]; mathematicians, physicists and the absent members of the old Circle were replaced, in growing numbers, by the pupils of Mr. and Mrs. [Karl and Charlotte] Bühler." Due to the absence of the so-called "left wing" of the Circle, most of the attendants and visitors of the Circle reported about the enormous and quasi-religious influence of Wittgenstein on the remainders.[22]

According to Neurath, the situation did not get any better over time. The "real founder" of the Circle, Hans Hahn, died unexpectedly after a surgery. His sister (who was Neurath's wife), wrote to Carnap almost desperately:

> Our philos. front has lost a good soldier & what the Vienna Circle was earlier ceased to exist. [sic] Only [Heinrich] Neider left to counteract the Schlick-W. Group, but he is short of the mathematical-physical qualifications & with his sociological-historical knowledge he is unable to do much in this Circle. (Olga Hahn-Neurath to Carnap, 1 August 1934. RC 029-10-41.)

In the same year, the *Verein Ernst Mach* was disbanded for political reasons. Carnap described Frank's visit to the Circle in 1935 to Neurath as follows: "Frank's visit to the Schlick Circle seems to bring its end. Waismann became so furious because Frank called him publicly a 'scholastic', that he does not want to appear in the public anymore."[23]

These reports do not show a harmonized or movement-like picture of the Circle: how did the unified "Vienna Circle" or "Logical Empiricism" narrative emerge, overcoming so much hatred, tension both in personalities, attitudes and theoretical commitments? Logical empiricism, as it eventually became known, especially its Viennese form, was a result of a massive, deliberate, and direct process of *brand building*. Logical empiricists had to fashion a common terminology and a seemingly shared stance on many questions such that their shared philosophical project would be easily recognizable: one should have been able to tell who is with them, and who is not. Mapping the "us" versus "them" approach (Creath 2012) was also useful to safeguard their projects and fields of publication. For example, Neurath was often concerned that Reichenbach was taking too much power as co-editor in the *Erkenntnis* (the other editor was Carnap from Prague). Once he even wrote to Carnap that he is always angry when he sees that Reichenbach's name appears on the front page of the journal with bigger characters.[24]

Neurath's goal was to make logical empiricism "identifiable" and to show a picture of a working enterprise. The "brand" should have also guaranteed the quality of the work, since as Neurath wrote, "[w]e are known in the world as the 'Vienna Circle.'"[25] Being a fellow traveler would come with many advantages, thus Carnap's

[22] This picture is drawn, for example, by Ernest Nagel, who participated at the meetings at the turn of 1934/1935. See Nagel to Carnap, 5 January 1935 (RC 029-05-16) and 6 March 1935 (RC 029-05-14).

[23] See Carnap to Neurath, 26 March 1935 (RC 029-09-66).

[24] See Neurath to Carnap, 26 October 1932 (RC 029-12-17).

[25] See Neurath to Carnap, 10 August 1932 (RC 029-12-36).

wife wrote to Ernest Nagel that Popper's way of overestimating the minor differences between himself and the Circle is "quite stupid, because he would have more success under the banners of the Wiener Kreis, quite well known and discussed now."[26] Because of the enormous, conscious and conscientious self-management, the intended readers were able to find the works of logical empiricists. Their unified appearance and purposeful cohesive strategy provided insurance for quality and sameness in appearance. By arranging their own translations, reports of their congresses, commenting on their own works, logical empiricists aimed at dispelling misunderstandings and misinterpretations of their works.[27]

5.3.1 Early Attempts at Practical Unification

Building a brand did come with a price: to smooth things out, members of the movement had to fight over the dominant themes, expressions, ideas, and goals. According to Carnap's diaries, when Neurath saw Reichenbach's invitation to the Prague Conference, he started to "inveigh against him" because Reichenbach used the word "philosophy."[28] Neurath took things into his own hands, and wrote a letter to Reichenbach, trying to persuade him not to use "Naturphilosophie," since it is misleading and points to a "philosophical," "internal," "introspective" conception of knowledge, which is not characteristic of the movement. Neurath did not suggest "exact philosophy" ("exakter Philosophie") either since it sounds like an "ersatz" for the old philosophy.

Neurath did suggest the moniker "scientific world-conception," but by 1932 he ceased to use the word "world." So, after he left behind "worldview" because of its irrational connotations, and "world-conception" because of metaphysical implications, he had to develop with a new term. His preferred replacement was "unified science," which indeed became internationally recognized and adopted.[29]

It is to be noted that Reichenbach persisted in using "Naturphilosophie": in October 1929 (just after Neurath's letter) he wrote a report on the Prague conference in which he discussed the tasks and aims of philosophy of nature. "[I]n addition to the individual sciences," said Reichenbach (1929/1978, 258), "there exists a mode of treating and evaluating its subjects matter that is itself not scientific but philosophical instead," which is, after all, epistemological, since it explores the nature and essence of knowledge-processes.

[26] Ina Carnap to Ernest Nagel, 23 January 1935 (RC 029-05-15).

[27] Whether Neurath was successful after all, is a further question: he wrote to Carnap, 21 September 1941, that "[s]uch is life, successively we [logical empiricists], the opposition become classics. Now we form a branch like others, as I learned from Laird's Introduction to modern philosophy" (RC 102-55-18). See letter 7 in the present volume.

[28] See Carnap's diary entry, 15 July 1929 (RC 025-73-03).

[29] Neurath to Reichenbach, 22 July 1929 (RC 029-15-15) and Neurath to Carnap, 9 October 1932 (RC 029-12-24). Cf. Neurath to Reichenbach and Carnap, 17 January 1935 (RC 029-09-97).

Despite the fact that Neurath was strongly opposed to the term "philosophy of nature," the expression surfaced again in the title of Reichenbach's 1931 brochure on the subject, "Aims and Methods of Modern Philosophy of Nature." Even though Reichenbach opposed the idea that philosophy shall be a "super science," or a science that holds the key to fundamental and infallible knowledge, he devoted still more space to talk about philosophy and how to deprive it of its classic character, than Neurath could have tolerated.

Reichenbach's brochure is an important document, not just because it reflects the core of a naturalization-debate, but because it was one of the first public and widely read attacks against what Reichenbach called "positivism." By marking off his Berlin-view from the Viennese outlook, Reichenbach might have impeded the unified brand-building process of Neurath, and thus sabotaged the common cause.[30] Nevertheless, for example, Carnap took Reichenbach's side and wrote in a letter that he "was especially happy because of the actual system-proposals" to be found in the "Wahrscheinlichkeitslehre." Though Carnap contrasted the "system-proposals" with Schlick's and Wittgenstein's philosophical considerations ("Erörterungen"), his conception opposed Neurath's attack on systems as well.[31]

Another example and form of brand building is provided by the series *Einheitswissenschaft*, edited by Neurath, Carnap, Philipp Frank, Hans Hahn, and Jörgen Jörgensen. In the usual historiographies and narratives, the series is depicted as a common effort in spreading the word, just like *Erkenntnis* and the Frank-Schlick series. Nonetheless, besides the formal editorial board, the origins of the series have an interesting lesson for us. In 1944, Neurath wrote the following to Carnap:

> Your letter on my articles hardly acceptable for the Erkenntnis immediately induced me to create the series EINHEITSWISSENSCHAFT, where I could publish my own stuff whenever I wanted to do it, without any consent from my strong teacher. Of course I did not insult you but invited you to be with me and to publish there, you by me highly admired giant of Logical analysis and a man who unveils the secrets of so many metaphysicians … (Neurath to Carnap, 18 November 1944. RC 102-55-06. See letter 28 in this volume.)

Given that the letter was written during an especially stormy period of their friendship, one might warn that we should take this passage with a grain of salt. Nevertheless, considering Neurath's intense and vehement character, it is not at all implausible to believe that he started the monograph series for himself, as an independent forum, as I would say, for brand building.[32] The fact that Neurath wrote two monographs (1933/1987; 1935/1987) and one article (1937/1987) for his new series seems to justify this idea.

The creation of *Einheitswissenschaft* was not enough in itself; Neurath left nothing to chance! He tried to arrange some positive reviews of his first brochure

[30] Neurath to Carnap, 12 April and 28 June 1930 (RC 029-14-18 and RC 029-13-10); see further Neurath's 14 April circular letter about his debate with Reichenbach (RC 029-14-17).

[31] Carnap to Reichenbach, 1 February 1935 (RC 102-64-05).

[32] On the history of the series see Hegselmann (1987, xiv–xviii). Cf. Neurath to Carnap, Frank and Hahn, 2 November 1932 (RC 029-12-15).

(1933/1987), Hahn's contribution (1933/1987) and Carnap's submission (1934/1987). He suggested that Hempel review his monograph, Frank review Hahn's paper, and Jörgensen review Carnap's work. But it was too late since Kurt Grelling (1935) had already started to review Hahn's work, and Carnap asked him to write about his and Neurath's brochures. Neurath's fears proved to be justified. He claimed that Grelling was "not standing entirely on their side" and his remarks in the review turned out to be quite negative – not helping the common cause.[33]

Building the brand of logical empiricism was a practical task, not a theoretical commitment for Neurath. This meant that after the early 1930s, he was willing to *overcome* philosophical conflicts, theoretical disagreements and oppositions for the greater practical good. A nice example of Neurath's "united by action" attitude is provided by the case of his relation to Karl Mannheim.

5.3.2 Neurath and Mannheim: From Rejection to Cooperation

Neurath and Mannheim's relation goes back to the turn of 1930 when the latter's *Ideology and Utopia* appeared in German. The book contained detailed discussions on themes such as how thinking is shaped by social positions, how collective conditions form the contents of experience, and how sociology and politics might become sciences. *Ideology and Utopia* is also considered to be one of the founding documents of Sociology of Knowledge as a discipline.

Neurath was happy in general with Mannheim's approach and with the idea of the social determinateness of knowledge. His behavioristic analysis of scientists and his sociology of sociology (1944/1970) shared many common themes with Mannheim's book. As Markus Seidel (2016) has recently pointed out that despite their different backgrounds, Neurath's and Mannheim's *politicized philosophy and sociology of science* have more in common than previously thought. Nevertheless, when Neurath (1930/1981) reviewed Mannheim's book, he was quite critical of it.

Neurath's criticism focused on a peculiar idea of Mannheim's. According to Mannheim (1929/1936, 104), knowledge has both social and political determinants, it is existentially related:

> It could be shown in all cases that not only do fundamental orientations, evaluations, and the content of ideas differ but that the manner of stating a problem, the sort of approach made, and even the categories in which experiences are subsumed, collected, and ordered vary according to the social position of the observer. (Mannheim 1929/1936, 130.)

Sociology of science aims at showing that not only our opinions, judgments, and commitments depend on our political and social status, but also what we recognize as *facts*, *problems*, or the *content* of experience! Thus "in the realm of politics," Mannheim claimed (1929/1936, 131), "the only knowledge that we have is a

[33] Neurath to Carnap, 5 and 16 March 1935 (RC 029-09-78 and RC 029-09-70). See further Carnap to Neurath, 13 March 1935 (RC 029-09-72).

knowledge which is limited by the position which we occupy." (Neurath and Frank will say later, however, that *all* our knowledge – including logical and natural scientific knowledge – has a substantial social and political dimension in Mannheim's sense.)[34]

From different political stances and social positions, different values and commitments arise, which always bring different perspectives into relief. Mannheim's asked how can the various political perspectives *synthetised* in order to avoid shoreless relativism and to enforce objectivity through the *scientific perspective*. In order to exceed individual and class interests, the intelligentsia will become "a relatively classless stratum which is not too firmly situated in the social order" – this is what Mannheim (1929/1936, 137) calls the "socially unattached intelligentsia."

In Neurath's eyes Mannheim was right in emphasizing the social embeddedness of thinking and arguing, but Neurath disagreed with Mannheim's view that the scientific perspective could be instantiated only by a "socially unattached" social stratum. In his review, "Bürgerlicher Marxismus" ("Bourgeois Marxism"), published in *Der Kampf*, Neurath wrote about the synthesizing process as follows:

> It is as if one thinks that 2 times 2 equals 7 since it is what the orbits say, the other that 2 times 2 equals 5 since this is God's decree, the third that 2 times 2 equals 8 since this corresponds to the cosmic view, whereas the scientist maintains that 2 times 2 equals 4. And now there is the one who synthesizes all four 'one-sided' viewpoints or takes the average viewpoint that consists in the claim that 2 times 2 equals 6. (Neurath 1930/1981, 352. Seidel's (2016) translation.)

By making the intelligentsia's *synthesizing process* equal with the mathematical *counting of average*, Neurath tried to ridicule Mannheim's idea and rhetorically played into his own hands the vantage point. But even Mannheim (1929/1936, 135) claimed that such a simple "additive synthesis" (as presented in Neurath's quote too) would not work.

So *that* critique of Neurath won't be relevant against Mannheim simply because he did not say anything like that of which Neurath later accused him in the review with regard the mathematical counting of average. Two years earlier, however, in his *Personal Life and Class Struggle*, Neurath (1928/1973, 283) emphasized that "the shaping of life and world-view is closely linked with the social and economic order within which they arise." That's why he stuck to the idea that "the philosophy of a class depends on the position of the class" (1928/1973, 289). In his review, Neurath emphasized again and again that Mannheim's considerations reflect only *his* bourgeois stance: *his* neutrality is only pseudo-neutrality.

In Neurath's eyes, Mannheim pursued metaphysics by talking about the socially unattached intelligentsia and its neutral perspective. Without going into the question

[34] It is quite probable that Mannheim's position and ideas were transferred to Frank via the Harvard-sociologist Robert Merton, who was a friend of Frank, and for a while also a collaborator on a "sociology of science" project (which included Ernest Nagel too). Frank discussed Mannheim, Lukács and other sociologists in an unpublished book manuscript, entitled *Science, Facts, and Values* (now in preparation for publication). On Frank's relation to the sociology of science and the mentioned project, see George Reisch's chapter in the present volume.

of who was right, we might stress two interesting points: (1) In this one-sided debate, Neurath defended that position which became quite dominant in the second half of the twentieth century in the strong program of the sociology of knowledge: no one can erupt from social determinateness,[35] and (2) Neurath's rhetorical moves allowed him to translate theoretical questions into social ones, as the closing lines of the review suggest:

> Mannheim looks for a comprehensive viewpoint, that aspect of the "world", which is called metaphysics! Marxism, on the contrary, looks for the adequate statements about social processes. Marxism predicts the fate of proletariat and other classes! Metaphysics against science! Mannheim against Marxism, after all against friendship: the bourgeois front against the proletarian front! That is the old, very well known song! (Neurath 1930/1981, 356.)[36]

It is not known whether Mannheim was aware of this review; though in his later writings he tried to answer these charges. However, criticisms of Mannheim's work that were similar to Neurath's were already widespread and common so he might have become aware of them from other sources.

Nonetheless, viewing things from Neurath's critical stance, some later events might be explained as well. In 1937, Charles Morris suggested doing a monograph about the sociology of science in the *International Encyclopedia of Unified Science*. He proposed the following authors: Louis Wirth (his colleague, the famous urban sociologist from Chicago), and Karl Mannheim.[37] While Neurath asked for some papers from Wirth (presumably thus he did not know his work), in the case of Mannheim he was able to formulate a stronger opinion:

> I have not read the last book of Mannheim. An older book of Mannheim is – what I would call – ful[l] of metaphysical formulations. But perhaps I would revise my opinion after my USA trip. I learned not always to act as a Torquemada. But I think we cannot decide this point in short time. (Neurath to Morris and Carnap, 12 March 1937. RC 102-51-60.)

Even though Neurath was exaggerating – Tomás de Torquemada was the first Spanish Grand Inquisitor in the fifteenth century, responsible for thousands of deaths that time – it is true that he was inclined towards harsh and extreme reactions.

By mentioning Hegel or Dilthey in a positive manner, Mannheim made an intolerable step towards metaphysics. In most cases, it is only Mannheim's language that reveals his philosophical past and tradition, but Neurath had a good eye to catch those supposedly dangerous words. Though Carnap's *Aufbau* does not belong to the tradition of *Geisteswissenschaften*, Neurath saw its allusion; Carnap wrote the followings in his diaries: "With Feigl to Neurath. Neurath grumbled over my presentation of the 'Geisteswissenschaften' in the 'Aufbau.' This is too idealistic for him; he

[35] On Neurath and the strong program, see Uebel (2000); on Frank and the sociology of science see Tuboly (2017).

[36] Recently Markus Seidel (2016) listed some reasons for thinking that Neurath's considerations did not apply to Mannheim. Even if he is right, it does not affect my main points, which are based on Neurath's internal (or so-called "emic") perspective.

[37] Morris to Neurath, copy to Carnap, 21 February 1937 (RC 102-51-63).

had some points to attack: he gets to name Dilthey: 'moral', 'state', 'manifestation.'"[38] Given Neurath's well-known opposition to that tradition, Mannheim's case was easily postponed for 2 years.

In 1939, however, something changed. With the increasing recognition of logical empiricism their congresses became more successful: after Prague, Paris, Copenhagen, and Cambridge, the next step was Harvard. Mannheim was invited to that Congress! He was happy to accept the invitation and proposed the following themes: (1) The Sociology of Valuations, (2) The Sociology of Knowledge and Logical Positivism, and (3) Specialization and Integration of Knowledge in the Social Sciences.[39]

Neurath liked all the three themes; the first could be related to John Dewey's (1939/1970) *Encyclopedia*-monograph on valuations that appeared the same year. The second session could have helped in understanding the context of the movement, and finally the third session would be a good match for the Social Sciences section. Mannheim promised, after all, to prepare a talk on the sociology of valuations. Even though at the congress there was a special section for the social sciences (with Neurath) and another for "Science and Society" (with Wirth and Zilsel), Mannheim did not deliver a lecture.[40]

Two years later Neurath approached Mannheim again to give a talk at the Sixth Congress, now in Chicago – again, unsuccessfully. But he did not give up. Because of the various war-restrictions, many European scientists were unable to travel, thus Neurath organized the aforementioned "Terminology" workshop for them in Oxford (October 1941). Due to his other commitments, however, Mannheim was not able to participate at the Oxford gathering either.[41]

After these conference-invitations, the line of events became blurry, but Neurath wrote the following to Mannheim 1 year later:

> I should like to meet you as far as our Encyclopedia is concerned. It is not urgent, you see we are now publishing the first 20 monographs, but on the other hand, we have to prepare carefully the next 60, too, and I should like to know a little, what, IN PRINCIPLE, of course, you would PERHAPS, assumed, you will have time, write on your own opinion as a representative one. (Neurath to Mannheim, 28 December 1942. In: Mannheim 1996, 183.)

Despite Neurath's idiosyncratic English, one might see that at the end of 1942, he brought himself to get Mannheim for the *Encyclopedia*. Actually, the final *Encyclopedia* would contain more than 100 monographs, some of them discussing the relation of science to society. By writing this one, Mannheim would indeed be a nice fit and a valuable contributor given his growing international reputation.[42] In Neurath's life, however, the ideal of the *Encyclopedia* was never realized: authors

---

[38] Carnap's diary entry, 19 December 1929 (RC 025-73-03). On the *Aufbau*'s relation to the Geisteswissenschaften, see Dewulf (2017) and Tuboly (2019)

[39] Mannheim to Neurath, 20 April 1939 (ONN).

[40] Neurath to Mannheim, 26 April 1939 and Mannheim to Neurath, 30 August 1939 (ONN).

[41] Neurath to Mannheim, 12 June and 15 September 1941 (ONN); Mannheim to Neurath, 19 September 1941, in: Mannheim (1996, 170).

[42] On the Encyclopedia see Reisch (1994), Dahms (1996).

were dropped, themes were changed, and monographs were delayed for years. Henceforth the discussions between Mannheim and Neurath were inconclusive. At the beginning Neurath planned for the *Encyclopedia* to have eight volumes, each volume consisting of ten monographs (that is eighty monographs), but only the first two volumes were published (twenty monographs in all), but only ten monographs in Neurath's life. Thus even if Mannheim had accept the invitation, he would have had to wait decades for his work to be published (his monograph would fit volumes seven or eight). The delays would have quickly become irrelevant as Mannheim died in 1947.

The story of Neurath and Mannheim's relation did not end here. Though Neurath's efforts to recruit Mannheim were unsuccessful, Mannheim approached Neurath too. Neurath wrote the following:

> I am also highly occupied, partly by writing books, partly by organizing our ISOTYPE INSTITUTE, which is now here at the British Institute of Visual Education. We are making educational Isotype Diagrams – Animated and illustrations for books and exhibitions. We think that the international educational situation after the war will need visual education much more than the prewar situation. And we think one should prepare something now. I do not speak of the visualization of my own ideas, but also of visualization of our contemporary knowledge as a whole. (Neurath to Mannheim, 28 December 1942. In: Mannheim 1996, 183.)

It is not known whether this letter kindled Mannheim's interest in Neurath's ISOTYPE (according to their letters, presumably they did not meet personally for months). However in March 1943 Mannheim wrote to Neurath to say that he is waiting for a detailed summary of Neurath's promised new book to close the deal. He also suggested a title for Neurath: "Humanisation and Popularization. New Ways of Visual Education and Humanisation of Scientific Knowledge." As Mannheim said, "[a] long title but provocative."[43]

What was the book, and where did Mannheim want to publish it? In 1942, Routledge & Kegan Paul started its *International Library of Sociology and Social Reconstruction* (ILSSR) series. The general editor was Mannheim, who held the position until his 1947 death. The series became quite well-known and successful: "its books were an important part of what there was to read" and their dark green dust-jacket "made what people in Britain thought a sociology book looked like" (Platt 2014, 236). ILSSR emerged as a fundamental tool in British sociological education and in the discipline's dissemination of knowledge. The particular books covered such themes as economic, town and country planning, sociology of crisis, art, and literature, criminology, foundations of thought, and, of course, sociology of knowledge.

Almost all of the ILSSR volumes were related to Neurath's earlier and present interests as well. Neurath advocated "planning" (the key category in British sociology) even before his active months in the Bavarian Soviet Republic in 1919.[44] The

[43] Mannheim to Neurath, 10 March 1943. In: Mannheim (1996, 187).

[44] For the significance and context of Neurath's ideas on planning, see George Reisch's and Thomas Uebel's chapters in the volume.

problems of *planning* and *reconstruction* turned out to be the two most important categories of British sociology during World War Two and even later. From that perspective, Neurath seemed to be a nice choice to publish a monograph in the series.

According to Mannheim's letter, Neurath indeed started to write a book for him. As Mannheim said, "[t]he longer I think about your proposed book the more I like its idea. I also feel that it would fit into the educational section of the International Library of Sociology, and Social Reconstruction admirably." Mannheim claimed that the book should be in agreement with the general idea of the series, namely that "from the platform of my Library the Educationalists and other people who will have to build up a new Europe should be informed about the best techniques." To that project, Neurath would contribute with his visual education and pictorial statistics.[45]

Since Neurath died unexpectedly in 1945, none of his works appeared in Mannheim's ILSSR, henceforth their common causes and aims surfaced only in their private correspondence. Nonetheless, a huge part of Neurath's manuscript turned up and was published partly in 1973, and wholly in 1996 – under the editorship of Juha Manninen – as "Visual Education: Humanisation versus Popularisation" (Neurath 1996).

One might raise the question: why and how did Neurath change his mind regarding Mannheim after the initial negative review? Since we cannot find any explicit trace of this in Neurath's (un)published works, we could only guess. One such guess may be Neurath's *brand building* as it was described in the previous two sections. In 1942, he wrote to Carnap:

> Collecting material on the second series of monographs I am preparing a survey on people who are related to us, but have some 'whims' – as it were – they should have an opportunity to express themselves with intensity (the first series of monographs stresses, what we have in common). (Neurath to Carnap, 29 July 1942. RC 115-07-58. See letter 13 in this volume.)

Neurath mentioned Feigl, Hempel, Tarski, Reichenbach, and the British psychologist Tom Hatherley Pear (who was the President of the British Psychological Society). But given the fact the Neurath refers to the second series of monographs (that is everything after volumes one and two), we can assume that the above remark in the letter applies to Mannheim as well. Although Mannheim's work fit into the tradition of *Geisteswissenschaften* by including metaphysical formulations, those differences (in theoretical commitments) could be overcome for the common goals. To give an example, see another letter by Neurath:

> Reichenbach [...] made me emotional, because I thought it not brotherly, how he treated you in the various cases, we had to discuss. And, I cannot deny, that his theoretical remarks looked perhaps more strange to me than they perhaps would do, if he were not so far away from brotherhood in a community, in principle. He is very often charming and I know very well how to go on with him in the nicest way ..... but I fear him, not as a scientist or debater,

[45] Mannheim to Neurath, 15 March and 18 August 1943; see further Neurath to Mannheim, 18 September 1943. In: Mannheim (1996, 188, 192–193).

> but as a person, who is not interested in creating brotherhood in this world. (Neurath's unsent letter to Carnap, 22 September 1945. RC 115-07-66. See letter 34 in the volume.)

Seemingly Neurath would get over some of the theoretical differences with Reichenbach if they are counterbalanced in *actions* and *decisions*, i.e. practical matters. Nevertheless, after Neurath charged Reichenbach with "sabotage,"[46] even the ideal of common action, based on brotherhood, was lost. Reichenbach did not help the cause (that is, showing a unified and functioning picture of logical empiricism) by being reluctant to write a monograph for the *Encyclopedia*.

Mannheim, however, never hindered Neurath, but helped him during the English period. Education, pedagogy, social planning, organization, and finally reconstruction were such common ideas and ideals for both of them that made common work possible in a time that called for *direct and planned action*. Thus the case of Neurath and Mannheim, diverging between accusations of metaphysics, negative criticism and helping each other by various invitations, exemplifies nicely how Neurath's brand-building project worked. Acting together for the greater good overcomes the theoretical differences.

5.4 The Story of an Unsuccessful Project

Erwin Dekker (2014, 116) was right in emphasizing that important changes occurred in Neurath's approach to philosophical and scientific cooperation: "throughout the 1930s Neurath himself moved away from [the] revolutionary idea of the reconstruction of society." The reason for downplaying much of his radical socialist overtones was actually in support of a positive common cause: building the brand of logical empiricism for the greater good, i.e. reconstructing society from the bottom up using language and science on *as many* fronts and with *as many* comrades as possible.

These tendencies were hindered in the 1930s by in-the-house debates, published controversies, and personal attacks (which were often made behind each other's backs). Even though Neurath was often ready for compromise, working out issues using compromises was simply too slow and often did not pay off for him. One such example was Herbert Feigl, a former member of the Circle, who emigrated to the United States in 1930. In 1943, Feigl published a paper about "Logical Empiricism" in Dagobert R. Runes' *Twentieth Century Philosophy: Living Schools of Thought*. In his paper, he aimed at summarizing the main tenets and original ideas of the movement, with some historical notes and a bibliography. Though he mentioned the Unity of Science Movement, the Congresses, and *Encyclopedia*, he never indicated that they were Neurath's core ideas. Neurath was mentioned only once, as one of the many non-philosophers who played an active role in the Circle. It is then not surprising that Neurath commented on this as follows:

[46] Neurath to Carnap, 22 July 1931 (RC 029-13-08).

> Feigl wrote a paper on Logical Empiricism. Do you think anybody reading this paper, will discover that I had a certain position in its history? [...] Please, look to the article by Feigl. He just put me into SOCIOLOGY – that is all. He does not even mention me as the initiator and editor-in-chief of the encyclopedia. DO YOU THINK HE WOULD TREAT ME SO IF I WERE [A] PROFESSOR, not to ask the question, what he did with all that if I were Schlick. (Neurath to Carnap, 24 September 1945. RC 102-55-14. See letter 36 in this volume.)

It is not simply that Feigl did not give sufficient credit to Neurath. What was even worse for Neurath was that Feigl acted against the ideal of brotherhood: "I never complaint about Feigl. I am not regarding him as very important or influential, but for a long time I regarded him as a potential friend. I FEEL IRRITATED AS A FRIEND" (ibid.).

Neurath also recalled some meetings with Feigl in Europe, which were influential for the young Feigl but were never properly acknowledged. Neurath suggested to Hannes Meyer, the director of the Bauhaus, for example, that he invite Feigl to Dessau. Neurath also suggested that Feigl should participate at a pedagogical conference in Geneva in order to increase his chances in finding a job – Feigl happily acknowledged this in a letter to Schlick.[47]

Even though Neurath did everything to become friends and close colleagues with Feigl, he was grieved by the (recurring) actions of the latter, especially by not getting proper credit or help in disseminating Neurath's point of view. Actually, when Feigl (1943, 401) listed those logical empiricists who worked on the social sciences, he mentioned only Hempel, Paul Oppenheim, Felix Kaufmann, Edgar Zilsel, and the American sociologist, George Andrew Lundberg. There was no mention of Neurath in that section at all.

The fact that Neurath was well-received in England should not surprise us, however. In his book on the cultural-political history of analytic philosophy, Thomas L. Akehurst (2010, 2), emphasized that "[t]here was a 'deep seated similarity of attitude and outlook' among the British analysts; [...] we can detect a clear pattern of cultural, political and philosophical beliefs shared by major British analytic philosophers." This shared worldview was centered partly on the idea that "a post-Kantian tradition of continental philosophy was the direct source of fascist ideology" (Akehurst 2010, 3). Relatedly, British philosophers had a positive program as well: the emerging analytic tradition was closely tied to liberalism and democracy. Neurath shared both of these characteristics.

In his last years, Neurath was busy with a larger project on Germany's education after the war, dealing with the sources, contexts, and future of Nazism and their interconnections. He thought that "in German literature, we find many elements, which fit together may form a pattern in harmony with the anti-democratic attitude."[48]

---

[47] Feigl to Schlick, 21 July 1929. Moritz Schlick Nachlass. The same event is recalled in Neurath's unsent letter to Carnap, 22 September 1945 (RC 115-07-66) and in his 24 September 1945 letter (RC 102-55-14). See letters 34 and 36 in this volume. Though it should be admitted that later Feigl (1969/1981, 62) noted his debt to Neurath regarding the Bauhaus.

[48] See Neurath and Lauwerys, "Plato's Republic, German Education and Human Brotherhood," in the Neurath Nachlass, ONN, K. 73. On Neurath and his fears of German education, see Antonia Soulez's chapter in the present volume.

Neurath tried to identify certain patterns and schemes in German literature, scientific studies and traditional ways of argumentations. Emphasizing the role of Plato's *Republic*, Kant's categorical imperative, the ideal of the genius who is above all rules and customs, the inclination towards something "the" (like "the state," "the rule," "the nation," "the race," "the great past"), Neurath separated the German climate (or atmosphere) from the British (or more generally, the Western European) one:

> The human climate of the western countries is different. Here a certain critical attitude plays its part, in Britain based more on certain traditional traits, in the continental West based more on certain rational arguments. But so or so, what grows up in the west are individuals, who think more of common happiness and of co-operation than of subordination under some super-human structure, THE STATE or THE RACE. Of course also here are collective traits of importance, but they never consummate the whole of a person. (Neurath, "Meeting, Belgium committee…," ONN, K. 79.)

In that sense – even though he had some intense debates about Plato's influence on the German climate and on German traditions after writing an article with Joseph A. Lauwerys (1945) – Neurath found eventually a cooperative, friendly and receptive attitude. He did not have to add much to the intellectual climate of British analysts with respect to their criticisms of continental thought and Nazism. He could add, however, his own experiences and practical works on how to re-educate people after the war. As he wrote to Carnap, "I like to be with the British in these hard days. Being in a new environment is rather thrilling and stimulating – I feel like a second youth here. On our life boat I thought of the future activity here."[49]

Without emphasizing it much: democracy was indeed a major concern for Neurath throughout his entire life, but most explicitly during his English years. He stressed England's own tradition in democracy (or at least its customs and ideas that created a supporting pattern for the democratic way of life):

> In Germany there is a whole ideology of regulations. Not only do you have to have regulations, but you have to admire them. In England, you admire your lack of regulations, your 'muddle'. […] Your Court procedure is based on the right of a single Judge to give a definite answer. Thus different solutions are admitted: you have more differences in local decisions. Differentiation is contrary to German feeling, because in Germany there are thirty different States, needing a common activity and a common basis. In England you have your common basis well-established, and therefore you do not fear differences. Thus you have a democracy. The muddle is related to democracy. (Neurath, "Contributing features in the emotional and intellectual isolation of the German," ONN, K.48.)

For Neurath, democracy came with pluralism, tolerance and 'muddle': both in theories and actions, an old theme, he issued already from his famous 1913 Vienna lecture: "the lost wanderers of Descartes."[50]

[49] Neurath to Carnap, 22 December 1942 (RC 115-07-61). See letter 16 in this volume. On the British analysts and Nazism see Akehurst (2010, 16–52).

[50] Actually Neurath emphasized this special continuity of his line of thought in "Universal Jargon and Terminology" (1941/1983, 216); but also in his "Argumentation and Action" manuscript (ONN, K39). See also Don Howard's chapter about the continuation of Neurath's thought regarding the practical relevance of philosophy, starting around the time of "The Lost Wanderers" article.

Neurath died on October 10, 1945, sitting, reading and laughing at his working desk after a usual walk in the streets around his house. In the recent literature, aiming at his rehabilitation, he emerged as an original thinker, often ahead of his time. Nonetheless, there was one thing he presumably did not see coming. Someone – according to Ayer's biographer, Ben Rogers (1999, 231), that someone was the journalist, Giles Romilly, who was the nephew of Churcill – under the pseudonym "Oxonian," described his experiences returning to Oxford in 1948:

> Since the end of the war, Professor Ayer's book [Language, Truth and Logic] has in Oxford acquired almost the status of a philosophic bible. [...] There is of course, no direct connection between Logical Positivism and Fascism. Yet an anti-aesthetic Philistinism is encouraged by it, since it reduces intellectual activity to a kind of artisanship, in which one either has "skill" or hasn't. And I would make this point, that a climate of opinion, in which a negative and rather arid philosophy of this type can rise to such pre-eminence is one more likely to favour Fascism than not; since Fascism steps into the vacuum left by an abeyance of concern with fundamental human values. ("A Visit to Oxford," *The New Statesman and Nation*, June 26, 1948.)

Logical empiricists have been accused of many things: relativism, nihilism, intellectualism, and dogmatism. In the United States, in a "Letter to the Editor" (*The Saturday Review*, 3 September 1949, p. 26) it was said that logical empiricists, as nihilists, are worse than communists since "communism (albeit in a perverted fashion) holds some things to be good and others to be bad." Neurath, running from fascism and Nazism, fighting them with all his efforts on the theoretical, educational and practical levels, could hardly imagine a worse outcome than his approach being accused of creating an atmosphere (in England!) supporting Fascism! He would have been even more surprised to find out that that C. E. M. Joad (1950) – who he invited to his Chicago Congress and Oxford conference in 1941 – has devoted an entire book of accusation, criticism, and disinformation to the "Fascism and Logical Empiricism" theme arguing in favor of the position first outlined by Oxonian.[51]

Throughout this debate A. J. Ayer's book and thoughts were under suspicion: in the philosophical circles of England, Neurath was already forgotten. By 1941, he claimed to Carnap that "I think, if universities should be interested in me, it would be in connection with particular problems of social sciences, unified science or visual education."[52] Though his name surfaced for a few more years in the context of visual education (thanks to the efforts of his third wife, Marie), his other activities fell victim to various misleading and oversimplified accounts of his works and intentions.

[51] It should be mentioned though, that in the United States Horace Kallen accused Neurath's unified idea of totalitarianism already in 1939, and later in 1945. See Reisch (2005, 167–190).

[52] See Neurath to Carnap, 21 September 1941 (See letter 7. in this volume.).

References

Akehurst, Thomas. 2010. *The Cultural Politics of Analytic Philosophy: Britishness and the Spectre of Europe*. New York/London: Bloomsbury.

Bergmann, Gustav. 1993. Memories of the Vienna Circle. Letter to Otto Neurath. In *Scientific Philosophy: Origins and Developments*, ed. Friedrich Stadler, 193–208. Dordrecht: Kluwer.

Carnap, Rudolf. 1934/1987. The Task of the Logic of Science. In *Unified Science*, ed. Brian McGuinnes, 46–66. Dordrecht: D. Reidel.

Cartwright, Nancy, Jordi Cat, Lola Fleck, and Thomas Uebel. 1996. *Otto Neurath: Philosophy between Science and Politics*. Cambridge: Cambridge University Press.

Creath, Richard. 2012. Before Explication. In *Carnap's Ideal of Explication and Naturalism*, ed. Pierre Wagner, 161–174. Basingstoke: Palgrave Macmillan.

Dahms, Hans-Joachim. 1996. Vienna Circle and French Enlightenment. A Comparison of Diderots Encylopédia with Neuraths International Encyclopedia of Unified Science. In *Encyclopedia and Utopia. The Life and Work of Otto Neurath (1882–1945)*, ed. Elisabeth Nemeth and Friedrich Stadler, 53–61. Dordrecht: Kluwer.

Dekker, Erwin. 2014. The Intellectual Networks of Otto Neurath: Between the Coffeehouse and Academia. *European Studies: A Journal of European Culture, History and Politics* 32 (1): 103–121.

Dewey, John. 1939/1970. Theory of Valuation. In *Foundations of the Unity of Science: Volume 1*, 379–447. Chicago: University of Chicago Press.

Dewulf, Fons. 2017. Carnap's Incorporation of the Geisteswissenschaften in the Aufbau. *HOPOS* 7 (2): 199–225.

Feigl, Herbert. 1943. Logical Empiricism. In *Twentieth Century Philosophy*, ed. D.D. Runes, 371–416. New York: Philosophical Library.

———. 1969/1981. The Wiener Kreis in America. In *Inquiries and Provocation. Selected Writings 1929–1974*, ed. Robert S. Cohen, 57–94. Dordrecht: D. Reidel.

Grelling, Kurt. 1935. Review of Einheitswissenschaft. *Erkenntnis* 5: 371–374.

Hahn, Hans. 1933/1987. Logic, Mathematics, and Knowledge of Nature. In *Unified Science*, ed. Brian McGuinnes, 24–45. Dordrecht: D. Reidel.

Hegselmann, Rainer. 1987. Introduction. In *Unified Science*, ed. Brian McGuinnes, ix–xxi. Dordrecht: D. Reidel.

Joad, C.E.M. 1950. *A Critique of Logical Positivism*. London: Victor Gollancz Ltd.

Kochan, Miriam. 1983. *Britain's Internees in the Second World War*. London: Macmillan.

Mannheim, Karl. 1929/1936. *Ideology and Utopia*. London: Kegan Paul.

———. 1996. *Mannheim Károly levelezése 1911–1946 [The Correspondence of Karl Mannheim 1911–1946]*. Budapest: Argumentum Kiadó.

Neurath, Marie. 1973. Marie Neurath: 1940–1945. In *Empiricism and Sociology*, ed. Robert S. Cohen and Marie Neurath, 68–75. Dordrecht: D. Reidel.

Neurath, Otto. 1928/1973. Personal Life and Class Struggle. In *Empiricism and Sociology*, ed. Robert S. Cohen and Marie Neurath, 249–298. Dordrecht: D. Reidel.

———. 1930/1981. Bürgerlicher Marxismus. In *Gesammelte philosophische and metodologische Schriften*, eds. Rudolf Haller and Heinrich Rutte, 345–356. Wien: Holder-Pichler-Tempsky.

———. 1930/1983. Ways of the Scientific World-Conception. In *Otto Neurath: Philosophical papers 1913–1946*, eds. Robert S. Cohen and Marie Neurath, 32–47. Dordrecht: D. Reidel.

———. 1931. Magie und Technik. *Erkenntnis* 2: 82–84.

———. 1931/1973. Empirical Sociology. In *Empiricism and Sociology*, ed. Robert S. Cohen and Marie Neurath, 319–421. Dordrecht: D. Reidel.

———. 1933/1987. Unified Science and Psychology. In *Unified Science*, ed. Brian McGuinnes, 1–23. Dordrecht: D. Reidel.

———. 1935/1987. What is Meant by a Rational Economic Theory? In *Unified Science*, ed. Brian McGuinnes, 67–109. Dordrecht: D. Reidel.

———. 1936. *International Picture Language*. London: Kegan Paul/Trench/Trubner & Co.

———. 1937. *Basic by Isotype*. London: Kegan Paul/Trench/Trubner & Co.
———. 1937/1987. The New Encyclopedia. In *Unified Science*, ed. Brian McGuinnes, 132–141. Dordrecht: D. Reidel.
———. 1939. *Modern Man in the Making*. New York: Alfred A. Knopf.
———. 1941. The Danger of Careless Terminology. *The New Era: The Journal of the World Education Fellowship* 22 (7): 145–150.
———. 1941/1983. Universal Jargon and Terminology. In *Otto Neurath: Philosophical Papers 1913–1946*, ed. Robert S. Cohen and Marie Neurath, 213–229. Dordrecht: D. Reidel.
———. 1944/1970. Foundations of the Social Sciences. In *Foundations of the Unity of Science: Volume 1*, 1–51. Chicago: University of Chicago Press.
———. 1996. Visual Education: Humanisation versus Popularisation. In *Encyclopedia and Utopia. The Life and Work of Otto Neurath (1882–1945)*, ed. Elisabeth Nemeth and Friedrich Stadler, 245–335. Dordrecht: Kluwer.
Neurath, Otto, and Joseph A. Lauwerys. 1945. Plato's Republic and German Education. *The Journal of Education* 77 (907): 57–59 (910): 222, 224; (913): 394.
Platt, Jennifer. 2014. The International Library of Sociology and Social Reconstruction and British Sociology. In *The Palgrave Handbook of Sociology in Britain*, ed. John Holwood and John Scott, 236–259. Basingstoke: Palgrave Macmillan.
Price, Henry H. 1940. *Hume's Theory of the External World*. Oxford: Clarendon Press.
Radcliff-Brown, Alfred Reginald. 1939. *Taboo: The Frazer Lecture 1939*. Cambridge: Cambridge University Press.
Reichenbach, Hans. 1929/1978. The New Philosophy of Science. In *Selected Writings: 1909–1953. Volume 1*, ed. Maria Reichenbach and Robert S. Cohen, 258–260. Dordrecht: D. Reidel.
Reisch, George. 1994. Planning Science: Otto Neurath and the International Encyclopedia of Unified Science. *British Journal for the History of Science* 27 (2): 153–175.
———. 1996. Terminology in Action: Neurath and the International Encyclopedia of Unified Science. In *Encyclopedia and Utopia. The Life and Work of Otto Neurath (1882–1945)*, ed. Elisabeth Nemeth and Friedrich Stadler, 79–86. Dordrecht: Kluwer.
———. 2005. *How the Cold War Transformed Philosophy of Science: To the Icy Slopes of Logic*. New York: Cambridge University Press.
Rogers, Ben. 1999. *A. J. Ayer. A Life*. London: Chatto & Windus.
Sandner, Günther. 2011. The German Climate and Its Opposite: Otto Neurath in England, 1940–45. In *Political Exile and Exile Politics in Britain after 1933*, ed. A. Grenville and A. Reiter, 67–85. Amsterdam/New York: Rodopi.
———. 2014. *Otto Neurath: Eine politische Biographie*. Wien: Zsolnay.
Seidel, Markus. 2016. Changing Society by Scientific Investigations? The Unexpected Shared Ground Between Early Sociology of Knowledge and the Vienna Circle. *Foundations of Science* 21: 117–128.
Stadler, Friedrich. 2001/2015. *The Vienna Circle. Studies in the Origins, Development, and Influence of Logical Empiricism*. 2nd ed. Dordrecht: Springer.
Tuboly, Adam Tamas. 2017. Philipp Frank's Decline and the Crisis of Logical Empiricism. *Studies in East European Thought* 69 (3): 257–276.
———. 2019. The Constitution of Geistige Gegenstände in Carnap's *Aufbau* and the Importance of Hans Freyer. In *Logical Empiricism, Life Reform, and the German Youth Movement*, ed. Christian Damböck, Günther Sandner, and Meike Werner. Cham: Springer (Forthcoming).
Uebel, Thomas. 2000. Logical Empiricism and the Sociology of Knowledge: The Case of Neurath and Frank. *Philosophy of Science* 67: 138–150.
———. 2007. *Empiricism at the Crossroads. The Vienna Circle's Protocol-Sentence Debate*. Chicago/La Salle: Open Court.

Part II
Science, Society, and Method

Chapter 6
Visualizing Relations in Society and Economics: Otto Neurath's Isotype-Method Against the Background of his Economic Thought

Elisabeth Nemeth

Abstract The article shows how two domains of Neurath's broad and multifaceted work are related to each other: the concepts and methods he wanted to implement in political economics, on the one hand, and the methods of visualization that he and his interdisciplinary team developed at the Social and Economic Museum of Vienna, on the other. Some of Neurath's suggestions in both domains are surprisingly modern even today.

The economist, philosopher and political activist Otto Neurath is a well known-figure in the history of philosophy of science.[1] He was a founding member of the "Vienna Circle," a group of philosophers and scientists who developed their ideas during the 1920s and 30s in Vienna. The views of the group – known as "Logical Positivism" or "Logical Empiricism" – shaped essential parts of the philosophy of science of the second half of the twentieth century. Since the 1980s, Neurath's unorthodox contributions to Logical Empiricism have been re-considered and

This article is a revised version of a paper that was first published as "Socially Enlightened Science. Neurath on Social Science and Visual Education," in Mélika Ouelbani (ed.), *Thèmes de Philosophie Analytique*, Université de Tunis, Tunis, 2006, pp. 83–112. A revised version was published as "Visualizing relations in society and economics: Otto Neurath's Isotype-method against the background of his economic thought," in Jean-Yves Béziau (ed.). *La pointure du symbole*, Paris: éditions Petra 2014. For this volume the text has been revised and amended. Throughout the paper all translations from German to English are mine, and Neurath's reprinted papers are quoted form Neurath (1981, 1991, 1998).

[1]On Neurath's life and work see: P. Neurath (1994), Neurath and Nemeth (1994), Cartwright, Cat, Fleck, Uebel (1996), Nemeth and Heinrich (1999), Stadler (2001), Uebel (2005b), Cat (2014), Sandner (2014), and the essays in Part 1 of this volume.

E. Nemeth (✉)
Department of Philosophy, University of Vienna, Vienna, Austria
e-mail: elisabeth.nemeth@univie.ac.at

J. Cat, A. T. Tuboly (eds.), *Neurath Reconsidered*,
Boston Studies in the Philosophy and History of Science 336,
https://doi.org/10.1007/978-3-030-02128-3_6

re-evaluated.[2] His activities in Red Vienna during the 1920s and 1930s have been studied, and some of them have been documented.[3] At present, Neurath's work on picture statistics and visual education is attracting ever more interest.[4] But so is his economical writing, which did not attract much interest after his death in 1945 but is now re-discussed by philosophers and economists. These discussions concentrate on Neurath's ecologico-economical arguments, the concept of "standard of living" [Lebenslage] as well as on his heterodox ideas on socialism.[5] In the following, I want to show how two domains of Neurath's broad and multifaceted work are related to each other: the concepts and methods he wanted to implement in political economics, on the one hand, and the methods of visualization that he and his interdisciplinary team developed at the Social and Economic Museum, on the other. Some of Neurath's ideas in both domains are surprisingly modern even today.

6.1 Re-considering the Basic Concepts of Economics

When Otto Neurath published his first scientific writings around 1910, he was a young political economist trying to convince the scholars of his time that their field of study should be redefined within a much broader conceptual framework than it had been until then. "In line with an ancient tradition," he wrote, "we define *wealth* as the *object of political economy*." While Neurath emphasized the venerated origin of this definition – Aristotle's *Nicomachean Ethics* – he went on to explain his own understanding of wealth:

---

[2] See, for instance: Nemeth (1981), Stadler (1982), Haller (1982/1991, 1985/1991), Uebel (1991, 1992, 2000b, 2007a), Cartwright, Cat, Fleck, Uebel (1996), Nemeth and Stadler (1996), and Stadler (2001).

[3] See, e.g., Stadler (1982), Novy (1983), Novy and Förster (1991), Blau (1999), Vossoughian (2008), and Sophie Hochhäusl's chapter in the volume.

[4] See Twyman (1975, 1982, 1985), Kinross (1981, 1984), Mueller (1991), Leonard (2001), Stadler (2011), Hartmann and Bauer (2002), Hartmann (2005), Blau (2006), Kraeutler (2008), Vossoughian (2008), Neurath and Kinross (2009), Burke (2009, 2010a, 2011), Kindel (2011), Hochhäusl (2011), Nikolow (2011), Heinrich et al. (2011), and the chapters of Angélique Groß and Silke Körber in the present volume. I want to draw special attention to two volumes: (1) the edition of a manuscript of the late Neurath which has not been published until 2010: *From hieroglyphics to Isotype* (see Neurath 2010) and to the excellent introduction by Christopher Burke (2010b). The volume includes the numerous illustrations intended by Neurath to accompany his text, and is completed by an extensive appendix showing examples from the rich variety of graphic material that he collected. (2) *Isotype. Design and contexts 1925-1971,* eds. Christopher Burke, Eric Kindel, and Sue Walker. London: Hyphen Press, 2013. The volume comprehends all periods of the development of Isotype. It presents both new material and new insights concerning the beginnings in Vienna, Neurath's time spent in the USSR (Izostat), developments in the Netherlands and America, the Isotype Institute in Oxford, Marie Neurath-Reidemeister's collaborations with African countries, and her children's books. It includes an enormous amount of new pictorial material. The quality of the reproduction of graphics and pictures is so high that looking at them feels almost like looking at the originals.

[5] See Martinez-Alier (1987), O'Neill (1993, 1998, 1999, 2007), Uebel (2004, 2005a, 2007b), and his chapter in the present volume, Lessmann (2007a, b), Nemeth et al. (2007), Nemeth (2007) and Stuchlik (2011).

> By 'wealth' we understand the totality of pleasure and displeasure that we find with individuals and groups of individuals. The term 'pleasure' has the advantage that in our use of language it comprehends complex and primitive facts at the same time. (Neurath 1911/1998, 471.)[6]

According to Neurath's definition, the phenomenon political economy attempts to explain should be understood as an irreducibly heterogeneous totality. In his view, the main intellectual challenge political economists face is finding a method to describe the components of that totality without reducing them to one unit of measurement. Calculation in monetary terms, Neurath argued, never gives an adequate account of the complex phenomenon called wealth.[7] It is important to note that the complexity Neurath identified manifests itself not only when we conceive of a group of individuals holistically, whose needs and preferences might differ, or even compete, but also when we think of an individual person.

> When talking of the total situation of a person, we must consider that many factors are significant for choosing a profession, a place of residence, even such factors as the opportunity for artistic or religious satisfaction (remember the emigrations for the latter reason) [...].
>
> In the end we have to consider a complex of pleasure and pain as *a whole*, if we want to characterise the entire situation of a person. The situation is the same if we want to describe the order of life [*Lebensordnung*] of a people, or of a temporal period, in order to infer from that its favourable or unfavourable conditions. Again we have to look at the entire situation. Here and at many other points as well, the calculus of value reaches its limits, because the value of a sum of goods is not derivable from the sum of the values of the individual goods. (Neurath 1909b/2004, 293.)

With these and similar reflections, Neurath contributed to the famous debate on method and values which had occupied German scholars in the social sciences from the eighties of the nineteenth century. This paper will present a very rough sketch of some of the assumptions and concerns Neurath derived from both the "Historical School" and the "Austrian School of Economics" in developing his own position, located somewhere in-between the two.[8]

At Ferdinand Tönnies's instigation, Neurath left Vienna after his first year at the university to study economics in Berlin. In Berlin, Neurath studied with Eduard Meyer and Gustav Schmoller – both well-known representatives of the "Historical School" – and received his doctoral degree in 1906. It is thus not surprising that many of his ideas pertaining to economics can be traced to these influences. First of

---

[6] In a footnote, Neurath refers his use of the term "pleasure" (as comprehending complex and primitive facts at the same time) to Gustav Fechner's *Vorschule der Ästhetik*.

[7] In a similar line of argument, Neurath also criticized Utilitarianism for (at least implicitly) suggesting a sort of a pleasure calculus (Neurath 1912a/1973) as well as Kautsky's idea to use "labor time units to calculate the cost and benefit of production and regulate distribution" (see Thomas Uebel's chapter in the present volume).

[8] The following rough sketch of Neurath's position in the *Methodenstreit* draws heavily on Uebel (2000a, 2002, 2004, 2005a) and Cartwright's, Cat's, and Uebel's detailed analysis of the debate in Cartwright, Cat, Fleck, Uebel (1996).

all, Neurath inherited their interest both in the history of economic systems and the history of economic thought.[9] Neurath also adopted the view that economic structures cannot be understood without taking into account political and cultural factors.[10] He also held that a systematic comparison of specific historical constellations of human social existence should be the central focus of scientific inquiry (Neurath 1911/1998, 471). Finally, for Neurath, the task of economics was to analyze in their totality the life conditions of a specific population.

However, Neurath's views owed something to the work of the "Austrians" too (Uebel 2004, 2007b). Two aspects of this work are especially important in the present context.

First, Neurath shared both the Austrians' concern for scientific standards and their view that economics was yet not a true science. But for Neurath this did not mean that economists should search for universal laws. On the contrary: he believed that certain features of the Austrian School's concept of true science were based on an anachronistic ideal of natural science that had reappeared in social science in a vague, misconceived form. One of the principal tenets of this ideal was that true science had to search for universal laws. Another was the assumption that without calculation (i.e., without measuring processes and relations on the basis of one single unit), no precise and reliable results could be obtained. Neurath held this ideal of scientific method and precision to be fundamentally mechanistic and therefore outdated, and wanted it to be replaced by the concept of knowledge that had emerged in modern physics and mathematics. The main points of reference in these advanced fields of modern science were Gregorius Itelson's definition of mathematics as the "science of ordered objects" on the one hand,[11] and Ernst Mach's descriptionist concept of science on the other.

Second, Neurath shared the Austrians' view that goods have no inherent value but are valuable only to the extent that human beings derive "pleasure" from them. When Neurath claimed that the "life conditions of the population in its totality" were to be the main focus of economics, he always added that this totality was to be seen as a complex of *individual* pleasure and well-being (see Neurath 1912a/1973). For Neurath, thinking of "society as a whole" never meant conceiving it as an entity in its own right, nor could the structuring principle of "society" be defined without reference to the states and actions of individuals. On the contrary, Neurath followed other modernists who defended individualism against the authoritarian tendencies of the time, like Ferdinand Tönnies, his father Wilhelm Neurath, and especially Josef Popper-Lynkeus (1923), who saw the unconditional respect for the individual as being the only possible principle of ethics and social philosophy.[12]

[9] See Neurath's *Antike Wirtschaftsgeschichte* (partly available in English as Neurath 1909a/2004), and his Ph.D. thesis on the history of interpretation of Cicero's *De Officiis* (1906/1998).

[10] See, for instance, Neurath's interest in the sociology of religion on the Balkan states: Neurath (1913/1998, 1914/1998), and in English as (1912b/2004).

[11] See Buek (1926). On Itelson's influence on Neurath see further Jordi Cat's chapter in the present volume.

[12] On Tönnies' individualism see Ringer (1969, Ch. 3.). On Wilhelm Neurath see Uebel (1995). On J. Popper-Lynkeus see Belke (1978).

Seen from this perspective, Neurath's project really seems to be an attempt to square the circle: On the one hand, he defended the "Historians'" view that "the concept of *life in its entirety*" should be the central focus of economic theory. He pointed out that one of the founders of the "Historical School" had stated that "the life of a nation is a whole, the manifestations of which are intimately connected" (Neurath 1909b/2004, 294). On the other hand, Neurath defined the object of economic theory as analyzing pleasure and displeasure as experienced by individual human beings. Hence, from Neurath's point of view, the "life of a nation as a whole" is the totality of experiences of a plurality of individuals. In this sense, Neurath's highly ambitious methodological project (called "calculation in kind") was to combine holistic and individualistic elements and thereby enlarge the theoretical framework of political economy. From the very outset, that theoretical project struck many of his contemporaries as rather strange. And when, after 1918, Neurath suggested several models of socialist economy, his social-democratic friends found them too radical and technocratic.[13] Nonetheless, during the twenties and thirties, Neurath's theoretical approach figured significantly in the debates on whether or not a socialist economy was possible – albeit mostly as a negative point of reference: "economy in kind" was referred to as a radical but illusory version of planned economy.[14] After 1945, Neurath's economic theory disappeared altogether. Yet, remarkably enough, since the 1980s this has changed. Some recent theoretical approaches – especially in the field of ecological and welfare economics – show surprising parallels with the questions Neurath sought to address, and even with the conceptual framework he developed. In order to show that Neurath's ideas are by no means as old-fashioned as had been assumed, we will say a few words about Martha Nussbaum's and Amartya Sen's concept of well-being.

> 'And he said, Now, this schoolroom is a nation. And in this nation, there are fifty millions of money. Isn't this a prosperous nation? Girl number twenty, isn't this a prosperous nation, and a'n't you in a thriving state?'
>
> 'What did you say?' asked Louisa.
>
> 'Miss Louisa, I said I didn't know. I thought I couldn't know whether it was a prosperous nation or not, and whether I was in a thriving state or not, unless I knew who had got the money, and whether any of it was mine. But that had nothing to do with it. It was not in the figures at all,' said Sissy, wiping her eyes.
>
> 'That was a great mistake of yours,' observed Louisa.

Nussbaum and Sen open their book *The Quality of Life* (1993) by quoting the above excerpt from Charles Dickens' *Hard Times*. And they stress that Sissy Jupe's problem has yet to be solved.

> When we inquire about the prosperity of a nation or a region of the world, and about the quality of life of its inhabitants, Sissy Jupe's problem still arises: How do we determine

[13] See, for instance, his debate with Helene Bauer in "Der Kampf" in 1923.

[14] See Uebel (2004, 51–7), and his chapter in the present volume; cf. Chaloupek (2007).

> this? What information do we require? Which criteria are truly relevant to human 'thriving'? Girl number twenty quickly discerns that just knowing how much money is available (the analogue of GNP per capita, still widely used as a measure of quality of life) will not take us very far. For we also need, at the very least, to ask about the distribution of these resources, and what they do to the people's lives. (Nussbaum and Sen 1993, 1.)

Nussbaum and Sen go on to emphasize that the problem is still more complex: "we need to know not only about the money they do have or do not have, but a great deal about how they are able to conduct their lives" (1993, 1). Information of a different kind is required: about life expectancy, healthcare, medical service, education ("not only about its availability but about its nature and quality"), about labor ("whether it is rewarding or grindingly monotonous, whether workers enjoy any measure of dignity and control"). We need to know about political and legal privileges, family relations and relations between the sexes.

> In short, to think well about Sissy's problem, we seem to need a kind of rich and complex description of what people are able to do and to be. [...]
>
> Economists, policy-makers, social scientists, and philosophers are still faced with this problem of measurement and assessment. They need to know how people are doing in many different parts of the world, and they need to know what is really involved in asking that question. When they face the problem well, they face it, so to speak, with wonder [...]; with a sense, that is, of the profound complexity of assessing human life, and with a desire to admit, at least initially, the widest possible range of accounts of how one might go about this, and what indicators one might trust.
>
> Of course, it is possible to wonder not at all – to stick to a mechanical formula that is easy to use and which has been used before. The unasked question does not have to be answered. (Nussbaum and Sen 1993, 2.)

Remarkably, Nussbaum and Sen (1993, 2) also believe that Aristotle's philosophy may help us "to face the problem well:" Aristotle gave a pluralistic account of what we mean by conducting our lives in a good way and thereby developed "a kind of rich and complex description of what people are able to do and to be." The sophisticated theoretical framework Sen developed in order to introduce rich descriptions of human life into economics shows strong affinities with Neurath's: first, the view that the concept of "well-being" is (and should be recognized as) a legitimate object of economic study; second the definition of well-being as a set of "functionings" which includes – to use Neurath's terms – "complex and primitive facts at the same time."[15] Thirdly, Sen's approach also integrates "holistic" and "individualistic" features. In the framework Sen suggested, "the total situation of a group of people" is conceived as a totality of unequally distributed social conditions of freedom in liv-

[15] In Sen's (1985, 197f.) terms: "The primary feature of well-being can be seen in terms of how a person can 'function,' taking that term in a very broad sense. I shall refer to various doings and beings that come into this assessment as functionings. These could be activities (like eating or reading or seeing), or states of existence or being, e.g., being well nourished, being free from malaria, not being ashamed by the poverty of one's clothing or shoes (to go back to a question that Adam Smith discussed in his Wealth of Nations). I shall refer to the set of functionings a person actually achieves as the functioning vector [...]."

ing one's own life. They determine the choices individual human beings are in a position to make. "A person's *capability set* can be defined as the set of functioning vectors within his or her reach" (Sen 1985, 200; emphasis added).[16] Hence, the "functioning and capability approach" suggests that a theory of social and economic development should describe to what extent people have (or achieve / lose) access to the social conditions under which they are able to conduct their lives in the way they choose – and how choices and opportunities are distributed between women and men, young and old people, urban regions and the country-side. Obviously, Sen's theory is much further elaborated than Neurath's and has the advantage of taking the social and political experiences of the twentieth century into account. Neurath paid less attention "to the freedom 'to do this', or 'to be that' that a person has," than Sen does. Nevertheless, the indicators of "well-being" Neurath suggested also include access to political and social freedom. He emphasized the right of each individual to take part in all human activities, to participate in shaping the "order of life" [*Lebensordnung*] concerning housing, nourishment, clothing, and medical care as well as law, morals, love, family, religion, and art (Neurath 1921/1981, 197; 1917a/2004, 1938).[17]

Let me give one more example to show that Neurath's views of human well-being and ill-being anticipated some approaches which became influential only at the end of the twentieth century. The *Human Development Report* (1997, 17) focused on poverty. In it we find the following list of "Criteria of ill-being."

The following criteria, drawn from various participatory studies, were used by local people in Asia and Sub-Saharan Africa for defining poverty and ill-being:

- Being disabled (for example, blind, crippled, mentally impaired, chronically sick).
- Lacking land, livestock, farm equipment, a grinding mill.
- Being unable to decently bury their dead.
- Being unable to send their children to school.
- Having more mouths to feed, fewer hands to help.
- Lacking able-bodied family members who can feed their families in a crisis.
- Having bad housing.
- Suffering the effects of destructive behaviors (for example, alcoholism).
- Having to put children in employment.
- Being single parents.
- Having to accept demeaning or low-status work.
- Having food security for only a few months each year.
- Being dependent on common property resources.

[16] "In examining the well-being of a person, attention can legitimately be paid to the capability set of the person and not just to the chosen functioning vector. This has the effect of taking note of the positive freedoms in a general sense (the freedom 'to do this', or 'to be that') that a person has" (Sen 1985, 200).

[17] Given the fact that Neurath was highly critical of Aristotelian *Metaphysics* (see for instance his letters to Carnap in the present volume), the Aristotelian features in his conception of the subject matter of economics might seem to be inconsistent. Yet, Neurath referred his approach to economics to the *Nicomachean Ethics* which is arguably independent of Aristotelian *Metaphysics*. Aristotle's *Ethics* is actually to a large extent empirically informed. Neurath could, in my view, quite easily be sympathetic to the empirical orientation of much of Aristotle's work and at the same time reject his *Metaphysics*.

In my opinion, this illustrates perfectly what Neurath regarded as the starting point of a truly empirical account of the life situation of a given population. The list presents a picture of the entire situation of a group of people, relating the individual components, which cannot be captured by a single unit of measurement, in different ways. It also asks social scientists and politicians to think about poverty by using the categories poor people themselves had used to describe their experience.[18]

Let us now return to Neurath's attempt to redefine the object of economics: according to Neurath, wealth was to be described as a heterogeneous complex of pleasure and displeasure of individuals and groups of individuals. Yet, this is only one aspect of what Neurath considered the object of scientific economics to be. We also have to ask how the changes in life situations are brought about. Neurath also stressed that what he was suggesting was by no means completely new. On the contrary: here, too, he wanted to recapture the complexity and richness of an ancient, though neglected scientific question.

> Political economists have always been interested in the processes that make people wealthy or poor. As long as this happened by cultivation of land or operating a plant it was treated as basically a technical question, but soon it was realised that it was the system of contracts, the system of taxes and duties, that were of decisive importance: thus the systems of organisations themselves became objects of inquiry. The classical school of economics has examined one unique form of such systems of organisation, free competition, and it has praised it just like the mercantilists praised theirs. In the course of examining the free market system as a cause of the growth of the population, one had to look into its structure in detail and so came across issues that did not have anything to do with wealth directly, e.g. one observed falling or rising prices entirely independently of whether this was conjoined with an increase or a decrease in wealth. Since it involved measurable quantities, just as with crop yields etc., that were easy to establish unambiguously, price theory soon became a discipline that was practised particularly eagerly. The question whether the system of organisation at issue would foster wealth or not receded in importance or was neglected altogether. Partly this was related to the idea that monetary calculation adequately reflected the distribution of wealth. (Neurath 1910b/2004, 272.)

Hence, Neurath called for a radical widening of the scope of economic studies on two levels: on the one hand, he assumed that there was a heterogeneous totality of the well-being and ill-being of individuals, and on the other hand, a heterogeneous totality of factors that both influence that complex structure of individual states and also transforms them ("the system of contracts, taxes, systems of organisations"). Economists, in Neurath's view, should describe and compare the effects of certain shifts on the individual states. (From an historical perspective they should examine shifts in the past. From a "purely theoretical" perspective, the effects of any possible shift should be investigated and compared with others.) Neurath tried to elaborate a method to address this highly complex situation in a precise way – the method he called "calculation in kind." In 1935 he described in a nutshell what this method was all about:

[18] See also Alkire (2002) for a comprehensive account of the concept of "multidimensional development" and Alkire and Santos (2010) and Alkire (2011) for the methodological issues involved in measuring multidimensional poverty.

> Economic theory deals with the influence particular institutions and actions have on the standard of living. (Neurath 1935/1987, 96.)

Note that this broad framework of "calculation in kind" includes analyses of markets, albeit from a specific point of view. Free competition is seen as a very specific type of organization. Its effects on life situations can and should be compared to the effects other "institutions," "actions" and their combination would have on the standard of living of a given population.

> In itself, calculation in kind does not represent any one socio-political or economic standpoint, it is merely a way of looking at things. Economic institutions and whole systems of economic organisations can be investigated by the in-kind calculus and it may be found, for instance, that under some circumstances the free market is more efficient than the planned economy. [...] What is essential is how we formulate the problem to be solved. The focus does not lie on the change of prices, of the interest rate, of wages, but on their influence on the satisfaction of needs. Even economic orders that make no use of these concepts may be examined on their efficiency. (Neurath 1917b/2004, 244.)

It is important to see the difference between the concept of "calculation in kind" and the concept of a "planned economy in kind." Before World War One, Neurath developed the calculation-in-kind-framework for *theoretically* exploring diverse forms of economic organization and their impact on the standard of living. He was however convinced that a socialist economy was a planned economy in kind and would therefore necessarily rely on some sort of calculation in kind. This was a crucial subject of the "calculation-debates" of the 1920s and 1930.[19] The highly political connotation that "calculation-in-kind" got after World War One should not make us forget that the concept was initially a theoretical concept and part of a highly ambitious academic project of research. Since Neurath's engagement in Munich in 1919, he was excluded from Academia and therefore not able to develop his ambitious methodological project further. Yet, even in its rather rudimentary form it enabled him to point at surprisingly modern theoretical topics of economics.[20]

---

[19] See Thomas Uebel's highly interesting reconstruction of Neurath's arguments concerning the possibility / unavoidability of in-kind-considerations in economics and politics (be it socialist or non-socialist) in the present volume.

[20] The economist Mooslechner emphasized an interesting example: Neurath who is till this day notorious among economists for his plead for a moneyless socialist economy, suggested in 1909 to investigate various monetary systems regarding their productivity: "[H]e did not rule out at that time that differences in the type of monetary organization will lead to corresponding differences in real productivity. 'The questions of productivity of monetary organization [...] are thus granted full legitimacy [...]' (Neurath 1909b/2004, 296)" (Mooslechner 2007, 105).

6.2 Making a Scientific Way of Looking at Society Accessible to the Public

Let us now make a big leap from the pre-World War One period, when Neurath still envisioned himself as a future university professor of political economy who wished to contribute to modernizing economic theory, to the mid-nineteen twenties, when Neurath was the director of the *Social and Economic Museum of Vienna*, one of the main institutions for adult education in *Red Vienna*. At the museum Neurath, together with an interdisciplinary team, developed the "Vienna method of picture statistics" which was meant to become a new and effective tool for "disseminating social enlightenment" (Stadler 2001, 700).

Enlightenment, in Neurath's definition, was "social" in the dual sense of the word. First, it was to show how the life conditions of broad masses of people depend on the social and economic order. Second, it was to give the broad masses of people access to that knowledge. In other words, it was to create the social and intellectual conditions under which knowledge about society would be produced for society.

> Most people first want to learn how it is that in periods of highest technological development, privation and poverty prevail. [...] When coffee is burnt or thrown into the sea, when cotton is destroyed, when machines are scrapped and millions of unemployed starve, enjoyment of life and health is restricted to a much greater extent than can be remedied by applying technical and hygienic methods. The social and economic order is the fate of broad masses of people who find it difficult to get an idea of how production, consumption, and the economic order are related to each other. (Neurath 1933/1991, 232.)

Note that the challenge, as Neurath saw it, was not only to present certain statistical data but to create special tools to visualize a type of relation we are not able to see as such with the naked eye. The challenge was to create special tools for discovering and revealing social facts.

> We need special tools to disseminate social enlightenment. In the *age of the visual,* primarily museums and exhibitions, pictures, and films should be the focus of our attention. While technical and hygienic correlations may be illustrated to some extent by photographs and models, social processes demand new special methods. We must try to show how the amount of production and consumption changes, how the reduction of employment is correlated to rationalization. The social engineer has to teach us how – as a result of birth and death, immigration and emigration – masses of human beings increase and decrease, how infant mortality and tuberculosis mortality of entire cities may be reduced by improvements in public housing. (Neurath 1933/1991, 234.)

Note that, in Neurath's view, the *Vienna Method of Picture Statistics* was a sophisticated tool for elaborating the "way of looking at things" he had advocated in his economic writings, and for making this "way of looking at things" accessible to the public. For Neurath, the museum was not a place where scientific truths about society and economy were conveyed to the layman. He rejected the idea of "popularization of knowledge," if by this is meant translating information "from the complicated to the simple" (Neurath 1996, 257; cf. Stadler 2001, 2011). The museum was to be a place where people – most of them without higher education – could learn to look at social issues in a new way and practice doing this.

What distinguished the "Vienna Method of Picture Statistics" from other ways of visualizing statistical data?[21] The most important principle is very nicely summed up in the term Marie Neurath[22], Otto Neurath's third wife, coined during their exile in the Netherlands: Isotype, which stands for "International System of Typographic Picture Education." Isotype combines the two Greek words "isos" (the same) and "typos" (type, symbol), indicating the methodological principle of visualization that Neurath (1931/2017, 110) thought was crucial: "*A larger quantity of things is to be represented by a larger quantity of speaking [sprechenden] signs*" (original emphasis) – and not, we should add, by symbols of different size. The Isotype representation repeats symbols of the same shape and size, thereby constructing visible quantities and visible differences between smaller and larger quantities. Very often the tables produced by the Isotype teams show a juxtaposition of longer and shorter rows of symbols (women, men, houses, ships, fruits, etc.). The most famous symbol is that of the unemployed (Fig. 6.2).

In this context it is most interesting to see how Neurath explained what he saw as the main advantages of his system of visual representation (Fig. 6.1).

Here Neurath presents four methods for visualizing quantitative data: squares, circles, rectangles, and figures. In all four cases, the same quantities are compared. And in all four cases there are two levels of comparison: 1 and 2, A and B. Neurath demonstrates in a most fascinating way that by using different methods of visualization we can make very different amounts of information accessible to the persons looking at them.[23] In the first case – the squares – we can only say: 2 is bigger than 1,

[21] On the development from the Vienna Method to Isotype see Neurath (1945/1973, 214–78) and the beautiful new book Neurath (2010). See also Angélique Groß's chapter in the present volume.

[22] Marie Neurath, born Reidemeister, played a crucial role in the whole process. She was the most important "transformer" of statistical data into pictorial graphics. (On the concept of the "transformer" see Neurath and Kinross 2009; see also her memories in Neurath and Cohen 1973, 56–64). She was also the author of numerous highly original books in picture language for younger readers (see, for instance, Kindel 2011). After Neurath's death she continued to develop the method further. It is due to her that the Isotype material came to the University of Reading (see Twyman 1982) where highly significant research on Isotype is going on. See: http://www.isotyperevisited.org/1981/01/isotype-and-the-university-of-reading.html

[23] Edward R. Tufte (1983, 61) stresses a very similar point: "The confounding of design variation with data variation over the surface of a graphic leads to ambiguity and deception, for the eye may mix up changes in the design with changes in data." The examples of bad graphics Tufte (1983, 69) gives are amazingly similar to those Neurath (1936, 75; 1991, 381) had in mind. It is a pity that Tufte does not seem to be aware of Neurath's work. I did not find any reference to Neurath in Tufte's writings. In any case, his conceptual framework seems strikingly similar. Remarkably enough, the subject of one of Tufte's early books is the political control of the economy. Here he described the "interplay between politics and macroeconomics in the United States and other capitalist democracies" and tried to "find specific links between political and economic life" (Tufte 1978, p. IX). At the end of his book he stressed that "those who write about national economics" bear a special responsibility. "That responsibility is to improve the level of public understanding so that voters can evaluate and repudiate corrupt economic policies" (1978, 154). Obviously, there is a link to graphic representation of quantitative data (and the "lie factor" – see Tufte (1983, 57) – that may be found there). In Neurath's early writings on national economics there is a strong emphasis on the impact public understanding of economic and social issues has on the advancement of democracy. See for instance Neurath (1908/1998, 1910a/1998), and Neurath and Schapire-Neurath (1910).

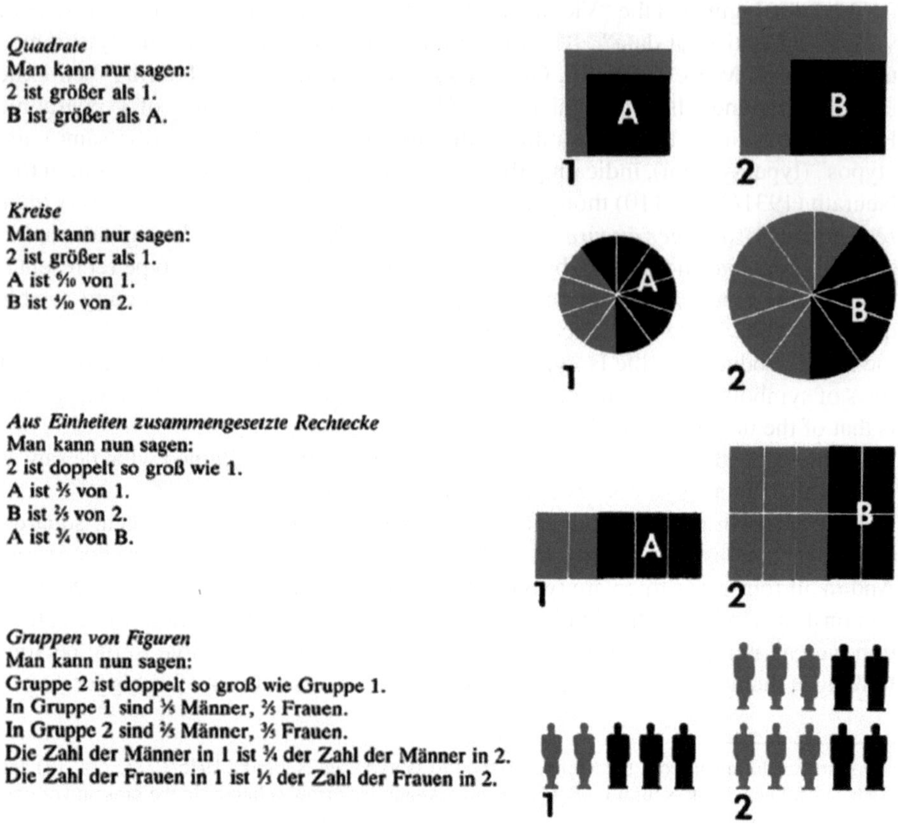

Fig. 6.1 Methods of Visualization

and B is bigger than A. Visualization by means of circles give us some more information – at least concerning the inner structure of 1 and 2. However, for the comparison between 1 and 2 we are still left with "2 is bigger than 1." In the third case, we use rectangles composed of unities, which enable us to see that 2 is twice 1 and gives us a more precise idea of the difference between A and B, namely A is ¾ of B. Finally, when we use groups of figures instead of rectangles, the information we receive is still richer, although not about the quantities. Now we see that the quantities represent women and men which makes it easier to remember what the comparison is all about.

Remember what we called Neurath's early attempts to square the circle between "holism" and "individualism": his search for a method of conceptualizing the life of a nation as a totality of individual human well-being. Already at that time, Neurath emphasized that symbolic representation would open up new avenues for comparing

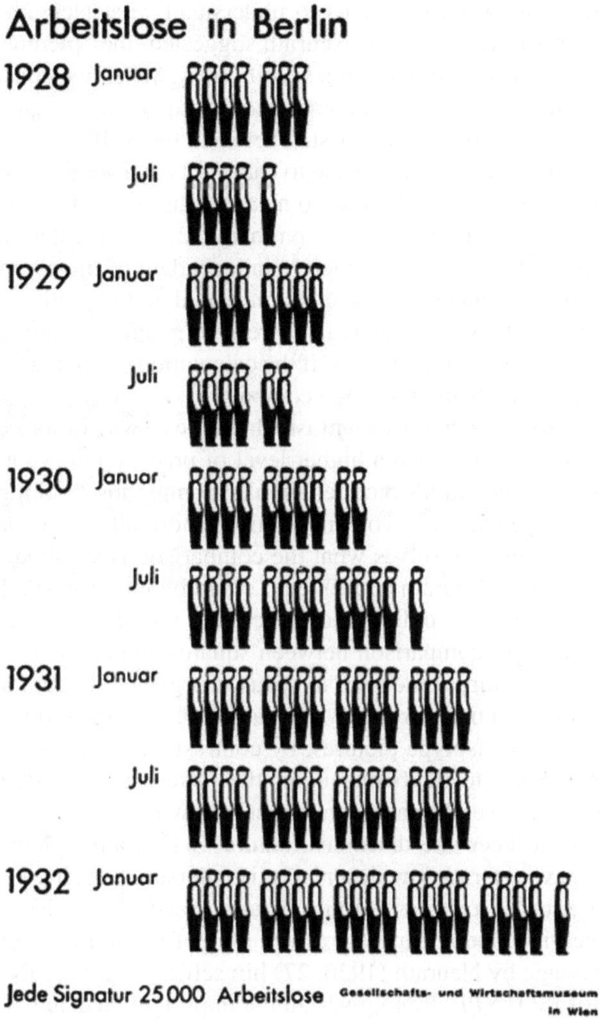

Fig. 6.2 Unemployed in Berlin

heterogeneous entities in a precise way without calculating them in units (either monetary or any other kind). And also at that time, Neurath described the procedures he had in mind by reminding us of how we proceed when we compare pictures.

> One cannot compare two states by comparing them bit by bit, say first the constitution, then the climate etc.; each of them has to be comprehended as a whole. After all, neither can we compare pictures in this way, nor can we do this in respect to the machines. The very idea of a calculus, however, consists of deriving a complex from the individual elements. (Neurath 1910b/2004, 280.)

Picture statistics make it easier for us to understand what Neurath meant in this early text. To compare two states as Neurath suggested, the "pictures" of the compared states have to be constructed in a specific way. When we construct the comparison as in the first example (as a comparison of squares), we can only conclude that there is an absolute difference of size between them. If we wished to attain a higher level of precision, we would have to carry out a calculation which would not be difficult in this case (we would have to measure the sides of the squares and calculate the areas and the difference between them). But the greater precision would be based on a procedure that is completely independent of the comparative procedures of our visual judgment. The calculation would end up with one very precise figure: the difference between the areas expressed in square centimeters. And in a more complicated case, it would be best if the calculations were made by an expert – a scientist trained in mathematics – or a computer.

Neurath's *Picture Statistics*, by contrast, elaborates a way of looking at things in which the steps we take to reach a higher level of precision are of a very different kind than calculation. Neurath's pictures should prompt those looking at them to go back and forth between at least two constellations (normally more than two) of elements, figuring out for themselves what the comparison is all about. Note that the visual representation as Neurath conceived it could by no means replace the verbal expression. On the contrary: only in the first case of visual representation – the one Neurath did not accept (comparison between squares and circles) – we are able to grasp the information immediately, so to speak at a glance: larger / smaller. And if we want to understand it more precisely, we have to change the level and the media – or trust in some expert. Isotype pictures, by contrast, are constructed in a "discursive" way: we are led to deliberate upon the different components, relate and compare them, so that we ultimately construct the "whole" we are looking at.[24]

It is important to keep the discursive nature of Neurath's picture language in mind in order to avoid a rather widespread misunderstanding. It consists in assuming that the criterion of a successful visualization according to Neurath lies in the speed with which its content can be grasped. One of the sources of this misunderstanding is a passage by Neurath (1936, 27) himself.[25] "A picture making good use of the system [of ISOTYPE, E.N.] gives all the important facts in the statement it is picturing. At the first look you see the most important points, at the second, the less important points, at the third, the details, at the forth, nothing more – if you see more, the teaching-picture is bad." Some interpreters deduce from this passage that an optimally visualized statement would be "recognizable at first glance" (Hartmann and Bauer 2002, 49). They overlook that the context of the passage is the *production* of a good teaching picture. Immediately after the passage quoted above Neurath

[24] This "discursive" procedure by Isotype forms the basis of what C. Burke (2011, 51) called "an early move away from 'mechanical objectivity' towards 'trained judgment' in scientific visualization." The "discursive" method is also a core element of what Neurath called "the scientific attitude" (see Nemeth 2011 and the quote at the end of this article).

[25] It is not the only case in which Neurath's all too catchy formulations were rather misleading. The same can be said of the terms "physicalism," "physicalist language," "unified science," "unified language," "index verborum prohibitorum."

(1936, 27) continues: "A good teacher is able to keep out all unnecessary details. For the selection, a clear sense of the needs of education is important, and a good teaching-picture may only be produced with the help of a good teacher." Thus the passage refers to the *selection* of information in the process of production of a teaching-picture. The "three glances" play the role of a checking procedure to be applied by the designers of the visualized table: Does the design include any bits of information that do not contribute to the content of the visualized statement (and would therefore distract the attention from the content that should be communicated)? The case of geographical representation might help to make the point clear.

See, for instance, the teaching-picture that visualizes the main export products of Latin America. (Fig. 6.3) It shows the shape of the continent and borders of countries but does not contain any additional information of physical geography (mountains, plains etc.). Only if it is the aim of the picture to show a relationship between specific physical geographical features and specific social or economic issues, the representation of mountains, plains etc. is justified. Otherwise the picture would invite a "forth" look and thereby distract the attention from the statement that the teaching-picture aims to visualize.

The table "Automobile Industry" (Fig. 6.4.) clearly illustrates the discursive nature of the Isotype pictures. At first glance we are struck by the enormous difference in car production between Europe and America. We almost automatically begin to count how many cars are produced in America and Europe, and then go on to do the same for the number of workers. On a third level, the table offers a means to help us think about the factors that lay behind this amazing difference: in the background of the American workers we see the assembly line.

The Vienna Method, Neurath (1996, 257) argued, could relieve less educated people of the humiliation and "inferiority complex" they often experience when they are confronted with written texts or the verbal presentation of abstract arguments. "Words separate – Pictures unite" was one of the slogans posted on the wall of Neurath's office at the museum. Again, this does not mean that we need no verbal deliberation when we try to figure out what a particular table represents. On the contrary: it is one of the advantages of visual representation, Neurath (1996, 259) argued, that it supports and initiates group discussion.[26] Yet, words *do* "separate," in Neurath's sense, in all the cases we use them in order to "hand on knowledge" from one person to the other. The "handing on of knowledge" orally presupposes that the person who wants to "tell or show" something speaks the same language as the person he or she addresses: the same national language, but also the language of the educated social strata. And this latter barrier might even be more difficult to overcome than the national one: the language of educated people not only includes class-specific jargon that can have an intimidating effect. The language of educated people is also based on a specific attitude to language and verbal communication. It is rooted in the teacher – student relation of secondary and tertiary education in

[26] See also Angélique Groß on the "activation of the addressee" in her contribution to the present volume.

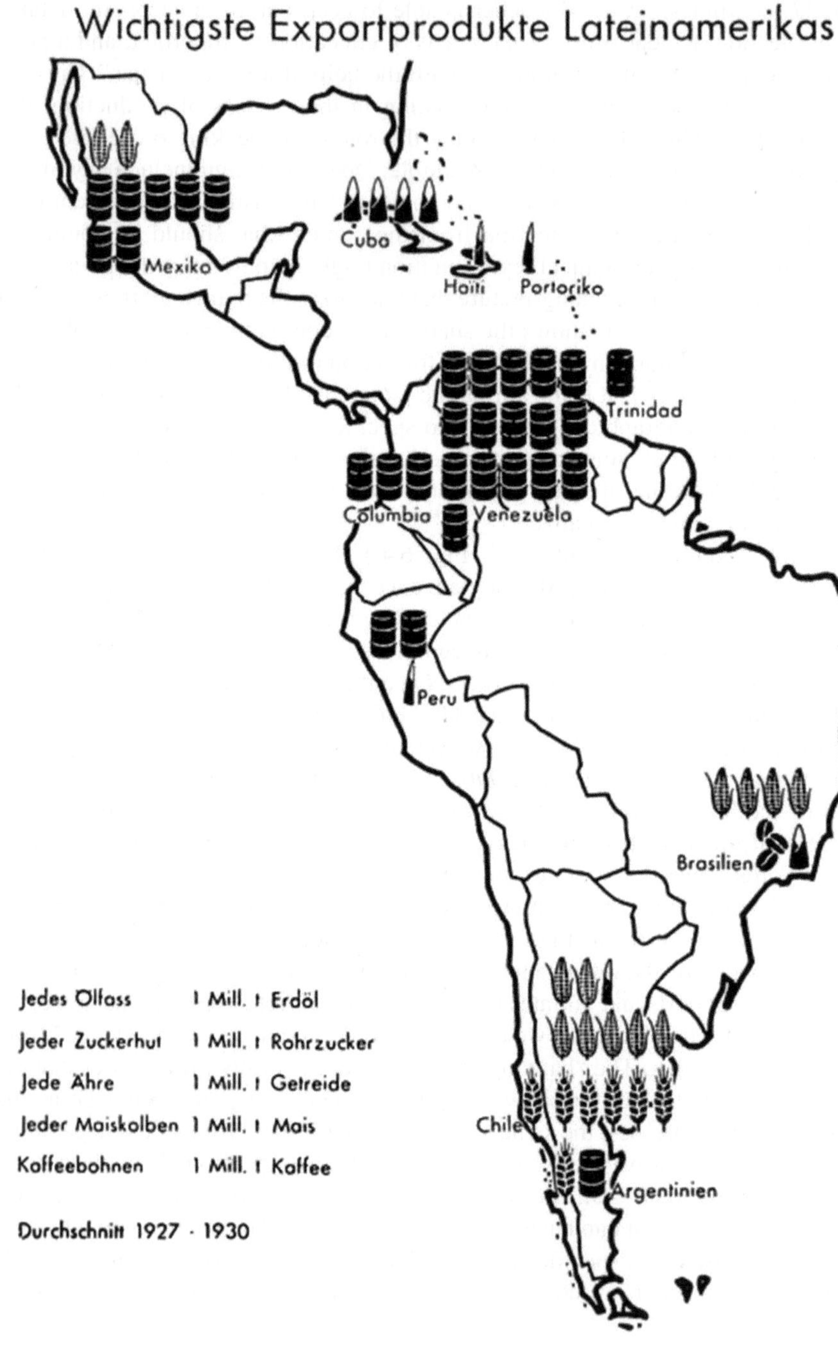

Fig. 6.3 Latin America's main products of export

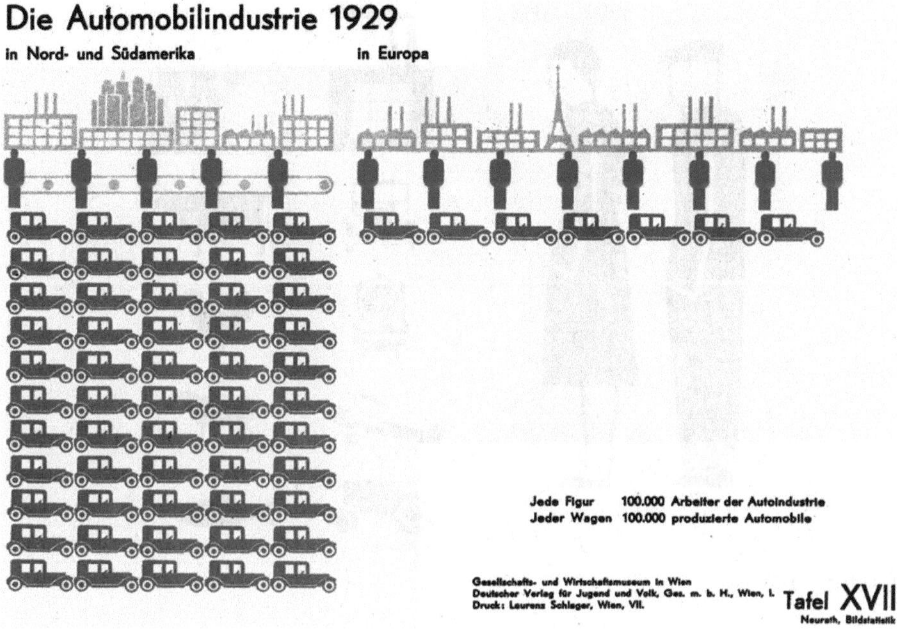

Fig. 6.4 Automobile industry

which the student is trained to follow the teacher's train of thought and to accept reproducing it in his or her mind.

In Neurath's conception of visual education, verbal communication assumes a different role.[27] Since Neurath's pictures cannot be grasped immediately but only by relating the components to each other and thereby constructing the entire complex step by step, the pictures even necessitate discourse (both subjective and inter-subjective discourse). But the person who looks at them is in a position to find her or his own words, to tell the story in his or her own language. She is also not obliged to "read" a picture by following a certain direction. There is no *right* direction. It was even one of Neurath's main concerns that the presentation of the material in a quantitative order in the pictures should not suggest the direction of reading and comprehending the data. On the contrary, these pictures should invite people to start at the point they find most striking or personally most interesting. Neurath (1996, 258) asked his team to use "narrative materials" and he pointed to the "relatively sovereign position" people who look at the pictures should be able to assume. Note again that it is not possible to assume a "relatively sovereign position" when the information presented can be grasped at a glance (Fig. 6.5).

However, Neurath wanted the individual symbols to be immediately accessible. He wanted them to "speak" to us directly. In the case of the unemployed, for

[27] Silke Körber shows in her chapter in the present volume how Neurath developed during his years in exile the "discursive" dimension of his visualization method further. She calls it "picture-text-style."

Fig. 6.5 Symbols

instance, the symbol speaks in an almost physical way: shoulders hanging down, hands in the pockets, we can almost imagine unemployed people standing in line. Many symbols are (thanks to Gerd Arntz and other artists) even aesthetically appealing.[28] They catch our eye and our attention. We understand immediately what they stand for. And this is why they are so well suited as public signs, labels, posters etc (Fig. 6.6).

However, as soon as these individual symbols are used to make social structures visible, our perception takes on a very different function. It becomes the starting point and the vehicle of a process of reading that we ourselves must initiate.

Let us consider another example – one that is relatively well known in Austria. (Fig. 6.7) It shows how the compiled "narrative materials" relate several dimensions and thereby try to make "social facts" visible. The table "Infant Mortality and Social Conditions in Vienna" juxtaposes two periods, the years from 1901 to 1905 and the period from 1925 to 1929. At the same time, two districts of Vienna are compared: a middle-class area and a well-known labor district. In all four pictures we see twenty babies and a certain number of small coffins covering over some of the babies. Each coffin represents one baby that died during the first year of its life. We see that during the first period, 3 of 20 babies died in the "better-off" district while 4 of 20 babies died in the poorer district. 25 years later, only 1 of 20 babies died in the better-off district, 2 of 20 in the poorer one. We see immediately that the situation has dramatically improved. Looking at both districts, we see that there had been a more than fifty percent decline in deaths. Yet on closer scrutiny we see that the two

[28] On the great importance of Gerd Arntz see Stadler (1982).

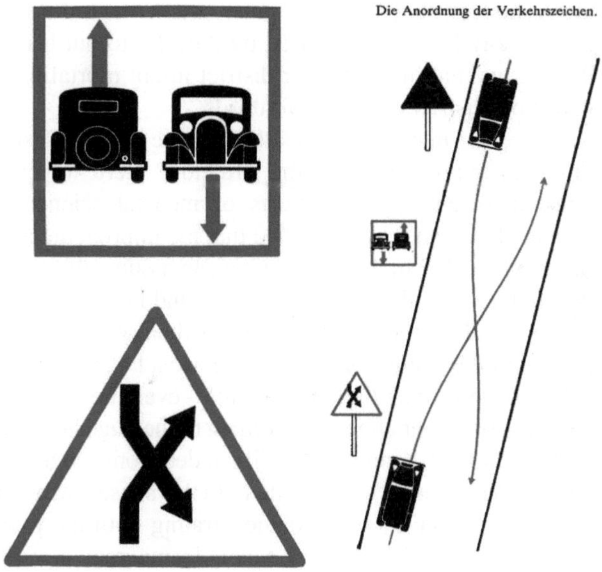

Figs. 6.6 Public signs (the one with the cars)

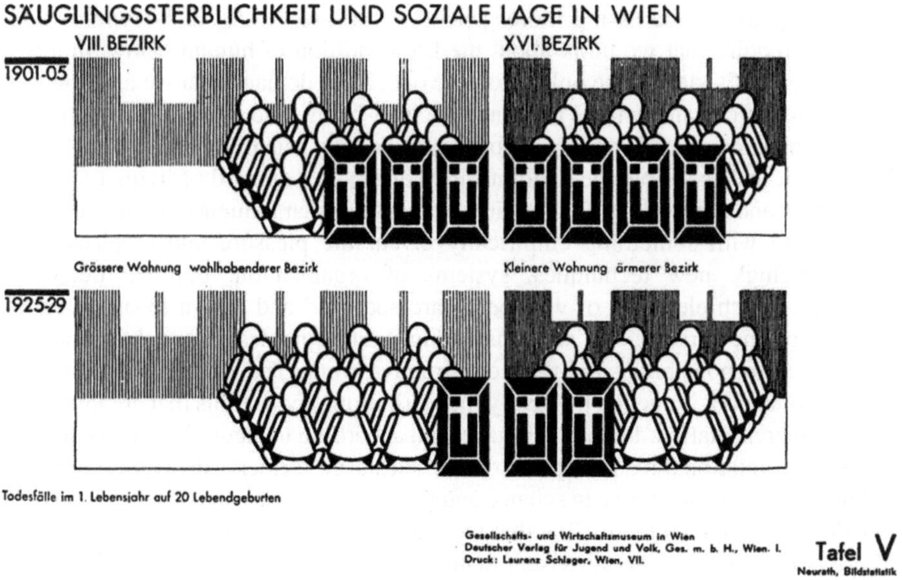

Fig. 6.7 Infant mortality

areas did not benefit to the same extent from the improvement. In both districts infant mortality has decreased by 2 of 20, to put it in absolute numbers. However, this means that in the poorer district infant mortality is reduced by half, and in the better-off district even by two thirds.

There are many questions that can be raised. First, what has brought about this dramatic improvement of the situation? Here we can imagine the visitors to the museum discussing the effects of medical science and hygienic measurements. Second, how can it be explained that the improvement affected the population so disproportionately? In this context we can assume that people think about differences in nutrition, availability of medical care and hospitals, about working hours of women, access to education etc. The pictures themselves provide several clues: in the background of the "better-off" babies we see a large, bright apartment whereas the poorer babies live in small, dark apartments – even the windows are shown to be smaller. At the same time, our attention is drawn to the huge housing problems Vienna faced after World War One – and to the political decisions the Social Democratic Government of Vienna took in order to solve them. In this sense, the pictures produced by the Museum were appropriate tools for demonstrating both the problems and the improvements that had been achieved by the Social Democratic government.

In this particular case, it is obvious that the deliberations and discussions the pictures initiate inevitably have some political impact. Yet, the same ultimately also holds for pictures addressing more general social problems. All Isotype pictures basically require that we think about the life condition of human beings in a way that in Neurath's view is the only scientific one. They demand that we ask how particular institutions and actions influence the well-being (and ill-being) of a group of people. Insofar as sociology is (or is becoming) a truly scientific enterprise, Neurath believed, it makes visible the relationship between the standard of living of a given population and the totality of institutions and actions that influence it. Of course, we are dealing with something empirically given: the pleasure and displeasure of human beings, new techniques, systems of organization, political decisions. However, which elements of well-being are specified and drawn to our attention depends on the persons who are considering these social issues, on their imagination, social responsibility and their constructive work. The same imaginative and constructive work is needed to specify the institutions and actions that are integrated into the representation. Picture language can, according to Neurath, help us develop this type of relational reasoning on social issues even further – in social science, in public discourse, and between science and the public: "If one starts with visual aids, one does not even get the feeling that there are two fields, science and non-science. There is a basis for common visual material" (Neurath 1996, 262). Finally, visual education can help social scientists, philosophers and the public become aware of the fact that scientific reasoning, as Neurath (in the tradition of Ernst Mach) conceived of it, has an enormous social and political impact.

In one of his last writings, Neurath argued that visual education could play a crucial part in what he called the "transfer of a scientific attitude," and that democracy would not be able to survive without an educational system that systematically supported the "transfer of a scientific attitude." As far as I can see, Neurath's insightful remarks on scientific attitude and democracy are no less significant today than they were in 1945:

> Education is not only the handing on of knowledge, it is also concerned with the ability to analyze observations and to find out something and contemplate all the matter under discussion from all sides. Let us use the expression "meditation". Education advocates a meditative mood. [...]
>
> The transfer of looking at more than one possibility, to be prepared to alter statements, is the principle of the scientific attitude. The social pattern, which permits more than one opinion etc. is the "democratic pattern".
>
> Part of education deals with the evolution of one's own judgement, of a "scientific attitude", a quality not restricted to scholars only; there are laymen who have it, and there are laymen who do not have it. The transfer of the scientific attitude is not mainly concerned with knowledge but also with the creation of certain habits, sincerity of research and integrity of arguing. That implies not only thinking of how to pass examinations but also of deliberating on various possibilities or arguments, in short how to become 'meditative' on certain subjects. (Neurath 1996, 260.)

References

Alkire, Sabina. 2002. Dimensions of Human Development. *World Development* 30 (2): 281–305.

———. 2011. *Multidimensional Poverty and its Discontents*. Oxford Poverty & Human Development Initiative (OPHI) Working Paper No. 46, Nov 2011, 1–30.

Alkire, Sabina, and Maria Emma Santos. 2010. *Acute Multidimensional Poverty: A New Index for Developing Countries*. Oxford Poverty & Human Development Initiative (OPHI) Working Paper No 38, 1–139.

Belke, Ingrid. 1978. *Die sozialrefomerischen Ideen von Josef Popper-Lynkeus (1838-1921)*. Tübingen: Mohr.

Blau, Eve. 1999. *The Architecture of Red Vienna*, 1919–1934. Cambridge: MIT Press.

———. 2006. Isotype and Architecture in Red Vienna: The Modern Projects of Otto Neurath and Josef Frank. In *Culture and Politics in Red Vienna*, Austrian Studies, ed. Judith Beniston and Robert Vilain, vol. 14, 227–259. Leeds: Maney Publishing for the Modern Humanities Research Association.

Buek, Otto. 1926. Gregorius Itelson. *Kant-Studien* 31: 428–430.

Burke, Christopher. 2009. Isotype. Representing Social Facts Pictorially. *Information Design Journal* 17 (3): 210–221.

———. 2010a. Back to Basics: Otto Neurath and Isotype. *Cambridge Literary Review* 1 (3): 221–235.

———. 2010b. Introduction. In Neurath (2010): vii–xvi.

———. 2011. The linguistic status of Isotype. In Heinrich et al. (2011): 31–57.

Burke, Christopher, Eric Kindel, and Sue Walker, eds. 2013. *Isotype. Design and Contexts 1925-1971*. London: Hyphen Press.

Cartwright, Nancy, Jordi Cat, Lola Fleck, and Thomas Uebel. 1996. *Otto Neurath: Philosophy Between Science and Politics*. Cambridge: Cambridge University Press.

Cat, Jordi. 2014. Otto Neurath. In *Stanford Encyclopedia of Philosophy*, ed. Edward N. Zalta. http://plato.stanford.edu/entries/neurath/

Chaloupek, Günther. 2007. Otto Neurath's concepts of socialization and economic calculation and his socialist critics. In Nemeth, Schmitz and Uebel (2007): 61–76.

Haller, Rudolf. 1982/1991. The Neurath-principle: Its grounds and consequences. In Uebel (1991) : 117–130.
———. 1985/1991. The first Vienna circle. In Uebel (1991): 95–108.
Hartmann, Frank. 2005. Humanization of Knowledge through the Eye. In *Making Things Public – Atmospheres of Democracy*, ed. Bruno Latour and Peter Weibel, 698–707. Cambridge, MA: MIT Press.
Hartmann, Frank, and Erwin K. Bauer. 2002. *Bildersprache, Otto Neurath, Visualisierungen*. Wien: Wiener Universitätsverlag.
Heinrich, Richard, Elisabeth Nemeth, Wolfram Pichler, and David Wagner, eds. 2011. *Image and imaging in philosophy, science, and the arts. Proceedings of the 33rd Wittgenstein-symposium in Kirchberg 2010*. Ontos-Verlag: Frankfurt.
Hochhäusl, Sophie. 2011. *Otto Neurath – City Planning. Proposing a Socio-political Map for Modenr Urbanism*. Innsbruck: Innsbruck University Press.
Human Development Report. 1997. Published for the United Nations Development Programme (UNDP). New York/Oxford: Oxford University Press.
Kindel, Eric. 2011. Reaching the people: Isotype beyond the west. In Heinrich et al. (2011): 175–193.
Kinross, Robin. 1981. On the Influence of Isotype. *Information Design Journal* 2 (21): 122–130.
———. 1984. The Work of Otto Neurath in Visual Communication. *Fundamentum Scientiae* 5 (2): 185–199.
Kraeutler, Hadwig. 2008. *Otto Neurath. Museum and Exhibition Work. Spaces (Designed) for Communication*. Frankfurt am Main: Peter Lang.
Leonard, Robert J. 2001. 'Seeing is believing'. Otto Neurath, graphic art, and social order. *History of Political Economy* 31: 452–478.
Lessmann, Ortrud. 2007a. *Konzeption und Erfassung von Armut. Vergleich des Lebenslage-Ansatzes mit Sens 'Capability'-Ansatz*. Berlin: Duncker und Humblot.
———. 2007b. A similar line of thought in Neurath and Sen: Interpersonal comparability. In Nemeth, Schmitz, and Uebel (2007): 115–130.
Martinez-Alier, Joan. 1987. *Ecological Economics*. Oxford: Blackwell.
Mooslechner, Peter. 2007. Neurath on money: Some reflections on Neurath's monetary thought in the historical context of the birth of modern monetary economics. In Nemeth, Schmitz, and Uebel (2007): 101–114.
Mueller, Karl H. 1991. Neurath's theory of pictorial-statistical representation. In Uebel (1991): 223–254.
Nemeth, Elisabeth. 1981. *Otto Neurath und der Wiener Kreis. Revolutionäre Wissenschaft als politischer Anspruch*. Frankfurt/New York: Campus.
———. 2007. 'Freeing up One's point of view': Neurath's Machian heritage compared with Schumpeter's. In Nemeth, Schmitz, and Uebel (2007): 13–36.
———. 2011. Scientific attitude and picture language. In Heinrich et al. (2011): 59–84.
Nemeth, Elisabeth, and Richard Heinrich, eds. 1999. *Otto Neurath: Rationalität, Planung, Vielfalt*. Wien-Berlin: Oldenbourg/Akademie.
Nemeth, Elisabeth, and Friedrich Stadler, eds. 1996. *Encyclopedia and Utopia. The Life and Work of Otto Neurath (1882-1945)*. Dordrecht: Kluwer.
Nemeth, Elisabeth, Stefan W. Schmitz, and Thomas Uebel, eds. 2007. *Otto Neurath's Economics in Context*. Dordrecht: Springer.
Neurath, Otto. 1906/1998. Zur Anschauung der Antike über Handel, Gewerbe und Landwirtschaft. In Neurath (1998): 25–109.
———. 1908/1998. Die allgemeine Einführung des volkswirtschaftlichen und staatsbürgerlichen Unterrichts. In Neurath (1998): 119–132.
———. 1909a/2004. Economic history of antiquity [excerpts]. In *Economic Writings. Selections 1904–1945*, ed. Thomas Uebel and Robert S. Cohen, 120–152. Dordrecht: Kluwer.
———. 1909b/2004. Remarks on the Productivity of Money. In *Economic Writings. Selections 1904-1945*, ed. Thomas Uebel and Robert S. Cohen, 292–296. Dordrecht: Kluwer.
———. 1910a/1998. Lehrbuch der Volkwirtschaftslehre. In Neurath (1998) : 239–396.

———. 1910b/2004. On the Theory of Social Science. In *Economic Writings. Selections 1904-1945*, ed. Thomas Uebel and Robert S. Cohen, 265–291. Dordrecht: Kluwer.

———. 1911/1998. Nationalökonomie und Wertlehre, eine systematische Untersuchung. In Neurath (1998): 470–518.

———. 1912a/1973. The problem of the pleasure maximum. In Neurath and Cohen (1973): 113–122.

———. 1912b/2004. Serbia's Successes in the Balkan War. In *Economic Writings. Selections 1904–1945*, ed. Thomas Uebel and Robert S. Cohen, 200–234. Dordrecht: Kluwer.

———. 1913/1998. Galizien und Bosnien während des Balkankrieges. In Neurath (1998): 297–340.

———. 1914/1998. Die konfessionelle Struktur Osteuropas und des näheren Ostens und ihre politisch-nationale Bedeutung. In Neurath (1998) :497–536.

———. 1917a/2004. The conceptual structure of economic theory and its foundations. In *Economic Writings. Selections 1904–1945*, ed. Thomas Uebel and Robert S. Cohen, 312–342. Dordrecht: Kluwer.

———. 1917b/2004. The Economic Order of the Future and the Economic Sciences. In *Economic Writings. Selections 1904-1945*, ed. Thomas Uebel and Robert S. Cohen, 241–262. Dordrecht: Kluwer.

———. 1921/1981. Menschheit. In Neurath (1981): 197–202.

———. 1931/2017. Pictorial Statistics Following the Vienna Method. *ARTMargins* 6(1): 108–118.

———. 1933/1991. Soziale Aufklärung nach Wiener Methode. In Neurath (1991): 231–239.

———. 1935/1987. What is meant by Rational Economic Theory? In *Unified Science*, ed. Brian McGuinness, 67-109. Dordrecht: Kluwer.

———. 1936. *International Picture Language*. London: Kegan Paul, Trench, Trubner & Co.

———. 1945/1973. From Vienna method to Isotype. In Neurath and Cohen (1973): 214–284.

———. 1981. *Gesammelte philosophische und methodologische Schriften*. Eds. Rudolf Haller and Heiner Rutte. Wien: Hölder-Pichler-Tempsky.

———. 1991. *Gesammelte bildpädagogische Schriften*. Eds. Rudolf Haller and Robin Kinross. Wien: Hölder-Pichler-Tempsky.

———. 1996. Visual education: Humanisation versus popularisation. In Nemeth and Stadler (1996): 245–335.

———. 1998. *Gesammelte ökonomische, soziologische und sozialpolitische Schriften*, Teil 1 und 2. Eds. Rudolf Haller and Ulf Höfer. Wien: Hölder-Pichler-Tempsky

———. 2010. *From Hieroglyphics to Isotype. A Visual Autobiography*. Eds. Matthew Eve and Christopher Burke. London: Hyphen Press

Neurath, Marie, and Robert S. Cohen, eds. 1973. *Empiricism and Sociology*. Dordrecht: D. Reidel.

Neurath, Marie, and Robin Kinross. 2009. *The Transformer. Principles of Making Isotype Charts*. London: Hyphen Press.

Neurath, Paul, and Elisabeth Nemeth, eds. 1994. *Otto Neurath oder die Einheit von Wissenschaft und Gesellschaft*. Wien-Köln-Weimar: Böhlau.

Neurath, Paul, 1994. Otto Neurath (1882–1945) Leben und Werk. In Neurath and Nemeth (1994): 13–96.

Neurath, Otto, and Anna Schapire-Neurath, eds. 1910. *Lesebuch der Volkswirtschaftslehre, 2 Bände*. Leipzig: G. A. Glöckner.

Nikolow, Sybilla. 2011. 'Words divide, pictures unite.' Otto Neurath's pictorial statistics in historical context. In Heinrich et al. (2011): 85–98.

Novy, Klaus. 1983. *Genossenschafts-Bewegung. Zur Geschichte und Zukunft der Wohnreform*. Berlin: Transit Verlag.

Novy, Klaus, and Wolfgang Förster. 1991. *Einfach bauen. Genossenschaftliche Selbsthilfe nach der Jahrhundertwende. Zur Rekonstruktion der Wiener Siedlerbewegung*. Wien: Picus.

Nussbaum, Martha, and Amartya Sen. 1993. Introduction. In *The Quality of Life*, ed. Martha Nussbaum and Amartya Sen, 1–6. Oxford: Clarendon.

O'Neill, John. 1993. *Ecology, Policy and Politics: Human Well-Being and the Natural World*. London: Routledge.

———. 1998. *The Market. Ethics, Knowledge and Politics*. London: Routledge.

———. 1999. Socialism, ecology and Austrian economics. In Nemeth and Heinrich (1999) : 123–145.

———. 2007. Pluralism and economic institutions. In Nemeth, Schmitz, and Uebel (2007): 77–100.

Popper-Lynkeus, Josef. 1923. *Die Allgemeine Naehrpflicht als Loesung der sozialen Frage*, 2. Auflage, hg. von Margit Ornstein, Wien/Leipzig/Muenchen: Rikola Verlag.

Ringer, Fritz K. 1969. *The Decline of the German Mandrines. The German Academic Community, 1890-1933*. Cambridge, MA: Harvard University Press.

Sandner, Günther. 2014. *Otto Neurath. Eine politische Biographie*. Wien: Zsolnay.

Sen, Amartya. 1985. Well-Being, Agency and Freedom: The Dewey Lectures 1984. *Journal of Philosophy* 82: 169–221.

Stadler, Friedrich, ed. 1982. *Arbeiterbildung in der Zwischenkriegszeit. Otto Neurath, Gerd Arntz. Ausstellungskatalog mit Forschungsteil.*, (ed.). Wien-München: Löcker.

———. 2001. *The Vienna Circle*. Vienna/New York: Springer.

———. 2011. Written language and picture language after Otto Neurath – Popularizing or humanizing of knowledge. In Heinrich et al. (2011): 1–30.

Stuchlik, Josuah. 2011. Felicitology: Neurath's naturalization of ethics. *HOPOS* 1 (2): 183–207.

Tufte, Eward R. 1978. *Political Control of the Economy*. Princeton, NJ: Princeton University Press.

———. 1983. *The Visual Display of Quantitative Information*. Cheshire, CT: Graphics Press.

Twyman, Michael. 1975. *The Significance of Isotype*. http://www.isotyperevisited.org/1975/01/the-significance-of-isotype.html

———. 1982. Isotype und die Universität Reading. In Stadler (1982): 185–188.

———. 1985. Using Pictorial Language: A Discussion of the Dimensions of the Problem. In *Designing Usable Texts*, ed. Thomas Duffy and Robert Walker, 245–312. Orlando: Academic Press.

Uebel, Thomas, ed. 1991. *Rediscovering the forgotten Vienna circle: Austrian studies on Otto Neurath and the Vienna circle*. Dordrecht: Kluwer.

———. 1992. *Overcoming logical positivism from within. The emergence of Neurath's naturalism in the Vienna Circle's protocol sentence debate*. Amsterdam-Atlanta, GA: Rodopi.

———. 1995. Otto Neurath's idealist inheritance. The social and economic thought of Wilhelm Neurath. *Synthese* 103: 87–121.

———. 2000a. Some scientism, some historicism, some critics: Hayek's and Popper's critiques revisited. In *The proper ambition of science*, eds. M.W.F. Stone and Jonathan Wolff, 151–173. London: Routledge.

———. 2000b. *Vernunftkritik und Wissenschaft: Otto Neurath und der erste Wiener Kreis*. Wien/New York: Springer.

———. 2002. Heterodoxer Neopositivismus als Antwort auf den Methodenstreit? – Die Philosophie der Sozialwissenschaften im Schatten einer langen Debatte. In *Gesellschaft denken. Eine erkenntnistheoretische Standortbestimmung der Sozialwissenschaften*, ed. Leonhard Bauer und Klaus Hamberger, 319–344. Wien/New York: Springer.

———. 2004. Neurath's economics in critical context. In *Economic writings. Selections 1904–1945*, eds. Thomas Uebel and Robert S. Cohen, 1–108. Dordrecht: Kluwer.

———. 2005a. Incommensurability, ecology and planning: Neurath in the socialist calculation debate, 1919–1928. *History of Political Economy* 37: 311–342.

———. 2005b. Otto Neurath: Leben und Werk. *In Internationale Bibliographie zur österreichischen Philosophie*, ed. Bearbeitet von T. Binder, R. Fabian, U. Höfer, and J. Valent, 7–51. Amsterdam/New York: Rodopi.

———. 2007a. *Empiricism at the crossroads. The Vienna Circle's protocol-sentence debate*. Chicago/Lassalle: Open Court.

———. 2007b. Otto Neurath as an Austrian economist: Behind the scenes of the early socialist calculation debate. In Nemeth, Schmitz, and Uebel (2007): 37–60.

Vossoughian, Nader. 2008. *Otto Neurath. The Language of the Global Polis*. Rotterdam: NAi Publishers.

Chapter 7
Traveling Exhibitions in the Field: Settlements, War-Economy, and the Collaborative Practice of Seeing, 1919–1925

Sophie Hochhäusl

Abstract Over more than four decades, the Austrian economist, sociologist, and philosopher, Otto Neurath made unique contributions to the fields of museology and curation, which culminated in the founding of the Social and Economic Museum of Vienna and its mobile exhibitions in the 1920s. But until today, Neurath's involvement in the organization of portable "field exhibitions," which predated those at the Social and Economic Museum by at least half a decade remains understudied. In this essay, I argue that field exhibitions, which were informed by Neurath's theories on war economy, are instructive in analyzing his overall curatorial ideas. Staged on the outskirts of the city in collaboration with allotment garden and settlement cooperatives, these exhibitions utilized plans and diagrams to convey social and political statements of facts through pictorial statistics and everyday objects. By pairing abstract graphic information with commonplace objects, they invited inhabitants into a conversation about the material world as well as the future by drawing on personal experience. As such, these field exhibitions created a communal environment for viewing and debating information and championed what I call "a collaborative practice of seeing."

7.1 Introduction

Surrounded by corn crops people are gathered to listen to a speaker who lectures from the balcony of a make-shift building. Just visible in the background are the tracks of a field train, rows of bricks, and stacks of wood. The distant view of bleak firewalls characteristic of Vienna, signals both the edge of the city and the presence

S. Hochhäusl (✉)
University of Pennsylvania, Philadelphia, PA, USA

Radcliffe Institute of Advanced Studies, Harvard University, Cambridge, MA, USA
e-mail: hochhaus@upenn.edu

J. Cat, A. T. Tuboly (eds.), *Neurath Reconsidered*,
Boston Studies in the Philosophy and History of Science 336,
https://doi.org/10.1007/978-3-030-02128-3_7

Fig. 7.1 Otto Neurath sitting on the balcony of a communal building at the Hoffingergasse settlement during a speech by Adolf Müller, Vienna, 1921. (Source: Archives of the Cooperative Altmannsdorf-Hetzendorf)

of the metropolis. The photograph's focus, however, is an architectural exhibit staged prominently along the front façade of the central building. With his arms folded over the top of the balcony, Otto Neurath observes the onlookers below and the scene before him. Though the panorama is invisible to the photograph's contemporary viewer, it would have revealed fields broken open by excavations, ditches, and foundations, from which rose modern buildings (Fig. 7.1).[1]

This image of construction at Siedlung Hoffingergasse was not uncharacteristic of social life on Vienna's outskirts in the early years of the 1920s, when gardening and settlement cooperatives banded together to construct homes with allotment

[1] For this research, I have relied on the visual archives of a number of still existing Viennese cooperatives and associations. I would like to extend my gratitude to Karl Sedlak, Chairman of the Altmannsdorf-Hetzendorf cooperative and Sylvia Wohatschek, clerk woman of the Austrian League of Allotment Gardeners for opening their archives to me and for revealing materials previously believed to be lost or missing. I would like to acknowledge the Graham Foundation's support for this research, which allowed me to digitize these rare materials and I thank Ádám Tuboly and Günther Sandner as well as my Harvard research partners Elizabeth Keto, Spencer Glesby, and Lara Teich for their commentary and feedback on this work.

plots to alleviate pressing resource and housing shortages after World War One.[2] Indeed, the photograph was representative of the endeavors of the Österreichischer Verband für Siedlungs- und Kleingartenwesen (ÖVSK), a centralized association of Austrian settlers and allotment gardeners, whose primary task became the cooperative construction of row houses throughout Vienna in the early 1920s.[3] Otto Neurath's presence in the photograph – although a rare visual document – was nothing out of the ordinary either, given that he had become the secretary of the ÖVSK in 1921. As Klaus Novy, Robert Hoffmann, Eve Blau, Nader Vossoughian and others have observed, Neurath was instrumental in the shaping of the ÖSVK along the lines of *Gemeinwirtschaft* or cooperative economy in Vienna.[4]

Yet the photograph and the staging of the architectural exhibition on the outskirts of the city reveal critical and little studied links between Neurath's work in the ÖVSK and his earlier theories on war economy, as well as his later conceptions of modern museology, that I set out to explore here. In fact, Neurath's early writing on labor and war economy, an area of study which he promoted as director of the *Deutsches Kriegswirtschaftsmuseum zu Leipzig* (German Museum of War Economy in Leipzig) during World War One (Sandner 2014, 81–89), defined his view on the production of housing and the making of community. In turn, field exhibitions, which were part of the production of housing, informed later curatorial practices at the *Gesellschafts- und Wirtschafts Museum in Wien* (Social and Economic Museum of Vienna) founded by Neurath between 1924 and 1925. Studying these field exhibitions, therefore, illuminates an understudied prehistory of the Social and Economic Museum that uncovers how small-scale self-organized displays came to reinforce and define a materialist theory of museology. This theory sought to synthesize depictions and objects of everyday life into legible assemblages of social and economic statements of facts.[5] Utilizing abstracted depictions of social statements such as charts, tables, and plans alongside quotidian objects, this distinct approach drew both on reproducible visualizations and concrete material things of and for the everyday to spur debate among viewers.[6] Based on the use of objects of the physical

[2] For the history of war-time resource shortages and alimentation in the city of Vienna see Healy (2004).

[3] Different acronyms and translations have been used for the Österreichicher Verband für Siedlungs- und Kleingartenwesen. Where possible, I have relied on Eve Blau's (1999) translations.

[4] Neurath became involved in the affairs of settlement and allotment garden cooperatives in 1920 and assumed leadership in the Hauptverband für Siedlungs- und Kleingartenwesen in January of 1921, representing the affairs of settlers and – though to a lesser. In October of 1921 Hauptverband für Siedlungs- und Kleingartenwesen was unified with Zentralverband für Kleingärtner und Siedlungsgenossenschaften, henceforth operating as ÖSVK with Neurath acting as Secretary. For Neurath's involvement in the ÖVSK see Novy (1981), Hoffmann (1982), Blau (1999, 89–133), Vossoughian (2008, 27–44), Sandner (2014, 165–171).

[5] By the mid-1920s, Neurath frequently used the term *Tatbestand*, which I have translated as statement of fact, to describe what was exhibited as ISOTYPE picture statistics at the Museum. See for example Neurath (1925, 18), (1929, 8), (1930, 29).

[6] There has been some previous interest in elucidating Neurath's engagement with the "everyday." See, for example, the work of McElvenny (2013) and Michelle Henning's chapter in the present volume.

world, this approach implied a theoretical alignment with a distinctly Austro-Marxist view on historical materialism that captured lived experience and everyday life.

Scholarship on Neurath has emphasized his commitment to the abstraction of social and economic information and has placed the invention of his International System of Typographic Picture Education or ISOTYPE firmly at the center of research.[7] Throughout the 1920s and 1930s, Neurath asserted that his approach of abstracting information transcended the museums of the past, which were "primarily cabinets of rarities and curiosities" and ushered in museums of the future, which "might be copied everywhere without loss of value."[8] He went on to declare that established museums "put on view separate things of special value or attraction, things of which there is only one in existence" (Neurath 1936, 70). Furthermore, he insisted that conventional curatorial strategies have "come down to us from the past, when rulers or churches got together works of art and strange things which were the property of one person only" (Neurath 1936, 69–70). What has been overlooked, still, is that in their stead Neurath not only placed abstracted information as ISOTYPE, but also quotidian objects that were of significance for particular groups of people including urban workers and citizens. At the Social and Economic Museum, Neurath facilitated communication about the importance of everyday life steeped in a materialist conception of history. This conception built on his experience in the ÖSVK where distinct constituents were invited to partake in debates about housing for the first time, prompted by field exhibitions. The basic premise of pairing reproducible charts with everyday objects aimed at arriving at individually and collectively interpretable statements about the material world. Such exchange of statements within the ÖSVK extended from conversation to collaborative events and unified actions.

Studying these exhibitions, then, contributes to a vast body of scholarship on Neurath's activities in the fields of architecture, exhibition making, and the theorization of Bildpädagogik or picture education. At the same time, this research expands and pushes against some established narratives by placing the exhibitions within the context of the cultural, political, and social objectives of settlement and allotment garden cooperatives.[9] Today, for example, it is commonly accepted that Neurath's ISOTYPE began to travel in the late 1920s when organizers at the Social and Economic Museum began to think about internationalism and new architectural display standards that allowed exhibitions to become mobile. Field exhibitions, which circulated within Vienna between 1921 and 1923, predate these "first" mobile

[7] Nader Vossoughian has suggested that "there was nothing 'real' or 'authentic' about the Museum of Society and Economy's collection," (2008, 79) and Hadwig Kraeutler (2010, 165) further showed that Neurath's ideas about museology allowed putting objects on view that escaped simplistic divisions low and high art as well as a long-valued culture of connoisseurship.

[8] Letter by Otto Neurath to Anton Weber, August 7 (1924), Papers of the Gesellschafts- und Wirtschaftsmuseum, Verein für Geschichte der Arbeiterbewegung; cf. Neurath (1936, 70).

[9] On ISOTYPE, see Angélique Groß's, Elisabeth Nemeth's, and Silke Körber's chapters in the present volume.

exhibitions at the Social and Economic Museum at least by half a decade and slightly dislocate the argument about internationalism. While exhibitions at the Social and Economic Museum in the 1930s sought to transcend borders and cultures (Vossoughian 2008, 90–110) and Neurath hoped that "the same charts [would] be utilized in different countries" (1933a, 209), earlier portable field exhibitions provided a local forum for conversation. These discussions drew allied constituents together to discuss practical questions and anxieties about housing, gardening, and urban food provisioning.[10] What is at stake here, is not an argument about the role of international versus local communication in Neurath's theoretical framework for mobile exhibitions – arguably one of the characteristic features was their reliance on local audiences while visualizing widely understandable statements of facts – but rather the central importance of facilitating conversation itself. The objective of the field exhibitions and the Social and Economic Museum by extension then was not the mere display of social realities, but their contemplation through picture statistics as well as objects of everyday life by involved audiences. The assemblage of these objects, abstract and material, was intricately linked to the goal of fostering discussion, or nurturing what I will call a *collaborative practice of seeing*.

I use the idea of a *collaborative practice of seeing* in dialogue with communication historian Fred Turner's (2013, 5) concept of the *democratic surround*, which showed that multi-media exhibitions in the twentieth century rehearsed "the political process of knitting oneself into a diverse and highly individuated society," by providing viewers with self-determined, "individuated experiences" of seeing. This, according to Turner (2013, 9), was "not only a way of organizing images and sounds; it was a way of thinking about organizing society." The claim that the organization of visuals can have societal analogs with real political consequences, is crucial for my analysis of field exhibitions as well. What I understand as the *collaborative practice of seeing*, however, relied heavily on simplified display strategies and easily-accessible information that placed humans at the center of viewing. A *collaborative practice of seeing* prompted individuals, groups, and societies to synthesize and make meaning while conversing about materials on display. Neurath thus championed a phenomenological or experiential reading of and engagement with the everyday – both individual and shared – much earlier than many of his twentieth century colleagues. ÖVSK members, furthermore, built connections and conversations around displays that reflected cooperative systems and economies. Neurath and his colleagues hoped that this *cooperative practice of seeing* invited involved viewers to debate the representations of the material world and its gaps, thus initiating possibilities to collaboratively envision alternative futures and perhaps even to resist dominant forms of power in action. Arriving at this approach, required engagement with diverse fields and led Neurath from theorizing war economy to politicizing education, and from organizing housing cooperatives to curating exhibitions (Fig. 7.2).

[10]The citation leads up to one of Neurath's most well cited aphorisms "words divide – pictures unite."

Fig. 7.2 Construction at Rosenhügel settlement designed by Hugo Mayer, Vienna, 1921. (Source: Archives of the Cooperative Altmannsdorf-Hetzendorf)

7.2 Human Forms of Organization: From War to Communal Economy

A diagram at the entrance of the 1918 exhibition "World Blockade and War Economy," Neurath's first and only exhibition as Director of the German Museum of War Economy, relayed graphically how a rabbit's fur could be utilized as substitute for cotton and how its meat, in the absence of beef and pork, could become a valuable source of food during wartime (Vossoughian 2008, 53).[11] Another chart in the exhibition explained how advanced industries could draw on still-available resources during periods of shortage and how workers employed in such industries could potentially adopt "nude labor" to contribute to saving stocks of clothing that were needed at the front (Vossoughian 2007). It is clear from the selection of themes, "World Blockade and War Economy" focused on military concerns, but the displays also described and sought to direct attention to the "circumstances of life" of ordinary people – anticipated as a department on its own merits in the museum (Sandner 2014, 96). To attain this goal, Neurath deployed two strategies in the exhibition: first, he built curatorial concepts around his theoretical texts on war economy (writ-

[11] The same idea of utilizing materials in various ways and of how museums shall represent these could be found also in "Museums of the Future:" "Isn't it curious: we are constantly told that we are living in the age of technology, and yet when we enter a modern museum of natural history, there is no sign of it. [...] A huge whale hangs in the middle of the hall; but we do not learn how the 'beard' is transformed into oldfashioned corsets, how the skin is transformed into shoes, or the fat into soap that finds its way to the dressing room of a beautiful woman" (Neurath 1933b, 459.)

ten as early as 1909) and, second, he popularized economic and scientific findings by rendering them legible and concrete (see e.g. Neurath 1918/1998).

To depict the commonplace rationing of foodstuffs and the use of substitute materials during wartime, Neurath gathered charts, diagrams, reliefs, and geographical and geological maps that showed statistically where resources came from and where they were needed. Whenever such abstractions were not used, Neurath made conscious decisions to use common items and objects including the role played by paper, pencils, or toys in wartime economies (Vossoughian 2008, 54). In fact, "World Blockade and War-Economy" synthesized material objects and their depictions into easily-legible displays of wartime scarcity. As a brochure stated, "movable maps and models will be efficient in illustrating as instructive presentation of increasing housing shortage, the quantity of food according to its official designation and real amount, and the use of accumulated resources of all sorts" (Das Deutsche Kriegswirtschaftsmuseum 1918, 8). This curatorial strategy set decisive precedents for Neurath's subsequent exhibitions, their pedagogical aims, and the materials they utilized. Displays, furthermore, illustrated his distinct belief that "war did not imply negative economic consequences." As Günther Sandner (2014, 96) has suggested, in Neurath's view, war could in come cases "become [...] an instrument for increasing economic prosperity. Such optimism might appear careless, considering the horrors of World War One, and Neurath's contemporaries indeed warned against hopeful interpretations. Yet his ideas reflected a dire need for narratives of new beginnings that could be foregrounded and forged from the conflict.

Another intellectual, who theorized the power of new beginnings after World War One, was Neurath's close colleague and friend, Josef Frank (Botstein 1996; Welzig 1998; Long 1996, 2002; Meder et al. 2008). "The war has brought us into the long-awaited situation of being able to start anew, and we are looking for a way," Frank (1931a, b, 1) wrote in an essay entitled "After Wars." This argument was not an unfamiliar one at the time, when figures at the Bauhaus among others, were grappling with new beginnings and how wartime technologies could be utilized for peaceful ends.[12] But Frank's argument quickly pivoted, introducing the potential of under-development into debates about modernization.

> The great experience was to realize that nothing has to be as it is and that everything can be different. That there must be no bread, that money has no particular value, and that we could find ourselves in a never imaginable situation. We have been wrenched out of our usual plans for the future, of our habitual calm and our regular doings, and we have lived an idle life for 4 years. We have met people of whose existence we had previously only vague suspicions and thus we have discovered a larger world. (Frank 1931a, b, 1.)

Over the next five years, Neurath, Frank, and many others set out to theorize, design, and eventually build this world, which brought them in contact with new people and with a debate about the material consequences of war economy. Others, such as fellow socialist, Emmy Freundlich (1921, 227), cautioned against overestimating the positive aspects of war economy. She signaled that "not because one wanted, but

[12] Peter Galison (1990) explored Neurath's involvement in Bauhaus debates.

because one had to, one was organized and took ration cards, circumventing the entire economy with all laws and regulations." Neurath, by contrast was optimistic, hoping that the lessons of war economy could be translated into peacetime measures.

In a series of texts written between 1910 and 1918, Neurath elucidated this view, positing that war allowed envisioning new economic orders. At stake was imagining economy beyond capitalism, as one that was planned or socialized (Hoffmann 1982, 140). Indeed, according to Neurath, the conflict had already led to a form of natural or barter economy, that, if planned, could give rise to full socialization. "The present urges us to create everywhere institutions for natural economy, which include the organization of natural exchange," Neurath (1916, 425–426) wrote in 1916 at the height of World War One. In a subsequent article published in *Der Volkswirt* in 1917, he envisioned a tiered system of natural income within a planned economy – *Naturaleinkommen statt Realeinkommen*, or natural income rather than real income – in which buyers with lower income would "pay a lower price than purchasers with a higher income for the same goods" (Neurath 1917b/2004, 249). Finally, in 1919, Neurath published the book *Durch die Kriegswirtschaft zur Naturalwirtschaft* (*Through War Economy to Natural Economy*), a feverish plea for the recognition of goods' use value (rather than their exchange value) and for the establishment of an economic system – i.e. socialization – based on it.[13]

With the end of World War One and the ensuing social and economic upheaval, it seemed possible that some of these ideas could come to fruition in countries such as Austria and Germany, where century-old monarchies had collapsed. After the overthrow of the Bavarian Soviet Republic in 1919 – and Neurath's brief imprisonment for his participation in it – it quickly became clear that full socialization would not become a reality in Weimar Germany or the newly founded Austrian Republic. But in Vienna the Social Democratic Party won an absolute majority in municipal the elections of 1920 making Austria's capital a fertile testing ground for left-wing ideas.[14] In this political context, Neurath advanced the idea, that individual sectors of the economy with strong cooperative organizations could provide unique avenues to prepare for future statewide socialization (Nemeth 1982/1991, 286) (Fig. 7.3).

At the time, the allotment garden movement represented one of the most potent sectors of the Austrian economy and its association *Zentralverband der Kleingaertner und Siedlergenossenschaften* counted 30,000 members and hundreds of existing clubs.[15] Throughout the war years this *Zentralverband* had formed a number of sig-

[13] For an analysis of Neurath's economic writings and the concept of a use-value centered economy see Neurath (1919), Uebel (2004, 9), and the chapters of Thomas Uebel and Elisabeth Nemeth in the present volume.

[14] In the federal Austrian legislative election of 1920, the Social Democratic Party lost their lead to the Christian Social Party taking 36 and 41.8% of the vote respectively. In the Austrian Constitutional Assembly of the previous year, 1919 the results had been almost in reverse with the Christian Social Part taking 35.9% of the vote and the Social Democratic Party 40.8%. In the municipal elections in Vienna, in 1919, Social Democrats won with an absolute majority of 54,2%, followed by the Christian Social Party with 27.1%. The stark difference between capital and the federal states, defined political rifts throughout the progressive period known as Red Vienna.

[15] These numbers slightly vary, but Adolf Müller (1923, 19) and Otto Neurath (1923a, 19) cited that there had been between 30,000 and 40,000 members of allotment garden communities in 1923.

Fig. 7.3 Organized allotment gardens on the outskirts, Vienna, 1919. (Source: Austrian Horticultural Society)

nificant bodies and committees, which were critical in shaping communal life in Vienna during a period of extreme shortage. From 1915 onwards *Zentralverband* maintained commissions for buildings, water, infrastructure, leases, and land, and a group publicizing lectures and events. It established an economy commission that not only acquired seedlings, building materials and tools, but also clothes and affordable household items. The allotment gardener's periodical, founded in 1915, frequently published on questions of food provisioning and the creation of cooperative stores and supermarkets. By 1920, the periodical prided itself that wholesale contracts with the city had made their goods "cheaper than anywhere else," although available items varied slightly from month to month due to persisting resource bottlenecks.

A critical tool in attaining municipal support already during wartime, was allotment gardeners' display of modest harvest festivals on the outskirts and, once annually, at *Neues Rathaus*, Vienna's City Hall. These exhibitions represented unique platforms to lobby the support of politicians and the Viennese public and to meet with other cooperatives and their representatives for organizing purposes. In annual produce competitions, municipal officials awarded allotment gardeners prizes for their self-sufficient production of foodstuffs in times of scarcity. At allotment garden fairs held in individual facilities on the outskirts, Viennese citizens purchased much needed fruit, vegetables, and honey products. Staged at allotment gardeners' so-called *Schutzhäuser* – buildings that functioned as communal gathering places for discussions, cultural activities, and cooperative affairs – these vegetable shows were not only commercial and promotional events, but venues that enabled conversation about collective achievement. By 1917, allotment gardeners also showcased modest abodes at the annual harvest fair at City Hall. After the end of the war, all of these established debates and institutions – allotment gardeners' cooperative organization, the existing discourse on alleviating food scarcity, and the pressing call to transform facilities into locales for more permanent housing – drew Neurath to the cause. In joining the effort, he sought to expand the organization by further developing strategic institutions and a central association based on communal economy.

What constituted communal economy? Principal among Neurath's achievements in founding the ÖSVK was the establishment of a new *Hauptverband für Siedlungs- und Kleingartenwesen* in January of 1921, representing the affairs of settlers and – though to a lesser degree – allotment gardeners to the city.[16] In October of 1921, the *Hauptverband* merged with allotment gardeners' longstanding centralized association, the *Zentralverband,* combining the interests of both allotment gardeners and settlers in the ÖSVK. This allowed the association to pool resources and begin to draw on a substantial network of institutions for cooperative work. A considerable attainment in establishing engines for a communal economy within and beyond the ÖSVK, was the creation of the *Gemeinwirtschaftliche Siedlungs- und Baustoffanstalt* (GESIBA) in 1921. The GESIBA was a collectively owned development company administered by the state, the city, and the ÖSVK and was charged with the production and distribution of materials for the construction of settlements.[17] By 1922 the ÖSVK also controlled its own Building Bureau alongside a *Warentreuhand*, an organization that selected furniture for settlers in local stores.[18] An organizer in the German Garden City movement and expert on its cooperative institutions, Hans Kampffmeyer, became the head of a newly founded municipal Settlement Office and the Viennese architect, Adolf Loos, its chief designer (Kampffmeyer 1926, Rukschcio and Schachel 1982, 229–297). The female architect, Margarete Lihotzky, worked in the ÖSVK's Building Bureau and headed the *Warentreuhand* (Lihotzky 1923, 9). By 1923, Josef Frank – the brother of Neurath's fellow logical empiricist, Philipp Frank – represented one of at least half a dozen architects working regularly in the ÖSVK. Neurath and the cooperative representative and city councilor, Adolf Müller (the man lecturing from the balcony in the photograph heading this text), moreover, helped to materialize thousands of housing units in

[16] The ÖSVK had emerged from the *Forschungsinstitut für Gemeinwirtschaft* (Research Institute for Social Economy), an organization founded by Neurath and socialist Käthe Pick, dedicated to translating the principles of war economy into socialization debates. At the time, Pick was involved in establishment of a *Betriebsräteschule* or work council school and a member of the Austrian Committee for the Socialization of Industry. The Research Institute's role was to support the committee and provide guidance and promotion in questions of communal economy (Hoffmann 1982, 142). The ÖSVK was from the beginning equally informed by debates about how to counteract the adverse economic circumstances through socialization as it was by socialist education discourse.

[17] For a history of the GESIBA see Banik-Schweitzer (1972) and Feller (1996).

[18] For primary commentary about the activities of Warentreuhand see Anonymous (1923a) and Lihotzky (1923, 9).

Fig. 7.4 Construction at Siedlung Hoffingergasse designed by Josef Frank, Vienna, 1921. (Source: Archives of the Cooperative Altmannsdorf-Hetzendorf)

dozens of sites between 1921 and 1923. These physical spaces as well as the organized exhibitions and demonstrations elicited the participation of up to 50,000 people, illustrating the success of the principles of communal economy (Anonymous 1921a, 5; Müller 1921, 261–263; Loos 1921, 10–11; Neurath 1921a) (Fig. 7.4).[19]

Economic planning in the context of the ÖVSK thus aimed not at theorizing lifeless organization, but at heightening the experience of life for all involved. Neurath had already pointedly made a similar argument in the essay "The Converse Taylor System" in 1917. In the text, he noted that many people feared that Taylorism – the scientific study of efficiency and the organization of work processes based on it – leads to the "general mechanization of being" or *Mechanisierung des Daseins* – and the subjugation of human labor to technology (Neurath 1917a/1973, 130). But then he proclaimed that the opposite could become a reality, since the Taylor System could transform itself into "a principal force of a new humanism" that would put people before all else (Neurath 1917a/1973, 131). Rhetorically turning the very idea of Taylorism on its head, Neurath stated:

> But all this is avoided if we also use the "Converse Taylor System", which, unlike the hitherto usual Taylor System, does not seek to regard the professions as something given, but puts men themselves in the foreground, and then examines the possible professions and forms of organization as to how far they correspond to men as we find them. Perhaps existing professions and forms of organizations are inadequate; perhaps in order to achieve full

[19] The largest ÖSVK sponsored protest staged in April of 1921 elicited the participation of 50,000 settlers.

humanity and the best use of all energies towards each envisaged goal, we must create new forms of profession and organization. (Neurath 1917a/1973, 131–132.)

In order to organize such a "full humanity," Neurath asserted that the Converse Taylor System would have to take into account the structural economic and social factors preventing people from entering the labor market. Any serious attempt to provide appropriate solutions would introduce systemic change, including the reduction of work hours or the very reconfiguration of occupations themselves (Neurath 1917a/1973, 133–134). Neurath was attentive to the need to adapt occupations to individuals' own circumstances. Of particular concern were the discouraging labor prospects for those with physical and mental disabilities as veterans returned to Vienna after World War One (Fig. 7.5).

War veterans had in fact founded many cooperatives. They were the ones who, based on their military service, had been the first to claim material contributions for the production of settlements from vast demobilization efforts in 1919. With the goal of building hundreds of housing units, by 1920, a cooperative of war-wounded veterans, the *Kriegsbeschädigte Lainzer Tiergarten*, began to collaborate on a settlement with Loos, Kampffmeyer, and Lihotzky at the former imperial hunting grounds at Lainz.[20] Although the plan was only partially realized, the work

Fig. 7.5 Construction at Siedlung at Lainz designed by Adolf Loos and Margarete Lihotzky, Vienna, 1921. (Source: Archives of the Horst Zecha, Siedlung Friedenstadt)

[20] For an analysis of the ÖSVK activities with war-veteran cooperatives see Blau (1999, 104–106), Klaus Novy, Wolfgang Förster & Verein für Moderne Kommunalpolitik (1991, 140–145).

Fig. 7.6 Construction with Pax bricks at Siedlung Hoffingergasse designed by Josef Frank, Vienna, 1921. (Source: Archives of the Cooperative Altmannsdorf-Hetzendorf)

and housing program for wounded workers was considered successful by its inhabitants. Built with materials on site, Franz Sekera, a new settler and a member of the *Hauptverband*, commented positively on the collective work completed by the disabled veterans. "A strong tie binds the members of the cooperative, both as settlers and as wounded soldiers," he wrote in the periodical *Der Siedler* (The Settler). "They hope to heal their bodies, wracked with heart, lung, and nervous diseases, in the fresh air, sun, and healthy, decent housing" (Sekera 1921, 92). Neurath's theory of the Converse Taylor System, which projected how impaired workers could be advanced in the labor process, provided a theoretical complement to the cooperatives' on-the-ground efforts to employ wounded veterans. In addition, his ideas raised important questions about the organization of labor in the context of the settlement movement overall (Figs. 7.6 and 7.7).

In the context of the ÖSVK, for example, Neurath developed a radical theory of a labor- and use-value-centered economy, following his hypothesis that "existing professions and forms of organization" were perhaps inadequate for human beings. Directly applying an aspect of his ideas on war economy to the realities of post-war settling, he advocated that a contribution of 1500–2000 hours of labor time should be counted towards the down payment on a house in the ÖVSK. In the absence of advanced building technologies and the abundance of work seeking settlers, he championed the implementation of labor-intensive construction practices such as the extraction of materials and the burning of bricks on site. These labor-intensive

Fig. 7.7 Construction of row houses at Siedlung Hoffingergasse with allotments in the foreground and background, designed by Josef Frank, Vienna, 1921. (Source: Archives of the Cooperative Altmannsdorf-Hetzendorf)

building practices soon became common for housing endeavors in Vienna more broadly, where until the 1930s even large-scale housing projects were built in brick. This system of organizing labor created and strengthened local community: ÖVSK representatives logged hours and familiarized settlers with self-help building techniques based on readily available materials.

One of the most important sites for developing this cooperative approach, was Siedlung Hoffingergasse, spearheaded, constructed, and later inhabited by the cooperative Altmannsdorf-Hetzendorf under the leadership of Müller. As its architect, this cooperative chose Josef Frank, who would later collaborate with Neurath at the Social and Economic Museum. Drawing on ideas from Neurath's "The Converse Taylor System," Frank avoided forms of construction that required heavy machinery and methods of prefabrication at Siedlung Hoffingergasse, since no technologically advanced construction industry existed in Vienna at the time. Instead, settlers completed basic bricklaying and carpentry work, excavated foundations, and utilized local cooperative workshops, for what would become more than 280 housing units. Furthermore, settlers chose to use regular bricks and so-called *Pax Ziegel*, for which materials were salvaged *in situ*.

For these practical purposes, row houses could not be tailored to each individual family's needs, but according to Neurath and Frank, the uniformity of design strengthened the collective identity of the settlement. As Neurath wrote about Siedlung Hoffingergasse,

> [t]he similarity of the units (types), the similarity of the building's parts (norms) is an expression of modesty, but also an expression of the sense for equality, which roots in both, fraternity and envy alike. Not one singular building is the subject of design, but the collec-

tivity of all houses. The singular building is like the brick within a house. A new community is created from the class solidarity of the labor-forces. (Neurath 1923a, 34)

Frank, too, insisted that settlements had to exhibit equality and similarity, rather than heterogeneity for the sake of picturesqueness.[21] According to Frank, the settlement's egalitarian objective distinguished it from the garden city, where architects had illustrated inhabitants' social and class differences through the design of communities that included "enhanced and disadvantaged houses" (Frank 1924, 26; Bojankin et al. 2012, 213). Siedlung Hoffingergasse's uniformity, by contrast, exemplified commonality, community, and cooperation.

Settlements' architecture indeed varied slightly from site to site, but row houses within each community adhered to a shared spatial organization and a common formal language. All settlements typically incorporated a number of social and communal spaces. Frank designated playgrounds, a cooperative supermarket, and a daycare center in Siedlung Hoffingergasse's general plan. A communal building, the *Genossenschaftshaus* (the equivalent to allotment gardens' *Schutzhaus* and the heart of the cooperative), inscribed community into social life of the settlement. These communal facilities were the center of cooperative life and the place for conversation. Neurath, who regarded Siedlung Hoffingergasse as one of the settlement movement's most critical achievements, stressed its capacity to symbolize the principles of cooperative economy. "Here, something like a new way of life is emerging, slowly and gradually [...]. The individual will no longer be lonely, but will feel carried and held by the whole to which he belongs," he wrote (1923a, 35). Using these methods with the support of the GESIBA, between 1920 and 1923 Neurath, Müller, Kampffmeyer, and others managed the construction of more than 20 settlement facilities, amounting to thousands of housing units.

Presenting the ÖSVK's achievements to a popular Viennese audience in a publication accompanying an exhibition in 1923, Neurath proudly highlighted the indivisible connection between economic order and the overall experience of life that had emerged from the settlement movement. "The Austrian organization of the allotment gardeners and settlers is based on the fundamental idea that the movement it serves is part of a tremendous upheaval, which [...] affects our entire life, our economic order as well as the feelings and thoughts of humans," he wrote (Neurath 1923a, 31). Such achievements – the possibility of rethinking the fundamental economic order through the construction of settlements and vice versa – were substantial, especially when considering that they emerged from adaptive responses to wartime deprivations and severe conditions of underdevelopment (Fig. 7.8).

[21] See Frank (1924), which was originally published in *Der Neubau*. I want to thank Christopher Long for sharing the text with me. In 2012 the text appeared in Tano Bojankin, Christopher Long, and Iris Meder's edited volume and translation *Josef Frank: Writings*. I have used both, this volume's as well as my own translations in some quotations above. Below I have indicated when my own translations have slightly deviated from official translations. For example, I translate the title of the essay "Die Wiener Siedlung," as "The Viennese Settlement" while it appears as "The Viennese Housing Settlement," in the translated volume. I have utilized this title, because it is clear from the context of this essay, that settlements are a distinct form of housing and the word "housing" did not appear in the original German title Cf. Bojankin et al. (2012, 211–222).

Fig. 7.8 Visitors at the allotment garden, settlement, and housing exposition in front of core houses with fully functioning allotment gardens. The crowd reaches back to the *Ringstraße*, Vienna, 1923. (Source: Austrian League of Allotment Gardeners)

7.3 Traveling Exhibitions as Education and Conversation: On the Limits of the City, at the Heart of the Metropolis

In the fall of 1923, the much-anticipated Allotment Garden, Settlement, and Housing Exposition opened at Vienna's City Hall, *Neues Rathaus*. With contributions from municipal offices, commercial exhibitors, the GESIBA, the ÖSVK and hundreds of cooperatives and individuals across the city, the exhibition featured planning documents, full model homes and visual data alongside vegetable and animal displays. This striking assemblage of objects – from data to living things – sought to illustrate that a modest culture of making-do, had made significant impacts on the urban economy in recent years. "When in 1918 more than 1,000 wagons of produce had yet to be imported, such importation by 1919 had become superfluous as a result of allotment work," Neurath (1923a, 10–11) wrote optimistically in a publication accompanying the exhibition. But exhibition organizers and participants were not content with solely advancing economic arguments. Charts, plans, and objects were carefully assembled to make educational statements about the material world and exhibited buildings put visions of social life on display and into circulation.

The question of communication and education had long played a key role in allotment garden and settlement debates. As discussed above, throughout the war years sharing horticultural and agricultural knowledge was critical for the survival of urban citizens, and expertise and insight were disseminated communally, in lectures, workshops, and in the periodical *Der Siedler*. Since 1918, the *Hauptverband* had organized hands-on activities open to all members in settlements and plant nurseries. The municipal Settlement Office at Parkring, where Loos and Kampffmeyer worked, handled dozens of individual queries and appointments daily, which consisted of personal and cooperative advice.[22] Starting in 1921, three times per week settlers could request conversations with more than half a dozen experts, including Frank, at the ÖSVK offices at Möringgasse in the fifteenth district (Anonymous 1921b, 117). Moreover, individual cooperatives could invite any of the ÖVSK experts to deliver instruction and presentations locally. International guests, such as the British architect Raymond Unwin and the German landscape architect Leberecht Migge, came to speak in Vienna on the invitation of Kampffmeyer, Loos, and Neurath.[23]

Lihotzky, who, together with Neurath was involved in numerous events, later recalled the dire necessity of these didactic activities. At prominent urban locations such as the *Urania* and *Universum* cinemas, she showed slideshows and, as of 1923, even short films to urban audiences. After lecturing in communities on the outskirts, she noted the devastating conditions of existing homes and gathering spaces for workers, where she gave lectures and provided advice (Anonymous 1923b, 1924). "By the light of candle stubs and dull oil lamps, I spoke in smoky taverns and far-off restaurants, and with the help of our drawings, explained to the people how they could create the buildings we had planned and typified, from the simplest alcove to the finished house, by means of self-help and mutual support and with our assistance, and thus how they could obtain a decent, humane framework for their lives," she remembered (Schütte-Lihotzky 2004, 87).

With the inauguration of a settlement school in the winter of 1921 and 1922, which offered 19 themed lectures at an adult education facility, the ÖSVK undertook the formalization of its educational programs (Anonymous 1921e, 125). Topics covered at the settler school targeted cooperative leaders and organizers and included planning and architecture, gardening and horticulture, and economic questions. Kampffmeyer, Loos, and Lihotzky lectured on interior design, and Frank provided advice in a series of talks dedicated to building with substitute materials.[24] Paul Vogt, the inspector of the ÖSVK's garden section, and Alois Zipfinger, the former head of the *Zentralverband*, spoke about the praxis of settlement and allotment gardening respectively, emphasizing the continuing need for food production. These

---

[22] The volume of inquiries and the intensity with which Kampffmeyer and Loos worked with individual settlers and cooperative representatives is reflected in the agenda ledgers of the municipal Settlement Office of the years 1921 and 1922, which have – if partially – survived in the Vienna City and County Archives.

[23] For a discussion of Leberecht Migge's lectures and his connections to Adolf Loos see Haney (2010, 121); Anonymous (1921c) and Anonymous (1921d).

[24] Kampffmeyer offered instruction about garden cities and the settlement, Adolf Loos lectured on "the settlement as educator," and Margarete Lihotzky on furniture and interior design. Josef Frank provided instruction in economic building techniques and a settler named Garder lectured on adobe construction.

very practical talks on design, construction, and urban agriculture were contextualized by instruction in economic topics: Otto Neurath lectured on the settlement movement's economy, and his colleague, Friedrich Bauermeister, presented findings on its finances. A band of legal and political experts, such as municipal politicians Max Ermers and Ludwig Neumann, taught courses on the culture and organization of the settlement movement. Adolf Müller provided instruction in the cooperative work of the settlement (Anonymous 1921e, 125). Overall, these courses offered material-theoretical instruction about the settlement movement's socio-economic character and potential. Moreover, the thematic breadth of the sessions illustrated the advantages provided by the association of the ÖVSK after 1922. While the settlement school was ongoing, individual experts continued to lecture around the city in settlers' clubs.[25]

It is important to note here that adult education at the settler school, which addressed cooperative representatives, was based on complex educational theory that Neurath and others had formulated since 1920. The settler school was in fact not unlike a *Betriebsräteschule* or work council school, which Neurath and the socialist Käthe Pick had founded concurrently to the creation of the ÖSVK (Leichter 1997).[26] In the 1921 essay "Educational Tasks of Socialism," Neurath posited

> [...] what can an economic plan affect and what does the creation of centralized associations of production matter, if they are not [...] controlled and influenced on all levels in a socialist sense? Effective socialist organizations today only exist in trade unions and works councils. Increasingly, however, they are becoming the bearers of the socialist order. (Neurath 1921b, 9.)

Delving fully into the finer points about processes of socialization that could occur through the transformation of individual sectors of industry, Neurath helped found a system of higher education in the work council schools. This system set out to grant access to solid professional development for people of all backgrounds according to their chosen occupation. In this manner, advanced instruction would not become exclusive to those with university degrees, but could extend to cooperative representatives, leaders of corporations and industries, as well as councilmen and women, regardless of their previous scholastic training. Thus, Neurath supported the organization of politics and labor in Vienna that relied on leadership by a cross-section of society, not on a few chosen functionaries. Writing about past eras, he stipulated:

> The firm framework of the school and curricula structure was predetermined by the goal, above all, to educate from a small class of urban intellectuals a leading elite. But now comes the new time. The masses do not want to recognize a previously preferred group whose power was inherited. Everyone who is suitable should be able to become an 'expert,' and the expert should be supported by the will of all. (Neurath 1921b, 9.)

[25] A number of articles in *Arbeiterzeitung* publicly announced ÖSVK experts' activities and lectures throughout the years of 1922. See for example Anonymous (1922).

[26] Käthe Pick and Neurath frequently wrote for the journal *Der Betriebsrat*, founded to promote the cause of the work councils. They also frequently lectured in work council schools.

According to Neurath, only when a group of work councilmen and women were granted access to such education, could a new order of planning and theorizing life come into being. Such a theortical study of life, however, necessitated a professional-philosophical inquiry into humanity, a general *Menschheitskunde* and *Lebenskunde*, as he understood it (Neurath 1921b, 11). This meant, that the organization of the economy was an attempt to understand and improve the organization of people's everyday life. Indeed, in 1921 and 1922 work council schools were established in Vienna, where Pick and others provided thematic instruction with the purpose of bridging the gap between economic theory and making change on the ground (see L. 1921). The settler school in this context was a related attempt to organize a didactic apparatus in the branch of housing and construction that would provide instruction about real-life concerns.

While work council and settler schools relied on relatively formalized structures of education and targeted the transformation of the movement from within, educational exhibitions served the double role of communicating to members internally and to the public externally. This function of communication had existed since the days of the allotment garden movement, when the staging of produce fairs had taken place in the *Schutzhäuser*. But *Schutzhäuser* in allotment garden facilities had deeper historical genealogies and were related to activities such as the *social hike*, a practice which had originated in turn-of-the-century Social Democratic activities. Groups such as the Austrian *Naturfreunde* posited then that the experience of nature should be enhanced by conversation and engagement with politics. Hikes and overnight stays in so-called *Schutzhütten* – alpine huts that functioned as communal centers – became the unlikely venue for the debate of urban politics, and the very idea of the *Schutzhütte* in turn became the model for settlers' cooperative *Genossenschaftshaus*.

To understand the political nuances undergirding the *Genossenschaftshaus*, it is helpful to consider Social Democratic politician Max Ermers's assertion that this communal space was "the heart and the brain of a settlement, simultaneously a town hall and a home for recreation [...]."[27] Ermers further explained the socio-political significance of the *Genossenschaftshaus*, stating that through occupation of the communal building, "the otherwise narrow mind of the allotment gardener and of the inhabitant of the single-family house grows into the social, the universal, and the important. Here the ideology of the settlement as a social category is born and spreads over the whole of its parts. Here is the seat of the freely elected administration, of political fights, of the diffusion of knowledge, or artistic experiences, of celebrations," Ermers (1924) argued. Events at *Genossenschaftshäuser* were incubators for conversation, activities, and cooperative life.

Therefore, when the architectural exhibition in the fields at Hoffingergasse was staged in front of the make-shift *Genossenschaftshaus*, the displays were not only meant to solicit viewers, but to stimulate conversation. Indeed, throughout the summer months of 1921 architectural displays were mounted regularly in settlements

[27] I have used Eve Blau's existing translation for this passage. For a careful analysis of the role of the *Genossenschaftshäuser* also see Blau (1999, 112).

and allotment gardens across the city on site. The workers' newspaper, *Arbeiterzeitung* and *Der Siedler* promoted these events, suggesting that they addressed local cooperatives, and the Viennese public as well.[28] While visual documentation of these exhibitions is extremely sparse, it can be inferred that by 1921 presentations were not standardized – neither in the depiction of their content nor in their display. Nonetheless, they did embody a distinct strategy of showcasing architectural and (to a lesser degree) economic and social relationships that were steeped in Neurath's theories of museology and communal economy. Crafted and exhibited with modest means, these diplays aimed at great communal impact and consisted of plans and photographs that were prepared by architects and artists working directly with cooperatives. They were commissioned through local organizations including the Settlement Office, and the ÖSVK's own Building Bureau. As the image in the introduction of this essay illustrates, they were slightly decorated with garlands and twigs (reminiscent of harvest celebrations) and arranged in an unassuming but clearly comprehensible manner. There were neither moving parts nor complex exhibition infrastructure, but the individual sections and displays were easily accessible for an audience of laypeople. With photographs serving as introductory images for technical drawings – "lead images," as Neurath would later term them – site plans, floor plans, and elevations became effortlessly legible, even to an illiterate reader. Photographs, moreover, were usually enlivened with people, and showcased buildings in the process of construction (rather than in their completed state), thus highlighting the communal ideas undergirding the projects. Finally, no new settlement was presented in isolation, but rather in an assemblage of graphic materials illustrating diverse facilities across the city, showcasing them in their greater social, economic, and cooperative context.[29] Staged alongside celebrations, speeches, and foundation stone ceremonies, these traveling exhibitions therefore were carefully curated within collective settings and as part of conversations between individual settlers, cooperative representatives, architects, and politicians, (as the presence of Neurath and Müller at Siedlung Hoffingergasse showed). This conversational quality and the possibility of debating cooperatives' futures based on visual information, embodied what I consider a *collaborative practice of seeing*.

Neurath's individual contribution in these exhibitions is hard to discern, but he was invested in them as a speaker, commentator, and organizer throughout the early 1920s. In 1922, for example, the ÖSVK secretariat under Neurath's leadership issued a promotional statement about the fourth Allotment Garden, Settlement, and Housing Exposition, an event directly expanding the tradition of produce competitions at Vienna's City Hall. The statement proudly highlighted the presence of the Settlement, Housing, and Construction Guild of Austria at the exhibition (another

[28] By 1922 *Der Siedler* featured a rubric that informed readers of lectures throughout Vienna as "News from the Organizations." See for example Anonymous (1921f). Also see Anonymous (1922).

[29] A potential fourth aspect of visual display was the use of lantern slides, which accompanied lectures starting in 1923. They were, however, predominantly utilized at more prominent urban locations, such as the *Urania* and *Universum* cinemas, or the Austrian Chamber of Labor, which commissioned Neurath and Lihotzky to hold lectures for their women's association in 1924.

one of the institutions Neurath helped found). It lauded the Building Bureau's prefabricated cooking niche, which Lihotzky had designed, and the work of the GESIBA, which had allocated one allotment garden hut as prize for a local tombola. Announcements in *Arbeiterzeitung* emphasized the participatory nature of the event, and solicited contributions from cooperatives to the exhibition in form of produce and graphic materials. "It is the duty of every allotment garden society and each settler cooperative to provide the best yields from the garden economy, in pictures, plans, sketches, and models of all kinds. It is the responsibility of every organization to donate a tribute [of produce]," the statement concluded (Anonymous 1922, 6).

It was in the fifth Allotment Garden, Settlement, and Housing Exposition of 1923 that all these curatorial strategies, conceived and tested on the outskirts by cooperatives, representatives, and individuals, finally converged in the center of the city on an unprecedented scale. "It has become practice to show what has been done on the limits of the city during the year in the field of allotment gardening, settlement and housing through a large-scale exhibition at the heart of the metropolis in City Hall when the autumn months begin," wrote the prominent Viennese mayor Jakob Reuman (1923, 3) in a guide explaining the 1923 exposition. But in this particular exhibition hundreds of allotment gardeners from dozens of organizations staged living things and objects for the first time. The exposition included fruit and vegetable displays, animal shows of rabbits and doves, and a wall of flowers, presented in a large open-air section of the City Hall arcades. Just like in earlier fairs, they were accompanied by prize ceremonies and public lectures, but now they were also complemented by musical events and performances. Due to the size of the presentation, the coordinating exhibition architect Ludwig Michael was mindful of the overall design of the open-air section and its unified layout. "Where will I exhibit?" he asked rhetorically in materials promoting participation. "This question has to be answered [...] in the sense that vegetables can only be accommodated in the vegetable group" (Michael 1923, 2). From the beginning, the exhibition targeted a coherent layout that built on the collective achievements of cooperatives and their representative institutions (Fig. 7.9).

The greatest attraction to popular audiences was shown on the main square in front of City Hall. Here seven full model homes could be viewed and visited. A collaboration between the GESIBA, the municipal Settlement Office, and the ÖSVK's own Building Bureau, these buildings showcased settler homes in all their varieties: from a tiny, but fully designed allotment hut of approximately two square meters, to a finished house of fifty-seven square meters (Neurath 1923b, c). Four of the buildings sponsored by GESIBA had the ability to be built in phases, "growing" from a small core outwards (Schütte-Lihotzky 1923b). These so-called core houses were planned by Lihotzky in collaboration with other Building Bureau architects and were intended to resolve settlers' still-substantial liquidity problems through design (Hochhäusl 2013). They extended previous questions of scarcity into a slowly recovering economy, while relying on conventional models of home financing. A main commentator on the "core house campaign," Neurath lauded the buildings' ability to accommodate still frail economic circumstances. "This difficult campaign

Fig. 7.9 Members of the exhibition committee in front of the flower exhibit at the allotment garden, settlement and housing exhibition. Neurath third on the left, back, Adolf Müller, second on the right, back, Kampffmeyer fourth on the right, front, Vienna, 1923. (Source: Austrian League of Allotment Gardeners)

could be tackled with the prospect of success because the ÖSVK's Building Bureau and the Settlement Office of Vienna's municipality have created types, which can be built with relatively small funds" (Neurath 1923c, 3).

Importantly, by enabling visitors to see these seven varying buildings with common principles, it was here that Viennese audiences were introduced not only to the visual, but the experiential qualities of settlements as *types*. This notion of the *type* had long occupied German and Austrian architects in a unified attempt to define "convincing architectural solutions" to the task of designing a small house – while avoiding both universalist tendencies and nativist nationalism.[30] In Vienna's settlement movement, *types* had been defined as row houses with standardized but slightly variable floor plans and measurements. While architects had established standardized sets of plans, many of them conceived from brick and wood, by 1923 the GESIBA had also made available standardized windows, door frames etc. But as visitors moved through GESIBA homes at City Hall, they witnessed for the first-time what *types* meant theoretically. With slight modifications of similar layouts, they saw not just how unified the allotment garden and settlement movement had become, but also how diverse it was. Viennese citizens experienced for themselves

[30] Josef Frank's and Otto Neurath's converging ideas on the type were first discussed by Eve Blau (2006, 256), which are part of the concluding remarks of this essay.

the benefits of utilizing the same windows, doors, and other basic building elements which brought down prices and communicated cooperation. And they could test *type* furniture from the cooperative store *Warentreuhand*, designed by Lihotzky while envisioning making a life in these homes.

From an experiential perspective, if visitors decided to rest on the sun porch of the core houses, they could enjoy the modest luxuries that the settlement movement afforded its residents. They could smell wet soil when moving through rows of vegetables with proper irrigation systems, and they could hear the sounds of chickens, doves, and other animals, all right in the center of the city. The display therefore constituted an immersive experience that exemplified the unifying principles as well as the individualizing tendencies that underwrote the practice of settling. Finally, on the outskirts, at Siedlung Heuberg, entire streets of differentiated row houses designed by Loos, Lihotzky, Frank and the municipal architect Adolf Meyer were open to visitors. There, in conjunction with the exhibit, Viennese citizens could witness communal life first hand.

Architectural critics and popular audiences embraced core houses, celebrating them for their experiential qualities and their "amazingly convenient" design (Wilson 1923, 6). Francesca Wilson (1923, 6) of *The Manchester Guardian* highlighted that "the great advantage of the Austrian movement has over similar movements in other countries, is that houses are being built by the people themselves for themselves, and not by contractors, whose first object is money. That is the reason why the convenience and comfort of the settlers is the first consideration, inside the house as well as outside." Neurath similarly attested that core houses embodied the possibility of alleviating housing shortages even in a state of emergency through large-scale cooperative organization. He also pointed out that a number of other architects had already adapted core house designs which made it a fundamentally collaborative endeavor. "None of these designs is the invention of an individual, but the result of many efforts," Neurath (1923c, 6) stated. By contrast, visitors at previous exhibitions had disappointedly remarked on the lack of sanitary conditions asking "How will I cook here?" and "What, to fetch water, I have to go outside?" (Altmann-Loos 2002, 106).

The space that most directly occasioned Neurath's curatorial involvement at the 1923 exposition, was the architectural and planning section in the ceremonial hall *Volkshalle*. Here GESIBA, the Settlement Office, and ÖSVK had collaborated to provide an immersive experience, showcasing a plethora of drawings, architectural models, and statistics. Unlike the produce displays, this section presented abstract planning materials, statistics, and data, appealing to viewers' intellects and imaginations. GESIBA materials featured an extensive selection of core house models and plans, which contextualized and expanded the experiential aspect of visiting the homes outside City Hall by providing financial and technical data points. Linking the construction of homes to urban economy, GESIBA also provided statistics on citywide building initiatives and how construction wages had been influenced by the introduction of the cooperative developer (Ermers 1923, 40–41). The municipal Settlement Office's contribution to the section focused on aerial plans, floor plans of buildings, and renderings of approximately twenty different settlements, thus illus-

Fig. 7.10 Displays by GESIBA (on the left) with core-house models in City Hall, Vienna, 1923. (Source: Austrian League of Allotment Gardeners)

trating how organizations across the city collaborated, shared information, and built collective expertise. A large site model of the Atmannsdorf-Hetzendorf's facility and illustrations of three of its settlements further invited viewers to picture communal life and networks of solidarity that had been built between these groups (Fig. 7.10).

Crucially, under Neurath's direct leadership the ÖSVK presented four aspects of its activities; the ÖSVK Building Bureau showcased the designs of housing and furniture and the ÖSVK's garden section highlighted schemes and charts about garden design and intensive produce production (Ermers 1923, 39). Under the rubrics "activities" and "movement," charts provided information about all of the ÖSVK's civic services (from Warentreuhand to educational and legal advice) alongside statistics. These illustrations were the first attempts at combining material-lived experiences into what Neurath would soon call "statements of social and economic facts."[31] Furthermore, all of these displays were crucial in facilitating a conversation with the broader Viennese public because they recorded numbers as picture statistics. Even illiterate people in the audiences at the exhibition could now compare the number of allotment garden huts in Vienna, the sizes of their plots, and the number of allotment gardeners who kept animals for the first time coherently in three charts. An additional organizational diagram of the entire settlement and allotment garden program synthesized the vast economic undertaking into a single drawing. This last

[31] See note 5. in this essay for Neurath's use of "statements of facts."

Fig. 7.11 Chart of allotment gardeners' material achievements compiled by the municipality's garden section in the midst of beekeeping display, Vienna, 1923. (Source: Austrian League of Allotment Gardeners)

display reaffirmed Neurath's approach from "World Blockade and War Economy" and foreshadowed consequent pictorial work in Vienna. Displays from the 1923 Allotment Garden, Settlement, and Housing exhibition overall became the primary collection for the emerging Social and Economic Museum (Fig. 7.11).

The resonances between the temporary Allotment Garden, Settlement, and Housing exhibition and what would become the permanent Social and Economic Museum of Vienna were multiple and manifold. Yet, I find their creative and intellectual underpinnings most clearly pronounced reading a chart by the municipal garden section at the exhibition in conjunction with a produce display of the allotment garden cooperative Rosenthal. Placed in the midst of the beekeeping display, the chart had been organized by cooperatives and apiary schools and was entitled "The Development of the Allotment Garden Movement and Its Material Results." It illustrated how many allotment gardeners operated in the city and how they had increased their production over the years. It did so by utilizing a language of picture statistics that was not yet systematized, as it deployed the multiplication and scaling of pictorial symbols at the same time. The chart captured the distinct material basis for the graphic displays, since it positioned visual representations of production in the midst of actual objects on display. It thus established real legible relationships between the conveyed information and the lived world while placing the viewer in a central role. Through this assemblage, abstraction and the material world were brought into conversation (Fig. 7.12).

Fig. 7.12 Grouping of various vegetables presented by allotment gardeners of the Rosental cooperative in the courtyard of City Hall, Vienna, 1923. (Source: Austrian League of Allotment Gardeners)

This idea was further exemplified in produce displays in the arcades of City Hall. Here, cooperative display strategies had always mandated that produce should not be exhibited as individualized items, but in groupings (cucumbers, tomatoes, cabbages, for example) that would render them into legible sections and products of common achievement. A photograph of such a cooperative produce section by Rosenthal allotment gardeners makes the leap from a *material* to a *materialist* conception of existence visible. It illustrates aggregations of things of the material world (groups of vegetables) and links them to their statistical depictions (pictorial statistics of vegetables).[32] In fact, the Allotment Garden, Settlement, and Housing exhibition of 1923 took place at a figurative moment when objects (living things included) were presented next to their scientific abstractions. Neurath would later coin the term physicalism (instead of materialism) for this very idea, which sought to describe "concrete events ordered in space and time processes" and which "emerge dialogically from everyday life rather than from a predetermined metaphysical system (Law 2015, 124)."[33] This way of seeing also presupposed an indi-

[32] These vegetable displays foreshadowed the idea of *transformation*, a process crucial for the creation of picture statistics, in which multiple of kind were translated into a single symbol.

[33] For Neurath's ideas on physcialism see Uebel (2007) and Cartwright, Cat, Fleck, Uebel (1996). Also see O'Neill (2007); Neurath (1931c/1973), (1931b/1973), "Physicalism, Planning and the Social Sciences: Bricks Prepared for a Discussion V.Hayek," July 26, 1945, Otto Neurath Papers, Haarlem, 202, K 56.

Fig. 7.13 Otto Neurath presenting the *Atlas for Society and Economy* to Viennese mayor Karl Seitz, Vienna, 1930. (Source: Otto and Marie Neurath Isotype Collection, University of Reading N-Files, N 958b)

visible connection between displays and their viewers, which had been so central in all ÖSVK exhibitions. Cooperative exhibitions were inherently social and they elicited not a need for viewing but for exchange, or seeing together. This centrality to study neither humans themselves nor their production, but rather the social relationships that bind them together, persisted at the Social and Economic Museum, even as Neurath shifted focus from illustrating cooperative economy to illuminating global and societal interconnections. So, too, would the culture of education and conversation: that *collaborative practice of seeing,* which relied on human cooperation (Fig. 7.13).

7.4 Conclusion

By the end of 1924, Neurath established the Museum für Siedlungs und Städtebau (Museum for Settlement and Town Planning) in Vienna which would be renamed and reconceived as the Social and Economic Museum in 1925. The lion's share of the museum's initial materials was secured from the Allotment Garden, Settlement, and Housing exhibition. Providing commentary in anticipation of the Allotment Garden, Settlement and Housing show in 1923, Neurath already remarked that the exhibited displays were worth preserving. "This will become an exhibition, that embodies a museum for 1 week in a certain sense," Neurath (1923b, 135) wrote in *Österreichische Städtezeitung* (Austrian Cities' Newspaper). It will be "educational and pleasant for the eye and the mind at the same time, because it outlines an image of a mass movement, to create [...] a freer and happier existence" (Neurath 1923b, 135). Didactic techniques and strategies based on human forms of organization had opened avenues towards more optimistic approaches of living life as a fuller humanity.

In conclusion, I want to offer some final points that warrant reflection and connect the 1923 Allotment Garden, Settlement, and Housing Exposition with more enduring displays at the Social and Economic Museum in the later years of the 1920s. As noted above, already by September 1923 Neurath conceived of the exhibition as the basis for a permanent collection, but throughout the fall he further outlined the focal themes, competences, and curatorial strategies the Social and Economic Museum would adopt in the following years. In the article "Permanent Retention of the Exhibition – A Settlement Museum" published in *Arbeiterzeitung* he wrote:

> The remarkable result, that the Fifth Allotment, Settlement and Housing Exhibition has to offer, created the desire to preserve the truly instructive and valuable exhibition material and to complement it in the future. While, so far, the laboriously designed panels, graphs, models and other displayed items were scattered in all directions at the end of an exhibition, this time, the ÖVSK will found a Settlement Museum to keep the most important objects and to make them accessible to the individual districts and federal states through traveling exhibitions. It should be shown everywhere, what the organized work of the municipalities and housing cooperatives can accomplish. (Neurath 1923d, 8.)

Such curatorial work on mobile exhibitions, he further asserted, was important, because it brought political work to specific engaged audiences in travel across regions.

Along with announcing the founding of the Museum, by the end of 1923, Neurath provided explicit instruction about the objects to be included in the new museum. He noted how such objects related to the institution's educational goals and to the urban visitors invited into conversation at the museum. One way to engage constituents was to provide them with the experiential qualities that had just been tested at the fair. Thus, not only abstractions of material objects should be shown. Full model homes and furniture – the material world in the making, now standardized and commonplace – would be incorporated as well. "Above all, models, charts, and overviews are exhibited. It is also planned to show fully furnished model houses" Neurath wrote (1923d, 8).

> A part of the museum will show in living rooms the objects that are conveyed by the Warentreuhand of the ÖSVK to those persons who want to supplement their household contents after expert advice. In this way, the Settlement Museum will serve general education to the widest extent and will certainly become within a short time a very popular institution. (Neurath 1923d, 8.)

This mission to provide experiential displays as a common educational platform within the context of a museum could be further strengthened by the use of a visual archive, Neurath argued. By the mid 1920s, the representation of information itself was reproduced by the Social and Economic Museum with the introduction of lantern slides and film already tested in ÖSVK lectures. The link between visual displays, conversation, and education, as well as the potential to transform politics through exhibitions, became ever more pronounced in the following years. It is no coincidence, for example, that after 1927 the Social and Economic Museum moved its permanent headquarters to City Hall. Exhibitions were staged in the same *Volkshalle* where the earlier settlement displays had been shown. Nader Vossoughian (2008, 72) has suggested the continued significance of this location, stating that City Hall constituted "the political heart of Vienna and could thus draw enormous crowds," a fact that the prized vegetable ceremonies foreshadowed.

Besides the central museum location at City Hall, Neurath organized displays at secondary locations, which reflected his belief that people wanted to see exhibitions "near their apartment," and "in connection with a lecture" (Neurath 1927, 3–4). The exhibition space located at the large communal housing project, Fuchsenfeld, provided a venue where workers, in particular, could think through information together within the context of a distinctly metropolitan community. The new site at Parkring, where Neurath opened the smaller "Zeitschau" exhibit in 1928, also fulfilled this goal by displaying graphic information in a storefront in one of the busiest transportation hubs at the center of the city. According to Neurath, the "Zeitschau" enabled people to converse about picture statistics while waiting for public transportation. By doing so, they would together consider the illustrated information and what it meant for their personal lives (Neurath 1936, 73). As Neurath would remark much later, the educational purpose of ISOTYPE was "the teaching of how to *argue*" (Blau 2006, 255). No longer was this the case only in a familiar, cooperative environment, but in one that, by extension, connected strangers as society. The main objective to encourage and preserve the culture of conversation that had been essential in ÖSVK exhibitions – learning how to *see* – had become teaching how to *argue* (Fig. 7.14).

In the following years, Neurath further theorized the meaning and purpose of *Bildung* or education in the context of the museum. Having permanently established the Social and Economic Museum in 1925, by 1927 he gave statistical courses in the Wiener Arbeiterhochschule, an extension school for adult workers administered by the Social Democratic Party (and similar in kind to the work council schools) (Stadler 1982, 231). He advocated bringing adult education to the museum, instating a program for work advice in the latter part of the 1920s (1931a, b, c, 125). The platform was initiated through a collaboration between the Work Advice Office of the City of Vienna and the Austrian Chamber for Labor. Although much larger in scope, adult education at the museum provided work advice and functioned similarly to the Settlement Office's capacity to consult on housing. For example, in a

Fig. 7.14 Worker making shoes, Vienna, 1925–1932. (Source: Otto and Marie Neurath Isotype Collection, University of Reading N-Files, N 175)

series of photographs the museum offered information about popular professions and was "dedicated to the enlightenment of students released from school and to their parents" (Neurath 1931a, b, c, 125). The program offered a local consultation and evaluation office where those seeking work were given information about the potentials and duties of various occupations. Because it was directed at young men and women who had bypassed higher education in favor of joining the work force, this program extended the goal of the work council schools to engage those Viennese who had previously been precluded from access to higher education.

This educational section relied more heavily on photographs, lantern slides, and film, while the majority of pictorial production in the Social and Economic Museum sill consisted of picture statistics. Neurath often remarked that social facts cannot be photographed (Blau, 2006, 256; Neurath 1933b, 462). As such, the education department continued to bring the material world into the museum and into its vast archive, classified in a special category, called n-files or "nature files." These n-files made up an almost equally expansive collection at the Social and Economic Museum as the t-files, *tafel* or "chart files," reserved for picture statistics and diagrams. Although often regarded as secondary or merely documentary, I argue that these "nature files" carried value, not only because they offered deeper insight into the physical and cultural landscapes of Vienna, but because they illuminated a dedication to the material world and the efforts Neurath and his colleagues undertook to render it legible in various ways in the museum (Fig. 7.15).

Fig. 7.15 Production of socks, Vienna, 1925–1932. (Source: Otto and Marie Neurath Isotype Collection, University of Reading N-Files, N 176)

Certainly, in conceiving ISOTYPE, Neurath extensively theorized processes of abstraction, and counted on citizens to debate the gaps between pictorial statements of facts. However, even as ISOTYPE evolved, he continued to bring depictions and objects of the everyday into the Museum, encouraging each visitor to draw on personal experience. One such experiential exhibition, depicted laborers while working; a series of photographs captured the goods they produced – shoes, tools, brushes and the like. But the photographs also conveyed the visceral experience of labor. They shed light on the daily tasks of cutting hair, the everyday engagement with a customer over the counter, and they even exemplified the heat of a workshop.

Speculating how picture statistics would lend themselves to other thematic museums in *International Picture Language*, Neurath (1936, 68) wrote as late as 1936, "What will an ISOTYPE museum of natural history look like? Certainly, it will have in it a number of birds, fishes, etc. But some account will be given of their living conditions, and other facts about them, such as their distribution over the earth, and their relations to man and society will be made clear." He continued,

> [a]n ISOTYPE museum of natural history will have not only animals and plants on view, but maps, number fact pictures, examples of things made from different animals and plants, their part in our existence, etc. So such a museum would be like a great ISOTYPE picture

> made up of natural things. The things have to be not separate, but in some relation to one another and in some relation to the experience and the knowledge of the on-looker. So the new knowledge will come into its right place. An ISOTYPE museum has to give the chance to every comer to make his selection himself. A great museum, specially, has to give a number of different chances. It is important to make this clear to everybody by the order of things, by the system of building. (Neurath 1936, 68–69.)

This last point elucidates not just the similarities between ISOTYPE and architecture, but also the parallel objectives of architecture and education. Indeed, in *International Picture Language* Neurath famously claimed that ISOTYPE museums were "nothing but a simple cover for simple teaching-material" (1936, 68–69). The relationship between architecture and ISOTYPE was similarly intertwined. Discussing the work of Frank and Neurath in her article "Isotype and Architecture in Red Vienna," Eve Blau (2006, 256) has shown that just as Neurath's picture statistics aimed at "giving a number of chances" to viewers to make their own connections and apply their own experience, Frank, particularly at Siedlung Hoffingergasse "provide a scaffold, a framework for dwelling." She also noted, that "modern architecture conceived in this way leaves enormous scope for agency and decision in the everyday life of the individual."

In this regard, it is relevant to highlight as a final point, that it was Josef Frank who at the Social and Economic Museum at City Hall developed a series of wooden panels that allowed the easy assemblage and rearrangement of existing displays. These panels and their frames relied on the standardized proportional system of measurements, which had become necessary due to the need for a unifying display strategy that fluidly facilitated pairing the ever-increasing amounts of charts. Frank's modular designs eventually also allowed the entire Museum to travel. More importantly, however, this display system was an extension of the standardized windows, frames, and parts that had been championed in settlements and by the GESIBA. Based on systematized, but not equal measurements, these panels created diversity and unifomity. They were, in addition, informed by the idea of the *type* in the settlement movement, which Frank and Neurath had so admantly studied and commented upon throughout the 1920s.

In conclusion, it is then important to emphasize that there was a deep, complex, and multifaceted connection between the production of housing and the practices of creating exhibitions. In fact, it would be no overstatement to say that the relationship between the staging of produce and Neurath's theories on war economy was not unlike the link between the construction of modern settlements in Vienna and the visualization of social and economic statements of facts in exhibitions at the Social and Economic Museum. It could even be argued that ISOTYPE itself was related to the settlement house as a type. Both ISOTYPE and *type* abstracted the old and built the new while underscoring heterogeneity and communality: statement by statement, earth heap by earth heap, brick by brick, house by house, and community by community (Fig. 7.16).

Fig. 7.16 Foundation stone setting ceremony, Siedlung Hoffingergasse, Müller, Neurath, and Kampffmeyer in the crowd, Vienna, 1921. (Source: Archives of the Cooperative Altmannsdorf-Hetzendorf)

References

Altmann-Loos, Elsie. 2002. *Adolf Loos, der Mensch*. Vienna: Prachner.

Anonymous. 1921a, April 5. Der Massenaufmarsch der Siedler. *Arbeiterzeitung*: 5–6.

———. 1921b, November. Auskunft und Beratung. *Der Siedler* 1 (8): 117.

———. 1921c, August. Unwin in Österreich. *Der Siedler* 1 (5): 69–71.

———. 1921d, August. Hauptverband für Siedlungswesen. *Der Siedler* 1 (5): 70–71.

———. 1921e, November. Siedlungsschuleschule. *Der Siedler* 1 (8): 125.

———. 1921f, July. Nachrichten aus den Organisationen. *Der Siedler* 1 (4): 62.

———. 1922, August 26. Kleingärtner, Mieter, Siedler, Bauarbeiter. *Arbeiterzeitung*: 6.

———. 1923a, August 8. Kleingärtner, Mieter, Siedler, Bauarbeiter – Warentreuhand. *Arbeiterzeitung*: 6.

———. 1923b, October 5. Film- und Lichtbildervortag über Siedlungs- und Kernhausbaus. *Arbeiterzeitung*: 9.

———. 1924, April 13. Vereine und Versammlungen. *Arbeiterzeitung*: 10.

Banik-Schweitzer, Renate. 1972. *Der staatlich geförderte, der kommunale und der gemeinnützige Wohnungs- und Siedlungsbau in Österreich bis 1945*. Vienna: Ph.D. dissertation.

Blau, Eve. 1999. *The Architecture of Red Vienna*. Cambridge: MIT Press.

———. 2006. Isotype and Architecture in Red Vienna: The Modern Projects of Otto Neurath and Josef Frank. In *Culture and Politics in Red Vienna*, ed. Judith Beniston and Robert Vilain, 227–259. (= Austrian Studies, 14).

Bojankin, Tano et al., eds. 2012. *Josef Frank: Writings* (Vol. 1). Vienna: Metroverlag.
Botstein, Leon. 1996. The Consequence of Catastrophe: Josef Frank and Post-World-War-I Vienna. In *Josef Frank, Architect and Designer: An Alternative Vision of the Modern Home*, ed. Nina Stritzler-Levine, 30-43. New Haven: Yale University Press.
Cartwright, Nancy, Jordi Cat, Lola Fleck, and Thomas Uebel. 1996. *Otto Neurath: Philosophy Between Science and Politics*. Cambridge: Cambridge University Press.
Das Deutsche Kriegswirtschaftsmuseum. 1918. *Das Deutsche Kriegswirtschaftsministerium*. (Leipzig: Veröffentlichungen des Deutschen Wirtschaftsmuseums zu Leipzig, Heft 3).
Ermers, Max. 1923. Gruppe II und IV: Siedlungswesen und Wirtschaftseinrichtungen. In *Führer durch die Wiener Kleingarten-, Siedlungs-, und Wohnbauausstellung, 2.-9. September 1923, Rathaus*, ed. Max Ermers. Vienna: Self-Published.
———. 1924. *Das Genossenschaftshaus der Wiener Rosenhügel-Siedlung und sein monumentaler Bilderschmuck*. Vienna: Hermes Buch- u. Kunstdruck.
Feller, Barbara. 1996. *75 Jahre bauen für Wien: die Geschichte der GESIBA*. Vienna: GESIBA – Gemeinnützige Siedlungs- und Bauaktiengesellschaft.
Frank, Josef. 1924. Die Wiener Siedlung. *Der Neubau* 6 (3): 25–29.
———. 1931a. Nach Kriegen. In *Architektur Als Symbol: Elemente Deutschen Neuen Bauens*, 1–6. Vienna: A. Scholl.
———. 1931b. Was ist modern? In *Architektur als Symbol*, 131–133. Wien: Löcker Verlag.
Freundlich, Emmy. 1921. Planwirtschaft, Kriegswirtschaft und Sozialismus. *Der Betriebsrat* 1 (15): 227–228.
Galison, Peter. 1990. Aufbau/Bauhaus: Logical Positivism and Architectural Modernism. *Critical Inquiry* 16 (4): 709–752.
Haney, David. 2010. *When Modern Was Green: Life and Work of Landscape Architect Leberecht Migge*. New York: Routledge.
Healy, Maureen. 2004. *Vienna and the Fall of the Habsburg Empire: Total War and Everyday Life in World War I*. Cambridge/New York: Cambridge University Press.
Hochhäusl, Sophie. 2013. From Vienna to Frankfurt Inside Core House Type 7: A History of Scarcity Through the Modern Kitchen. *Architectural Histories* 1 (1): 1–19.
Hoffmann, Robert. 1982. Proletarisches Siedeln: Otto Neuraths Engagement für die Wiener Siedlerbewegung und den Gildensozialismus 1920 bis 1925. In Stadler (1982): 140–148.
Kampffmeyer, Hans. 1926. *Siedlung und Kleingarten*. Wien: Springer.
Kraeutler, Hadwig. 2010. *Rondom Rembrandt* and Beyond: On Otto Neurath's Isotype Contributions to Visual Communication and Museum Learning. In *Image and Imaging in Philosophy, Science and the Arts*, ed. Richard Heinrich, Elisabeth Nemeth, Wolfram Pichler, and David Wagner, vol. 2, 137–173. Frankfurt: Ontos.
L. 1921. Bildung und Erziehung: Zweite Betriebsräteschule in der Tschechoslowakei. *Der Betriebsrat* 1 (12): 192.
Law, Alex. 2015. *Social Theory for Today: Making Sense of Social Worlds*. Los Angeles: SAGE.
Leichter, Käthe. 1997. *Käthe Leichter: Leben, Werk Und Sterben Einer österreichischen Sozialdemokratin*. Wien: Ibera & Molden.
Lihotzky, Margarete. 1923. Neues Wohnen. Der Kampf gegen den Möbelschund. *Arbeiterzeitung* 9.
Long, Christopher. 1996. The Wayward Heir: Josef Frank's Vienna Years, 1885–1933. In *Josef Frank, Architect and Designer: An Alternative Vision of the Modern Home*, ed. Nina Stritzler-Levine, 44-61. New Haven: Yale University Press.
———. 2002. *Josef Frank: Life and Work*. Chicago: University of Chicago Press.
Loos, Adolf. 1921. Der Tag der Siedler. *Neue Freie Presse*: 10–11.
McElvenny, James. 2013. International Language and the Everyday: Contact and Collaboration Between C.K. Ogden, Rudolf Carnap and Otto Neurath. *British Journal for the History of Philosophy* 21 (6): 1194–1218.
Meder, Iris, et al. 2008. *Josef Frank 1885–1967: Eine Moderne der Unordnung*. Salzburg: Anton Pustet.
Michael, Gottlieb. 1923. Über Zweckmäßigkeit und Schönheit der Ausstellung. *Kleingärtner und Siedler* 1 (7): 2.

Müller, Adolf. 1921. Die Siedlerbewegung. *Der Betriebsrat* 17: 261–263.

Nemeth, Elisabeth. 1982/1991. Otto Neurath's Utopias – The Will to Hope. In *Rediscovering the Forgotten Vienna Circle: Austrian Studies on Otto Neurath and the Vienna Circle*, ed. Thomas Uebel, 285–292. Drodrecht: Kluwer.

Neurath, Otto. 1916. Naturaltausch und Ernährungsproblem. *Der Volkswirt* 1: 425–426.

———. 1917a/1973. The Converse Taylor System. In *Empiricism and Sociology*, ed. Marie Neurath and Robert S. Cohen, 130–135. Dordrecht: D. Reidel.

———. 1917b/2004. The Economic Order of the Future and the Economic Sciences. In *Otto Neurath: Economic Writings. Selections 1904–1945*, ed. Thomas Uebel and Robert S. Cohen, 241–261. Dordrecht: Kluwer.

———. 1918/1998. Die Kriegswirtschaftslehre und ihre Bedeutung für die Zukunft. In *Otto Neurath: Gesammelte Ökonomische, Soziologische, und Sozialpolitische Schriften, Band II*, ed. Rudolf Haller and Ulf Höfer. Wien: Hölder, Pichler, Tempsky.

———. 1919. *Durch Die Kriegswirtschaft Zur Naturalwirtschaft*. Munich: G. D. W. Callwey.

———. 1921a. Die Österreichische Siedlerbewegung. *Neue Freie Presse (Evening Edition)* April 19, 1921: 3.

———. 1921b. Erziehungsaufgaben des Sozialismus. *Bildungsarbeit* 8 (2/3): 9.

———. 1923a. *Österreichisches Kleingärtner und Siedler Organisation*. Wien: Kommissionsverlag der Wiener Volksbuchhandlung.

———. 1923b. Die Vorbereitung der Kleingarten- Siedlungs- und Wohnbauaustellung. *Österreichische Städtezeitung* 9, September 1923: 131–135.

———. 1923c. Kernhausaktion der Gemeinde Wien. *Österreichische Städte Zeitung* 7, July 1923: 1–8.

———. 1923d. Dauernde Erhaltung des Ausstellungsmaterials, Ein Siedlungsmuseum. *Arbeiterzeitung*, September 6 (1923): 8.

———. 1925. Gesellschafts- und Wirtschaftsmuseum in Wien. *Österreichische Gemeindezeitung*: 1–12.

———. 1927. Der Weg des Gesellschafts- und Wirtschaftsmuseums. *Österreichische Gemeindezeitung* 4: 2–4.

———. 1929. Bildstatistik und Arbeiterbildung. *Bildungsarbeit* 1: 8.

———. 1930. Das Sachbild. *Die Form* 2: 219–225.

———. 1931a. Bildhafte Pädagogik im Gesellschafts- und Wirtschaftsmuseum in Wien. *Museumskunde, Neue Folge III*: 125–129.

———. 1931b/1973. Physicalism. The Philosophy of the Viennese Circle. In *Empiricism and Sociology*, ed. Marie Neurath and Robert S. Cohen, 48–51. Dordrecht: D. Reidel.

———. 1931c/1973. Sociology in the Framework of Physicalism. In *Empiricism and Sociology*, ed. Marie Neurath and Robert S. Cohen, 58–90. Dordrecht: D. Reidel.

———. 1933a. Die pädagogische Weltbedeutung der Bildstatistik nach Wiener Methode. *Die Quelle* 3: 209–212.

———. 1933b. Museums of the Future. *Survey Graphic* 22 (9): 458–463.

———. 1936. *International Picture Language – The First Rules of Isotype*. London: Kegan Paul, Trench, Trubner & Co.

Novy, Klaus. 1981. Die Pioniere vom Rosenhügel: Zur wirklichen Revolution des Arbeiterwohnens. *UmBau* 4: 43–60.

Novy, Klaus, Wolfgang Förster, and Verein für Moderne Kommunalpolitik. 1991. *Einfach Bauen: Genossenschaftliche Selbsthilfe nach der Jahrhundertwende: Zur Rekonstruktion der Wiener Siedlerbewegung*. Vienna: Picus.

O'Neill, John F. 2007. Pluralism and economic institutions. In *Otto Neurath's Economics in Context*, ed. Elisabeth Nemeth, Stefan Schmitz, and Thomas Uebel, 77–100. Dordrecht: Springer.

Reumann, Jakob. 1923. Geleitwort. In *Führer durch die Wiener Kleingarten-, Siedlungs-, und Wohnbauausstellung, 2.-9. September 1923, Rathaus*, ed. Max Ermers, vol. 3. Wien: Self-Published.

Rukschcio, Burkhardt, and Roland Schachel. 1982. *Adolf Loos: Leben Und Werk*. Wien: Residenz Verlag.
Sandner, Günther. 2014. *Otto Neurath: Eine politische Biographie*. Wien: Zsolnay.
Schütte-Lihotzky, Margarete. 1923b. Wiener Kleingarten- und Siedlerhüttenaktion. *Schlesisches Heim* 3 (4): 23–85.
———. 2004. *Warum ich Architektin wurde*. Salzburg: Residenz.
Sekera, Franz. 1921, September. Invalide als Siedler. *Der Siedler* 1(6): 92.
Stadler, Friedrich, ed. 1982. *Arbeiterbildung in der Zwischenkriegszeit. Otto Neurath und Gerd Arntz*. Wien: Löcker.
Turner, Fred. 2013. *The Democratic Surround: Multimedia & American Liberalism from World War II to the Psychedelic Sixties*. London: University of Chicago Press.
Uebel, Thomas. 2004. Neurath's Economy in Critical Context. In *Otto Neurath: Economic Writings. Selections 1904–1945*, ed. Thomas Uebel and Robert S. Cohen, 1–108. Dordrecht: Kluwer.
———. 2007. *Empiricism at the Crossroads: The Vienna Circle's Protocol-Sentence Debate*. Chicago: Open Court.
Vossoughian, Nader. 2007. The War Economy and the War Museum: Otto Neurath and the Museum of War Economy in Leipzig, c. 1918. In *Otto Neurath's Economics in Context*, ed. Elisabeth Nemeth, Stefan Schmitz, and Thomas Uebel, 131–140. Dordrecht: Springer.
———. 2008. *Otto Neurath: The Language of the Global Polis*. Rotterdam: Nai Publishers.
Welzig, Maria. 1998. *Josef Frank (1885–1967): Das Architektonische Werk*. Wien: Böhlau.
Wilson, Francesca. 1923. The resurrection of Vienna, houses on the land settlement. *The Manchester guardian*, October 26: 6.

Chapter 8
Generating Cognitive Tools: Neurath's Educational Ideal and the Concept of ISOTYPE

Angélique Groß

Abstract Neurath was a practitioner and organizer in many ways. In relation to education, he did not express a philosophy, a theory of education. Neurath just practiced education, he arranged it for employees institutionally, personally, didactically and methodically. In a similar way, his methodical procedure of transformation, which is manifested in the pictorial presentation system of ISOTYPE, was not based on an explicit theory of depiction. He rather developed a concept, whose realization evolved due to a successful practice of transmission. However, in the practice of educational depiction, Neurath resorts to constants – instead of all gradual evolution and instead of all crudities – which are describable analytically as theoretical implications. In this respect, the present contribution points out the context of the educational and descriptive practice and its evolution out of Neurath's ideal texture on the one hand, out of the Wissenschaftliche Weltauffassung. In comparison to the educational landscape of Vienna on the other hand, commonalities and demarcations are illustrated to expatiate on the novelty of Neurath's legislating. In addition, the evolution of ISOTYPE will be retraced to define different levels of progression and to elaborate its principles and structure. In conclusion, there are factors described with regard to his educational ideal that illustrate the aim of education, the didactics used, the impact of the transmission method and the addressed form of reception. With regard to his concept of depiction, there are factors described that illustrate the principles of ISOTYPE, the addressed form of cognitive pattern and the intended field of validity.

A. Groß (✉)
Adult Education Centre, Lebach, Germany
e-mail: angelique.gross@gmx.de

J. Cat, A. T. Tuboly (eds.), *Neurath Reconsidered*,
Boston Studies in the Philosophy and History of Science 336,
https://doi.org/10.1007/978-3-030-02128-3_8

8.1 Introduction[1]

The manifesto published by members of the *Vienna Circle (Wiener Kreis)* as *Wissenschaftliche Weltauffassung* in 1929, was partly drafted by Otto Neurath in cooperation with his fellow campaigners and partly quite specifically further developed[2]: for him, it is a comprehensive structure of thought which frames his entire creative work as an educator of the people and of the working population and from which his enlightenment concept, his educational idea and his transmission approach are derived.

Wissenschaftliche Weltauffassung is based on a socio-political question, namely the question of the ways and means to achieve a socio-political way of life which fulfills the expectations of the people who inhabit it and in which the people are therefore happy (Neurath 1929a/1991, 139; 1912/1973, 122). In this respect, it stands for a configuration program for social and political life. It links science, on the one hand, with society on the other and uses enlightening education as the mediating authority for the two dimensional and reflexive connections: It distributes scientific findings and makes them exploitable for improving living conditions The way of life in society therefore becomes scientific. At the same time, science is assigned a constitutive role in social life (see Fig. 8.1).

Enlightenment carries science with its structural characteristics as far as possible into society and, at the same time, it does justice to society's educational needs. In this context, it becomes relevant, on the one hand, to basically open science for society and, on the other, to reduce obstacles within society and create access to science.

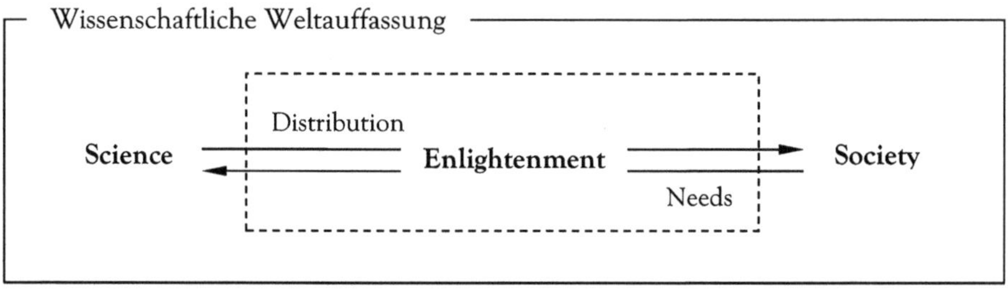

Fig. 8.1 Wissenschaftliche Weltauffassung

[1] The paper was translated from German by Morag Paul. Neurath's German papers are quoted from his collected works with the original dates and the page numbers of Haller and Rutte (1981), Haller and Kinross (1991), and Haller and Höfer (1998).

[2] On the production history of the Vienna Circle's manifesto see Uebel (2008).

8.2 Neurath's Concept of Enlightenment and His Educational Ideal

Enlightenment constitutes unity of thought and action which integrates the primacy of reason into the field of practice. The result is a holistic structure, which, with the aid of the utopian idea – the alternative reality concept which has been captured in thought (Nemeth 1994, 111) – creates a dialectic connection between theory and practice (Haller 1993a, 156; Uebel 2000, 27, 332). Enlightenment seeks to promote *independent thinking*, that is the intellectual handling of scientifically generated knowledge and classification of knowledge within the individual value system, sovereign processing and reflection (Neurath 1933b/1991, 273; 1936/1991, 361; 1928a/1973, 251–252). It therefore applies "[...] to fashion cognitive tools for everyday life, [...] for the daily life of all of those who in some way join in working at the conscious re-shaping of life" (Neurath et al. 1929/1973, 305, translation modified). Since the recipient himself decides the extent to which content becomes relevant for action or not, the moment of independent thought is coherent with *independently responsible action.*

Enlightening education therefore only provides the informational basis and leaves it up to the individual if and how he deals with this information from a judgmental and behavioral point of view. It merely stands for the creation of individual cognitive tools and only influences the will of the individual indirectly via the effect on the individual's insight (Neurath 1914/1998, 423). In this respect, although utopias may be intersubjectively scientifically founded constructions, they are very different from one individual to another (see Fig. 8.2). They mark the boundary between conveying the intersubjective, empirically generated finding and the subjective, judgmental and individually specific decision to act, between abstract, scientific information and concrete, committed practice. The unity of thinking and acting can therefore only be a subjective one.

Enlightenment becomes socially relevant when decisions to take action within a political context become effective. Because a *democratically structured social system* allows and requires that all citizens, no matter from which social level, participate in the development processes, creating them according to their utopias, Neurath's enlightenment aims are democratic in their implementation (Neurath 1929a/1991, 139; 1912/1973, 122). Vice-versa, democratic decisions on creative action taken by the enlightened and self-determined individual are substantiated by a scientific-utopian basis.

The *content* of such enlightening education can only be value-free information which logical-empirical science generates as findings with its strict, standardized processes (Neurath 1926a/1991, 56; 1936/1991, 396).[3] In order to come close to

[3] For this reason, the educational content is material, intersubjectively verifiable, formulated language-critically, antimetaphysical and relativistic. Detailed explanations of the logico-empirical scientific theory of Neurath and the Vienna Circle can be found, among other sources, in Haller (1993b), Stadler (1993), Nemeth (1982) and Uebel (2000).

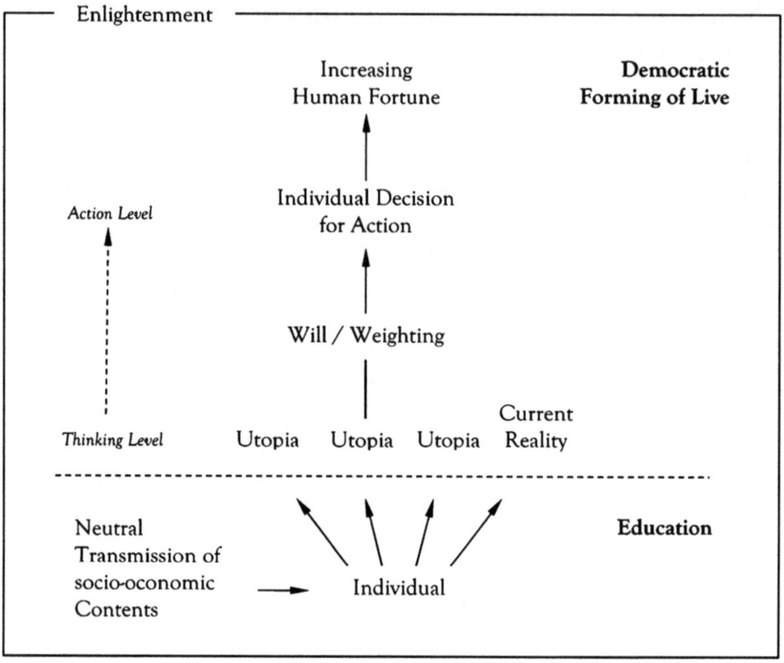

Fig. 8.2 Structure of the concept of enlightenment

taking account of the connection within democratic decision-making environments, it is primarily socio-scientific statistical content which is to be conveyed (Neurath 1908/1998, 120; 1910/1998, 423f).

Of *those to be addressed*, it is largely the workers who come into the focus of Neurath's enlightenment addresses, insofar as the disadvantage related to their social level needs to be compensated on several educationally (and therefore configuration-) relevant levels (Neurath 1928a/1973, 248, 258).

Both orientations not only lead to specification of the enlightenment concept but also to specification of the *transmission method* within this enlightenment concept. Because the enlightenment technique to be used is regarded as a moment which decides on the success or failure of the appropriation processes, it is assigned a constitutive role. Neurath therefore generates a pictorial pedagogy which, due to its special form, illustrates the structure of scientific knowledge schematically and meets the requirements of those who are addressees (Neurath 1928b/1981, 280).

One aspect of educational theory which is closely linked to the aspiration to reach the workers as the group of addressees is that of *Activation* of the addressee. This point is dealt with on several levels, firstly that of motivation, of attracting the addressee in order to initiate a process of appropriation and further pursue it. The moment of activation is furthermore connected to the promotion of an (active) questioning attitude in the recipient: Through schematic shortening of what is presented, there are multiple capabilities for connecting the received content which opens up a random number of interlinks and thus induces independent cognitive treatment of the material conveyed (see Sect. 8.3.2). If interlinks are now created which go beyond what is directly shown, the creation of a questioning attitude is promoted in the recipient which can lead to the continuation of educational interests. The third form of activation is that of association: since socio-scientific quantities are presented where the connection is only generated by correlation on behalf of the recipient, it is the recipient himself who creates meanings through his cognitive constructional efforts (Neurath 1925a/1991).[4]

In this respect, Neurath's *enlightenment concept* presents itself as the intellectual education of workers in which socio-economic content with the aim of self-activated appropriation is neutrally demonstrated (Groß 2015, 73).

Derived from this, it can be seen in the interlinking of materials and formal educational principles that *education* according to Neurath is the methodically structured, independent appropriation of empirical-rational content for the purpose of intellectual development of the personality (Groß 2015, 269).

8.2.1 Neurath and the Educational Landscape of Vienna

But why does Neurath see a need to develop his 'own' special practice of enlightenment when, at the time, there are already two established *Viennese Educational Movements* in Vienna, the popular education by the Bourgeoisie and the socialist education of workers?[5] So how do Neurath's theory and practice differ from the other educational concepts?

When comparing educational theory, there are similarities and differences: the popular education by the Bourgeoisie, for example, – institutionalized in the popular university lectures, in the Volksheim (house of the people), popular education association and the Urania – promotes the neutral transmission of scientific content in much the same way as Neurath in order to support intellectual abilities and competence in democratic organizational activities (Uebel 2000, 295f). The socialist worker education movement, on the other hand – institutionalized in the educational center, the Viennese Workers' School, the Viennese Union School and the Workers' University – besides scientifically orientated education of functionaries, in view of the masses, constitutes education in workers' culture, above all. The proportion of culture and festival and vacation culture shows a significant predominance towards the education of the masses in conveying of content (Weidenholzer 1981, 71) to correspond to the theoretical significance of culture as a tool for creating a proletarian

[4] Here, basic formal human capacities are addressed: Since, as anthropologically dispositioned constants, curiosity, visual perception and the formation of associations are used as tools of awareness and in their activation function of cognitive structures (see also Riedl and Parey 1980).

[5] Even if there are close interdependencies in Vienna between national civil education and social education of workers, both movements are dealt with separately here in order to underline their different basic constructs.

class consciousness and a class-conscious competence to act. For Neurath, however, instead of the formulation of party political educational aims, neutrality is indispensable as a category of mediation: Only where judgmental dealings with the content conveyed are not laid down in advance is independent thinking promoted.[6]

In formal respects, Neurath's educational concept therefore integrates educational aims of both Viennese educational movements. From the point of view of material, on the other hand, he tends rather to conform to the popular education by the Bourgeoisie (Altenhuber 1999, 74; Filla 1992, 93) and to exclude judgmental content with a party political perspective as well as cultural content.

Also in a comparison with the concepts of the established Viennese educational structures, there are various similarities and differences. In the practice of popular education by the Bourgeoisie, in spite of the aspiration to be scientific, the conveying of cultural content plays almost equally significant a role as scientific content (Filla 1998, 83; 1996, 91; Glaser 1981, 335).[7] Neurath, on the other hand, exclusively teaches philosophy of science and scientific findings at the *Verein Ernst Mach* (Ernst Mach Society) and at the *Gesellschafts- und Wirtschaftsmuseum* (GWM, Social and Economic Museum) and consistently realizes his claim of scientific integrity on a didactical level.

The conceptional efforts to involve the working community in educational matters are common to both educational movements and to Neurath. However, Neurath's primary practical focus is on the education of the proletarian masses (and not on the education of party functionaries), so that he obtains a differentiated perception of the needs of the working masses and, in connection with this, the identification of a problem regarding transmission. Due to the fact that the two Viennese educational movements fail to question their transmission methods (Neurath 1929a/1991, 139), it can be assumed that they could not reach anyone other than German-speaking and literate workers and that this failure constitutes one of the reasons why they are hardly able to integrate the workers.[8] The identification of this mediation problem

[6]Although Neurath also emphasized the necessity of a class-conscious, democratic competence to act, the socialist worker education for Neurath is only defined via the addressee and not, for example, via a close ideological guidance of educational aims and their didactical or methodical implementation: in 1908, he claimed that the current enlightenment was not objective enough because it was "clouded by the vigour of the party political fight" (Neurath 1908/1998, 120). Neurath's own utopia, however, is indeed socialist-revolutionary, the desire, as it were, to determine how enlightened workers are to deal with the knowledge conveyed to them. Detailed treatment of the ideas and biography of the political Neurath is provided by Günther Sandner (2014). See also Sandner's chapter in the present volume.

[7]The restriction to scientifically generated content only takes place in lectures of the popular university which increasingly lose significance, however.

[8]According to Langewiesche (1979, 265) 1901/02, of a total audience of 33,221 at the popular university lectures, 30% were worker. Filla (2001, 87) notes that, in the 1920's, 40,000–50,000 of participants in courses offered by the adult education center "Volkshochschule" were also recruited to one third from the group of the working population. Consequently, education at the Volkshochschule is dominated by the middle classes in spite of the relatively large high proportion of workers and is only accessible for an active minority of the total of 580,000 workers (Filla 1999, 107; 1998, 85; 2001, 87). Petrasch (2007, 85, 94), on the other hand, speaks of Urania reaching

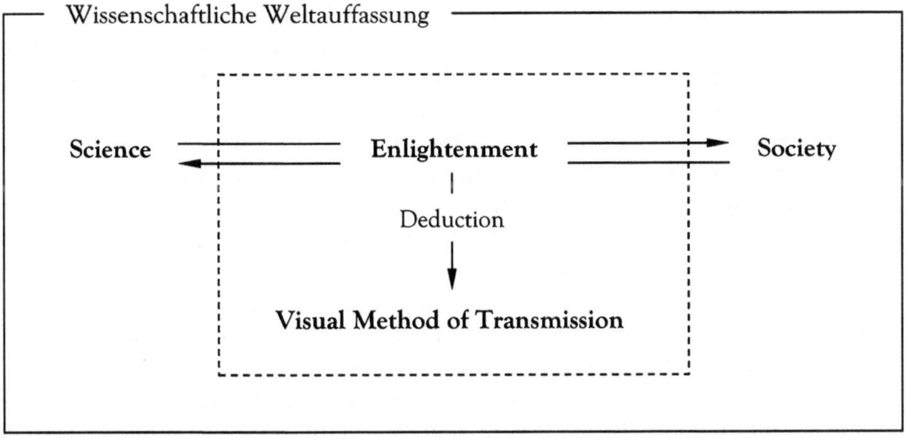

Fig. 8.3 Specification of Wissenschaftliche Weltauffassung

leads Neurath to inevitably seek a radically different approach in spite of a similar aim being formulated by the Viennese educational sphere: the education of workers cannot be carried out in the same way as civil education has been conducted until now. The popular methods of mediation using word and letter as passed down from civil education are replaced by the pictorial pedagogy which was not passed down from popular education (see Fig. 8.3).

Neurath's theoretical principles are therefore partially to be found in both of the established Viennese educational movements. However, the methodological procedures used differ greatly from the conventional ones and close a gap which Viennese educational practice makes apparent. For this reason, the novelty in the structure of enlightenment according to Neurath is rather more to be found in practice than in theory.

8.3 Neurath's Educational Practice. The Concept of ISOTYPE

Neurath makes various demands of the transmission method of pictorial pedagogy if it is to fulfill its purpose in the context described.

440,561 adult visitors in 1917/18. Since Urania does not try to attract workers and there are no social statistics with information on the proportion of employees, it can be assumed that, in this connection, too, one can hardly speak of broad inclusion of workers.

However, in socialist worker education, too, with the institutional focus on the education of functionaries, only a small proportion of workers participate in educational seminars. A survey from 1931 suggests that only 9.6% of the worker population frequents educational programs (Kuleman 1979, 28f).

On the one hand, these requirements concern methodical moments of the preservation of the structure of scientific knowledge, namely correctness (Neurath 1944b/1991, 601; 1932/1991, 211; 1925b/1991, 3; 1927/1991, 118; 1937/1991, 401) and neutrality i.e. lack of association and suggestion (Neurath 1929b/1991, 134; 1930a/1991, 153; 1931b/1991, 195; 1933a/1991, 260).[9] Furthermore, they contain a moment of opening of an ordered, scientifically logical handling of the knowledge conveyed (Neurath 1933b/1991, 273; 1936/1991, 361). The depiction of correlations therefore takes on a double function here in which, on the one hand, it allows scientifically correct presentation of factual relationships and, on the other, ensures the order of a network-type, logical and flexible way of thinking (Neurath et al. 1929/1973, 305; and Neurath 1944b/1991, 601). An essential requirement of Neurath of the educational method is therefore that of its *scientificness*.

Further standards of Neurath aim at the direct, cognitive effect of using pictures to convey information. The aim is to achieve optimization of reception through activating the recipient (Neurath 1929a/1991, 143; 1931a/1991, 185) and, at the same time, by using anthropological dispositions with only few prerequisites (Neurath 1933c/1991, 235; 1933b/1991, 298). In addition, the content to be presented is to result in optimization of storage by using recognized cognitive storage mechanisms (Neurath 1931a/1991, 184; 1935/1991, 342). Therefore, a further normative principle is the *efficiency of transmission*.

Finally, there are requirements by Neurath of pictorial pedagogy to provide emotional, psychological incentives. As an initial impetus, these incentives serve to awaken human curiosity and attract attention in order to gain access to education. Furthermore, they serve to maintain an intrinsic interest in education which remains effective in the long-term (Neurath 1936/1991, 378). Neurath's last normative principle is that of *motivation*.

Since it is not possible to fall back on a functioning pictorial pedagogy with scientific character, efficiency of transmission and motivation, it becomes necessary for Neurath to generate this. These individual principles are too abstract, however, and not meaningful as concrete instructions for generating pictorial pedagogy: "Therefore: Pictures! But this insight alone is not enough, it is necessary to know how to make correct use of the pictures" (Neurath 1926b, 57). Here, normative principles (only) serve as target factors for the construct of pictorial pedagogy and they are bound by the derivation from specific operational rules for construction.

8.3.1 The Levels of Progression of ISOTYPE

The creation of the equivalent of abstract target principles and concrete conventions does not take place as a theoretical construction and also not as a direct, stringent process or even as a one-off invention. Neurath rather conducts this performance of translation within the practice of transmission without a methodical-theoretical

[9] For more considerations in the secondary literature see Runggaldier (1979, 246).

foundation according to the principle of "trial and error." At the beginning there is only the idea of pictorial pedagogy and its implementation means years of improvisation, evaluation and modification so that only little by little is improvisation replaced by a reflected method of transmission (Hartmann 2002, 58; Neurath 1931c/1991, 205; 1926d/1991, 74). In the overall developed enlightening context, this protracted genesis of pictorial pedagogy applies as an approach to a solution for the recognized problem of educating workers which, itself, is full of deviations.

Henceforth the efforts regarding the development of conventions for pictorial pedagogy are at the center of the enlightenment concept and are correspondingly institutionalized: The GWM subscribes to being collectively responsible for the whole process of evolvement and presentation. It bundles several functions of the network of experts within itself and is (1) the institution for generating the technique for enlightenment using pictorial pedagogy, (2) the institution for enlightenment using this technique and (3) the institution for international distribution of the concept of enlightenment using pictorial pedagogy (Neurath 1931b/1991, 195).[10]

Since 1923, there have been pictorial representations whose connection is referred to in 1925 as the Viennese Method of Pictorial Statistics and which is renamed *ISOTYPE* in 1934 thus shortening *International System of TYpographic Picture Education* as an acronym.[11] Whereas many of the concrete construction rules must first find themselves within the graphic and draft-related field, in the statistical context, there are however already two conventions that are fixed from the start: statistical quantities are represented by "speaking," i.e. self-explanatory symbols which represent a defined quantity (Neurath 1931a, 180). And: these symbols are repeated in rows as often as the statistics require "verlangt" (Neurath 19250/1991a, 22).

Many other methods of implementation are just starting to evolve: where the symbols are three-dimensional at first and naturalistic, they later soon become two-dimensional and more and more stylized into types.[12] Only with growing experience is it possible to integrate the isomorphy and reduction of the subject information. The symbols slowly become module-type elements which are conventionally put together. The tables are two-dimensional from the start in order to create a construct which emphasizes the information by doing without three-dimensional aspect. Whereas the pictures are still arranged at the beginning using grid lines, later only axes are explicitly drawn in until these too are eliminated and it is apparent which letters belong to which group only from the smaller or greater distance between them.

---

[10] Due to the required emigrations, several institutions in sequence make efforts at generating pictorial pedagogy: the GWM in Vienna (1925–1934), the International Foundation for Visual Education in Den Haag (1934–1940) and finally the Isotype Institute in Oxford (1940–1945). Cf. Neurath (1931b/1991, 113). Neurath's museum pedagogy is described by Kraeutler (2008) and Groß (2015, 65ff).

[11] An extensive description of the genesis with regard to transmission methodology in institutional and personal interrelationships has been written by Groß (2015, 91ff). A graphic perspective of the genesis is detailed by Kinross (2013) and Groß (2015, 91ff).

[12] In semiotic terminology, the symbols constructed by Neurath are icons. Neurath himself uses various terms for the symbols: signs, signatures, types, isotypes and letters.

Guiding images are often only used at the beginning so that it can be recognized what a table refers to even without knowledge of language. They are then used less and less frequently and then finally only used if they are a reference quantity as part of the statistical representation. While the pictures of the first few years are extremely focused on writing, with increasing certainty, the proportion of writing in the symbol and image constructions becomes significantly less and only more again at the start of experimentation with abstraction. At the beginning, the number of symbols and rows of symbols is very large and can only be considerably reduced after the first few years. In an attempt to convey scientific content factually and logically correctly, at the beginning the detailed factual description dominates. This is only restricted by slowly developing courage to reduce this in favor of the reception capability of the pictures.[13] The diversity of the first few years on almost all levels of presentation is therefore successively suppressed by increasing uniformity.

The limits of the material presented become increasingly broader with regard to content: from concrete, matter-of-fact topics such as civic education and health, becoming partially supplemented by more abstract subjects such as machine power, the power of resistance and history of art and later being extended to include abstract natural sciences such as chemistry. In the tables, which are mainly social science-related, over the years various types of picture are developed such as statistical tables, statistically utilized cartograms, map tables and organizational charts.[14] In health education, statistical tables and multiple step progression schemes are used.[15]

In the context of constant evolvement of ISOTYPE, the standard works can be used to determine individual watersheds and stages of development. *The early days* of the method (from 1923) are determined by a simultaneous, diverse juxtaposition of tables, each of which characterizes different stages of experimentation (see Fig. 8.4).

In the children's book *Die bunte Welt* (*The Colorful World*) from 1929, significant changes can be seen in that there is a uniform concept on a rough level that implies some contradiction in concrete terms (see Fig. 8.5).

The Atlas *Gesellschaft und Wirtschaft* (*Society and the Economy*) from 1930 marks a significant watershed for the first time: experience gained until now has led to a relatively closed system and to the sure mastery of the trade (see Fig. 8.6).

[13] Whereas at the beginning, correct scientific knowledge which is accurate in every detail and a reduction in the diversity of statements seem to be contrasting elements which are difficult to incorporate, they become ever more compatible in the course of development.

[14] Kinross (2013) illustrates which different forms of statistical tables are used.

[15] Within the context of presentation, the tables are given priority. However, because their illustrative performance is as special as it is limited, other media are used in a presentation. These, too, are not used without reflection and adherence to rules: it is attempted to use models, basic English, picture-text-style and film for reflecting and arranging in similar moments as with the tables. Extension of the rules of pictorial pedagogical tables, in the sense of transferability to a more comprehensive system of transmission, are therefore included in the thought process and the beginnings of the implementation are realized. Nonetheless, within Neurath's lifetime, the main focus remains on the pictorial pedagogy created using Isotype.

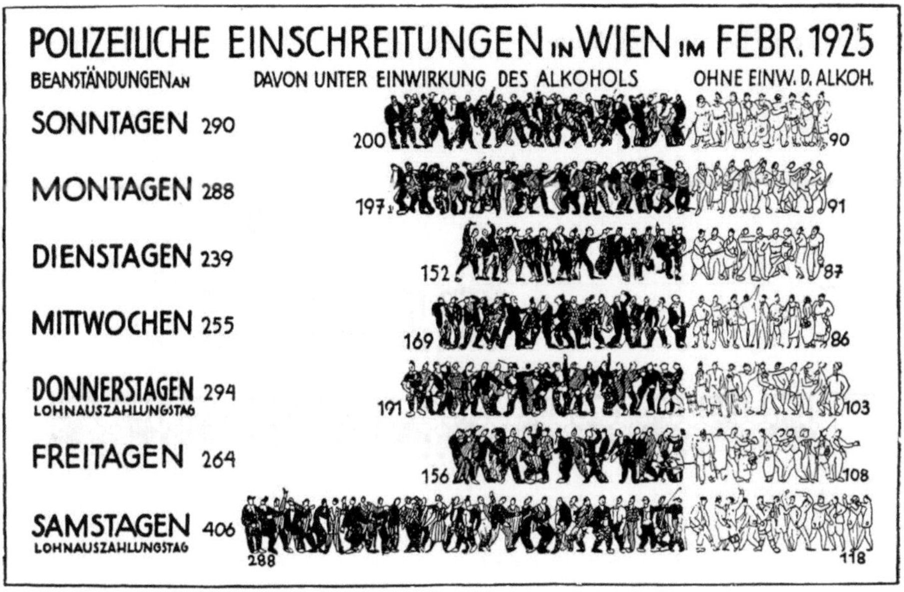

Fig. 8.4 Polizeiliche Einschreitungen in Wien im Febr. 1925. (© Otto and Marie Neurath Isotype Collection, University of Reading)

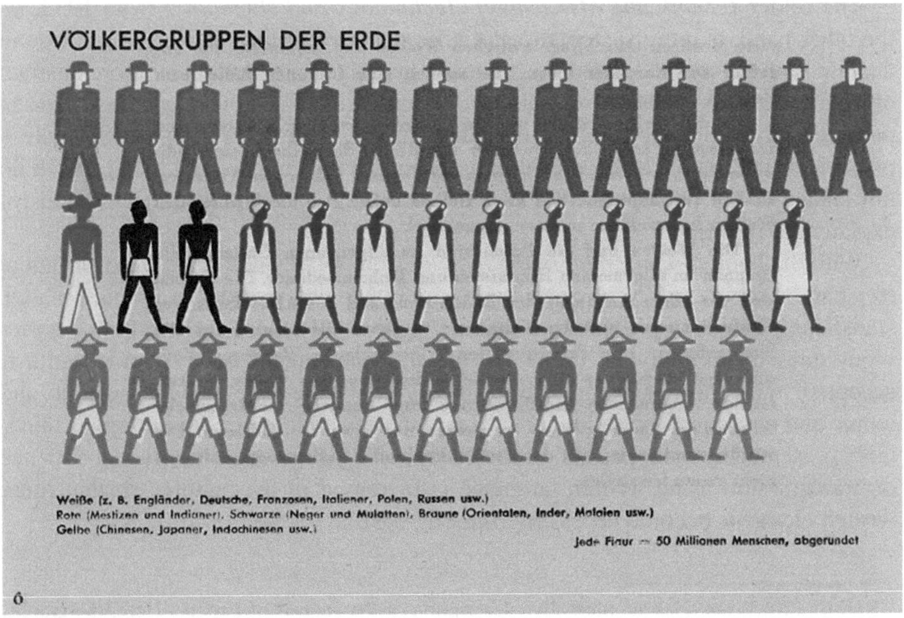

Fig. 8.5 Völkergruppen der Erde. (© Otto and Marie Neurath Isotype Collection, University of Reading)

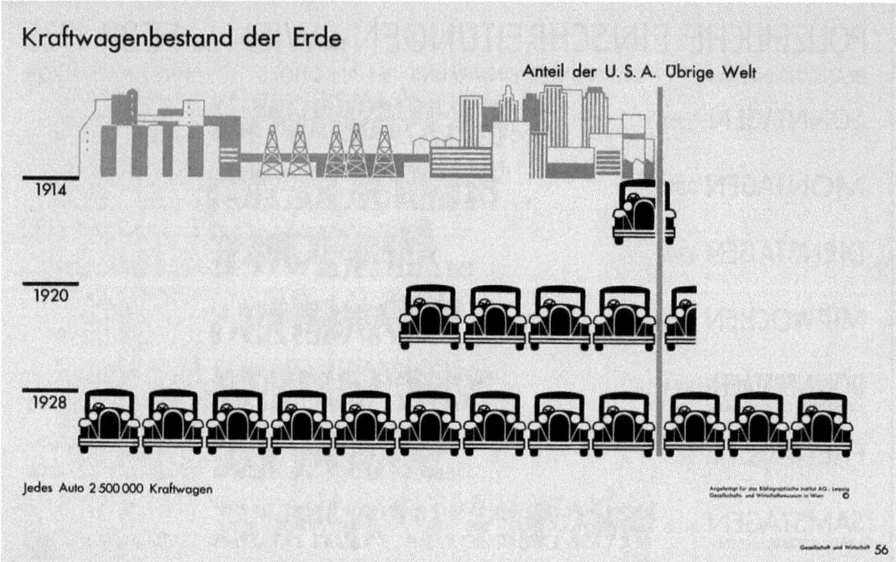

Fig. 8.6 Kraftwagenbestand der Erde. (© Österreichisches Gesellschafts- und Wirtschaftsmuseum, Wien)

The folder *Technik und Menschheit* (*Technology and Humanity*) from 1932, on the other hand, is more systematic and goes one step further: due to the routine in dealing with the method, attempts are made here to create pictures of more abstract objects and to free oneself from hitherto strictly adhered-to pictorial constructs. In particular, the tendency to grade the tables in stages of perception which make it possible to integrate a further level of assertion: details are now included which do not endanger the perception of the main assertions and which hold potential for explanation (see Fig. 8.7).

With the book *Modern man in the making* from 1939, once again a development step can be registered. Now experiments are conducted in order to find alternative statistical methods of presentation that are significantly more abstract than the previous ones. Admittedly, in doing so, the infringement of basic requirements is accepted: with increased abstraction, the symbols used are less isomorphic and iconic and consequently no longer self-explanatory. Quantities are no longer inevitably presented by the repetition of these symbols. At the same time, the dependence on transmission using written language is increased in the pictures so that more written elements become necessary (see Fig. 8.8).[16]

[16] Kinross (2013) provides an alternative description of the genesis of Isotype. Here, the graphic developments are clarified in detail so that, within formal setting of topics and their content, pictures are described in chronological comparison. As watersheds in the development, the periods "1925," "1926," "1927," "1928," "1928–29," "1929–30" and "from 1930" are deduced so that, on the one hand, there is a differentiated description of the early years and, on the other hand, an undifferentiated description of the later years.

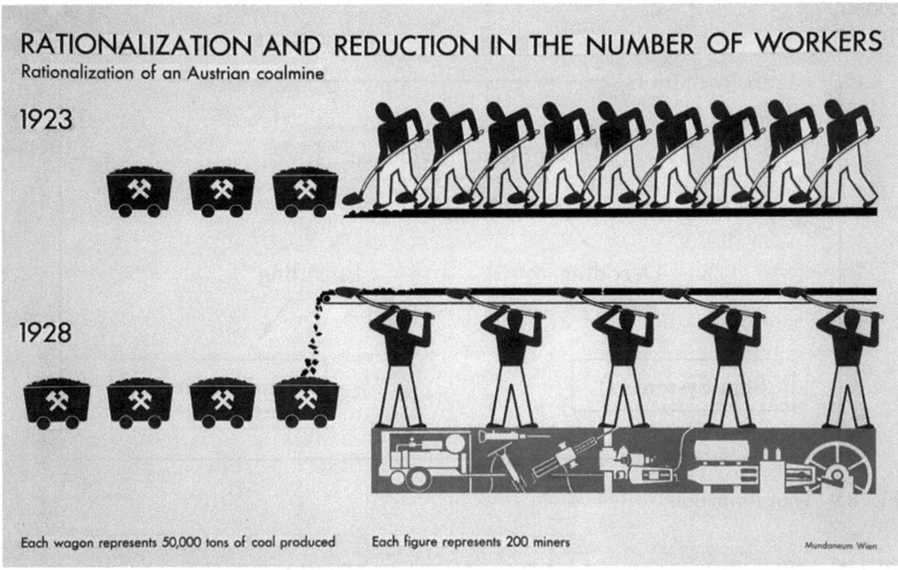

Fig. 8.7 Rationalization and reduction in the number of workers. (© Otto and Marie Neurath Isotype Collection, University of Reading)

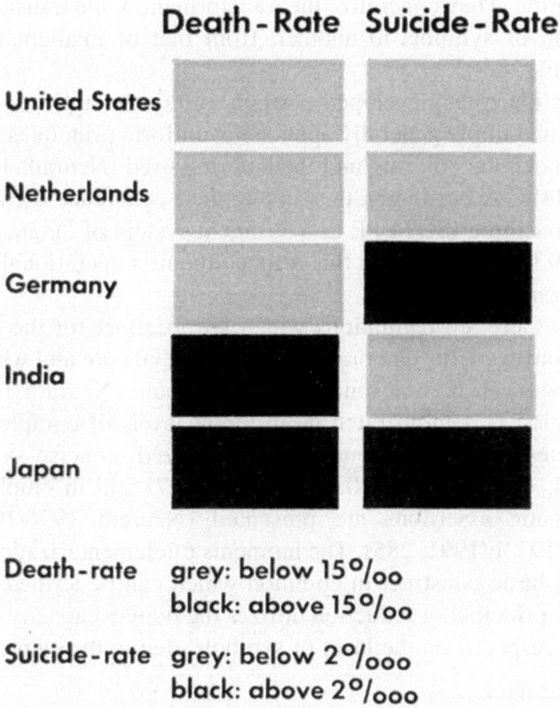

Fig. 8.8 Death-rate/Suicide-rate. (© Otto and Marie Neurath Isotype Collection, University of Reading)

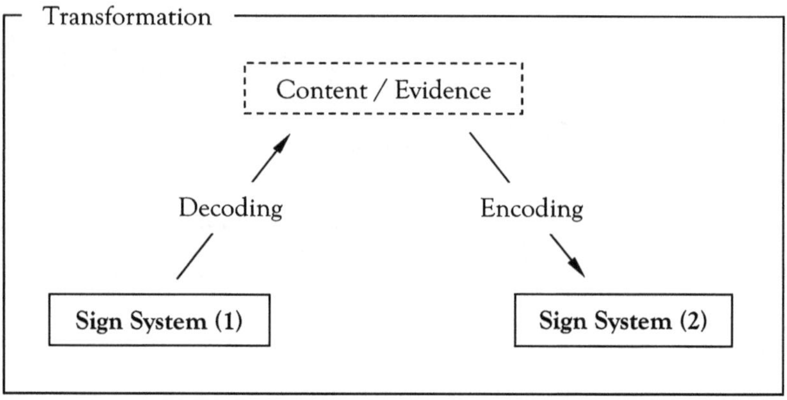

Fig. 8.9 Transformation

8.3.2 The Principles and Structure of ISOTYPE

In the course of the genesis of ISOTYPE, operational construction rules are derived from the formulated normative principles of scientificness, efficiency of transmission and motivation. They concretize the transformation, the translation of one content of a system of symbols to another, from that of mathematics into that of ISOTYPE (see Fig. 8.9).

The construction rules developed partially affect the aspiration of methodical standardization and imply generally applicable, uniform principles for the presentation of the symbols, the coloring and the lettering used (Neurath 1942/1991, 591f; 1930b/1991, 144).[17] Accordingly, they create their coherence dictionary, grammar and style and constitute ISOTYPE as a stringent system of language (Neurath and Kleinschmidt 1939/1991, 423). In this respect, the first operational principle is that of *systematization*.

Furthermore, there are regulations which are in effect for the emphasis of the material to be conveyed, the elaboration of its essential core and which characterize the difference between the relevant and the irrelevant (Neurath 1935/1991, 343). This differentiation is demonstrated on different levels of complexity in that elementary characteristics are presented in emphasized, concise symbols (Neurath 1935/1991, 344; 1933b/1991, 270; 1926c/1991, 67) and in emphasized, concise pictures, schematic assertions are presented (Neurath 1936/1991, 377, 381; 1945/2010, 54; 1933b/1991, 285). The moments of elementarization and schematization have one basic construct in common which can be termed the principle of *reduction*.[18] The principle of reduction utilizes the human capacity to form associations in various respects: on the level of symbols, due to the elementarization, that

[17] For more details in the secondary literature see Hartmann (2002, 83f).

[18] Since reduction refers to the formal aspect of the content, it is not a didactical reduction alone but also and above all a methodological reduction.

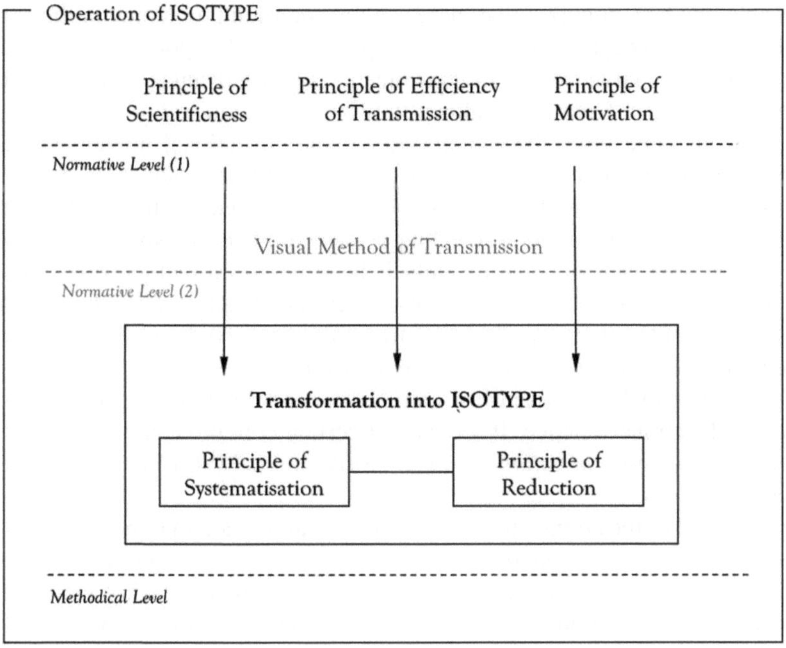

Fig. 8.10 Operation of ISOTYPE

which is presented is as concise as possible in order to invoke an unambiguous association with the known subject. On the level of the picture, that is the assembly of symbols, the schematization allows an active linking of information to form a cognitive network of knowledge (Hartmann 2002, 45).

In the sense of a unique distinguishing feature, the two elementary principles of systematization and reduction constitute the structure of ISOTYPE in their synthesis.

The *presentation concept* of ISOTYPE therefore corresponds to a methodical structuration which, using the principle of systematization, provides clear allocations of visualization and, using the principle of reduction, provides concentration on the essential and, based on the human capacity for creating association, is valid for a normative set field of conditions consisting of efficiency of transmission, motivation, a scientific way of thinking and statistically correlative content (see Fig. 8.10).

8.4 Summary

Based on the normative overall content of *Wissenschaftliche Weltauffassung*, Neurath realizes a concept of enlightenment, an educational idea and a presentation concept which he does not grasp theoretically. However, they can be analytically derived and explained from the specific realization. It becomes particularly noticeable that he recognizes that there is a transmission problem with this pictorial enlightenment concept for workers and he places central importance on the method of transmission: a large number of the educationally disadvantaged working population is not literate or German-speaking, they have no command of this cultural technique. At the same time, the established Vienna educational movements use the German language both orally and in writing and, for this reason, they do not reach the target group formulated. However, there is no adequate method of transmission which does not require a cultural technique. Since this is decisive Neurath decides on a radical approach to solve it: Only if he generates a method of transmission himself which follows on from an anthropological constant and which can therefore directly be adopted, is the problem of transmission solved and the aspirations of *Wissenschaftliche Weltauffassung* can be realized.

The extent to which he and his team are successful in doing so remains to be tested in educational science. It seems that certain conditions are emerging which differentiate the *scope of validity* of the pictorial method ISOTYPE and therefore of the whole endeavor.

With regard to the *content* to be presented, it can be said that the visualization of indirect content such as quantitative correlations, scientific knowledge content and thoughts lead us to assume a rather more abstract scope of validity for ISOTYPE. However, here the method can obviously connect the aspiration of consistent visuality with that of reduction whereby the thoughts to be presented relate to physical objects with natural figurativeness.[19]

With regard to the *addressees,* the limits of performance seem to exceed Neurath's original one by far. Although initially it was exclusively conceived as a method for the disadvantaged working class, trials in nursery schools and schools involve the children from an early age. The educated bourgeoisie are also soon integrated into the efforts of the *Volksbildungsinstitut für soziale Aufklärung*, (*Adult Education Centre for Social Enlightenment*) as the GWM was subtitled. Thus the partially conceived starting point for the genesis of ISOTYPE, which was based on the principle of enlightenment, becomes subject to an increasing degree of generalization: "The educated and the uneducated are both capable to roughly the same degree of grasping the main points of visual information and argumentation" (Neurath 1944a/1991, 596).

A comprehensive modification of the ISOTYPE concept seems to accompany this. Deduced from the specific, socially pedagogical enlightenment context,

[19] For purposes of relativization, it should be emphasized that ISOTYPE functions for scientific knowledge, as Neurath defines it, namely for statistical-correlative knowledge. For example Bühl (1984) and Mittelstraß (2001) define scientific knowledge as being more complex than hypothetical and open and able to take criticism. In this case, the structure of scientific knowledge cannot be formed as correlation alone.

The striving for abstraction which has been formulated cannot mean the distancing from or detachment from figurativeness through presenting complex scientific knowledge. It can only imply the formal separation of the relevant out of the entirety. In this respect, the concept of abstract as used by Neurath is not to be understood as didactic but rather as closely linked to the formal principle of reduction.

ISOTYPE becomes more and more a tool for international understanding and for 'mass information and mass education' (cf. Vossoughian 2008). In this context, it can be stated that, even in Neurath's time, the possibilities for the use of ISOTYPE become increasingly independent and lead beyond the context of its generation.

References

Altenhuber, Hans. 1999. Staat und Volksbildung in Österreich 1918–1939. In *Erwachsenenbildung in der Zwischenkriegszeit*, ed. Wilhelm Filla, Elke Gruber, and Jurij Jug, 72–82. Innsbruck/Wien: Studien-Verlag.

Bühl, Walter. 1984. *Die Ordnung des Wissens*. Berlin: Duncker & Humblot.

Filla, Wilhelm. 1992. Ludo Moritz Hartmann: Wissenschaftler in der Volksbildung. In *Aufklärer und Organisator. Der Wissenschaftler, Volksbildner und Politiker Ludo Moritz Hartmann*, ed. Wilhelm Filla, Michaela Judy, and Ursula Knittler-Lux, 67–100. Wien: Picus.

———. 1996. Aufklärung als Vorbotin und Element der Moderne in der österreichischen Volksbildung. In *Erwachsenenbildung in der Aufklärung*, ed. Wilhelm Filla, Elke Gruber, and Jurij Jug, 79–97. Wien: Promedia.

———. 1998. Entstehung der modernen Volksbildung in Wien. In *Erwachsenenbildung von 1848–1900*, ed. Wilhelm Filla, Elke Gruber, and Jurij Jug, 78–94. Innsbruck/Wien: Studien-Verlag.

———. 1999. Zwischen Arbeiterbewegung und Bürgertum. In *Erwachsenenbildung in der Zwischenkriegszeit*, ed. Wilhelm Filla, 97–110. Innsbruck/Wien: Studien-Verlag.

———. 2001. *Wissenschaft für alle – ein Widerspruch? Bevölkerungsnaher Wissenstransfer in der Wiener Moderne. Ein historisches Volkshochschulmodell*. Innsbruck/Wien/München: Studien-Verlag.

Glaser, Ernst. 1981. *Im Umfeld des Austromarxismus*. Wien/München/Zürich: Europa-Verlag.

Groß, Angélique. 2015. *Die Bildpädagogik Otto Neuraths. Methodische Prinzipien der Darstellung von Wissen*. Cham: Springer.

Haller, Rudolf. 1993a. *Neopositivismus. Eine historische Einführung in die Philosophie des Wiener Kreises*. Darmstadt: Wissenschaftliche Buchgesellschaft.

———. 1993b. Marksteine und Grundlagen der wissenschaftlichen Philosophie. Zur Neubewertung der Philosophie des logischen Empirismus. In *Wien – Berlin – Prag. Der Aufstieg der wissenschaftlichen Philosophie*, ed. Rudolf Haller and Friedrich Stadler, 38–55. Wien: Hölder-Pichler-Tempsky.

Haller, Rudolf, and Ulf Höfer, eds. 1998. *Otto Neurath. Gesammelte ökonomische, soziologische und sozialpolitische Schriften*. Wien: Hölder-Pichler-Tempsky.

Haller, Rudolf, and Robin Kinross, eds. 1991. *Otto Neurath. Gesammelte bildpädagogische Schriften*. Wien: Hölder-Pichler-Tempsky.

Haller, Rudolf, and Heiner Rutte, eds. 1981. *Otto Neurath. Gesammelte philosophische und methodologische Schriften*. Wien: Hölder-Pichler-Tempsky.

Hartmann, Frank. 2002. Bildersprache. In *Otto Neurath. Visualisierungen. Bildersprache*, ed. Frank Hartmann and Erwin K. Bauer, 15–105. Wien: WUV Universitätsverlag.

Kinross, Robin. 2013. The graphic formation of Isotype. 1925–40. In *Isotype. Design and contexts. 1925–1971*, ed. Christopher Burke, Eric Kindel, and Sue Walker, 107–177. London: Hyphen Press.

Kraeutler, Hadwig. 2008. *Otto Neurath. Museum and Exhibition Work. Spaces (Designed) for Communication*. Frankfurt: Peter Lang.

Kulemann, Peter. 1979. *Am Beispiel des Austromarxismus*. Hamburg: Junius.

Langewiesche, Dieter. 1979. *Zur Freizeit des Arbeiters*. Stuttgart: Klett-Cotta.

Mittelstraß, Jürgen. 2001. *Wissen und Grenzen. Philosophische Studien*. Frankfurt a.M.: Suhrkamp.

Nemeth, Elisabeth. 1982. Die Einheit der Planwirtschaft und die Einheit der Wissenschaft. In *Schlick und Neurath. Ein symposium*, ed. Rudolf Haller, 437–450. Wien: Rodopi.
———. 1994. Utopien für eine wissenschaftliche Sicht der Welt und des Wissens. In *Otto Neurath oder die Einheit von Wissenschaft und Gesellschaft*, ed. Elisabeth Nemeth and Paul Neurath, 97–132. Wien/Köln/Weimar: Böhlau Verlag.
Neurath, Otto. 1908/1998. Die allgemeine Einführung des volkswirtschaftlichen und staatsbürgerlichen Unterrichts. In Haller and Höfer (1998): 119–132.
———. 1910/1998. Volkswirtschaftslehre für die Allgemeinheit. In Haller and Höfer (1998): 411–470.
———. 1912/1973. The problem of the pleasure maximum. In *Empiricism and sociology*, ed. Marie Neurath and Robert S. Cohen, 113–122. Dordrecht: D. Reidel.
———. 1914/1998. Zum österreichischen Auswanderungsgesetzentwurf. In Haller and Höfer (1998): 411–470.
———. 1925a/1991. Darstellungsmethoden des Gesellschafts- u. Wirtschaftsmuseums. In Haller and Kinross (1991): 18–27.
———. 1925b/1991. Gesellschafts- und Wirtschaftsmuseum in Wien. In Haller and Kinross (1991): 1–17.
———. 1926a/1991. Aufgaben des Gesellschafts- und Wirtschaftsmuseums in Wien. In Haller and Kinross (1991): 56.
———. 1926b/1991. Bildliche Darstellung sozialer Tatbestände. In Haller and Kinross (1991): 57–62.
———. 1926c/1991. Das Gesellschafts- und Wirtschaftsmuseum auf der internationalen Städtebauaustellung in Wien. In Haller and Kinross (1991): 63–71.
———. 1926d/1991. Statistische Bildertafeln zur Sozialversicherung. In Haller and Kinross (1991): 72–77.
———. 1927/1991. Bildliche Darstellung gesellschaftlicher Tatbestände. In Haller and Kinross (1991): 118–125.
———. 1928a/1973. Personal life and class struggle. [excerpts]. In *Empiricism and sociology*, ed. Marie Neurath and Robert S. Cohen, 249–298. Dordrecht: D. Reidel.
———. 1928b/1981. Lebensgestaltung und Klassenkampf. In Haller and Rutte (1981): 227–293.
———. 1929a/1991. Bildstatistik und Arbeiterbildung. In Haller and Kinross (1991): 139–143.
———. 1929b/1991. Die neue Zeit, Köln 1932. In Haller and Kinross (1991): 133–138.
———. 1930a/1991. Das Sachbild. In Haller and Kinross (1991): 153–171.
———. 1930b/1991. Gesellschaft und Wirtschaft. In Haller and Kinross (1991): 144–149.
———. 1931a/1991. Bildstatistik nach Wiener Methode. In Haller and Kinross (1991): 180–191.
———. 1931b/1991. Das Gesellschafts- und Wirtschaftsmuseum in Wien. In Haller and Kinross (1991): 190–196.
———. 1931c/1991. Bildhafte Pädagogik im Gesellschafts- und Wirtschaftsmuseum in Wien. In Haller and Kinross (1991): 197–206.
———. 1932/1991. Bildstatistiken des Gesellschafts- und Wirtschaftsmuseums in Wien. In Haller and Kinross (1991): 210–213.
———. 1933a/1991. Bildstatistik – ein internationales Problem. In Haller and Kinross (1991): 258–264.
———. 1933b/1991. Bildstatistik nach Wiener Methode in der Schule. In Haller and Kinross (1991): 265–336.
———. 1933c/1991. Soziale Aufklärung nach Wiener Methode. In Haller and Kinross (1991): 231–239.
———. 1935/1991. Isotype und die Graphik. In Haller and Kinross (1991): 342–354.
———. 1936/1991. Internationale Bildersprache. In Haller and Kinross (1991): 355–398.
———. 1937/1991. Aufklärung über Tuberkulose durch Isotype. In Haller and Kinross (1991): 399–402.
———. 1942/1991. Isotype-Institut und Erwachsenenbildung. In Haller and Kinross (1991): 591–595.

———. 1944a/1991. Visuelle Hilfsmittel in der Erwachsenenbildung. In Haller and Kinross (1991): 596–599.

———. 1944b/1991. Visuelle Hilfsmittel und Argumentieren. In Haller and Kinross (1991): 600–617.

———. 1945/2010. *From hieroglyphics to Isotype. A visual autobiography*. London: Hyphen Press.

Neurath, Otto, and Harry Edwin Kleinschmidt. 1939/1991. Gesundheitserziehung mit Isotype. In Haller and Kinross (1991): 421–448.

Neurath, Otto, Hans Hahn, and Rudolf Carnap. 1929/1973. The scientific conception of the world: The Vienna circle. In *Empiricism and sociology*, ed. Marie Neurath and Robert S. Cohen, 299–318. Dordrecht: D. Reidel.

Petrasch, Wilhelm. 2007. *Die Wiener Urania. Von den Wurzeln der Erwachsenenbildung zum lebenslangen Lernen*. Wien/Köln/Weimar: Böhlau.

Riedl, Rupert, and Paul Parey. 1980. *Biologie der Erkenntnis*. Berlin/Hamburg: Parey.

Runggaldier, Edmund. 1979. Das Problem der 'Darstellung' bei Neurath und Carnap. In *Wittgenstein, der Wiener Kreis und der kritische Rationalismus*, ed. Hal Berghel, Adolf Hübner, and Eckehart Köhler, 245–247. Wien: Hölder-Pichler-Tempsky.

Sandner, Günther. 2014. *Otto Neurath. Eine politische Biographie*. Wien: Zsolnay.

Stadler, Friedrich. 1993. Wien – Berlin – Prag. Zum Aufstieg der wissenschaftlichen philosophie. In *Wien – Berlin – Prag. Der Aufstieg der wissenschaftlichen Philosophie*, ed. Rudolf Haller and Friedrich Stadler, 11–37. Wien: Hölder-Pichler-Tempsky.

Uebel, Thomas. 2000. *Vernunftkritik und Wissenschaft. Otto Neurath und der erste Wiener Kreis*. Wien/New York: Springer.

———. 2008. Writing a Revolution: On the Production and Early Reception of the Vienna Circle's Manifesto. *Perspectives on Science* 16: 70–102.

Vossoughian, Nader. 2008. *Otto Neurath. The Language of the Global Polis*. Rotterdam: NAi Publishers.

Weidenholzer, Josef. 1981. *Auf dem Weg zum, Neuen Menschen*. Wien: Europaverlag.

Chapter 9
Rationality and Pseudo-Rationality in Political Economy: Neurath, Mises, Weber

Thomas Uebel

Abstract In this chapter Neurath's controversial contributions to the socialist calculation debate are confronted with the criticism of two well-known opponents, Ludwig von Mises and Max Weber. Each side's arguments are considered at a certain level of abstraction so as to allow what are lasting points of significance in Neurath's proposals to shine through more clearly. It is argued that while these points are closely interwoven in their presentation with his schemes for marketless socialism, they are conceptually independent of them. Suitably so, they have proven influential in just this independent capacity.

Many philosophers are likely to consider Otto Neurath's work in economics, if they are aware of it at all, to be of specialist interest only, like his work on visual education or in museology.[1] To show that matters are otherwise, this paper will review Neurath's contribution to the socialist calculation debate and bring into focus what broadens its significance and makes it relevant to political economy generally. In doing so we cannot, however, rely on the received wisdom of mainstream economics for in this field too Neurath is controversial and remembered mainly as the advocate of a particularly radical form of socialism. What makes worthwhile our doubly uphill struggle is the view we will get from the top. That is not a detailed working out but, just as important, a clear enough outline of the fundamentals of Neurath's conception of how to think about an alternative socio-economic order (and some of the struggles it takes to reach them). This conception also, I believe, informed other members of the left Vienna Circle in their shared efforts to free philosophical thought from its fruitless preoccupation of double-guessing science as to the nature of reality and to reorient it towards the provision of tools to make its surface more humanly bearable. Reconceptualizing

[1] For work that shows, contrary to the first impressions one might have, the implications of Neurath's work in visual education and museology for his comprehensive modernist enlightenment agenda, see, e.g., Burke et al. (2013) and Groß (2015) and her chapter in this volume on the former and Kraeutler (2008) for the latter.

T. Uebel (✉)
Philosophy, School of Social Sciences, University of Manchester, Manchester, UK
e-mail: thomas.uebel@manchester.ac.uk

J. Cat, A. T. Tuboly (eds.), *Neurath Reconsidered*,
Boston Studies in the Philosophy and History of Science 336,
https://doi.org/10.1007/978-3-030-02128-3_9

significant stretches of economics was an instance of the task they chose – "to fashion intellectual tools [...] also for the daily life of all those who in some way join in working at the conscious reshaping of life" and "to rationally transform the social and economic order" (Carnap et al. 1929/2012, 81).[2]

There are two aspects of Neurath's contribution that we must consider for present purposes. Both concern the seemingly devastating criticisms directed at his proposals for the total socialization of the national economies of Germany and Austria after their defeat in World War One by Ludwig von Mises and Max Weber. Independently but along similar lines they argued that Neurath's proposals were simply unworkable in developed complex industrial societies. What was long lost in the subsequent calculation debate that turned into a wholly different direction and focused on the provision of prices in the absence of a market, was that while Mises and Weber had important points to make, so did Neurath. Nowadays this is no longer news to those ecological economists who oppose as fundamentally mistaken the methodology employed by neoclassical or Austrian economics to deal with issues of sustainability and environmental degradation with arguments pioneered by Neurath.[3] What therefore can be saved from his original argument for the ages at an abstract level is perhaps more important, certainly of greater practical relevance than what his interventions had originally intended. The implications of Neurath's insights for economic thought about human welfare are, in any case, radical and give the lie to the long-standing myth, promulgated for generations by self-styled critical theorists, that social science informed by logical empiricist sensibilities is bound to conform to "positive" fact and cannot challenge the existing order.[4]

9.1 Neurath's Argument

The socialist calculation debate was a long-running dispute over the economic foundations of socialism that stretched from the last third of the nineteenth century to the last quarter of the twentieth.[5] The possibility of economic calculation under social-

[2] Compare Carnap's remarks on "scientific humanism" and socialism in his intellectual autobiography (1963, 82–4). Neurath's views on economics do not appear to have had much influence outside of the left wing, not even on Karl Menger, the son of the originator of Austrian economics, who edited, much to the displeasure of that tradition's third generation, a second expanded edition of his father's foundational text (1923).

[3] For discussions of Neurath's role in the socialist calculation debate and the development of ecological economics see, the pioneering work of Martinez-Alier (1987/1990, 1995) and since then O'Neill (1996, 1998, 2004, 2006), Uebel (2005, 2007a, 2008, 2018), O'Neill and Uebel (2015); for an overview of the development of Neurath's own thought in and on economics, see Uebel (2004) and of logical empiricist philosophy of social science in general, Uebel (2007b).

[4] See Dahms (1994) and (1997) with reference particularly to Horkheimer (1937/1972) and Adorno et al (1969/1976). For a critical assessment of Horkheimer's original charge see O'Neill and Uebel (2004).

[5] For accounts of the debate see Lavoie (1985) and Steele (1992), albeit with an emphasis on the later stages when the debate had moved to Anglophone publications. For supplementation with regard to the Austrian debate in the 1920s, see Chaloupek (1990).

ism, of accounting for the best possible use of given resources, was put into question already in Albert Schäffle's *Die Quintessenz des Sozialismus* (which also warned of reduced productivity due to motivation deficits and infringements of the sovereignty of labor and consumption). With the labor theory of value unable to allow for the proper mediation of supply and demand and no replacement in sight, supply and demand "would fall into a hopeless quantitative and qualitative discrepancy" (1875/1892, 87).[6] Schäffle's book – large parts of which set out to clarify the socialist doctrines against their distortion by opponents – was widely discussed at the time.[7] Leading theorists of German Social Democracy tended to ignore his criticism under the guise of rejecting utopianism, like Karl Kautsky who continued to merely gesture towards the unspecified use of labor time units to calculate the cost and benefit of production and regulate its distribution "on the morrow of the revolution" in (1902/1907).[8] By contrast, Neurath, as we shall see, took the bull by the horns.[9] While the goal of immediate implementation of market- and moneyless economy was also announced by the Bolsheviks in the years of "war communism" but not systematically implemented and later discarded, it was Neurath's writings in 1919 on how new forms of planned administrative economies were about to emerge from the German and Austro-Hungarian war experience that prompted Mises' and Weber's interventions (which provoked many more responses in turn).[10]

Neurath's motivating reasoning possessed undeniable force and was effectively presented in numerous pamphlets. During World War One planned production for perceived need was seen to have replaced production for profit alone; what if the goals set by war aims were replaced by the goal of satisfying the needs of the population in peacetime?

> Let us ask first: is this economy as efficient, as "productive" as it could be in light of the available raw materials and energies? That it is not! The supporters of the free economy do not even in principle want to produce as much as possible. They rather want to earn as much as possible, according to the declared principle of this form of economy, they want to achieve maximum "net profit". [...] Moreover we ask whether the free market economy is

[6] Schäffle's criticism concerned the practice of socialism and so differed from the critique of the consistency of Marx's system launched by Böhm-Bawerk (1896/1949), a former student of his.

[7] According to an obituary in *Berliner Volkszeitung* (quoted in Small 1904) one banker bought 10,000 copies of *Quintessenz* for distribution amongst the educated public to stimulate discussion. Schäffle, it may be added, was not an apologist for the existing capitalist order: the second edition of his *Bau und Leben des sozialen Körpers* (1896) outlined numerous measures of state intervention to ameliorate "the social problem" that were curtly dismissed by Mises (1925) in the course of his review of Cohn (1920) who also showed sympathies for Neurath's position; for an informed appreciation of Schäffle's work, see Hodgson (2013).

[8] Kautsky's (often reprinted) Delft address stirred Nicholaas Pierson, an economist and former Dutch prime minister, to publish a riposte (1902/1935) which was lionized some 30 years later in Hayek (1935) but until then had remained untranslated and so was much less known than Schäffle's three main criticisms which soon became standard warhorses.

[9] For remarks on Neurath's "scientific utopianism" (see his 1919b/1973) in different contexts, see Uebel (1996) and (2008).

[10] Prior to the publication of Neurath's socialization plans Mises' and Weber's arguments against socialism lacked the argument(s) they were deploying against Neurath; see Uebel (2007a).

> as economical as an economy should be in the light of the limited resources of the earth and our limited lifespans – and as it could be given at least the same output. That it is not! In the first place, it wastes raw materials. [...] We do not even stop to think whether we should conserve materials for the sake of future generations, but let the decision be made by the desire for net profit on the part of a few of our contemporaries. Secondly, the free market economy also wastes human powers. [...] But machines would be "more expensive", human labor "cheaper", so net profit decides that humans will suffer, fall ill and die early for the entrepreneur would be less well-off with machines. Thus prompted we ask, thirdly, whether the free market economy is rational, that means, whether it employs technological inventions wherever possible, whether it produces the greatest yield for the least effort. [...] It is rational only to that degree precisely which is conducive for net profits. (Neurath and Schumann 1919, 13–18.)

Neurath thus argued that the dysfunctional "anarchy" of the market could be overcome by a "planned economy" that would be organized along entirely different principles. According to his proposal for a marketless socialism, not only power relations with regard to economic decisions had to change from the existing order but also the very way economic decisions were being taken. By rejecting the free market one rejects the organizing principle of production for profit and thus the "rule of money." Neurath's "economy in kind" (or economy *in natura*) required a "calculation in kind" (calculation *in natura*) so as to develop an appropriate "economic plan." In this way alone, Neurath argued, was it possible to realize an economy the direction of which was determined by planned social need satisfaction.

Let us clarify Neurath's terms. "*Economy in kind*" means "a large-scale economy in kind, [...] a socialized economy, [where] money no longer is a driving force. No longer is there a 'net profit' for which production occurs. Money could remain at best as a token for a claim on all sorts of goods and services which the individual consumer is given to enable him to arrange his consumption" (Neurath 1919a/1973, 145). "*Calculation in kind*" means "there are no units that can be used as the basis of such a [production] decision, neither units of money nor hours of work. One must directly judge the desirability of the two possibilities" (1919a/1973, 145). Needless to say, a great challenge hides in this conception. But Neurath was adamant:

> To many it seems impossible to proceed in this manner, and yet it is only in this field that we are not used to it. For even in the past one has not started from units of teaching or sickness in order to decide whether new schools or hospitals should be built; rather one directly set over against one another, even if only at general outlines, the totality of changes caused by schools and those caused by hospitals. (Neurath 1919a/1973, 146.)

As we can see, it was the incommensurability of the values involved (health, education, etc.) that called for both calculation in kind and economy in kind. What was rejected was not just calculation in units of specific goods or services but any universally applicable unit of a means of measurement. Calculation in kind had to be developed more or less from scratch. The replacement of "capitalist profit calculation" with "socialist utility calculation" by an economic plan required the development of a new and comprehensively organized form of statistics in kind. It had to assess, on the one hand, the social need in food, housing, clothing, health and education provision, etc., *in natura* and, on the other hand, it had to assess, again in specific quantities, the available or required raw materials and machinery etc., as well as the amounts of labor it would take to transform the raw materials into the desired goods, etc. On the basis of such a "*universal statistics*" (Neurath's term) a variety of economic plans were then to be drawn up which specified what social needs could be satisfied at what cost in terms of other social needs remaining unsatisfied etc. It was then up to the populace to choose, either directly or through its representatives, which of the plans was to be implemented as that which satisfied the most pressing social needs etc.

9.2 Mises' Response and Neurath's Rejoinders

Mises pointed out that use values were subjective and did not allow for an objective measure of the economic efficiency of actions (or at least not one that was intersubjectively intelligible) and so did not allow for the appropriate coordination of individuals' actions. Such a measure was only afforded by money prices for goods and services. Since he also assumed it to be the essence of economic thinking to maximize the utility of expended effort, Mises is easily seen to have argued that without a commensurate cardinal measure of value like money, objective economic calculation was impossible. Since socialism precluded markets and money, no such calculation was possible.

Against this it is immediately and rightly objected that it is often possible to make rational decisions about how to direct one's expenditure of effort even if no exact commensurate measure is available but only comparative assessments of the satisfaction of incommensurable values can be attempted, however vague. Against this counter, Mises had no argument but he had never intended to dispute the point. Mises' argument was rather that the distinction between lower- and higher -order goods, between consumption and production goods, is of crucial importance. (Consumption goods can be consumed right away, production goods are inputs into processes making consumption goods.) Concerning lower-order, consumption goods it was indeed possible to arrive at a reasoned judgment concerning the merit of their relative utility by ranking and comparing them with the effort required for their procurement without reference to a neutral measure like money. Mises' point concerned higher-order, production goods: modern capitalist economies are simply too complex to allow the economical use of production goods to be assessed in terms directly related to the use-values of consumption goods.

> Calculation *in natura*, in an economy without exchange, can embrace consumption-goods only; it completely fails when it comes to deal with goods of a higher order. And as soon as one gives up the conception of freely established monetary price for goods of a higher order, rational production becomes completely impossible. Every step that takes us away from private ownership of the means of production and from the use of money also takes us away from rational economics. (Mises 1920/1935, 104; cf. Mises 1922/1951, 119.)

Once the relevance of the distinction consumption/production goods is observed, it seems that Neurath's plans can only make sense within the circumference of a directly surveyable household economy. Rational calculation in modern capitalist economies required the medium of a universally commensurate measure: money. Since without a market, it is meaningless to speak of prices, what is needed therefore is a market in production goods – just what socialism precludes.

By the end of the year in which Mises' counter appeared, Neurath had published his plans in numerous other places.[11] His response to Mises' objection seemed to remain opaque for a long time. Neurath reiterated that it would be possible to assess the efficiency of production goods in non-monetary, namely "technical" terms, by a comparative statistics establishing average yields, labor expenditure and production times for given quantities of goods. And he also pointed again to areas in which in-kind calculation was long in use in the administration of public health, security and education. But did these sectors face the very same problems a moneyless economy faced? To start with, public health, security and education were clearly demarcated areas reserved for the state to deal with; as long as they remained embedded within a market economy, the means to achieve their goals carried a price it was thought worthwhile to be paid. Whether this in-kind reasoning could be extended to the whole economy was just the issue.[12]

So what did Neurath's "official" argument establish? That money calculation was not necessary for economic rationality, that no market was needed at all, does not follow from the mere fact that in some cases production decisions are undertaken for social need, not profit. Call the idea that the rational employment of production goods could be ensured entirely by in-kind considerations the "*strong in-kind calculability assumption*." It says that alternative uses of production goods can be assessed as fully as is required for rational decision making by quantitative in-kind labor and production technology statistics – *money calculation is not necessary for any rational economic decision making*. Did Neurath have any good argument for the strong in-kind calculability assumption? It would appear that, at best, his universal statistics was an ongoing research program.

But this was not Neurath's only argument. In the same year which saw publication of his *Wirtschaftsplan und Naturalrechnung* (*Economic Plan and Calculation in Kind*, 1925a/2004), a remarkable small book that combined a restatement of his conception of a socialized economy with some not wholly convincing asides against Mises, an equally contentious Marx-exegesis and a very moving humanist pathos, Neurath also published a brief article in the theoretical journal of the Austro-Marxists which was either missed or disregarded by his critics, but which spelt out in greater detail his fundamental reasoning.[13]

[11] See, e.g., besides Neurath and Schumann (1919) also Neurath (1919, 1920a/2004, 1920b/2004).

[12] That Neurath (1925a/2004, 429) was able to point to a passage where Mises (1922/1951, 389) himself adverted to concept of wealth divorced from money and prices does not establish this point either.

[13] Incidentally, neither Neurath's reading of Marx nor his socialization plans remained uncontroversial among Austro-Marxists: see, e.g., Helene Bauer (1923), Käthe Leichter (1923a) and Otto Leichter (1923b).

> The socialist economy [...] is concerned with "utility", with the interest of the social whole and the welfare of all of its members with regard to housing, food, clothing, health, entertainment, etc. To this end it seeks to employ the given sources of raw materials, the extant machines and labor power etc. Right at the start it must be determined what this is, the "interest of the social whole". [...] One has to find the best way to achieve a non-wasteful exploitation of [natural resources], to ensure the health of the next generation, etc. Now how can this "best use" be calculated in a socialist economic order? For a socialist calculation there does not exist a unit of the sort which capitalism finds in "money". (Neurath 1925b/2004, 468.)

To convince us that commensuration by a single unit will not work, Neurath asked three rhetorical questions: "Some had the idea to introduce a certain amount of labor as a unit. But how could this make it possible for the excessive exploitation of a coalmine to figure as a negative entry in the balance? How could a quantity of electricity which a river provides us with be entered as an increase in amounts of labour units? Or the increase in wind power used in the running wind mills?" (Neurath 1925b/2004, 468.) Neurath addressed all candidates for that single commensurating unit: any such candidate is meant to be ruled out.

Consider how ecological incommensurabilities figures here: "the excessive exploitation of a coal mine" involves the "intergenerational context" with the crucial problem of having to determine "the rate of discount needed to weigh future costs and benefits". Inevitably, "[w]e need a political decision, therefore, on the rate of discount and the time-horizon" (Martinez-Alier 1987/1990, xxi). No objective facts are available that could determine this rate of discount. Inasmuch as the second and third examples trade on assessing present-day but counterfactual costs and benefits, the argument seems weaker: why could "shadow-pricing" or "contingent valuation" (two terms for the monetization of non-monetary goods used by environmental economics) not work here? The reasoning is similar: here too shadow-pricing "from above" requires a political decision about the value of the type of labor at issue, and, more generally, a hierarchy of needs. However, such an assignment of value to labor is neither objectively determined nor is there anything timeless about the hierarchy of needs. On the other hand, contingent valuations "from below" are subject to distorting framing effects and differential resource constraints, and so also cannot count as objective.

Neurath drew the obvious conclusion: the automaticism of the market does not work here. Indeed the point can be generalized: against the claim that monetary algorithms could overrule incommensurability stands the realization that they would require a prior non-algorithmic weighing up of competing needs etc. and their determination in monetary values. It is this thought that justified for Neurath the view that socialist calculation is moneyless: socialist economic plans are designed by multi-criterial evaluation. So when it came to showing that the very logic of capitalism consists in the wrongful absolutization of the profit motive as definitive of economic action, Neurath invoked ecological incommensurability. Call this reason the *'weak in-kind calculability assumption'*: *monetary calculation is not sufficient in all cases for rational economic decisions.*[14] So Neurath's best response to Mises' counter was all along that in-kind calculation was unavoidable in any responsible economic decision making concerning non-renewable resources, the allocation of exhaustible resources – whether a market remained in place or not. On pain of failing this ecological dimension of economic reasoning altogether, economic rationality was bound to do without or go beyond calculation with monetary units – even when it came to production goods.

9.3 The Neurath-Mises Debate Reassessed (with Hayek in the Mix)

Mises never engaged with Neurath's incommensurability argument. His defenders may point out that he too rejected the idea of assigning money prices to extra-economic values. He wrote that monetary calculation "can never obtain as a measure for the calculation of those value-determining elements which stand outside the domain of exchange transactions," like the "beauty of the waterfall which the scheme [for erecting a waterworks] might impair" (Mises 1920/1935, 98–9). Of course, one could try to capture some of the value of the beauty of the waterfall in terms of the monetary value of "the diminution of tourist traffic or similar changes" its removal would bring about (ibid.), but that would not be valuing its aesthetic appeal. Instead these extra-economic values can "be embraced straightaway within the ambit of our judgement of values" because "all those ideal goods are goods of a lower order." So when Mises went on to declare: "Once we see clearly how highly we value beauty, health, honor and pride, surely nothing can prevent us from paying a corresponding regard to them" (1920/1935, 100), he suggested that comparative in-kind valuations will determine for us personally the opportunity cost we would be willing to incur so as to enjoy the extra-economic values in question. Mises's waterfall example illustrates "extra-economic' elements" that are "not substitutable against each other on the market and therefore do not enter into exchange-relationships" (1920/1935, 99).

But this admission does not work against Neurath's ecological argument for that deals in the impact the ecological gain or loss has on the proper deployment of *production goods*. To allow for rational calculation concerning environmental goods at all Mises must insist that they be brought into the market more directly and for real (as by the assignment and enforcement of property rights over them in "free market environmentalism").[15] Since Austrian economists reject the equilibrium assumptions of neoclassical economics, make-believe markets can never replace real markets and non-market valuations via shadow-pricing etc. can never play the

[14] Note that this argument was already implicit in the passage quoted above from Neurath and Schumann (1919).

[15] As latter-day proponents of Austrian economics indeed suggest: see Cordato (1992) or Sagoff (2008).

indispensible signaling role of prices. This strategy remains barred for Mises also at the level of production goods. (The same holds for Hayek, who also overlooked Neurath's ecological argument.) In sum: Neurath's ecological incommensurability argument *supports* his "*weak in-kind calculability assumption*" by establishing the insufficiency of capitalist monetary calculation for purposes of economic rationality.

What about Neurath's "*strong in-kind calculability assumption*?" For Mises, at least as he is naturally read, the central point is that rationality in complex economies requires commensuration of all values onto a common one. The possibility of just that, of course, is what is denied by Neurath's ecological argument: rationality is not renounced just because we have to make do with merely partial orderings of alternatives, moreover, with a plurality of those. What was a presupposition of rational action for Mises was for Neurath but a recipe for "pseudo-rationalism."[16] The very determinacy of action that Mises appeared to insist on, was anathema to Neurath in that it set an illusory standard that was rarely if ever met. This is a good counter-charge, but does it save the *strong in-kind calculability assumption*? Here Hayek's development of Mises' argument becomes important.[17]

Hayek stressed that comprehensive rational planning requires the concentration of what is dispersed and often only tacit knowledge amongst many agents in the head of the planner and that this is an unrealistic assumption.

> In a centrally planned society the selection of the most appropriate among the known technical methods will be possible only if all that knowledge can be used in the calculations of the central authority. This means in practice that this knowledge will have to be concentrated in the heads of one or at best a very few people who actually formulate the equations to be worked out. It is hardly necessary to emphasize that this is an absurd idea even in so far as that knowledge is concerned which can properly be said to 'exist' at any one time. (Hayek 1935/1948, 155; cf. Hayek 1940/1948, 202.)

> To assume all the knowledge to be given to a single mind in the same manner in which we assume it to be given to us as the explaining economists is to assume the problem away and to disregard everything that is important and significant in the real world. (Hayek 1945/1948, 91.)

This "information problem" also refutes Neurath's appeal to the future "universal statistics" in the light of which planning decisions were to be made. With it falls his strong in-kind calculability assumption and his hope that for rational economic calculation it is not even necessary to have money calculation.[18] But Hayek's argument does not overcome the weak in-kind calculability assumption.

---

[16] For Neurath's argument against pseudo-rationalism in various guises see (1913/1973) and (1935b/1983).

[17] For discussion of another methodological difference between Mises's and Hayek's argument that matters for empirical social science, see Uebel (2017).

[18] Note that this conclusion is established only as far as Neurath's proposals for calculation in kind are concerned. Whether Hayek's argument holds against more recent elaborations of suggestions first made by Oskar Lange that computing machines should be able to effect the calculations needed, namely that V. Kantorovich's method of linear programming be employed for the task (Cockshott 2008), or that Hayek's own argument has limitations that speak against the across-the-board applicability that it is often credited with (O'Neill 2012), cannot be discussed here.

What can be concluded as a result so far then is this. Neurath claimed that monetary calculation (not only private ownership of the means of production) was *not necessary* in order to run a complex economy. Mises countered that monetary calculation (and private ownership etc.) was not only necessary but *sufficient* for running a complex economy. But both Neurath and Mises seem to be mistaken. Monetary calculation and markets *are necessary but not sufficient* for running a complex economy in a "rational" fashion.

9.4 Weber vs. Neurath

Now let us turn to the debate between Neurath and Max Weber.[19] This will highlight an important feature of Neurath's perspective that complements his partial success over Mises. That is the claim that not only is *monetary calculation not sufficient* for a well-ordered economy, but also that *calculation in kind* is *necessary* for such a well-ordered economy.

Recall that calculation in kind considers how perceived needs of a population in real terms (food, housing) could be met by determining the cost of production in terms of input quantities of material labor, and also provides comparisons of alternative uses in terms of quantities of different outputs. Calculation in kind is essential for an economy in kind. Unlike monetary calculation in a market economy which guides, even directs investment decisions by means of a cost-benefit analysis and the criterion of maximum profit, calculation in kind *provides the data* (by a material input-output analysis of production possibilities) for decisions about the allocation of resources in an economy in kind. But there is also a still further dimension of calculation in kind: that it provides criteria for assessing economic outcomes. Calculation in kind not only deals in quantities of raw materials and products and so requires inventories thereof, but also deals in all aspects of human "life conditions" and so requires "inventories of standards of living." Neurath (1931/1973, 401) mentioned "shelter, food, clothing, health, books, theatres, friendly human surroundings, all this belongs to the conditions of life, even the quantity of malaria germs that menace […].")[20] It is this further use of calculation in kind that becomes of great importance in the debate with Weber.

Weber's criticism of Neurath paralleled Mises' only in part: Weber recognized a dimension of Neurath's argument that Mises missed altogether. Weber (1921/1978, 101) shared Mises' overall conclusion, in his own words: "the comparison of different kinds of processes of production, with the use of different kinds of raw materials and different ways of treating them, is carried out today by making a calculation of comparative profitability in terms of money costs. For accounting in kind, on the other hand, there are formidable problems involved here which are incapable of

[19] Some of this debate is virtual, as we'll see, as Weber died in 1920. For some live interaction between them see Cartwright, Cat, Fleck, Uebel (1996, 54) and Neurath (1910b/2004, 295).

[20] For elaboration see Neurath (1917/2004) and (1937/2004).

objective solution." Weber stressed particularly that calculation in kind was unable to deal with what had become known as the "imputation" problem: how to assign differential value to individual parts of the production process. Thus Weber (1921/1978, 103) concluded: "We cannot speak of a rational 'planned economy' so long as in this decisive aspect we have no instrument for elaborating a rational 'plan.'"

Weber and Mises strongly differed, however, in recognizing different kinds of rationalities. Unlike Mises who only recognized instrumentally rational action, Weber recognized four types of social action: instrumentally rational, value-rational, affective and traditional.[21] "Instrumentally rational" (*zweckrational*) action is determined by the purpose, the means and consequences of acting, with satisfaction of one's goals of overriding importance. "Value-rational" (*wertrational*) action is determined by belief in the unconditional intrinsic value of certain behaviors, independently of any instrumental success attained.[22] "Affective" action is determined by emotions and feelings and "traditional" action is determined by lived custom or habit. Closely related to these four types of social action are two types rationality of economic action.

> The term "*formal rationality* of economic action" will be used to designate the extent of quantitative calculation or accounting which is technically possible and which is actually applied. The "*substantive rationality*", on the other hand, is the degree to which the provisioning of given groups or persons (no matter how delimited) with goods is shaped by economically oriented social action *under some criterion* (past, present or potential) *of ultimate values*, regardless of the nature of these ends. (Weber 1921/1978, 63; emphases added.)

Note that what is substantively rational in the specific context of economic action responds to what is value-rational in the general context of social action – resulting in a mixed form. The substantive rationality of economic action represents one of the mixed forms of instrumental rationality Weber allowed for: here pure means-ends reasoning is modified by the recognition of intrinsic values that need to be observed.

With these categories in hand consider Weber's charge that one "cannot speak of a rational 'planned economy'" as long as the imputation problem has not been addressed and solved. What follows is that what renders calculation in kind deficient – less than fully "rational" – as an instrument of economic action is *not* that it is not pure instrumental rationality. Given "both calculation in kind and in money are rational techniques," calculation in kind fails because it falls short in its *formal* prowess. Now this is not a minor problem: the "fact that the problem of imputation

---

[21] Mises (1929/1960, 85) wrote that "everything that we regard as human action […] is instrumentally rational: it chooses between given possibilities in order to attain the most ardently desired goal." In this quotation "instrumentally" is restored from the original German: Mises evidently considered it universal and exclusive and explicitly rejected Weber's distinctions.

[22] Note that the goals and means of an instrumentally analyzed situation could be subjected to consideration from the perspective of value-rationality and the decision made on purely instrumental grounds could be modified – but from instrumentalist perspective, Weber (1921/1978, 26) noted, "value-rationality is always irrational."

of the part contributed to the total output of an economic unit by the different factors of production and by different executive decisions is not capable of the kind of solution which is at present attained by calculations of profitability in terms of money" is *fatal* for Weber (1921/1978, 104). Calculation in kind fails to account for "precisely the process of provision for mass demand by mass production" which is central for modern industrial economies.

Importantly, however, Weber (1921/1978, 105) conceded that formal rationality has shortcomings too. "It is naturally entirely correct that mere money accounts [...] tell us nothing whatever about the nature of the real provision of a given group with what it needs; namely, real articles of consumption." Weber thus concluded that in this "complete indifference of even the formally most perfect rationality of capital accounting towards all substantive postulates, an indifference which is absolute if the market is perfectly free," lies "the *ultimate limitation*, inherent in its very structure, of the rationality of monetary economic calculation" (1921/1978, 108; emphasis added). Weber went even further and issued what can only be regarded as a tragic diagnosis of modernity. "Substantive and formal (in the sense of *exact* calculation) rationality are, it should be stated again, after all largely distinct problems. This *fundamental and, in the last analysis, unavoidable element of irrationality in economic systems* is one of the important sources of all 'social' problems, and above all, of the problems of socialism" (Weber 1921/1978, 111; emphasis added). This is surely a stunning twist to Weber's argumentation: having declared a method of economic calculation lacking in rationality that fails to solve the imputation problem, he now implicated the very method that can solve it in an irrationality that apparently pervades economics as whole.

9.5 A Neurathian Response

A defender of Neurath can reject Weber's parallel between monetary and in-kind calculation (that both aim for a kind of profit). True, both aim to determine what are better and worse investment decisions. But there are differences great enough to discount any talk of a straightforward analogy: monetary cost-benefit calculation will offer an optimal solution, it provides an algorithm, an automatic decision procedure. By contrast, calculation in kind only produces proposals for what could be done with given resources for different ends or for how different possible resources could be employed in pursuit of a given end. *Calculation in kind was an aid for deliberation, not a determinative algorithm.* Since calculation in kind was never meant as a tool for formal economic rationality as Weber understood it, calling it inadequate as formal rationality is to miss its very point. This is not, of course, to say that it is adequate as formal rationality in Weber's sense: at least in market economies the imputation problem remains. But noting the point missed by Weber allows bringing into focus what it is that calculation in kind does. And that is precisely what monetary calculation cannot do: *serve considerations of substantive rationality.* So while Weber claimed its insufficiency for implementing in full one type of economic rationality, Neurath can stress its indispensability for another type.

In other words, whereas Weber charged that calculation in kind was *insufficient* for formal rationality (incapable of achieving the formal rationality of economic action required in modern economies), Neurath claimed that it was *necessary* for assessing substantive rationality (required to establish the substantive rationality of economic action). Three things are notable about this response. First, that Weber could not really have disagreed; second, that Neurath not only could have answered as envisaged, but actually did so; third, that the Neurathian response did not go unnoticed either (his influence on ecological economics worked largely via K.W. Kapp – but that's another story).[23] To go straight to the second point. Neurath's *Was bedeutet rationale Wirtschaftsbetrachtung?* ("What is Meant by Rational Economic Theory?") signals by its title his (re-)engagement with Mises' and Weber's old charge. But readers are not told so. Instead Neurath wrote that he decided, "in the interest of calm and tranquility" to forego the "discussion of particular theses by particular authors" (1935a/1987, 71): the title is the only, still rather oblique reference to the opposition he addressed.[24] Like his better response to Mises, his considered reply to Weber found itself handicapped, coming some 15 years after the original exchange, by a distinct lack of visibility.

To be sure, readers expecting to find an explanation of how calculation in kind can deal, after all, with the problem of imputation and capital accounting that Mises and Weber claimed it could not, are bound to be disappointed. That was precisely the kind of problem Neurath did *not* solve. Instead, not for the first time, Neurath pressed the point that the basic categories of economic thought require rethinking. Note also that his title speaks of "Wirtschafts*betrachtung*" – not "*Theorie* der Wirtschaft": his expression designates more of a reflective evaluative assessment of an economy than a set of universal conditionals, let alone of mathematical equations. (Likewise, when he spoke of "*Erfolgsbetrachtung*" he did not mean a "theory of success" but a consideration of what makes for success.) So when Neurath asked "*wie man eine logisch konsequente Wirtschaftsbetrachtung aufbauen kann*," he did not ask "how to construct a consistent economic theory" but rather: how can we develop a logically rigorous way of rationally considering an economy? Everything now depends on what is considered "rational" – and Neurath was fully aware of the contested nature of term.

Neurath began his paper by noting one agreement between defenders and detractors of the free market: "that we can ascribe to certain institutions an influence on human living standards; and this already seems to give us the logical foundation we need for a comparative *Wirtschaftsbetrachtung*" (1935a/1987, 67–8). By contrast, he bemoaned the commensurabilist consensus shared even by fellow socialists and he rebuked them for failing to clarify "that this widely touted monetary calculation cannot be advocated on theoretical grounds as a form of economic calculation, even if it may have to be used in practice as a socio-technical expedient" (1935a/1987,

[23] For elaboration of the third point, see O'Neill and Uebel (2015) and Uebel (2018).

[24] Note, incidentally, that the talk of "rationality" with regard to economic calculation was initiated by Mises and Weber and that Neurath here reacted to it.

68).[25] His aim by contrast was "to show that the *Erfolgsbetrachtung* relevant to our field cannot be constructed on the basis of a single unit." What then was the criterion of "success" that Neurath used? It was the institution's "influence on human living standards." Rather than aim for an imaginary convergent value of assessments of components of conditions of life Neurath insisted to speak of "silhouettes" of conditions of life or standards of living to which each component made a distinctive and discernible contribution. Different silhouettes, representing the projected outcome of alternative economic plans, could be compared with regard to individual components but had to be assessed holistically.[26]

Consider now Neurath's position on the imputation problem (which so bothered Weber and Mises):

> The *calculations in kind* of an in-kind macro-economy proceed from the economy as a whole, and the *economic value* of an individual concern can be determined only within the framework of an economic consideration of the whole. If we want to find out whether a certain way of managing a concern would be preferable to another, we must examine ultimately how it would change the output of the whole economy. Not so in a monetary economy. Here the balance sheet of each concern is autonomous but it states something only about the profits and *nothing at all about economic value*. We cannot therefore say that a money economy comes equipped with an instrument for estimating the value of individual concerns, whereas an in-kind macro-economy has to do without one. (Neurath 1935a/1987, 104–5. Terminology slightly altered; last emphasis in the original.)

Add to this that Neurath took it as established that a money economy comes unequipped with an instrument for measuring economic value for society as whole and we do not go too far in regarding Neurath's talk of "economic value" here as engaging with Weber's distinction of formal and substantive rationality. Neither do we go too far, I believe, in having Neurath say that *Wirtschaftsbetrachtung* by means of monetary calculation would miss the point. It is calculation in kind that is *necessary* to assess whether certain criteria of substantive rationality have been met – here, the raising of human living standards.

Like his 1925 journal article, Neurath's 1935 monograph strikes me as a neglected classic of the calculation debate. To be sure, there is much in it that he had com-

[25] Neurath made this comment still before he could have known of the market-socialist model of Lange (1935–36) which subsequently dominated the socialist calculation debate; whether he knew of Taylor (1929/1964) is unclear as well, but he certainly was acquainted with the German-language discussion following Mises' intervention in which virtually all contributors adopted money for its commensurating role in economic calculation, including Polanyi (1922) and (1924).

[26] Already in 1909 Neurath (1910a/2004, 294) stated: "If we want to compare the orders of life of two nations with each other, we cannot describe them as the sum of some elementary constituents and compare these individually. We cannot reach a sum by saying: more meat is eaten in the one country, fewer clothes are worn in the other. Neither do we compare the artistic achievements of architecture so as to say: this hall is more functional than that one, but less beautiful; let us add up advantages and disadvantages. In comparing two works of art we look at one as a whole and look at the other as a whole." Some 28 years later he stated about the concept of standard of living: "We cannot regard it as a weight made up of the sum of the weights of the various parts. We cannot even specifically enumerate all the things which might be counted in the standard of living. Nevertheless, it can be shown that this concept suffices for both our theory and practice" (Neurath 1937/2004, 516).

mented on before. Indeed, substantive rationality itself was what Neurath had understood economics to be concerned with all along – albeit not under this heading. As he had stated already in 1910, so in 1935 he reiterated that "we go back to the oldest tradition if we try to delimit economics as a discipline concerned with 'welfare', 'wealth', and 'happiness' as the product of certain institutions" (Neurath 1935a/1987, 71; cf. 1910b/2004, 272). But just as his earliest arguments for his socialization proposals contained an incommensurability argument that became foundational for ecological economics, so his more or less last return to the debate in print engaged with Weber's reflections about rationality in economics.[27]

The confrontation with Weber shows that the criterion by which Neurath measured the adequacy of solutions to economic problems was all along what Weber called substantive rationality. This constitutes a notable, if partial, agreement with Weber and encourages an instructive parallel between Neurath's positions in the socialist calculation debate vis-à-vis Mises and Weber. Just as Neurath's argument against Mises' and Hayek's market fundamentalism succeeds *independently of his argument for marketless socialism*, so does his argument for calculation in kind. *Calculation in kind* is *necessary* to assess whether in running an economy certain criteria of substantive rationality have been met. This complements his argument that *monetary calculation* is *not sufficient* for running a well-ordered economy. Both points are surely of major methodological significance for political economy. That these points are nowadays discussed mainly in the debates between ecological and environmental economists does not mean that their relevance is limited to these applications: they generalize to all of welfare economics. Complementing each other, Weber's and Neurath's concepts of *substantive rationality* and *calculation in kind* provide a framework for attempts to enrich the vocabulary and concern of welfare economics generally. On just this count Neurath's approach joins forces, perhaps unexpectedly, with the campaign waged by Amartya Sen and Vivian Walsh and Hilary Putnam, albeit in its substantive methodological, not its rhetorical respects.[28]

9.6 Conclusion

Specifically with regard to Neurath we can conclude that neither Mises nor Weber show his alternative approach to economic calculation irrational *in toto*. They are correct that marketless socialism is highly problematic, at least given the

[27] In very his last monograph Neurath (1944/1970, 40) wryly commented on his views of monetary calculation and calculation in kind being regarded as "left deviation" in the Soviet Union.

[28] For discussion of the substantive convergence see O'Neill and Uebel (2008) and a highly interesting comparison see Lessmann (2007). Needless to say, still beyond Putnam and Walsh's anti-positivist rhetoric there are some substantive disagreements about Weberian methodology and meta-ethical matters that remain which must, however, be resolved on another occasion.

arguments provided by Neurath. But Mises was and his followers are wrong in holding monetary calculation and market rationality to be sufficient for a well-ordered economy. Indeed, Weber himself appears to have suspected this by detecting a paradox at the heart of formal economic rationality. Neurath's own concept of calculation in kind – however much in need of further elaboration[29] – provides a common denomination for the tools required to assess substantive economic rationality when environmental or common welfare provision issues are at stake. To be sure, the limitations of calculation in kind are clear: it provides data, not decisions. This contrasts with the algorithmic rationality of formally rational money calculation. But this limitation also has a clear benefit for, with a view to wider political-institutional implications, we may add: whether criteria of substantive rationality are adequately met, indeed which criteria of substantive rationality are appropriate, cannot be done by algorithm but requires deliberation.

References

Adorno, Theodor et al. 1969/1976. *The Positivist Dispute in German Sociology*. London: Heinemann.

Bauer, Helene. 1923. Geld, Sozialismus und Otto Neurath. *Der Kampf* 16: 195–202.

Böhm-Bawerk, Eugen von. 1896/1949. *Karl Marx and the Close of his System*. Ed. P. M. Sweezy. New York: Augustus M. Kelley.

Burke, Christopher, Eric Kindel, and Sue Walker, eds. 2013. *Isotype. Design and Contexts 1925-1971*. London: Hyphen Press.

Carnap, Rudolf. 1963. Intellectual Autobiography. In *The Philosophy of Rudolf Carnap*, ed. Paul A. Schilpp, 3–84. LaSalle, Ill: Open Court.

Carnap, Rudolf, Hans Hahn, and Otto Neurath. 1929/2012. The Scientific World-Conception. In *Wissenschaftliche Weltauffassung. Der Wiener Kreis*, eds. Friedrich Stadler and Thomas Uebel, 75-115. Wien: Springer.

Cartwright, Nancy, Jordi Cat, Lola Fleck, and Thomas Uebel. 1996. *Otto Neurath: Philosophy Between Science and Politics*. Cambridge: Cambridge University Press.

Chaloupek, Günther K. 1990. The Austrian Debate on Economic Calculation in a Socialist Economy. *History of Political Economy* 22: 659–675.

Cockshott, Paul. 2008. *Calculation in natura from Neurath to Kantorovich*. https://pdfs.semanticscholar.org/0e3a/443d6fb314eb8b160576faa9928aa151d6fb.pdf. Accessed 17/03/2018.

Cohn, Arthur W. 1920. *Kann das Geld abgeschafft werden?* Jena: Fischer.

Cordato, Roy. 1992. *Welfare Economics and Externalities in an Open-Ended Universe: A Modern Austrian Perspective*. Dordrecht: Kluwer.

Dahms, Hans-Joachim. 1994. *Positivismusstreit. Die Auseinandersetzungen der Frankfurter Schule mit dem logischen Positivismus, dem amerikanischen Pragmatismus und dem kritischen Rationalismus*. Frankfurt a.M.: Suhrkamp.

———. 1997. Der Positivismusstreit. Ein kritischer Rückblick. In *Bausteine wissenschaftlicher Weltauffassung*, ed. Friedrich Stadler, 75–90. Wien: Springer.

Groß, Angélique. 2015. *Die Bildpädagogik Otto Neuraths. Methodische Prinzipien der Darstellung von Wissen*. Cham: Springer.

Hayek, Friedrich A., ed. 1935. *Collectivist Economic Planning. Critical Studies on the Possibilities of Socialism*. London: Routledge.

[29] For reflection on the possibilities see, e.g., Martinez-Alier et al. (1998) and Sarkar (2019).

———. 1935/1948. The Present State of the Debate. In *Individualism and Economic Order*, 148–180. Chicago: University of Chicago Press.

———. 1940/1948. Socialist Calculation: The Competitive Solution. In *Individualism and Economic Order*, 181–208. Chicago: University of Chicago Press.

———. 1945/1948. The Use of Knowledge in Society. In *Individualism and Economic Order*, 77–91. Chicago: University of Chicago Press.

Hodgson, Geoffrey. 2013. Albert Schäffe's Critique of Socialism. In *Economic Theory and Economic Thought: Essays in Honor of Ian Steedman*, ed. J. Vint, J.S. Metcalfe, H.D. Kurz, N. Salvadori, and P. Samuelson, 296–315. London: Routledge.

Horkheimer, Max. 1937/1972. The Latest Attack on Metaphysics. In *Critical Theory: Selected Essays*, ed. M. O'Connell, 132–187. New York: Continuum.

Kautsky, Karl. 1902/1907. *The Social Revolution and On the Morrow of the Revolution*. London.

Kraeutler, Hedwig. 2008. *Otto Neurath: Museum and Exhibition Work. Spaces Designed for Communication*. Frankfurt: Peter Lang.

Lange, Oskar. 1935–36. On the Economic Theory of Socialism. *Review of Economic Studies* 4 (2): 123-142.

Lavoie, Don. 1985. *Rivalry and Central Planning. The Socialist Calculation Debate Reconsidered*. Cambridge: Cambridge University Press.

Leichter, Käthe. 1923a. [Review] O. Neurath, Gildensozialismus, Klassenkampf, Vollsozialisierung, 1922. *Der Kampf* 16: 119–120.

Leichter, Otto. 1923b. *Die Wirtschaftsrechnung in der sozialistische Gesellschaft*. Wien: Wiener Volksbuchhandlung.

Lessmann, Ortrud. 2007. A Similar Line of Thought in Neurath and Sen: Interpersonal Comparability. In *Otto Neurath's Economics in Context*, ed. Elisabeth Nemeth, Stefan W. Schmitz, and Thomas Uebel, 115–130. Dordrecht: Springer.

Martinez-Alier, Juan. 1987/1990. *Ecological Economics. Energy, Environment and Society*. 2nd edition. Oxford: Blackwell/Oxford.

———. 1995. Political Ecology, Distributional Conflicts and Economic Incommensurability. *New Left Review* 21: 70–88.

Martinez-Alier, Juan, Giuseppe Munda, and John O'Neill. 1998. Weak Comparability of Values as a Foundation for Ecological Economics. *Ecological Economics* 26: 277–286.

Menger, Carl. 1923. *Grundsätze der Volkswirtschaftslehre. Aus dem Nachlass herausgegeben von Karl Menger*. Vienna: Hölder.

Mises, Ludwig von. 1920/1935. *Economic Calculation in the Socialist Commonwealth*. In Hayek (1935): 89–130.

———. 1922/1951. *Socialism: An Economic and Sociological Analysis*. Enlarged edition. London/New Haven: Jonathan Cape/Yale University Press.

———. 1925. Neue Beiträge zum Problem der sozialistischen Wirtschaftsrechnung. *Archiv für Sozialwissenschaft und Sozialpolitik* 51: 488–500.

———. 1929/1960. Sociology and History. In *Epistemological Problems of Economics*, 68–129. Princeton: Van Nostrand.

Neurath, Otto. 1910a/2004. On the Theory of Social Science. In *Economic Writings. Selections 1904-1945*, ed. Thomas Uebel and Robert S. Cohen, 265–291. Dordrecht: Kluwer.

———. 1910b/2004. Remarks on the Productivity of Money. In *Economic Writings. Selections 1904-1945*, ed. Thomas Uebel and Robert S. Cohen, 292–296. Dordrecht: Kluwer.

———. 1913/1983. The Lost Wanderers and the Auxiliary Motive (On the Psychology of Decision). In *Otto Neurath: Philosophical Papers 1913-1946*, ed. Robert S. Cohen and Marie Neurath, 1–12. Dordrecht: D. Reidel.

———. 1917/2004. The Conceptual Structure of Economics and its Foundation. In *Economic Writings. Selections 1904–1945*, ed. Thomas Uebel and Robert S. Cohen, 312–341. Dordrecht: Kluwer.

———. 1919a/1973. Character and Course of Socialisation. In *Empiricism and Sociology*, ed. Robert S. Cohen and Marie Neurath, 135–150. Dordrecht: D. Reidel.

———. 1919b/1973. Utopia as a Social Engineer's Construction. In *Empiricism and Sociology*, ed. Robert S. Cohen and Marie Neurath, 150–156. Dordrecht: Reidel.

———. 1919. *Die Sozialisierung Sachsens. Drei Vorträge*. Chemnitz: Verlag des Arbeiter- und Soldatenrats im Industriebezirk Chemnitz.

———. 1920a/2004. A System of Socialisation. In *Economic Writings. Selections 1904-1945*, ed. Thomas Uebel and Robert S. Cohen, 345–370. Dordrecht: Kluwer.

———. 1920b/2004. Total Socialisation. In *Economic Writings. Selections 1904-1945*, ed. Thomas Uebel and Robert S. Cohen, 371–404. Dordrecht: Kluwer.

———. 1925a/2004. Economic Plan and Calculation in Kind. In *Economic Writings. Selections 1904-1945*, ed. Thomas Uebel and Robert S. Cohen, 405–465. Dordrecht: Kluwer.

———. 1925b/2004. Socialist Utility Calculation and Capitalist Profit Calculation. In *Economic Writings. Selections 1904-1945*, ed. Thomas Uebel and Robert S. Cohen, 466–472. Dordrecht: Kluwer.

———. 1931/1973. Empirical Sociology. The Scientific Content of History and Political Economy. In *Empiricism and Sociology*, ed. Robert S. Cohen and Marie Neurath, 319–421. Dordrecht: D. Reidel.

———. 1935a/1987. What is Meant by Rational Economic Theory? In *Unified Science*, ed. Brian McGuinness, 67–109. Dordrecht: Kluwer.

———. 1935b/1983. Pseudorationalism of Falsification. In *Otto Neurath: Philosophical Papers 1913-1946*, ed. Marie Neurath and Robert S. Cohen, 121–131. Dordrecht: D. Reidel.

———. 1937/2004. Inventory of the Standard of Living. In *Economic Writings. Selections 1904-1945*, ed. Thomas Uebel and Robert S. Cohen, 513–527. Dordrecht: Kluwer.

———. 1944/1970. Foundations of the Social Sciences. In *Foundations of the Unity of Science. Towards an International Encyclopedia of Unified Science*. Vol. 2, 1–52. Chicago: University of Chicago.

Neurath, Otto, and Wolfgang Schumann. 1919. *Können wir heute sozialisieren? Eine Darstellung der sozialistischen Lebensordnung und ihres Werdens*. Leipzig: Klinkhardt.

O'Neill, John. 1996. Who won the Socialist Calculation Debate? *History of Political Thought* 17: 431–442.

———. 1998. *The Market. Ethics, Knowledge and Politics*. London: Routledge.

———. 2004. Ecological Economics and the Politics of Knowledge. The Debate Between Hayek and Neurath. *Cambridge Journal of Economics* 28: 431–447.

———. 2006. Knowledge, Planning and Markets. A Missing Chapter in the Socialist Calculation Debate. *Economics and Philosophy* 22: 1–24.

———. 2012. Austrian Economics and the Limits of the Market. *Cambridge Journal of Economics* 36: 1073–1090.

O'Neill, John, and Thomas Uebel. 2004. Horkheimer and Neurath: Restarting a Disrupted Debate. *European Journal of Philosophy* 12: 75–105.

———. 2008. Logical Empiricism as Critical Theory? The Debate Continues. *Analyse & Kritik* 40: 379–398.

———. 2015. Analytical Philosophy and Ecological Economics. In *Handbook of Ecological Economics*, ed. J. Martinez-Alier and R. Muradian, 48–73. Cheltenham: Elgar.

Pierson, Nicholaas. 1902/1935. *The Problem of Value in the Socialist Community*. In Hayek (1935): 41–85.

Polanyi, Karl. 1922. Sozialistische Rechnungslegung. *Archiv für Sozialwissenschaft und Sozialpolitik* 49: 377–420.

———. 1924. Die funktionelle Theorie der Gesellschaft und das Problem der sozialistischen Rechnungslegung. *Archiv für Sozialwissenschaft und Sozialpolitik* 52: 218–227.

Sagoff, Mark. 2008. *The Economy of the Earth*. 2nd ed. Cambridge: Cambridge University Press.

Sarkar, Sahotra. 2019. Deliberative Decision and Formal Multicriteria Analysis. In *A Sustainable Philosophy. The Work of Bryan Norton*, ed. S. Sahotra and B. Minteer. Cambridge: Cambridge University Press. Forthcoming.

Schäffle, Albert. 1875/1892. *The Quintessence of Socialism*. London: Sonnenschein.
———. 1896. *Bau und Leben des socialen Körpers*. 2nd ed. 2 vols. Tübingen: Laupp'sche Buchhandlung.
Small, Albion W. 1904. Albert Schäffle. *American Journal of Sociology* 9: 708–709.
Steele, Daniel. 1992. *From Marx to Mises. Post-Capitalist Society and the Challenge of Economic Calculation*. LaSalle, Ill: Open Court.
Taylor, Fred M. 1929/1964. The Guidance of Production in a Socialist State. In *On the Economic Theory of Socialism*, ed. B.E. Lippincott, 39–54. New York: McGraw-Hill.
Uebel, Thomas. 1996. The Enlightenment Ambition of Epistemic Utopianism. Otto Neurath's Theory of Science in a Historical Perspective. In *Origins of Logical Empiricism*, ed. Alan Richardson and Ronald Giere, 91–112. Minneapolis: University of Minnesota Press.
———. 2004. Neurath's Economics in Critical Context. In *Economic Writings. Selections 1904-1945*, ed. Thomas Uebel and Robert S. Cohen, 1–108. Dordrecht: Kluwer.
———. 2005. Incommensurability, Ecology and Planning. Neurath in the Socialist Calculation Debate, 1919-1928. *History of Political Economy* 37: 311–342.
———. 2007a. Neurath as an Austrian Economist. Behind the Scenes of the Early Socialist Calculation Debate. In *Otto Neurath's Economics in Context*, ed. Elisabeth Nemeth, Stefan W. Schmitz, and Thomas Uebel, 37–59. Dordrecht: Springer.
———. 2007b. Philosophy of Social Science in Early Logical Empiricism: The Case of Radical Physicalism. In *The Cambridge Companion to Logical Empiricism*, ed. Thomas Uebel and Alan Richardson, 250–277. Cambridge: Cambridge University Press.
———. 2008. Calculation in Kind and Marketless Socialism: On Otto Neurath's Utopian Economics. *European Journal of Economic Thought* 15: 475–501.
———. 2017. Values, Facts and Methodologies. A Case Study in Philosophy of Economics. In *Integrated History and Philosophy of Science*, ed. Friedrich Stadler, 93–107. Cham: Springer.
———. 2018. Calculation in Kind and Substantive Rationality: Neurath, Weber, Kapp. *History of Political Economy* 50: 289–310.
Weber, Max. 1921/1978. *Economy and Society. An Outline of Interpretive Sociology*. Ed. G. Roth and C. Wittich. Berkeley: University of California Press.

Chapter 10
Neurath's Theory of Theory Classification: History, Optics & Epistemology

Gábor Á. Zemplén

Abstract Otto Neurath's early work on the classification of systems of hypotheses in optics provided some of the key insights of Neurath's later philosophy of science. The chapter investigates how Neurath developed his theory of theory-classification in response to inconsistencies he stumbled upon while studying the historical theories. Neurath's empiricism and thoroughgoing fallibilism informed his mapping of the group of theories, locating "elementary notions" of theories and taking into account the "blurred margins" of theories. To replace false dichotomies the project provided a finer-grained analysis of theories and could be utilized to locate unconceived alternatives of a domain. The first sections discuss the close links between Neurath's optical essays, his notion of an "auxiliary motive," and his attack on pseudorationalism in the "Lost Wanderers of Descartes" paper. The last sections provide a comparison of the two essays, with an extended table of the elementary notions Neurath listed, and discuss Neurath's two-tier methodology for historical reconstruction.

10.1 Neurath's Historicism: Balancing the Empirical and the Normative

After the formative University years in Berlin the young Neurath embarked on a number of research projects as an assistant teacher at the *Neue Wiener Handelsakademie*. Of these his early economic work has received significant

G. Á. Zemplén (✉)
Department of Argumentation and Marketing Research, Institute of Business Economics, Eötvös University (ELTE), Budapest, Hungary

Morals and Science Research Group (MTA Lendület), Hungarian Academy of Sciences, Budapest, Hungary

Department of Philosophy and History of Science, Budapest University of Technology and Economics (BME), Budapest, Hungary
e-mail: zemplen@gti.elte.hu

J. Cat, A. T. Tuboly (eds.), *Neurath Reconsidered*,
Boston Studies in the Philosophy and History of Science 336,
https://doi.org/10.1007/978-3-030-02128-3_10

attention, while some, like the research on logical calculus with his second wife, Olga Hahn (on this see Jordi Cat's chapter in the present volume), or his lectures and articles on the history of optics have been less studied. Neurath's two papers discussing the history and classification of optical theories were published after his work on war economics, the papers on the pleasure maxim, and on Schröder's logic. The most important precursor to the historical study of optical theories was the paper on the auxiliary motive (the "lost wanderers" of Descartes, discussed in more detail in Sect. 10.2).

Before the outbreak of the war Neurath delivered a lecture at the Philosophical Society of the University of Vienna on 2nd March, 1914 that was printed in the yearbook of the Society ("Zur Klassifikation von Hypothesensystemen," Neurath 1916/1983). In July 1914 he finished and submitted another manuscript to the young but prestigious journal of the Berlin Society for the History of Science and Medicine, the *Archiv für die Geschichte der Naturwissenschaften und der Technik.* This paper appeared in the fifth volume of the journal as "Prinzipielles zur Geschichte der Optik" (Neurath 1915/1973). I will quote from the English translations, the longer Viennese lecture "On the Classification of Systems of Hypotheses' (With Special Reference to Optics)" printed in (Neurath and Cohen 1983: 13–31) and the latter article in the Berliner journal "On the Foundations of the History of Optics" printed in (Neurath and Cohen 1973: 101–112). I sometimes add in square brackets key notions in German, and refer to the two essays as KH and PO (see Neurath 1981, pp. 71–84 for PO and pp. 85–101 for KH).

This chapter provides an analysis of the optical essays and outlines Neurath's particular method of (partial) reconstruction of theories, one that is acutely sensitive historically and thoroughly naturalistic. The significance of these investigations for Neurath's developing views on scientific theories is easily overlooked, even though some of the important insights gained during his intense period of study often reappear in later years, providing stock examples, 'Neurathian tropes' in many key papers, closely linked to analysis of economic theories and the continuing quest for a framework for a unified science. The choice of topic and the tool-kit developed for the analysis of optical theories shows a conscious effort to construct a normative historicist agenda. The fallible and communitarian epistemological approach combines an anti-philosophical attitude rejecting a priori considerations as much as possible with an empirical stance towards the history of scientific developments, and proposes a minimal formal framework, some form of logical analysis. The approach is directed against certain undesirable social processes standing in the way of scientism, e.g. epistemically detrimental polarisation of opinions, and certain philosophical stances, like types of foundationalism.

The optical essays are intimately linked to the famous 1913 paper investigating situations in which two or more alternatives are equally rational, ("Die Verirrten Des Cartesius Und Das Auxiliarmotiv (Zur Psychologie des Entschlusses)," Neurath 1913/1983). The essay on "The Lost Wanderers of Descartes and the Auxiliary Motive (On the Psychology of Decision)" extended a classical problem (choice without preference) to the domains of ordinary decision making, and has received considerably more attention than the optical essays. The paper (henceforth VC) already used an example from the history of optics, but, as Michael Stöltzner (2000, 25) notes: "it is most surprising that in VC Neurath does not provide any example [of] how the auxiliary motive is applied within science. Rather, VC is concerned with the various aspects of the social implementation of the auxiliary motive. To my mind, the missing examples can be found in his subsequent works on the history of optics (PO, KH)."

What exactly Neurath does in these early works is not a trivial question, since the texts contain elements from many traditions and, thus, are not typical representatives of any style or field of investigation. The papers from this early period are at once celebrated early examples of theoretical pluralism (Cartwright, Cat, Fleck, and Uebel 1996) and a starting point to reconsider the role of values in science (Howard 2006), or an attempt to develop an anti-Cartesian – epistemological as opposed to methodological – programme (Mormann 1996). Rheinberger (2010) reckons the optical essays among the early examples of pre- World War One historical epistemology, outlining a method of reconstruction and interpretation in an attempt to plug epistemology into history instead of adding an historical dimension to epistemology.

I first revisit the "Lost Wanderers" paper (Sect. 10.2) and Neurath's take on pseudorationalism via endorsing a thoroughgoing fallibilism, as well as empiricism (Sect. 10.3). I investigate how Neurath developed his theory of theory-classification in response to inconsistencies he stumbled upon while studying the historical theories. His approach took into account the "blurred margins" of theories, and developed a method of locating "elementary notions" of theories. Neurath considered optics as a model discipline (Sect. 10.4), and his project was aimed at providing a finer-grained analysis of theories to replace false dichotomies. At the same time his mapping could locate unconceived alternative theories of a domain.

The last sections compare the two essays in detail. They provide an extended table of the elementary notions that Neurath discussed to show the development of Neurath's mapping of the group of theories (Sect. 10.5), and introduce Neurath's two-tier methodology of historical reconstruction (Sect. 10.6) As many themes reappear in the 1930s, some citations from the mature works of Neurath are included to show the long-lasting influence of the constrained period and topic, his attempt to map optical theories and to develop a historicist epistemology.

10.2 Pseudorationalism, the Auxiliary Motive, and False Dichotomies

In his recent review of the discussions on transient underdetermination and values in science, Justin Biddle (2013, 131) describes Neurath's auxiliary motive as "any factor that fills the gap between 'insight' and decision; some of these factors might properly be characterized as values, while others clearly are not." For Neurath, a proper attitude for rationalism is to recognize the limits of actual insight, and he develops a minimal framework where "'insight' is insufficient to determine uniquely what to believe; some other factors, which he called 'the auxiliary motive,' will inevitably play a role" (Biddle 2013, 131). As Biddle notes, Neurath's point is to emphasize the presence of an auxiliary motive, as ignoring it amounts to "pseudorationalism":

> One of the lessons that can be drawn from Neurath's work for the current discussion of "science and values" is that there are a wide range of different factors that can fill the gap between "insight" (i.e., logic, evidence, and epistemic values broadly construed) and decision making in science. These factors might be sociological in nature (e.g., in one particular scientific sub-discipline, one set of norms are typically employed for evaluating research, as opposed to some other set, which are employed in some other sub-disciplines); they might be consciously adopted ethical or political values; they might be unconsciously held subjective preferences or ideological assumptions, and so on. Some of these factors, again, are properly described as values, while others are not. (Biddle 2013, 131–132.)

An important characteristic of Neurath's early work is that it develops a fallibilist framework bracketing the question of values, but leaving room for them. The attack on "pseudorationalism" points to scenarios, where we are "lost in the forest," that is, we need practical action without proper justification of the decision: "[t]here is not the slightest reason to doubt that a great military leader like Napoleon is frequently incapable of deciding by means of reflexion exactly what he should do" (Neurath 1913/1983, 5). In these cases a rationalist has to admit that the only rational strategy is to resort to a decision-maker, and to find resolution via an auxiliary motive that might be just instinctive or even random, a substitute for button-counting [Surrogate des Knöpfeabzählens].

Neurath's naturalistic approach connects the problem of choices without preference to social, economical, or psychological considerations, acknowledging all these as potential contributing factors playing a role in theory-choice. His early anti-foundationalism tackles the problem of "logical" foundations of empirical science from a very particular point of view, that is, ultimately considering coin-tossing as (practically) rational. Theory-choice in philosophy of science is therefore seen as having much in common with real-life decisions under time-constraints, in need to be optimized for an economical use of resources. For Neurath (1913/1983, 11) the coming stage of development of the world, after the ages of "Instinct," "Authority," and "Pseudorationalism," is the dawning era of the "Auxiliary motive" (see Don Howard's chapter in the present volume).

Mormann (1996, 93–95, see also 1999) triangulated Neurath's anti-Cartesian approach to accounts of language as calculus and as universal medium, describing the position as treating humans neither as hostages of language, nor as fully able to master language. And Stöltzner (1996, 2001) provided an analysis of Neurath's paper on the auxiliary motive with an in depth study of the Machian roots of the project, and the novelties of Neurath's approach as well as their relation to the optical essays. He traced the lineage that connects Neurath's lost wanderer all the way back to Buridan's Ass, and noted that in a pragmatic sense Neurath's auxiliary motive represented a significant addition to scientific methodology:

> Thus, for Neurath, decisions between principles and elementary experiences could be made not just on the ground of economy as Mach had recommended but by any rational criterion, though ultimately through an auxiliary motive. The same holds for the comparison between already established systems of hypotheses or parts of them. Here the difference from Mach's concept of a theoretical system becomes evident. If a theory were only an economical arrangement of facts, there would be only one rational criterion deciding between two compelling theories: their respective economy.
>
> Though unspecified, the auxiliary motive is a formal procedure that can be used in such a way that mathematical hierarchies are respected as long as they are not elevated into an ontology. But if the ontology is "smeared out" throughout the system, then the comparability condition can be well met and human decisions made. (Stöltzner 2001, 110.)

The 1913 "Lost Wanderers" outlines Neurath's radical pragmatist starting point, and the optical essays work out the application for historical analysis and reconstruction of scientific theories, with policy-relevant observations. As VC had its particular target, "pseudorationalism," the optical papers KH and PO target the warlike spirit behind spurious "dichotomies" that are pervasive in the controversies of scientists and standard histories. Neurath recognized that many of the classifications based on dichotomies, like particle vs. wave theories are crude, that is, a way science is often portrayed and taught hampers scientific development. Just as the auxiliary motive extended a traditional problem of choice to the sphere of scientific theory-choice, the optical essays localize the problem in the natural sciences:

> Dichotomies, however, are not only crude intellectually, but also mostly the product of scientific pugnacity. One characterises the opponent as pungently as possible for the purpose of beating him down as forcefully as possible. At such occasions transitions are only troublesome. Thus dichotomies are a result of a warlike spirit. I do not want to examine in detail here how far dichotomies, precisely through their deficiencies, have a stimulating effect on scientific life, as pointed out by Vaihinger. Even if that were the case, they would be useful for science perhaps, but themselves unscientific. (Neurath 1916/1983, 15.)

Inconsistency in grouping theories is what struck Neurath first, as he read years before these papers were written the German translation of William Whewell's (1840) history of the inductive sciences. The author classified Descartes' theory as a first form of emission-theory but the translator Joseph Littrow (director of the Observatory in Vienna) added a note that corrected Whewell, and stated that the theory is a wave theory in a "vague way." The contradictory classification of the same theory in a single volume indicated a problem that Neurath could resolve: Whewell's reading has support in the *Dioptrics* and the *Meteora*, and Littrow probably means the *Principia philosphiae*. Yet the enigma remains, as both sides of the divide are present to some extent in the works of one individual.

To evade fiascos like this one, the ideas of the scientists should be presented in a more fine-grained manner. Instead of classification based on one symbolic property, following a (potentially false) dichotomy in ordering rival theories, Neurath proposed a less crude system, where a number of aspects, a whole set of important properties was investigated. The need to develop a theory to classify theories was also connected to problems of the public understanding of science, as talking of past theories using the corpuscle/wave and right/wrong dichotomies was widespread (and still is). To solve these problems Neurath developed a calculus for science without ontological dead weight. He provided a radical perspective in the spirit of the Vienna Circle, namely, developing a novel conception of the social role and task of (social) science, working towards the radical democratization of science, connecting science, education, and everyday life (see Ibarra and Mormann 2003, 245).

If in the "Lost Wanderers" Neurath discussed the auxiliary motive *in abstracto*, in the optical essays he did *in concreto*. The "auxiliary motive" of the VC paper allowed Neurath to criticize Cartesian foundationalism, but also to endorse the Cartesian recipe for practical action in the *Principles of Philosophy*: follow the usual laws and customs, "act energetically even if insight is insufficient," and change yourself "rather than the world – a view which is, on the whole, of a stoical character" (Neurath 1913/1983, 2; see also Stöltzner 1996, 114; and Koterski 2018).

In these essays Neurath fused a probabilistic and mathematical approach with a praxis-centered one concerning the psychology of decision, and introduced an extra-logical motive relevant for real-life decision-processes. And he worked himself into a field, where the individual agendas of researchers can be studied, and a procedure for theory-classification can be developed and tested. The application mapped disunity in the field, a set of competing driving ideas, combinable in many forms, and showed that historically only some combinations were actualized.

Neurath's continuing fight against pseudorationalism went hand in hand with his communitarian epistemological perspective to help underdogs, the theories not unambigously classified in the primitive system [Übergangsanschauung] and that are generally appreciated less. The first exemplar of the field, the "Cartesian optical theory" already displayed the problem of "mixed theories" [Mischtheorien], and Neurath (1916/1983, 14) saw here a general character of science, as we "frequently encounter situations like this." He did not condemn Descartes, and took ambiguity to be a characteristic of other fruitful theories, like Newton's.

Before investigating in detail the analysis of elementary notions of optical theories, first Neurath's theoretical position is revisited. Rejecting the Cartesian tradition, the only sane choice for Neurath was anti-foundationalism:

> Whoever wants to create a world-view or a scientific system must operate with doubtful premises. Each attempt to create world-picture by starting from a tabula rasa and making a series of statements which are recognised as definitively true, is necessarily full of trickeries. [...] We can vary the world of concepts present in us, but we cannot discard it. Each attempt to renew it from the bottom up is by its very nature a child of the concepts at hand. (Neurath 1913/1983, 3.)

But how to build an empiricist project in the time-constrained world of researchers acknowledging the plurality of views typical of "science in the making," and also the interpretative difficulties of theory-reconstruction? His essays on the history of optics construct the outlines of a descriptive and normative project with an "empty hand."

10.3 Neurath's Cool Hand: Empiricism in a Fallibilist Framework

In the influential re-appreciation over twenty years ago (Cartwright, Cat, Fleck and Uebel 1996, 4) the authors suggested that the relatively late debates on data-reporting and on the proper form for protocols (and a Marxist debate on the materialist conception of history) gave rise to Neurath's radical position and his rejection of scientific method as well as his anti-foundationalism. In contrast to this appreciation of the development of Neurath's views, by now we have more evidence that science needed no ideal method and no secure foundations for Neurath already before World War One. To recall the line from the later prison drama, "sometimes, nothing can be a real cool hand," as pervasive fallibilism was not a weakness of Neurath's position, but rather an asset in his fight against "apriorism" and "pseudorationalism."

Neurath did not focus on justification (that will at one point become pseudorational), but on the potentials of the inductive sciences. He built his case not from the icy slopes of logic, but rather the murky waters of history, bracketing the question of foundations and acknowledging that natural science is always surrounded by the richness of experience. This stance is echoed in the "Scientific World-Conception" paper, where he contrasts our historically shaped means of expression and our rich language and script with logic as a doctrine of tautological transformations (Carnap et al. 1929/1973, see also Jordi Cat's chapter in the present volume for integrating the sources). His position is at once radical, and at the same time it endorses the creative potential of the non-formal aspects of theories. For the progress-bound inquiry development was a paramount concern, and Neurath devised a framework exploiting analogies, novel concepts and creativity.

Even an inconsistently used piece of fiction [Phantasiebild] can be a driving idea, for Neurath admitted not just inductive reasoning as his auxiliary motive, but any creative burst. Empirical science can develop by figments of mind [Phantasiebilder] or analogies, some of which, if pursued intensely can have benefits, like leading to novel formalisms later. The potential of the "elementary notions" wane with gradual formalization, so Neurath did not want to limit the investigation to abstract symbolic aspects of theories. A view of theories that takes them to be equations or proposition leaves no room for the "blurred margins of analogies, which leads the scientist to further assumptions" (Neurath 1916/1983, 29).

In this world any actor can only meaningfully follow a few insights, so world-view differences and our own agenda mean that any researcher can only go his own way: "only the belief in the correctness of a definite analogy or a group of analogies can create the energy that is needed to overcome all difficulties" (Neurath 1916/1983, 29). To develop a partial insight a whole life may be needed, and therefore a certain diversity of opinions is recognized as a typically beneficial characteristic of disciplinary development. As a result, Neurath started off with disunity, and prescribed energetic action, as an exploitable partial insight or potential analogy can best be justified by pursuing it.

As Neurath went "bottom up" when looking at the theories, he deliberately relied on a broad conception of what a theory can be. It might be simple to claim, with Heinrich Hertz, that Maxwell's theory is just his system of equations, an example he explicitly cites, but this clearly leaves out many of the important aspects of a theory that can have an impact on our world-view. It is a mistaken assumption that theories could generally be equated with equations or propositions. Theories are, instead, complexes with abstract symbolic, visual, analogical aspects. Neurath's project tackles heads on the complexity of science (including visuals) and all the interpretative difficulties:

> We must try to see clearly how a physical theory hinges on the images used, and how far on those features that actually carry the argument. Perhaps we cannot grasp some developments unless we consider the images and pictures; in other uses we must rely on what governs the mathematical treatment of phenomena; or, maybe, both ways of looking at it are steps. At all events, the latter analysis has not yet become current among historians of physics. (Neurath 1915/1973, 102.)

> Some modern physicists, who, like Poincaré or Duhem, are reckoned among conventionalists, allow that the mathematically important features are relevant to classification and analysis. But this leaves open the philosophical question. Those who wish to give more weight to the imagery of hypotheses (as I believe one must in some cases), may without contradiction add this to the analysis. (Neurath 1915/1973, 102–103.)

Neurath analyzed some optical theories in detail, most notably Newton's *Opticks*, both with respect to language use, and the pictorial means of theory-propagation, his diagrams. The papers show his expertise in historical research and optics, many examples display knowledge and sensitivity to details of the theories in question. He recognized that the analyzability of these historical theories is limited. For Neurath not just the analogies, but also the theories were "blurred:"

> Each system of hypotheses, even if its formulations are of the utmost precision, has, to use this expression, a blurred margin. This always and necessarily exists. The amount of difficulties can grow through new insight; at best we can approach it asymptotically. A complete mastery of the whole multiplicity seems an impossibility to us. (Neurath 1916/1983, 24.)

Historians of science face ambiguities and potential inconsistencies when attempting the classification of systems of hypotheses. This is a problem that has more or less been disregarded by philosophies of science after the linguistic turn (see Meheus 2002, where Kitcher's paper mentions Neurath, and Cat 2005). Neurath's project assumes that as these theories have proven their merit, the theory *in toto* was some achievement, but also that the theories are unsharply bounded entities so justification of the content is both unnecessary and full of trickeries. The interpretation and method of comparison pose real challenges, as already a simple *classification* can be problematic. Neurath discussed techniques of appropriation: slight rephrasing can make a view look more akin to our current notions (Malus speaks of "emitted magnet-like corpuscules," stressing "emitted corpuscles" distances him from modern notions, while "magnet-like" approximates him, Neurath 1916/1983, 30). His attempt is directly linked to the historical observation that inconsistency can be stimulating:

> We now have to deal with Newton (1642–1727); as he does not have as consistent a central idea as Huyghens, he therefore operates with a much greater wealth of elementary notions. It was precisely his inconsistency that was highly stimulating and gave posterity an opportunity to form hypotheses of many kinds, many of which have proved fertile. According to his words he attaches little weight to the character of light, but in fact he is very dependent on the notions that he forms of it. Actually he expresses them several times. (Neurath 1916/1983, 20.)

Neurath talks about systems of hypotheses, but does not even differentiate between hypothesis and reality [was man als Hypothese, was man als Realität bezeichnet], as in the case of Optics hypothesis and experience [Erfahrung] are colorfully intertwined. The difficulty of their separation is shown in Goethe's polemic against Newton. Goethe notes that the description of observational facts in Newton's text shows a mismatch with the depiction of the experiment. The diagram of an observation in Newton's *Opticks* shows the fuzzy edges of the displaced image, and Goethe notes that an explanation is offered by Newton later, but the text neglects features of the visual representation of the observations at this experimental level of description. In the example of the displaced colored strips the description of the "brute facts" is different for Newton and Goethe. Underlying it, we find a different mathematical idealization: for Newton two displaced rectangles, and for Goethe one (Zemplén 2018). Neurath (1916/1983, 24) discusses the blurred margins interpreted differently, and reproduces part of Newton's diagram (Fig. 10.1):

For Neurath this is a clear case where different languages neglected different aspects of the observations, and the selection of 'facts' was driven by theoretical findings. The Polemical part of Goethe's *Farbenlehre* provides an example[1] for the

Goethe points out that in Newton's illustration (Figure 8) the red strip has fringes of which Newton makes no mention at all. "Why does he not mention this phenomenon in his text of which he has a careful, though not quite correct, engraving made in copper? A Newtonian will probably answer: this is just a residue of the decomposed light which we never can get rid of entirely, and that still plays its tricks here."

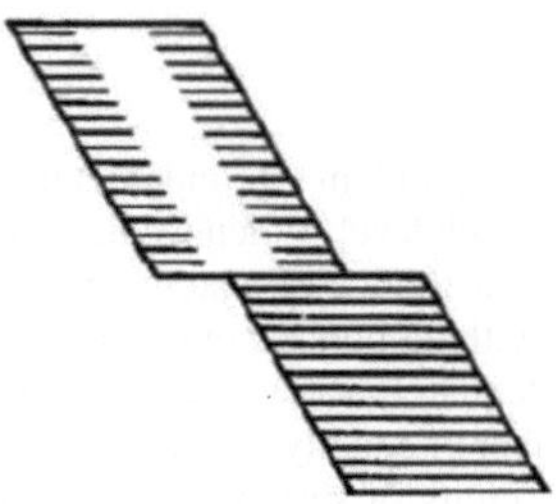

Fig. 10.1 Neurath's (1916/1983, 24) diagram of and commentary on blurred margins

[1] This example has been analysed in detail in Zemplén (2006). Recently the problem of underdetermination with respect to Newton and Goethe's competing views on color has been taken up by Olaf Müller (2015) in detail. His decidedly Quinean approach (Müller 2016) has also been critically scrutinized (Lampert 2017).

introduction of the Neurath Principle,[2] the extension of Duhemian holism to all sciences, 'observation statements' included (Haller 1982/1991; Zolo 1989). In case of conflict between statements of facts and theories we are unable to tell which part of the theory is to be blamed, and cannot even tell whether to revise the theory *or* reject the new statement as useless or even false (Neurath 1932/1983). In his "The Unity of Science as a Task" (Neurath 1935/1983), the mature Neurath again cites an example from the history of optics:

> But even the initial statements of successful science are not fixed, since one could begin at the beginning with different unified languages that cannot be translated into each other straight away. And even if the unified languages were more or less fixed – in fact the statements of yesterday and today, appearing at the beginning and at the end of a book, belong to often slightly differing languages – nevertheless, to make good predictions, we could set out from different observation statements that we select from the large number at our disposal that can be steadily increased. What one person neglects as unimportant – and then he shapes his concepts accordingly – may seem essential to another for the predictions. For example, Goethe strongly criticised Newton for omitting certain blurred margins of the spectrums as unimportant, whereas he himself started from this very point.
>
> This is how matters stand in *every 'layer'* of scientific work, not only in the narrower sphere of systems of hypotheses, as Poincaré and Duhem have pointed out with such intensity. (Neurath 1935/1983, 116–117; original emphasis.)

Historical research had a decisive impact on the young Neurath's views on the building-blocks of empirical science, and his thoroughgoing anti-foundationalism, stressing that any description of phenomena fail at "capturing the whole multiplicity." In these papers Neurath cares little about constraining any of the potentials of empirical science. Instead, he focuses on mapping the conceptual space of alignments and misalignments, in a quest to explore unconceived possibilities.[3] For Neurath the "exploratory" aspect is closely connected to recognizing that various languages can be developed to talk about the world. To carry out the task of the historian it is important to find the facts *that are neglected and disregarded* in the descriptions. Systems of hypotheses connect and also select the facts:

> One should always indicate which facts have been neglected, which favoured. The systems of hypotheses of physics, like all other systems of hypotheses are *an instruction directing not only the connectedness, but also the selection of* facts. (Neurath 1916/1983, 24; original emphasis.)

[2] The Neurath Principle extended the problem of underdetermination to observation reports: "Science is ambiguous – and is so on each level. When we have removed the contradictory groups of statements, there still remain several groups of statements with differing protocol statements that are equally applicable; that are without contradictions in themselves but exclude each other. Poincaré, Duhem and others have adequately shown that even if we have agreed on the protocol statements, there is an unlimited number of equally applicable, possible systems of hypotheses. We have extended this tenet of the uncertainty of systems of hypotheses to all statements, including protocol statements that are alterable in principle. (Neurath 1934/1983, 105, translation from Howard 2006, 11)

[3] Although coming from very different angles, an impressive amount of commonality can be established between the early Carnap and Neurath. Before Carnap was working on logically conceivable models of the world (Stone 2010), Neurath worked out a method to localize logically conceivable theories of a domain (for the sources see Jordi Cat's chapter in the present volume).

For a phenomenal domain well-confirmed alternatives can be conceived, in fact are conceived in a given period, and Neurath's mapping project helps identify further alternatives. He thinks of a way to make "artificial" theories, utilizing the "natural," historical ones as assets. This approach has numerous overlaps with Kyle Stanford's more recent take on the historical record of scientific inquiry and on the problem of unconceived alternatives (Stanford 2006).

Neurath's history of the optical theories is sensitive to *variation* as a resource for *selection*. The perspective is intimately tied to an evolutionary view. Heterogeneity is ever-present where living organisms are in pursuits of goals, both on the level of individual preferences, and on the level of communities, like the pursuing influence of scholasticism and other large scale differences in culture. The comparison (a theme also important for Mach) provides *some* mapping of the investigated theories based on properties with yes/no values, but with no clear ontology and no precise identification of the theories "proper."

10.4 Optics as a Model for the Mosaic of Science

Why did Neurath study the historical theories of optics? His interest in optics was first sparked by the pictures of crystals in polarized light, found in a mineralogical atlas (Neurath 2010, 33).[4] While many members of the Vienna Circle, like Carnap, Reichenbach, and Moritz Schlick had an early interest in general relativity, Neurath focused on optics (and, to some extent, mechanics), subjects very dear to Mach (and, to some extent Schlick). With some exaggeration, the optical essays could be seen as a token of appreciation towards Mach, written by an enthusiastic apprentice. As Neurath's letters testify, he approached Mach with intent to work further on the subject,[5] and promised sending him KH. Mach finished his book on the subject in June 1913, and manuscripts were circulated, but his treatment is significantly different. After Mach's *Prinzipien der Physikalischen Optik* appeared (1921), Neurath returned to the topic in a 1922 lecture on January 21. It is probable that the area intrigued him for decades to come. For example, as Carnap left for the USA in 1936, Neurath mentions in a letter that he would be happy to continue working on the history of optics, to develop their position [Betrachtungsweise].[6]

The history of optics as a field of investigation could provide Neurath a testing ground for a model of empirical science and for his theory of theory-classification. Several cues suggest that he had carefully picked his domain of investigation. He did not want to lose the complexity of the subject (science as a whole), or the historical perspective, so picked a field that utilized mathematical abstractions and

[4] I thank Ádám Tamás Tuboly for pointing out this passage to me, as well as the later cited letter to Carnap.

[5] One of the questions raised in the optical essays and also addressed to Mach is the history of the sound-light analogy, one of the main themes of the informed history of optics by Darrigol (2012).

[6] See Neurath to Carnap, 30 April 1936 (RC 102-52-29). Rudolf Carnap Papers, 1905–1970, ASP.1974.01, Special Collections Department, University of Pittsburgh) All rights reserved.

symbolic representations as well as analogies, and played a role in shaping our scientific world-view. When he delineates the domain, he stresses the partial independence as well as the embeddedness of the field:

> The subject I choose for my comments is the history of optics from the beginning of the seventeenth to the beginning of the nineteenth century. The reasons for the selection of this discipline and this period are mainly the following: the subject of optics was rather clearly fixed at a relatively early time. Advances were made successively by representatives of the most different trends. At one time this trend, at another time that trend achieved important results. The wealth of elementary notions and elementary hypotheses is large enough to make a sufficient number of combinations possible.
>
> Acoustics, for example, would have been all too simple; its most important principles were already stated in final form early on, and everything that came later was an improvement. On the other hand, the theory of electricity displays too much confusion. […] The history of optics is suited for our comments only up to the middle of the nineteenth century, before the union of electrical and optical theories. After that, one has to take all physical notions into account to do justice to optics; its relative isolation reached its end. (Neurath 1916/1983, 16–17.)

Neurath created a simple framework to ameliorate the problem of "primitive" analyses, stressing a single characteristic of theories. His optical essays developed a methodology to classify theories according to a number of "elementary notions," their lack or presence in a theory. The framework gives us a mapping-possibility for the not-fully-rational choices in science that nevertheless yielded significant gains in a scientific domain.[7] For Neurath, this is a more scientific way of talking about theories, than the common practice (Fig. 10.2):

> This already shows what difficulties arise from dividing existing views into two types. Such dichotomies occur in all disciplines. They really belong to a primitive form of concept formation, which any theory bent on perfection must try to shed as fast as it can. Dichotomies might perhaps do, if the two types could be described as **A** and non-A, but this is almost never so. Usually each outlook received its hallmark separately, without examination whether the two might not be compatible. Suppose a theory rests on three elementary notions, ***a, b, c,*** with contradictories ***a*, b*,*** **c***. The following constellations may occur: (Neurath 1915/1973, 104; original emphases.)

Fig. 10.2 Constellations of theories

a b c
a b c*
a b* c
a b* c*
a* b c
a* b c*
a* b* c
a* b* c*

[7] "[In KH and PO] Neurath asserts that the historian of science as well as the scientist should attempt a classification of systems of hypotheses arranged according to logical alternatives which – or so I shall argue – could ultimately be resolved by an auxiliary motive. Within Neurath's radical pragmatist conception of science, 'on board of his boat', the auxiliary motive also plays a role in establishing such a system of hypotheses. In the latter case, the auxiliary motive acts in the same way as inductive or abductive modes of inference. It replaces them if there is no basis at all for their standard application due to lack of relevant information or if they only yield equally probable alternatives" (Stöltzner 2000, 25).

Neurath picks a neutral stance, a multi-element charting of the theories to decrease hostility and to dissolve tensions between factions. He does not assume that the theories necessarily have a definitive structure, only that they can be classified. He notes that some "elementary" notions are linked, nevertheless he takes care to show how decoupling is possible for some of the items (e.g. for periodicity and interference, linked in modern optical theories). His scheme is coarse, but his eye is that of a trained philologist. He starts from *some* mapping of the theories, never wants a full redescription of any of the theories. The complexity (long texts combining diagrams, words and mathematical abstractions) is only tackled to reach the depth where *some* significant results can be obtained.

The analysis is open-ended, and the two essays show how he approximated the model to "historical reality." The tables offer much less than the text of the papers discussing the theories in more detail, but they offer enough to show the applicability of the scheme and help Neurath highlight its availability for general use. His solution draws on a chemical analogy for the history of ideas: the old chemistry was characterized by single elements, but with the development of modern chemistry we now use names that are typical for the composition (KH 14). Finding the "composition," however, is only an analogy, Neurath does not fully embrace atomism, his charting of the theories is only a tentative approximation.

It is easy to read later passages as continuing the same research project that was first developed and tested in Neurath's short-time dabbling with the history of optics. In his "Unified Science as Encyclopedic Integration" Neurath (1937a, 4) gives an example from optics (Huyghens' principle and Euler's theory) before moving on to economics (Smith, Ricardo, and Sismondi), and finally concluding: "[o] nly a complicated comparative scheme could show the amalgamation of various elements, the common and different features of various theories."

What became known as "Logic of science" was for Neurath at first a "History of Ideas," but all the same carried out by a social engineer, taking into account the practical difficulties, including the lack of impartial procedures. To optimize the functioning of the scientific community Neurath developed a fundamentally communitarian epistemological framework and outlined a project for future historians of science to help society by a more nuanced way of talking about theories, their merits, and their potentials. Scientific progress requires a co-ordination of multiple composite perspectives, an enormous number of elements that he later described as a "Mosaic."

10.5 Comparison of the Optical Essays KH and PO

The essays show commitment to a number of theoretical views characteristic of the early Neurath, like the shared ideals of the natural and the human sciences. They stress the need for autonomous concept-formation for the History of Science, and several similes and metaphors point towards the non-reductive naturalization of the social sciences. The texts employ a wide range of examples, and not just from optics. The analysis minimizes metaphysical ballast as it should not be biased towards any particular philosophical stance. Neurath does not want to introduce a distinction between hypothesis and reality, and the method for analyzing historical theories as composites of elementary notions acknowledges a hermeneutic difficulty that is linked to vagueness, the "blurred margins" on every level of the investigation (later congestions or cluster-concepts [Ballungen]). Neurath discusses the early thinkers generally in isolation, at time linking some of these individuals, and for the nineteenth century he tends to group them (Brewster, Biot, Malus; Young and Fresnel), alluding to factions of the opinion-polarization of the field (emission- and wave-theories). So the framework is not *narrowly* individualistic.

The KH and the PO contain many overlaps, but the stresses are placed differently. The same key figures are investigated in both texts (see Table 10.1), and there are differences in accent. The analysis proceeds according to the same order, but with different excursions and historical details in the two papers. The shifts in Neurath's take between the two versions are probably due to both the historical development of his views, and to the different positioning of the enterprise for different audiences (in Vienna and Berlin). Before the analysis of content, it is instructive to compare the beginnings of the papers, easily portrayable as an example of twentieth century *Naturgeschichte* (KH) and *Naturphilosophie* (PO):

> Everyone who takes up theory of science or the history of the sciences feels oppressed by the profusion of facts. Early on, a start was already made with the classification of stones, plants, and other objects. [...] Of course groupings of complexes of ideas were formed, but this was not always preceded by sufficient analysis. There were wild growths of new classifications when the traditional ones were altogether abandoned. There was no continual cooperation of scholars in the field of the history of science. (Neurath 1916/1983, 13.)

> History of science, if seen as more than a mere chronicle of findings and biographies, is a young discipline. It can aim much higher: like the history of any field of enquiry, it *may try to shed light on the psychology of the enquirer*; besides, it *may exhibit the logical structure of theories*, and from it derive how they may develop. To follow how such possibilities happen to be realized by this or that enquirer is an especially engaging task. (Neurath 1915/1973, 101; emphasis added.)

In the earlier finalized KH the paper builds the argument from the historian's perspective, and there is an outline of a two-tier theory of reconstruction, introduced in more detail in the next section. KH discusses the "pugnacity" of scientific debates and false dichotomies (meaning-polarization) in more detail, and offers a clear example for the extension of Duhemian holism that has come to be known as the Neurath-principle, and, presumably linking the discussion to Mach in the Viennese context, has more on the role of analogies. In the *natural history* focus of the KH, the historian has the same problem of classification as the Biologist with plants, the Chemist with molecules. For Neurath the "Views of Physics" [die Anschauungen der Physik] are objects for contemplation, analysis, and grouping [in Gruppen bringen]. Just as not all chemical combinations are found among minerals, similarly not all possible theories show up in the world. To override passion, the project takes its start with the enumeration of neutral logical possibilities, acknowledging that there is a need for the impartiality of a *scientists gaze* on its object, unfortunately lacking

Table 10.1 Neurath's elementary notions discussed in KH and PO

	KH elementary notions	PO elementary notions
– – Descartes	+Wave (?) +Emission (?)	+Wave + Emission
	–Polarisability	–Polarisability
	–Periodicity, (+?)	–Periodicity
	+Diffraction	+Diffraction (theoretically)
– – Grimaldi	+Sound	+Sound
	+Diffraction	+Diffraction
	+Interference	+Interference
		+Periodicity (vibration)
		+Fluidatio (current)
		+Undulatio (wave motion), etc.
– – Malebranche	+Sound (–Interference)	+Sound (appr. = HP)
	+Periodicity	+Periodicity
	+Diffraction (physical model)	
+ + Huyghens	+Sound (–Interference) +HP	+Sound +HP
	–Periodicity	(–Diffraction)
	+ Polarisability	–Periodicity
	+/– Diffraction (theory-experience tension)	+Polarisability
	–Emission	–Interference (like X-rays)
+ + Newton	+Periodicity (theory of "fits," like X-rays)	+Periodicity (like gamma rays)
	+Emission –Sound –Wave (–HP) (+ether)	+Diffraction
	+Diffraction	–Interference
	–Interference	+Polarisability
	+Polarisability (to support +Emission)	–HP (light particle has poles)
+ – Euler	+Wave + Sound [+HP]	+ Periodicity
	–Emission (today redundant objection)	+HP
	+Periodicity (abandons HP)	
+ +Young Fresnel	+Periodicity +Sound +Wave	+Interference +Wave
	+Interference (both light and sound)	+Diffraction
	[+HP, – Emission]	[+Periodicity, +Polarisability, +HP]
– – Biot Brewster Malus	–Wave + Emission	+Periodicity (surges)
	+Polarisability	–Interference
	(+using statistics)	+ Emission +Polarisability
	+Periodicity (for both emission and wave)	(+using statistics) (magnetic analogy) (light particle has poles)

in the history of science, and even impacting our public understanding of science via the way science is taught.

In the PO Neurath (1915/1973, 105) introduces natural combinations, which occur in nature (i.e in history), and artificial ones, potential but nonexistent outlooks derivable from a set of elementary notions. He knows that a *fully impartial* scheme is unrealistic, the analysis has to pick some perspectives that are not self-evident, and any abstract scheme will have its limitations in a world of real, "mixed" theories that resist easy classification (Jordi Cat in this volume discusses sources for the issues of classification and imprecision). He is cautious, and is not trying to disentangle the intertwining of facts and hypotheses:

> The variety of views in the history of optics stems partly from the different ranges of findings and partly from different ways of accounting for them. For a clear picture, we must group the elementary observations on which the various outlooks rest. In practice this is rather difficult, because scientists often omit important things; sometimes actual findings are deliberately or otherwise neglected. Our analysis would aim less at finding what scientists knew but rather at stating, for each theory, the essential parts of experience.
>
> To show on what kind of classification of outlooks a systematic historical treatment would have to rest, let us confine ourselves to periodicity, polarization, interference and diffraction, leaving aside whether in any instance these are elementary findings or elementary notions. Findings and notions are usually equal in number. Most often hypotheses are richer than the known range of facts. (Neurath 1915/1973, 105.)

After introducing the abstract mapping scheme, the KH lists elementary notions that are phenomena known via observation and experiment, and are linked to conceptual developments in the field,[8] starting with periodicity (Newton's rings and the colors of thin plates), interference (split rays by reflection), polarisability (tourmaline-crystal), and diffraction (light sent through a hole). The text then contrasts the individual theories, introducing a number of other elementary notions ("fruitful analogies," like wave, emission), while finally in the table three aspects are investigated: periodicity, Huygens' principle, and emission, with four physicists, Huygens (n,y,n), Newton (y,n,y), Euler (y,n,n), and Young (y,y,n). In PO, in contrast, a scheme is printed with four properties and three thinkers: periodicity, polarisability, Huygens' principle, and interference; Huygens (n,y,y,n), Newton (y,y,n,n), Young (y,y,y,y).

Similar classificatory schemes, or, as he came to call them "chessboards" abound in Neurath's oeuvre, several times on topics connected to pseudorationalism and mapping models of social order, so it is quite safe to assume that the classificatory pattern developed here became a general device for organizing theoretical complexes for Neurath. His chessboards are flat, the list-items of the schemes have no direct ontological implications, and the "structure" of the comparison has no hierar-

[8] A first list of "elementary notions" could simply have been derived from the table of contents in (Whewell 1840) that Neurath consulted.

chies.[9] Later tables usually also end with a line containing only "yes," so by Neurathian standards the PO table is final (enough), it shows Young as endpoint incorporating all three mentioned (clear) examples of elementary notions, displaying the progress in the field.

Neurath's history is revisionistic, driven by an Enlightenment agenda (see Angélique Groß's chapter in the present volume). The historical data, however, obstruct the execution of the project. Both papers publish tables, but these are different, and the textual analysis shows that as the research progressed, Neurath refined his take on the individual theories. Assignments of theory-properties change between the versions, some theories have multiple and/or contradictory values. Some of the items can be linked, and this suggests coupling and hierarchical relations with respect to some "elementary" notions, like the 'analogy with sound' and the "employment of Huygens' Principle." Table 10.1 gives a more detailed comparison of the texts of KH and PO. This extended table of Neurath's analysis shows the mentioned properties and some of the classificatory problems and the links he notes (in brackets), and items that are only in his tables, not discussed in the text [square brackets]. Before the names +/– signs show presence in the published chessboards in KH and PO. HP refers to Huyghens' principle.

The comparison shows that Neurath's analysis improves, and his historiographical input gradually refines the systemized output. Importantly the elementary notions link not just to parts of experience, but to models, methods, ontologies. Neurath obviously consulted some available histories of the subject, as well as many of the main texts of his protagonists. A few of his historical observations are odd, and it is curious, for example, how he did not exploit the deep tension between Newton's and Huygens' theories that his analysis reveals. Some elementary notions that were crucial for the development such as "Emission" and analogy between light and sound (waves) do not appear in the tables, but are used for the broader categorization of factions.[10] Neurath makes interesting observations tracing these analogies, like "the wave theorists displayed a greater uniformity of hypothesis formation than the emission theorists, who needed a special hypothesis for almost each new phenomenon treated by wave theories" (Neurath 1916/1983, 21). The distances of views can show surprising connections (Brewster is closer to Fresnel than to Euler, see Neurath 1916/1983, 30). The later PO more directly points to development, a fusion of the two research programs organized along the 'wave' and 'particle' party-lines.

---

[9] See Neurath's unpublished essay, "Argumentation and Action," in the Otto Neurath Nachlass (K.39–41), and the "CHESSBOARD OF ATTITUDES" and the "CHESSBOARD OF THREE SOCIAL QUALITIES" there. Otto Neurath Nachlass, Wiener Kreis Archiv, Rijksarchief in Noord-Holland, Haarlem, The Netherlands. All rights reserved.

[10] Recognizing Newton's vagueness on the issue could also have been a reason for dropping the "Emission" rubric in the second table. Neurath in later works often revisited the role of contradictions: they do not destroy the *entirety* of a system, and might be useful asset for a later (consistent) set.

10.6 Neurath's Two-Tier Historicist Epistemology: Naturalizing the Genius

In the early period, history was an important (possibly nomothetic) discipline for Neurath, his revisionist and historicist project offers an empirical and pragmatic approach to epistemological problems (for overviews see Cat 2014, Uebel 2000). Classifying theories based on too crude dichotomies is deemed problematic, and so Neurath provides a mathematical framework to do more justice to the plurality of positions. Neurath's program is agnostic about "the world of facts," and fallibilism allows for no secure foundations either in the theoretical domain or in the world of elementary experiences. Varieties breed new ideas more quickly, the previously unknown implication of a combination of basic ideas *is* progress, as fruitful science is an incorporation of more and more unknown in the community of knowers.

The optical essays show the young Neurath's acute sensitivity to social processes (like the formation of research traditions, opposing scientific camps), highlighting and criticizing scientific "pugnacity." The essays pay special attention to the epistemic relevance of visuals, and cite several examples of 'blurred' representational content. The optical essays touch on many of the core issues of his epistemology. They fuse Duhem's holism with Machian considerations, but the agenda has no ontological preference for sensations over theories (like Mach's sensory intuition), and no containment of the set of theoretical statements like Duhem. Later justificatory approaches aimed at raising confidence in a theory, but Neurath is deflationist, as Reisch (1997, 449) noted. His target in the optical essays is to decrease the overconfidence in any single solution – and the confidence that it is easy to grasp a theory.

He develops the systematics of unsharply bounded theories, an analysis sensitive to structural relations but one that has no mathematical hierarchies and no clear metaphysics/ontology. Without the need to say exactly what the historical theories are, Neurath develops a way to compare them, as he re-describes a group of theories as conglomerates of elementary notions. Neurath at the end of the first paper notes that an actual scheme cannot be impartial, and the analyst has to prioritize some of the historical theories:

> Our comments have shown that a comparative study of systems of hypotheses, which has to be regarded as the basis of all historical research, can proceed in two stages. As a first stage a pure analysis of the system has to be proposed that leads to a grouping of elementary views. We saw that in this way, without special general considerations, some useful insight can be gained already within the specific field of research, within physics, chemistry or any other science. Schemes of this kind allow us to indicate what the contributions of individual scholars have been. By not giving preference at first to any elementary notion, all groupings of hypotheses are taken into account with the same intensity; the usual, and often unjustified, neglect of transitory systems – called like that mostly on the basis of crude dichotomies – is avoided. For a best possible mastery of historical development it is desirable to make a preparatory survey of all groupings of individual views that are possible in principle. Maybe the start can be made from the theory of greatest power, if there is one of this character. (Neurath 1916/1983, 30.)

Neurath separates two stages of the historical research. The methodology first prepares the ground for a more refined appreciation of the scientific theories in question, theories that are messy, and not like Maxwell's. The systematic completion of the first stage would lead to a greater insight into the character of systems of hypotheses through preparatory historical work, and could be used to create a routine through which many could be qualified for profitable work. The first stage of analysis, the charting of the elementary notions of a field, is for the craftsmen, promoting a fuller democratization.

The second phase of analysis is for the inquisitive specialist, willing to commit significant resources to studying individual creativity.[11] This stage does not culminate in a justificatory enterprise, but rather in an inquiry to find the driving ideas of an individual [die treibenden Ideen aufzufinden], the fuller exploration of the insight suspecting order, where before no one suspected:

> The second stage needs a premise of general considerations. In it one can no longer recognise the significance of certain hypotheses with the help of reflections within the individual sciences. Our reference to a total world-view becomes a duty. Only with special gifts and after comprehensive previous studies can a man satisfy such demands. Historians of physics in this sense will therefore be either philosophically trained physicists or philosophers trained in physics. (Neurath 1916/1983, 31.)

An important feature of progress in both the natural and the social sciences is that some people with special gifts can grasp hidden connections. Although Neurath attacked non-empirical 'empathic' practices, his pragmatic project is not hostile to idealist thinking.[12] His early historiography reserves a place for the agent of surprisingly swift progress both in natural science and in the history of ideas, and his hermeneutics can naturalize the problem:

> In my cursory remarks about the history of optics, I have already emphasised a significant analogy. A number of physicists, such as Malebranche, Huyghens, Euler, etc., expressed the opinion that light is something like sound. In connection with this analogy it is remarkable that the authors mentioned did not know anything about interference of sound or of light. The thinkers who analysed light were influenced by the advances of acoustics, just as those occupied with acoustics were by the advances of optics. Only Young was simultaneously concerned with interference of both light and sound. The theory of interference of sound was later developed by W. E. Weber in the first third of the nineteenth century. Such occurrences in the field of scientific thinking have given rise to the idea that a principal task of genius consists of acts of empathy with the world. The genius is supposed to be able to recognise whole complexes of facts as related, even if he is not consciously aware of all elements of the two analogised views. It is just the non-formulated part of the analogy that contains, as it were, a driving force. (Neurath 1916/1983, 26.)

[11] "By philosophical assumptions of a very general kind about the structure of the world we can possibly obtain foundations for the classification of systems of hypotheses but, as can be seen at once, this assumes much intellectual preparation" (Neurath 1916/1983, 27).

[12] The revival of early nineteenth century German Idealism, and Weimar Classicism has not only impacted Neurath from the Vienna Circle. For example, on Schlick and Schiller see Ambrus (forthcoming).

The "scientific attitude" of Neurath in the 1930s can be roughly described as a compromise between empiricism and materialism (Mach and Marx), and in the optical essays as a reconciliation of empiricism and idealism (Mach and Goethe). The Romantic connection was tuned down in later years as Neurath saw how regimes and ideologies can subvert the ideals. But all in all there is much continuity in his work. As he writes in the 1937 papers on Unity of Science and its Encyclopedia:

> The program of the unification of scientific language requires a logical analysis of science. The history of science shows the importance of such an analysis for the progress of scientific work.
>
> The technique of such logical analysis was gradually perfected and employed consciously. As a consequence, feats of logical analysis which were heretofore possible only to men of genius may now be taught systematically to scientists of ordinary attainments. This method of analysis is the subject matter of the new discipline called the *Logic of Science*. (Neurath 1937a, 5, 1937b, 268.)

His critique of stable structures as timeless models of science is closely linked to his historicism, and the early recognition that the blurred margins are a necessary part of science and of scientific progress (see e.g. Neurath 1941). Neurath's study of the history of optics provided a rich resource for his fight against "a priorism" and "pseudorationalism" just as for his later physicalism.

References

Ambrus, Gergely. forthcoming. Schlick on the Meaning of Good. In *The Socio-Ethical Dimension of Knowledge: The Mission of Logical Empiricism*, ed. Christian Damböck and Adam Tamas Tuboly. Cham: Springer. Forthcoming.

Biddle, Justin. 2013. State of the Field: Transient Underdetermination and Values in Science. *Studies in History and Philosophy of Science Part A* 44: 124–133.

Carnap, Rudolf, Hans Hahn, and Otto Neurath. 1929/1973. The Scientific Conception of the Word: The Vienna Circle. In *Empiricism and Sociology*, ed. Marie Neurath and Robert S. Cohen, 299–318. Dordrecht: D. Reidel.

Cartwright, Nancy, Jordi Cat, Lola Fleck, and Thomas Uebel. 1996. *Otto Neurath: Philosophy Between Science and Politics*. Cambridge: Cambridge University Press.

Cat, Jordi. 2005. Modeling Cracks and Cracking Models: Structures, Mechanisms, Boundary Conditions, Constraints, Inconsistencies and the Proper Domains of Natural Laws. *Synthese* 146 (3): 447–487.

———. 2014. *Otto Neurath*. In Stanford Encyclopedia of Philosophy, ed. Edward N. Zalta. URL = https://plato.stanford.edu/cgi-bin/encyclopedia/archinfo.cgi?entry=neurath

Darrigol, Olivier. 2012. *A History of Optics from Greek Antiquity to the Nineteenth Century*. Oxford/New York: Oxford University Press.

Haller, Rudolf. 1982/1991. The Neurath-Principle: Its Grounds and Consequences. In *Rediscovering the Forgotten Vienna Circle: Austrian Studies on Otto Neurath and the Vienna Circle*, ed. Thomas Uebel, 117–129. Dordrecht: Kluwer.

Howard, Don. 2006. Lost Wanderers in the Forest of Knowledge: Some Thoughts on the Discovery-Justification Distinction. In *Revisiting Discovery and Justification. Historical and Philosophical Perspectives on the Context Distinction*, ed. Jutta Schickore and Friedrich Steinle, 3–22. Berlin: Springer.

Ibarra, Andoni, and Thomas Mormann. 2003. Engaged scientific philosophy in the Vienna Circle: The case of Otto Neurath. *Technology in Society* 25: 235–247.

Koterski, Artur. 2018. Neurath's Decisionism and the Earliest Reviews of Logical Empiricism. In *Rationality and Decision Making: From Normative Rules to Heuristics*, ed. Marek Hetmański, 314–336. Leiden: Brill Rodopi.

Lampert, Timm. 2017. Underdetermination and provability: A reply to Olaf Müller. *British Journal for the History of Philosophy* 25 (2): 389–400.

Mach, Ernst. 1921. *Die Prinzipien der physikalischen Optik*. Leipzig: Verlag von Johann Ambrosius Barth.

Meheus, Joke, ed. 2002. *Inconsistency in Science - Studies in the Sources of Scientific Creativity*. Vol. 2. Dordrecht: Springer.

Mormann, Thomas. 1996. Encyclopedism as an Anti-Cartesian Account of Language and Science. In *Encyclopedia and Utopia the Life and Work of Otto Neurath (1882–1945)*, ed. Elisabeth Nemeth and Friedrich Stadler, 87–97. Dordrecht: Kluwer.

———. 1999. Neurath's Opposition To Tarskian Semantics. In *Alfred Tarski and the Vienna Circle*, ed. Jan Woleński and Eckehart Köhler, 165–178. Dordrecht: Springer.

Müller, Olaf. 2015. *Goethe mit Newton im Streit um die Farben*. Frankfurt am Main: S. Fisher.

———. 2016. Prismatic Equivalence – A New Case of Underdetermination: Goethe vs. Newton on the Prism Experiments. *British Journal for the History of Philosophy* 24: 322–346.

Neurath, Otto. 1913/1983. The Lost Wanderers of Descartes and the Auxiliary Motive (On the Psychology of Decision). In Neurath and Cohen (1983): 1–12.

———. 1915/1973. On the Foundations of the History of Optics. In Neurath and Cohen (1973): 101–112.

———. 1916/1983. On the Classification of Systems of Hypotheses (With Special Reference to Optics). In Neurath and Cohen (1983): 13–31.

———. 1932/1983. Protocol Statements. In Neurath and Cohen (1983): 91–99.

———. 1934/1983. Radical Physicalism and the 'Real World'. In Neurath and Cohen (1983): 100–114.

———. 1935/1983. The Unity of Science as a Task. In Neurath and Cohen (1983): 115–120.

———. 1937a. Unified science as Encyclopedic Integration. In *International Encyclopedia of Unified Science*, ed. Rudolf Carnap, Charles Morris, and Otto Neurath, 1–27. Chicago: University of Chicago Press.

———. 1937b. Unified Science and Its Encyclopedia. *Philosophy of Science* 4 (2): 265–277.

———. 1941. The danger of careless terminology. *The New Era* 22 (7): 145–150.

———. 1981. *Gesammelte philosophische und methodologische Schriften, Gesamtausgabe der Schriften Neuraths*. ed. Rudolf Haller and Heinrich Rutte. Wien: Hölder-Pichler-Tempsky. Band 1.

———. 2010. *From Hieroglyphics to Isotype: A Visual Autobiography*, ed. Matthew Eve and Christopher Burke. London: Hyphen Press.

Neurath, Marie, and Robert S. Cohen, eds. 1973. *Empiricism and Sociology*. Dordrecht: D. Reidel.

———. eds. 1983. *Otto Neurath: Philosophical Papers 1913–1946*. Dordrecht: D. Reidel.

Reisch, George A. 1997. How Postmodern was Neurath's Idea of Unity of Science? *Studies in History and Philosophy of Science Part A* 28 (3): 439–451.

Rheinberger, Hans-Jörg. 2010. *Historical Epistemology*. Palo Alto: Stanford University Press.

Stanford, P. Kyle. 2006. *Exceeding Our Grasp: Science, History, and the Problem of Unconceived Alternatives*. Oxford: Oxford University Press.

Stöltzner, Michael. 1996. The Auxiliary Motive in the Forest and in Optics. In *Encyclopedia and Utopia the Life and Work of Otto Neurath (1882–1945)*, ed. Elisabeth Nemeth and Friedrich Stadler, 113–126. Dordrecht: Kluwer.

———. 2000. An Auxiliary Motive for Buridan's Ass: Otto Neurath on Choice without Preference in Science and Society. *Conceptus* 33 (82): 23–44.

———. 2001. Otto Neurath. In *Ernst Mach's Vienna 1895–1930. Or Phenomenalism as Philosophy of Science*, ed. John Blackmore, Ryoichi Itagaki, and Setsuko Tanaka, 105–111. Dordrecht: Kluwer.

Stone, Abraham D. 2010. On the Sources and Implications of Carnap's *Der Raum*. *Studies in History and Philosophy of Science Part A* 41: 65–74.
Uebel, Thomas. 2000. *Vernunftkritik und Wissenschaft: Otto Neurath und der erste Wiener Kreis*. Wien: Springer.
Whewell, William. 1840. *Geschichte der inductiven Wissenschaften*. Vol. 3. Stuttgart: Hoffmann'sche Verlags-Buchhandlung.
Zemplén, Gábor Á. 2006. The development of the Neurath-principle: Unearthing the Romantic link. *Studies in History and Philosophy of Science Part A* 37: 585–609.
———. 2018. Theory-Containment in Controversies: Neurath & Müller on Newton, Goethe, and Underdetermination. *Journal for General Philosophy of Science* 49 (4): 533–549.
Zolo, Danilo. 1989. *Reflexive Epistemology: The Philosophical Legacy of Otto Neurath*. Dordrecht: Kluwer.

Part III
Logic and Semantics

Chapter 11
Neurath and the Legacy of Algebraic Logic

Jordi Cat

Abstract In this paper I introduce a broader context, and sketch an integrated account with the purpose of examining the significance of Neurath's attention to logic in early works and subsequent positions. The specific attention to algebraic logic is important in integrating his own interest in mathematics and combining, since Leibniz, the ideals of a universal language and of a calculus of reasoning. The interest in universal languages constitutes a much broader, so-called tradition of pasigraphy that extended beyond philosophical projects. I argue (1) that Neurath's works can be embedded in a richer intellectual landscape that includes developments in logic and their local reception in Vienna, and that his attention to logic developed a sustained symbolic standpoint – with semiotic and typographic expressions –; (2) that specific aspects of the work in algebraic logic became a standard and a resource in subsequent work often thought independent, while its value was steadily challenged by the separate goal of empirical theorizing and practical application in social domains – including in the areas of economics, history and visual communication –; that (3), in particular, the presentation of systems of algebraic logic by Neurath's sources such as Stanley Jevons and Schröder was not isolated from discussions of political economy; and finally (4) that some of his positions in matters of language, unity and epistemology in the articulation of logical empiricism and its debates are better understood in terms of shared but diversified acquaintance with pasigraphy, formal standards and logical projects.

J. Cat (✉)
Department of History and Philosophy of Science and Medicine, Indiana University Bloomington, Bloomington, IN, USA
e-mail: jcat@indiana.edu

J. Cat, A. T. Tuboly (eds.), *Neurath Reconsidered*,
Boston Studies in the Philosophy and History of Science 336,
https://doi.org/10.1007/978-3-030-02128-3_11

11.1 Introduction

The historical construction of Neurath's intellectual figure has long pitted him against the background, preferences, standards and positions of more canonical figures.[1] Against traditional philosophy, it has pitted a declared antagonism to metaphysics shared with fellow logical empiricists, also their related goal of a unified demarcation of the sciences. Against other logical empiricists' penchant for philosophical theorizing, the construction has pitted his social activism; against other scientific philosophers' commitment to the conceptual and methodological standards of the exact natural sciences, his theoretical and practical interest in the social sciences; against their theoretically normative project, his empirical version of representation and practical rationality; against their reductionistic theoretical model of unification of the sciences, his anti-reductionistic and social encycopedia model; against their commitment to the value of precision and organization exemplified in symbolic logic and its application in logicism, his pictorialism, his informalism and attention to the consequences of unavoidable imprecision of expression. The dialectic is also borne out by records of Neurath's own acrimonious personal exchanges, in a permanent tension between building consensus and momentum – through associations and movements – and introducing critical dissent.

This historical construction has been extremely valuable in its task of differentiating and integrating Neurath's views and activities in broader intellectual and social contexts and projects. But it has succeeded, as any model necessarily does, at the expense of features that require alternative accounts. In particular, it has left out the differentiating and integrating role of Neurath's symbolic standpoint and his sustained attention to symbolic logic.[2]

With this in mind, the established historical construct renders positively surprising a number of notable records and testimonies. For instance, Popper (1973, 51) begins his reminiscences of Neurath as a pre-1919 public figure with a reference to economics papers in his father's library, education papers read to the Social-pedagogical Society of Vienna and, last but not least, a paper read to the Philosophical Society at the University of Vienna: "I think it was on Schröder's *Algebra of Logic*." Similarly, the mathematician Karl Menger (1994, 61) – who led the Mathematical Colloquium in parallel to the meetings of the Vienna Circle – began his intellectual portrait of Neurath by referring to the fact that "Neurath had written a few papers on Boolean algebra."

If those records suggest idiosyncratic personal references, others, attempting to establish a new project and consensus in the history of logic, hardly can. The Harvard logician C.I. Lewis (1918, 389) listed in his survey of 1918 "titles of all the positive contributions to symbolic logic and logistic in the strict sense" and

---

[1] I have in mind, for instance, Uebel (1992) and (2007), Cartwright et al. (1996), Nemeth and Stadler (1996), Stadler (2001), Reisch (2005), Vossoughian (2008), Burke et al. (2013) and Sandner (2014).

[2] An early valuable discussion by Eckehart Köhler (1991) is otherwise cursory, dismissive and, equally significantly, isolated from his discussion of other work by Neurath.

supplemented works of his acquaintance with some he borrowed from bibliographies of Venn and Peano. Alongside works by the likes of Boole, Cantor, Cayle, Dedekind, De Morgan, Frege, the Grassmanns, Hilbert, Jevons, Peano, Peirce, Russell, Whitehead and Venn, among others, we find listed Neurath's and Olga Hahn's papers. In fact, two of Hahn's papers, her dissertation on the logical significance of coefficients in systems of logical equations and one with Neurath on dualism, Lewis indicated with an asterisk, marking those "considered the most important contributions to symbolic logic" (ibid.). References to Neurath's and Hahn's papers were included also by Alonzo Church (1936, 165) in his equally canonical bibliography of symbolic logic in *The Journal of Symbolic Logic*.

In this paper I introduce a broader context, and sketch an integrated account with the purpose of examining the significance of Neurath's attention to logic in early works and subsequent views. The specific attention to algebraic logic is important in integrating his own interest in mathematics and combining, since Leibniz, the ideals of a universal language and of a calculus of reasoning. The interest in universal languages constitutes a much broader, so-called tradition of pasigraphy that extended beyond philosophical projects. I argue (1) that Neurath's works can be embedded in a richer intellectual landscape that includes developments in logic and their local reception in Vienna, and that his attention to logic developed a sustained symbolic standpoint – with semiotic and typographic expressions –; (2) that specific aspects of the work in algebraic logic became a standard and a resource in subsequent work often thought independent, while its value was steadily challenged by the separate goal of empirical theorizing and practical application in social domains – including in the areas of economics, history and visual communication –; that (3), in particular, the presentation of systems of algebraic logic by Neurath's sources such as Stanley Jevons and Schröder was not isolated from discussions of political economy; and finally (4) that some of his positions in matters of language, unity and epistemology in the articulation of logical empiricism and its debates are better understood in terms of shared but diversified acquaintance with pasigraphy, formal standards and logical projects.

In Sect. 11.2., I introduce the project of algebraic logic as the specific project and set of resources that Neurath engaged most directly and explicitly. In Sect. 11.3., I examine Ernst Schröder's own project and its relation to mathematics and political economy, with attention also to Jevons. In Sect. 11.4., I focus on the culture of logic and psychologism in Germany and Austria, especially around the University of Vienna. In Sect. 11.5., I introduce Neurath and Olga Hahn, their education and more personal circumstances and Neurath's pre-1909 work and projects. In Sect. 11.6., I survey the products of their collaboration in the study of algebraic logic, especially Schröder's works. In Sect. 11.7., I track the occurrence and changing role that the specifics of Neurath's engagement in symbolic logic played in subsequent work in the human and social sciences and their philosophy. Finally, in Sect. 11.8., I trace the occurrence and changing role in his contributions to the inseparable projects of logical empiricism and the unity of science.

11.2 History of the Project of Algebraic Logic: The Path to Schröder

The late nineteenth century and the turn of the twentieth offered logicians and mathematicians developments and innovations that placed formalism at the heart of both disciplines but also pitted one against the other in those terms, probing their own and each other's foundations. Not surprisingly, we find the first presentations of symbolic logic and logicism, by, among others, Louis Couturat (1905/1914), C.I. Lewis (1918), Susan Stebbing (1930/1942) and Rudolf Carnap (1929), with a performative agenda to document and establish a new consensus and to emphasize its modernity.[3] And they do so by looking to the history of logic since Aristotle as a history of the discovery of the symbolic and formal essence of reasoning and a history of logic's relation to the parallel evolution of mathematics.

A cursory selection of highlights, mainly from Lewis' substantial survey, might prove helpful to introduce the reader to a legacy of problems and resources and the standards adopted from it. Carnap and Neurath themselves would not be alien to the same rhetorical and pedagogical use of historical narrative and, on logic, they would follow suit and refer to some of the key historical figures.

The validity of Aristotle's syllogistic arguments, it is noted, depended on the symbolic expression of the form of subject-predicate propositions and of the form of the argument, that is, independently of the specific words describing subjects or predicates. Similarly, mathematical knowledge required axiomatic systems that rested on definitions presenting necessary truths about certain objects in intuition (see especially Stebbing 1930/1942).

Within such a tradition, Leibniz featured prominently as a turning point and the source of the algebraic project, although not without occasional references to the medieval project of universal science, *mathesis universalis* – a project conceived in the Middle Ages by Ramon Llull and given a rationalist articulation by René Descartes. Leibniz noted the similarity between definitions and algebraic formula, that is, mathematical equations introducing quantities and fundamental concepts or else identity statements. He upheld the goal of the *characteristica universalis*, in the image of algebra, as the cornerstone of the possibility of knowledge of the world, as both a rational and universal symbolic language and the basis for a calculus of reasoning, *calculus ratiocinator*. Symbols for fundamental simple concepts (the 'alphabet of human thought') instantiated in reality could be combined according to rules of combination, in the *ars combinatoria*, to form symbols for complex concepts. In the universal mathematics, the intellect understands reality, since, for Leibniz, truths of thought (reason) capture truths of reality as conceived intellectually and created by God. Together, then, the constitution of language and the calculus formed a program for a general science.

[3] Lewis, expounding on Couturat's narrative – and his narrow focus on Huntington's recent formulation –, provided the main source for other authors.

Leibniz's language and calculus admitted of two equivalent interpretations: relations of intension, such as inclusion, between concepts and relations of extension, also such as inclusion, between individuals and classes. Thus A + B expressed both 'A and B' in intension, and 'either A or B' in extension. One calculational challenge consisted in interpreting addition (+) and subtraction (−) as also inverse logical operations. To introduce negative predicates, Leibniz had to take advantage of the Aristotelian model.

The adopted resulting standard was effectively this: Algebra provided a mathematical model of symbolic language and of reasoning; according to the model, the construction by symbols would take place, as Llull had imagined, according to mechanical rules of combination by operations, a 'universal calculus.' As a matter of application, two old methods and goals were equally preserved and presented in the new logical terms, namely, analysis and synthesis: the analysis of predicates in the intension of subjects and their synthesis, or art of invention (Lewis 1918, 5–9).

Other researchers such as Gottfried Ploucquet, Fréderic de Castillon and, especially, Johann Heinrich Lambert investigated a calculus of logic, unaware of most of Leibniz's ideas. They explored analogies with arithmetical relations that often lacked any logical interpretation such as the use of coefficients other than 0 and 1 – at least within an intensional perspective that considers things within the broader domain of possibility.[4]

The consolidation of the algebraic project was initiated in English-speaking nations visibly led by William Hamilton in Britain, alongside George Bentham, followed by Augustus de Morgan in the 1840s, George Boole in the 1850s and Stanley Jevons in the 1860s along with Hamilton's students William Thomson and Thomas Baynes. By then, British mathematics had undergone a major change with the introduction of the Leibnizian notation of the calculus through the early 1810s – by John Herschel, George Peacock and Charles Babbage, founders of the undergraduate Cambridge Analytical Society –, followed by the general formulation of algebra as a science of combination of symbols by Peacock in 1830, and a general calculus of operations by Duncan Gregory in 1840 (see Peckhaus 1999, 430–440).

This British revival of studies in logic sought to work within the tradition of Aristotelian logic, adopting its linguistic model of subject-predicate statements and can be characterized by the contribution of new symbolic notations (symbolic creativity, also diagrammatic, tracking old and new conceptions), the quantification of the predicate (extending the applicability of mathematical symbols and operations) and generally an extensional interpretation. Without adopting algebraic symbols, de Morgan included a treatment of the equivalence (and distribution) of operations of conjunction and disjunction.

Boole's mathematical recasting of logical reasoning rested on replacing the Aristotelian tradition of non-symbolic logic with a more psychologistic project, both more empirical and more formal, and still both descriptive and normative. This project consisted in determining and expressing laws of thought in terms of rules for operations on so-called elective symbols, symbols for selections from the universe

[4] Lambert was the most explicit in his analysis; see Lambert (1764) and Lewis (1918, 23).

of conceivable objects. From a formal standpoint, the rules corresponded to the algebra of operations on numbers 0 and 1 (see Boole 1854 and Lewis 1918, 51–52): '=' symbolized the relation of equality between classes with the same membership, or, equivalently, concepts with the same extension; 'x + y' represented strict disjunction, symbolizing the class of things members either of x or of y, but not both; '+' and '-' could be interpreted as strictly inverse operations; 'x-y' represented the class of members of x that are not members of y; 'x.y' or 'xy' represented the class of things members of both x and y. The operations, and their logical interpretation, satisfied properties such as commutativity and distributivity. Elective symbols, the variables, but not any coefficients, satisfied the so-called index law, $x^2 = x$. For Boole, then, logic provided the interpretation of algebraic relations that enter the system or calculus.

Without a relation of inclusion, from the extensional perspective the application of the equality sign to represent the 'is' connecting predicates and subjects required expressions for the quantification of the predicate, x = vy (see Lewis 1918, 56–57). Similarly, the logical interpretation of more complex functions of x, y or both, suggested the use of coefficients in the terms of the function's perturbative expansion. Algebraic methods allowed Boole to use logical equations to express one class as a function of the any relevant others (Lewis 1918, 68). Applied to propositions, symbols x or y represented the times when some proposition X or Y is true, but without any symbol for implication relation. The only option was to quantify the equation between times of truth for different propositions, x = vy.

It is worth noting that de Morgan and Boole used the algebraic treatment of logical relations of classes to express a further parallelism, namely, to the numerical relations between probabilities. Boole (1854, 259) presented the statement of parallelism between the two kinds of relations in a dual display of parallel columns of algebraic relations. With a new meaning, this format would become a graphic tool of algebraic thinking especially in Schröder and Neurath (see below).

Whereas Boole's interests lay mainly in mathematics, Jevons was interested in economics and mental and moral philosophy (see Sect. 11.3., below). Closer to those more psychological disciplines, thus, Jevons intended his logical calculus to be a calculus of terms in intension, that is, of meanings. Also in contrast to Boole, he interpreted the sum of terms as a disjunction of non-exclusive classes or meanings. The interpretation allowed Jevons to retain the relation a + a = a, which for Boole held no logical meaning (constrained by the arithmetical meaning). Jevons symbolized the negation as –A, rather than as Boole's (1-a). In these terms he introduced a principle of duality: a thing must either be or not be (Jevons 1869). In the relation a = b, a means the same as b (extensionally, a is identical with b). The relation supports his first principle of reasoning, the procedure of 'substitution of similars' (it should not be confused with the substitution for variables of their values). The product ab expressed the conjunction of meanings and it was commutative, ab = ba. The sum and the multiplication satisfied also a distributive relation, a(b + c) = ab+ac (see Jevons 1864 and Lewis 1918, 73–75). Jevons also discarded inverse operations such as subtraction and division.

Peirce, son of the mathematician Benjamin Peirce, integrated his interest in reasoning with the mathematical techniques of probability and statistics he applied in his work for the US Coast Survey and the Department of Weights and Measures. Accordingly, he followed Leibniz in conceiving symbolic logic as the general science of mathematical form, but focused its application on relations characteristic of logical classes or else on the calculus of probabilities (for instance, arithmetical operations multiplication and division). He extended the mathematical formalism to relative terms (from an extensional standpoint, an ordered pairs or triads, etc. of individuals), turning a calculus introduced by de Morgan into a systematic mathematical theory. He also introduced the relation of inclusion, being as small as, or material implication, equivalent to the meaning of $\leq$ and symbolized by $-\!\!<$ so that x $-\!\!<$ y reads 'if x is true, y is true.' With it, Peirce introduced nine principles about it and its relation to the operations of sum and multiplication (see Peirce 1867, 1880 and 1885; Lewis 1918, 79–83).

Following Boole, he also noted the algebraic parallelism between the logical and probabilistic relations, which he introduced by means of the two-column display. The distinction rested on the distinction between logical identity and numerical equation. The former, represented as a ≡ b, states that classes a and b have the same membership, all a's and all b's; the latter, a = b, states that the number of members is the same. The relation of proportion or frequencies of members is the basis for the calculus of probabilities (Peirce 1880).

In his survey, Lewis noted that, according to William James, Peirce's series of articles "Some Illustrations of the Logic of Science" (1877–78) are the sources of pragmatism.[5] While this connection might seem paradoxical, it might also make less surprising the relation in Neurath between attention to formal reasoning and the practical rationality of individual and social decision-making and calculation.

Schröder integrated his Leibnizean interests in algebra, reasoning and universal language into a neo-Leibnizean system of algebraic logic. He developed his system in the footsteps of not just Boole's, but Grassmann's efforts, linked to Grassmann's more familiar contribution to arithmetic, and to Peirce's developments. By the second decade of the twentieth century, only a few years after the posthumous publication in 1909 of the *Abriss der Algebra der Logik*, a helpful summary his compendious earlier 1890s *Vorlesungen über die Algebra der Logik*, Schröder's work (1890a, 1891, 1895, 1909) constituted to logicians such as Couturat (1905/1914) and Lewis (1918) the culmination of the algebraic tradition. Lewis (1918, 111) referred to it as the Boole-Schröder algebra, systematizing and extending earlier contributions on three fronts: the algebra of classes, the algebra of propositions and, after Peirce, a calculus of relations or relative concepts. The third development would provide the grounds for Russell's logicism with its alternative relation between mathematics and logic.[6]

[5] Lewis (1918, 79 n. 121). Lewis cited the title incorrectly as "Some illustrations of the Science of Logic;" the series consists of the essays "The Fixation of Belief" and "How to Make our Ideas Clear," "The Doctrine of Chances," "The Probability of Induction," "The Order of Nature" and "Deduction, Induction and Hypotheses."

[6] I am grateful to Volker Peckhaus for a helpful discussion of the significance of Schröder's contributions.

Like Peirce, and Jevons before him, also Schröder introduced his discussion of logic in the idealized language of mathematics and presented it as a critical development of Boole's ideas. Whereas Boole and Jevons had introduced their calculus of reasoning in the context of mental philosophy, or psychology normatively understood, Schröder introduced his ideas first in the context of mathematics, in particular, in his pedagogical efforts to systematize the science of numbers in general (see Sect. 11.3., below). He organized his compendium around seven algebraic operations. With them in mind he criticized Boole's use of algebraic relations to represent logical ones. With Jevons he criticized Boole's strict interpretation of addition, +, and the logical use of subtraction and division.

Central to Schröder's development of the logical calculus, in Boole's tradition, is the extensional interpretation of numbers and concepts, in such a way that the logical interpretation of the calculus reduces to the algebraic by adopting numbers as classes (with parts and wholes) and then concepts as classes to which they apply and propositions in relation to classes that render them true. The two logical calculi, of concepts and propositions, are reduced to one (Couturat 1905/1914, 3–4). The same approach led him to contribute to set theory (the so-called Schröder-Bernstein theorem about mappings between sets and the ordering of their respective cardinality; see Carnap 1929, 51).

Contributors and commentators alike have actively constructed the project of algebraic logic as a historically connected project. This is how commentators have assessed the significance of Schröder's work, as he did himself, only adding future-oriented claims about precedence and inspiration (for instance in O'Connor and Robertson 2009). Specifically, the two main dimensions of significance of algebraic logic are the relation to Leibniz's project and the evolving history of logic in relation to mathematics. This framework applies not just to Schröder's own assessment of the significance of his project, but also to Carnap's and Neurath's, including the project articulated in the Vienna Circle Manifesto and subsequent exchanges between them.

In this historical narrative, the project of a universal characteristic had preceded – that is, anticipated – logicism, while the universal calculus had preceded symbolic logic. Leibniz's project had anticipated also the analysis of relations and the arbitrary character of the choice of primitive concepts (Lewis 1918, 11).

Leibniz constituted the source reference for situating subsequent work, including Schröder's, in relation to either of his two connected symbolic projects forming the organon for a general science: the *calculus ratiocinator* and the *lingua characteristica* (or *characteristica universalis*). Thus, Jourdain (1914, v) and Couturat (1905/1914, 93) declared that the calculus of reasoning had been developed by Boole, De Morgan, Jevons, Venn, Peirce, Ladd-Franklin and, most systematically, by Schröder, while the universal *lingua characteristica* had been developed by Frege, Peano, Russell and Whitehead. They also admitted the caveat that Frege himself – although the lesser-known figure with little acknowledgment from his contemporaries prior to Russell – had explicitly pointed to Leibniz's precedent and intended his own symbolism to contribute to both projects, playing the connected roles of universal language and calculus of reasoning.

Subsequently, Jean van Heijenoort (1967a) developed the theme as a matter of a linking two strands in the history of logic, logic as calculus and logic as language.

Replying to Schröder's criticism in 1880 of Frege's *Begriffsschrift*, Frege himself pointed to Leibniz and Boole, and contrasted Boole's project of a *calculus ratiocinator* with his own project, closer to Leibniz, aiming also at a universal *lingua characteristica*.[7] Not surprisingly, Frege's new analytic method, philosophy of mathematics, logical calculus, logical notation and philosophy of language are inseparable. Russell's case and even Wittgenstein's are arguably similar, partly due to Frege's contribution. The notion of logic as a calculus expressed by Frege's propositional calculus rests on the conception of logic as a language, not just the mathematical standard of a formal system from axiomatics, with a regulated system of signs that enables his theory of quantification. It is along similar lines that Schröder's extensional calculus would bridge the gap between the models of axiomatics and set theory before Löwenheim's contributions – in the wake of the projects of axiomatization of set theory and the logicism of the *Principia Mathematica*.[8]

Below, I will emphasize the same connection between both projects in Schröder and then Neurath, who will place a sustained and stronger emphasis on the linguistic one. More broadly, Schröder's accomplishments and the significance of algebraic logic have been appreciated further in connection with the evolving and creative relation between logic and mathematics.[9] From this conceptual and historiographical perspective, logic is no longer an exclusive branch of philosophy developed by philosophers or rhetoricians (Peckhaus 1999).

In this regard, Grattan-Guinness (2004) has distinguished between two traditions within symbolic logic: algebraic logic and mathematical logic. Each evolved by engaging different developments in mathematics, or else by engaging them differently. And since the adopted ideal of mathematics was itself informed by logical virtues – e.g., in axiomatics –, the relations between logic and mathematics, whether in the pursuit of logic or the foundations of mathematics, were meta-logical.

The algebraic tradition, the one to which Venn had originally applied the label 'symbolic logic,' originated with the project a powerful representation of syllogisms in terms of symbols regulated according to mathematical theory. And mathematical theory extended to evolving algebraic theory, from arithmetic operations to differential operators, functional equations and group theory.

Operating with the Aristotelian focus on syllogistic reasoning, the algebraic project paid attention to the linguistic elements associated with the roles of subjects and predicates, and the copula binding them in propositions. But it also extended its

[7] See Frege (1879), Schröder (1880), van Heijenoort (1967a) and (1967b, 1–2), and Peckhaus (2004, 599).

[8] In *Principia Mathematica*, Whitehead and Russell pointed to both dimensions of Schröder's contribution: his notation as one, alongside those of Frege and Peano, which they will adopt and modify as convenient, and to his calculus, for instance, his explanation of the proposition that 'p and not-q imply r' is equivalent to 'p implies q or r' (Whitehead and Russell 1910, vol 1, viii and 123). In the wake of reviews by Venn, Husserl, Peano and Peirce's students, for instance, Whitehead and Russell's mention might have drawn an even wider international attention to the work of the isolated teacher at a provincial German technical college.

[9] See Couturat (1914), Lewis (1918), Dipert (1991), Grattan-Guinness (2000), Hailperin (2004) and Peckhaus (2004).

scope with the introduction of relations, or relatives, and the quantification of the predicate. In addition, it settled on their extensional classification in terms of classes, parts and wholes, eventually reaching out to set theory proper. On such interpretation, together with the algebraic focus on equations, logical connectives centered on relations of conjunction, disjunction, equality and logical equivalence, eventually extended by Peirce and Schröder with the asymmetric relations of inclusion and its logical expression in implication.

The central role of mathematicians in this tradition tracks the ascent in Europe of the intellectual authority of scientific practices through the nineteenth century, without its reduction to the new experimental status of psychology (Peckhaus 1999, 438–39). By the end of the nineteenth century, Schröder, a mathematician, had come to be considered both the leading representative of algebraic logic in Germany and the mathematician responsible for completing the Boolean project.

If algebraic logic was concerned with mathematics at the service of logic, mathematical logic was, by contrast, concerned with logic at the service of mathematics. The latter project emerged out of concurrent efforts by Frege and by Peano's Italian school to craft a new and more precise logical notation to further the logical rigorization of mathematics, in its concepts, its proofs and the systematic (axiomatic) organization of its statements. Alongside the new treatment of arithmetic and set theory, the inspiration came from the project of rigorization of analysis, primarily in the hands of Cauchy and Weierstrass. It was the rigorous and precise new language and rules of logic, at least the considered desirability and possibility of crafting and applying such standards, that enabled the project of logicism, initiated philosophically by Frege in the late 1870s and early 1880 and developed systematically in the 1910s by Russell and Whitehead. With their attention focused on proofs, they placed implication at the center of the logical project applied in logicism. It is worth noting that in 1913, during the same period, Norbert Wiener was awarded his doctoral degree for a dissertation comparing Schröder's and Russell's projects, especially on relations (Grattan-Guinness 1975). Schröder died in 1902, when Hilbert and Russell's ideas were setting new grounds. Renewed interest in his project was briefly facilitated by the posthumous edition of the relatively reduced survey of his compendium in 1909.

11.3 Schröder's Algebra of Logic: Calculus of Reasoning and Universal Language

11.3.1 Schröder from Algebra of Numbers to Algebra of Logic

In this Section I draw attention to Schröder's specific place in the history of symbolic logic by drawing attention to the following features: (1) his earlier foundational interest in arithmetic and algebra; (2) his historical self-awareness placing his project within the algebraic tradition he traced back to Descartes and especially

Leibniz and more immediately to the British project of Boole and Jevons; (3) his commitment to the two connected ideals of that tradition, namely, a universal language (pasigraphy) and a rational calculus; (4) his Machian naturalistic approach straddling a line between psychologism and logicism; (5) the close relation to mathematics strengthening the relation of logic to a more scientific philosophy and to exact empirical sciences; (6) the surprising but sustained role since Boole and Jevons of economics in the presentation and significance of algebraic logic; and (7) more technical aspects such as the quasi-axiomatic approach, the extensional interpetation, the centrality of relations and subsumption, status and interpretation of identities, the significance of univocality, and the principle of duality, or dualism.

In this subsection I emphasize the role of the foundational attention to mathematics as the context for the introduction of concepts such as equality, duality and univocality – that is, determinateness or precision – and the ideal of pure form that would play a role in the project of algebraic logic. I begin with a brief account of Schröder's academic life (see Dipert 1991 and O'Connor and Robertson 2009). Schröder's father, Heinrich, was born and educated in Munich and, after graduating from the University, became a physics teacher in different high schools in the Southwest of Germany, first in Munich and, when his son Ernst was born in 1841, in Mannheim. In 1860 Ernst graduated from the University of Heidelberg, where he had studied mathematics with Otto Hesse, physics with Gustav Kirchhoff and chemistry with Robert Bunsen. Two years later he received a doctorate for work under Hesse on the extension of the application of fractional powers in algebra to geometry, e.g., p/q-sided polygons. Next, after spending two more years studying mathematics and physics, in Königsberg, he qualified to teach mathematics and natural sciences in German gymnasiums and to lecture at the Eidgenössische Technische Hochschule in Zürich, where taught as Privatdozent until 1869. Back in Germany, he was appointed to teach mathematics and natural sciences at the Realgymnasium in Bade-Baden.

In 1874 he was appointed full professor at the Technische Hochschule in Darmstadt and in 1876 in Karlsruhe, where he stayed till his death. It was in the processes of developing his lectures in mathematics that by 1873 he had completed his comprehensive school textbook *Lehrbuch der Arithmetik und Algebra für Lehre und Studirende* (1873) and, prior his move to Darmstadt, a corresponding brief survey, *Abriss der Arithmetik und Algebra für Schüler an Gymnasien und Realschulen* (1874a) and a more abstract programmatic text, also for students, *Üeber die Formale Elemente der absoluten Algebra* (1874b). In Karlsruhe, after his move to the local Polytechnische Schule, he wrote *Operationskreis des Logikkalkuls* (1877), his directorial speech *Über das Zeichen* (1890b), the equally criticized and hailed monumental *Vorlesungen über die Algebra der Logik (exakte Logik)* (3 vols.,1890–1895) and "Über Pasigraphie" (1898a).[10] The brief ouline of his system, *Abriss der Algebra der Logik* (2 vols.) was edited by Eugen Müller and published posthumously in 1909 and 1910.

[10] See Schröder (1877, 1890a, 1890b/1892, 1891, 1892, 1895 and 1898a, b).

Schröder's evolution from work on the calculus of algebra to work on the calculus of logic has not been sufficiently appreciated in a way that exhibits the role of mathematics[11]: it's the path from *Lehrbuch der Arithmetik und Algebra* (1873), *Abriss der Arithmetik und Algebra* (1874a) and *Über die Formalen Elemente der absoluten Algebra* (1874b) to *Operationskreis des Logikkalkuls* (1877).

Schröder intended the texts on arithmetic to serve a pedagogical function associated with his teaching positions. Yet, they are distinctively ambitious in documenting higher-level modern developments and in attending to foundational issues such as the principled and sophisticated conceptual systematization of the discipline, its distinctive scope, its scientific status and its ideals (partly in the axiomatic tradition of geometry set by the Euclidean system as a model of science and education).

In *Lehrbuch der Arithmetik und Algebra* his sources featured more advanced recent textbooks on arithmetic and number theory by H. Grassmann (1844), by Bertrand and Lejeune-Dirichlet and by Hankel. Grassmann, in particular, had introduced in 1844, in *Die lineale Ausdehnungslehre*, two connections between extensive magnitudes, a synthetic connection ($a \cap b$) and an analytic connection ($a \cup b$). Schröder's early extension of the discussion of mathematical operations to rules of logical reasoning was accompanied by another work, by another Grassmann. In 1872, Hermann's brother Robert introduced in a series of short books under the title *Die Formenlehre oder Mathematik* the idea that any object of thought was composed of (mental) pegs satisfying relations of identity, =, non-identity, ≥, and subordination, < or >, which Schröder adopted as well. In the second booklet, *Die Begrieffslehre oder Logik*, Grassmann introduced the algebraic notation applied to concepts, not classes (in the Aristotelian German tradition since Leibniz and especially Lambert, whom he cited; see R. Grassmann 1872, and Grattan-Guinness 2000, 157–58).

Schröder (1873, 1) advanced his conception of pure mathematics as the science of numbers in general ('Lehre von den Zahlen'), not of magnitudes ('Grössenlehre'). Algebra is general arithmetic and an instance of the presumed more general and abstract discipline he called absolute algebra. In this regard, Schröder's interests were clearly formal and systematic, well within the tradition of axiomatics, which by this time was starting to expand beyond geometry.

In *On the Formal Elements of the Absolute Algebra* (1874b, 26) Schröder addressed the ideal of a complete system of formal algebra introduced in the handbook of 1873, such that we can determine directly and with certainty whether for two arbitrary formulas, one implies the other, they are equivalent or they are mutually independent.[12] Logic didn't meet the formal standard and could not address the formal decision problem without adopting a semantic indirect path through a real domain. Schröder's view was that absolute algebra must be strictly formal. In the handbook he already distinguished the domain of the formal – or absolute –

[11] But see relevant remarks in Grattan-Guinness (2000) and Peckhaus (2004).

[12] Yet again, this text was surprisingly appended as a supplement on recent algebra to the Baden-Baden gymnasium program.

from the domain of the real; and formal algebra is a theory of pure connections (Schröder 1873, 180–181). From that standpoint, for Schröder (1874b, 27) the logical calculus was only an interpretation of absolute algebra. And within this framework he would develop Peirce's calculus of relatives (ordered n-tuples), that is, for relative concepts, and introduced matrices to tabulate them. It was ultimately absolute algebra, the formal theory of connections that makes logic and mathematics possible, that remained his ultimate goal (Peckhaus 2004, 566).

As a remedy for the problem of completeness, he suggested the project of completing the methods of logic so that the consequences of any set of premises may be proven to be derivable. The project of contributing to algebraic logic, then, grew out of the formal considerations of algebra he had initiated in his handbook of 1873. To pursue it, he had to follow Boole in the assumption that equations were premises in a deductive chain and operations were logical operations acting on concepts (Schröder 1874b, 6). Thus, he explicitly points to the road paved by Robert Grassmann in 1872 and, especially, Boole 1854 (see Schröder 1874b, 7), and to the goal, he declared again in 1877, of offering the foundations and the technique of a completed logical calculus.

Central to his presentation of the logical system was the conception of signs. The fundamental role of the concept of number, rather than extensive magnitudes, as proper object of mathematics pointed to the level of representation. Schröder focused on the expressive elements. According to Schröder, signs expressing numbers may be either numerical ('numerisch'), or digital (his own English translation of *ziffrig*), – such as 1874, 5 or 6.13/2–7– or literal (*literal*) – such as a, 5x + 3, ab-2c, containing at least one letter (Schröder 1874b, 6; 1873, 357).

In turn, according to Schröder, the emphasis on signs, including letters, placed his conception of algebra, and the connective, formal ideal of absolute algebra, under the Leibnizian ideal of the scientific *characteristica universalis*, the universal symbolic language of *ars combinatoria*. In his review of Frege's *Begrisffschrift* and the introduction to the *Vorlesungen*, he would call it the logical ideal of a pasigraphy and the linguistic ideal of a world language (such as the popular Volapük) (see Schröder 1880, 81 and 1890a, 93–94).

The linguistic and intellectual (rational) dimensions of the calculus were linked to the focus on operations; and he subtitled the arithmetic books accordingly, *Die Sieben algebraischen Operationen* (*The Seven Algebraic Operations*). In particular, he focused on the intellectual manipulation of symbols through the regulated application of operations (the rules of application and definition were effectively the same). When he considered the commutativity of operations such as multiplication, he analyzed it into two non-commutative operations, left-multiplication (or composition) and right-multiplication (Schröder 1890a, 139). In the footsteps of Galois, commutativity expressed the place of group theory in algebra, and the assumption that groups were groups of permutations and their compositions.

One fundamental principle regulating the universal symbolic language concerned the univocality of the relation between numerical signs, and the operations establishing it: "two units can be connected to each other in only one way" (Schröder 1873, 17). More specifically, he added, "[w]e call the units of two quantities unambiguously connected to each other if each unit on one quantity is uniquely assigned to a unit of the other" (1873, 19). Univocality identified the rigor or precision that rendered the algebraic language a standard for others and it served the calculus by facilitating the use of operations as rules of inference and substitution – equations. Schröder's extensional approach is clear from the application of the criterion to collections or sets of objects ('Menge von Objecten'): "we call the units of two sets univocally connected if each unit in one set is uniquely assigned to a unit of the other" (1873, 19). He contrasted the univocal and the ambiguous mappings in two diagrams paired together side by side (1873, 20), which will set the visual example for the parallel, double-column presentation of dual sets of expressions. From the notational and presentation standpoint, the graphic application of duality expressed the Jevonsian, contrasting meaning.

The heart of Schröder's system lay in the extension from univocal equations to numerical comparisons and inclusions (subsumption, 'Subsumtion,' or subordination, 'Unterordnung') – after Robert Grassmann's (1872) development in 1872 of his brother's doctrine of extension of 1844. Schröder took mathematics as fundamental, and, with it, despite the formal ideal of absolute algebra, the primacy of domains of objects and the part-whole relation. This establishes his extensional – "real" – interpretation of concepts, and thereby its application to connections between propositions as a calculus of logical conclusions. Univocal relations of equality between concepts, he then objected, may be misleading and lead to false conclusions: by the algebraic properties of the relation of equality symbolized by '=', the equations 'metal = silver' and 'gold = metal' imply 'silver = gold' (Schröder 1873, 26). Just like numbers may satisfy relations of inequality and ambiguity at the conceptual level, $\sqrt{9} = +/-3$, identity must be distinguished from what he called logical subordination, or subsumption (1873, 26). Equivalently, substitution (operation or rule) must be distinguished from inclusion (1873, 32).

The extension of algebra to a logical calculus of reasoning required, then, extending relations of numerical comparisons to relations of logical subordination. For this purpose Schröder introduced the corresponding sign, €, so that a symmetric relation of coordination, like the case of composition and multiplication, may be decomposed into asymmetric relations of subordination. For instance, in the relation between concepts of metal and silver, silver € metal, or between numbers $\sqrt{9}$ € −3 (Schröder 1873, 27–29). As a mode of conceptual analysis, this is the meaning of the copula connecting subject and predicate in propositions or judgments of the form 'A is B' (1873, 30).

The relation is the basis for representing thought processes in a logical calculus. Among the basic principles he listed characterizing the relation of inclusion or subordination was the relation of transitivity that modeled after a classic syllogism: if A € B, and B € C, then from these premises it followed that A € (ibid.). Another important one connected the coordination with equality: if A is subordinated to a univocal, single-valued term B (B = b), then A = B. Similarly, if A € B and B € A, then A = B (1873, 30). The binary relation of equality was, then, clearly derivative. Note that Schröder introduced the symmetric sign without reversing the order of the terms A and B, hence expressing the asymmetric nature of the relation.

In his treatment of algebra in the mathematics handbook the relation of subordination already emerged as the fundamental relation in a calculus, to which he would devote first a small book mapping algebraic operations to logical operations, *Der Operationskreis des Logikkalküls* (1877), and then the celebrated comprehensive systematization and development of Boole's and Peirce's systems in the *Vorlesungen*. An entire system of logic, thus, emerged from the presentation of algebra in the *Lehrbuch* of 1873 by means of the application of operations and relations extended from numbers to propositions and inferences.

A related central idea was his conceptions of equations: two numbers are equal "if they represent the sum of one" (1873, 22), and "the sign = (read: equal) between two expressions was thus to be used only to say that these are (different) names for the same number" (1873, 23). Elsewhere he expressed the idea in more algebraic terms: "if one of the numbers can be matched with the other without residues" or "if they are 'identical', i.e., only (different) names for the sum total of one" (Schröder 1874a, 4). He gave the example of two hands each finger in one matching a finger in the other; a fundamental principle of number theory is that the number is independent of the order in the counting (1874a, 4). Different names, different hands. Each number is equal to itself.

Next, Schröder offered the following classification of equations: numerical, either correct, in accordance with the definitions of the numbers – $2 + 3 = 5$ – or false – $2 + 3 = 4$ –; literal, either synthetic, stipulated for some values – $x + 1 = x$ –, or analytic (formal), applying to all values – $a(b + c) = ab+ac$. He distinguished the relation of equality, as identical intension and extension, from the relation of correlation, symbolized by '(=),' to indicate in the identity of a value of domain member.

Synoptic tables and two-column presentations appeared first in the discussion of mathematics. Schröder introduced the dual format generally to present a synoptic classification into mutually exclusive kinds, for instance a nested classification of kinds of numbers (determinate/indeterminate, and then known/unknown and partially/fully, respectively) and a nested classification of kinds of equations (numeric/literal, and then right/false and synthetic/analytic, see Schröder 1874a, 358–359).

He introduced the two-column dual format to present a particular dichotomy, two sets of algebraic operations, from among the seven he identified in total, in an inverse relation: to the set formed by addition, multiplication and exponentiation he opposed the corresponding set formed by subtraction, division and root/logarithm (1873, 119). This is the generalized algebraic expression of duality, or dualism, that would provide a distinctive organizing principle in his systematic formulation of algebraic logic.

The total number of operations assumed another duality, between non-commuting operations such as left- and right-multiplication that compose each commutative operation. The inverse relation was important for the purpose of calculation and inference when the goal was to express an unknown value in terms of known ones. Like substitution for Jevons, the elimination of unknowns provided the central methodological correspondence, with its limitations, that Boole and Schröder postulated between the method of reasoning and the algebraic method of solving equations for unknown variables. Even despite the omission of degrees and the logical formulas of tautology and absorption that challenged the fundamental role of algebra, we can see at least how the determinateness or univocality of the solution was central to the project.

Having completed his school handbooks on mathematics, he turned to logic. Recent notices on the algebraic calculus of logic by Arthur Cayley in 1871 and James Ellis in 1873 came to Schröder's attention and enabled him to put his project of absolute algebra first and foremost within what was effectively Boole's British tradition of mathematical logic. First in *On Formal Elements of Absolute Algebra* (1874), Schröder embraced his ideal of a formal and complete system of logic and announced the plan to complete Boole's calculus. Next, in connection to the system of algebra in the handbook, he laid out the first definitions and axioms of his system of logic (without inclusion). In the 37-page monograph *Der Operationskreis des Logikkalküls* (1877) (*The Circle of Operations of the Logical Calculus*) he made sure to draw a connection also to the older and German precedent, Leibniz. Afterward, he would become aware of Peirce's own work in the British tradition and even corresponded with him. Peirce actually used Schröder's *Operationskreis* to teach at Johns Hopkins (Grattan-Guinness 2000, 161).

In the three volumes of *Vorlesungen üeber die Algebra der Logik* that took up the rest of his life, Schröder pursued the project of algebraic logic in the footsteps of Leibniz, the Grassmanns, Boole and Peirce. In line with the growing adoption in mathematical research of the axiomatic ideal, already in *Operationskreis* Schröder extended its application to the logical calculus. In addition, joining the English-speaking tradition led him to introduce English terminology and English translations of his own terms.

11.3.2 Schröder's Algebra of Logic

Schröder's hailed systematization and extension of a certain constructed tradition were based on the fundamental role he gave the relation of subsumption and the law of duality. Along with the basic definitions and axioms, in *Operationskreis* Schröder introduced the principle of duality ('Dualismus'), which he considered empirical ('empirische') in nature. Subsequent discussions suggest he meant a synthetic general statement based on instances. It is also a constructive instrument, in the classical conception of postulates. He replaced as the fundamental binary relation in the system that of equality with the asymmetric relation of inequality for numbers, inclusion for classes, subsumption for concepts and implication for propositions. The sequence tracks the roots of his system of algebra of logic to the relations and operations in algebra (see below). Notice that also in the *Lehrbuch*, the main axiom, of inherence of signs, tracking countable things as units, was inductive (Schröder 1873, 16–17). The law of duality, or dualism, established the systematic correspondence between theorems in terms of + and 1 and theorems in terms of x and 0; and it serves the methodological task of deriving new theorems.[13]

Within mathematics, especially geometry, a tradition of attention to formal duality, reciprocity, and higher-level formal analogies was developed in projective geometry, especially by French mathematician J.V. Poncelet, J.D. Gergonne and M. Chasles in the 1820 (with reciprocal theorems about points and lines). The inspiration for a broader methodological dual perspective has been traced in the case of the physicist James Clerk Maxwell (see Harman 1998, 159–161). Moreover, Grassmann's *Die lineale Ausdehnungslehre* (1844) was a key source of Schröder's treatment of arithmetic and algebra and provided a more philosophical context for the introduction of duality. Grassmann's conceptual framework included the dualist philosophy of polarity in the Kantian tradition and the post-Kantian tradition of *Naturphilosophie*, along with Schelling, Ritter or Schleiermacher. But dualism ran through the so-called algebraic tradition in the more specific notion of inverse operations such as addition and subtraction, multiplication and division. For instance, Jevons' calculus, after Grassmann's and even Boole's, exhibited more explicitly a duality between logical addition and multiplication (Peckhaus 2004, 570); but Jevon's work came to Schröder's attention only after his original introduction of the law of duality in 1877.

Schröder had then two mathematical sources, geometric and algebraic, but it is the algebraic one exhibited in Peirce's logic that Schröder emphasizes (also as part of his own conception of the historical significance of his project; see Grattan-Guinness 2000, 164). From a conceptual standpoint, the distinctive philosophical consideration that Schröder attached to the principle includes its "empirical" or inductive status; it is not proven a priori but derived from its application in restricted (synthetic) cases.

Commentators such as Couturat (1905/1914, 20) and Lewis noted, after Neurath and Hahn, the methodological value of the formal symmetry in the law of duality for the derivation of theorems. According to the principle of duality, from each valid general formula in a system of logic, another true one may be derived by exchanging signs for operations of addition and subtraction with signs for multiplication and division, and the symbol 1 with 0 (see Schröder 1877, 3).[14] Based on the principle, the axioms, definitions and theorems are organized in dual columns. Despite the nod to axiomatic standard, the application is rather loose, especially by contrast with formalizations by Peano, borrowing Schröder's symbolic notation, and then Hilbert.[15] Schröder would give the parallel presentation of dual statements a central place in the *Vorlesungen* (Fig. 11.1.).

[13] Peirce (1867, 251) had noted in passing such parallelism between the laws of union, or disjunction, and intersection, or conjunction, and was stressed by his student Christine Ladd (1880).

[14] Schröder (1873, 146) noted that the key notion was contained in the statement about the general character of the product of indeterminate quantities, in the *Lehrbuch*.

[15] Peckhaus (2004, 585–588) calls Schröder's geometry-inspired approach quasi-axiomatics; unlike derived principles, especially inductive ones such as duality, axioms are formal to the extent that they are derived from self-evident intuitions.

Definition ($3_\times$).	**Definition (3_+).**

Wenn es für gegebene Gebiete a, b und c zutrifft, dass zugleich

$c \mathrel{=\!\!\!\!\in} a$ *und* $c \mathrel{=\!\!\!\!\in} b$	$a \mathrel{=\!\!\!\!\in} c$ *und* $b \mathrel{=\!\!\!\!\in} c$

ist, so soll — kürzer — gesagt werden, es sei:

$c \mathrel{=\!\!\!\!\in} ab.$	$a + b \mathrel{=\!\!\!\!\in} c.$

Fig. 11.1 Dual statements in *Vorlesungen* (1890a, 196), vol. 1

Among the axioms and definitions three are fundamental (Schröder 1873, 5):

- Definition of equality (between two or more class symbols): two symbols are equal if they are different names for the same class.
- Axiom (of reflexivity): every symbol is equal to itself.
- Axiom (of transitivity of equality): if two symbols are equal to a third, they are also equal to each other.

Other axioms introduced the operations of addition and multiplication, their separate commutativity and associativity and their joint distributivity, the identity operation ($aa = a$, $a + a = a$) and the complementary or negation symbol ($aa_1 = 0$ and $a + a_1 = 1$). Schröder's axiomatic approach to establishing the independence and primitiveness of propositions appeared clearly in his proof of Peirce's law of distributivity for (logical) addition and multiplication, $a(b + c) = ab+ac$, arguing that the reverse direction cannot be proven and hence must be introduced as a separate axiom.[16]

Among the theorems is the so-called Schröder's theorem: $xa + ya_1 = 0$ is equivalent to $xy = 0$ and $a = ux_1 + y$, with u an arbitrary class. Alternatively, for $ax+bx_1 = 0$, the indeterminate concept x lies between a_1 and b (see Couturat 1905/1914, 39).

This system derived directly, but only partially, from the system of algebra he would call in the *Vorlesungen* the 'identical calculus' (Schröder 1890a, 157–167). On the algebraic model that had inspired the system in *Operationskreis*, equality by definition – nominal definition – became reduced to equalities between terms as rules of substitution that, like the kinds of equations in the *Lehrbuch* he had classified as formal, analytic equations, in the extensional interpretation meant they are just equalities valid for all values (Schröder 1873, 359).

Along with the identical calculus and the principle of duality – and the dual presentation for operations + and x –, in the *Vorlesungen* Schröder reintroduced the relation of subsumption, the asymmetric meaning of the copula in 'a is b' – 'gold = metal' – he had introduced in the *Lehrbuch* along with the symbol for it, € (and its mirror image, for 'metal = gold'). The relation completed the foundation of the system of logical calculus. Like the relation of equality, it was grounded in the

[16] Schröder presented the proof first at the 1883 meeting of the British Association of the Advancement of Science along with a summary of the laws for algebraic operations he had established in 1877 (see Schröder 1884a and 1884b).

following principles Schroder introduced as theorems (see also Peckhaus 2004, 573):

- Principle of identity (his term) (or of reflexivity): a € a.
- Principle (of transitivity): if a € b, and b € c, then a € c.
- Principle (of distributivity): if bc = 0, then a(b + c) € ab +ac.

The extensional relation of equality became derivative in the algebraic way he had introduced in the *Lehrbuch*: if a € b and b € a, then a = b (Schröder 1890a, 184). The algebraic roots introduce a tension, noted by Husserl in his extensive review and more recently Grattan-Guinness, insomuch as Schröder insisted again that the subsumption symbol denotes both inclusion and equality, thus challenging the independence of subsumption from equality, if not its primacy. But this is only a notational inadequacy rooted in the analogy with Grasmmann's general relation of non-identity, symbolized by $\leq$. In fact, in the *Vorlesungen* Schröder provided frequent examples of propositions and their relations in terms of inequalities. From the logical, not linguistic standpoint, subsumption became a model of inference between valid propositions, or in Schröder's innovative term, the foundation of the propositional calculus.

After devoting the second volume of the *Vorlesungen* to the logical calculus of propositions (Schröder 1891), in the third volume Schröder (1895) developed the algebra of relatives, that is, relational concepts or elements in the domain of thought. For instance, the first domain of thought is represented by a logical sum, or disjunction, over all it elements. Then a second domain can be constructed out of pairs of elements, i/j from the first in a certain ordered relation and represented by the sum over the pairs. The general form includes coefficients, a_{ij}, with values of either 0 or 1 that, for the purpose of solving a problem involving systems of equations for a system of relatives, Schröder represented with matrices.[17]

The representation of relatives would allow the formal and systematic representation and calculation of basic concepts not just in mathematics, but also in the logical analysis – and calculus – of human relations in, for instance, political economy and jurisprudence.[18]

Based on its relevance to two connected projects, the significance of relatives is twofold. First, attention to the formal representation and calculation of relations was the basis for strengthening the relevance of (modern) logic to (modern) mathematics, especially, as Frege, Russell and Whitehead showed, in the project of logicism. For Schröder, the algebra of relatives would ultimately support a logic of relatives; both were interpretations of the formal general theory of connections or absolute algebra as a foundation for all of science. In line with Frege's own project, Schröder extended the application of the calculus of relatives to arithmetic and declared as a goal the logical definition of number as a relative – 'number of' – and the deduction of all propositions about the concept. Pure mathematics would become "only one branch of general Logic" (Schröder 1898b, 46; see Peckhaus 2004, 597–598).

[17] See Dipert (1978, 287–318), Grattan-Guinness (2000, 170–72), and Peckhaus (2004, 591–96).

[18] By this time, discussions of logic included examples of human relationships and even a calculus such as Alexander Macfarlane's, extending Leibniz's combinatorial discussions; see Peckhaus (2004, 596, n.54).

Second, also in relation to Frege and more important for this paper, there is the relation to the tradition of artificial universal languages. Interest in such languages was revived in the seventeenth century mainly for religious and intellectual reasons and especially popular in Europe at the turn of the twentieth century for intellectual and political ones. For Schröder, the symbolic representation and formal understanding of basic relative concepts in the different sciences aimed at the connected ideals of an absolute algebra and a scientific universal language.

11.3.3 Logic Meets Political Economy. From Boole and Jevons to Schröder's Algebra and Pasigraphy.

In this section I examine how Jevons' and Schröder's attention to language and reasoning set a standard for the connection of mathematics, logic and economics so that Neurath's own case will become more persuasive and also more distinctive.

In relation to Leibniz's ideal of general science and theory of signs (*ars combinatoria*), with the combination of universal rational language (*characteristica universalis*) and the mechanical calculus of reasoning (*calculus ratiocinator*), Schröder's theory of relatives sought to provide both the semantical part, the symbolic language, and the syntactical part, the calculus (Peckhaus 2004, 598–599).

Already in the introduction to his school handbook, *Lehrbuch der Arithmetik und Algebra* (1873), Schröder had introduced the science of number in terms of the construction and manipulation of a sign language ('Zeichensprache'). Then, in *Der Operationskreis* (1877), the emphasis on the logic of operations associated Leibniz – and Boole – with the logical calculus; he was relying on an account of Leibniz's theory of signs by the philosopher and philologist Adolf Trendelenburg (1867).

Also Frege (1879) appealed to Trendelenburg's account of Leibniz's ideal in the preface to the *Begriffsschrift*. Frege pointed to Leibniz's ideal of a general *calculus ratiocinator*, noting that it depended on a general system of clear and precise notation. The language would exhibit how the structure of propositions could express meaning (van Heijenoort 1967a). It is from Trelendenburg's text that Frege borrowed the term 'Begriffsschrift,' or conceptual script or notation, for his more limited version, for arithmetic, of Leibniz's project.[19]

In 1880, Schröder (1880, 81) published a review of Frege's book in which he shared Frege's Leibnizian ideals but criticized him for concentrating on the logical calculus at the expense of the universal characteristic, the universal theory of signs expressing the universal classification that included algebraic signs and he now referred to as Leibniz's ideal of pasigraphy (general script). Frege replied defending the language component of his project in terms of the Leibnizian distinction, this time borrowing Trendelenburg's term *lingua characterica* (see van Heijenoort 1967b, 1, and Peckhaus 2004, 599).

[19] See van Heijenoort's introduction to the translation of the *Begriffsschrift* in van Heijenoort (1967b, 1 n. b).

In the first volume of the *Vorlesungen*, Schröder emphasized, again in relation to Leibniz – and now also Descartes –, quoting Trendelenburg, the significance of a symbolic language for algebra and logic as a sign language that constituted a scientific universal language and expressed Leibniz's the logical ideal of pasigraphy. The latter he distinguished from the linguistic ideal of a world language such as the popular Volapük (Schröder 1890a, 93–95).

The same year, he gave a formal address at the technical school in Baden-Baden titled "Über das Zeichen."[20] There he combined Trendelenburg's history of philosophy with accounts from philology, psychology and biology (Darwin's among others) to defend the general claim that human intelligence and civilization, with the reach of intellectual and practical goals, depended on the designative or symbolizing human activity, their use of signs standing for things. In particular, Schröder claimed, language and reason are inseparable.[21] This capacity, he claimed, animals possess in a much lesser degree.[22] In particular, the valuable use of linguistic signs stems from their introduction by the mind through the exercise of the capacities of association (similarity) and abstraction (selection or isolation) (see Schröder 1892, 3442). In the ideal language, borrowing words from Trendelenburg quoted in the *Vorlesungen*, signs stand in a more perfect and law-governed connection with things:

> A designation of this character, if extended over the whole field of the objects of thought, will, in contrast to the verbal sign, in its present greater or less indifference to the contents of mental images, be a *figurative* language of the ideas of the mind, an ideographic language, and, as opposed to the special languages of the nations of the world, a universal language of the thing, a pasigraphy. (Schröder 1892, 3444; original emphasis.)

For Schröder, however, Leibniz's ideal of pasigraphic science remained that, an ideal. He made the point in relation to Frege's *Begriffsschrift* and in the same terms in reply to Peano's optimistic statement, introducing his *Formulaire de Mathématique* in 1894, that "le problème proposé par Leibniz est (donc) resolu" (quoted in Schröder 1898b, 45). While, interestingly, Frege and Peano present their respective projects in relation to Leibniz's, Leibniz's problem, according to Schröder, remained the following:

> [E]xpressing *all* the notions which it comprises, adequately and in the concisest possible way, through a minimum of *primitive notions*, say "categories," by means of purely logical operations of general applicability, thus remaining the same for every branch of science and being subject to the laws of ordinary Logic, but which latter [sic] will present themselves in the shape of a "calculus ratiocinator." (Schröder 1898b, 46; original emphasis.)

In terms of language, he added, for "the categories and operations of this 'lingua characteristica' or 'scriptura universalis' easy signs and simple symbols, such as

[20] See Schröder (1890b), which appeared translated in *The Open Court* in 1892 as "Signs and Symbols," see Schröder (1892).

[21] Here he was following both Lambert and the philologist Max Müller.

[22] The essay deserves close comparison with the symbolic, or semiotic, approach in both Mach's *Contributions to the Analysis of Sensations*, of 1886, and Peirce's subsequent writings on semiotics.

letters, are to be employed, and – unlike the 'words' of common language – they are to be used with absolute consistency" (Schröder 1898b, 46; see also Schröder 1880, 81–82).

My purpose in this subsection is not to highlight only the way Schröder's project of algebraic logic was inseparable from his commitment to Leibniz's (and Descartes') philosophical ideal of a universal language. I want to highlight the neglected relation of Schröder's presentation of his language and calculus of logic to ideas in political economy.[23] Both dimensions set an enabling precedent for Neurath's own projects and, in particular, for the significance and limitations of ideals of universal language and rational calculation.

The relation of algebraic logic to economic theory preceded Schröder, as he himself noted. In *An Investigation of the Laws of Thought* (1854), Boole had stated that propositions connecting universal terms, as in the case of scientific definitions, their symbolic expressions can be connected by the equality sign; in other words, the proposition can be expressed by an equation. Then mathematical operations help explore conceptual relations. According to Nassau William Senior, wealth "consists of things transferable, limited in supply, and either productive of pleasure or preventive of pain" (Boole 1854, 59). If w = wealth, t = things transferable, s = limited in supply, p = productive of pleasure and r = preventive of pain, then w can be expressed as $w = st (p + r(1-p))$ (see Boole 1854, 60, 106 and ff.). The mathematical manipulation of variables allowed Boole to derive different conclusions about w.

The case of Jevons is more telling and would find a place both in Schröder's and Neurath's projects.[24] Jevons' interest in logic was inseparable from the question of its relation to mathematics, as he had learned from Boole and De Morgan. They had looked to algebra and the differential calculus and attempted to make the former the basis for the logical calculus of reasoning. In the same algebraic tradition of Leibniz, Jevons also linked the project of logic to the linguistic project of a universal symbolic language that would underpin logical reasoning and its application in the empirical sciences. He called this universal system of symbols first, in 1874, the *logical abecedarium* and subsequently, in 1877, the *logical alphabet* (Jevons 1874 and 1877).

Also after Leibniz's example, the universal language could be manipulated through permutation and combination in an *ars combinatoria* that rendered logic an objective mechanical science (Jevons 1874, vol 1., p. 198). With the right rules, this would in turn contribute to the mechanization of logical operations in the model of previous attempts to mechanize mathematical operations in the form of calculating machines.[25] Thus he referred to a *logical abacus* and more generally the *logical*

[23] Also in this respect Schröder's writings should be compared to Mach's and his appeal to a principle of economy of thought, rejected by philosophers such as Husserl and Jerusalem as an expression of psychologism, especially in matters of logic, the paradigmatic case of a priori; see Kusch (1995).

[24] I am grateful to Margaret Schabas for drawing my attention to the role of logic in connection to mathematics and its application in economics.

[25] The project was much in line with the standard set by the optimistic industrial culture of machines and engines prompted by the Industrial Revolution and placed at the heart of the economy. Jevons mentioned attempts by Blaise Pascal, Arthur Colmar, Charles Babbage and Alfred Smee (Jevons 1874, 123–124).

machine (Jevons 1874, 119–129). Schröder himself followed Jevons on many points, emphasized the mechanical nature of logical and mathematical calculations and the value of substitutions and, especially, elimination for the sake of solving logical problems by analogy with solving mathematical equations; and he explicitly discussed and applied Jevon's combinatorial method.[26]

What Jevons sought was a mechanical method of acquiring scientific knowledge from experience that stood between the informal inductive logic defended differently by Mill and Whewell and the formal deductive logic defended and articulated by Boole and De Morgan (Schabas 1990, 54). The difficulty stemmed from adopting the algebraic formulation of logic and the assumption that mathematics – and the other sciences – was somehow founded on logic (Jevons 1874, vol. 1, p. 154)[27]: "The mathematician is only strong and true as long as he is logical, and if number rules the world, it is logic which rules number." Quality precedes quantity, he added.

Logic was for Jevons both normative, and, after Leibniz, a precondition of reasoning, and also, after Boole and physical psychologists David Hartley and Alexander Bain, a branch of psychology grounded in the empirical study of the laws of thought. After Leibniz's precedent, the normative character of logic, and of mathematics, was inseparable from its connection to reality and the possibility of its knowledge. The relation is mediated by the linguistic dimension and rests on the respective empirical connections between signs and thoughts and thoughts and things, so that Jevons (1874, vol. 1, p. 9) could claim that the immediate object of logic, and of mathematics, are signs and the indirect ones the others.

For Jevons, the psychological laws of thought stem from the exercise of powers of the mind, after Bain, namely, identity, discrimination and retention. Accordingly, Jevons (1869) would introduce as basic the laws of identity (whatever is, is), contradiction (a thing cannot be and not be) and duality (a thing must either be or not be). The central place for the concept of equality – whether in the form of similarity or identity – Jevons expressed symbolically, after Boole, with the equality sign from algebraic equations.

The role of equations placed as a central principle of logic and mathematics the principle of substitution of similars in thought and language: reasoning in general, proceeds by the substitution of logically equivalent terms – and thoughts – as in mathematical calculation equations between signs substitute equivalent numbers, variables or functions. The principle also served the process of reasoning by deduction understood by Boole as a process of elimination of known terms, which in the analogy to mathematics, implied the elimination of known quantities in order to determine the value of the unknown quantity.

Logic and mathematics here rely on an extensional interpretation. And Jevons could apply the law of duality in a series of combinatorial substitutions to generate

[26] See, for instance, Schröder (1890a), Lesson 40, sect 26 and especially Appendix 6.

[27] The tension between the roles of mathematics and logic is discussed in Schabas (1990, Ch. 4.), and in Mosselmans (2007, Ch. 4.). The value of Boole's system, according to Jevons, lied only in an analogy between mathematics and logic, so that logic is the algebra of two terms, 0 and 1 (Jevons 1874, vol. 2, 293).

mechanically the logical space fixed by 2^n combinations of logically possible terms, positive and negative – the logical abacus. The result, Jevons hoped, was a 'natural classification' that supported and unified proper empirical knowledge.

The path from quality to quantity, from logic to mathematics, depended on the path from intension to extension through his problematic analysis of the notion of number. Also here the symbolic or linguistic dimension becomes prominent as is the psychology of experience. The application of signs alongside the extensional interpretation of terms provides the framework for his doctrine of units distinguished, psychologically, in space and time and named accordingly. Unity is an abstraction that a general term applies to each object in the extension. Schröder would adopt a similar view. And Frege, seeking a rigorous articulation of logicism, criticized them both accordingly:

> If we use 1 to stand for each of the objects to be numbered, we make he mistake of assigning the same symbol to different things. But if we provide the 1 with differentiating strokes, it becomes unusable for arithmetic. (Frege 1884/1968, 50.)

Frege criticized Schröder explicitly also on account of the particular psychological role of symbols in his account of mathematics Schröder introduced in his so-called axiom of Symbolic Stability: "it guarantees us that throughout all our arguments and deductions the symbols remain constant in our memory – or preferably on paper" (Frege 1884/1968, xxi).[28] Nevertheless, the criticism stood alongside Frege's own commitment to the ideal of universal language, the *characteristica universalis*, of the Leibnizian project of the logical calculus. In particular, it must be considered in relation to his own project of a logical notation, the *Begriffschrift* (Frege 1879), and, finally, in his own philosophy of language, with a distinction between sense and reference that sought to prevent relations of identity from collapsing into relations of symbolic equality and rules for substitution (Frege 1892/1980). Instead, it was Neurath who would allow for such a collapse from a psychological, typographic standpoint (see below).

The relevance of logic to Jevons' political economy followed the relation of logic to mathematics and the unity of the sciences: "logic is the superior science, the general basis of mathematics as well as of all the other sciences" (Jevons 1874, vol. 1, 156). In order to think of economics as founded in logic, Jevons sought first to understand economics as a mathematical and not just a logical science, that is, based on considerations of occurrence and absence and of reasoning that are quantitative, and not just qualitative, matters of degree (Jevons 1871, 8). Therefore, a theory of economic quantities required a mathematical science. Mathematics, then, facilitated the logical foundation of economics as a rational calculus.

Even if Jevons intended economics as an empirical science, a logical foundation would justify the application of algebra (besides the calculus), which, in turn, facilitated deductive reasoning and a quasi-axiomatic construction after the

[28] Before his Leibnizian commitment to pasigraphy, Schröder had embraced this symbolic and typographic standpoint in the *Lehrbuch der Arithmetik und Algebra* (1873).

application of the geometrical standard of deductive structure to mechanics (that is, along the lines of Mill's methodological monism). Jevons' concerns clearly followed in the footsteps of British methodologists such as John Herschel, J.S. Mill and William Whewell. Both pure mathematics and its application in physics and economics are based on the mechanical nature of objective formal logic and, in particular, on the ideal of a natural classification, which for Jevons rested on the application of the general principle of substitution of similars through the use of equations (see Schabas 1990, Ch. 5. and especially Mosselmans 2007, Ch. 4.). The application of substitution was relevant because it was general, and it was general because it was ultimately a general conceptual, qualitative activity underpinning the very method of inquiry, acquisition and application of general scientific knowledge: "The whole value of science consists in the power which it confers upon us of applying to one object the knowledge acquired from like objects" (Jevons 1874, 1).

Then the logical framework licensed the use of analogies between physics and economics as well the bridge between qualities and quantities by adopting units of economic experience: pleasure and pain. The extensional interpretation of terms raised the additional issue of relating variables representing quantitative features of individuals to features of groups that embody general quantities and general relations between them. The solution involved the application of concepts and techniques involving probabilities and statistics such as the so-called fictitious mean and the law of large numbers.

Indeed, economics, wrote Jevons, is the 'calculus of pleasure and pain' quantified along the dimensions of intensity and duration of feeling – the psychology of utility from Bentham. And utility was the net outcome of pleasure minus the pain in the relation between a commodity and its consumer. Jevons formulated the theory of economic exchange, the fundamental phenomenon in his theory, in terms of relations of inequality between utility values of something to someone, which he expressed by means of the symbols of equality and inequality (like Fechner expressed the psychophysical relation of stimulus intensities in similar terms). Prices for a certain quantity express the relation of balance and equality in the mind, and, as a result, economics, according to Jevons, was a branch of mathematical psychology. Still, the methodology of economics, its so-called logic, was based, as it was for Mill, on the physical, or concrete deductive method. For its application, Jevons (1871) relied on indirect measurement of utility and analogies to phenomena of equilibrium and motion in classical mechanics (see also Schabas 1990, Ch. 5. and Mosselmans 2007, Ch 4.).

Schröder introduced both an economic standpoint on the role of symbolic language and economic illustrations of the application of his algebraic formalism. As I have suggested above, he was partly following Mach's bio-economic analysis of cognition. With intelligence, Schröder, declared, comes the possibility of dishonesty and fraud. And the use of a sign, he added, is 'an artful act of dishonesty,' namely, we replace the thing with a name or sign, much like we exchange the thing for a bank note (Schröder 1892, 3441). In order to avoid fraud, he continued, we need to establish a form of equivalent compensation, a form of mental equation so that the sign "is equalised by its user substituting *in thought* for the sign at the proper time the thing itself; by his constantly associating with the sign, with absolute logical consequence, the notion of the thing" (ibid.). Any departure from logical consequence is, in that sense, "like a declaration of insolvency by a bank or by a merchant who does not cash on presentation the notes or drafts issued by him." In other words, it is a form of "intellectual bankruptcy" (ibid.). Once the reliable mechanism of coordination (denomination) is in place, the art of describing reality can benefit from an increase in the number of signs or terminology. In a later essay, "On Pasigraphy," he called it 'capital of denotation' (in the original, 'Bezeichnungskapital'), with the added note in the English version stating that "'Capital' is here to be taken in the sense of Adam Smith and Political Economy," namely, the part of someone's stock expected to afford him revenue (Schröder 1898a, b, 55).

Economic terminology appeared also throughout the *Vorlesungen*. There he used again the terms 'Beweiskapital' (Schröder 1890a, 310), 'wissenschaftliches Kapital' (1890a, 378), 'Wort-Kapital' (1895, 172) and 'Erkenntniskapital' (1895, 217), since the logical benefit relies on investing the capital of signs ("Kapital order Vorrat an Zeichen, über welche der Kalkul verfügt," ibid. 196). With an appropriate system of signs, correct logical inference and separating truth from error become best exercised with "the greatest possible saving of mental labor' and 'provident economy'" (Schröder 1892, 3443).

In the *Vorlesungen*, Schröder also engaged in the application of the symbolic universal language and its rules to economic concepts, even for the purpose of rational calculation and potential decision-making. But this is hardly alien to his own predecessors in his constructed historical tradition of pasigraphy and algebraic logic.

Explaining the practical value of signs, he quoted Leibniz at length describing the social value of signs in communication and the individual value in the convenient internal management of a large number of thoughts, or mental images of things. Leibniz presented the situation as a sort of mental economy, by analogy with the practice in large trading cities of not exchanging even money ('Geld'), but symbolic tokens such as stamps or records (Schröder 1890a, 40).

With the use of the symbolic language of mathematical signs, the calculus of reasoning aims to formulate conceptual problems in logic as algebraic problems to be solved through the application of rules governing algebraic operations. Among his examples, Schröder (1890a, 557–558) included Boole's analysis of the concept of wealth discussed above.

Schröder contributed problems of his own that sought to show how the manipulation of symbolic language could clarify or simplify instructions in order to enforce them. One of them concerns a practical matter of political economy in relation to wartime: "On a strategic railway, it is forbidden for a certain time to transport all goods except those which may serve war purposes if they are explosive or not intended for the coal and steel industry, and those which are intended for the coal and steel industry if they are not explosive or not useful for war purposes. The transport ban is to be simplified" (Schröder 1890a, 376–377). His solution involved the introduction of variables a = useful for war purposes, b = explosive and c = intended for the mining industry. Then, applying Peirce's law of distributivity, the transportation rule is $a(b + c_{|}) + c(b_{|} + a_{|})$, where '|' indicates negation. Applying a dual theorem, he then derived the expression $a(b + c_{|}) + c(b_{|} + a_{|}) = a + c$ (Schröder 1890a, 377).

11.4 Vienna, on the Foothills of the Icy Slopes of Logic

In this section I examine briefly and more broadly the rising relation of logic to psychology in the German-speaking academic landscape, especially in Austria. My purpose is to extend the record of intellectual sources with authors such as Mach, Petzoldt, Wundt and Stöhr and of academic opportunities such as the philosophy curriculum at the University of Vienna all of which paved the way for Neurath's acquaintance with logic, his research and his sense for its empirical application – especially challenging in the human sciences –, also the rhetorical use of its historical protagonists and its broader methodological value – especially in relation to the role of logical analysis in the context of logical empiricism.

In the wake of Hegel's death in 1831, the rise of psychologism (on its path to experimental psychology) marked the rise of scientific philosophy in opposition to speculative philosophy in mental sciences such as logic, epistemology, ethics and aesthetics. This form of naturalism replaced more generic forms of materialism in philosophy, urging that philosophical questions and answers be engaged on the grounds of "results" from empirical psychology and cultural history. For instance, an explanatory, genetic method aimed to replace the normative, critical, transcendental method in the Kantian tradition and to reformulate its problems accordingly: Helmholtz, Sigwart, Mach, Brentano, Wundt, etc. (see Kusch 1995). Perhaps the most contentious case was the application of this program to psychology. Psychologism, as critics labeled it, concerned the controversial process of an increasingly problematized relationship between experimental psychology and the discipline of philosophy.

It was Wilhelm Wundt who, besides having been recognized as responsible for the institutionalization of experimental psychology, has been considered the most important figure in the institutionalization of psychologism in relation to philosophy (Kusch 1995, 125–134). Out of his laboratory came, for instance, most experimental psychologists holding chairs in philosophy at German universities. For Wundt (1906), philosophy was devoted to the study of the genesis and systematic structure of scientific knowledge. The first volume of his *Logik* begins precisely engaging the late-nineteenth century debate over the unity of the sciences, natural and human – 'Naturwissenschaften' and 'Geisteswissenschaften' –, now extended to their own relation to philosophy. Wundt adopted a monistic perspective related to Mach's, claiming that the interrelation of different forms of knowledge simple expresses the diversity of perspectives on the common domain of experience. It is worth noting that to the third edition of *Logik* he added a third volume devoted to the logic of the human sciences, which would be decisive in Neurath's methodological thought (Wundt 1908).

Unlike Mach, whose principle of economy of thought Wundt (1907, 301) presented in the second volume, Wundt considered psychology as an empirical human science, coordinated with the natural sciences such as physiology, but not reducible to them. From that standpoint, psychology provided the grounds for establishing of general laws of thought from psychological facts and logic, part of the same inductive process, presents them in their regulative objective form: "Logic has to account for the laws of thought which are effective in scientific knowledge" (Wundt 1906, 1; quoted in Kusch 1995, 128).

As quantification was central to his defense of the scientific empirical nature of psychology, mathematics, especially algebra, provided a model that extended to logic itself. It is not surprising, then, that in this spirit he devoted a chapter to the algebraic tradition, focused on the quantification of concepts and logical equations. Along the way he mentioned its main protagonists, from Leibniz to Boole, Jevons, Venn, Peirce and Schröder. In the first edition, Wundt (1880, 221) mentioned Schröder's *Logikkalkuls*, while in the second and third, he cited the *Vorlesungen* (Wundt 1893, 250 and 1906, 249). Schröder himself had followed partly in Wundt's footsteps, consistent with the psychological underpinnings of Boole's project, and cited the first edition of the *Logik*.[29]

In Austria, as well as in Germany, the end of the nineteenth century marked the consolidation of experimental psychology as a discipline and its intellectual visibility and authority grew with the proliferation of alternative schools and projects such as psychophysics, child psychology, Gestalt psychology and psychoanalysis. Authors labeled – or denounced as – psychologists up to 1910 included Brentano and his school (Stumpf, Meinong, Höfler and others), members of the empiriocritical school (Mach, Avenarius, Petzoldt) and even the object theorists (Meinong, Ehrenfels, Höfler).

Philosophy in Vienna became taught and promoted in the areas of metaphysical, historical and scientific philosophy. The central academic positions became chairs in natural philosophy, history of philosophy and psychology. Stadler (2001, 78) has stressed the intellectual and academic roles of a fundamental polarization between scientific and metaphysical philosophies. The rise of scientific philosophy took place alongside the rise in prestige and success of scientific disciplines such as biology, physics, experimental psychology and the social sciences. From a methodological standpoint, this approach to philosophy combined different empiricist standards and formal standards, with attention to logic, mathematics and language, especially linguistic analysis and criticism.

In addition, on the philosophical front, the rise of scientific philosophy took place alongside the Catholic intellectual tradition emphasizing metaphysics, scholasticism and history, free from a (Protestant) sustained tradition of idealism and its later expression in neo-Kantianism, Herbatianism – or 'psycho-philosophy' – and radical versions of phenomenology.

It was the interest and precedents in the form of subjectivism and speculative metaphysics that would prompt the application of scientific philosophy with new,

[29] Partly in association with Mill; see Schröder (1890a, 177).

critical forms of objectivism and realism, most visibly first by the rationalist Bernard Bolzano inspired by Leibniz and grounded in exact logical and linguistic standards, and then his followers such as Robert Zimmermann and Kasimir Twardowski. Zimmermann applied the formalist standpoint to aesthetics and Twardowski applied it to phenomenology and, after leaving Vienna for Lwów, encouraging the development of logic (and the formation of the influential Lwów-Warsaw school of logic). Subsequently, the new objectivism became gradually grounded on more empirical scientific standards, visibly first by Brentano, in the footsteps of Bolzano and Leibniz, and his followers, and then by Mach and his followers in Vienna and also in Berlin such as Joseph Petzoldt. By the end of the century, in Vienna, other influential thinkers in those constellations included Alois Höfler, Friedrich Jodl and Adolf Stöhr.[30]

At the University of Vienna scientific philosophy gained a stable position. Franz Brentano taught at the University of Vienna from 1874 to 1895 (demoted to Privatdozent in 1880 on account of his marriage). His work was launched and informed by traditional intellectualist interest in Aristotle and Leibniz, combining metaphysics and logic, alongside interest in the new experimental philosophy. Thus, *Psychology from an Empirical Standpoint* (1874) was a leading contribution to the new psychologism, especially through his doctrine of intentionality of mental acts intending objects that might lack existence, which was perceived as a critical deflation of metaphysics. For Brentano, philosophy's intellectual role was fundamentally of guidance and synthesis, in a formal Leibnizian sense. This position set a legacy of replacing subjectivism, scholastic speculative metaphysics with a synthesis of formal and empirical, traditional and modern, philosophical and scientific, specifically, through attention to Leibniz and experimental philosophy.

Meanwhile, educational reform around the mid-nineteenth century paved the way for the academic autonomy of philosophy and its social significance. For instance, the possibility of writing doctoral dissertations on philosophical topics since 1872 and their requirement as a criterion of academic appointment freed the discipline from clerical control. In addition, the new autonomy facilitated in turn the development of scientific philosophy and its role in social and political reform. Here education became particularly salient. The growing perception of the instrumental value of education beyond the growth of disciplinary autonomy and disciplinary resources led to a growing interest in education in relation to formal tools for thinking and as a subject matter in relation to experimental psychology. It led also to the proliferation of education projects and associations with shifting intellectual and social significance, e.g., the increasing emphasis on its use for social reform the distinctive culminating, in the wake of the Soviet Revolution, in the Red Vienna period, with its cultural associationism, child psychology and adult education.

Besides formal teaching and publications, scientific philosophy was established and popularized also through the activities of the Philosophical Society of the University of Vienna. It was founded in 1888 around followers of Franz Brentano

[30] For a general overview see Johnston (1972), Schorske (1979), Janik and Toulmin (1973), Blackmore et al. (2001) and Stadler (2001).

such as Twardowski, a student of Brentano and Zimmermann, and Höfler, who trained as a physicist under Boltzmann, as a philosopher with Brentano and as a psychologist with Brentano's student Alexius Meinong, and eventually occupied academic posts in Prague and since 1907 in Vienna as professor of science education. For instance, his book on logic was a psychological introduction to logic, originally titled *Grundlehren der Logik* (1896), but as a subtitle indicated, it was actually intended as an introductory textbook to philosophy, the so-called propaedeutics in philosophy ('philosophischen Propädeutik'); a later edition was titled *Grundlehren der Logik und Psychologie* (1906). He nevertheless defended the notion that logic and mathematics were a priori forms of knowledge and, as a matter of psychology, only their application, but not their exact formulations or justification, might be determined by experience. Similarly, he defended the intellectual value of metaphysics to address unsolved scientific problems. As part of his educational efforts, Höfler also penned, for instance, a number of textbooks in physics. He considered it a discipline whose practice involved the empirical application of mathematical knowledge and was not informed by philosophy (see Blackmore et al. 2001, 237–276).

The Philosophical Society was recurrently led by Höfler, until 1922, with periods of leadership by Zimmermann (1890–1898) and Friedrich Jodl (1903–1912). Höfler and Jodl shared a naturalistic emphasis on the philosophical value of science, especially the role of psychology, which for Höfler extended to the science of pedagogy and pedagogy in the sciences, and for Jodl extended to a monistic and evolutionary view of intellectual, ethical and religious life (Blackmore et al. 2001, especially 276–314). Höfler's discussion of a lecture on comparison in physics by Mach in 1895 would help secure Mach's appointment to Brentano's chair. Later, after 1907, Hahn, Frank and Neurath, all likely to have been instructed by Höfler at some point, would be regular speakers.[31]

Despite the Society's enabling role if the development of scientific philosophy, there was a sustained interest in idealism is exhibited by the fact that by 1927, under the neo-Kantian Robert Reininger, the Philosophical Society acted as part of the German Kant Society (Blackmore et al. 2001, 281; Stadler 2001, 79).

In 1895 the physicist Ernst Mach was appointed to Brentano's old philosophy chair, renamed chair of 'philosophy, in particular history and theory of the inductive sciences' (Stadler 2001, 120). Mach stopped teaching in 1898 following a disabling stroke. In *Analysis of Sensations* (1886), *Popular lectures* (1896) and *Knowledge and Error* (1905), Mach introduced an avowedly anti-metaphysical, anti-subjectivist general scientific philosophy, centered on cognition, including science itself. His influential view postulated the existence of certain natural elements of psychological life (apprehended as sensations) and their representation by signs – with the distinctive role for language, logic and mathematics –, and was framed within a combination of energeticist, economic and evolutionary biological perspectives. This framework was widely shared by Schröder, Stöhr and others, facilitating further Neurath's adoption of similar perspectives and an appreciation for logic.

[31] For a record of titles for lecture delivered at the Philosophical Society see Reininger (1938) and Blackmore et al. (2001, 283–98).

The physicist Ludwig Boltzmann returned to the physics professorship he had obtained at the University of Vienna in 1873, and left in 1876, eventually succeeding his teacher Joseph Stefan in 1894; leaving again in 1890 to return in 1902 to take over also Mach' philosophy course, and effectively taking over Mach's chair until the year of his suicide, in 1906. For Boltzmann philosophy was ultimately philosophy of science to the extent that it was already part and parcel of scientific practice, for instance, in physics, the attitude towards atomism as an issue that separated physicists themselves. He adopted a psychological attitude to the concept of atoms as pictures and fictions rather than mere signs or real entities (Planck's position).

In the footsteps of Mach and Boltzmann, Adolf Stöhr, at the University of Vienna since 1885, was appointed to Mach's chair in 1911 and occupied it until 1921. In the new tradition of scientific philosophy, psychology was central to Stöhr's work, especially Mach's psychology and anti-metaphysical critique of language. Perception was a matter of psychology grounded in physiology and physics; philosophy, including ethics and logic, rested on psychology and philosophy of language. From this standpoint, Schröder, Höfler and Stöhr placed the analysis of names at the heart of formal knowledge. Like Schröder, Stöhr also adopted the example set by the use of symbols and operations in algebra and, without referring to Schröder, used the title *Algebra of Grammar* (1898).

He collected his lectures in a textbook with the emblematic title, in the footsteps of Brentano and Höfler's own, *Lehrbuch der Logik in Psychologisiender Darstellung* (1910). Stöhr's text benefited from the posthumous publication in 1909 of the first part of Schröder's *Abriss*, but he also referred to earlier texts and introduced key notions and symbols of Schröder's formulation, including the extensional interpretation of logical equation and identity equation in terms of the application of different names to identical individuals or classes (Stöhr 1910, 128–130). In a special chapter on logical calculus he introduced principles and operations of algebraic logic such as mathematical substitution and places it within the history of a symbolic tradition in deductive logic, with references to Leibniz, Boole, Jevons, Peano and Venn (Stöhr 1910, Ch. VI.). Stöhr, we may say, was closely following Wundt's *Logik*, which he referred to.

In Austria, as well as in Germany, then, also logic fell within the scope of psychologism. The commitment to the methods and concepts of the sciences as intellectual standards was reaching through the nineteenth century beyond the scientistic commitment to the resources of mathematics in Boole and others: logic as applied psychology or a branch thereof. The influence in this case of British empiricism became more powerful than the British appeal to mathematics in algebraic logic. From this standpoint, logicism, from Frege to Russell and Whitehead, would be considered a form of anti-psychologism.

Despite his teacher Bolzano's attention to mathematical logic without psychology, Brentano continued in the tradition of Aristotelianism and the psychological foundations of logic (both present partly in Boole and Schröder). Thus, Brentano introduced a phenomenological approach to judgments in logic. It included a theory of existential judgments with symbolic representation after Boole ('+A', A exists, and '–A', A doesn't exist), as well as his central concepts of mental presentation and intentionality.

As I mentioned above, both Höfler and Stöhr followed Brentano's example with different treatments, especially associating psychology and the more formal analysis of language. Logic found its renewed place in the development and the teaching of scientific philosophy, that is, as experimental and formal, associated with scientific concepts, methods and results. The emphasis on science and language connected doctrines old and new, from Aristotle to Leibniz, Boole and Schröder. In Höfler's case, it is emblematic that he intended the teaching of logic and psychology to constitute the introduction to the discipline of philosophy. In Stöhr's case, the references to Schröder's works prior to 1909 suggests he would have been teaching ideas of algebraic logic in his logic lectures prior to the publication of the textbook. But his bibliography aimed also to include recent work, and thus it included, for instance, the paper by Olga Hahn and Neurath on Schröder's discussion of Dualism (see below, and Stöhr 1910, 425 and 427; Hahn and Neurath 1909). Both Neurath and Hahn had likely already been by then in contact with Stöhr (see below).

From the standpoint of the visibility of logical treatments in relation to algebra and psychology, the first decade of the twentieth century closed with several landmark publications: the first volume of the third edition of Wundt's *Logik* appeared in 1906, the second in 1907, the third in 1908; then Schröder's posthumous *Abriss* was published in 1909–1910 and in 1910, Stöhr's *Lehrbuch*.

The teaching of logic became part of curriculum that reflected the new intellectual landscape, in which discussions of logic, mathematics, psychology and philosophy were inseparably entwined. The register of philosophy lectures for the same decade shows that logic was taught through a system of rotation among the faculty that included Höfler, Jodl, Stöhr and Laurenz Müller.[32] Müller was a Catholic priest teaching and writing in theology and philosophy of religion, but also interested in the philosophical significance of science, especially religious controversies such as the cases involving Galileo and Darwin's theory of evolution. Among them four professors, logic was taught regularly, almost every year:

- Winter semester 1901/02: Logic and Epistemology ('Erkenntnistheorie'), by Höfler.
- Summer semester 1902: Logic and Philosophy of Science ('Wissenschaftslehre'), by Jodl.
- Summer semester 1903: Logic with Attention New Reform Attempts, by Müllner (and History of Philosophy, by Jodl).
- Summer semester 1904: Logic with Attention New Reform Attempts, by Müllner.
- Summer semester 1905: Logic and General Theory of Method ('allgemeine Methodenlehre'), by Jodl, and Natural Philosophy, by Boltzmann.
- Winter semester 1905/06: Natural Philosophy, by Boltzmann and Stöhr.
- Summer semester 1906: Logic for Teacher Candidates ('Logik für Lehramtskandidaten'), by Stöhr.
- Summer semester 1907: Logic and Theory of Method, by Jodl.

32 I am indebted to Professor Karl Sigmund for providing the information; see Öffentliche Vorlesungen an der k.k. Universität zu Wien, Archiv der Universität Wien.

- Summer semester 1908: Logic with Attention New Reform Attempts, by Müllner, and Logic for Teacher Candidates ('Logik für Lehramtskandidaten'), by Stöhr.
- Winter semester 1908/9: Logic with Didactics ('Logik mit Didaktik'), by Höfler.
- Summer semester 1909: Jodl on Logic and General Philosophy of Science ('allgemeine Wissenschaftslehre'), by Jodl.
- Summer semester 1910: Logic for Teacher Candidates, by Stöhr.

11.5 Enter Neurath and Olga Hahn

In this section I place Neurath and Olga Hahn mainly in the academic context outlined above and lay out Neurath's early intellectual evolution prior to their work in logic in 1909–10. Besides the role of his personal and intellectual relation with Olga Hahn and her gradual loss of sight, I draw attention to Neurath's acquaintance with logic as part of his polymathic education alongside interests in mathematics, natural sciences, languages and history, and his participation in more foundational and methodological debates about the sciences, especially in relation to the human or social sciences. Besides the question of the distinction between natural and human sciences, his work in political economy placed him in a conflict between, on the one hand, historical approaches to the human sciences and, on the other, formal – exact – and psychological approaches. His attempts and reasons to resolve the conflict would set the stage for his incipient considerations of the methodological roles of both history and scientific logic and their synthesis at the service of rational empirical theory and practical decision-making.

Neurath's polymathic intellectual life and his inseparable commitment to social reform began at home. Otto Neurath's father, Wilhelm Neurath (1840–1901), was a Hungarian Jew that, like so many Jews living within the confines of the diverse Austro-Hungarian Empire, sought a new life in to the more liberal capital. He earned two doctorates, one in philosophy at Vienna and the second in Staatswissenschaften at Tübingen, turning into a prolific author of works in political economy and teaching back in Vienna first at the Technische Hochschule and later at the Hochschule für Bodenkultur. His scholarship guided his social activism – which included the value of education. Most prominently, he proposed economic reforms that were grounded in an ethical and social criticism of the neo-classical concept of marginal utility, which he understood to undermine the realizability of an economy that prioritized the use-value of goods and aim at social welfare (M. Neurath and Cohen 1973, 1–4; Johnston 1972, 192–193; and Uebel 1995).

His intellectual and academic outlooks are reflected in a letter submitted to the University of Tübingen with his doctoral dissertation where he suggested references from the social historian Wilhelm Roscher, the museologist and advisor to the Ministry of Trade Franz Migerka,[33] and the Rector of the University of Prague,

[33] Migerka was the organizer of a special exhibition on women at work at the Vienna World Fair of 1873 and married Otto Neurath's future maternal aunt.

Ernst Mach (see M. Neurath and Cohen 1973, 1). Wilhelm Neurath's ethical standpoint in his theoretical work was much in the spirit of an anti-religious emphasis on personal and social values urged by the new Ethical Movement of Wilhelm Jerusalem and Friedrich Jodl. All in all, his son Otto had in his father a model of polymathic versatility, encyclopedic learning and reforming zeal (Johnston 1972, 192–193). These features characterized his father's 13,000-volume library and father-son interactions (Neurath and Cohen 1973, 5).

Otto Neurath (1882–1945) began University studies in the fall of 1902. In fact, Neurath spent only two semesters at the University of Vienna, the 1902–3 winter semester and the 1903 summer semester. From the proportion of courses he enrolled in, it could be stated that his interests were mainly in mathematics and philosophy; but his voracious curiosity reached further.[34] In his choice of subjects he replicated the polymathic education of his recently deceased father.[35]

During the first semester, his more than 40 h per week included lectures in mathematics, history of philosophy, moral theology, physics, chemistry, forensic psychiatry, Greek mythology, comparative grammar (both in Indo-European languages and Arabic), and nineteenth-century literature.[36] During the summer semester he took Müllner's Logic with Attention New Reform Attempts, Jodl's History of Philosophy, Müllner's history of Philosophy of the Middle Ages, moral theology, experimental chemistry, world geography, European geography and an advanced course in Arab grammar.

During this period he became close with fellow students Hans Hahn and his sister Olga, already friends, Philip Frank and Anna Schapire, a student of history of German literature who would become Neurath's first wife in 1907 (Neurath and Cohen 1973, 6).

He would soon be moving to Berlin after the summer of 1903 at the suggestion of his father's friend the sociologist Ferdinand Tönnies, and added political economy to his subjects in a combination with history characteristic of the so-called Historical School represented by Gustav Schmöller, Eduard Meyer and others (see Uebel 2004, 12–30). The encounter with Tönnies and the decision that followed took place at the summer academy in Salzburg in 1903.

A report on the lectures and discussions was, significantly, Neurath's first publication and an expression of his interest, in his father's footsteps, in political economy, with distinctive combination of features that he considers interrelated: a historical perspective, a focus on "the connectedness of the social," an expectation of logically consistent foundations and the attribution of technical value in the solution of social problems.[37]

[34] The emphasis on mathematics is Marie Neurath's in her note to Neurath's memories (Neurath and Cohen 1973, 7). In a letter to his son Paul, Neurath placed the emphasis on the natural sciences to tell of his additional interest in the social sciences; see Uebel (2004, 16).

[35] See W. Neurath's autobiographical sketch, in Neurath and Cohen (1973, 2). W. Neurath died in March 1901.

[36] I am indebted for this information to Karl Sigmund; see Otto Neurath's *Nationale* for winter 1902–3 and summer 1903, Archiv der Universität Wien.

[37] Neurath (1903); see especially the translation of the first and last paragraphs in Uebel (2004, 16).

Tönnies' advice that Neurath study under Meyer and Schmöller in Berlin was prompted by a seminar essay that Neurath wrote around the same time and succeeded in publishing early in 1904. The essay applied a historical interest in antiquity to the question, which had interested his father, of interest on money, that is, the price of credit.[38] From a methodological standpoint, Neurath adopted a familiar approach: he declared that questions about the spread in the adoption of the practice of demanding a fixed percentage of borrowed capital required both a comparative study and the use of several sciences (Neurath 1904/2004, 112). His attention did not focus only on different relevant social conditions for the different introductions of interest practices, including Egyptian "provision in kind." Another aspect stands out: the very topic concerns the decision to set a certain percentage as the calculation of a measure of value, loosely combines his growing focus on issues in economics with the longstanding interest in mathematical calculation and reasoning – the focus of algebraic logic.

In the same essay, Neurath also hinted at the broader issue of language from the standpoint of the textual analysis of historical sources and risk of anachronism in comparative analyses. The representation of facts required the sort of interpretive activity that philogists such as Augustus Böckh had introduced in historical method (Neurath 1904/2004, 111). By then German historicism had fully followed in the footsteps of critical history in the hands of philologists and Biblical interpreters. A decade later he would further remark that the task was no mere deductive matter of inferring facts and intentions from an author's principles and, once in involved in logical empiricism defended the values, would continue defending the methodological values of interpretive methods such Weber's *Verstehen*, although understood within the framework of empirical physicalism, free from metaphysical connotations.[39] Of course, this was another example in which informed judgment had to tackle ambiguity and supplement mechanical calculation.

In 1906 in Berlin he wrote a dissertation under Eduard Meyer – and secondarily, Gustav Schmoller – on the economic history of antiquity, with an emphasis on Greek and Roman economies, and a separate discussion of Cicero *De Officiis*, his treatise on moral obligations, especially of politicians. Already here Neurath introduced another methodological resource: having noticed that systematic patterns of economic reasoning emerge beyond legal reasoning and philosophical inquiries, he settled on the prior identification of possible forms of economic organization, not unlike the introduction of definition in the foundations of an axiom system, and then tracked their realization through antiquity. Any such possibilities might be actualized in historical recurrence, yet without controversial historiographical pitfalls such as anachronism or the search for general laws of development or cycles.

His Berlin period provided the opportunity for strengthening his appreciation of formal tools in addition to historicism. In the *Lebenslauf* submitted along with his 1906 dissertation in Berlin, Neurath mentioned three teachers and thinkers: Tönnies,

[38] See Wilhelm Neurath's textbook, W. Neurath (1896).

[39] Uebel (forthcoming) elaborates on this apparent inconsistency in Neurath's position and its conflict with the established accounts of logical positivism in Uebel.

Simony and Itelson. Oskar Simony was a colleague of his father's at the School of Agriculture engaged in research on topics as varied as algebraic topology, mathematical forestry and radiation. Gregorius Itelson was, according to Marie Neurath, "probably the strongest influence upon him at that time" (in Neurath and Cohen 1973, 7).

A Russian-born Socratic and encyclopedic figure, Itelson was the earliest translator of Einstein into Russian and a philosophy teacher to Russian immigrants in Berlin, tragically killed by the same anti-semitic violence that he fled after Russian pogroms of the 1880s (Freudenthal and Karachentsev 2011). Neurath described Itelson in a letter to Tönnies of 1906 praising his stimulating dialectical, Socratic attitude and his multidisciplinary scholarship, while lamenting his isolating bachelorhood (quoted in Freudenthal and Karachentsev 2011, 113). His dialectical approach to philosophizing opposed the practice of solipsistic reflection through monologues.

Concomitantly, Itelson adopted an objective, universal and aprioristic conception of logic and mathematics. Logic is the science of all objects, actual and possible, and not a theory of empirical laws or normative rules of thought, lacking in necessity in one case and truth in the other. Even Husserl, he declared, ended up in the hold of psychologism. He expressed these anti-psychologistic views publicly in two lectures, "The Reform of Logic" and "Logic and Mathematics," before the attendants of the section "Logic and Philosophy of Science" at the International Congress of Philosophy of 1904 in Geneva. Mathematics, Itelson declared, is the science of ordered set-like objects. With these lectures, and his position on the scope of mathematics and logic, Itelson entered the debates on psychologism and the unity of science, and was later criticized accordingly by Husserl and Windelband (Freudenthal and Karachentsev 2011, 115–117). To mark his view, he introduced, along with Couturat and others, the term 'logistic' to designate technical symbolic logic – and also misleadingly suggesting logicism.

Itelson's apriorism extended to the interpretation of relativity theory, in which space-time and causality involve the application of logical elements and relations to an empirical world. He assumed that empirical science would be unified by psycho-physics, albeit subjected to a neo-Kantian idealist critique. In the same vein, metaphysics could be criticized through a logical analysis of language within the framework of science, a critical development of Leibniz's rationalism. For his doctrine he introduced the term "empirical rationalism."

It's not surprising that Neurath declared Itelson's philosophy a close forerunner of the tenets of logical empiricism, at least as presented in the Vienna Circle's Manifesto. For Neurath (1937b/1983, 191) it represented the dual role of experience and formalism in the critique of rationalist metaphysics. With more general and less formal developments, it reinforced Neurath's preoccupation with the terms of the unity of science debate, especially through the critique of metaphysics and this by means of a regimented universal language – that is, engineered and controlled in the pragmatic, technological attitude that for him connected thought and intervention –, something like his later *index verborum prohibitorum* (M. Neurath in Neurath and Cohen 1973, 7).

It is significant that Neurath would choose to refer to Itelson to acknowledge the value of logic as well as mathematics. For instance, in relation to the unity of science in the scientific worldview, he noted that "it was important to develop an account of all the sciences using only one kind of 'style'" and he "became convinced of the possibility of speaking about stars and about men with the same logical techniques and with the same scientific dispassionateness" (Neurath 1937a/1983, 178–179).

In relation to logical empiricism, more generally, Neurath emphasized the value of Itelson's logical and mathematical critique of rationalism:

> The "rationalism" that we rejected as metaphysical principle, as a supreme judge in Leibniz, is descending as it were to the level of science. The extent to which the auxiliary means of logic and mathematics are applicable when we wish to make predictions is shown to us precisely by experience. "Formal logic," which is mocked so much, will now become a major tool of committed empiricists who, what is more, are setting out to conquer the whole domain of science and reserve no propositions for that which one once called "metaphysics." Gregorius Itelson aptly named this attitude "Empirical Rationalism" in contradistinction to former "Metaphysical Rationalism." (Quoted in Freudenthal and Karachentsev 2011, 119.)

In Berlin Neurath might have become acquainted also with the work of Mach's follower Joseph Petzoldt, later founder of Berlin's Positivist Society. Petzoldt had written in 1895 an influential essay defending Mach's empiricist principle of causality in mechanics from Wundt's critique. Like Itelson later, and Einstein himself after both of them, Petzoldt adopted a formal approach in the tradition that associated causality with determinism, namely, in terms of a law of univocality (*Eindeutigkeit*).[40] In fact, this was part of a broader discussion of the application of mathematics through different uses and meanings of univocal equations, just as for Schröder it had been a formal matter of mathematical calculation and symbolic reasoning (see above). In the determination of physical events, according to Petzoldt, there is no ambiguity.

As Neurath's intellectual perspectives became densely enriched, so were his experiences of empirical conditions and the decision to represent them for theoretical and practical uses. On his return to Vienna, in 1906 Neurath acquired his first military experience serving in the voluntary army corps (see Neurath and Cohen 1973, 7–10). It would be his first exposure to military organization and its relation to war conditions, at least prior to his study of the Balkan Wars and the subsequent experience of World War One. This acquaintance added wartime situations to the historical laboratory of economic organizations and famously led him to the study and defense of economies in kind, that is, without recourse to universal monetary units for the calculation of economic equivalences and rational decisions. Meanwhile he undertook postdoctoral studies with Friedrich von Wieser and Eugen von Böhm-Bawerk, representatives of the Austrian School, on the formal marginalist doctrines of Johannes Heinrich von Thünen (see Uebel 2004, 28). On 5 March, 1908, Neurath gave his first lecture at the University of Vienna's Philosophical Society: "War and Moral Principles" (Reininger 1938, 29; Blackmore et al. 2001, 289). The title

[40] Petzoldt (1895) and (1900); on its influence on Einstein's development of the general theory of relativity see Howard (1996).

indicates the abstract and universalist perspective on the war situation consistent with his technical economic interest.

Through his early economic writings, Neurath was progressively but rapidly engaging the three foundational debates animating the development of the social sciences: the fin-de-siècle debates over method, or *Methodenstreit*, – inductive historicism of the Historical School vs the a priori realism of the Austrian School – and over the unity of the sciences – natural vs human sciences –, and the brewing debate over the role of values, or *Werturteilsstreit*.[41] About method, Neurath was aiming at a holistic synthesis of inductive historicism and Austrian a priori decisionism. The latter was based on abstractions – whether Aristotelian intelligible essences or ideal types – in abstract, deductive theories. Here, I suggest that his interest in mathematics and logic informed his technical focus on calculation and decision-making, and was reinforced by it in turn. More distinctively, the same interest played a role in how Neurath promptly acknowledged limitations in the empirical and practical applicability of formal rationality, especially in a theoretical and practical setting that was characterized also by his attention to historical conditions. The latter consideration, I want to suggest, would become the source for his most distinctive contribution to the Vienna Circle debates, but, I suggest, it cannot be separated from his sustained attention to projects based on formal calculation and reasoning.

In 1907 Neurath married Anna Schapire and during the following 7 years he took a job as teacher of political economy at the New Vienna Academy of Commerce, *Neue Wiener Handelsakademie*. It is during this period when he began writing his intellectually original and socially ambitious works in economics, beginning with three historical textbooks, in his father's footsteps, serving both pedagogical and social purposes: *Antike Wirtschaftsgeschichte* (Neurath 1909c), *Lehrbuch der Volkswirtschaftslehre* (Neurath 1910d) and the collection of readings co-edited with his wife, *Lesebuch der Volkswirtschaftslehre* (Neurath and Schapire-Neurath 1910). The historical framework considered and used history as a laboratory of possibilities represented by what in his dissertation he had first called "ideal limit forms," which could be multiply instantiated, with contextual differences, in different geographic areas and historical periods. We can note here aspects of both historicism and the Austrian-School abstractionism. With Anna he also edited a translation of Francis Galton's *Hereditary Genius*, a historicist, evolutionary and naturalistic exploration of both social Darwinism and planned interventionism, much in line with Neurath's own social and economic outlook.

During the same period, on his return to Vienna, Neurath began assisting his old friend Olga Hahn (1882–1937), who had recently lost her sight. According to Neurath's son, Neurath had already met the Hahn siblings at their summer villa around 1898, while accompanying on of his Gymnasium teachers who was a family friend (Neurath and Cohen 1973, 29). Olga lost her sight during the period Neurath lived in Berlin (ibid. 29–30). As many of her friends began neglecting her, Neurath

[41] Uebel develops this point in Uebel (2004, 12–15).

stepped in. It is after 1906 that his assistance and their companionship intensified, especially over shared intellectual interests and soon led to another fruitful intellectual collaboration. Her blindness would only have intensified the intellectual focus on their more abstract shared interest in mathematical and logical subjects she was already familiar with, far removed from the visual world of her painter sister Louise. Either Neurath himself or someone else would have been responsible for reading out new texts, identifying the symbolism, and typing up the papers resulting from their discussions.

Upon graduation from a Viennese Gymnasium, Olga registered as auditor at the Faculty of Philosophy of the University of Vienna for the winter 1902–3 semester – also Neurath's first term. Also according to her curriculum vitae, submitted with her application for a doctoral degree, her subjects were philosophy and mathematics; then she engaged in private studies and left the university in 1910 after receiving an Absolutorium in April (the Absolutorium certified having attended classes, without passing any final exam).

11.6 The Logic Papers

In this Section I introduce Neurath's papers on algebraic logic and his collaboration with Olga Hahn. I will show how the papers are related and draw attention to their focus on the themes of axiomatics, equality, univocality and dualism that I consider significant in at least three ways: they exhibit Neurath's technical engagement with the tradition of algebraic logic, also his adoption of a symbolic standpoint that includes semiotic and typographic dimensions, and they constitute some of the resources that are to play a future role in subsequent work, empirical, historical and methodological – his "philosophy."

Algebraic logic fell precisely in the overlap of Olga's subjects and in 1909 the posthumous publication of Schröder's *Abriss* followed the release in 1908 of Wundt's third and last volume of the third edition of his *Logik*. Stöhr's own logic text came out in 1910 for his semester teaching. Between 1909 and 1910 her studies led Neurath and Olga to collaborate in the joint writing of one paper, while each published others separately: three by Neurath and two by Olga. The following are the titles in the complete series, in order of publication:

*"Ernst Schröders Beweis des 12. Theorems: Für die identischen Operationen gilt das 'Kommutationsgesetz'" ("Ernst Schröder's Proof of Theorem 12, that the Law of Commutativity holds for Identical Operations") (Neurath 1909a).

*"Zum Dualismus in der Logik" ("On Dualism in Logic") (O. Hahn and O. Neurath 1909).

*"Eindeutigkeit und Kommutativität des logischen Produktes ab" ("Univocality and Commutativity of the Logical Product ab") (Neurath 1909b).

*"Zur Axiomatik des logischen Gebietkalkuls" ("On the Axiomatics of the Logical Calculus of Domains") (Hahn 1909).

"Definitionsgleichheit und symbolische Gleichheit' ('Definition Equality and Symbolic Equality") (Neurath 1910a).

"Über die Koeffizienten einer logischen Gleichung und ihre Beziehung zur Lehre von den Schlüssen" ("On the Coefficients of a Logical Equation and their Relation to Theory of Conclusions") (Hahn 1910).

They all addressed primarily Schröder's results and their interrelation is marked by corresponding citations. For instance, in the last of his papers, "Definitionsgleicheit…," Neurath cited not only his first two, but also quoted from Hahn's own "Zur Axiomatik…," which in turn cited Neurath's first, "Ernst Schröders…," and their joint paper "Zum Dualismus…."

In February 1911, Olga submitted her last paper as the thesis required to earn a doctoral degree. The work was evaluated and praised by Stöhr, who had included a survey of algebraic logic, including Schröder's system and was also that year's logic instructor. Olga's contribution was a method for solving systems of (logical) equations for connecting relatives, extending the results in Schröder's treatment, also from a historical standpoint, considering earlier presentations by Lambert and more recent work by Peirce's student Ladd-Franklin (Hahn 1910).[42]

One might assume that Olga's interests might have been somehow linked to work by her mathematician brother Hans, but during the same period, Hans Hahn was working on problems in the variational calculus, instead. The timing and subject of her research can be best understood in relation to her and Neurath's shared interest in logic and to an academic context in which algebraic logic resonated with the standards and interests of mathematicians as well as philosophers (even from the standpoint of psychologism). The academic context reflected the interest of German-speaking mathematicians in Schröder's work, and not work by the philosophers Frege and later Russell – both defenders of logicism. Similarly, I believe, the timing of Neurath's own attention to algebraic logic amidst his intense dedication to economics teaching, research and writing, can be better understood in relation to the timing of the new publications and Olga Hahn's studies.

What I am drawing attention to is ultimately the broad and connective character of Neurath's logic work, that is, over and above the technical meaning and value of his particular results. The articles place Neurath's intellectual life and work within the shifting mathematical and philosophical cultures of the turn of the century: (1) mathematics deepening and extending its claim to rigorous foundations and philosophy developing a new relation to the emerging experimental psychology; and (2) a specific tradition that integrated both, since Leibniz, around the linked projects of algebraic logic and universal language.

On the one hand, logic work tracks his personal relation to these cultures through his own education and his personal relation to the Hahns – part of a lifelong pattern of serial integration of intellectual and erotic bonds, first with Anna Schapire, then

[42] For the supporting documents, see "Ansuchen um Zulassung zu den Rigorosen", Rigorosenakt Olga Hahn, Archiv der Universität Wien, Sig. PH RA 3111 and "Beurteilung der Dissertation Olga Hahn (von Adolf Stöhr)", Rigorosenakt Olga Hahn, Archiv der Universität Wien, Sig. PH RA 3111. I am grateful to Karl Sigmund, Johannes Friedl and Christoph Limbeck for assistance.

Olga Hahn and last Marie Reidemeister. On the other, the articles on logic establish the disciplinary connection to the German-speaking mathematical and psychological communities, especially in relation to logic. What was at stake there is the application of the symbolic language and calculus of mathematics, embodying a certain standard of representation of the empirical world and of rationality of thought and action. Indeed, the disciplines of mathematics, logic and psychology provided the medium of intellectual production as much as its social content and context.

Neurath's intellectual activity was informed, I suggest, by a path that would take him rapidly from the icy slopes of logic to its foothills. In the next sections, below, I will argue that far from being an isolated episode, working in algebraic logic proved significant for his subsequent work and relation to fellow logical empiricists. The themes can be traced in the domains of economics and politics, including the project of pictorial languages, and in his shifting endorsement, positions and critical attitude within the projects of logical empiricism and unity of science.

Neurath's own three papers addressed the symbolic relation of equality and the property of commutativity of the logical operations sum and product applied to terms representing classes.[43]

In the first paper, 'Ernst's Schröder's Proof of Theorem 12, that the Law of Commutativity holds for Identical Operations,'[44] Neurath considered Schröder's dual proofs – equivalent according to the principle of duality – that ab = ba and a + b = b + a. From the extensional perspective, Schröder's Venn diagrams make clear that the logical product represents the intersection of classes, or domains, and the sum represents the union. Neurath is particularly concerned with Schröder's symbolic notation representing the product, 'ab', which omits a multiplication sign.[45] The products 'ab' and 'ba' are defined identically and are to be considered the "same" product.

The key to Schröder's proof is the introduction of both the logical (Boolean) operations and the identity relation in terms of the extensional relation of subsumption (see translation in the Appendix).

For instance, for the definition of identity:

If a € b and simultaneously b € a, it is said that $a = b$; and, if $a = b$ holds, so must a € b and b € a.

For the product:

If c € a, c € b, so holds c € ab; and, if c € ab, so hold c € a and c € b.

And for the sum:

If a € c, b € c, so holds $a + b$ € c; and, if $a + b$ € c, so hold a € c and b € c.

---

[43] In his brief and dismissively critical overview, Köhler (1991, 109–112) somewhat misleadingly refers to sets and set-theoretic operations, whereas Schröder's extensional approach, following Boole, is based on part-whole relations rather than membership.

[44] Neurath (1909a/1981, 1–3) and Appendix in this volume, 489–492.

[45] Here Köhler (1991, 110) speculates that the absence of the sign motivated Neurath to declare that one can "equate symbolically" 'ab' and 'ba.'

Unlike Schröder and Neurath, one can, with Köhler (1991, 111), introduce the definition of the left and right products (and sums) in terms of subsumption. Then, Schröder and Neurath applied the expressions above to the dual theorem

$$ab \in a, ab \in b \qquad \text{and} \qquad a \in a+b, b \in a+b.$$

Neurath was thus assuming the commutativity of the conjunction, and concluded that ab = ba, independently of the resulting commutativity of the product operation.[46] To conclude that the two products are identical, Neurath effectively proved that they are logically equivalent. But, are they logically identical? Neurath made this metatheoretical assumption, but effectively he adopted an explicit notational (semiotic), symbolic standpoint, stating that the asymmetric order of signs is an artifact of a linear, one-dimensional notation.[47] Then he concluded that the sign 'ab' for the product rendered redundant the 'symbolically equal' sign 'ba.' The equality 'ab = ba' is precisely what he called in the third paper, "Identity of Definition and Symbolic Identity," a symbolic identity, a notational relation.[48]

Following Schröder in his discussion of identity and equality (see above), Neurath distinguished the two interdependent elements: expression and denotation. Identity is based on the doubling up of symbolic representations – signs – of the same object or domain. The redundancy, and hence the difference between the terms, is expressive, symbolic, notational. From that standpoint, Neurath was in a position to advance an explanation for the asymmetry in the order of signs – the one-dimensionality of the typographic mode of construction of expressions also marked by the equality sign. Redundancy rendered only one forced choice notationally helpful and semantically meaningful. He was also in a similar position to critique Schröder for conflating mathematical and logical standards for the manipulation of signs, namely, an inadequate "reliance on mathematical nomenclature in logic."[49]

Neurath upheld the role of symbolic identity in elimination and substitution, that is, a regimentation of a symbolic logical language to form rules at work in its associated calculus. The focus on the symbolic level, however, undermined the logical, and algebraic, utility of equations as expressions of properties of operations such as commutativity.[50] From a calculational standpoint, and its application to the

[46] Köhler (1991, 111) introduces the symbol '&' for conjunction, which Schröder and Neurath, after him, didn't use, and he himself has noted the notational role in Neurath's argument.

[47] This typographic standpoint is of course a matter of both types and concrete visible token marks.

[48] See Neurath (1910a/1981, 19–21), and this volume 509–511. Köhler (1991, 111) notes, despite the asymmetry in the definitions of ab and ba, that Neurath seems to think that the equality 'ab = ba' is the same as 'ab = a.b'.

[49] Neurath (1909a/1981, 3) and this volume, 492.

[50] According to Köhler (1991, 111), it undermines the expressive power, namely, for representing the properties of operations. As a criticism, he also points to alternative treatments of the question of equality he considers more enlightening and influential such as Frege's (1892) "On Sense and Reference." Later, Carnap (1947), committed to his teacher Frege and to logicism, would offer also an intensional account of asymmetric identities in *Meaning and Necessity*. A different source is Waissmann's (1936/1977) Wittgensteinian defense of redundancy in language and the treatment of identity.

axiomatic construction of a system, the symbolic standpoint undermined the old conception, from Boole to Schröder, of equations as rules of substitution.

Still, the expressive, symbolic standpoint in algebra and logic belongs squarely in the Leibnizian tradition of a privileged universal language for the dual purposes of reasoning and knowledge of reality (with the demand that symbols express elementary real concepts). However, unlike for Schröder and Frege, Neurath's notational or semiotic focus on signs pitted against each other the two faces of the Leibnizian algebraic project: the language against the calculus. In Neurath, this standpoint would reappear in subsequent linguistic projects: the semantics of a picture language for statistical information and the structural public nature of his physicalism and syntacticism about language, especially the basic universal language for the unity of the sciences.[51]

Now, what's so significant about the property of commutativity in the application of a logical operation? The case of commutativity provided the occasion for addressing the complex nature of the relation of equality as the meaning of equations, and, in particular, of equality signs. It's easy to see in relation to both developments in logic and mathematics the broader significance of identities at the heart of the philosophical projects concerning language and knowledge, from idealism to logicism, namely, in relation to the notion of analyticity – not aprioricity – and the method of analysis. The latter would play such a central role in the debates among logical empiricists, especially as figures such as Carnap, Schlick and even Waismann developed earlier contributions of Frege, Russell, Hilbert and Wittgenstein. Schröder himself took up the issue, even before Frege did, in relation to the general and formal nature of general algebra in his *Lehrbuch* of 1873 and again throughout the first volume of the *Vorlesungen*.[52] Neurath's discussion continued in two follow-up papers.

In the second paper, "Univocality and Commutativity of the Logical Product ab," Neurath (1909b) turned to Schröder's more recent statements in the *Abriss*. On this occasion he insisted on the contrast between the arithmetical asymmetry between numerical factors from the mathematical operation of multiplication and the logical symmetry. The latter, he noted, is masked by the accidental "one-dimensional descriptive ordering" in the sequence of signs forming "the two names" for equally valid statements – that is, semantically equivalent, after Schröder, in terms of shared domain, whether this is operating as referent or truth-maker.[53] Neurath's pursued the question at the symbolic level with a proposal reported by the mathematician Karl Menger (1994, 61): the proposal consisted in an alternative, symmetrical expression of the product with a vertical arrangement of the symbols for each term, one above and the other below the symbol representing the product.

Neurath's attention to commutativity led him, beyond the purely symbolic level of presentation and representation of logical identity, to probe the relation in Schröder's system also at the level of conceptual structure. For Neurath, Schröder effectively relied on the more fundamental relation of univocality (see my discussion

[51] On the relation between syntacticism and physicalism, see Derek Anderson's chapter in the present volume.

[52] The first volume of the *Vorlesungen* includes a full section on the topic.

[53] Neurath (1909b, 18), and this volume, 506.

of Schröder, above). Commutativity implied that two products have the same value, while univocality, the uniqueness condition for determinate values and meaning of symbols and operations, implied that two domains have only one product.[54]

To the extent that Schröder tried to offer a proof of commutativity, Neurath noted, it did not rely on the definition of the products. In the case of signs standing for statements, their valid product represented simply their joint validity, their logical product without the algebraic distinctions.[55] Instead, then, Schröder's demonstration relied on the proof of univocality: that for any second value, (ab),' of the product, it would have to be identical to any other value (ab), (ab)' = (ab).[56] Schröder used the substitution of values in the axiom of products of subordinations[57] to do the same in the univocality condition for the product and replace the value (ab)' with the value (ba) and generalize to the equation ab = ba (Schröder 1909, 32).

In Schröder's system, Neurath concluded, univocality and its proof occupy a more fundamental place. This standard of formal meaning, central to mathematical calculation as well, would prove central to Neurath's subsequent scientific and metascientific (philosophical) explorations. In particular, many of his views would rely on judgments about the success of its applicability as a standard of clarity and precision to the linguistic representation of empirical cases, including the representation of any language of practical and theoretical social significance.

In the third and last paper on the identity of logical products, "Definition Equality and Symbolic Equality,"[58] Neurath insisted on the distinction between two interpretations of the equality relation that expresses the property of commutativity of the product: (1) ab and ba are defined simultaneously by denoting the same relation between two objects – a and b –; and (2) ab and ba are either two relations separately defined or separate instances of the same relation with the same results as the separate mathematical products (ab) and (ba).[59] The first is the case of symbolic equality, represented as ab == ba. The second is a relation of identity.

Nevertheless, with Axiom VI_x', above, and its reciprocal, Schröder could also define the products asymmetrically in terms of the more fundamental relation of subordination (see also Köhler 1991, 111).

Neurath found support elsewhere for his project of drawing distinctions between uses and interpretations of equality signs: in particular, in Petzoldt's and Wundt's attention to the exact representation of empirical causal relations in physical theory and psychology.[60] Neurath (1910b/2004) had been reading Wundt's *Logik* for a

---

[54] Neurath (1909b, 18), and this volume, 506.

[55] Nevertheless, with Axiom VI_x', below, for instance, Schröder could also define the products asymmetrically in terms of the more fundamental relation of subordination.

[56] Neurath (1909b, 17), and this volume, 505. Schröder's (1909, 30) more general introduction used the example of putatively different null domains.

[57] Axiom VI_x' is $(x \text{ € } a)(x \text{ € } b) \text{ € } (x \text{ € } ab)$ (Schröder 1909, 23).

[58] Neurath (1910a/1981, 19–21), and in this volume, 509–11.

[59] Neurath (1910a/1981, 19), and this volume, 509.

[60] Neurath (1910a/1981, 19), and this volume, 509. See Petzoldt (1895) and Wundt (1894) and (1893/1895).

review in the context of the debate over the disunity between the natural and the human sciences (more below).

For Wundt the issue of identity relations was a conceptual issue in logic and, relatedly, also in psychology, not just as a source of logic, but in relation to the problem of psychophysical parallelism, namely, the relation between (claims about) mind and matter. As mentioned above, Petzoldt sought an interpretation of mathematical physics in line with Mach's Humean empiricist standpoint. Relations in mechanics, then, could be causal only in the deterministic sense that the relations between mechanical quantities are univocal. From Wundt he borrowed the distinction between definitional equality, represented by the equation $c = s/t$, and causal equality, as in $v = gt$. As Neurath noted, in the physical case, Petzoldt considered definitions to be definitional equalities and, ultimately, identities.[61]

To accommodate the formal case in algebraic logic Neurath turned to Petzoldt's mathematical examples and introduced an extended formal classification.[62] In addition to symbolic asymmetries, as in symbolic equalities, Neurath pointed out that definitions provided asymmetries in the conceptual relation between terms on different sides of the equality sign, that is, one side defined the other, but not vice versa. Otherwise, definition equalities might also be entangled with symbolic equalities – type (1), above – and computational equalities – type (2). Neurath settled on the following classification:[63]

I. Equality: $(a + b)(a - b) = a^2 - b^2$.
II. Definition equality: $a + a + a + a + \ldots + a$ (b times) $== ab$.
III. Symbolic equality: $\sqrt[b]{a} == a^{1/b}$.

Neurath's linguistic standpoint, unlike Frege's, left no cognitive significance for the expressive difference in symbolic presentation. The symbolic approach paved the way for the syntacticism and physicalism that would come to characterize his social and empirical language and its unifying role.

In the same vein as the previous article, Neurath also broadened the scope of his discussion beyond the linguistic and conceptual to the structural. The place of the law of commutation in an axiomatic system he now addressed quoting a recent paper by Olga Hahn on the axiomatic dimension of systems of algebraic logic such as Schröder:

> It differs from O. Hahn's axiomatization (*Archiv f. system. Phil.* XV, 1909, p. 347):
>
> "Similarly, one can prove that $a + b$ € $b + a$, whereby the law of commutation for addition and multiplication is proven. While in Schröder's presentation there is no reason for the introduction of the commutation law, here it becomes necessary due to the asymmetric expression of the definitions of addition and multiplication."
>
> The interesting fact that the commutation law in logic depends on the axiom system suggests the question of the degree to which commutation laws may be eliminated by

[61] Neurath (1910a/1981, 20), and this volume, 510.

[62] Note that as theoretical approach, Neurath had already adopted empirical classifications in his work in the history of economics, namely, with the classification of systems of organization of production and distribution of goods.

[63] Neurath (1910a/1981, 21), and this volume, 511.

> variations of the axiom system. Today a systematic approach to axiomatics is becoming more and more a requirement. (Neurath 1910a/1981, 20, my translation in this volume, p. 510.)

This article was Neurath's final attempt at exploring identity relations and dismissing the substantive significance of the law of commutation – e.g., over the law of univocality, as in the second paper – in a logical system. And it echoed the history of axiomatics in its formalist unmooring from absolute foundations, in intuition or otherwise. The role of the commutation law is, like the postulate of parallels in systems of geometry, relative not just to symbolic, typographic, conventions, but also to each specific axiomatization. Here Neurath might have been relying on his readings of Poincaré and Helmholtz, for instance, with the Hahns and other fellow members of their informal discussion group.

Not only was Neurath exploring the regimentation of symbolic language through the analysis of kinds of identities and the role of univocal determination. He was also exploring the relativity of logical systems much like his historical models of economic systems, also acting as systems of calculation and representation. It is the application of such standards, the absolute system, absolute precision, and univocal determination of meaning and calculation, that his scientific and metascientific – or philosophical – work would challenge. Language, calculation and organization stand, fall, limit, multiply or evolve together. In this case, the analysis of the algebraic standard is the analysis in the Leibnizian tradition of an ideal language and rational calculation, and unified systematization as well.[64]

Neurath's attention to axiomatic structure as the standard for a (logical) system points to Olga Hahn's ongoing own explorations of algebraic logic. Her own first article, "On the Axiomatics of the Logical Calculus of Domains," followed Neurath's second in the same issue of *Archiv für systematische Philosophie*, the journal edited by Wilhelm Dilthey, Benno Erdmann, Paul Natorp, Ludwig Stein and Eduard Zeller and reflecting rather ecumenically the diversity of their interests. It was likely written around the same time in the course of their ongoing discussions of the literature, including Schröder's posthumous *Abriss*, published the same year.[65]

In the article Hahn addressed Schröder's loose attempt at an axiomatization of his logical calculus. Schröder proceeded as one might have expected in this regard, by identifying first principles and definitions and adding several others throughout the presentation. Schröder (1909, 25) discussed also in the *Abriss* the distinction between principles, definitions and postulates and adopted axioms as general and self-evident principles. Taking the relation of subsumption as fundamental, Schröder's (1890, 169) first principle is

Principle I. a € a.

The subsumption relation expresses the copula 'is' of predication and connects concepts – that is, as extensions over domains of elements – that may be distinguished

[64] Walter Dubislav (1931) devoted an entire monograph to definitions in 1931, in a book, *Die Definition*, published in the *Erkenntnis* series edited by Carnap and Reichenbach, a collaboration between the Berlin and Vienna groups. He distinguished, following Leibniz, the following kinds of definitions: nominal, real, causal and essential (1931, 24). He also listed Neurath's article in the bibliography, but, as in other authors' bibliographical mentions, didn't discuss it (1931, 155).

[65] O. Hahn (1909), and this volume, 507–9.

as subject and predicate (see my exposition of Schröder's work, above). Principle I represents the most general, hence 'formal' principle of identity, '*a* is *a*,' as in 'gold is gold' and 'white is white' (Schörder's examples, ibid.). This is different from the symmetric relation he identifies in Definition 1:

Definition 1
If (a € b) and (b € a), then $a = b$, and vice versa (Schröder 1890a, b, 184).

Nevertheless, also here the identity relation is based on denoting one and the same domain (Schröder 1890a, b, 185). Schröder (1890a, b, 186) derived then the theorem $a = a$. Taken as an extensional principle of individuation, the identity recovers its Leibizian rationalist metaphysical roots.

In the spirit that Neurath would echo in his third article, Hahn's declared that "it may be of some interest to discuss the various possibilities for an axiom system for this calculus."[66] It was the same heuristic – adopting pluralism about axiom systems as units of scientific knowledge – that had been greatly valued in the new formal development of mathematics, even mathematical physics, through the second half of the nineteenth century in the wake of its success in non-Euclidean geometry.

Her approach consisted first in adopting Schröder's basic identity statements:

Principle I. a € a.
Principle II. If a € b and at the same time b € c, then a € c.
Definition 1. If (a € b) and (b € a), then $a = b$, and vice versa.

Then she added an additional principle suggested by Schröder himself concerning the negation of a logical element:

Principle III. $(a_|)_| = a$.[67]

Hahn emphasized the fact that the relation of negation indicated by the stroke was univocal and invertible. Then she proceeded to derive a number of theorems in addition to the ones that followed from Schröder's original principles of identity. She concluded that the alternative axiomatic basis was sufficient to derive the entire calculus, as presented by Schröder.

For instance, she proved relations between the negation of the product and the sum of negations, and commutation laws for the operations of addition and multiplication. She did so echoing Neurath's position: "While in Schröder's presentation there is no reason for the introduction of the commutation law, here it becomes necessary due to the asymmetric expression of the definitions of addition and multiplication."[68] She also proved the theorem that ($b_|$ € $a_|$) = (a € b), whose significance lies in its heuristic value in the derivation of dual theorems.

The relevance of negation to duality was precisely in the focus of her earlier article, co-authored with Neurath at the very onset of their collaboration, "On

[66] O. Hahn (1909, 345), and this volume, 507.

[67] See Theorem 31, $(a_|)_| = a$, Schröder (1890a, b, 305).

[68] O. Hahn (1909, 347), and this volume, 509. However, Schröder's introduction of the operations in terms of subsumption is in fact based on asymmetric expressions.

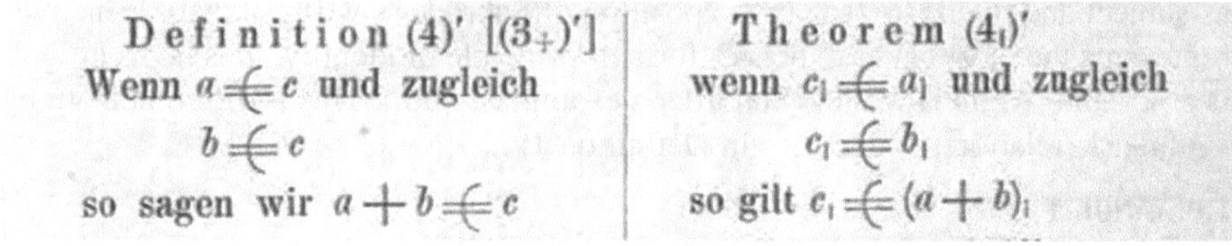

Fig. 11.2 Example of O. Hahn and Neurath's notation

Duality in Logic."[69] As in the case of equalities, the methodological and organizational roles of the principle of duality derive form its nature as a methodological principle of symmetry (see my discussion, above).

From a purely formal standpoint, in 1905 Louis Couturat (1905/1914, 20) had offered a pragmatic defense of Schröder's principle of duality: "Hence a true formula may be deduced from another true formula by transforming it by the principle of duality; that is, by following the rule given above [of interchanges and transpositions]. In its application the law of duality makes it possible to replace two demonstrations by one."

In the paper on dualism Hahn and Neurath set out to exploit the heuristic value of axiomatics to elucidate the origin and nature of the principle of duality in the context of Schröder's system. Schröder required the exchange of operation symbols – multiplication and addition – and of 1 and 0, and reversing the relation of inclusion. Instead, Hahn and Neurath require that classes be exchanged for their complements, which applies to 1 and 0. Negation plays a key role. This is precisely the main original contribution of their paper, relating the symmetry of duality to the complementarity of domains (Fig. 11.2).

For Schröder, complementary classes served a more distant, unrelated role. Hahn and Neurath objected that for Schröder the law of duality is an empirical – that is, inductive – principle suggested by a number of specific dual theorems about different operations. In their article, they nevertheless proceeded in the same extensional spirit, without seeking a proof.[70] It is worth noting that in a passing footnote, no doubt by Neurath, pointed out that the more precise and formal presentation of the Schröder's calculus in the *Abriss* came close to Itelson's conception of logic, the science of objects in general (see my discussion, above).

One question remains: in light of such contributions to the literature on algebraic logic, and the fact that they were hardly cited and discussed by any logicians, why were there included in literature reviews and bibliographies such as Couturat's or Church's? I suggest that the one should consider the function of such bibliographies and the logic surveys they accompanied. They played the role of identifying symbolic or algebraic logic as an intellectual discipline or area of research and track

[69] O. Hahn and Neurath (1909), in this volume, 493–505, and in Haller and Rutte (1981, 5–16).

[70] Köhler (1991, 112) objects that Hahn and Neurath fare no better by appealing to a list of theorems too, but that is just as well, since the principle is not an axiom or a theorem, but a metatheoretic statement operating as a proof procedure. For Köhler, again, implicit in their argument is yet another redeeming metatheoretic intuition.

its developments. With such a purpose in mind, their authors would have aimed at a comprehensive account of available sources, which might include minor contributions as well as more significant ones as well as works in larger formats with similar ambitions.

To conclude this section: in addition to the considerations of language and symbolic expression, also considerations of duality and complementarity would play immediately a role in Neurath's scientific and meta-scientific works. A cursory survey of examples of such roles is the subject of the final section.

11.7 Applications and Challenges of the Algebraic-Logic Standard of Language and Reasoning in the Social Sciences

In this Section I begin arguing that Neurath's attention to logic as language and calculus was hardly an intellectually isolated episode and I show, in particular, that elements of his own work and the literature he had become acquainted with played a significant role in his more familiar works in different areas in the human sciences – psychology, political economy and history – and related social projects.

11.7.1 Algebraic Tools for a Rational Economic Calculation In Kind: From War Economics to the Calculus of Pleasure

In Sect. 11.5., above, I have described how during the few years prior to 1909 Neurath had been undertaking projects that involved a certain affinity between different disciplines such as economics, history, mathematics and logic. He was also involved in debates over the unity of the sciences and the methods and role of values in the social sciences. A central element in those debates was the role of precise description and reasoning as a standard for scientific rationality and objectivity; logic and mathematics, synthesized in algebraic logic, embodied just that standard.

Then, his activities in 1909 gave way to a period of integration of different specific elements from the work in logic in (1) the pursuit of specific empirical and social projects with historical dimensions, and (2) the participation in debates of methodological, philosophical, or meta-scientific nature that continued through changing circumstances into the period of logical empiricism, in Vienna and exile. In fact, the two were inseparable, each constraining the other in varying ways.

In Sect. 11.3.3, above, I have argued that Neurath's detailed acquaintance with algebraic logic, especially in the works of Schröder and, either through it or separately, of Jevons and others, didn't drive a wedge separating his interests in disciplines such as political economy. Economic thinking appeared also in illustrations and applications through the algebraic tradition: in Boole's logic writings, especially in Jevons' works in logic, methodology and political economy, and finally in Schröder's, which also recounted some of the previous instances.

The form and purpose of algebraic logic were hardly independent of the project of economics. In particular, they resonated with the themes and terms of the debates over method – the so-called *Methodenstreit* – that confronted Neurath's own teachers and local authors. While the historical school of Schmöller, Meyer and others defended empirical research and inductive generalizations, the Austrian school of Carl Menger and others defended an a priori approach based on definitions and abstractions and the focus on rational decision. If Neurath rejected Menger's Aristotelian essentialism, Schmöller's inductivism and Weber's ideal types (Uebel 2004, 7), the resources and broader perspective of formal logic expressed in mathematical symbolism, I suggest, played a role in enabling Neurath to articulate a synthesis of inductive historicism and abstract decisionism.

Interest in the algebraic standard of representation and reasoning through a universal symbolic language and calculus provided Neurath with a standard of economic theory. Similarly, his review of Wundt's *Logik* placed his interest in logic and the social sciences on a higher level in relation to psychology and the debates over the unity of the sciences. A productive resource turned out to be the growing ties between logic and economics, between psychology and economic theory and between logic and psychology. And yet, in the process of exploring such connections in social phenomena, he soon identified both value and limitations in their application. And it was this realization that would prompt decisive and distinctive contributions to the debates over logical empiricism: For instance, his familiar defense of a limited kind of rationalism, with limited univocality, he distinguished from pseudorationalism.

In the post-1909 economics writings, Neurath sought to articulate a theory of economy in kind already identified in his historical analysis of 1906. The challenge consisted in integrating the goals of the decisionism of the Austrian school while defending a standard of rationality without abstract universal units of calculation relative to universal standard of profit. Neurath's alternative involved a comparison of multiple indicators – individual physical health, intellectual satisfaction, etc. – combined into different possible orders of life and correlated to subjective preferences although without the conditions for comparing individual or social utility. The role of utility was instrumental and formally objective, unmoored from any reductive psychologism or psychological realism beyond individual statements of pleasure attributions; and it focused on representations of actual and possible distributions of goods, that is, any goods with use value, or real income.

What Neurath explored and proposed was a so-called calculation in kind, without the market standard of decision such as profit and price as abstract and universal unit of measure in monetary calculations; and without a socialist, objective measure of value such as labor either. It was at odds with the Austrian school's rational choice theory, market capitalism, neo-classical marginal utility theory, even welfare economics.

Reasoning in kind was nevertheless both a consequentialist and a comparative approach much in the way the dominating models of practical rationality in economics. It assumed only the possibility of an ordinal ranking of pleasures and a comparisons of indicators, only with incomplete mechanical calculations to be performed in relation to market-free, non-monetary alternative considerations: use-value, productivity, economic efficiency and welfare goals. The multi-dimensional decisions were to be ultimately based on the exercise of judgment – and decision.

Wartime cases of economic administration illustrated how the model could be realized (Neurath 1910c). In the *Lesebuch* edited with his then-wife Anna Schapire, Neurath drew attention to the American economist Henry George for having noted the economic significance of wars.[71] Around the same time, Neurath objected that economists he looked up to such as Jevons and Pareto had hardly paid any theoretical attention to different kinds of economic crises.[72] It is worth noting that George's work presented familiar aspects of graphic interest: in a book on the case of industrial depression, *Progress and Poverty*, George (1882, 197) introduced a dual presentation reminiscent of Schröder's, except with a contrastive meaning, namely, displaying a contrast between 'the current statement' and 'the true statement' about rent, wages and interest which was reprinted in translation (Fig. 11.3.).

War economy provided the model of a possible peacetime economic order, the administrative organization of the economy or economy in kind based on calculations in kind. The theory of war was a representation of war economy as a whole: an overall picture of the multiple changes of an entire complex situation in terms of the

(S. 194)

Die gewöhnliche Darstellung.

Die Grundrente hängt von der Grenze des Anbaues ab, steigt, wie die letztere sinkt, und sinkt, wie jene steigt.

Der Lohn hängt von dem Verhältnis zwischen der Arbeiterzahl und dem Betrage des ihrer Beschäftigung gewidmeten Kapitals ab.

Der Zins hängt von der Ausgleichung zwischen Angebot und Nachfrage des Kapitals ab, oder, wie vom Gewinn behauptet wird, vom Arbeitslohn (oder dem Preis der Arbeit) steigt, wie der Lohn sinkt, und sinkt, wie der Lohn steigt.

Die richtige Darstellung.

Die Grundrente hängt von der Grenze des Anbaues ab, steigt wie die letztere sinkt und sinkt, wie jene steigt.

Der Lohn hängt von der Grenze des Anbaues ab, sinkt, wie letztere sinkt und steigt, wie jene steigt.

Der Zins hängt (da sein Verhältnis zum Lohn durch die dem Kapital innewohnende Netto-Zunahmefähigkeit bestimmt wird) von der Grenze des Anbaues ab, sinkt wie letztere sinkt und steigt, wie jene steigt.

Fig. 11.3 Dual presentation of contrasted statements about rent, wages and interest in Henry George's *Progress and Poverty* (1882, 197), reprinted in Neurath and Schapire's *Lesebuch* (1910), vol. 2, p. 275

[71] Neurath and Schapire-Neurath (1910, vol. 2, p. 168). See also Neurath (1910a/2004, 162).

[72] Neurath (1910b/2004, 290 n. 25). He devoted an entire chapter in his textbook, Neurath (1910b, 40–47).

distribution of pleasures and displeasures tracked by real incomes (all goods used, consumed, even if not purchased) (Neurath 1910a/2004, 153–154).

As the conceptual laboratory of history had shown Neurath (1906–1907 and 1909c), there has been, and therefore is, a plurality of possibilities of economic organization bringing about the circulation of goods: the money economy, based on the instrument of prices and the goal profitability, is "only one of the possible ways" (Neurath 1910a/2004, 172). It's characterized by disorder and intentional waste, and from the standpoint of maximizing productivity and consumption, inefficient. The administration of war suggests the value of an alternative way, an organized economy in kind, which from the standpoint of alternative goals, distribution of goods maximizing pleasure through "unrestricted production and consumption" and a "full utilisation of all energies," (Neurath 1910a/2004, 194) proved more efficient. It is a technical matter of practical rationality in matters of collective choices and valued goals. Independely of its practical value, the theoretical model, like the historical framework, shared with the algebraic logical framework in Jevons and Schröder the combinatorial exploration of possibilities.

The kind of reasoning involved required a calculation in kind, without the comparison of prices (Neurath 1910a/2004 162), and, as we have seen, reasoning rested on means of representation such as language. For the sake of the alternative sort of economic theory, and its associated economic plan and organization, Neurath (1910a/2004, 154) called for a new "means of representation" opposed to the "traditional method of description." In particular, he insisted on a symbolic notation in order to comprehend the overall change in complexes in terms of symbolic formulas (Neurath 1910a/2004, 153). Neurath suggested also the use of synoptic tables arranging the distribution of real income across groups at a certain time side by side with the inventory at a later time (ibid.). On the one hand, then, Neurath's symbolic proposals incorporated accounting standards and the modes of algebraic logic. On the other, the economic counterparts to the forms of logical reasoning and calculation Neurath was introducing were representing and planning change: duality now met history, contrast represented change.

Seven years later, he would acknowledge the further limitations of formal reasoning in the understanding and planning of causal influence, and the role of judgment and will: "Even if one were able to determine the effect of one or another of these economic orders precisely and indisputably in some intricate chain or reasoning, this would not be enough to influence the will in any unique way" (Neurath 1917a/2004, 242). The measures of individual and collective happiness associated with real income were now operationalized in terms of comparable indicators he called qualities of life. They are linked to the set of material and social conditions that enabled them at any given time. As in formal algebraic theory, Neurath (1917b/2004, 315) required that the two sets be correlated univocally.

The same themes appeared in is review of Wundt's *Logik*, of 1910. In the third volume Wundt appended to the third edition, logic and psychology gave way to the discussion of the logic of the social sciences in the debate of unity. Neurath's review synthesizes too the intellectual interests he was pursuing in 1909: mathematical logic and economics, disciplines grounded in the standard of a formal calculus and a formal, abstract psychological models of rationality – as well as empirical, inductive history.

In Wundt's discussion of the social sciences, Neurath identified the theme of a universal science and taxonomy, which became a clear and specific idea in Leibniz. But the project was beset by a problem of univocality: the search was on for a determinate fundamental level, whose "objects are more and more determined," (Neurath 1910b/2004, 267) including, physics, mathematics and logic ("the title 'Logic,' so ambiguous today" [*so vieldeutigen*]) (ibid.).

For Neurath logical precision was not sufficient on additional grounds. While Wundt followed the logical standard and endorsed abstract economics based on the definition of concepts, Neurath pointed to the problem of the particular approach to language reform, namely, by eliminating terms with an ambiguous designation of concepts without general empirical research. In general, Neurath objected, "logically correct connections" when "inferences are valid" leave conclusions empirically irrefutable because the conditional prevents them from asserting directly anything about actual economic reality (Neurath 1910b/2004, 270).

Wundt grounded economic abstraction such as ideal types on simplified psychological constructions that had led to utility theory (Neurath 1910b/2004, 274). Economic theory relied on an exact treatment of economic phenomena, that is, abstract but measurable quantities such as prices. But there were problems about transfers of goods "involving non-measurable but comparable quantities" that can be treated with exactness (Neurath 1910b/2004, 276).

This was the calculus of pleasure, in Jevons' quantitative sense. Neurath valued that the formal emphasis as a tool for representation and decision avoided a deeper reductive psychological analysis of individual experiences motives.[73] What was compared is the (utility) value that different goods (a, b, c, …) have for different individuals (A, B, C…) (see also Neurath's textbook, Neurath 1910b, 52 and 64). Thus, if for A: c > a > b, for B: a > b > c, and for C: b > c > a, the initial state of economic distribution [(Aa), (Bb), (Cc)] may be replaced with the more satisfying one [(Ac), (Ba), Cb)], but economic theory cannot provide a logical derivation or calculation of the latter state from the former.

For the mathematical representation of relations Neurath could rely on the example of prior uses of inequality relations. He had encountered them in Schröder's multiple examples of algebraic order and inclusion, in Jevons' economic calculus of pleasure and pain, especially in the formulation of the theory of exchange at its heart (Jevons 1874, vol. 1, p. 121) and in Fechner's (1888, 106) psychophysics comparison of stimuli and sensations, also present in Wundt's (1906, 185–186) review of psychophysics more generally.

---

[73] In a note Neurath (1910b/2004, 290, n. 27) points out that "the above account required no examination of the motives" and, more radically, that if we ignore moral pleasure or pain, the transition from one distribution of wealth to a more satisfying one represents an improvement since it doesn't constrain the particular kind of transfer of goods involved and takes into account, or at least makes room for, different possibilities.

Another concept connecting logic and economics was complementarity. In Pareto's own criticism of the availability of cardinal utility measures Neurath discovered the challenge posed by the pervasive relation of complementarity between different goods (Neurath 1910b/2004, 280).[74] Our personal satisfaction from consuming one cannot be established independently of the other (Pareto 1906, 239–240). However indirectly, this relation suggests a purely formal resemblance to the relation of complementarity of domains underpinning Schröder's logical calculus and which Neurath and Olga Hahn (1909) had recently addressed in their discussion of dualism.[75] The empirical application of the formal calculus here began to challenge the very idea of a calculus, which "consists in deriving a complex from the individual elements" (Neurath 1910b/2004, 280). The implied holism prevents the mechanical calculation of the maximum collective pleasure, the theoretical and political cornerstone of utilitarianism.

As a science, economic theory aimed, according to Neurath, at systematicity, rationality and exactness, but not measurability. Neurath stressed the point that the relation of inequality expressing the difference in personal valuation didn't have to relate measurable quantities: in general, "exact relations are also possible between non-measurable quantities" (1910b/2004, 277). This fact expanded the scope of economic theory and decisions; still, Neurath (1910b/2004, 277) remarked, they were grounded on formal logic rather than quantitative mathematics: the "calculus of logic represents an example of how far such systems of relations can be developed."[76]

Rather than pointing to Peirce and Schröder, in the economic context Neurath pointed to Jevons:

> It is noteworthy that even a man like Jevons, who after all was equally significant as an exact logician as he was as an exact political economist, only considered relations of [measurable] quantities and thus deprived his speculations in political economy of the opportunity to enter the fruitful field of a goods transfer in which quantities do not matter. (Neurath 1910b/2004, 277.)

On similar grounds he criticized also Wundt, in his double capacity as logician and empirical scientist, for failing to distinguish between two standards of exactness, the logical or symbolic standard and the mathematical, quantitative one, and for failing, like Jevons, to uphold in the social sciences the more general logic one: "Even Wundt, who after all does also consider exact logic and the parts of mathematics that do not deal with measurable quantities, does not indicate that the method of *symbolic-exact* representation and that of *quantitative-exact* representation do not have to coincide at all" (Neurath 1910b/2004, 277; original emphasis).

---

[74] He introduced the notion of complementary goods as a fundamental concept also in his textbook on economics of the same year (Neurath 1910b, 1).

[75] For Schröder (1909, 26), the exhaustive conjunction of two domains of objects (including propositions) may be represented as a + b = 1 (totality, universe), so that negation allows the definition of a complementary (or reciprocal) domain b = not-a.

[76] Notice that Schröder's *Vorlesungen* bore the subtitle *Exakte Logik.*

Economic theory seems then, for Neurath, to fall within the Leibnizian tradition, both a language of representation and a calculus of reasoning, but its symbolic standard is logical, not merely mathematical, if only partially. The standard of a universal language standard remained in place, but Neurath rejected a quantitative one, e.g., the kind of universal measure provided by monetary prices. In the logical space, however constructed – for instance, combinatorially, as in Jevons and also Schröder – or constrained by attention to empirical cases, logical possibilities could then be laid out and economic theory become scientific in the way Neurath had already associated with the historical approach: "*True science consists in systematically examining all possible cases*" (Neurath 1910b/2004, 277; original emphasis). Language led to classification. As in Jevons's own project, the logical framework provided a combinatorial approach to seeking a useful system of classification.

There can be little doubt that the logical standpoint pervaded Neurath's analysis of economic theory in particular and his view of the sciences more generally, and both in terms of a universal language for representation and a calculus of reasoning, also to support decision-making – that is, policy-making. The application extended well beyond Jevon's logicism and was enabled by Neurath's joint education in the related fields of logic and political economy, related by symbolic formalism, its rational value and its expression in general a priori abstraction and particular empirical measurement. It was prompted further by Neurath's simultaneous participation in the debates over method, unity and also values during the same pre-war period.

His commitment to empirical science was, in this early sense, inseparable from a broader logical standpoint, found also in Jevons, Itelson, Schröder, Wundt and other scientists and philosophers, with a commitment to exact thinking. The synthesis of disciplines and epistemological standards is consistent also with his interest during the same period in Mach, Duhem and Poincaré, shared by his friends Hans Hahn and Philip Frank.

The group shared with so many scientists and philosophers metascientific, foundational preoccupations that led them to the shift in intellectual perspectives taking place through the second half of the nineteenth century and expressed, for instance, in the debates over psychologism. Moreover, adopting philosophical and historical perspectives to address broader or foundational questions, whether by Helmholtz, Mach, Schröder, Duhem, Poincaré or Neurath himself, seems gradually led by a twofold mistrust of naïve inductive empiricism and of speculative philosophy and by a commitment to a growing sense of scientific standards, closer to the disciplines that sought both foundation and reform, including the social sciences. Logic, partly in the image of mathematics, became the last scientific refuge of philosophy.

In fact, after his return to Vienna, Neurath, Hahn and Frank joined the Philosophical Society of the University of Vienna, led by Friedrich Jodl between 1903 and 1912. Their talks were hardly empirical in content. Thus, on November 23, 1906, Hahn had lectured on empiricism in geometry, presumably also on Poincaré's conventionalism, with the title "Does Geometry Ultimately Rest on Facts?" (Reininger 1938, 28; Blackmore et al. 2001, 289). And Frank's first lecture, on December 4, 1909, bore the title "Is there Absolute Motion?" (Reininger 1938, 30). Neurath's topics, while in line with his researches and debates engaged during 1908–10, were further removed. On March 5, 1908, Neurath gave the lecture "War and Moral Principles" (ibid., 29). Then, on January 17, 1910, he lectured on the a priori under the title "The concept and range of validity of the a priori" (ibid., 30).

In his subsequent lecture on June 1st, 1912, Neurath returned to the calculus of pleasure in political economy he had addressed first in the review of Wundt. This time he developed the earlier argument and presented his conclusions from the theoretical and political perspective of the British tradition of utilitarianism, only criticizing the possibility of applying its first principle, of maximum collective happiness: "The Problem of Pleasure Maximum" (Reininger 1938, 31; Neurath 1912/1973). The lasting and often unacknowledged contribution of this article was the introduction of a classification of measurement scales typically attributed to the Harvard psychologist S.S. Stevens.[77] I want to focus on how Neurath argued in more detail that the total pleasure of a human collective cannot be derived univocally from the mere sum of individual pleasures.

The problem of univocal determination appeared in the formal context of solving an algebraic problem. Neurath's algebraic-logical treatment relied on the introduction of two relations, pleasure and inequality (greater-than), and the sum operation. Pleasure, as Jevons had already noted, is the relation between a person, M, and a good, m, which Neurath (1912/1973, 114) denoted by (Mm). Different pleasure relations may then be ordered by means of the inequality relation, (Aa) > (Ba) or (Aa) > (Ab). The relation is of course transitive. By duality of reasoning, comparison could be extended to displeasures and take the same form, only modifying the interpretation of the attribution: (Bb) > (Ab) will read "B is less upset by b than A" (1912/1973, 121).[78]

Neurath introduced the additive property of pleasures (Aa) + (Bb) and explored the limits of the distributive property of addition in composition with inequalities. Both are necessary to the utilitarian use of the calculus: Additivity is the condition for calculating collective pleasure; inequality provides the comparative relation between collective distributions of pleasure that point to a desirable choice. But indeterminacy appears in cases in which individual comparisons fail to determine collective comparisons (the policy choices): (Aa) > (Ab) and (Ba) > (Bb) and yet (Aa) + (Bb) ? (Ab) + (Ba) (ibid. 115). The failure extends to cases of complementary goods such that (Aa) + (Ab)? (A(a + b)) (ibid. 116).[79]

To the algebraic model of reasoning by calculation, Neurath added a combination of Jevon's and Schröder's notations for the logical elements: capital letters, A, in Jevons and small letters, a, in Schröder. Jevons introduced capital letters to refer

[77] On the influence of Neurath and Carnap on Stevens see Hardcastle's (2003) discussion of Carnap and Stevens at the head of the so-called Harvard Science of Science Discussion Group, in 1940–41.

[78] On reasoning with inequalities, see Jevons (1874, vol. 1, 186–188).

[79] Neurath's notation for the indeterminate relation when the inequality sign doesn't apply was the question mark "?".

to negations of terms and thus a different interpretation than Neurath's for combinations in the logical alphabet of possibilities such as A = Ab ('A is not B').[80] Once again, this suggests that algebraic logic, as a standard of symbolic language and reasoning, was not for Neurath the subject of a parallel interest, but a resource.

11.7.2 Algebraic Tools for Rational Decision-Making and Theorizing Through Classification

Several months later, on January, 27, 1913, Neurath gave what may be considered a follow-up lecture, "The Lost Wanderers of Descartes and the Auxiliary Motive (On the Psychology of Decision)" (Reininger 1938, 31, and Neurath 1913/1983). The continuity with the meta-theoretical project of logic and the theoretical and practical projects of political economy is evident. It is also continuous with his earlier participation in the debate over values in the social sciences and in relation to war. His position remained a consistently technical one: science is an instrument, even if judgment must be exercised to reach any conclusions or choose among possible ones, but the valued goals, even if they can be partly determined empirically in relation to, say, society's values, are nevertheless external to the economic solution. From a theoretical and practical standpoint, the expanded focus on psychology is also significant, especially in an intellectual landscape in which a variety of views and local authors were identified as psychologistic and so many others were simply turning to psychology as an emerging scientific discipline in the footsteps of, for instance, Fechner and Wundt.

The essay addressed the broader limits to the empirical application of formal reasoning that Neurath had exposed earlier – in relation to the challenge of determining univocal solutions to the economics problem of maximum pleasure, and especially in cases of entangled pleasure attributions. Defending the value of rationality also in ordinary psychology of practical thinking, Neurath pointed to the case of Descartes's rationalism to distinguish between rationalism and pseudo-rationalism. Rationalism properly understood acknowledged, according to Neurath, the limits of insight and theoretical argument in the process of reaching conclusions to make choices among hypotheses or decisions. Pseudorationalism, by contrast, failed to acknowledge that thought processes, as matter of empirical psychology, are not reducible to "a system of logical relationships" (1913/1983, 3) and that missing premises would fail to determine a unique conclusion (ibid., 7). Instead, "one finds oneself very often in the position of having to choose one of several hypotheses of equal probability" (ibid., 3) and is unable to mechanically reach a decision or make a principled choice.

The problem, theoretically and practically, takes the form of a kind of holism that extended to scientific reasoning beyond Duhem's well-known problem of auxiliary hypotheses (Cartwright et al. 1996). Thinking processes are temporal unities in the

[80] See, for instance, Jevons (1874, vol. 1, 122).

way are sciences are; and in the theoretical case, Neurath (1913/1983, 3) noted, "the phenomena we encounter are so much interconnected that they cannot be described by a one-dimensional chain of statements."[81] From an epistemological standpoint, "the correctness of each statement is related to that of all others" (ibid.). And this led Neurath, without recourse to a boat metaphor, to a position of anti-foundationalism about out system of concepts: "Each attempt to renew it from the bottom up is by its very nature a child of the concepts at hand" (ibid.). Neurath's solution was the theoretical and practical necessity of provisional rules, or auxiliary motives. He would later call them extra-logical terms or factors (Neurath 1934/1983, 104, 106). Their job was, at the most theoretical, to help ever provisionally construct and apply a scientific system. Rather than a rejection of logical formalism, or rationality, against opponents, or in defense of radical empiricism, his persistent argument since 1910 consisted in exposing the limits of empirical applications of the formal, normative, standard with which he himself was familiar.

About a year later, on March 2nd, 1914, Neurath developed the themes further in a new lecture, "On the Classification of Systems of Hypotheses (With Special Reference to Optics)" (1916/1983) and a second essay to be published first, "On the Foundations of the History of Optics" (1915/1973). The focus on the application of the logical standard remained in the human sciences, but went beyond the earlier cases of economics and psychology, in this case extending to the historical sciences, with a historical theory of scientific theories. The focus is however similar in the attention to the formulation of possible theoretical cases and the problem of empirical selection among them. And temporality was relevant again, only now as the historical dimension of the social phenomena at hand. In addition, Neurath gave the challenge of capturing an intellectual totality a disciplinary dimension that would sustain into the future the theme of unity in a science and among the sciences.

In the case of "fields of knowledge" such as psychology, Neurath (1916/1983, 13) asked "what there is in common." His answer was the notion of theory in term of possibilities and the notion of methodology in terms of multiple choices (underdetermination). The unified representation of the scientific field, then, took the form of a classification as a comparative study of possible systems of hypotheses. In general, Neurath (1916/1983, 24) noted that any system of classification selection and connectedness of facts or hypotheses. The selection would involve auxiliary assumptions such as analogies that played a key role in the construction and selection of systems of hypotheses and in the selection of hypotheses within it.[82] They link a particular hypothesis to a whole of concepts and assumptions at work at a given time (Neurath 1915/1973, 101). Neurath had learned it about physics from Duhem.[83]

[81] Note that in the case of symbolic equalities, Neurath also pointed to the typographic artifact of one-dimensionality in our symbolic representations.

[82] On the role of analogies and their relation to Mach's work also on optics, see Stöltzner (1996/1983).

[83] See Duhem (1906). Duhem was also concerned with classification as the proper aim of science, without deeper explanations.

On a related Duhemian theme, symbolic precision and exactness, Neurath (1910b/2004) had previously characterized theories after the standard of algebraic logic (and language): as a systematic representation and comparison of (logical, conceptual) possibilities with symbolic exactness (see above; the use is ambiguous between exactness and precision). Again, now Neurath (1916/1983, 24) pointed to the challenges of empirical application and noted the unavoidable limitations in the symbolic precision of classified hypotheses: systems of classification cannot avoid blurred margins and reach clarity only asymptotically. The notion of blurred margins is Neurath's representation of the gap between the artificial distinctness of the formal construction and 'the whole fullness' of a phenomenon (remember here his previous example in rational psychology of decisions of the logical, one-dimensional train of thought picked out in a psychological complex whole). As a source of blurred margins, Neurath (1916/1983, 25) pointed to the role of empirical data. Duhem (1906) had noted a similar semantic gap, the indeterminacy in the relation between the symbolically precise theoretical facts (quantities) and practical facts (qualities) (with the implication that laws are neither true nor false).

The theme would evolve into Neurath's metalinguistic doctrine of the inevitably limited precision of our scientific language in (linguistic) contact with empirical contents and ordinary contexts. Neurath (1916/1983, 24) illustrated the idea in terms of the optical phenomenon of refracted light rays of different colors with blurred edges.[84]

In a later optical illustration Neurath pointed to the application of language as writing and thinking "arrayed one-dimensionally," and of artificially precise networks of mathematical concepts with functional and logical connections. He referred to the blurred edges of a projected beam of light from a lamp on the walls of a mine, representing the inexhaustible whole, which, from the analytical standpoint of precise elements, he represented as the manifoldness of experience and complexes of representations (Neurath 1921/1973, 198).[85]

How did the classification come about? Following a logical analysis of known cases into constituents, "precisely expressed notions of the systems of hypotheses can be systematically registered" and then "their 'combinations' can be classified according to different characteristics" (Neurath 1916/1983, 24). The classification, then, consisted in combinations of "all imaginable relations" that "can be produced by purely logical operations" (ibid., 28). The combination could then be "fixed by a kind of formula" (ibid., 15). This inquiry into the "logical structure of theories" leads to a theory in the form of a taxonomy in the form of a network for exploring "how they may develop" (Neurath 1915/1973, 101).

One might begin, like Jevons, with a basic dichotomy, A-theories and non-A-theories. But the choice, Neurath (1916/1983, 15) remarked, had logical but no practical value (as in Schröder's complementary domains). A more complex set of elements had the practical value of accommodating a plurality of cases while

[84] Neurath's linguistic treatment of the conceptual blur is the doctrine of *Ballungen*, see Cat (1995) and Cartwright et al. (1996).

[85] On the optical illustration, see Gábor Zemplén's chapter in the present volume.

constructing a unifying totality. Then would come the matter of empirical relevance: identifying natural combinations according to some accepted theory.

The new challenges were only the outcome of upholding the (algebraic) logical standard. Unlike other interpretations of Neurath's different works, my focus on one running theme and resource also suggests a new interpretation of the essays on history and classification. Neurath was following here Schröder and Jevons too (alongside Mach, Duhem, Poincaré and others) as he embraced the notion of a logical classification as a calculus of combinations into classes taken as unities (this is the extensional interpretation of terms as domains of logical elements). It was the Jevons of the logical alphabet and natural classification, including a heuristic role for analogies and analogical reasoning.

A few brief remarks developing the discussion of Jevon's project in Sect. 11.3.3., above, will make clear the connection to Neurath. Jevons (1874, vol. 2, pp. 346–48) adopted the view that the practice and products of classification were based on identifying analogies that exhibited laws of union or correlations of properties.[86] He further attempted to distinguish imprecise distinction between natural and artificial systems[87]: artificial classifications are based on arbitrarily selected properties of objects and natural analogies involving a concurrence of essential characteristics of objects grouped by closest degree of analogy (Jevons 1874, vol. 2, pp. 351–352). Still, he admitted of a multiplicity of possible systems of classification and that any choice among them would be based on pragmatic considerations (ibid., pp. 348–349).

Of particular relevance here is Jevon's interest in the symbolic presentation of the theory of classification (ibid., 367–371). Jevons sought to generalize on the so-called bifurcate classifications resulting from a dichotomy of the universe and the application of the principle of duality with considerations of negatives; such was the mechanical combinatorial source, after Leibniz, of the logical abecedarium – later renamed 'alphabet' – (Fig. 11.4.) and logical machine (Fig. 11.5.) (ibid., 367 and 371–374, and 1874 vol. 1, p. 109).

At the symbolic level, then, classification "may be explained most precisely by the use of letter combinations" (Jevons 1874, vol. 1, p. 376). For instance, he added: "If we pay regard only to three qualities or circumstances in which things may resemble each other, namely the qualities a, B, C, then there are according to the laws of thought eight possible classes of objects" (ibid.). The combinations include the negations of the qualities, a, b, c (Fig. 11.6.).

This was precisely the combinatorial scheme Neurath adopted (1916/1983, 15): "from three elementary notions and their negations we get eight combinations in total." Nevertheless, despite his close acquaintance with both Jevons and, following Jevons, Schröder, Neurath adopted Schröder's notation (Figs. 11.7 and 11.8) – with small or lowercase letters for the positives and the vertical stroke sign for negations.

[86] In the plural scene of nineteenth-century British new empiricism, the view can be traced back especially to Herschel and Mill, also Whewell.

[87] He was following Ampère, and indirectly a long literature going back to the seventeenth century. Duhem would adopt a more regulative and metaphysical notion of natural classification (Duhem 1906).

THE LOGICAL ALPHABET.

I.	II.	III.	IV.	V.	VI.	VII.
X	AX	AB	ABC	ABCD	ABCDE	ABCDEF
	aX	Ab	ABc	ABCd	ABCDe	ABCDEf
		aB	AbC	ABcD	ABCdE	ABCDeF
		ab	Abc	ABcd	ABCde	ABCDef
			aBC	AbCD	ABcDE	ABCdEF
			aBc	AbCd	ABcDe	ABCdEf
			abC	AbcD	ABcdE	ABCdeF
			abc	Abcd	ABcde	ABCdef
				aBCD	AbCDE	ABcDEF
				aBCd	AbCDe	ABcDEf
				aBcD	AbCdE	ABcDeF
				aBcd	AbCde	ABcDef
				abCD	AbcDE	ABcdEF
				abCd	AbcDe	ABcdEf
				abcD	AbcdE	ABcdeF
				abcd	Abcde	ABcdef
					aBCDE	AbCDEF
					aBCDe	AbCDEf
					aBCdE	AbCDeF
					aBCde	AbCDef
					aBcDE	AbCdEF
					aBcDe	AbCdEf
					aBcdE	AbCdeF
					aBcde	AbCdef
					abCDE	AbcDEF
					abCDe	AbcDEf
					abCdE	AbcDeF
					abCde	AbcDef
					abcDE	AbcdEF
					abcDe	AbcdEf
					abcdE	AbcdeF
					abcde	Abcdef
						aBCDEF
						aBCDEf
						aBCDeF
						aBCDef
						aBCdEF
						aBCdEf
						aBCdeF
						aBCdef
						aBcDEF
						aBcDEf
						aBcDeF
						aBcDef
						aBcdEF
						aBcdEf
						aBcdeF
						aBcdef
						abCDEF
						abCDEf
						abCDeF
						abCDef
						abCdEF
						abCdEf
						abCdeF
						abCdef
						abcDEF
						abcDEf
						abcDeF
						abcDef
						abcdEF
						abcdEf
						abcdeF
						abcdef

Fig. 11.4 Jevons' logical abecedarium, Jevons 1874, vol. 1, p. 109

Fig. 11.5 Jevons' logical machine, Jevons 1874, vol.1, frontispice

Fig. 11.6 Symbolic representation of the eight combinations of three classes (properties) and their negations, Jevons 1874, p. 367

ABC	***a*BC**
AB*c*	***a*B*c***
A*b*C	***ab*C**
A*bc*	***abc***

Neurath had introduced Schröder's notation already in the paper on dualism with Olga Hahn and here he was applying it to the historical classification of systems of elementary concepts or hypotheses: $a_{|}$, $b_{|}$, $c_{|}$ (the stroke is sometimes replaced by a typographic choice of the visually analogous 'l' or '1') (Fig. 11.9.). In the 1983 English translation of "On the Foundations of the History of Optics" the stroke sign has been replaced by asterisks, obscuring the typographic genealogy of the conventions in the logic literature (Neurath 1915/1973, 104).

Neurath's (1916/1983, 15) choices for the three elementary characteristics were periodicity, polarizability and interference; he added a fourth, diffraction. What determined the choice of elementary terms is some combination of conceptual conjecture, analysis and decision.

Kombinationen.

1)	$abcde - \gamma_3')$	17)	$a_{,}bcde$
2)	$\beta)$ $abcde_{,} - \gamma_1')$	18)	$a_{,}bcde_{,}$
3)	$\gamma_3)\ \gamma_1'')\ abcd_{,}e$	19)	$a_{,}bcd_{,}e - \gamma_1'')$
4)	$\gamma_1')\ \gamma_1'')\ abcd_{,}e_{,}$	20)	$a_{,}bcd_{,}e_{,} - \gamma_1'')$

Fig. 11.7 Fragment of table of symbolic representations of logical combinations of four classes, Schröder (1890a, b, 563)

Tabelle:

3) $abcd_{,}e$

4) $abcd_{,}e_{,}$

5) $abc_{,}de$

Fig. 11.8 Fragment of table of symbolic representations of logical combinations of four elements, Schröder (1890a, b, 564)

a	b	c
a	b	c_1
a	b_1	c
a	b_1	c_1
a_1	b	c
a_1	b	c_1
a_1	b_1	c
a_1	b_1	c_1

Fig. 11.9 Neurath's symbolic presentation of the eight combinations of three theoretical elements and their negations forming a classification of systems of hypotheses (original publication in Neurath 1915, 376)

Finally, at the level of presentation, the choice for the graphic expression of logical combinations of elementary terms were the classification tables (Fig. 11.10.). Their use was established in the taxonomical tradition, alongside synoptic tables introduced by Petrus Ramus in the sixteenth century, which Jevons and Schröder also used to organize logical combinations. They were also common in a widespread graphic tradition of presentation of numerical information – e.g., Luca Pacioli's double-entry bookkeeping in the fifteenth century. Such kinds of grids mark out a visible spatial arrangement to map out conceptual relations in an efficient way, at a glance – synoptically. Neurath's symbolic standpoint in logic was not just consistent with his later structural, physicalist formulation of empiricism, it was another visual example of it. Before he became concerned with the logic of image, and its social value, he was concerned with the image of logic.

Physiker	Periodizität	Polarisierbarkeit	Huyghensches Prinzip	Interferenz
Huyghens	nein	ja	ja	nein
Newton	ja	ja	nein	nein
Young	ja	ja	ja	ja

Fig. 11.10 Neurath's table of optical theories, or systems of hypotheses, Neurath (1915, 385)

11.7.3 Universal Symbolic Language for Social Representation and Intervention, from Numbers to Pictures (Pictorial Pasigraphy)

Neurath returned to Vienna sometime in 1919 in the aftermath of World War One; the time corresponded also to the wake of the turmoil of his involvement in German political projects of Soviet-inspired central planning, which landed him briefly in prison. In "Anti-Spengler," drafted during his imprisonment, Neurath appealed to the principle of duality in the form introduced in projective geometry in order to represent the structural objective that made sense of human communication and shared objective content of experience, so that two speakers "always manage to coordinate their statements unambiguously" (Neurath 1921/1973, 200).[88] In symbolic terms from geometry: "There could be a far-reaching parallelism or 'dualism' such as in projective geometry, where we can enunciate two sets of statements that have quite the same structure and can be represented by one single notation and yet describe totally different facts, depending on how symbols are interpreted" (ibid.). The pasigraphic ideal of a universal language had to accommodate, formally, the possibility of different languages or interpretations.

It is important to note that, whereas for Jevons the role of logic in economic theory was mainly foundational, Neurath was clear that the role of the symbolic standard was both theoretical and practical, as was the role of social theories. Pure logic could have anti-metaphysical, practical value: "It is precisely Marxism that uncovers indirect relations and detours, and thus might ascertain that cultivating pure logic and the most general problems of mathematics and physics is especially favorable to revolutionary thinking" (Neurath 1928/1973, 295).

His post-war jobs in the Museum for Settlement and Town Planning, in 1923, and in Social and Economic Museum of Vienna after 1924, and his involvement in adult education extended the emphasis on the social and practical dimension of social information. Still, before settling on the new emphasis, Neurath delivered a talk on economics at the Philosophical Society on 26 February 1923 still with the theoretical focus included in his pre-war writings and with an explicit reference to

[88] This is similar to the structural description Carnap would introduce in his *Aufbau* later in the same decade.

the role of logic. The lecture bore the title “The Conceptual Building of Economic Theory. A Logical Treatment” (Reininger 1938, 37).

While in previous discussion his focus was decision-makes at the level of economic government, now the emphasis extended to citizens as democratic participants. Neurath (1925/1973, 214) defended, acccordingly, the practical value of a universal visual pasigraphy, also portrayed in the spirit of internationalism, linked to the universal language of statistical information (information in the form of social correlations was of special practical value). While Neurath (1933/1973, 222) acknowledged that “only quantitative facts are socially significant,” a schematic pictorial representation would have cognitive mnemonic value, at the expense of quantitative exactness (ibid., 220; see also Elisabeth Nemeth’s chapter in the present volume).

The pasigraphic dimension was explicit, including Schröder’s own example of a (failed) project of international languages of cognitive and social value, Volapük. His old interest in pictures, also, like economic organizations in kind, can be traced to Egyptian culture, feeding into the project of a universal language beyond mathematics: “The problem of an international language attracted me fairly early, Volapük had come and gone; Esperanto reigned uneasily in its place. Scientists tried to improve on it by evolving Ido as a more consistent language. Basic English and Interglossa were yet to come” (Neurath 1946/1983, 96). The very project of a universal Visual Thesaurus in the 1930s connected his encyclopedia project to Leibniz’s old ambitions by extending the symbolic to the pictorial. Indeed, Neurath (1936/1983, 142–143) believed his encyclopedia would drive the unification of scientific language and graphic representation; and he attributed to Leibniz the value of both unifying encyclopedic compendia and visual representations.

Again, Neurath was extending symbolic representation beyond the quantitative. Pictures are symbols, and their role depends on the representation of quantitative information according to the rule “one symbol means a given number.” Then Neurath’s displays would introduce combinations of symbols by juxtaposition in order to represent different amounts based on the interpretation of a picture as a unit. If the initial concern with the image of a language of logic, he soon extended it to the logic of a language of images. As in his earlier discussion of Schröder’s identities, especially in the case of commutativity of logical products, Neurath’s standpoint was explicitly typographic. The communication technique, the pictorial symbolic language, was first known as the Vienna method and from exile was named, accordingly, International System of TYpographic Education (ISOTYPE). It is worth noting that Neurath chose to emphasize the technical, instrumental, or functional dimension of visual communication in the same way he was conceiving both language and knowledge as tools, and referring to it accordingly, as a method – the Vienna Method – or a technique.[89] The linguistic standard was still very much at the heart of its conception: “The important point is that there must be a system of rules, a sort of grammar of picture language; that, and a dictionary of symbols” (Neurath 1933/1973, 222). The universality of the pictorial language as a unified method of representation was as constructed as were its symbols, or those of the

[89] See Burke et al. (2013), also Angélique Groß’s chapter in the present volume.

artificial languages Carnap would prefer; it was a uniformity that would require "a unified, planned, central control of all museums and educational institutions" (ibid.).

Visual communication was not divorced from reasoning. In a Socratic fashion, Neurath's interest in the social value of education focused on the practice of arguing – "teaching how to argue" (Neurath 1937/1973, 239) – and the presentation of arguments. Part of their role was the parsing of essential and incidental aspects of issues; in this sense, the notion of argument was close to the narrative notion of plotline, and the selective task is one of the conditions of storytelling. In the case of child education, Neurath (1996, 284) stressed the fact that symbolic drawings were natural tools for storytelling, and that only as children grow up, they lose such symbolic habit in favor of more detailed, realistic pictures.

While arguing takes place in general through the use of symbols, for Neurath it also had a dual character, narrative and logical, so that its public expression would facilitate encouraging the latter, at least in some a functional sense; hence its alleged social and scientific significance: "Sincere arguing is something of international importance because it creates a common basis for discussions and decisions" (Neurath 1996, 285). Its social function was to allow some critical distance from tradition and authority in the task of reaching one's own judgments and making decisions with their help but within the limits of univocal inference he had long drawn attention to: "any decisive action has to be based on a decision which cannot be derived from any unequivocal result of arguing" (ibid., 286). In this way, in light of the broad shared goals, he could conclude the close connection between logical empiricism and visual education. But it is the film experience that could educate most effectively, Neurath suggested, in this critical form of communication. In his presentation, Neurath introduced the notion of a visual argument. The use of film allowed the temporal dimension to be integrated with the pictorial medium, combining visual and verbal aids to exhibit and facilitate argument, presumably in both the narrative and logical senses (Neurath 1945/1973, 240 and 1946/1983).

11.7.4 Logic at the Service of the Integration of the Human Sciences

Neurath would continue to uphold the role of logic, especially Jevons', in later discussions of the social sciences, and especially in relation to goal of unification – something Jevons himself had done. In the earlier phase of the unity project within the Vienna Circle, Neurath had followed Carnap in trying to address at once the issues of (1) unity and demarcation, (2) empirical grounds – that is, alongside the formal analytic truths of mathematics –, (3) a logical-linguistic framework and (4) objectivity. His earlier solution consisted in endorsing the universal language of physicalism across the sciences (a metalinguistic and epistemological standard of testability without reductionism to the concepts and generalizations of physics).

Addressing the anti-metaphysical imperative to compare statements only with other statements in relation to human perception and behavior, he pointed to the practical value of using Jevon's logical machine:

> All this could be developed experimentally with the help of a 'thinking machine' as suggested by Jevons. Syntax would be expressed by means of the construction of the machine, and through its use, logical mistakes would be avoided automatically. The machine would not be able to write the sentence: "two times red is hard." (Neurath 1931b/1983, 67).[90]

Jevons' logical machine, then, provided the material, physicalist, coordination of the logical – and the syntactical – and the empirical.

Neurath also pointed to Jevons in the context of the encyclopedia movement, later in the 1930s, this time in relation to the unifying method of empiricism for both the natural and the social sciences: "Mill, Jevons, and Pearson were all very interested in the social sciences and tried to apply the empirical procedure to all questions without distinction" (Neurath 1937/1987, 133). Neurath's purpose then was to endorse Jevons and the others as historical exemplars in his encyclopedic approach to unification. The project aimed to arrive at "a synthesis on an empirical basis" (ibid., 134).

With the emphasis on empiricism, Neurath was not rejecting Jevons' emphasis on logic; he was countering, instead, a rationalist, even metaphysical, alternative to unity based on a grand system of logical relations alone:

> The all-embracing view of the scholastics is to be found again in a somewhat weakened form among the leading rationalists who, like Descartes and Leibniz, took part in the expansion of the individual sciences. The striving for logical derivations which we encounter among the scholastics was to revive again in Leibniz in a thoroughly modern form. He was to become the forerunner of logicism. But empiricism does not look kindly upon such synthesis, and any rationalism a priori appears suspect to it. (Neurath 1937/1987, 133–134.)

Like social engineering, logical engineering had its limitations too; neither science nor society could be fully mechanized.

The indictment against (pseudo) rationalism, a priori metaphysics, cannot be mistaken for an indictment of the use of logic. What Neurath was defending was after all a logical form of empiricism whose scientific goal involved the empirical application and testing of formal linguistic structures and whose philosophical goal involved the demarcation, by means of a logical analysis of language, of tautological, identity statements from empirically testable statements (Neurath 1937/1987, 135). In this respect, he noted that mathematicians and logicians were contributing new creative opportunities, pointing to Frege, Peano, Schröder, Peirce and Russell and noting that their "ideas are influencing an increasing number of scientists from every discipline" (ibid.). He singled out Peirce and Russell as having "found a most intimate way of combining their interest in empiricism with their interest in logic," Dewey, whose own interested in logic didn't prevent him from guiding American thinkers "in the direction of empiricism," and Schlick, at the center of a "circle for the development of *logical empiricism*" (ibid.; original emphasis).

Neurath picked up again the question of the logical unity of the sciences and the value of symbolic logic in relation to language and classification, now in his review of Carl Hempel and Paul Oppenheim's book *Der Typusbegriff im Lichte der neuen*

---

[90] Mosselmans (2007, 50) quotes Neurath merely to identify in Jevons a "logical positivist attitude."

Logik (Hempel and Oppenheim 1936; Neurath 1937c/1983). Even though concerned especially with psychology, Hempel and Oppenheim adopted a formal approach to typology as a method cutting across differences between natural and human sciences. In fact, the authors concluded the book pointing explicitly to the goal of "die logische Einheit der Wissenschaft" (Hempel and Oppenheim 1936, 125). To contribute to this project, they replaced the standard of classification in terms of single-term designations attached to objects or phenomena of scientific interest with a general theory of order with special attention to concepts that admit of distinctions of degree.

The authors sought to establish the method of typology on the notion of type based not on the notion of class but of series. The fundamental idea was the topological order that replaces the quantitative standard of a metric with a more general notion of ordered series and the relation of betweenness. In turn, the fundamental relation also made fundamental in typological language the role of two-term predicates: if, in a series m, l, h, i, k, a quality symbolized as g is placed between i and k, the single-term (monadic) predicate 'g' is replaced by the two-term predicate '(i, k)' (Neurath 1937c/1983, 186). Neurath objected, besides to the relativity of series-location, to the emphasis on mixes of extremes, after the Aristotelian model, in the same way he had rejected as simplistic the classification of hypotheses based in the binary opposition between A and non-A theories, favoring, instead a combinatorial approach he now associated with chemistry (Neurath 1916/1983, 15).

In addition to unification and symbolic logic, the book spoke also to Neurath's earlier specific interest in classification, surveys, scales and general symbolic precision – beyond quantitative measurement for the relations between pleasures or life quality indicators.[91] Thus Neurath connected the book's focus to his earlier discussions of empirical representation in the social sciences and offered the following meta-classification:

> One could now start from cases in which object are characterized by names, given to themselves and to their properties [...]. In all cases of this kind one either ascribes or denies a property to an object. And it seems to depend on the richness of our "catalogue of qualities" – if one can use this expression – whether we have enough names at our disposal, with the understanding that in the case of classification through measurement one will use cardinal numbers, in the case of a topological order, ordinal numbers, and in the case of non-ordered qualities, perhaps numbers that, by their value of position, indicate neither quantitative nor topological order. (Neurath 1937c/1983, p. 185.)

Next he distinguished between the catalog of qualities – with the single-term labels (the categorical scale) providing a uniform syntactical treatment –, an atlas of qualities – with a topological arrangement – and a table of qualities – with formulas (1937c/1983, 185). The order of the different modes of representation also sets the terms for a historical evolution in empirical sciences, even if the transformations of one mode into another will always be only partial (ibid.).

[91] See, for instance, his review of Wundt in Neurath (1910b/2004) and the additional discussion of aggregate pleasure in Neurath (1912/1973).

Instead of classifications from above, he returned to the logical model of 'many small scientific units as a logical start' reminiscent of the logical combinatorial approach based on elementary units: "a great many scientific units (many of them very small) might be assembled step by step as systematically as possible. Such an increased of assembling is closely connected with the actual increased of scientific investigation and comprehensive logicalization: chemistry and optics could not be really joined by means of a mere classification" (Neurath 1937/38/1983, 200).

The symbolic, notational standpoint and the combinatorial method inherited from Neurath's early work in algebraic logic and subsequent applications found use through to the end of his life. Tables of classification appeared, for instance, in his prehistory of logical empiricism, where philosophers (and their doctrines) are compared in terms of the categories anti-metaphysics, empiricism, logic and mathematics; Leibniz is the closest to scientism, the sole difference between them Neurath (1936/1981, 694) identified on that count is only the former's commitment to metaphysics. To the extent that he applied any standard of duality in his classification tables, it was Jevon's logical principle of duality in the polar sense of A and its complementary not-A, e.g., often represented in each taxonomical category by the symbols + or – (ibid.). It is not the sense of correspondence and equivalence in Schröder's principle discussed by Neurath. The same format, with an explicit mention of duality and the +/– expression of its application, made a final appearance in the context of his concern with education and politics in handwritten notes on his copy of proofs of an article by the educator C.G. Field on Plato's use in education (Fig. 11.11.).

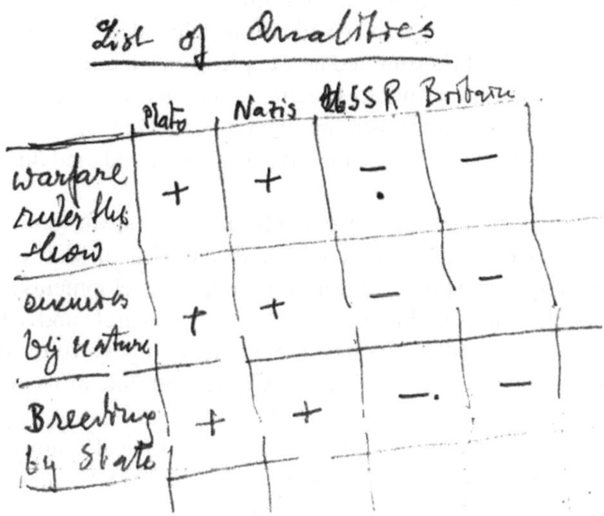

Fig. 11.11 Neurath's handwritten "List of Dualities" on a copy of C.G. Field's article "Plato's 'Republic' and Its Use in Education," *The Journal of Education*, 77 (909), April 1945, pp. 161–162. (I'm grateful to Adam Tamas Tuboly for drawing my attention to this file)

11.8 Applications and Challenges of the Algebraic Logical Standard in Neurath's Participation in the Movements of Logical Empiricism and the Unity of Science

11.8.1 Differences Within a Shared History of Interest in Logic and Language

In the opening, I pointed out that, for decades now, Neurath's activities and contributions have been duly assessed, even vindicated, in a dialectical manner, roughly in terms of at least three oppositions: (1) intellectual theory/social-political action, (2) formalism/antiformalism in language and method and (3) natural/social sciences. His distinctive historical role has been singled out – that is, classified and explained – primarily in terms of an emphasis on each three of the second elements: social-political action, antiformalism (or informalism) and the social sciences. In fact, his alleged informalism has been explained, in turn, by reference to the other, social factors (and this in addition to references to the social nature, not just application, of standards of objectivity and rationality).[92]

This story is too narrow to take into account Neurath's attention to logic both before and during his logical empiricist militancy; it's also too narrow to take into account the role of the formal, symbolic standards in his own social theorizing. I have been telling a long but different story, introducing a larger context that places Neurath in an active and substantive relation to logic and mathematics and to their application in social thought and action (in relation to both decision theory and its application in social practice).

From the same standpoint, to conclude this argument, I suggest that the new periods of intellectual involvement provided an expanded context in which new circumstances and occasions prompted new expressions of attention to the symbolic standard, only this time sensitive to specific situations. In other words, I suggest that, in relation to logical empiricism and the unification of the sciences, Neurath's proposals and criticisms stemmed from a shared background, and differences within it, not from it.

The second half of the 1920s represented Neurath's postwar return to philosophy of science with his participation in theoretical discussions of Marxism, in activities of the Monist Society, in the private discussions of new developments the foundations of physics and mathematics within the so-called Schlick Circle and in the foundation and activities of the Ernst Mach Society, the public springboard for the constitution at the end of the decade of the Vienna Circle and the movement of logical empiricism.[93]

Neurath public involvement in the movement of logical empiricism included, broadly speaking, two periods and contexts, the 1929–33 Vienna Circle and the

[92] See, among others, Uebel (1992) and (2007) and Cartwright et al. (1996).

[93] For an overview of the intellectual and associationist Viennese landscape, see Stadler (2001).

post-1933 exile. The first was geographically and intellectually more localized with an emphasis on the Vienna Circle and the issues and doctrines of logical empiricism, while the second was consistently international in perspective, with an emphasis on the Unity of Science movement and the Encylopedia project.

I have been documenting that different ways and times Neurath had long been paying attention to standards of logical calculus and universal language, that is, the rational and the pasigraphic sides of the algebraic logic tradition. His interest survived the turmoils of the 1910s and the social opportunities of Red Vienna through the 1920s, and found expression, even if a diverse and shifting one, also in his efforts at articulating a program for an intellectual and social movement around the Vienna Circle. This includes his communications with fellow Circle members such as Carnap and in his proposals for formulating an account of scientific method and unification.

Neurath's proposals were throughout anti-Cartesian in the more general sense of opposing an unrestrained standard of rationalism he had labeled pseudorationalism, of which he accused also Leibniz. According to a number of charcaterizations, it involved a precise language and rules of calculation, first principles and the assumption that any problem had a uniquely determined solution (see above). Neurath considered it incompatible with the application of the algebraic-logical standards of reasoning and representation in empirical science and human rational action, of economic policy-making. His more critical rhetoric can be associated with the Encyclopedia period in exile, but, for the same intellectual and practical reasons at work in the writing of the Vienna Circle's Manifesto, in 1929, it never amounted to a blanket condemnation.

We may say that also Carnap had entered logical empiricism with a philosophical outlook informed primarily by active interests in logic and mathematical science. While Neurath's preferred science was economics, Carnap's was physics. What about logic? As I have argued in previous sections, Neurath's differences with Carnap, and other philosophers or Circle members, cannot be reduced to a dismissal or avoidance of formal logic and language. Rather, it was a matter of Neurath's putting forward critical qualifications of the significance of symbolic languages and calculi that may be traced to (1) his different kind of attention to actual scientific practice and its social management and (2) differences in logic background and attitude towards it.[94]

Having attended Frege's lectures at Jena between 1910 and 1914, after the First World War Carnap turned to Whitehead and Russell's *Principia Mathematica* (1910–1913), which Frege had referred to in his lectures.[95] Soon he would endorse unreservedly their logicism, more comprehensive and notationally convenient than

[94] Rather than subscribing to overdrawn contrasts between pre-1930, pre-1935 and post-1935 Neurath, drawn for instance by Mormann (1999), and equally between Neurath and Carnap, usefully drawn by Uebel and, especially after 1935, by Mormann, I want to emphasize any differences only in relation to a shared background and interest in symbolic language and logic.

[95] For a detailed discussion of Carnap's involvement in logic in the 1920s, see Reck (2004) and Reck and Awodey (2004).

Frege's own, in 1920 in his teacher certificate dissertation, "Of What Philosophical Significance is the Problem of the 'Foundation of Philosophy'?": geometry "can be derived completely without use of intuitive elements from arithmetic, and the latter in turn from deductive logic" (quoted in Carus 2007, 97).

By the end of the decade, Carnap had recently published *Die logische Aufbau der Welt* (1928/1967) where he adopted Russell's formal methods and so-called supreme maxim of philosophizing, with which the text opens: "Wherever possible logical constructions are to be substituted for inferred entities" (Carnap 1928/1967, 1). In early September 1930, he participated in a session devoted to the foundational crisis in mathematics at the Second Conference on the Epistemology of the Exact Sciences held in Königsberg in which von Neumann discussed Hilbert's formalism in axiomatics, Arend Heyting discussed Brouwer's intuitionism and Carnap discussed logicism, its challenges and some of the affinities with the constructive element in intuitionism and the pure formal calculus of the formalist method (Stadler 2001, 352).

In 1920 Carnap had emphasized that the starting point of logicism was precisely deductive logic "represented in algebraic form – on the basis if an idea of Leibniz – by Peano, Russell, Couturat, Frege and Schröder, among others" (quoted in Carus 2007, 97). By the end of the decade Carnap had also separated older from new symbolic logic, or logistic: In the *Abriss der Logistik* (1929) he introduced a distinction between older and modern works. Older systems he associated with De Morgan, Boole, Jevons, Peirce, and older works with Frege, Peano, Schröder, Whitehead and Couturat, the last two on algebraic logic. In particular, he listed Schröder's *Vorlesungen* and even mentioned the Schröder-Bernstein theorem. Modern logistics Carnap (1929, 107) associated resolutely with Whitehead and Russell. And, to mark the change, he noted that the term 'algebraic logic' had been replaced by 'symbolic logic' and 'logistic' (1929, 1–2). The term 'logistic,' Carnap (1929, 2) also noted, was introduced in 1904 by Meinong, Itelson, Couturat and Lalande.[96]

In his article 'Die alte und die neue Logik' (1930), we find a similar distinction: The most important older works: by Frege, Peano and Schröder (1930, 26). In both cases, Carnap emphasized how the new symbolic logic developed from the older (1930, 14). He had been clear that in this development, "all of mathematics becomes a branch of logic" (Carnap 1929, 2). The role of symbols is to facilitate the precision of concepts (language) and exactness of demonstrations (1929, 1).

When Neurath, in the course of his studies of mathematics and philosophy in Vienna, turned his attention to logic, the more science-oriented scholars were paying attention to algebraic logic, and, according to contemporaries, its complete systematization was Schröder's. Whitehead's own *A Treatise on Universal Algebra* (1898) had recently come out and it would still be over a decade before the publication of the *Principia*. Schröder was critical of Frege's *Begriffschrift* as a competing attempt at realizing Leibniz's ideal of universal conceptual script, the

[96] Itelson, Couturat and Lalande made the proposal at the 1904 at the International Congress of Philosophy in Geneva.

characteristica, and its application in rational calculus (see Sect. 11.3., above).[97] He was also sensitive, like Carnap would be, to history in both directions, the historical identity of the project since Leibniz and a sense of progress and modernity. Thus, while finding merit in symbolic work by Peano and the Italian school as representatives of an Italian pasigraphic movement, he also criticized them for not having caught on to progress made first in America by Peirce with the algebra of relatives, so that Peano's notation lacked an expression for the relation 'of' and the corresponding operations (Peckhaus 2014, 222). Like Jevons, only metaphorically, Schröder illustrated progress in logic with progress in technology and borrowing from Minkowski's introduction of their Section in the Congress of Mathematicians in Zurich on 10 August 1897. Schröder used a boat simile that placed Peano with "those who persist in still using sailing ships whilst steamboats have already been invented, constructed and are waiting at their service" (1897, 161 and 1898, 61).

The differences between Carnap and Neurath over logic, then, was hardly just the difference at the heart of the usual contrast between naturalism and rationalism, or between social and natural sciences. It was also a difference in generations of practitioners in modern symbolic logic, with Frege acting as bridge figure. Neurath belonged in the pre-1910, older logic (Schröder); Carnap – Frege's student – belonged in the post-1910 new logic, which developed on the back of an explicit program of logicism (Russell). But the difference cannot be considered outside the shared historical and cultural background, for instance, the Leibnizean genealogy of the different stages in the progress in symbolic logic, the appeal of modern formalism, the value of the relation to mathematics, the broader social value of pasigraphy, and the symbolic standpoint.

Carnap's challenge in the *Aufbau* consisted in reconstructing rationally the unified and objective nature of scientific knowledge and its basis in experience. For the logical construction of scientific objects urged by Russell Carnap adopted the method of applying purely structural concepts and relations and the step-by-step derivation of higher-level objects – concepts – and general relations between them through rules and definitions. But to model the empirical foundation of scientific knowledge he simply turned to a modern, extensional interpretation of formal terms in set theory and postulated sets of Gestalt-like complex elementary experiences.[98]

For Carnap, the modernity of logicism and his own project was also rooted in history of logic and more ordinary projects of pasigraphy. In this he shared Neurath's narrative and culture. Carnap had already been interested in universal constructed languages such as Esperanto, independently of any rationalist – Cartesian or Leibnizean – ideal. Carnap also quoted the semiotic work of the psychologist Richard Gätschenberger, in *Symbola* (1920): "'All philosophers are correct, but they express themselves in with varying degrees of ineptness, and they cannot help this, since they use the *available* language and consequently speak in a hundred

[97] About the debate between them, see Couturat (1905/1914) and Peckhaus (2004).

[98] In all the techniques Carnap was relying, also rhetorically, on modern scientific developments, both empirical, e.g., Gestalt psychology, and formal and foundational, e.g., set theory, axiomatics and logicism.

sublanguages, instead of inventing one pasigraphy.' This neutral language is the goal of construction theory" (Carnap 1928/1967, 287).

As part of the history of logic, the application of modern logistics in that neutral and universal sense was based on concepts of the theory of relations that, as Carnap (1928/1967, 8) noted in the opening sections of the *Aufbau*, could be found in "Leibniz's ideas of a *mathesis universalis* and of an *ars combinatorial*." In particular, he added, the "application of the theory of relations to the formulation of a constructional system is closely related to Leibniz' idea of a *characteristica universalis* and of a *scientia generalis*" (ibid.).

The Leibnizean genealogy of the project of logical unification wasn't just Frege's own conception (see above), or even Schröder's.[99] Also Russell had discussed Leibniz at length, especially in his early *Critical Exposition of the Philosophy of Leibniz* (1900), a work that marked his transition out of idealism and monism into logical analysis and construction. The selection of Leibniz texts reprinted there includes Leibniz's fragments on the *characteristica universalis* – the universal Alphabet of Thought – and the *ars combinatorial* (Russell 1900, 282–283).

Carnap also adopted for the same purpose the duality of material and formal modes of speech, where pseudo-questions (and metaphysical assumptions) are prompted by the material mode as "the more usual formulation." To express the correspondence and the contrast, Carnap adopted the dual format introduced by Schröder and adopted also by Neurath: "In the rest of the paper we shall at all times help the reader by using both modes of expression and write the formal, and strictly speaking, only correct, expression of our thought in a parallel column on the left of the more usual formulation" (Carnap 1932/1995, 41).

11.8.2 Old and Modern Symbolic Logic, Neurath and Carnap, Meet in the Manifesto for Logical Empiricism

By the time of the so-called Vienna Circle Manifesto, in August 1929, immediately after the publication of the Aufbau at least three different projects were in progress: Carnap's interest in explicative rational reconstruction, Schlick's interest in the conditions of coordination of empirical foundations and Neurath's interest in the social sciences and naturalism about the conditions of scientific practice (Uebel 1992 and 2007). The debates posed the challenge of constructing a unified umbrella formulation of the program that could identify the movement's viewpoint, if not doctrine.

It should be easier to accept an integrated account of the authorship of the Manifesto, "The Scientific Conception of the World: The Vienna Circle" (Carnap et al. 1929/1973). Signed by Hahn, Neurath and Carnap, and aimed at prompting Schlick to turned down outside offers and stay in Vienna, it may be assumed that statements about logic and mathematics were penned by Hahn and Carnap, while

[99] Carus (2007, 103) stresses the Fregean connection.

Neurath added the references to the social sciences and social goals. If in fact, as Carnap's diaries suggest, Carnap only revised Neurath's original draft (Stadler 2001, 335), Neurath had not been thinking just about formal logical and mathematical interests and standards as Schlick's, or even Hahn and Carnap's alone, but also his own.

The intellectual core of the Vienna Circle grew primarily out of the informal discussions of the Schlick Circle (1924–34). Its members included Hahn, Neurath, Carnap, Waismann, Kaufmann and Reidemeister and their discussions focused on issues in the philosophy of language and mathematics prompted first by Schlick's interest in Hilbert's program of formal axiomatics, followed by Wittgenstein's *Tractatus* and later Carnap's *Aufbau*.

A broader pool of members and social interests surrounded the foundation of the Ernst Mach Society (1928–34) in November 1928 (Stadler 2001, 328–334). Neurath himself supported scientific adult education, as long it included new ideas and results from both natural and social sciences. But, over that distinction, he was responsible drawing a boundary around them with the descriptor Society for the Promotion of Knowledge of the Exact Sciences, along with an opening lecture about Mach's exact world conception,[100] and adopting the broad category of empiricist rationalism, close to the term introduced by Itelson, "empirical rationalism" also associated with the empirical application of doctrines such as critical positivism, conventionalism, logicism, free-thinking and dialectical materialism and, later, with French experimental rationalism – note the primacy of the rational and formal, even as Hahn was defending their value in experimental science.

Carnap, Hahn and Neurath signed the Manifesto on behalf of the Ernst Mach Society. The Manifesto mixed up two overlapping strands, the Schlick's Circle's discussions of foundations of the exact sciences – mainly mathematics and logic – and the Mach Society's social and educational goals. Tension followed, pitting Neurath against Schlick, to whom the Manifesto was addressed. Only Neurath, more actively than Carnap, could straddle any divide. Thus, while mentions of logistic and a commitment to logicism and figures such as Frege, Russell, Whitehead and Wittgenstein, fall squarely within the scope of Carnap's public statements (and private revisions to the draft?), other references to symbolism and logic and historical sources and declared "precedents" such as Leibniz and Schröder fall, on the other hand, also well within the purview of Neurath's own intellectual biography. Only his history of qualifications to the otherwise worthy empirical application of the formal standards of symbolism and rationality he omitted. Instead, he chose the rhetoric of a public unity of the program and movement under the rubric of a scientific world conception.[101]

[100] Neurath (1937a/1983, 174) later wrote, in relation to the new discipline of the logic of science, a replacement of philosophy, and linguistic (anti-metaphysical and social) unification, that "progress of the unification of scientific language requires a logical analysis of the sciences" and that "Mach's far-reaching conclusions did not depend upon new experimental data, but simply upon a rigorous logical analysis of the traditional formulations."

[101] My discussion of this aspect supplements the account in Stadler (2001) and the detailed reconstruction of the manifesto's writing in Uebel (2008).

The Manifesto listed the five main strands that 'came together' in the scientific world conception, aimed at unified science against metaphysics ('the common goal of all'): the more empirical and social strands are (1) positivism and empiricism, (2) foundations, aims and methods of empirical science, and (5) hedonism and positive sociology; then we find the formal categories of (4) axiomatics – a modernized mathematical standard for Schlick as well as Carnap and, less so, Neurath – and (3) 'logistic and its application to reality' and listed as main representatives of the latter Leibniz, Peano, Frege, Schröder, Russell, Whitehead and Wittgenstein (Carnap et al. 1929/1973, 304).[102]

The issue of the relation of logic to reality was central to its methodological relevance to the possibility of objective empirical theorizing for Schröder and Neurath, and central to Itelson's conception of logic, also to the relation of mathematics to empirical reality for Neurath (see above), Schlick (especially in his *General Theory of Knowledge*) and Hahn. The same year, Hahn wrote from the symbolic standpoint that, from logic and mathematics had permeated theories of scientific knowledge of scientist-philosophers such as Helmholtz, Mach, Peirce, Schröder, Poincaré and Duhem, before Neurath, and characterized the Circle's new interpretation of mathematics, as built on convention and tautology (and the interpretation of philosophy as well)[103]:

> If logic were to be conceived – as it has actually been conceived – as a theory of the most general theory of objects, as a theory of objects as such, then empiricism would in fact be confronted with an insuperable difficulty. But in reality logic does not say anything at all about objects; logic is not something to be found in the world; rather, logic first comes into being when – using a symbolism – people *talk about the world*, and in particular, when they use a symbolism whose signs do *not* (as might at first be supposed) stand in an isomorphic one-one relation to what is signified. (Hahn 1929/1980, 40; original emphasis.)

Hahn's concern was precisely with the question, "how is pure empiricism compatible with the existence of mathematics?" If empiricism is compatible with logic, the answer is only a matter of commitment to logicism, that "mathematics is part of logic, and hence, that the propositions of mathematics too do not say anything about the world, but are merely directions for saying what has been said in another way" (Hahn 1929/1980, 41–42).

Logic was at the heart of the exact world-conception, the source of its exactness; and its role would be discharged, following Russell, through the method of logical analysis. Its key role was to unify formal and empirical sciences and distinguish the latter both from older forms of empiricism and from philosophy as a kind of

---

[102] Carnap had done the same in the *Abriss*, of the same year.

[103] An important question that deserves attention is the distinction between sign and symbol. These authors often lack a systematic distinction, or a consistent statement of their equivalence: they sometimes use the terms interchangeably, or else their use is inconsistent in taking 'sign' as the more general term, almost in Peirce's sense, or else as a concrete token of a symbol, or as an index, in contrast with the more conventional character of a symbol, restricted further by the rules of a specific system of symbols, or language, including mathematics.

knowledge (Carnap et al. 1929/1973, 306). This was still the Carnap of the *Aufbau* and the constitutive system: "The aim of scientific effort is to reach the goal, unified science, by applying logical analysis to the empirical material" (ibid., 309).

The reform was at its heart a symbolic project of language reform: only the control of language and linguistic statements would rid the scientific conception of pseudo-problems that could be traced to the reliance on a priori speculation and ordinary language alike (ibid., 307–308).[104] And since no symbolic language would do, the ideal is a modern one, that is, one regulated by modern logic: "Only modern symbolic logic ('logistic') succeeds in gaining the required precision of concept definitions and of statements, and in formalizing the intuitive process of inference of ordinary thought, that is to bring into a rigorous automatically controlled form by means of a symbolic mechanism" (ibid.). The automated mechanism provides also the social and formal conditions for effective communication – "what unites men in language" – and for objective description through structural formulae. Unity, objectivity and freedom from metaphysics would have to rely, then, on symbolic form determined by modern symbolic logic.

So, reform of language and reform of logic went hand in hand, and the Manifesto emphasized the role of logic's shifting relation to mathematics in a history of mutual aid in establishing each other's foundations – axiomatic structure – and in eliminating each other's ambiguities and contradictions. And that is precisely the history of symbolic logic from algebraic logic to logicism, from Leibniz to Russell. The Manifesto made a point of including the goal of a practical relation of logic to reality. First, what Carnap called the old logic: "Since Leibniz and Lambert, the idea had come up again and again to master reality through a greater precision of concepts and inferential processes, and to obtain this precision by means of a symbolism fashioned after mathematics. After Boole, Venn and others, especially Frege (1884), Schröder (1890a, b) and Peano (1895) worked on this problem" (Carnap et al. 1929/1973, 310). Then came the new logic: "On the basis of these preparatory efforts Whitehead and Russell were able to establish a coherent system of logic in symbolic form ('logistic'), not only avoiding the contradictions of traditional logic, but far exceeding that logic in intellectual wealth and practical applicability" (ibid.). Here logicism constituted an application to, also an outcome of, confronting conceptual and logical challenges in the foundations of mathematics, and, in particular, of arithmetic. In general, more practical applications extended to social reality, closer to Neurath's sustained attention to the value and challenges of symbolic representation and reasoning: the scientific would conception would serve life, personal and public, from education to architecture to "the shaping economic and social life according to rational principles" (ibid., 317–318).

[104] Carnap would soon add to the list of sources the material mode of expression.

11.8.3 Neurath's Physicalism and Syntacticism from the Semiotic Standpoint. The Role of the Old Algebraic Standards

More traces of the early symbolic standards I am tracking in Neurath's later works appeared in a more personal follow-up piece on the scientific world-conception (Neurath 1930/1983). In the essay, Neurath set out to explore the historical conditions of the new view in connection to "other spheres of life and science" (ibid., 46). He insisted on the instrumental nature of language and thought and their historically and socially holistic dimensions: "Our thinking is a tool, it depends on social and historical conditions." Our own use of language in thought is historical in ways that develop his later linguistic – social – doctrine of empiricism in the use of protocol statements: "We confront our present thinking with earlier thinking, but we have no possibility of taking a judge's stand on a point outside" (ibid.). The same must apply to linguistic expression at the concrete symbolic level: "We owe our means of expression, our rich language and script, to definite historical premises" (ibid.). Neurath's concern was the same he had pursued in the context of economic theory, using historical knowledge to determine logical possibilities and effective empirical options. In this case, the constructive goal was the reform of language at the service of intellectual, scientific and social goals.

Neurath was delving into his long own acquaintance, going back to at least his University courses, with the diversity and history of languages and into the linguistic interest that permeated Viennese culture and academia. Alfred Stöhr, for instance, was one such scholar, successor to Mach and Boltzmann's University Chair, interested both in symbolic logic and language much in the psychological and anti-metaphysical way of Mach and the mathematical way of Schröder (see above), and the author of *The Algebra of Grammar*. From his own symbolic, semiotic, standpoint, Neurath warned against the conceptual risk of the combinatorial capacity for "freely moving symbolism" and "unlimited word formation" in alphabetic writing, the sort of "tongue acrobatics" that Stöhr had called *glossurgy* (ibid.). This capacity might have been a source of artistic creativity, but, at least in ordinary language, it was also a source of philosophical problems – that is, what Carnap had declared pseudo-problems (ibid.). Those problems, in Neurath's familiar diagnostic, were "much dependent on language" (ibid.).

The remedy, as indicated in the Manifesto, consisted, much in the pasigraphic tradition and renewed interest in constructing universal artificial languages, in applying a combinatorial approach to more strictly regimented symbolisms without fixed uses and meanings: "The letters as signs without conceptual meaning are, however, well suited for a strictly scientific symbolism" (ibid.). Then, "one combines signs whose meaning one can define at will" (ibid.). To the extent that modern science exemplified the scientific world-conception, "it owes its success partly to the new symbolism that can be used for the purification of language" (ibid.).[105] The

[105] Neurath drew attention to the scientific value of general names (variables).

declared goal was "building up a system of symbols with the aid of logic and mathematics" (Neurath 1931a/1983, 48).

Neurath's remark on the importance for science of general names (symbolically, variables) was reminiscent of Schröder's claim, in his essay on signs, that signs are handles on ideas that free the latter from their association with concrete impressions and give them the generality and control that makes human reason possible (Neurath 1930/1983, 46; Schröder 1890a, 5 and 1890b/1892, 3431). Schröder here as well as in the *Vorselungen* was following Johann Heinrich Lambert, whose work on logic, *Neues Organon* (1764) included a second volume on a theory of signs or *Semiotik*, 'the theory of the designation of thoughts and things,' as a study of Leibniz's universal language or characteristic at the foundation of the sciences. Neurath was familiar with the work of both authors, Schröder and Lambert as well, who were the subjects of Olga Hahn's doctoral dissertation and were mentioned in the Manifesto. The independent introduction into German of the term 'semiotik' in relation to the study of linguistic signs – that is, independently of Peirce's, not of Locke's and the medical tradition – suggests that Neurath's symbolic, notational standpoint adopted already in algebraic logic may be called also a semiotic standpoint.[106] Indeed, Neurath drew these connections in relation to the term 'semiotik' in a letter to Carnap of April 9, 1932:

> Semeion is called the sign, of which semiotics – the doctrine of the characteristics of the disease. As a theory of signs par excellence in Leibnitz, Lambert, but in a very narrow and unpleasant sense. (Neurath to Carnap, 9 April 1932; RC 029-12-55. My translation.)[107]

His first significant contribution to logical empiricism was precisely in the form of the universal physicalist language, an application of the pasigraphic ideal to a philosophical problem of the scientific world-conception. Physicalism entered the picture around 1930 when it was adopted, albeit with different conceptions, by Neurath and Carnap as a solution to the problem of objectivity of empirical knowledge that Carnap had undertaken in the *Aufbau*.

Neurath had objected to Carnap's proposal of a system, a logical construction, founded on the subjective language of phenomenology. Carnap, from a theoretical, scientific perspective, adopted the language of physics, the basis of the construction system that reconstructed the unity of sciences. For Neurath, instead, physicalism echoed the social nature of materialism in the sense associated with socialism and preserved some degree of disciplinary autonomy for the social sciences. Its scope was, then, broader, consisting, instead, in descriptions of publicly accessible objects and events in space and time.[108] Neurath (1931a/1983, 49 and 1931b/1983, 55) spoke of spatio-temporal data expressing space-time linkages. This proposal would

[106] Note also that for Peirce (1884), algebraic logic was a contribution to the 'philosophy of notation.'

[107] In the followings sections, the Neurath-Carnap letters are quoted from the Carnap Archive at Pittsburgh (Rudolf Carnap Papers, 1905–1970, ASP.1974.01, Special Collections Department, University of Pittsburgh). All rights reserved.

[108] For a discussion of the first stage of the so-called protocol-sentence debate see, Uebel (1992) and (2007).

be the core of the account of the protocol language and the protocol sentences that exhibit the objective empirical basis of scientific knowledge.

Aside from an implicit reference to the cental status of spacetime in Einstein's theory of relativity, Neurath placed an explicit emphasis on two inseparable features of his proposal, the anti-metaphysical (against Wittgenstein's picture theory of meaning) and anti-subjective (against Carnap's phenomenological reduction). Both depended on a commitment to more formal relations, at three different levels: (1) the semantic commitment to the (meta)linguistic framework, without references to any extra-linguistic world or reality, with the methodological implication that statements are only compared with statements; (2) the syntactical aspect of language – as in Russell and Carnap's emphasis on structural descriptions –; and (3) the particular focus on relations of order, arrangement or linkages in space and time (Neurath 1931a/1983, 1931b/1983 and 1932/1933/1983).

The latter feature connects Neurath's syntacticism with his physicalism. In relation to unity, the universality of the objective (and empirical) language suggests a requirement of completeness and reflexivity (completeness extended to include itself) without a semantic dependence on the suspicious extra-linguistic facts. Language, then, and metalanguage, must be of this world, just like any other physicalistically describable fact. They are a kind of spatio-temporal physical structure – of public objects and events – such as speech behavior. Statements in speech or thought are physical events. As a matter of meaning, physical correlations could be both expressive of a speaker's physical (material) states and indicative of physical (material) states of the world (see Uebel 1992, 2007; and Derek Anderson's chapter in the present volume).

Neurath (1931b/1983, 53) used an analogy to ornaments as physical arrangements, which indirectly integrated the system of pictorial symbols, and telegraphic signals: "What is at all scientifically expressible is not richer in fundamental relations than the symbols on a Morse tape which the telegrapher reads as they are sounded by his apparatus" (Neurath 1931a/1983, 49). Or more generally: "In language nothing but order is essential, and that is already represented by a sequence of signs in Morse code" (Neurath 1931b/1983, 62).

The structural emphasis was consistent with the typographic approach in visual education through pictorial symbols and, similarly, with the formal standards of algebraic logic – even when steeped in psychologism or assuming the dominating extensional interpretation. The issue of symbolic order is what was at stake, from the symbolic, or semiotic, standpoint, in Neurath's discussion of symbolic equality and the law of commutation (see Sect. 11.6., above).

By contrast, Carnap's syntacticism about language in the same period evolved out of his logicism and the problems of unity and objectivity he had followed Neurath in solving in terms of some notion of physicalism. The roots of his turn to logical syntax included Gödel's communication in 1931 of his incompleteness theorem. If symbolic logic had been inspired by algebra and Schlick's epistemology and Hilbert's formalism on axiomatics in geometry, now Carnap was modeling a metalogic based on arithmetic. The system of arithmetic, Carnap learned, was consistent but incomplete.... except about its own formal features (Carnap 1934/1937).

Very succinctly put, he promptly turned to the philosophical project of constructing a metalanguage in terms of syntax, derivation rules, rather than truth, testability or meaning. Content, semantics, is a formal matter of syntax or metalogic. In fact, until 1935 and his next fatal encounter, with Tarski, "semantics" meant syntax. Meaning was introduced by formal logical equivalence or equality (e.g., ostensive and nominal definitions within the same language system, and translations between languages) (ibid.). Out of this development came the distinction between the material and formal modes of speech, the principle of tolerance in matter of logic and the centrality of analyticity problem of standard of definitions and identities described in the metalanguage. In particular, Carnap declared that identity or definition cannot depend on extensional matters of truth or designation of objects (equality of truth value or denotation); if syntax depended on semantic sameness of descriptive meaning, it must be an intensional matter (ibid.).[109]

For Carnap, as well as Neurath, the syntactic turn was part of the linguistic turn in science and philosophy, and of the culture of language criticism in Austria and Germany. It had a marked normative significance in line with the terms and goals laid out in the Manifesto and expressed in different ways by kindred groups: "The aim of logical syntax is to provide a system of concepts, a language, by the help of which the results of logical analysis will be exactly formulable" (Carnap 1934/1937, xiii). The project was meant to replace philosophy with the logic of science, which "nothing other than the logical syntax of the language of science" (ibid.). This is a standpoint from which to study scientifically what otherwise may be the object of sociology in Neurath's sense, language as a "historically given method of communication, and thus of mutual influence, within a particular group of human beings" (ibid., 5). Other scientific standpoints, according to Carnap, are the syntactical, or formal, that is, his own, the psychological and the semasiological (the study of meaning of expressions or semiotics; Carnap borrowed the term 'semasiology' adopted in psychology by Bühler, Külpe and Gätschenberger) (ibid.).[110] Carnap's formal standpoint considered languages as calculi, sets of rules or conventions for the formation and transformation of finite series of symbols, or linguistic expressions (ibid., 4).[111]

The contrast between Carnap's and Neurath's respective doctrine of physicalism rests on differences over the linguistic basis of empirical, objective language in, at least, scientific practice and for effective social communication, more generally. Distinctively, Neurath disagreed with Carnap over the normative – instrumental or ideal – value of precise symbolic language (in the mathematical model of algebraic

[109] The literature on Carnap's syntactic turn is vast, but to stay within the context of the Circle, the reader may do well to start with Uebel (1992) and (2007).

[110] As he had done in the *Aufbau*, in the bibliography Carnap listed Gätschenberger's *Symbola* (1920).

[111] In the tradition of Leibniz, Lambert and turn-of-the-century symbolic logic, Gätschenberger (1920, 126, 133, and 454) declared that sematology contributed to algebraic logic, especially to the calculus of relations sought by Peirce and Schröder, and from that standpoint, it was the mother of mathematics. On Gätschenberger's relevance to Carnap's early works, see Haller (1959); on his place in German semiotics, see Eschbach (1987).

logic and subsequent symbolic logic) over ordinary language (source of no less confusion than metaphysics). The limitations in the value of the former, as I have discussed (see Sects. 11.5 and 11.7., above), extended, in addition, to its empirical application, as a methodological matter of representation, reasoning and decision-making.

Neurath adopted his semiotic standpoint against Carnap's use of 'semantics' after 1931. In a letter to Carnap of April 9, 1932, he wrote as follows:

> Neither I nor Olga can befriend semantics. A faint picture of words, a "mantic", reminiscent of visionary art. Strange and scholarly in an unfair sense.
>
> In addition, as the learned woman knows, semantics is moreover still wrong, it is called "semiotics", Semeion is called the sign, of which semiotics – the doctrine of the characteristics of the disease. As a theory of signs par excellence in Leibnitz, Lambert, but in a very narrow and unpleasant sense. Sema the character knows no semantics.
>
> We both think that SYNTAX – Logical Syntax – General syntax or the like sounds much better and can be popularized for emergencies. (Neurath to Carnap, 9 April 1932; RC 029-12-55. My translation.)

Neurath also pressed Carnap on the distinction between empirical, logical and terminological, or glossological, normative standards for the construction and use of words and statements. The construction of language was a normative task at the very level of elementary expressions. Not any group of sounds can make up an acceptable, meaningful word in a given language, e.g., 'nots' (Heidegger's *Nichte*).[112]

Similarly, it is not surprising that in the wake of Carnap's syntactic turn, Neurath wrote to him to remind him of Neurath's early essay on symbolic equality and request Carnap's opinion.[113]

Standards of symbolic language and reasoning were at the heart of Neurath's, and the Circle's, project of unification, even with the shifting limits he set on the application of aims driving the program of algebraic logic. Noticeable changes in his statements responded to changing circumstances, including responses to positions by other authors. Some were more pointed such as the explicit opposition of the systematic dimension of the ideals of knowledge, the system-encyclopedia model opposition; others were more a matter of explicit emphasis, such as declaring the only possible universal terminology a jargon.

For Neurath, unlike for Carnap – after Frege –, (see Carus 2007, 102–103) ordinary language was unavoidable in empirical science in general, and, in particular, the social sciences and their practical social role. Rational reform had, then, its limits. Already in "On Protocol Statements" Neurath had departed further from the unqualified generalizations in the Manifesto and had stressed how empirical science required a universal jargon that reflected what above I have called its historical and social holistic character, that is, the thickly and locally embedded character of scientific practices in social and historical conditions, and in which, accordingly, we "combine terms of ordinary and advanced scientific languages, since in practice, the

112 Neurath to Carnap, 3 November, 1932; RC 029-12-13. In the same letter Neurath also analyzed cases of empirical falsehood as forms of logical contradiction or nonsense.

113 Neurath to Carnap, 5 April 1934; RC 029-10-77.

terms of both languages overlap" (Neurath 1932/33/1983, 92). Protocol statements are an unavoidable source of imprecise cluster expressions that belie the one-dimensional linear formal ideals: predictions are "checked in turn by common language statements" endowed with a "relatively great stability" that "is not linked with a precision equivalent to that of scientific formulas" (Neurath 1936/1983, 151). Instead, in the actual cases and in Neurath's linguistic version of the epistemic reconstruction, they operate as "manifestly imprecise, equivocal, indeterminate formulas" (ibid., 152).

Still, in an early letter to Neurath of October 7, 1928, Carnap himself had considered a similar situation, although only as a default, which was for Neurath the only situation:

> [A] logic, a method of concept formation must be established which takes into account the fact that we always have crystals and dirt mixed before us, which therefore specifies the demands on science. Concepts and statements are to be provided as long as the "ideal language" is not available. (Carnap to Neurath, 7 October 1928; RC 029-16-01. My translation.)

When in 1935 Carnap turned to semantics in Tarski's sense, Neurath perceived the move as metaphysical and pseudorational and at odds with the scientific attitude (Uebel 1992, 2007). Without mentioning the Leibnizian language-calculus standard or its fortunes on the back of the development of symbolic logic, it has been argued that the members of the Vienna Circle, including Neurath, embraced the universalism of the linguistic framework, but not a calculus, without distinguishing between Cartesian and anti-Cartesian possible versions, and only when Carnap embraced semantics, he singly left the universal-language camp to join a Cartesian version of the calculus one (Mormann 1999, 170–171).[114] Carus, in the present volume, questions elements of this portrayal of Carnap. My elaboration on the theme of logic and its associated standards emphasizes the way in which (1) Neurath's early (anti-Cartesian) anti-pseudorationalistic commitment evolved in expression and argumentation, without reversals, as he entered new debates, while (2) its includes a sustained, although restricted, appeal to the dual standard of both language and calculus the earlier tradition of symbolic logic picked up also by Schröder and Frege, if not by Russell and Whitehead as well.

I have argued above that limitations on language preceded Neurath's warning about restrictions implied by the unavoidable presence of *Ballungen* terms after his attention had turned to the broader philosophical picture prompted by Schlick and Carnap's works, especially the latter's *Aufbau* (Uebel 1992, Cat 1995, Cartwright et al. 1996). Neurath also introduced the idea of a universal jargon along with the presence of Ballungen and the absence of an epistemological tabula rasa – the historical holism expressed by the boat metaphor – as early as in "On Protocol Sentences" (Neurath 1932/33/1983, 92). Physicalism, even if purified from metaphysical terms, can be only a universal jargon. As a result, scientific language, from this Neurath's perspective, despite its "growing equipment of systematic

[114] Mormann adopts Hintikka's terminology, paraphrasing van Heijenort's focus on logic, language as universal medium and language as calculus.

symbol formation, can by no means be regarded as an approximation to such an ideal language" (ibid., 91).

Syntacticism might have been compatible with physicalistic jargon, but it could not have provided a consistent enough metalanguage (Mormann 1999, 174). The view preceded, then, also the reaction to Carnap's semantic turn and the subsequent attention to the encyclopedia as the anti-systemic model of unification.[115] And yet, the anti-pseudorationalist limitations didn't prevent Neurath from endorsing, in however limited a form, the calculus standard. In both cases, the ideal universal language and its calculus, the logical-mathematical background he shared with others, prompted him to keep formulating his alterantive views in relation to those standards. Only the occasions and expressions varied.

11.8.4 Logical and Terminological Unity of the Encyclopedia in Exile, and Back to Symbolic Equality

Neurath's preoccupation with the empirical conditions of scientific practice, especially in the empirical and, in particular, the social sciences, led him to an interest in their methodological and social management. With this approach to empiricism, and mainly in this regard, it should be acknowledged that his participation in debates over logical empiricism or with fellow Circle members became complicated by a tendency to substitute, and conflate, epistemological questions for logical ones.

At the same time, he also declared that logic and (empirical) behavioristics constitute alternative perspectives on scientific practice with their respective sets of concepts, or terms, such as contradiction and deduction – logical – and acceptance – behavioristic. So that when discussing "the problem of unified science and encyclopedia," he could, if he wished, "separate the behaviouristic considerations strictly from those about the logic of science" (Neurath 1936c/1983, 170).

Also, both logic and universal language, the two sides of the Leibnizian algebraic project, were relevant to unification. Neurath's commitment to the value of symbolic logic as an ideal was sustained but growingly critical and distant: notably since the position laid out in 1913 in "The Lost Wanderers of Descartes," first as calculus (1913/1983, 1932/33/1983 and 1934/1983), then explicitly also as language (1932/33/1983), and as two sides of pseudorationalism (1936c/1983, 1935a/1983 and 1935b/1983). Still, Neurath pointed to a similar standard in a joint extension of both methodological holism and logicism, echoing some of the programmatic (and propagandistic) statements of the Manifesto and Carnap's Fregean commitment to the logical foundations of the sciences, including the empirical ones.

[115] In a number of places, Neurath adopted the characaterization of the encyclopedia as a model, in a Duhemian sense, part of the opposition to the characaterization of unity as a system (Neurath 1936/1983, 145); but he used the term "model" also for a broader category to contrast the encyclopedia with the system as models of knowledge (ibid., 156). In relation to Duhem, it must be noted that he adopted the notion of logical unity in the sense of oneness, as a fundamental feature of natural classification.

In 1937, after the International Congress of Philosophy and in preparation for the Third International Congress for the Unity of Science in Paris at the end July, Neurath felt compelled to talk up the value of logic in logical empiricism, retaking in exile the international propagandizing tone in Manifesto on behalf of the movement as a whole, now detached from its original geographical and cultural location. That is, he would restate the early intellectual endorsement of the connection of science to technical logical development in philosophy in a manner consistent with the Circle's joint projects of intellectual and social reform. In particular, he would insist on the desirable meta-scientific (or scientific?) synthesis of philosophy and science as recapitulated in the synthesis of formalism (in symbolic logic, in foundations of mathematics and in mathematics) and empiricism. In the Circle's scientific philosophy formalism was represented primarily in two approaches to the foundations of mathematics, formalism in Schlick's Hilbertian picture and logicism in Carnap's.

In "Unified Science and its Encyclopedia," already in exile and in his Encyclopedia phase, Neurath wrote:

> The program of the unification of scientific language requires a logical analysis of the sciences. The history of science during the last decades shows the importance of such an analysis for the progress of concrete scientific work. (Neurath 1937a/1983, 174.)

Examples of such scientific relevance were for Neurath scientist-philosophers such as Mach, Duhem, Einstein, Poincaré, Boltzmann and Russell. The logic of science is the study of the scientific application of the method of logical analysis.

In the task of unification we find, again, the two faces of the Leibnizian project: the calculus and the universal language. On the calculus, Neurath is again explicit: "The logical calculus in its widest sense becomes an essential apparatus of a unified science, and all logistic research acquires very great direct and indirect importance for the evolution of the logic of science" (1937a/1983, 175). In particular, the task "of modern logical analysis of science is to build up a more consistent framework for the special science and for a unified science" (ibid.). The two parts were inseparable: "The process of the logical organization of a single science cannot be divorced from the process of building up bridges or connections between the different sciences." The calculus, as the combinatorial calculus to construct logical classification, allows for systematization, in the deductive sense, but only locally and from below, from the connection between available small units and not as a single closed deductive system (Neurath 1936b/1983, 145 and 153).

And here is where the linguistic dimension plays a key technical, constructive role in the form of his older proposal of physicalism:

> As has already been indicated, the unification of scientific language is a special and technical task. The fundamental thesis of our movement is that terms similar to those employed in physics and in our everyday language are sufficient for constructing all sciences. This thesis, known as Physicalism, has been progressively confirmed by special investigations in recent years. (Neurath 1937a/1983, 175-176.)

As in the case of logical analysis, Neurath pointed to empirical evidence from actual historical cases of applications in the sciences or about them. Both, then, the evidence and the goal are historical and constructive, and this sets out the contrast in the role of logic with the "[pseudo] rationalistic anticipation of the system of the sciences' based on the notion of 'a philosophical system which is to legislate for the sciences" (ibid., 176–177).

He had made the point elsewhere with a broader historical focus so as to distinguish between older and modern projects, just like Carnap had done about older – Schröder and Neurath – and modern – Russell and Carnap – symbolic logic: "The striving for logical derivations which we encounter among the scholastics was to revive again in Leibniz in a thoroughly modern form. He was to become the forerunner of logicism. But empiricism does not look kindly upon such synthesis, and any rationalism a priori appears suspect to it" (Neurath 1937/1987, 133–134).

It is this more liberal sense, also narrowly informed by attention to empirical science, that he could endorse a restricted application of the standards in algebraic logic, since 1909, and could later insist on the Manifesto's theme and standard. Writing to Susan Stebbing in England on April, 8 1939, he distinguished between "unification by logicalisation" and "unification by visualization" (quoted in Körber in this volume) Once in exile in England, Neurath acknowledged the two perspectives, ideal and restricted, when the logical met the empirical, that is, in the project of logical empiricism. On the one hand, he noted, "Carnap's work shows how symbolism helps us to overcome some difficulties which arise when we seriously try to logicalize' and 'empiricalize' our scientific enterprise" (1941/1983, 213). On the other, he warned, "formalization is no magic sieve, saving empiricism, saving empiricism" (ibid.); no only language may be unavoidably imprecise, but sometimes "a symbolism may even conceal the ambiguity of certain explanation" (ibid.).

Neurath made the point in vivid terms in a letter of August 8, 1944 to Hans Reichenbach's brother Bernhard, also exiled in England:

> LOGIC deals with ARGUING, and therefore NEVER CAN LOGIC TELL YOU ANYTHING WITHOUT APPLYING IT TO FACTUAL STATEMENTS. Fechner said, you cannot get out from a roasted turkey more boiled apples than you put into it before .....
>
> Logic is only PLAYING WITH APPLES, and now listen:
>
> ALLL, ALLLLLLLLL, ALLLLLLLLLLLLLL factual statements, which tell about some stories, earthquakes or weather, forests or mountains, human beings or comets are based on SELECTED material, with well KNOWN remaining details and others we shall perhaps know later on .... THEREFORE ALL LOGIC APPLIED TO SUCH FACTUAL STATEMENTS CAN NEVER PRESENT YOU WITH UNAMBIGUITY, SINCE THE START IS AMBIGUOUS, EVERY ARGUMENT IS EITHER LOGICAL OR ILLOGICAL, to say it is logical OK, but things in history do not happen according to the logic, sounds to me Bushman language, which I do not understand. (Neurath to Bernhard Reichenbach, 8 August 1944; original emphasis; Otto Neurath Nachlass.)[116]

As a matter of empirical method, not just meaning, Neurath repeatedly concluded, the selection of scientific statements was a matter of decision, the outcome of a process that involved a limited use of reasoning and limited benefit from logical determination (same as the general empirical psychology of decision making 1913):

---

[116] I'm grateful to Adam Tamas Tuboly for drawing my attention to the letter.

> *Multiplicity* and *uncertainty* are essential. From data at our disposal we can, in more than one way, deduce predictions that are in harmony with science; the multiplicity of predicting cannot be excluded by any method; no degree of systematic procedure can alter this. One can, so to speak, not agree on a 'machine' that unambiguously produces 'inductions' in the wider sense. The progress of science consists, as it were, in constantly changing the machine and in advancing on the basis of new decisions. Still, the result in fact is far-reaching unity that can *not* be deduced logically. (Neurath 1935a/1983, 116; original emphasis.)

Against the Jevonsian mechanical ideal, here judgment, purpose, extra-logical factors distinguish the rationality of actual process from the pseudorationality of the ideal. The well-known implications for an epistemological picture are holism and antifoundationalism.

In this sense, Neurath distinguished the conditions for accepting content statements, synthetic statements assessed in relation to protocol statements, from logical statements. In their case the presence of uncertainty and role of decision is possible but not inevitable: "While to make progress at all, of course, we have to choose between several equally possible groups of content statements and do this on the basis of a decision, such a decision is unnecessary in a logical statement (that, for example, a system of statements in a given language is consistent)" (Neurath 1934/1983, 103). And he added, echoing the Jevonsian ideal, that "such a decision is discarded in the sphere of combinatorial analysis" (ibid.).

The pseudoscientific and pseudorationalist position is the belief that there is one solution and a calculus and a precise language with the power to reach it, so that decision, as he attributed to several Polish logicians, is replaced by "the calculus of the logic of science" can replace decision in the practice of science (Neurath 1936a/1983, 136). Instead, "we have to reach a 'decision' not based on a calculus" (Neurath 1946/1983, 235). The difference in uncertainty stems from a difference in modal multiplicity: "In logic and mathematics we are dealing each time with *one* possibility *in principle*, in the sciences with *several* possibilities *in principle* that are in competition with each other" (ibid., 104; original emphasis). The uncertainty consists in the inability to conclude that we know whether a statement is analytic or synthetic (ibid.). For Carnap this was a matter of logical tolerance and the pragmatic challenge of choosing a linguistic framework. It was also a matter of convention about the way we use symbols in relation to the same content. This was also how Neurath had discussed symbolic equalities as a matter of practical or conventional constraints on the use of symbols and the dual pattern of presentation (see Sect. 11.5., above).

In the last decade of his life, Neurath devoted his time and energy in exile to the coordination, first out The Hague and later from Oxford, of Isotype projects such as the pictorial contributions to educational booklets and films, and the coordination, and of Unity of Science projects such the Encyclopedia of Unified Science and annual International Congresses for the Unity of Science. To carry things out, the core group of patient collaborators included his wife Marie in Oxford, Carnap and Charles Morris in Chicago and Jörgen Jörgensen in Copenhagen.

The projects had a clear pragmatic character, as had the Manifesto before them, aiming, for instance, at actual expressions of linguistic unification and social collaboration. The three main fronts of such collaborations were the sciences, science education, and scientific philosophy as philosophy of science, and, in particular, the logic of science. For practical reasons, then, Neurath would have to straddle the line between touting the goals of unity and universality and clarifying the actual limitations that challenged the empirical and practical applications of ideal formal standards of language and logic – that is, of representation and reasoning. In this way he remained engaged in projects of symbolic standardization and reform harking back to his work in the first decade of the century.

The focus on language as a social tool for communication and community, for unity, remained extended to logic and to its application in linguistic analysis – a philosophical project pursued, among others, by Russell, Tarski and Carnap. Thus one of the results of the First Congress for the Unity of Science, held in Paris in 1935, was an international commission to explore the standardization of logical symbolism in the face of growing numbers of suggestions and inconsistent usage. The committee was formed by Carnap, Heinrich Behmann, Heinrich Scholz, Adolf Lindenbaum and Paul Bernays.[117] Neurath himself drafted the proposal, a six-page document titled "Unification of Logical Symbolism" ("Vereinheitlichung der logischen Symbolik"), stressing the goal of unity and the condition of enough continuity with tradition (Otto Neurath Nachlass K13.). The final text was discussed at special meetings first during the Third Congress, held again in Paris the last week of July 1937 and again during the Fourth Congress, in Cambridge, England, the third week of July the following year.

The Sixth International Congress for the Unity of Science took place in Chicago the first week of September 1941. Since many Europen participants were unable to cross the Atlantic because of the war outbreak, Neurath organized in England a smaller conference on the first week of October (with Susan Stebbing and the education expert Joseph A. Lauwerys). The topic was terminology. I have introduced his position, which he published in "Universal Jargon and Terminology:" to the extent that unification requires universality, the universal language can only be a universal jargon. That is, it can function as a tool for the encyclopedia model of unification only if based on unavoidably uncertain and imprecise expression (Neurath 1941/1983).[118]

In his correspondence with Carnap, Neurath persisted in the bitter attacks on the latter's semantic turn and its association with Polish philosophy. On one front, Neurath insisted on distinguishing the idea of precise formula in a formal calculus from empirical expressions, holistically embedded "aggregations," and it was in the empirical application of the calculus that he sensed the danger of incurring in metaphysical commitments.[119] He pointed to the specific risk of "calculus absolutism," which he resisted, but required some form of integration within the alternative proj-

[117] These are names listed by Neurath on the cover letter from 10 May 1937 with the proposal sent out for circulation. Nachlass K13. The Congress Report included a mention of the committee that listed Alfred Tarski and Olaf Helmer (Erkenntnis 7, 63).

[118] See also Adam Tamas Tuboly's chapter in the present volume.

[119] Neurath to Carnap, 25 September, 1943 (RC 102-55-03). See letter 22 in the present volume.

ect, of "orchestration."[120] On another front, Neurath pointed to a Catholic Austria-Poland axis in which a genealogy that originated with Brentano led to Twardowski and then Kotarbinski.[121] He associated their work in logic with metaphysical doctrines and approaches of neo-Scholasticism, Thomism and Aristotelianism, even in the case of Lukasiewicz and his "free-will tendency in multi-valued logic."[122] By association with Tarski and the alleged metaphysics of semantic talk of truth, Neurath extended his indictment to Carnap himself, who in his final letters he would accuse of embodying also the rigid values of Platonism, Puritanism and Prussianism.[123]

Neurath remained defiantly committed to his physicalism and syntacticism, which he considered authentically anti-metaphysical, and to its ensuing decisionistic methodology, which he considered authentically rational. It's hardly surprising, then, that when Russell took on the question of protocol sentences in the late 1930s and targeted especially Neurath's views, and his anti-metaphysical criticisms of Wittgenstein and Schlick, Neurath would adopt the same attitude and react to Russell the same way he had reacted to the others and more recently to Carnap's semantic turn.

Russell's presented his effectively own contribution to the protocol sentence debate in an essay, "Basic Propositions," included in the book *An Inquiry into Meaning and Truth* (Russell 1940, 137–149). It could hardly have been an intellectually less welcoming reception on Neurath's arrival in Britain. Russell's book contained a series of lectures delivered at Oxford in 1938; the book was subsequently presented at seminars attended by Carnap and Morris – and Hempel? – in 1938–9 at the University of Chicago and most likely also by Reichenbach at UCLA in 1939–40, also by Carnap at Harvard in 1940.

In the essay on basic propositions Russell sought, as had Popper before him, to identify the source of the epistemic authority of the distinctively empirical evidence associated with protocol statements. Russell was concerned throughout the book with perceptual knowledge and empirical evidence, and argued that such an empirical claim on our beliefs could not be available within a self-enclosed social linguistic world in which the syntactical standpoint reduced empirical truth to formal, logical truth (Russell 1940, 140). According to Russell, the corresponding refusal to acknowledge a semantic concept of truth, in relation to some extra-linguistic reality, rendered Neurath's empiricism, inter alia, completely unintelligible. It was mere "ultra-empirical" verbalism. Russell was joining forces with Schlick and Wittgenstein and, indirectly, with the empiricism of the Carnap of the *Aufbau* and the more recent semantic turn. Empirical knowledge, the epistemic force of basic statements, rested on individual perceptual access to extra-linguistic reality.

---

[120] Neurath to Carnap, 16 June, 1945 (RC 102-55-11). See letter 31 in the present volume.

[121] Neurath to Carnap, 15 January, 1943 (RC 102-55-02). See letter 17 in the present volume.

[122] See letter to Carnap, 22 September 1945, unsent (RC 115-07-66). See letter 34 in the present volume.

[123] See letter to Carnap, 22 September 1945, unsent (RC 115-07-66). See letter 34 in the present volume.

Neurath included in "Universal Jargon and Terminology" a reply to Russell's criticism marshalling the combined retorts he had wielded against Carnap's metaphysical subjectivism, or idealism, in 1928–31 and against Schlick's and Wittgenstein's metaphysical realism in 1934, even denouncing, for good measure, Russell's talk of "absolute truth" (Neurath 1941/1983, 226–228, and 1934/1983).

The following year, Neurath's British intellectual ally and collaborator Susan Stebbing published the third edition of her *A Modern Introduction to Logic* (1930). To his satisfaction, it contained a section on definition and analysis with criticism of Russell's own logical method in philosophy of language central to the both symbolic logic and the methodology of the logic of science – symbolic language, logical analysis and definitions (Stebbing 1930/1942, 439–442). Stebbing attributed to Russell, in the *Principia Mathematica*, a contradiction between definition as purely symbolic equation (simplifying substitution rule) and conceptual analysis:

> It is obvious that Mr. Russell has been led into contradiction because he wanted to do two things at once, viz. to define *definition* as concerned with symbols, and to point out that the *analysis of a concept*, which may be most suitable *expressed* in a definition, constitutes an advance in knowledge.
>
> The analytic definition of a symbol does indeed entail the analysis of a concept which the symbol expresses. A definition is an equation; it asserts that a given set of symbols is equivalent to some other symbol, or set of symbols. Both the definiendum and the definiens express the *same* referend, the latter being an analysis of the former. It is for this reason that an analysis of a symbolic expression may constitute an advance in knowledge. The definition enables us to see what it was we were *meaning* when we used the expression that is now defined. We sometimes use words to refer to a referend with regard to which we have no clear idea. Then we may come to see that another set of words, or symbols, much more clearly expresses what the referend is. It is in this way that the analytic definition of a symbol entails an analysis of what is expressed by *both* sets of symbols, although the analytic definition is *of* the symbol, not *of* the concept which the symbol expresses. [...] But in logical analysis there are not two *things* but two *expressions* which mean the same. [...] The correct analysis of a symbolic expression is enlightening since it shows what we meant when we used the defined expression. (Stebbing 1930/1942, 440; original emphasis.)

For Stebbing, logical symbols, in particular, served the dual purpose of abbreviation and abstraction, and as a consequence, they revealed form (see, for instance, Stebbing 1930/1942, Ch. XXIII). Her remarks dovetailed with Neurath heated exchanges with Carnap over the latter's "lapse" into the metaphysical, semantic turn, which included the issue of symbolic definitions and the relevance of Neurath's early work.

After reading Russell's new book, Neurath had promptly approached Carnap in April 1941 asking about his opinion not without sharing initial reactions: "What do you think about Russell's last book? It looks somewhat strange? Neo-neo-Platonic Mysticism – I think it is full of misunderstandings. He seems not to grasp our purpose. His position – an interesting coincidence – seems to be very similar to Schlick's conception in his latest days."[124]

Carnap replied in conciliatory tone: "Yes, Russell's book is somewhat disappointing. But I think one can still call it empiricism though – as you say – it is

[124] Neurath to Carnap 4 April 1941 (RC 102-55-20), see letter 5 in the present volume.

the Schlick-denomination of empiricism."[125] The following year, in a letter of 17 July 1942, Neurath supplied Carnap with a list of detailed replies to Russell's statements.[126] In line with Russell's argument, Neurath's criticisms extended to semantics, entering the ongoing debate with Carnap himself, and to the interpretation of logical symbols. Neurath's sustained but turbulent sailing through symbolic matters of language and logic retained its early original references:

> You answer, that all my remarks touch even Russell's existence symbol in his logic. Of course, it does. I always knew that and I always looked with some suspicion to that. But as long as such symbols remain within calculus not much may happen, we should not be too pedantic, but if they did enter the empiricist sphere through semantics, we have to be careful. But that is another long story, connected with the problems of introducing symbols, an action not very properly done by Russell, as you may see, where he introduces letter combinations. How to say that ab and ba are the same or vice versa, only ab may be used. Russell's remark is a very weak one and only an additional one – Stebbing made the right remark on that in her book. This whole symmetry business is a difficult problem in itself. I am not sure, what to do in detail with it. My remarks on monogrammatic writing touch only one point, and my remarks on symbolically equivalence etc. (Neurath to Carnap, 25 September 1943, RC 102-55-03. See letter 22 in the present volume. See Neurath 1910a.)

In a follow-up letter of July 20, Neurath returned to the frequent topic of his satisfying efforts rebuilding his lost library, especially reading and buying scientific texts in physics, chemistry and mathematics. He concluded by listing Schröder: "I always hope I shall find a Schroeder, I have now a VENN with his nice ellipses (I had them only in my Schroeder)."[127]

11.9 Conclusion

If I am right, tracking Neurath's attention to symbolism and symbolic logic introduces a broader perspective and a new object of study that significantly supplement current accounts of his intellectual life and works; they are also integrative in how they yield a richer understanding of Neurath's more familiar interests and views in economics, pictorial language and logical empiricism. For these reasons, they also require further and more detailed attention.

References

Blackmore, John, Ryoichi Itagaki, and Sezsuko Tanaka, eds. 2001. *Ernst Mach's Vienna 1895–1930*. Dordrecht: Kluwer.

Boole, George. 1854. *An Investigation of the Laws of Thought*. Cambridge: Macmillan.

Burke, Christopher, Eric Kindel, and Sue Walker, eds. 2013. *Isotype: Design and Contexts 1925–1971*. London: Hyphen Press.

[125] Carnap to Neurath, 5 June 1941 (RC 102-55-19), see letter 6 in the present volume.

[126] Neurath to Carnap, 17 July 1942 (RC 102-56-04), see letter 11 in the present volume.

[127] Neurath to Carnap, 20 July 1942 (RC 115-07-57). See letter 12 in the present volume.

Carnap, Rudolf. 1928/1967. *The Logical Structure of the World.* Trans. R. George. Berkeley: University of California Press.
———. 1929. *Abriss der Logistik.* Wien: Springer.
———. 1930. Die alte und die neue Logik. *Erkenntnis* 1: 12–26.
———. 1932/1995. *The Unity of Science.* London: Thoemmes.
———. 1934/1937. *The Logical Syntax of Language.* London: Kegan Paul, Trench.
———. 1947. *Meaning and Necessity.* Chicago: University of Chicago Press.
Carnap, Rudolf, Hans Hahn, and Otto Neurath. 1929/1973. The Scientific Conception of the World: The Vienna Circle. In Neurath and Cohen (1973): 299–318.
Cartwright, Nancy, Jordi Cat, Lola Fleck, and Thomas Uebel. 1996. *Otto Neurath: Philosophy Between Science and Politics.* Cambridge: Cambridge University Press.
Carus, André. 2007. *Carnap and Twentieth-Century Thought. Explication as Enlightenment.* Cambridge: Cambridge University Press.
Cat, Jordi. 1995. The Popper-Neurath Debate and Neurath's Attack on Scientific Method. *Studies in History and Philosophy of Science Part A* 26 (2): 219–250.
Church, Alonzo. 1936. A Bibliography of Symbolic Logic. *Journal of Symbolic Logic* 1 (4): 121–216.
Couturat, Louis. 1905/1914. *The Algebra of Logic.* Chicago: Open Court.
Dipert, Randall. 1978. *Development and Crisis in Late Boolean Logic: The Deductive Logics of Peirce, Jevons and Schröder.* Bloomington, IN: Indiana University Ph.D.
———. 1991. The Life and Work of Ernst Schröder. *Modern Logic* 1: 119–139.
Dubislav, Walter. 1931. *Die Definition.* Leipzig: Meiner.
Duhem, Pierre. 1906. *La Théorie Physique: Son Object, et Sa Structure.* Paris: Chevalier & Rivière.
Eschbach, Achim. 1987. Sematology as basic science. In *R. Gätschenberger, Foundations of semiotics*, i-xxi. Philadelphia: John Benjamin.
Fechner, Gustav. 1888. *Elemente der Psychophysik.* Vol. 1. Leipzig: Breitkopf and Härtel.
Frege, Gottlob. 1879. *Begriffsschrift, eine der arithmetischen nachgebildete Formelsprache des reinen Denkens.* Halle: Louis Nebert.
———. 1884/1968. *The Foundations of Arithmetic.* Trans. J.L. Austin. Oxford: Basil Blackwell.
———. 1892/1980. On Sense and Reference. In *Translations from the Philosophical Writings of Gottlob Frege.* Trans. Max Black. ed. Peter Geach and Max Black, 56–78. Oxford: Blackwell.
Freudenthal, Gideon, and Tatiana Karachentsev. 2011. G. Itelson – A Socratic Philosopher. In *Otto Neurath and the Unity of Science*, ed. John Symons, Olga Pombo, and Juan Manuel Torres, 95–108. Dordrecht: Springer.
Gätschenberger, Richard. 1920. *Symbola.* Karlsruhe: G. Braun.
George, Henry. 1882. *Progress and Poverty.* London: Kegan Paul, Trench & Co.
Grassmann, Hermann. 1844. *Die lineale Ausdehnungslehre.* Leipzig: Wiegand.
Grassmann, Robert. 1872. *Die Formenlehre oder Mathematik.* Stettin: R. Grassmann.
Grattan-Guinness, Ivor. 1975. Wiener on the Logics of Russell and Schröder. An Account of his Doctoral Thesis, and of his Subsequent Discussion of it with Russell. *Annals of Science* 32: 103–132.
———. 2000. *The Search for Mathematical Roots, 1870–1940. Logics, Set Theories and the Foundations of Mathematics from Cantor Through Russell to Gödel.* Princeton: Princeton University Press.
———. 2004. The Mathematical Turns in Logic. In *Handbook of the History of Logic*, ed. Dov M. Gabbay and John Woods, vol. 3, 545–556. Amsterdam: Elsevier.
Hahn, Olga. 1909. Zur Axiomatik des logischen Gebietkalkuls. *Archiv für Philosophie*, 2. Abteilung: *Archiv für systematische Philosophie*, Neue Folge, Bd. 15: 345–347. See also Appendix 1 in the volume.
———. 1910. Über die Koeffizienten einer logischen Gleichung und ihre Beziehung zur Lehre von den Schlüssen. *Archiv für Philosophie*, 2. Abteilung: *Archiv für systematische Philosophie*, Neue Folge, Bd. 16: 149–176.
Hahn, Olga, and Otto Neurath. 1909. Zum Dualismus in der Logik. *Archiv für Philosophie*, 2. Abteilung: *Archiv für systematische Philosophie*, Neue Folge, Bd. 15: 149–162. Reprinted in Haller and Rutte (1981): 5–16. See also Appendix 1 in the volume.

Hahn, Hans. 1929/1980. Empiricism, Mathematics, and Logic. In *Hans Hahn: Empiricism, Logic, and Mathematics*, ed. Brian McGuinness, 39–42. Dordrecht: D. Reidel.

Hailperin, Theodore. 2004. Algebraical Logic. In *Handbook of the History of Logic*, ed. Dov M. Gabbay and John Woods, vol. 3, 323–388. Amsterdam: Elsevier.

Haller, Rudolf. 1959. Das 'Zeichen' und die 'Zeichenlehre' in der Philosophie der Neuzeit. *Archiv für Begriffsgeschichte* 4: 113–157.

Haller, Rudolf, and Heiner Rutte, eds. 1981. *Gesammelte philosophische und methodologische Schriften*. Vienna: Hölder-Pichler-Tempsky.

Hardcastle, Gary. 2003. Debabelizing Science: The Harvard Science of Science Discussion Group, 1940–41. In *Logical Empiricism in North America*, ed. Gary Hardcastle and Alan W. Richardson, 170–196. Minneapolis: University of Minnesota Press.

Harman, Peter. 1998. *The Natural Philosophy of James Clerk Maxwell*. Cambridge: Cambridge University Press.

Hempel, Carl G., and Paul Oppenheim. 1936. *Der Typusbegriff im Lichte der neun Logik: Wissenschaftstheorische Untersuchungen zu Konstitutionsforschung und Psychologie*. Leiden: Sijthoff.

Howard, Don. 1996. Relativity, *Eindeutigkeit*, and Monomorphism: Rudolf Carnap and the Development of the Categoricity Concept in Formal Semantics. In *Origins of Logical Empiricism*, ed. Ronald N. Giere and Alan W. Richardson, 115–164. Minneapolis: University of Minnesota Press.

Janik, Allan, and Stephen Toulmin. 1973. *Wittgenstein's Vienna*. New York: Simon and Schuster.

Jevons, William S. 1864. *Pure Logic, or The Logic of Quality Apart from Quantity*. Stanford: London.

———. 1869. *The Substitution of Similars, the True Principle of Reasoning*. London: Macmillan.

———. 1871. *The Theory of Political Economy*. London: Macmillan.

———. 1874. *The Principles of Science. A Treatise on Logic and Scientific Method*. 1st ed. 2 vols. London: Macmillan.

———. 1877. *The Principles of Science. A Treatise on Logic and Scientific Method*. 2nd ed. 2 vols. London: Macmillan.

Johnston, William M. 1972. *The Austrian Mind. An Intellectual and Social History 1848–1938*. Berkeley: University of California Press.

Jourdain, Philip E. B. 1914. Preface. In Couturat (1905/1914): iii-iix.

Köhler, Eckehart. 1991. On Neurath's Writings on Logic, Ethics and Physics. In *Rediscovering the Forgotten Vienna Circle*, ed. Thomas Uebel, 109–116. Dordrecht: Kluwer.

Kusch, Martin. 1995. *Psychologism. A Study in the Sociology of Philosophical Knowledge*. London: Routledge.

Ladd, Christine. 1880. On De Morgan's Extension of the Algebraic Logic. *American Journal of Mathematics* 3: 210–225.

Lambert, Johann H. 1764. *Neues Organon*. Leipzig: Wendler.

Lewis, Clarence I. 1918. *A Survey of Symbolic Logic*. Berkeley: University of California Press.

Menger, Karl. 1994. *Reminiscences of the Vienna Circle and the Mathematical Colloquium*. Dordrecht: Kluwer.

Mormann, Thomas. 1999. Neurath's Opposition to Tarskian Semantics. In *Alfred Tarski and the Vienna Circle: Austro-Polish Connections in Logical Empiricism*, ed. Jan Woleński and Eckehart Köhler, 165–178. Dordrecht: Kluwer.

Mosselmans, Bert. 2007. *William Stanley Jevons and the Cutting Edge of Economics*. New York: Routledge.

Nemeth, Elisabeth, and Friedrich Stadler, eds. 1996. *Encyclopedia and Utopia: The Life and Work of Otto Neurath (1882–1945)*. Dordrecht: Kluwer.

Neurath, Wilhelm. 1896. *Elemente der Volkswirtschaftslehre*. 3rd ed. Leipzig: Klinkhardt.

Neurath, Otto. 1904/2004. Interest on Money in Antiquity. In *Otto Neurath. Economic Writings. Selections 1904–1945*, ed. Thomas Uebel and Robert S. Cohen, 111–119. Dordrecht: Kluwer.

———. 1906–1907. Zur Anschauung der Antike über Handel, Gewerbe und Landwirtschaft (Cicero, De Officii, I 42). *Jahrbücher für Nationalökonomie und Statistik* 3, Folger 32 and 34.
———. 1909a. Ernst Schröders Beweis des 12. Theorems: Für die identischen Operationen gilt das 'Kommutationsgesetz,' *Archiv für Philosophie*, 2. Abteilung: *Archiv für systematische Philosophie*, Neue Folge, Bd. 15: 104–106. Reprinted in Haller and Rutte (1981): 1–3. See also Appendix 1 in the volume.
———. 1909b. Eindeutigkeit und Kommutativität des logischen Produktes ab. *Archiv für Philosophie, 2. Abteilung: Archiv für systematische Philosophie*, Neue Folge, Bd. 15: 342–344. Reprinted in Haller and Rutte (1981): 17–18. See also Appendix 1 in the volume.
———. 1909c. *Antike Wirtschaftsgeschichte*. Leipzig: Teubner.
———. 1910a. On the Theory of Social Science. In *Otto Neurath. Economic Writings. Selections 1904–1945*, ed. Thomas Uebel and Robert S. Cohen, 265–291. Dordrecht: Kluwer.
———. 1910b. Definitionsgleichheit und symbolische Gleichheit. *Archiv für Philosophie*, 2. Abteilung: *Archiv für systematische Philosophie*, Neue Folge, Band 16: 142–144. Reprinted in Haller and Rutte (1981): 19–21. See also Appendix 1 in the volume.
———. 1910c/2004. War economy. In Otto Neurath. *Economic writings. Selections 1904–1945*, ed. Thomas Uebel and Robert S. Cohen, 153–199. Dordrecht: Kluwer.
———. 1910d. *Lehrbuch der Volkswirtschaftslehre*. Wien: Alfred Hölder.
———. 1912/1973. The Problem of the Pleasure Maximum. In Neurath and Cohen (1973): 113–122.
———. 1913/1983. The Lost Wanderers of Descartes and the Auxiliary Motive (On the Psychology of Decision). In *Otto Neurath: Philosophical Papers 1913–1946*, ed. Robert S. Cohen and Marie Neurath, 1–12. Dordrecht: D. Reidel.
———. 1915/1973. On the Foundation of the History of Optics. In Neurath and Cohen (1973): 101–112.
———. 1916/1983. On the Classification of Systems of Hypotheses (With Special Reference to Optics). In *Otto Neurath: Philosophical Papers 1913–1946*, ed. Robert S. Cohen and Marie Neurath, 13–31. Dordrecht: D. Reidel.
———. 1917a/2004. The economic order of the future and the economic sciences. In *Otto Neurath. Economic writings. Selections 1904–1945*, ed. Thomas Uebel and Robert S. Cohen, 241–261. Dordrecht: Kluwer.
———. 1917b/2004. On the theory of social science. In *Otto Neurath. Economic writings. Selections 1904–1945*, ed. Thomas Uebel and Robert S. Cohen, 265–291. Dordrecht: Kluwer.
———. 1921/1973. Anti-Spengler. In Neurath and Cohen (1973): 158–213.
———. 1925/1973. The Social and Economic Museum in Vienna. In Neurath and Cohen (1973): 214.
———. 1928/1973. Personal Life and Class Struggle. In Neurath and Cohen (1973): 249–298.
———. 1930/1983. Ways of the Scientific World-Conception. In *Otto Neurath: Philosophical Papers 1913–1946*, ed. Robert S. Cohen and Marie Neurath, 32–47. Dordrecht: D. Reidel.
———. 1931a/1983. Physicalism: The Philosophy of the Vienna Circle. In *Otto Neurath: Philosophical Papers 1913–1946*, ed. Robert S. Cohen and Marie Neurath, 48–51. Dordrecht: D. Reidel.
———. 1931b/1983. Sociology in the Framework of Physicalism. In *Otto Neurath: Philosophical Papers 1913–1946*, ed. Robert S. Cohen and Marie Neurath, 58–90. Dordrecht: D. Reidel.
———. 1932/33/1983. On Protocol Statements. In *Otto Neurath: Philosophical Papers 1913–1946*, ed. Robert S. Cohen and Marie Neurath, 91–99. Dordrecht: D. Reidel.
———. 1933/1973. Museums of the Future. In Neurath and Cohen (1973): 218–223.
———. 1934/1983. Radical Physicalism and the 'Real World.' In *Otto Neurath: Philosophical Papers 1913–1946*, ed. Robert S. Cohen and Marie Neurath, 100–114. Dordrecht: D. Reidel.
———. 1935a/1983. The Unity of Science as a Task. In *Otto Neurath: Philosophical Papers 1913–1946*, ed. Robert S. Cohen and Marie Neurath, 115–120. Dordrecht: D. Reidel.
———. 1935b/1983. Pseudorationalism of Falsification. In *Otto Neurath: Philosophical Papers 1913–1946*, ed. Robert S. Cohen and Marie Neurath, 121–131. Dordrecht: D. Reidel.

———. 1936/1981. Die Entwicklung des Wiener Kreises und die Zukunft des Logischen Empirismus. In Haller and Rutte (1981): 673–702.
———. 1936a/1983. Individual Sciences, Unified Science, Pseudorationalis. In *Otto Neurath: Philosophical Papers 1913–1946*, ed. Robert S. Cohen and Marie Neurath, 132–138. Dordrecht: D. Reidel.
———. 1936b/1983. Encyclopedia as 'Model.' In *Otto Neurath: Philosophical Papers 1913–1946*, ed. Robert S. Cohen and Marie Neurath, 145–158. Dordrecht: D. Reidel.
———. 1936c/1983. Physicalism and the Investigation of Knowledge. In *Otto Neurath: Philosophical Papers 1913–1946*, ed. Robert S. Cohen and Marie Neurath, 159–171. Dordrecht: D. Reidel.
———. 1937/1987. Towards an Encyclopedia of Unified Science. In *Unified Science*, ed. Brian McGuinness, 130–165. Dordrecht: D. Reidel.
———. 1937a/1983. Unified Science and its Encyclopedia. In *Otto Neurath: Philosophical Papers 1913–1946*, ed. Robert S. Cohen and Marie Neurath, 172–182. Dordrecht: D. Reidel.
———. 1937b/1983. The New Encyclopedia of Scientific Empiricism. In *Otto Neurath: Philosophical Papers 1913–1946*, ed. Robert S. Cohen and Marie Neurath, 189–199. Dordrecht: D. Reidel.
———. 1937c/1983. The Concept of 'Type' in the Light of modern Logic. In *Otto Neurath: Philosophical Papers 1913–1946*, ed. Robert S. Cohen and Marie Neurath, 183–188. Dordrecht: D. Reidel.
———. 1941/1983. Universal Jargon and Terminology. In *Otto Neurath: Philosophical Papers 1913–1946*, ed. Robert S. Cohen and Marie Neurath, 213–229. Dordrecht: D. Reidel.
———. 1945/1973. Visual Education: Humanisation versus Popularisation. [Excerpts]. In Neurath and Cohen (1973): 227–248.
———. 1946/1983. The Orchestration of the Sciences by the Encyclopedism of Logical Empiricism. In *Otto Neurath: Philosophical Papers 1913–1946*, ed. Robert S. Cohen and Marie Neurath, 230–242. Dordrecht: D. Reidel.
———. 1996.Visual Education: Humanisation versus Popularisation. In Nemeth and Stadler (1996): 245–335.
Neurath, Marie, and Robert S. Cohen, eds. 1973. *Empiricism and Sociology*. Dordrecht: D. Reidel.
Neurath, Otto, and Anne Schapire-Neurath, eds. 1910. *Lesebuch der Volkswirtschaftslehre*. Leipzig: Klinkhart Second edition: 1913, Leipzig: Glöckner.
O'Connor, J.J., and E.F. Robertson. 2009. Ernst Schröder. *MacTutor History of Mathematics Archive*, University of St. Andrews. http://www-history.mcs.st-andrews.ac.uk/Biographies/Schroder.html
Pareto, Vilfredo. 1906. *Manuale di Economia Politica*. Milan: Società Editrice Libraria.
Peckhaus, Volker. 1999. 19th Century Logic Between Philosophy and Mathematics. *Bulletin of Symbolic Logic* 5 (4): 433–450.
———. 2004. Schröder's Logic. In *Handbook of the History of Logic*, ed. Dov M. Gabbay and John Woods, vol. 3, 557–609. Amsterdam: Elsevier.
———. 2014. Ernst Schröder on Pasigraphy. *Revue d'Historie des Sciences* 67 (2): 207–230.
Peirce, Charles S. 1867. On an Improvement in Boole's Calculus of Logic. *Proceedings of the American Academy of Arts and Sciences* 7: 250–261.
———. 1880. On the Algebra of Logic. *American Journal of Mathematics* 3: 15–57.
———. 1885. On the Algebra of Logic: A Contribution to the Philosophy of Notation. *American Journal of Mathematics* 7: 180–202.
Petzoldt, Joseph. 1895. Das Gesetz der Eindeutigkeit. *Vierteljahrsschrift für wissentschaftliche Philosophie und Sociologie* 19: 146–203.
———. 1900. *Einführung in die Philosophie der reinen Erfahrung*. Vol. 1. Leipzig: Teubner.
Popper, Karl R. 1973. Memories of Otto Neurath. In Neurath and Cohen (1973): 51–56.
Reck, Erich H. 2004. From Frege and Russell to Carnap: Logic and Logicism in the 1920s. In *Carnap Brought Home. The View from Jena*, ed. Steve Awodey and Carsten Klein, 151–180. Chicago: Open Court.

Reck, Erich H., and Steve Awodey. 2004. Frege's Lectures on Logic and Their Influence. In *Frege's Lectures on Logic. Carnap's Student Notes, 1910–1914*, ed. Erich H. Reck and Steve Awodey, 17–42. Chicago: Open Court.

Reininger, Robert, ed. 1938. *50 Jahre Philosophische Gesellschaft an der Universität Wien 1888–1938*. Vienna: Verlag Philosophische Gesellschaft an der Universität Wien, E. Kainz and J.B. Wallishausser.

Reisch, George. 2005. *How the Cold War Transformed Philosophy of Science: To the Icy Slopes of Logic*. New York: Cambridge University Press.

Russell, Bertrand. 1900. *A Critical Exposition of the Philosophy of Leibniz*. Cambridge: Cambridge University Press.

———. 1940. *An Inquiry into Meaning and Truth*. London: Allen & Enwin.

Sandner, Günther. 2014. *Otto Neurath. Eine politische Biographie*. Wien: Zsolnay.

Schabas, Margaret. 1990. *A World Ruled by Numbers. W. Stanley Jevons and the Rise of Mathematical Economics*. Princeton: Princeton University Press.

Schorske, Carl E. 1979. *Fin-de-Siècle Vienna. Politics and Culture*. New York: Alfred Knopf.

Schröder, Ernst. 1873. *Lehrbuch der Arithmetik und Algebra für Lehrer und Studirende* (pt. 1 and only). Leipzig: Teubner.

———. 1874a. *Abriss der Arithmetik und Algebra für Schüler an Gymnasien und Realschulen* (pt. 1 and only), Leipzig: Teubner.

———. 1874b. *Über die Formalen Elemente der absoluten Algebra, appended to Beilage zum Programm des Pro- und Real-Gymnasiums in Baden-Baden für 1873–74*. Stuttgart: Schweizbart.

———. 1877. *Operationskreis des Logikkalkuls*. Leipzig: Teubner.

———. 1880. Review of Frege's *Begriffsschrift. Zeitschrift für Mathematik und Physik, Hist.-literarische* 25: 81–94.

———. 1884a. On the Most Commodious and Comprehensive Calculus. In *Report of the Fifty-Third Meeting of the British Association for the Advancement of Science*, 411–412. London: John Murray.

———. 1884b. Exposition of a Logical Principle, as Disclosed by the Algebra of Logic but Overlooked by the Ancient Logicians. In *Report of the Fifty-Third Meeting of the British Association for the Advancement of Science*, 411. London: John Murray.

———. 1890a. *Vorlesungen über die Algebra der Logik (exakte Logik)*. Vol. 1. Leipzig: Teubner.

———. 1890b. *Über das Zeichen*. Leipzig: Fock.

———. 1891. *Vorlesungen über die Algebra der Logik (exakte Logik)*, vol. 2, pt. 1, Leipzig: Teubner.

———. 1892. Signs and Symbols. *The Open Court* 6: 3431–3434, 3441–3444 and 3463–3466.

———. 1895. *Vorlesungen über die Algebra der Logik (exakte Logik)*, vol. 3, pt. 1: *Algebra und Logik der Relative*. Leipzig: Teubner

———. 1898a. Über Pasigraphie, ihren gegenwärtigen Stand und die pasigraphische Bewegung in Italien. In *Verhandlugen des Ersten Internationalen Mathematiker-Kongresses in Zürich vom 9. Bis 11. August 1897*, ed. F. Rudio, 147–162. Leipzig: Teubner.

———. 1898b. On Pasigraphy. *The Monist* 9: 44–62 (corrigenda p. 320).

———. 1909. *Abriss der Algebra der Logik*. ed. K.E. Müller, pt. 1: Elementarlehre, Leipzig: Teubner.

Stadler, Friedrich. 2001. *The Vienna Circle: Studies in the Origins, Development, and Influence of Logical Empiricism*. Vienna/New York: Springer.

Stebbing, Susan. 1930/1942. *A Modern Introduction to Logic*. 3rd ed. London: Methuen.

Stöhr, Adolph. 1910. *Lehrbuch der Logik in Psychologisiender Darstellung*. Wien: Franz Deuticke.

Stöltzner, Michael. 1996. The auxiliary motive in the Forest and in optics. In Nemeth and Stadler (1996): 113–126.

Trendelenburg, Adolf. 1867. Über Leibnizens Entwurf einer allgemeinen Charakteristik. In *Historische Beiträge zur Philosophie*, vol. 3, 1–47. Berlin: Bethge.

Uebel, Thomas. 1992. *Overcoming Logical Positivism from Within*. Amsterdam: Rodopi.

———. 1995. Otto Neurath's Idealist Inheritance: The Social and Economic Thought of Wilhelm Neurath. *Synthese* 103: 87–121.

———. 2004. Neurath's economics in critical context. In *Otto Neurath. Economic writings. Selections 1904–1945*, ed. Thomas Uebel and Robert S. Cohen, 1–108. Dordrecht: Kluwer.

———. 2007. *Empiricism at the Crossroads*. La Salle: Open Court.

———. 2008. Writing a Revolution: On the Production and Early Reception of the Vienna Circle's Manifesto. *Perspectives on Science* 16 (1): 70–102.

———. Forthcoming. Neurath on *Verstehen*. Forthcoming manuscript.

van Heijenoort, Jean, ed. 1967a. Logic as Calculus and Logic as Language. *Synthese* 17: 324–330.

———, ed. 1967b. *From Frege to Gödel. A Source Book in Mathematical Logic*. Cambridge, MA: Harvard University Press.

Vossoughian, Nader. 2008. *Otto Neurath: The Language of the Global Polis*. Rotterdam: NAi.

Waismann, Friedrich. 1936/1977. The Concept of Identity. In *Friedrich Waismann: Philosophical Papers*, ed. Brian McGuinness, 22–29. Dordrecht: D. Reidel.

Whitehead, Alfred N. 1898. *A Treatise on Universal Algebra*. Cambridge: Cambridge University Press.

Whitehead, Alfred N., and Bertrand Russell. 1910–1913. *Principia Mathematica*. 3 vols. Cambridge: Cambridge University Press.

Wundt, Wilhelm. 1880. *Logik: Einte Untersuchung der Prinzipen der Erkenntnis und der Methoden wissenschaftlicher Forschung*. 1st ed, vol. 1, *Erkenntnislehre*. Stuttgart: Enke.

———. 1893. *Grundzüge der physiologischen Psychologie*. Vol. 1. 2nd ed. Leipzig: Wilhelm Engelmann.

———. 1906. *Logik: Einte Untersuchung der Prinzipen der Erkenntnis und der Methoden wissenschaftlicher Forschung*. Vol. 1. 3rd ed. *Erkenntnislehre*. Stuttgart: Enke.

———. 1907. *Logik*. Vol. 2, 3rd ed. *Logik der Exakten Wissenschaften*. Stuttgart: Enke.

———. 1908. *Logik*. Vol. 3. 3rd ed. *Logik der Geisteswissenschaften*. Stuttgart: Enke.

Chapter 12
Neurath and Carnap on Semantics

A. W. Carus

Abstract Carnap is still often portrayed as a "representationalist." While the genealogy of this prejudice may not actually go back to Neurath's response to Carnap's embrace of Tarskian semantics, there is a continuity of motivation and rhetoric. However, based on a reading of the later Neurath-Carnap correspondence reproduced in this volume, it would appear that the apparent dispute between them over semantics really was largely terminological, with certain differences of emphasis amplified by personality differences and the long interruption of personal contact due to the war. Their conceptions of a language of science can be reconciled. Carnap was neither a representationalist nor an anti-representationalist nor an inferentialist (though it may appear that he can legitimately be portrayed as any of these), since ultimately to embrace one of these positions is to endorse an "order of explanation" or ontological primacy, and Carnap rejected ontology.

Some of the cruder stereotypes about logical empiricism, and especially about Carnap, have faded into relative obscurity since the bright light of historical scholarship has been trained on them in the past few years. But others survive unscathed, and seem almost ineradicable. One such hardy perennial is the idea that Carnap was a doctrinaire "representationalist," as Robert Brandom puts it. What he has in mind is Carnap's supposed adherence to the "myth of the given" castigated by Wilfrid Sellars (1962). Brandom, following Rorty, regards "the idea of things known simply by our grasp of our own meanings" – by which he means analyticity – as another, different kind of "given" from that discussed by Sellars, "the idea of things known simply by being in some sensory state." Parenthetically, Brandom (2013, 92) adds "Carnap, of course, embraced both forms of givenness."

I am grateful to the editors, Jordi Cat and Adam Tuboly, for suggesting that I contribute to this volume and for helpful comments. I also thank Thomas Uebel for some valuable suggestions.

A. W. Carus (✉)
Munich Center for Mathematical Philosophy, Ludwig-Maximilians Universität, Munich, Germany
e-mail: awcarus@mac.com

J. Cat, A. T. Tuboly (eds.), *Neurath Reconsidered*,
Boston Studies in the Philosophy and History of Science 336,
https://doi.org/10.1007/978-3-030-02128-3_12

Those who know their Carnap will be aware that he did not, of course, embrace *either* "form of givenness." But Brandom's offhand remark is by no means original; it hardly expresses a view that is peculiar to him. He probably just thought he was repeating what everyone knows (hence the "of course"); certainly his teacher Rorty would have conveyed such an understanding of Carnap to him. But so would the major names of the generation in which Rorty grew up, including both Quine and Sellars, who each, according to Rorty (and Brandom) were able to overcome one of the two kinds of givenness while still clinging to the other (ibid.).

But this view of Carnap as a "representationalist" goes back further than the 1950s. Otto Neurath, Carnap's own friend and comrade on the "left wing" of the Vienna Circle, leveled the same accusation at Carnap from the moment Carnap first incorporated semantics into his conception of meta-discourse in the mid-1930s. Those who currently perpetuate the notion of Carnap as a "representationalist" may not consciously be influenced by Neurath, but they are repeating the same patterns, and some of the same motivations appear to be involved.

So it is not too surprising that efforts have been made to rehabilitate Neurath's critique of semantics (Mormann 1999), and to argue that it was not rooted in misunderstanding at all, but in a deeply rooted antipathy to Carnap's "Cartesianism" and to Neurath's own development of an anti-Cartesian alternative to the Cartesian mainstream of the "language as universal medium" school of thought, as portrayed by Hintikka (1988, going back to van Heijenoort 1967), and contrasted with the "language as calculus" school. Carnap, according to Mormann, had – just as Hintikka (1992) portrays him – adhered until 1935 to the "language as universal medium" view, but with his adoption of semantics, defected to the "language as calculus" side, while remaining "Cartesian" (see Sect. 12.1. below for details). Mormann's defense of Neurath may be regarded, then, as a way of shoring up the "obvious" interpretation of Carnap that Brandom inherited from Rorty and from Sellars, Quine, and ultimately from Neurath himself.

Thomas Uebel (2001) suggests a different interpretation of Neurath, in which his rejection of semantics was *not* in fact rooted in a principled and fundamental difference between himself and Carnap (as Mormann contends), but mostly in emotional and relatively superficial overreactions to Carnap's unimaginative, wooden-sounding terminology. In principle, at least, Uebel contends, it should have been possible for Carnap and Neurath to overcome their disagreement over semantics. Uebel's view, then, is not so much the traditional "misunderstanding" view, but it also rejects Mormann's contention that Carnap and Neurath misleadingly papered over deep and indeed unbridgeable philosophical differences.[1]

I have repeatedly endorsed something like Uebel's view of the matter (Carus 2007, Ch. 11., 2017), and would see the difference between Neurath and Carnap regarding the form of the language of science as less a practical disagreement (even less a theoretical one) than a matter of emphasis. Each accepted the other's concep-

[1] There have also been other interpretations of Neurath's rejection of semantics that are less programmatic than Mormann's or Uebel's but seek merely to clarify Neurath's position while holding out the possibility that it could be worked out more coherently as an alternative to the "mainstream" view of Tarski and Carnap; the best example of this is Mancosu (2008).

tion of the language of science, in principle, but each thought his own conception of greater importance and the other's of subsidiary interest, and each had a different conception of how the two fit together, which neither spelled out very clearly, if at all. I have tried to reconstruct what their joint idea of how the two fit together might have looked like if Neurath had lived a bit longer and they had had a chance to talk in person after the war had separated them for so long. But this looks dangerously like mere idle speculation if it is not at least consistent with what we know about Carnap and Neurath. If Mormann's picture (buttressing the traditional interpretation of Carnap handed down from Rorty, Sellars, Quine, and others) is right, then of course my attempts over the years to synthesize Carnap's and Neurath's conceptions of the language of science into a single coherent picture are going nowhere. So I had better be able to show that there is something wrong with Mormann's picture of Carnap. But the present paper tries also to give an alternative interpretation of Neurath's motivation on the basis of the Neurath-Carnap correspondence reproduced in this volume.

The first section reviews Carnap's transition from the *Syntax* to his incorporation of semantics into meta-discourse. Section 12.2. surveys Neurath's response to semantics from 1935 to his death (1945), especially the last few years. Section 12.3. addresses Carnap's response, and argues that Carnap and Neurath are in fundamentals not so far apart, i.e. that their correspondence supports Uebel's interpretation rather than Mormann's. Section 12.4. goes on to argue that not only representationalism (as well as its supposed opposite, inferentialism) but the whole idea of a preferred "order of explanation," so often invoked by Brandom, is irrelevant and out of place in Carnap's perspective.

12.1 From Syntax to Semantics

The reception of Carnap's *Logical Syntax* was significantly distorted by the timing and circumstances of its publication.[2] The Vienna Circle was in the process of breaking up and fleeing from fascist regimes; Carnap moved to Chicago soon after the 1934 publication of the original, and by the time the English translation appeared in 1937, Carnap had already signaled in "Wahrheit und Bewährung" (1936a) that he had now left behind the "formal mode of speech" highlighted so prominently in the final ("philosophical") section V of the *Syntax* (the only section, it seems, that most readers actually attended to). As discussed elsewhere in detail (Awodey and Carus 2009; Carus 2007, Chap. 10.), this section was written in mid-1932, before Carnap had arrived at the principle of tolerance, which is therefore hardly in evidence there. Since this section appeared to most readers to articulate the philosophical point of the book, it is not surprising that they (a) missed the principle of tolerance entirely and (b) thought that with the incorporation of semantics, the *Syntax* had been left behind.

[2] As I have discussed in much more detail in Carus (2007, 33–37).

The consensus view in the current Carnap literature regards the principle of tolerance as the major turning point in Carnap's career (as well as the most important contribution of the *Syntax*), and no longer sees the incorporation of semantics into meta-discourse in 1935 as a major discontinuity at all (Creath 1991; Ricketts 1996; Carus 1999; Awodey 2007; Awodey and Carus 2009). The rejection of "meaning" (specifically of Wittgenstein's picture theory of meaning in January 1931 – in Carnap's famous "sleepless night" – and thus of any substantial account of meaning) was not taken back. There was no restoration of meaning in 1935. The new designation relation is not a relation between language and world (some pre-existing or objectively existing world) but between designators and an explicitly specified universe of discourse whose elements have only the content that is explicitly specified, nothing beyond. The designation relation is, in other words, entirely schematic, and Carnap's early semantics is in spirit (if not yet quite in practice) model-theoretic in the modern sense.[3] The "Carnapian linguistic turn" (Carus 2015a) and the language pluralism of the principle of tolerance remained in place.

Why then did Neurath (who during Carnap's last Vienna years was perhaps closer to him than anyone else) see the incorporation of semantics into meta-discourse as such a fundamental break with the *Wissenschaftslogik* of the *Syntax*? The more recent literature on logical empiricism, insofar as it addresses this question, largely accepts the consensus view on the relative insignificance of this step, so tends to regard Neurath's objections as based on misunderstandings. Thomas Uebel, the foremost interpreter of Neurath over the past twenty years, essentially takes this view, while also showing how given the context, Neurath's misunderstandings were not entirely arbitrary or unmotivated. It is certainly true, for instance, that Tarski emerged from a Polish philosophical school some of whose members did indeed adhere to ideas of "truth" in a much more absolute and substantial sense than anything suggested by Tarski's famous monograph of 1935. And Carnap hardly bothered to absolve himself from such guilt by association, since he thought it blindingly obvious that he himself could not possibly be considered to entertain such ideas.

Mormann, in contrast, argues that this consensus view is mistaken, and that there really was a major discontinuity in Carnap's development in 1935, which Neurath accurately discerned. The Vienna Circle, up to then, had all (though to varying degrees) been unthinking adherents of the Cartesianism of the Frege-Russell tradition of "language as universal medium" (in Hintikka's (1988) classification), according to Mormann, but in 1935 Carnap and Neurath moved in different directions: Carnap remained Cartesian, but changed sides to the "language as calculus" view, while Neurath went on to develop an anti-Cartesian version of the "language as universal medium" idea (Mormann 1999, 171). In Mormann's view, this move of Neurath's to a more anti-Cartesian position was simply a return to Neurath's original anti-Cartesianism as expressed e.g. in an early paper "Die Verirrten des Cartesius und das Auxiliarmotiv" (Neurath 1913/1983). By "Cartesianism" in Neurath's sense, Mormann (1999, 166) means this:

[3] Hintikka (1991) questions this. But Schiemer (2013) takes a different view, based on a close reading of the hitherto unstudied notes and drafts for the second part of the *Untersuchungen zur allgemeinen Axiomatik*.

(1) Infallibilist foundationalism: Starting from indubitable true statements an all-embracing final system of philosophical and scientific knowledge can be erected.
(2) Deductivism: Science is a deductive system of general statements ensured by proofs.
(3) Transparentism: The system of (scientific and philosophical) knowledge can be expressed in a language whose concepts refer to clear and distinct ideas.

Mormann regards (3) as the "hard core" of the Cartesian program, in Neurath's sense, since (1) and (2) depend on (3) for their realization. And Mormann regards Neurath's rejection of Tarskian semantics to have been motivated by his opposition not only to (1) and (2), but most especially to (3) – as expressed, for instance, in his insistence that the scientific vernacular generally still retains unclear concepts inherited from ordinary language, which he referred to as "clots" ("Ballungen"). In Sect. 12.3. below, we consider the question whether Carnap can be seen as representing a "Cartesian" position in Mormann's sense.

But first, we consider the degree to which Neurath's own position against semantics in 1935–45 can accurately be portrayed as motivated by "anti-Cartesian" impulses.

12.2 Why Did Neurath Reject Semantics?

Neurath's rejection of semantics at the 1935 Paris conference, in response to Carnap's (1936a) "Wahrheit und Bewährung," is well known, not least from Carnap's own report in his autobiography (Carnap 1963, 61; see also the detailed and well-documented account in Mancosu 2008). Two years later, Carnap and Neurath had a private meeting when both were attending another conference in Paris,[4] and from Neurath's prepared document for that meeting, we at least get a little more detail about Neurath's motivations. Dated 12 July 1937, it begins with a section entitled "I. Self-knowledge is the first step toward improvement" that begins as follows:

> Two good lessons from the past years: CARNAP: Better not to disguise proposals as dogmas. NESS: Never forget that every way of putting things is only fully understood when the target is specified. These wise lessons can be articulated much more crisply, and then they become something more than pedagogical maxims.
>
> I admit that in my paper "Physicalism" (Scientia, November 1931, p. 299) and in other papers that partly occasioned the current discussion about truth (SCHLICK, HEMPEL, CARNAP, etc etc) I didn't speak of PROPOSALS where actually it was proposals I was putting forward.
>
> On the other hand, in contrast to many others currently participating in this discussion about truth, I have clearly specified my target: WITTGENSTEIN and those around him. (UCLA Box 5/CM20, section 10, item 2.)

[4]This 1937 meeting is also discussed by Mancosu (2008, 215–19), with documentation from the Neurath Nachlass.

The main burden of the discussion document, however, is once again to suggest (this time perhaps in a bit more detail) using the model of Neurath's protocol sentence form (which avoids any mention of correspondence to anything external to language) to avoid the locutions Carnap and Tarski have introduced in semantics, especially truth itself (but also the distinction between "Bezeichnetes" and "Bezeichnendes"). So in substance nothing much has changed. Neurath is still worried primarily by the rhetoric, and points to the fact that even at the 1935 Paris conference, Tarski was misunderstood by some people in precisely the ways he, Neurath, has feared.[5] Also he points to the incriminating historical background of semantics in Polish Catholicism and Thomism, and wants no one to associate logical empiricism with that stuff. We do learn, though, perhaps more unambiguously than before, that the fear of a return to Wittgenstein and the (correspondence-theoretical) picture theory of meaning was an important motivation for Neurath in opposing the adoption of semantics.

And this is still where we are in Neurath's first extended letter pertaining to semantics after those discussions of the mid-1930s (among those reprinted here, Letter 18 of 29 January 1943), Neurath's objection is still to the supposed Aristotelian connotations of Carnap's terminology in the *Introduction to Semantics* (Carnap 1942). While Neurath certainly complains that his old friend has always had a weakness for formal systems, and that this weakness allowed him to be seduced by the Polish Catholic obscurantists, there is otherwise no reference to long-running deep philosophical disagreements. On the contrary, Neurath says he is merely defending "Carnap One" (the Carnap of the *Syntax*) against "Carnap Two" (of the *Semantics*). Again, nothing has changed since 1937.

The next relevant letter is the long, meandering Letter 22 of 25 September 1943.[6] One can certainly share Carnap's frustration at its vagueness and loose free association, but it seems clear that Neurath's main priority here is something he calls "pluralism," by which he means something like the denial that there could be a single authoritative system, in science or in human values – or even a single criterion for choosing one theory over another; he objects to Carnap's mention that he is working on defining degree of confirmation; as he put it a bit later, "[w]e do not have any reason to assume a one-dimensional ranking of theories."[7] He repeatedly expresses admiration for the permissive, pluralistic English or Anglo-Saxon attitude in both cognitive and practical matters, contrasting it with German absolutism, idealism, or authoritarianism. There is certainly an implication, at least, that Carnap has tendencies of the latter sort. The immediate object of his scorn (and the antithesis to his

[5] Mancosu (2008, 209) documents that it was Louis Rougier, in particular, who in his presentation at the Paris conference took Tarski's account of truth to vindicate Schlick's position in the controversy between the left and right wings of the Vienna Circle about protocol sentences and "Konstatierungen" (affirmations), and thereby to shift the boundary between metaphysical and scientific language.

[6] See Neurath to Carnap, 25 September 1943 (RC 102-55-03); see letter 22. in the present volume.

[7] Neurath to Carnap, 1 April 1944 (RC 102-55-05); see letter 26. in the present volume.

vision of pluralism) is not Carnap, though, or even Wittgenstein, but Popper – perhaps as a proxy for Carnap. Again and again, Popper is castigated for suggesting a decision method to identify the single theory that has at a given time withstood all tests. In this way, Neurath suggests – rather more stridently even than in his review article of Popper's *Logik der Forschung* (Neurath 1935/1983) – that Popper promotes a cognitive (and by implication a practical, political) authoritarianism.[8] (It is of course ironic that at this very moment Popper was at work in New Zealand on *The Open Society and Its Enemies* (Popper 1945), where he was himself advocating pluralism and fallibilism, by his own lights, against the authoritarianism inherent in the philosophical tradition, especially Platonic and German idealism, in both knowledge and social organization. It is doubly ironic that insofar as Popper does indeed exhibit the tendencies toward verisimilitude and "the *one* true system" that so annoyed Neurath, this was due to his incipient (later much more pronounced) realism – which Neurath himself, according to Carnap's (1963, 51) autobiography, had promoted as superior to idealism in Vienna Circle discussions. In any case, if Popper is the antithesis to the "pluralism" Neurath is stressing to Carnap, then "anti-Cartesianism" is hardly an adequate description of his position; Popper would even in 1940 have rejected at least (1) and (3) of Mormann's three criteria for Cartesianism, and his "deductivism" (i.e. his partial or potential support for (2)) extended only to the generation of testable instances of theories.

In one part of his letter Neurath even harks back to the extreme positivism of the early Vienna Circle, suggesting that science should not entertain universal laws that inherently go far beyond what we could possibly know, but should restrict itself to limited generalizations in the way Mach had suggested. Carnap had certainly been attracted by such ideas during the late 1920s, and more or less deprecated the legitimacy of all theoretical language – and though they left little trace in his actual publications, he echoed such ideas e.g. in his 1929 Bauhaus lectures (Carus 2007, 218–19). In "Testability and Meaning" and certainly in his Encyclopedia volume *Foundations of Logic and Mathematics* (Carnap 1939), however, he had returned to accepting theoretical language, without Neurath having raised much of an objection at the time. Perhaps he retrospectively wondered whether this hadn't been a slippery slope, and now regretted not having objected sooner.

Neurath's letter is less an expression of anti-Cartesianism, then, than of a far more generalized anti-authoritarianism. Mormann's three criteria for Cartesianism were, after all, rejected by Popper as well, and yet Popper is obsessively held up as the antithesis to the "pluralism" Neurath advocates. He not only resists the idea that we can construct or identify a single consistent system of knowledge and emphasizes that our knowledge even at its best remains fragmentary and incomplete. But we are not even able, as Popper thinks we are, to identify or pick out one theory as cognitively better than its rivals. An obvious analogy (*not* suggested by Neurath himself) is to the granddaddy of the English empiricism he had recently come to appreciate (and whose grandchildren, including Mill, he *does* mention): Locke, like Neurath, had been very concerned in his *Essay* to limit the claims of knowledge, and

[8] See the discussion on Neurath and Popper in Cartwright, Cat, Fleck, and Uebel (1996).

it seems clear that this was motivated by the same anti-authoritarian impulse that underlay his *Two Treatises* and his support for the revolution of 1688.

These themes continue through Letter 26 (1 April 1944), now with even more on Popper, including many quotations and page references to the *Logik der Forschung*, and once again the spirit of late-1920s Vienna radical positivism is invoked, with Neurath reminding Carnap of their discussions of "our" Gustav Kirchhoff and his rejection of explanation in favor of mere description. However, things now also begin to get more personal, as well; "I am really sorry that you agree to such an extent with Schlick, Tarski, and Popper that you feel more with them than with me. [...] I feel that what you like is a tendency towards CALCULUS, connected with [...] what I would call a mixture of crude realism and metaphysical absolutism. You very seldom speak of EMPIRICIST problems as such. I feel very uneasy."[9] Here we can indeed begin to detect an indication of long-suppressed differences, perhaps fundamental differences of the kind Mormann suggests.

But it must also be kept in mind that by this time, Carnap and Neurath had not seen each other for nearly five years, and had relied only on written communication. As Neurath himself admits, this was not the ideal mode of interaction for him; in Letter 28 he suspects he may get on Carnap's and the others' nerves for presenting his case more forcefully in person than they might like, since unlike them he doesn't publish his ideas but rather "I [...] used the discussion as my medium."[10] It seems obvious three-quarters of a century later, as it seemed obvious even at the time to Ina Carnap (who eventually intervened with a letter to Neurath to try to calm things down)[11] that personal contact in conversation could have sorted out all the differences in short order. Carnap himself readily admitted that his personality was more a north German, perhaps Prussian, protestant, somewhat repressed and introverted one, while Neurath was the voluble, loud, excitable, but basically generous and affectionate Viennese extravert. One should not give way too easily to the temptation to read genuinely doctrinal differences into the last year or two of the Carnap-Neurath correspondence when there are mostly much simpler explanations.

This applies especially to the following long Letter 28, where Neurath gives anguished and elaborate expression to his feelings of having been insulted by Carnap, who had inserted a note at the end of Neurath's encyclopedia monograph (then just appearing) to say that he had been unable to exercise editorial responsibilities for this volume (which he in fact hadn't). This gives Neurath an occasion to open the floodgates of recrimination about all sorts of wrongs over the years, real and imagined, inflicted on him not only by Carnap, but also by Reichenbach, by Schlick, and the rest of the Vienna group. First he recounts his humiliation by Carnap over an article he had submitted to *Erkenntnis* a decade previously. "One day you wrote me a letter, in a similar teacherlike tone, that you can hardly accept an article of mine for the *Erkenntnis* since my style of presentation did not reach the level wanted there etc. etc." But actually, Neurath continues, the article was no

[9] Neurath to Carnap, 1 April 1944 (RC 102-55-05); see letter 26 in the present volume.

[10] Neurath to Carnap, 18 November 1944 (RC 102-55-06); see letter 28 in the present volume.

[11] Ina Carnap to Neurath, 24 August 1945 (RC 102-55-10); see letter 33 in the present volume.

worse than many others published there, so why, Carnap, are you only critical toward me? "You are usually not very critical towards other people, but particularly irritated by me."[12]

He then develops an elaborate speculative story line (freely admitting that it is "guesswork") about his role in the Circle as the outsider who always annoys the others by giving them a bad conscience about straying from strict empiricism, and hence makes them uncomfortable and incurs their resentment. First it was Wittgenstein, he says; he still doesn't understand what not only Carnap, but also Schlick, Waismann, and Hahn saw in that "antiscientific thinker full of metaphysics," and remembers with resentment how Feigl and Carnap forced him, Neurath, to go along with a positive mention of Wittgenstein in the "Manifesto." Then it was Popper: "you preferred his attitude to mine. I think today what I thought then, that you like the strong antithesis of YES and NO in his looking at positive and negative instances, and then his tendency to start from ONE world system as the most complete one." And now, in the third stage of Carnap's progressive alienation from his loyalty to empiricism (and to Neurath), it is of course Tarski, "whose Aristotelian metaphysics seems useful to you in building up semantics. What may be helpful in building up a calculus, [though,] can become very dangerous in building up an instrument for empiricism." Admit it, Neurath urges Carnap, admit "that you FEEL A LITTLE THAT YOU ARE IN SOME WAY UNEMPIRICIST, then I should understand that I myself as your bad conscience irritate you enormously." Then, for a moment, Neurath steps back from his tirade to put himself in perspective:

> Let me come to an end. All that is guesswork, but I think I should write it to you since I would say it to you. Of course writing and speaking are different, scripta litera manet, but I tried to write as peacefully as possible, whereas in a conversation I should call you names, as usual. (Neurath to Carnap, 18 November 1944. RC 102-55-06. See letter 28 in the present volume.)

Finally he admits that despite all the insults and his hypersensitivity toward them, he is quite resilient, and holds out that prospect for the present incident as well, even managing a bit of irony at his own expense:

> In spite of the fact that all such events touch me very much, I usually overcome them and discover a way how to go on better than before. You see, your treating me badly by not mentioning me induced me to start with publishing my main ideas regularly – I like that now. Your letter on my articles [being] hardly acceptable for the *Erkenntnis* immediately induced me to create the series *Einheitswissenschaft*, where I could publish my own stuff whenever I wanted to do it, without any consent from my strong teacher. Of course I did not insult you but invited you to be with me and to publish there, you by me, highly admired giant of logical analysis and a man who unveils the secrets of so many metaphysicians. …… And so life could go on in a rather smooth way, as far as we are concerned. (Neurath to Carnap, 18 November 1944. RC 102-55-06. See letter 28 in the present volume.)

And so when Neurath near the beginning of the letter accuses Carnap (along with Hempel, Popper, Reichenbach, Morris, and others) of closing off "pluralism" from the outset (the immediate context is once again Carnap's project of an inductive

[12] Neurath to Carnap, 18 November 1944 (RC 102-55-06); see letter 28 in the present volume.

logic), and then follows this with "I think this antagonism between us is an old one," it is hard to see this as more than a reiteration of the previous accusations about Carnap's supposed hostility to "pluralism," but now charged with personal resentment about Carnap's new affront. And sure enough, immediately after this sentence about the old antagonism, Neurath launches into his story about getting on everyone's nerves and therefore provoking their hostility and their desire to hurt him (as in the case of the rejected *Erkenntnis* paper, and now again). So it is once again hard to see this as evidence of a fundamental philosophical or *doctrinal* divergence between Neurath and Carnap traceable back to Vienna or before.

The final letters descend even further into the personal. Letter 32 of 23 August 1945 complains at length of the condescending behavior toward himself of Schlick, Hempel, and Carnap himself during the Vienna years and since then, but even of his old school friend Hans Hahn, and then letter 33 of 24 August is Ina's intervention. She is well-placed to mediate, as a Viennese who understands very well how the cerebral north-German Carnap can rub people the wrong way. The Viennese Philipp Frank, she tells Neurath, would never have gone to the trouble of inserting a note into the back of Neurath's encyclopedia volume as Carnap had done:

> [...] he does not have Carnap's earnest righteousness and also he has a certain mellowness, half wisdom, half "Wurstigkeit"; it would never have occurred to him to drop his name even if he had felt like Carnap did, for the simple reason that he would not have thought it important enough and that he would have been too lazy to write the attending letters. Ah, but that's the difference between the Viennese temperament and the zealous Lutheran from Prussia! I do not try to excuse Carnap, I am just trying to point out the way he is made. Yes, he could be induced to treat you with greater circumspection but that would not improve the friendship. He does not have the saving grace of a light touch and of a felicitous formulation which might soften the blows which he is striking in the name of science, impartiality, and other suchlike gods. But then again, perhaps you cannot have the charm and the reliability in one person! (Ina Carnap to Neurath, 24 August 1945 (RC 102-55-10); see letter 33 in the present volume.)

Not surprisingly, Ina's intervention was largely successful; Neurath's final letters to both Ina and Carnap himself are much more conciliatory than before, and insofar as they are still antagonistic, they focus entirely on the personal; there are no further accusations of Carnap's failure to adhere sufficiently to "pluralism." Indeed, Neurath had already, in one of his most hostile letters, fully accepted that the issue was mainly personal for him:

> I personally always fear that I do queer things and therefore I like people like Frank who try to reduce my overestimating protocol-statements etc. without grieving me. And you see, that is the point. My letters dealt with YOUR UNKINDNESS. That is the main point. One has no "right" to ask for kindness, but one can object to unkindness. Look at [Carnap's] letter to Morris [about Neurath's encyclopedia volume]. You only tell of my bad qualities and that you drop ONLY your name, – as a concession – because he is such a violent boy ..... No kind word about me in this letter. You see, it could e.g. come into your wise brain that even strong elephants sometimes need much of their energies to overcome the obstacles of life. You see, to reach the shores in shoes only is not just the best start for a new life in a foreign country and an internment is also not the best introduction. (Neurath to Carnap, 16 June 1945. RC 102-55-11. See letter 31 in the present volume.)

All in all, by the final letters there seems to be every reason for optimism about the future of the Carnap-Neurath friendship *and* their philosophical collaboration; Uebel (2001) is on the right track – if anything he understates the prospects for reconciliation. When in the earlier letters of the 40s Neurath accuses Carnap of insufficient "pluralism" it seems very likely that he is talking about an attitude, not a doctrine, just as in his very last letters to Carnap he focuses on trying to get Carnap to admit to a certain over-zealous Platonism (by which he means something more like Lutheran Prussianism or punctiliousness) in his demeanor and behavior – which Carnap would presumably have admitted without hesitation! Certainly there are differences between Neurath's and Carnap's priorities and attitudes, just as there had been before 1935; but from the correspondence of 1935–45, at least, there is little evidence of a fundamental philosophical difference of the kind suggested by Mormann.

12.3 Carnap's Cartesianism

Now we come to Carnap's responses to Neurath's critique of semantics, and to the larger question in the background whether Mormann's hypothesis of "Cartesianism" (in Neurath's – or Mormann's conception of Neurath's – sense) applies to Carnap. Here there are many fewer documents to draw on, both because Carnap's letters are not nearly as long as Neurath's, but also because Carnap was initially at a loss to understand what Neurath was saying, and kept asking for clarification.

Carnap's initial response to Neurath's doubts about semantics in the Paris meetings of 1935 and 1937 was to show that the concept "true" could not be replaced everywhere by the concept "accepted by scientists" or something of the kind, but also to point out that the semantic use of "true" had no implications whatever regarding the "truth" in some language-transcendent sense of the sentences in question; "'F' is true" meant simply "F." To Neurath's concern that sentences be compared only with sentences, as in syntactic *Wissenschaftslogik*, rather than (as it appeared to be the case in semantics) with matters of fact in the world, Carnap answered that in that respect, nothing had changed, but that in the end there is still the question (as there was before) *how* sentences can represent matters of fact (which he regarded as an empirical question; by the 1940s it would become a question of pragmatics). To Neurath's insistence that this could be misunderstood, and to Neurath's insinuations of guilt by association (pointing to the provenance of the Polish logicians and the philosophical backgrounds of some of their colleagues), Carnap imprudently did not respond.

His response to Neurath's Letter 17 is, accordingly (in Letter 18) simply to ask for clarification about where Neurath discerns Aristotelian metaphysics in the *Introduction to Semantics* and what in particular he has in mind. In his response to Neurath's Letter 22 (Letter 24 of 4 February 1944) he goes into more detail. Regarding semantics itself, he essentially repeats what he had to say in 1935 and 1937, as Neurath still seems to be conflating the epistemological question of justifying or establishing the truth of a sentence with the logical or linguistic question of the semantic truth definition:

> On Semantics. If I see it correctly, you raise objections or doubts in two points:
>
> 1. Semantics may be a[l]right as a mere calculus, but if it is applied to the language of empirical science, it seems doubtful or at least an open question whether it is fruitful and useful.
> 2. Some of the chief concepts of semantics are metaphysical; therefore empiricists have to reject them as meaningless.
>
> I am not quite clear how you intend to combine these two objections; they do not seem to me to fit well together. (Carnap to Neurath, 4 February 1944. RC 102-55-04. See letter 24 in the present volume.)

In response to the first point, Carnap assures Neurath that he has little interest in semantics as a mere calculus, but is concerned to apply it to the analysis of questions about science. He exhorts Neurath to patience and tolerance while the first steps are taken; we can't know whether a new tool is useful until we've had a chance to develop it and try it out for a few years. But he points out that Neurath himself is constantly making semantic statements in everyday life, and points once again to the scientific examples he gives in §14 of *Introduction to Semantics*. He expresses exasperation that Neurath keeps asking for examples but does not engage with the many examples he has given him previously. In any case, he says that the question of usefulness and applications can only be put off if we feel reasonably sure that the concepts we are using make sense and are not metaphysical, which brings us to Neurath's *second* point. On this, Carnap again repeats essentially what he had said in 1935 and 1937:

> On (2). That the semantical concept of truth is not metaphysical can very easily be shown by the following translation: "The sentence 'this tree is green' is true" means not more and not less than "This tree is green". (If the latter sentence does not occur in your strangely restricted language you may take instead any other sentence which you regard as meaningful.) (By the way, if the term 'true' were to occur only in connection with a direct quotation as in the example given, the term would indeed hardly be useful; but that is another question.) This translation shows that the concept of truth is not metaphysical but scientific. Furthermore, the translation makes it clear that the term 'true' is not at all meant in the sense of 'absolutely certain', 'indubitable' or anything like that as you sometimes seem to believe. And the translation also shows that 'true' has nothing to do with 'accepted'; you make time and again the mistake of demanding that I should translate my semantical sentences into sentences with the term 'accepted'. (Carnap to Neurath, 4 February 1944. RC 102-55-04. See letter 24 in the present volume.)

Further to Neurath's continuing tendency to conflate the epistemological question of the grounds for the truth of a sentence with the logical question of defining truth, Carnap addresses Neurath's attempt to cast doubt on semantics by an elaborate discussion of what scientists actually do:

> To your letter p.11, the last two paragraphs. I am in complete agreement with your description of the scientific procedure. I should classify this as belonging to the methodology of science. I do not see what it has to do with semantics. Which assertion of mine concerning semantics seems to you to be in contradiction to your description? – Likewise p.12 you see[m] to believe that semantics intends to propose a new conception concerning scientific procedure; that is certainly not the case. (Ibid.)

So far, this is all covering familiar ground. Now we come, though, to Neurath's new objection – that Carnap does not share his "pluralism." Carnap responds by rejecting Neurath's accusation, but with the reservation that he is not clear about it and would like Neurath to give a more precise formulation:

> On pluralism. I believe that in this point I am in agreement with your attitude. I say "I believe" and "attitude", not "opinion" because your formulation of pluralism (p. 9) is so vague that I am not able to see in it any clear thesis. As you formulate it now, nearly everybody would agree, including Schlick, Popper, Russell. Since pluralism seems to you very important, and in particular also the question whether I agree with you in this point, please give me a more precise formulation of what is asserted by it. – Perhaps the difference here is not a difference of opinion but of emphasis. We emphasize the importance of the task of systematization in science; you, on the other hand, emphasize the fact that the statements accepted by scientists at a certain time do not form a well connected system and you point to the dangers involved in overlooking this fact. I think you are right in both points. (Ibid.)

Is Carnap being disingenuous here? If we consider the radical implications of what Neurath appears to mean by "pluralism" (see the discussion in Sect. 12.2. above), then perhaps Carnap's summary of Neurath's views is somewhat narrow and careful. Certainly Carnap would not at this point have signed on for a return to the radical positivism of late-1920s Vienna with its almost complete banishment of theoretical language (Carus 2007, 218–9), nor would he have been willing to *give up* on the idea of a deductive unity of the sciences (in the very long run), while certainly admitting, as he later did explicitly (Carnap 1963, 883) that it was at best a distant goal. But as he actually paraphrased Neurath's pluralism, in the penultimate sentence of the above quotation, his concurrence with Neurath's two points is hardly controversial. In fact, he could have gone substantially further. He was *fundamentally* on Neurath's wavelength regarding the *voluntarism* that Neurath is perhaps trying to articulate in his Letter 22. From quite early on (Carus 2007, Ch. 1.) he thought that human conceptual systems were a matter of choice and under human control. Though constrained by the facts of the world, frameworks have broad leeway to articulate these facts differently and represent them differently (Jeffrey 1994); this impulse goes back at least to *Der Raum* (Carnap 1922), though it only found its adequate philosophical expression in the principle of tolerance. The incorporation of semantics into meta-discourse did not change this.

But with this voluntarism came the responsibility of imposing appropriate constraints (empiricism, logicism, etc.) on this broad range of choice (Ricketts 1994). Without any constraints whatever, voluntarism could be interpreted to mean that we could make up any reality we liked, and that of course was *not* what Carnap had in mind. So he wanted to be sure Neurath wasn't going too far; this was why he wanted a more precise articulation of what Neurath had in mind when he demanded "pluralism," and also thought Neurath shouldn't go too far in deprecating the efforts of systematizers like himself:

> I should be still more inclined to agree with you were it not for the impression that you exaggerate very much laying more stress on the dangers of systematization than on its usefulness, importance and indispensability; I doubt whether your overstressed warning is fruitful. (Carnap to Neurath, 4 February 1944. RC 102-55-04. See letter 24 in the present volume.)

In fact he ends this part of his letter with a plea for mutual tolerance between people of different priorities and temperaments:

> I believe [that] if we are careful to avoid exaggeration on both sides, we shall easily be able to come to an agreement on this point; "we" includes Popper, Tarski, etc. For here there is really no serious difference of opinion; it is more a difference in temperament and therefore in the direction of interest. I think it would be best for the development of science if the people on the one side who see more the turbulent whirl of material in all its colorfulness and vagueness, and those on the other side who love nice structural schemata would not polemicize against each other but rather realize that the work of both is necessary for science. (Ibid.)

Carnap then goes on to respond to Neurath's tirade against Popper. In their previous discussion of Russell's *Inquiry into Meaning and Truth*, Carnap – though responding in great detail (Letter 19) to Neurath's detailed comments (Letter 11), eventually tires of the exchange and asks Neurath why he goes on harping on this book, which they agree is wrong and not very interesting. Popper, on the other hand, he is willing to defend (against Neurath, at least) up to a point. A remark from an earlier letter about this casts an interesting light on his general attitude toward discussions within the Vienna Circle (and now among those he considers part of the "movement"):

> In the case of Popper, I believe your reaction is chiefly caused by the fact that he criticized the Vienna Circle quite unnecessarily. He was overcritical and so you are now. Even when he wrote the book he was in agreement with us on most fundamental points. When later he came into personal contact with us the agreement became even more strong and conscious to him. Some of his views which you criticize, e.g. the refutability of hypotheses, have the same defects – and, I think, the same as many of our earlier views: they might be taken as first approximations but closer inspection shows that they are not entirely adequate but must be replaced by better approximations. I suppose that the same holds for many of our present views, including mine, where we do not see today how they should and can be improved. In the case of some of your views which are not shared by most of the people in our movement I often defend them in just the same way: I admit to those who criticize them that the formulations are not quite adequate but I point to the fact that they may be taken as first approximations. I like to defend your views but sometimes you make it, by Jove, hard for me to do so when you stubbornly stuck to your old formulations years after they have been shown to you to be inadequate (e.g. "No facts, only statements", your form of protocol sentences, "the semantical concept of truth is only applicable to calculi not to the language of science"). The views of all of us within the movement of empiricism do of course differ more or less from each other. I think for the sake of the movement it would be much better if we were more tolerant towards each other. If your intolerance would become the general custom, then I am afraid you would be among the first to be declared a heretic and excommunicated. By tolerance, of course, I do not mean acceptance of each others' views. The differences of opinion should and will be discussed. But this discussion is not helped by labelling the views of the others as nonempiricist and metaphysical. (Carnap to Neurath, 29 January 1943. RC 115-07-62. See letter 18 in the present volume.)

Now, since Neurath continues harping on Popper and using him as a proxy in his complaints about Carnap's supposed failure to share his "pluralism," Carnap goes into a little more detail:

> Popper's position [...] seems to me good as a first approximation. I think in the question of asymmetry we have to distinguish two assertions of Popper's:

> 1. In testing a universal law, there is an asymmetry between favorable (confirming) and unfavorable (disconfirming) cases. If somebody proposes a law hypothetically then he usually has already a number of favorable cases, and he is looking around for further cases. The asymmetry is this: an unfavorable case has a much stronger influence upon our judgment concerning the law than an additional favorable case.
> 2. One single unfavorable case refutes the law. (Carnap to Neurath, 4 February 1944. RC 102-55-04. See letter 24 in the present volume.)

The discussion of these two points that follows is very interesting, because it further illustrates Carnap's ethos of rational discussion, as already spelled out in the previous quotation (from Letter 18):

> Some objections might perhaps be raised against (2). But still I would think that it is acceptable as a first approximation. On the other hand, your rejection of (1) in your last letter is surprising to me. I think that all empiricists and all good scientists agree in this; Who, in your opinion, does not? (Carnap to Neurath, 4 February 1944. RC 102-55-04. See letter 24 in the present volume.)

The interesting thing about this passage is not what Carnap says, but what he does *not* say. When he says that some objections may be raised against (2), he does not mention that he himself, in §§25-6 of "Testability and Meaning" (Carnap 1936b–7; summarized in Carnap 1963, 879), pointed out a fatal flaw in just this thesis (2) of Popper's – an objection that Popper never responded to and to this day remains unanswered by his followers (Carus 2015c). He could have made this discussion with Neurath easy for himself by going along with Neurath and saying "Yes, you're right, Popper is wrong, and I've shown myself that his falsification criterion doesn't do what he claims, so I dissociate myself from him and agree with you." But he does not want Neurath to reject Popper for the wrong reasons; he wants him to understand and appreciate the insight of genuine value in Popper before focusing on the limitations of that insight. His ethos of rational discussion rejected not only a premature rush to judgment, but also the exclusion of viewpoints in a sweeping and generic, undifferentiated sort of way.

He also points out that Neurath can't have it both ways; he can't *both* say that Popper is *wrong* about universal laws, falsifying instances, etc. but then *also* say that these assertions of Popper's are *metaphysical*:

> You remember how our opponents called us solipsistic because we rejected the thesis of realism as meaningless? We always emphasized that if a sentence is metaphysical and hence meaningless, then the same holds for its negation. Now Popper makes certain assertions concerning universal laws, negative instances etc. You deny these assertions. This is your good right. But then you go on to say that Popper's assertions are metaphysical. If so, however, your counter-assertions are likewise metaphysical and meaningless. Seriously, I think it would be better if you would at long last abandon your habit of calling people who are empiricists and antimetaphysicians metaphysicians if you do not share their opinions. First, it is absurd; and second it does not help a successful discussion. (Carnap to Neurath, 4 February 1944. RC 102-55-04. See letter 24 in the present volume.)

There are a few further remarks in the following letters, but there was hardly an opportunity for a proper debate of the issues themselves (as opposed to personal animosities and reconciliations) before Neurath's death in late 1945. There are no grounds in this correspondence, in any case, for Mormann's hypothesis of a deep philosophical fault line between Carnap and Neurath that they both were in denial about and papered over for the sake of a harmonious friendship. For one thing, we can see that the friendship was not particularly harmonious; both were too honest to sacrifice things they believed in for personal harmony. And second, it is pretty obvious that the single *substantive* (or possibly doctrinal) bone of contention, the issue of "pluralism," was less a thesis of Neurath's than an attitude. In this it resembled "physicalism," which was one of the earlier foci of conflict and debate between Carnap and Neurath (cf. Carus 2007, 239–50). But *unlike* physicalism, which was a more articulate and explicit (i.e. more objective) attitude or value, "pluralism" remained somewhat vague, and may in the end have been a way for Neurath to represent *personal* characteristics of Carnap's (as acknowledged in Ina's letter) rather than anything that could serve as an orientation for *Wissenschaftslogik*. Even if we want to interpret it as something more objective, though, it was evidently not a matter of dispute between Carnap and Neurath, at most one regarding which (as Carnap suggested) their emphases differed.

There remains the question of Carnap's supposed Cartesianism. Going back to Mormann's three criteria of (Neurathian) Cartesianism in Sect. 12.1. above, it is obvious that Carnap rejected (1); his above-quoted response to Neurath about "pluralism" makes this very clear. (2) and (3) are different. Carnap obviously recognized that both were false, of science as it currently exists; so in that sense he does not qualify as "Cartesian" by Mormann's Neurathian criteria. On the other hand, both (2) and (3) represented aspirations for him, to some degree. He thought both were goals worth working toward, though as he grew older he became less and less confident they could ever be attained. He also left open the possibility that any eventual deductive unity of our knowledge could be undertaken in different, mutually incompatible ways, and encouraged the pursuit of different programs for such a unity. In his review of Popper's *Logik der Forschung* in *Erkenntnis* (Carnap 1935), for instance, he acknowledged Reichenbach's extremely negative review of Popper's book (in the same issue), but didn't think that we needed to *decide* between Popper's and Reichenbach's conceptions of probability; we should view them as two different *proposals* for the structure of the language of science. He hoped both authors would go on to develop their respective proposals in more detail, and apply them to actual scientific problems, so that eventually we could see which one was working out better. This pluralism is obviously also not very "Cartesian." It was Mormann (2001) himself, after all, who first suggested that Carnap saw philosophy as a "science of possibilities" – a *Möglichkeitswissenschaft* in the sense of Robert Musil.[13]

Carnap realized, then, that current science was far removed from anything like a deductive system, and was aware that the scientific vernacular used for regular communication among sciences, and even within the most advanced sciences, could hardly count as transparent, clear, or distinct. And he may well have realized that this would always remain the case, however far the deductive system-builders like

[13] A suggestion strongly endorsed by Carus (2007); see the discussion, including an illustrative quotation from Musil, on p. 64. The fundamental kinship between Carnap and Musil has also been addressed by Jacques Bouveresse (e.g. 2001, 2012).

himself advanced their projects.[14] But this would not have discouraged him; he would have continued to think it was better to extend the islands of clarity in the murky sea of vagueness in which the sciences communicate with each other than to wallow complacently in vagueness.

And here is where it seems quite reasonable to think he and Neurath could have come to a kind of rapprochement regarding the language of science. Carnap strove for clarity and, ultimately, some sort of deductive unity, while Neurath strove for pluralism and openness within the scientific vernacular envisaged as a tidied-up version of ordinary language, one in which all the terms that have explications within any of the sciences *defer* to those explications (in cases when pressure is put on them and they need to be more precise than they are in ordinary language). Neurath was very concerned to emphasize that this vernacular was not, and could not be, a perfect deductive system, and would continue to have "clots" (Ballungen) of vagueness in it. Carnap could readily have agreed to this as a matter of fact in science as it now is, but one could also see him, ever the forward-looking optimist, proposing that precisely these *Ballungen* be regarded as the most necessary and promising target concepts for explication (in his sense; see Carus 2007, 273–384). Neurath might have seen this as a threat, worrying that the overzealous Carnaps of this world would set about removing *all* the *Ballungen* from the scientific vernacular and thus constraining or eliminating "pluralism." But he might have been coaxed into accepting such a conception of it in view of Mark Wilson's many examples showing that the *Ballungen* never really go away; they only shift their position under the rug, or ramify further into more branches of meaning (Wilson 2006, 2018). On the assumption (or reassurance) that the supply of *Ballungen* is inexhaustible, Neurath might well have been willing to concede that some might be targeted for explication, and might actually succumb. In any case, it is easy to imagine a scenario whereby Carnap and Neurath could have converged on a conception for the iterative and continuous *improvement* of the scientific vernacular along such lines.

12.4 Carnap's Representationalism and Inferentialism

By way of conclusion, I return to the question of Carnap as a "representationalist" that, as we saw, Robert Brandom took over unquestioningly from tradition. In Brandom's scheme of things, everyone is either a representationalist (like Carnap) or an "inferentialist" (like Brandom). Representationalists come in many varieties,

[14] It has been known since Reisch (1991) that in his capacity as the editor of the *Encyclopedia of Unified Science*, Carnap oversaw Thomas Kuhn's *The Structure of Scientific Revolutions*, which interested him very much and which he found very compelling; see also Friedman (2003). It is thus hard to believe he would not have been interested in other well researched pieces of recent "descriptive pragmatics" (as he would have classified it) that tended to undermine his default view of a gradual convergence toward deductive unity and transparency, such as Mark Wilson's work, as I suggest (Carus 2012), where I also suggest how he might have accommodated such news; see also Wilson's reply (Wilson 2012) in the same volume.

but they have in common that they regard language as representing some pre-existing, extra-linguistic given, while inferentialists regard the system of representations as ultimately internal to language; representation is re-created or explained (away) by means of intra-linguistic inferences.

Of course, not everyone sees Carnap this way; the wave of new research over the past two or three decades[15] has begun to turn the tide, and while consensus has been reached on some issues (as on the transition from syntax to semantics; see above, Sect. 12.1.), the degree to which Carnap can be regarded a representationalist is in flux, under discussion. This has gone so far, in fact, that some have even gone to the other extreme and begun to talk about "Carnap's Inferentialism" (Chalmers 2012; Peregrin 2014, 2019) in the *Syntax*, by which approximately the same thing is meant as Neurath's aspiration to compare sentences only with sentences, in *Wissenschaftslogik*, and never sentences with facts. And the mistake still made by those who now think of Carnap's *Syntax* period as inferentialist is also essentially the same as Neurath's mistake; they are misled by the semantic terminology to think that meaning (in the pre-1931 sense of correspondence with some pictured complex of matters of fact) was reinstated in 1935, and was no longer a matter of interrelations among sentences.

Even before 1931, if one wants to get picky, Carnap could be regarded as a "structural" inferentialist. The metaphor of the railway map in the *Aufbau* indicates how he wants to go about referring to a set of concepts purely structurally, without ever having to employ ostension or direct matching up of a node in a structure with a particular space-time point in the world or a particular object. This structuralist inferentialism is not widely appreciated, perhaps because Carnap's attempt there (§§153–5) to do away with reference altogether and thus to make all knowledge purely structural failed. But this failure hardly dilutes the fundamentally "inferentialist" character of the project of picking out even the basic difference among human senses not by ostension but by their structural characteristics, e.g. the number of (experiential) dimensions they exhibit (Carnap 1928; Friedman 1999, Chs. 5 and 6, 2007). This inferentialism is rather different in tendency from Brandom's, of course, but it certainly shares Brandom's goal of explaining reference away (or rather accounting for reference without requiring that there be anything independently accessible for linguistic reference to refer to).

So it is entirely legitimate to talk about Carnap's inferentialism (cf. also Tuboly 2017). But we also saw in Sect. 12.3. above that Carnap aspired, at least, to a kind of Cartesian representationalism – admittedly a rather off-beat, pluralistic sort of Cartesianism that Descartes himself would not have been very happy with, and a very schematic representationalism. Still, how can that be? Are representationalism and inferentialism really opposites, as Brandom has always claimed? Are they mutually exclusive? Some have argued compellingly that they are not, and that both are indispensable, so there is not only no need to choose between them, but either choice would leave an incomplete and one-sided conception of meaning (e.g. Kremer 2010). Carnap's case, though, was different.

[15] Of which an overview up to 2015 is provided in Carus (2015b).

For both representationalism and inferentialism as commonly understood (or understood by Brandom, anyway), a question of *reduction* is at stake. Brandom refers to this as the "order of explanation." His preferred order of explanation is that from inference to representation, rather than putting representation first and then deriving inference from it (the order he attributes to the mainstream of analytic philosophy). In other words, he wants to *reduce* representation to inference. But there is no point to reduction of this sort outside an *ontological* context – one is only motivated to reduce this to that if one wants to make a case that really or ultimately *there are* only thats, while thises are constructions from thats. That seems to be Brandom's motivation as well; he wants to argue that the kind of reasoning we employ in constructed languages of science (which he calls "algebraic reasoning") is ultimately parasitic on a more primitive kind of practical reasoning (exemplified by his "material rules of inference"; cf. Brandom 1994, 97–107; Carus 2004, 319–27) that he calls "hermeneutic reasoning" (Brandom 2009).

Now Carnap actually once argued specifically against this idea that constructed languages are parasitic on colloquial or evolved languages, in his reply to Strawson (Carnap 1963, 933–40), who had relied on the intuitive plausibility of the idea that colloquial languages are learned first, and that we learn constructed languages later, through the medium of a colloquial one. In his reply Carnap sought to undermine the plausibility of this apparently obvious consideration by pointing out that while it is true that we learn colloquial languages first, this is purely contingent. Strawson wrongly infers from this contingent fact that no other way is possible. But we know from the cases of Esperanto and other such made-up languages that we can *create* languages for everyday communication, and there is no reason to think that children couldn't grow up speaking those instead of their usual mother tongues.[16] How far could we vary the parameters of such artificial languages, how precise could we make them, before it became impossible for children to learn them as first languages? We have no idea (this is an empirical question that for good reason hasn't been studied),[17] but we do know that adolescent and adult humans are in fact able to learn complex artificial languages (e.g. algebraic topology, musical scores, programming languages) to a point of amazing fluency, and to build up extensive networks of intuitions around them. So we can hardly make dogmatic pronouncements on this issue, which remains an empirical question.

In Carnap's view there is no room for anything *but* an empirical question here. There is no way of answering the question "what is language *really*?" other than (a) empirical research; (b) stipulation, i.e. construction of a new language – in which case what that language "is" is exhaustively specified in the given rules. There are no grounds other than empirical (or stipulative) ones on which one could claim that

---

[16] Children have in fact been brought up Esperanto-speaking, e.g. George Soros, but apparently there are no known cases of monolingual upbringing in Esperanto; native Esperanto speakers have also, so far, always learned another language from childhood; see the article "Native Esperanto speakers" in Wikipedia.

[17] Which is also one reason why the learnability literature (e.g. Steven Pinker) isn't very empirical.

one kind of language is primary in some sense to others, in the sense that the others must always or necessarily be dependent on it. It is certainly true that the first constructed languages (the language of the law, the language of geometry) were set up using ordinary colloquial language as a meta-language, and that has not changed over the millennia, but it is also true that our colloquial languages contain many more concepts from constructed languages than they did when this process first started in ancient Greece. And the practical need to use a colloquial language as the metalanguage for setting up constructed languages does not imply there is something inherent in human cognition that requires this "order of explanation."

For Carnap, then, neither inferentialism nor representationalism is of interest as a reductive *doctrine*. Each can be of interest as a conceptual exploration, to find out what different strategies and raw materials can be used to define logical operators and set up elementary logical or arithmetical languages or theories, and how these different approaches relate to each other (cf. Murzi and Steinberger 2017; Hjortland and Standefer 2018). But these are technical questions in logic and mathematics like any questions in those fields, and have little bearing on the correctness or the preferability of one "order of explanation" over another. About an "order of explanation" Carnap would have said exactly (and for the same reasons) what he said about questions of existence in "Empiricism, Semantics, and Ontology" (Carnap 1950/1956). They can be understood in an *internal* sense, then they can be answered, usually trivially; the order of explanation internal to a framework depends on what universe of discourse comes with the framework. Whatever doesn't come with the framework has to be constructed, where that is possible, out of the materials that *do* come with the framework. If inference comes with a framework but not representation, then representation has to be constructed from inference, as Brandom prefers. If only representation comes as a primitive concept within the framework, though, then inference must be constructed from representation. What does not make sense, for Carnap, is to ask the *external* question "which order of explanation is *right* or *correct*?" in some general or absolute sense, without indexing this question to a language framework. In response to this question Carnap would ask "for what?" – just as he does in the case of existence. The question itself, when stated without a reference language, is strictly speaking meaningless, but it can be turned into a practical question regarding the *preferability* of this or that order of explanation.

In that case, though, it has to be indexed to a *purpose*, relative to which one order of explanation can be judged to be preferable to (i.e. more conducive than) another. Carnap and Neurath would have been in complete agreement – against Brandom, Rorty, and perhaps many or most other philosophers of recent decades – that an order of explanation purporting to explain what language really or essentially is by reducing one of its functions to another (e.g. by reducing representation to inference or vice versa) is of no great interest.

12.5 Note on Sources and Quotations

The location of an unpublished document from the Carnap papers (Manuscript collection No. 1029) in the Young Research Library, University of California at Los Angeles, is specified by the abbreviation (UCLA. . .). Translations from German are my own. In the quotations from the correspondence between Neurath and Carnap, I have corrected minor errors of spelling, grammar, punctuation, and usage to avoid distraction. References are to the letter numbers in the part of the correspondence that is reproduced in this volume.

References

Awodey, Steve. 2007. Carnap's Quest for Analyticity: The *Studies in Semantics*. In *The Cambridge Companion to Carnap*, ed. Michael Friedman and Richard Creath, 226–247. Cambridge: Cambridge University Press.

Awodey, Steve, and A.W. Carus. 2009. From Wittgenstein's Prison to the Boundless Ocean: Carnap's Dream of Logical Syntax. In *Carnap's Logical Syntax of Language*, ed. Pierre Wagner, 79–106. Basingstoke: Palgrave Macmillan.

Bouveresse, Jacques. 2001. *La voix de l'âme et les chemins de l'esprit: Dix études sur Robert Musil*. Paris: Seuil.

———. 2012. Rudolf Carnap and the Legacy of *Aufklärung*. In *Carnap's Ideal of Explication and Naturalism*, ed. Pierre Wagner, 47–62. Basingstoke: Palgrave Macmillan.

Brandom, Robert. 1994. *Making It Explicit: Reasoning, Representing, and Discursive Commitment*. Cambridge: Harvard University Press.

———. 2009. Metaphilosophical Reflections on the Idea of Metaphysics. *Harvard Review of Philosophy* 16: 44–57.

———. 2013. Global Anti-representationalism? In *Expressivism, Pragmatism and Representationalism*, ed. Huw Price et al., 85–111. Cambridge: Cambridge University Press.

Carnap, Rudolf. 1922. *Der Raum. Ein Beitrag zur Wissenschaftslehre. Kant-Studien*, Ergänzungshefte Nr. 56. Berlin: Verlag von Reuther & Reichard.

———. 1928. *Der logische Aufbau der Welt*. Berlin: Weltkreis.

———. 1935. Besprechung: Karl Popper, Logik der Forschung. *Erkenntnis* 5 (4): 290–294.

———. 1936a. Wahrheit und Bewährung. In *Actes du Congrès international de philosophie scientifique, Sorbonne, Paris 1935, fasc. 4*. Unité de la science, 18–23. Paris: Hermann.

———. 1936b–7. Testability and Meaning. *Philosophy of Science* 3 (4): 419–471; and 4 (1): 1–40.

———. 1939. *Foundations of Logic and Mathematics*. Chicago: University of Chicago Press.

———. 1942. *Introduction to Semantics*. Cambridge, MA: Harvard University Press.

———. 1950/1956. Empiricism, Semantics, and Ontology. In *Meaning and Necessity*, 2nd ed., 205–221. Chicago: University of Chicago Press.

———. 1963. 'Intellectual Autobiography' and 'Replies and Systematic Expositions.' In *The Philosophy of Rudolf Carnap*, ed. Paul A. Schilpp, 1–84, 859–1013, LaSalle, IL: Open Court.

Cartwright, Nancy, Jordi Cat, Lola Fleck, and Thomas Uebel. 1996. *Otto Neurath: Philosophy Between Science and Politics*. Cambridge: Cambridge University Press.

Carus, A.W. 1999. Carnap, Syntax, and Truth. In *Truth and its Nature (If any)*, ed. Jaroslav Peregrin, 15–35. Dordrecht: Springer.

———. 2004. Sellars, Carnap, and the Logical Space of Reasons. In *Carnap Brought Home: The View from Jena*, ed. Steve Awodey and Carsten Klein, 317–355. LaSalle, IL: Open Court.

———. 2007. *Carnap and Twentieth-Century Thought: Explication as Enlightenment*. Cambridge: Cambridge University Press.

———. 2012. Engineers and Drifters: The Ideal of Explication and its Critics. In *Carnap's Ideal of Explication and Naturalism*, ed. Pierre Wagner, 229–243. Basingstoke: Palgrave Macmillan.

———. 2015a. Williamson on undoing the linguistic turn. *Carnap Blog*. http://awcarus.com/2015/04/williamson-on-undoing-the-linguistic-turn/

———. 2015b. Rudolf Carnap. *Oxford Bibliographies Online*. http://www.oxfordbibliographies.com/view/document/obo-9780195396577/obo-9780195396577-0248.xml

———. 2015c. Popper and Carnap again. *Carnap Blog*. http://awcarus.com/2015/05/257/

———. 2017. Carnapian Rationality. *Synthese* 194: 163–184.

Chalmers, David. 2012. Inferentialism and Analyticity. http://consc.net/books/ctw/excursus19.pdf (Excursus 19 of the online "Extended Edition" of Constructing the World, Oxford: Oxford University Press).

Creath, Richard. 1991. Every Dogma Has Its Day. *Erkenntnis* 35 (1/3): 347–389.

Friedman, Michael. 1999. *Logical positivism reconsidered*. Cambridge: Cambridge University Press.

———. 2003. Kuhn and logical empiricism. In *Thomas Kuhn*, ed. Thomas Nickles, 19–44. Cambridge: Cambridge University Press.

———. 2007. The *Aufbau* and the Rejection of Metaphysics. In *The Cambridge Companion to Carnap*, ed. Michael Friedman and Richard Creath, 129–152. Cambridge: Cambridge University Press.

Hintikka, Jaakko. 1988. On the Development of the Model-Theoretic Viewpoint in Logical Theory. *Synthese* 77: 1–36.

———. 1991. Carnap, the Universality of Language, and Extremality Axioms. *Erkenntnis* 35: 325–336.

———. 1992. Carnap's Work in the Foundations of Logic and Mathematics in a Historical Perspective. *Synthese* 93: 167–189.

Hjortland, Ole, and Shawn Standefer. 2018. Inferentialism, Structure, and Conservativeness. In *From Rules to Meanings: New Essays on Inferentialism*, ed. Ondřej Beran, Vojtěch Kolman, and Ladislav Koreň, 115–140. New York/London: Routledge.

Jeffrey, Richard. 1994. Carnap's Voluntarism. In *Logic, Methodology, and Philosophy of Science IX*, ed. D. Prawitz, B. Skyrms, and D. Westerståhl, 847–866. Amsterdam: Elsevier.

Kremer, Michael. 2010. Representation or Inference: Must We Choose? Should We? In *Reading Brandom: On Making It Explicit*, ed. Bernhard Weiss and Jeremy Wanderer, 227–246. Abdington: Routledge.

Mancosu, Paolo. 2008. Tarski, Neurath, and Kokoszynska on the Semantic Conception of Truth. In *New Essays on Tarski and Philosophy*, ed. Douglas Patterson, 192–224. Oxford: Oxford University Press.

Mormann, Thomas. 1999. Neurath's Opposition to Tarskian Semantics. In *Alfred Tarski and the Vienna Circle: Austro-Polish Connections in Logical Empiricism*, ed. Jan Woleński and Eckehart Köhler, 165–178. Dordrecht: Kluwer.

———. 2001. Carnaps Philosophie als Möglichkeitswissenschaft. *Zeitschrift für Philosophische Forschung* 55: 80–101.

Murzi, Julien, and Florian Steinberger. 2017. Inferentialism. In *Blackwell Companion to Philosophy of Language*, ed. Bob Hale, Alexander Miller, and Crispin Wright, 2nd ed., 197–224. London: Wiley Blackwell.

Neurath, Otto. 1913/1983. Die Verirrten des Cartesius und das Auxiliarmotiv. Repr. in Otto Neurath, *Gesammelte philosophische und methodologische Schriften*, vol. 1 (Wien: Hölder-Pichler-Tempski), 57–67. Translated as: The Lost Wanderers of Descartes and the Auxiliary Motive (On the Psychology of Decision). In *Otto Neurath: Philosophical Papers 1913-1946*, ed. Robert S. Cohen and Marie Neurath, 1–12. Dordrecht: D. Reidel.

———. 1935/1983. Pseudorationalismus der Falsifikation. Erkenntnis 5: 353–365. Translated as: Pseudorationalism of Falsification. In *Otto Neurath: Philosophical Papers 1913–1946*, ed. Robert S. Cohen and Marie Neurath, 132–138. Dordrecht: D. Reidel.

Peregrin, Jaroslav. 2014. *Inferentialism: Why Rules Matter*. Basingstoke: Palgrave Macmillan.

———. 2019. Carnap's inferentialism. In *The Vienna Circle in Czechoslovakia*, ed. R. Schuster. Dordrecht: Springer. Forthcoming.

Popper, Karl. 1945. *The Open Society and Its Enemies*. London: Routledge.

Reisch, George. 1991. Did Kuhn Kill Logical Empiricism? *Philosophy of Science* 58: 264–277.

Ricketts, Thomas. 1994. Carnap's Principle of Tolerance, Empiricism, and Conventionalism. In *Reading Putnam*, ed. Peter Clark and Bob Hale, 176–200. Oxford: Blackwell.

———. 1996. Carnap: From Logical Syntax to Semantic. In *Origins of Logical Empiricism*, ed. Ronald Giere and Alan Richardson, 231–250. Minneapolis: University of Minnesota Press.

Schiemer, Georg. 2013. Carnap's Early Semantics. *Erkenntnis* 78: 487–522.

Sellars, Wilfrid. 1962. Philosophy and the Scientific Image of Man. In *Frontiers of Science and Philosophy*, ed. Robert Colodny, 35–78. Pittsburgh: University of Pittsburgh Press.

Tuboly, Adam Tamas. 2017. From 'Syntax' to 'Semantik' – Carnap's Inferentialism and Its Prospects. *Polish Journal of Philosophy* 11 (1): 57–78.

Uebel, Thomas. 2001. Carnap and Neurath in Exile – Can their Disputes be Resolved? *International Studies in the Philosophy of Science* 15: 211–220.

van Heijenoort, Jean. 1967. Logic as Calculus and Logic as Language. *Synthese* 17: 324–330.

Wilson, Mark. 2006. *Wandering Significance: An Essay on Conceptual Behavior*. Oxford: Oxford University Press.

———. 2012. The Perils of Polyanna. In *Carnap's Ideal of Explication and Naturalism*, ed. Pierre Wagner, 205–224. Basingstoke: Palgrave Macmillan.

———. 2018. *Physics Avoidance: Essays in Conceptual Strategy*. Oxford: Oxford University Press.

Chapter 13
Rejecting Semantic Truth: On the Significance of Neurath's Syntacticism

Derek Anderson

Abstract This chapter argues for the centrality of Neurath's syntacticism to his physicalism, encyclopedism, and the unity of science program. I defend the intelligibility of Neurath's opposition to semantics and underline the role syntacticism plays in his anti-metaphysical empiricism. I argue that Neurath was correct to fear that the semantic turn would spell the end of logical empiricism. Many of the most influential metaphysical arguments of the twentieth century turn on premises that would be undermined by a Neurathian rejection of semantics.

Neurath notoriously opposed Carnap's semantic turn and insisted that a semantic theory of truth would undermine the logical empiricist agenda. Many scholars, including Carnap, have regarded this opposition as either unintelligible or misguided.[1] Here I take up a defense of Neurath. I argue that his opposition to semantics was intelligible and his concerns prescient.

Section 13.1 outlines Neurath's syntacticism and its relationship with both his physicalism and his Unity of Science Movement. Here I defend the intelligibility of Neurath's opposition to semantics and the place his syntacticism occupies in his broader political and scientific projects. Section 13.2 argues that Neurath's fear of semantics was warranted. The semantic turn contributed decisively to the demise of logical empiricism and many of the most influential developments in metaphysics from the second half of the twentieth century were spurred by semantics and would have been undermined by Neurathian syntacticism.

---

[1] See Mancosu (2008) for discussion of Carnap's reaction to Neurath, as well as Tarski's and Kokoszyńska's, all of who ultimately found Neurath's resistance to semantics unfathomable. See also Uebel (2001, 2007) for a contemporary discussion of whether Neurath's syntacticism is acceptable. Other criticisms of syntacticism may be found in the contemporary literature, although their target is Carnap's (1934/1937) *Logical Syntax of Language* (LSL). Neurath's project is sufficiently close to Carnap's that we may take criticism of LSL to bear on Neurath's syntacticism, which be found in Coffa (1987), Creath (1990), Oberdan (1992), Uebel (2007), Friedman (1999), Tuboly (2017) and André Carus's chapter in the present volume.

D. Anderson (✉)
Philosophy Department, Boston University, Boston, MA, USA
e-mail: deanders@bu.edu

J. Cat, A. T. Tuboly (eds.), *Neurath Reconsidered*,
Boston Studies in the Philosophy and History of Science 336,
https://doi.org/10.1007/978-3-030-02128-3_13

13.1 Physicalist Syntacticism

Syntacticism entails a constrained approach to theorizing about language that limits the scope of investigation to syntactic entities and relations that hold between expressions in virtue of their syntactic properties. Semantic entities such as propositions, contents, and truth-values, semantic properties such as being true and being false, and semantic relations such as reference and satisfaction are systematically omitted.

Neurath's conception of syntacticism flows from his physicalism: the treatment of linguistic entities as physical particulars, empirically observable and located at particular places in space and time. The language of science is a material tool that is constructed for purposes that accord with the goals of society. For Neurath, the relevant goals involve implementing radical socialism at a global scale (Cartwright Cat, Fleck, Uebel 1996). Part of this socialism is the effort to organize our scientific and linguistic practices in a way that promotes general knowledge and welfare. Language can be fashioned into a more effective empirical tool through considered and directed manipulations. We can organize exact rules for producing new sentences on the basis of sentences we have already accepted. In this way, we can make sense of scientific theories themselves and make decisions about which theories to accept.

13.1.1 Neurath's Physicalism: First Aspect

Neurath's physicalism is a thesis about language. Differences in his formulations of this thesis indicate that there are two aspects of physicalism. The first aspect concerns the nature of language as a physical system. The second concerns the proper organization of language for use in the program of unified science. Neurath does not explicitly distinguish these two aspects, indicating that he regards them as closely connected.

In its first aspect, physicalism is the thesis that language is a physical formation that cannot be treated as different in kind from the rest of the physical world. As such, it has no special properties beyond those of other physical systems. Most notably, the language of science does not "express" anything beyond its own physical structure. The study of language is the study of physical processes. Any act of saying or writing is a spatio-temporal arrangement. This spatio-temporal arrangement exhausts what is expressed by language. Neurath writes,

> [w]hat is scientifically expressible is no richer in fundamental relations than the symbols on a Morse tape which the telegrapher reads as they are sounded by his apparatus. In a sense unified science is physics in its largest aspect, a tissue of laws expressing space-time linkages – let us call it: *Physicalism*. (Neurath 1931a/1983, 49.)

Note that the physical system of language is here identified with unified science itself for reasons that will become clear shortly.

This passage poses a small puzzle. In the first sentence, Neurath gives a statement of his physicalist syntacticism: what is scientifically expressible is only the physical structure of symbols. An expression of the language of science only "expresses" its own physical structure. Immediately after giving this ultra-deflationist picture of scientific language, Neurath says the language of unified science expresses all physical space-time linkages! The laws of physics would seem to be more than physical relations between symbols, but somehow Neurath disagrees. The puzzle is resolved when we recognize that Neurath takes laws of physics to be syntactic strings constructed in a physical medium. The laws of physics are physical items, e.g. ink patterns on a piece of paper, and they "express" connections between other physical items, e.g. other ink patterns on other pieces of paper. How these connections are implemented will be discussed later.

The unity of language as a physical system is part of what Neurath has in mind when he talks about the unification of science. "All these disciplines [geology, botany, zoology, and the science of science itself] are constructed of the same bricks, as it were" (Neurath 1931a/1983, 48). These bricks are expressions of the physical language. Neurath (1931b/1983, 52) emphasizes the same point again when he writes, "[f]or 'physicalism' it is essential that *one* kind of *order* is the foundation of all laws, whichever science is concerned, geology, chemistry, or sociology" (original emphases). The unity of order is uniformity in the physical nature of the expressions of the language of science.

Neurath's physicalism does *not* entail that science is unified in virtue of capturing the unity of reality as expressed in the subject of physics. It would be thoroughly confused to think of Neurath as holding that all scientifically investigated phenomena are grounded in the properties of microphysical systems and that this unity of reality is the metaphysical ground of the unity of science. Physicalism provides the foundation of all laws in the sense that the physicalist language is the physical medium for the expression of every law, each of which is a syntactic object.

Because language is physical, part of the world like anything else, unified science can engage with it as a subject of inquiry. Hence, language can engage with itself. In explaining the physical nature of language, Neurath (1931b/1983, 53) writes, "scientific language itself is a physical formation whose structure, as physical arrangement (ornament), can be discussed by means of the very same language without contradictions." We use syntactic language systems to study the physical order and language itself is part of that order, so language can be used to discuss itself. The physicality of language guarantees there is no methodological problem with using language to discuss language. "The statements themselves also form part of other statements as physicalist elements" (1931b/1983, 55). There is no physical restriction on constructing statements within statements, so there is no principled reason why language can't engage with itself as object (in the sense of "object" as argument position of a syntactic structure).

It follows immediately that the project of modeling language within language, pursued by Gödel (1931) and Carnap (1934/1937), is possible. Where Gödel and Carnap argued for this possibility by showing how to encode a language using its own syntax within an abstract mathematical framework, the argument that language can engage with itself because of its physical nature is unique to Neurath. Neurath's project precedes the project of syntactically modeling language within language pursued by Carnap and Gödel. Further, Neurath's physicalism provides an independent argument for the possibility of this project.

This aspect of Neurath's physicalism, viz. the way it enables language to talk about language, undercuts the theoretical role of a metaphysically robust semantic theory. Once Neurath's physicalism is accepted as a fundamental postulate, what could semantics be except more physical syntactic strings? There cannot be a domain of semantic meta-objects expressed by a semantic theory, since that theory is merely another collection of physical strings that "expresses" only linkages between syntactic strings. Semantic theorizing, if we were to engage in that kind of thing, would itself have no semantic content; it would only express syntactic relationships. So there is clearly a sense in which Neurath's physicalism renders semantics impossible.

13.1.2 Neurath's Physicalism: Second Aspect

In its second aspect, Neurath's physicalism is a thesis about the proper composition of the language of unified science, viz. that our adaptation of language for the purpose of unified science should start with ordinary common language and transform it so that incorporated expressions can be controlled by empirical evidence. In explaining this second aspect, Neurath (1931b/1983, 54) writes, "[w]hat matters is that all statements contain references to the spatio-temporal order, the order that we know from physics. Therefore this view is to be called 'physicalism.' Unified science contains only physicalist formulations."[2]

According to Neurath's physicalism, the language of unified science is not restricted to terms that belong to the technical jargon of physics. It can and must include words used to describe ordinary objects such as tables and people. These terms do not need to be completely precise. Neurath (1936b/1983, 162) says, "[f]or example, we do not use completely precise terms when we say: 'Man A formulates: in the room was a table perceived by A.' But this kind of formulation as known in everyday language is always needed where predictions are empirically checked by confronting predictions with protocol statements." What is essentially physicalist about the protocol statement in this example is that it contains no terms that cannot be placed under the control of empirical evidence concerning things that are part of the spatio-temporal order.

The notion that expressions can be placed under the control of the senses plays a central role in Neurath's physicalist criterion for curating the language of unified science. Physicalism is meant to rule out metaphysical statements by excluding

[2] Here "references" must be understood in a syntactic way, hence the spatio-temporal order itself must be comprised of syntactic items and 'reference' must be connection to such items either as parts of the statements themselves or parts of other statements reached via transformation rules.

certain expressions from the language altogether, those that are not under the control of the senses. Such expressions are to be labeled "devoid of sense:"

> Our knowledge of phenomena is controlled by sight, hearing, tasting – our sense organs. [...] A statement which cannot be controlled is a thesis devoid of sense. Those who thus succeed in formulating a system of laws which they apply in *predicting events* were best regarded as "*representatives of a scientific* conception of the universe." (Neurath 1931a/1983, 48; original emphases.)

Neurath is not interested here in providing a theory of empirical justification, but rather in specifying what it is for a system of laws to count as admissible into the system of unified science. A necessary condition for incorporating a statement into the system is that deployments of that statement be appropriately controlled by our sense organs. This is a requirement concerned with how to properly engineer a language for unified science. An acceptable statement must be connected (in some indeterminately complex way or other) to sense experience through social practices.

This is where Neurath's formulation of physicalism comes closest to what Carnap claims was the Vienna Circle's official doctrine of physicalism in his intellectual autobiography. Carnap (1963, 59) writes, "[t]he thesis of physicalism, as originally accepted in the Vienna Circle, says roughly: Every concept of the language of science can be explicitly defined in terms of observables; therefore every sentence of the language of science is translatable into a sentence concerning observable properties." However, Neurath's physicalism is not committed to this strong form of translational reductionism. Neurath's program does not entail that every theoretical term of a scientific theory must be translated into a complex of observation predicates, nor that every sentence that expresses a scientific concept can be translated into an observation sentence. Neurath only requires that expressions connect up with observation statements in some way, but this way is intentionally left unspecified.

The difference is worth emphasizing. At the time in question, Carnap (1936) maintained that every viable concept of science could be defined through a logical relation to what he called "observable predicates." An observable predicate is defined as a predicate P such that a person can come to a decision between "P(a)" or "not-P(a)" with the help of only a few observations. The logical relation Carnap proposed was given by reduction sentences. What it is for a theoretical predicate Q to be reducible to observable predicates is for Q to be uniformly replaceable by a new predicate Q* explicitly introduced by a reduction sentence of the form $(x)(O_1x \rightarrow (O_2x \rightarrow Q^*x))$ or $(x)(O_1x \rightarrow (O_2x \leftrightarrow Q^*x))$, where O_1 and O_2 are observable predicates. O_1 specifies the directly observable experimental setup. O_2 specifies the directly observable outcome of an experiment that would confirm a predication of Q. Reductions could involve chains of such reduction sentences. What this means is that, for Carnap, every significant scientific expression must admit of explicit definition via a chain of reduction sentences connecting the use of that expression to some set of observations made under conditions that can be specified using observable predicates.

Neurath's program could make use of Carnap's reduction sentences in controlling the use of predicates, but Neurath does not call for explicit reductive definitions of this or any other kind. In fact, Neurath explicitly rejects the call for a definitive criterion of significance for acceptable language. "As scientific people, we are prepared to check all our tenets by observation statements, but also – far removed from every absolutism – to alter the principles on which the checking is based" (1935/1983, 115). Neurath only insists that we engage in a general practice of critically evaluating the language of unified science, but even in this project "everything remains ambiguous and in many ways uncertain" (1935/1983, 116). Neurath's approach is to treat the criterion of significance as necessarily vague, ambiguous, indeterminate, and never absolutely and eternally acceptable in any particular form.

Our best practice for evaluating the significance of language changes as our science and social organization develops. The best practice is a function of socio-historical circumstances. There is no rational way to predict a priori how our theoretical apparatus should be connected with our observation statements as science and society develop. Neurath says this practice of seeking out new ways to clarify our terms must (for best results) produce a uniform, coordinated effort at any given time, but he says, "[i]t is not [a logical consequence of our program]; I stress this again and again; I see it as a *historical fact* in a sociological sense" (1935/1983, 115; original emphasis). The best practice must be uniformly enacted, but it is a matter of historical sociological fact which practice should be enacted at a given time and place.

Rather than insisting that some specific principle for inclusion in the physicalist language be developed, Neurath's physicalism relies on an open-ended exclusionary principle. "One can hope to remove only some coarse errors and certain coarse nonsense; much remains uncertain at first, though one cannot do without it" (1935/1983, 118). Science cannot proceed while simultaneously and instantaneously removing every term that might be suspect. Neurath maintained that this reliance on expressions riddled with uncertainty is unavoidable in science.[3] It is both epistemically and practically intractable to replace our whole language with something perfectly precise and clearly connected with observation.

Understanding Neurath's physicalist criterion of significance in this negative and open-ended way makes his approach much more feasible than the reductionist picture advocated by Carnap. Neurath's approach is much closer to the holistic assessment of scientific theories famously advocated by Quine (1951). Neurath (1935/1983, 118) says, "[t]he whole of science is basically always under discussion." Because we are not forced to evaluate the significance of scientific concepts on a term-by-term basis as Carnap required, Neurath's criterion of significance allows us to judge whole theories at once.[4] A theory can be accepted if it makes

[3] For more in-depth discussion see Cat (1995).

[4] As Adam Tuboly has pointed out to me, Carnap (1934/1937, §82) also maintains a Duhemian perspective on the confirmation of theories, writing: "*Thus the test applies, at bottom, not to a single hypothesis but to the whole system of physics as a system of hypotheses* (Duhem, Poincaré)" (original emphasis). We should take care to distinguish this holistic view of theory confirmation from Carnap's (1936) criterion of cognitive significance for theoretical terms.

good predictions, even if the contribution of individual expressions of that theory cannot be traced directly to particular observations. Yet where Quine (1951, 20) saw this holism as resulting in "a blurring of the supposed boundary between speculative metaphysics and natural science," Neurath's approach avoids this result by strictly adhering to physicalist syntacticism. The holistic, vague, ambiguous, indeterminate, and non-absolute physicalist criterion of significance is, after all, only concerned with whether certain physical strings should be incorporated into the body of unified science. These decisions have no bearing on metaphysics or ontology of any kind. We cannot always tell with certainty whether a given term is directly definable in terms of observation or controlled more indirectly by observation. Perhaps we even accidentally use a term that is completely beyond control of the senses. But these facts have no metaphysical implications.

13.1.3 Physicalism as Social Program

Neurath (1935/1983, 115) speaks of implementing a "program of unified science" and a "program of empiricism." To invoke an anachronistic metaphor, unified science can be thought of as a kind of computer program. The language of unified science – the code in which the program is written – is specified in physical syntactic terms. Expressions are identified by their physical properties. These physical properties are coordinated with syntactic types. The computations the program runs are given by transformation rules – statements within the body of unified science that specify in syntactic terms inferences that can be made. Syntactic types thus codify relations between physical expressions according to their syntactic classifications. Semantic properties such as truth or reference are fully left out of account.

The purpose of the program is to connect statements generated by the sense organs with predictions generated by laws of various sciences. These predictions are then used as an aid to the social planning of action. The code must be constructed in a physical medium since there is nothing else to construct it from. Its functioning must be established through social intervention, since the relevant operations on statements can only be carried out by groups of people within a language-using community. The community as a whole implements the program of science.

The picture of science-as-computer-program is borne in Neurath's discussion of scientific laws. Neurath (1931c/1983, 62) writes: "Unified science contains all scientific *laws*; these can be connected without exception. Laws are not statements; they are directions for obtaining predictions from observation statements (Schlick)" (original emphasis). Here we find two ideas that are connected with the computer program metaphor. Scientific laws – mere syntactic strings – function as parts of inferential chains connecting observation statements with prediction statements. Theoretical statements play an important role in the code that generates predictions on the basis of observation, but they do not stand for facts. The idea that all scientific laws "can be connected without exception" means that laws from any set of scientific domains can occur in a single inferential chain.

The conception of laws as inference rules sheds light on the earlier puzzle of how Neurath could maintain that laws conceived as mere syntactic strings express space-time linkages. When these strings are part of a system of inferences that is implemented within the social fabric of a community, they "connect" the events that are observed with events that are predicted. An event occurs. It is observed by the sense organs of some individual who, as part of the program of unified science, records the event with an observation sentence. Scientific laws are then used to infer prediction sentences, which are used in guiding interactions with future events.

Neurath's treatment of laws as derivational rules is exemplified in Carnap's (1934/1937) account of the logic of science as logical syntax.[5] According to Carnap, we introduce L-Rules (constructed from purely logical vocabulary) and P-Rules (constructed from logical vocabulary together with physical predicates) and these function together "as transformation rules of the physical language" (1934/1937, §82, p. 316). The L-Rules assign measurable magnitudes and determine valid inferences that are used to test the P-sentences that serve as primitive laws. These P-sentences "contain no constants as interior arguments" and so are treated as valid for all arguments, e.g. for all space-time points. Valid P-sentences are "tested by deducing consequences on the basis of the transformation rules of the language, until finally sentences of the form of protocol sentences are reached" (ibid. p. 317). We use the laws to derive observation statements that would confirm them. Carnap gives the example of Maxwell's equations, which are formulated as P-primitive statements. "[S]entences of protocol form can be deduced from the Maxwell equations in conjunction with the other primitive sentences of classical physics; in this way, the Maxwell theory is empirically tested" (ibid. p. 319). When a valid P-sentence is not confirmed by observation, we can delete that sentence from our theory and formulate a new one. Carnap's approach provides a clear example of the kind of syntactic treatment of laws Neurath advances. But Neurath never says Carnap's analysis of the logic of science is the best we can expect to achieve. He advocates a continuing analysis of the logic of science.

Neurath (1937/1983, 175) maintains, "[t]he process of the logical organization of a single science cannot be divorced from the process of building up bridges or connections between the different sciences." The idea that all scientific laws can be logically connected with one another reflects Neurath's (1931b/1983, 53) conviction that "under certain circumstances it must be possible to link the laws of all sciences with each other to make *one* definite prediction." Predictions about a town we plan to build must take account of the geography, the weather, potential natural hazards, economic factors, sociological factors, and so on. Neurath (1931b/1983, 54) writes, "[f]or 'physicalism' it is essential that *one* kind of *order* is the foundation of all laws, whichever science is concerned, geology, chemistry, or sociology" (original emphases). The order Neurath is speaking of is the syntactic medium of

[5] See the discussion in *The Logical Syntax of Language* (1934/1937, 315–333). Carnap explicitly acknowledges Neurath's influence in the development of his account on pp. 320–321.

scientific prediction. What is required is that the various laws can be made to *physically work together*.

Scientific laws are not the only components of the language of unified science that are used for deriving predictions from observations. Neurath also grants a central place to the implementation of symbolic logic.

> The evolution of modern logic makes possible the organization and utilization of all research as upon the foundations of mathematics; and at the same time it clarifies the application of mathematical and other calculi to concrete subject matter. We thus become clearer concerning such matters as the relation between what are called pure and physical geometry, or the logical structure of the probability calculus and its relevance to all the concrete sciences. (Neurath 1937/1983, 175.)

Here, Neurath makes oblique reference to the sorts of logical investigations Carnap (1934/1937) is engaged in. Neurath's endorsement of symbolic logic as central to the program of unified science shows the close affinity he sees between his own work and Carnap's work on logical syntax.

Making statements physically work together is a social engineering project. Statements don't interact with each other all by themselves. They only interact through the actions of groups of people. Neurath should be understood as proposing and advocating a social engineering project that would enforce a pattern of language use within a global community, a pattern that Neurath hopes would facilitate the project of unified science and ultimately provide epistemic power for the masses.

Neurath indicates in several places (1915/1973, 1916/1983, 1930/1983, 1934/1983) that he accepts a Duhemian view of science according to which observations can be equally well accounted for by indefinitely many different theories. What distinguishes theories from one another is that they are composed of different and potentially incompatible sets of laws and/or formalisms. For Neurath, even the protocol statements that serve as our evidential basis can vary from theory to theory.[6] Hence, on Neurath's picture, social coordination is required for establishing our formally specified derivational laws of nature, the rules of logical inference, and the format of the program's input, i.e. the protocol statements. These syntactic items can take a myriad of different forms. There can be no a priori consensus about what the program of unified science should or must look like.

13.1.4 Syntacticism: Comparing Statements with Statements

"An unblemished syntax is the foundation of an unblemished unified science. Language is essential for science; within language all transformations of science take place, not by confrontation of language with a 'world', a totality of 'things' whose variety language is supposed to reflect. An attempt like that would be metaphysics" (Neurath 1931b/1983, 54). The transformations of science are processes

[6] See Uebel (1996) for discussion of how Neurath's principle is a distinctive development of Duhem's thesis on this point.

that transform statements – physical structures – into other statements. "Thinking in terms of language as physical process is the starting point of all science" (Neurath 1931b/1983, 54). But these transformations are specified independently of any relation between the expressions of the language of science and the world apart from those expressions. The transformations of statements are governed by principles that are fully syntactic. They only relate statements with other statements.

The idea of a mechanized program for knowledge production helps us understand what Neurath (1931b/1983, 53) means when he says that "statements are always compared with statements, certainly not with some 'reality', nor with 'things.'" The program proceeds from observation statements to prediction statements. The inputs to the system are statements, the internal procedures are defined over statements, and the outputs are statements. To go beyond this picture and speak of a reality that corresponds to the statements of science is to pass into the realm of pseudo-rationalistic metaphysics. Syntacticism blocks this passage.

When Neurath says that statements are always compared with statements and never with reality, the phrase "compared with" bears some scrutiny. Expressions are not to be compared with reality, as a picture might be, and assessed for truth or falsehood on the grounds of such comparison. Neurath (1934/1983) qualifies this position by pointing out that statements can be compared with other physical objects in the way that physical objects can be compared with other physical objects. For example, we can say that the statement "This chair has four legs" has more words than the chair has legs – the chair has four legs, the statement has five words. We can describe the physical properties of a statement and compare those properties with the physical properties of other systems.

When the impossibility of comparing statements with reality is contrasted with the necessity of comparing statements with other statements, how should we understand the special way in which sentences can be compared with other sentences? The scientifically relevant relations between statements are given by a comparison of their physical structure as coded by syntactic properties. The syntactic properties of an expression are determined by their physical properties. They are shapes or patterns in some physical medium.

Neurath's physical syntacticism is a weapon of unified science against any metaphysics or ontology rooted in a priori reflection. In the first instance, Neurath (1931a/1983, 53) rejects the idea that statements can be assessed for truth or falsehood through comparison with reality: "Statements are compared with statements, not with 'experiences', not with a 'world' nor with anything else. All these meaningless duplications belong to a more or less refined metaphysis and are therefore to be rejected." Neurath takes semantic relations themselves to belong to metaphysics and thus thinks they should be rejected. Moreover, Neurath advocates Carnap's idea that a syntactic approach to language reveals broad swaths of philosophical discourse about metaphysics and ontology as meaningless: "Precisely for the purpose of evading such idealistic metaphysics, physicalism tries to replace pseudo-content statements (Carnap's 'content language') by statements about language conventions (Carnap's 'formal language')" (Neurath 1934/1983, 101).

Sentences that purport to state metaphysical theses are excluded from the language of unified science and replaced with metalinguistic discourse: more statements about statements. This appeal to Carnap's methodology of demonstrating the impossibility of translating sentences of metaphysics into the formal mode indicates that Neurath embraced syntacticism for the purpose of doing away with metaphysics. Hence, one of the goals of physicalism is that it should render metaphysical statements meaningless: "everything that was put forward as philosophy by scholastics, Kantians, phenomenologists, is meaningless except that part of their formulations that can be translated into scientific, that is physicalist, statements" (Neurath 1931b/1983, 57).

13.1.5 Encyclopedism

The output of the program of unified science is a collection of statements, observations and predictions, together with laws and procedures for going forward. Many of these laws and procedures will be constructed through the use of logical and mathematical techniques. They can be refined through investigation. Hence, part of the program's output may be an updated set of instructions about how the program should proceed.

Neurath is opposed to characterizing this output as a "system" of the kind that he attributes as the goal of science according to Descartes. The purpose of unified science is not "to reach an absolute point from which all particular things should somehow radiate" (Neurath 1936a/1983, 153). The output of the program of unified science is never to be regarded as absolute truth; nor is it to be regarded as complete. "*'The' system is the great scientific lie*. Not even as an anticipated goal is it a useful guiding thought" (Neurath 1935/1983, 116; original emphasis). Moreover, none of the sentences that compose the output at a given time are to be regarded as certain or as irrevocable. For these reasons (among others) Neurath advocates the choice of the term "encyclopedia" for this output.

An encyclopedia is a physical structure. It is a product of a particular society at a particular place in history. There may be more than one at a time even within a single society. These multiple encyclopedias need not coincide with one another at all, although they may need to confront one another. This picture is not very different from the actual sociology of science in which different groups of scientists with competing theories present their data and their competing explanations and try to work things out.

According to Neurath (1936a/1983, 146), scientific progress is characterized as the transformation of one encyclopedia into another. "The march of science progresses from encyclopedias to encyclopedias. It is this conception that we call *encyclopedism*" (original emphasis). We continue to improve and change our encyclopedias as we are confronted with new data, new desires, and new decisions about the directions our societies should take.

Neurath sometimes suggests that the word "true" might be used as a shorthand for indicating that a claim is included in the encyclopedia one currently accepts (although he is very wary of using the word "true" at all; see the next section). "One can completely renounce the use of these terms ['true' and 'false'], but one can also try to redefine them appropriately. It would, for example, be perfectly expedient to use the term 'true' for all statements that are 'valid' for us in the sense given above, that is, are either part of our encyclopedia or can be deduced from it" (Neurath 1936b/1983, 161).

The idea of treating truth in this hyper-deflationary way naturally raises the question: what about the possibility that encyclopedias might contain contradictions? Neurath (1936b/1983, 160) acknowledges this possibility. "In the actual encyclopedias that we use, there are often theories in contradiction with each other, which, if restricted to certain areas, however, produce good predictions." Whether by design or by accident, our best encyclopedia (collections of physical strings) may contain contradictions. But Neurath is explicitly unconcerned with this possibility on the assumption that we can design ways to limit our inferential chains to non-contradictory fragments of our encyclopedia. In effect, this means implementing practical procedures to prevent logical explosion where two theories we accept are incompatible. This is very much in the spirit of how scientists actually proceed given the incompatibility of various scientific theories in good standing.

Neurath's encyclopedism has a close affinity with Carnap's logical syntax in this regard. Sarkar (2013) argues that Carnap's (1934/1937, §17) principle of tolerance does not require a language to be consistent. That is to say, the principle of tolerance implies that there are no logical constraints whatsoever imposed on our choice of logical syntax. For all Carnap says, it is possible to adopt an inconsistent language. Sarkar regards this as an objection that needs to be responded to, and cites Beth (1963) and Gödel (1995) as raising the objection against Carnap. But for a proponent of the principle of tolerance, there really is no reason to believe that logic itself imposes sanctions against the use of inconsistent languages, especially once it is recognized that inconsistent languages can be safeguarded against explosion. In principle we may discover reasons for adopting inconsistent languages for certain purposes. This is what Neurath recommends. Moreover, the practical and empirical contexts in which our encyclopedias are developed will tend to safeguard them from containing contradictions.[7]

Yet the possibility of unwanted contradictions appearing within what we take to be a consistent fragment of our encyclopedia is unavoidable. When we find these contradictions, we fix them. The mere possibility of accepting contradictions is not a deep objection to Neurath's syntactic encyclopedism. It is a real problem that must be dealt with in practical ways.

[7] Sarkar argues this is why Carnap was not overly concerned with the objections raised by Beth and Gödel.

13.1.6 Rejecting "True"

As noted above, Neurath sometimes advocated excluding the use of the word "true" altogether from the language of unified science.

One source of uneasiness concerning the word "true," noted by Mormann (1999), is that Neurath rejects the idealization of a perfect body of scientific knowledge. This rejection is connected with his encyclopedism. Neurath writes:

> I propose that one no longer use the term 'the system of science' or any other similar terms, and that one equally avoid all expressions that sound as if they supported the absolutism of the 'system'. We should never say that certain formulas are 'unshakable', 'definitely free from contradiction', 'absolutely true', nor that they 'approximate' such a state more and more, as if this were something determined or determinable. (Neurath 1936a/1983, 145.)

The fact that there is no ultimate encyclopedia is one reason that we should not speak of "absolute truth" or, what usually comes to the same thing, "truth."

Another reason for Neurath's rejection of "true" is that we cannot be *certain* any of our statements will not be revised in the future. Neurath's argument against truth concerning certainty is tricky and has confused several philosophers, including Carnap. What does Neurath mean by "certainty?"

Neurath writes:

> When we say that one statement is more certain than another, we maintain something concerning our 'conduct' in this respect; for example that we do not intend to spend more time and effort in order to test its truth; moreover, that we do not foresee that the development of science must soon change it, in other words, what would be necessary to do in this case, we do not feel obliged to do. (Neurath 1936a/1983, 146.)

Certainty is thus construed as a psychological inclination and a physical process, but one that is governed by the program of unified science insofar as that program dictates our conduct regarding how we spend our time and effort and how we update our encyclopedias. The necessary lack of certainty is connected with the necessary incompleteness of every encyclopedia and the fact that we can have no justification for refusing absolutely to revise some part of any encyclopedia. Because we can always gather more data, we can never regard an encyclopedia as ultimate. Thus, Neurath's argument about certainty *does not* turn on an identification of the concept of truth with the concept of certainty, pace Carnap who ascribed this position to Neurath during their debate at Congrès Descartes in 1937 (see Mancosu 2008). Neurath's argument turns only on the fact that we cannot regard any physical repository of knowledge as complete. Neurath extends this point even to logico-mathematical statements, arguing that what we now regard as analytic we might soon regard as contradictory.

Another reason Neurath regards truth with suspicion is the fact that the language of unified science must, as a matter of historical necessity, be constructed using inexact terms – what Neurath calls *Ballungen*, terms from common language that have no precise use.[8] *Ballungen* cannot be given exact rules of interface with the

[8] For a detailed discussion of *Ballungen* see Cartwright Cat, Fleck, Uebel (1996).

program of unified science but are nevertheless essential to that program because they form the basis of the common language with which we must speak when doing science. To abandon these terms altogether would be to abandon the connection history has established between our senses and our observation statements. Since *Ballungen* cannot be given an exact analysis, nor can they be removed from our encyclopedia, an encyclopedia cannot be regarded as an exact or perfect reflection of reality.

The reasons presented thus far are consequences of treating knowledge as a syntactic product of the program of unified science. None of these considerations gives the deepest reason for Neurath's rejection of truth. The deepest reason follows from physicalism itself. In this connection, Neurath writes:

> The study of language can perfectly well be combined with the study of physical processes; for one always stays in the same field. In staying within the closed area of language one can express everything. Thus *statements are always compared with statements*, certainly not with some 'reality', nor with 'things' [...]. [Idealistic and realistic elements] can be completely eliminated if the transition is made to pure unified science. [...] If a statement is made, it is to be confronted with the totality of existing statements. If it agrees with them, it is joined to them; if it does not agree, it is called 'untrue' and rejected; or the existing complex of statements of science is modified so that the new statements can be incorporated [...]. *There can be no other concept of 'truth' for science.* (Neurath 1931b/1983, 53; original emphases.)

It is because language is to be treated as a physical medium, and because the use of language is to be specified with regard to its physical syntax only, that the notion of semantic truth should be to be rejected by proponents of Neurath's program of unified science. A strict syntacticism simply does not admit of a substantive semantic theory of truth.

Neurath allows that "truth" could function as a tool in inference, as part of the inner workings of the program of unified science, i.e. as a piece of syntax. As mentioned above, it could function as everyday shorthand for indicating which sentences belong to or follow from the accepted encyclopedia. Neurath also accepts that a Tarski-type truth definition can be given as part of the purely formal fragment of the language of unified science. He does not object to Tarski's (1936/1956) theory of truth for formalized languages *qua* piece of mathematics. Neurath only objects to applying Tarski's definition in a way that would make statements of science count as true or false as a consequence of how things are in reality (see Mancosu 2008). Treating "true" as part of an acceptable logical formalism is consistent with treating logical formalism as a purely physicalistic component of the mechanized production of knowledge.

13.2 Neurath's Prophecy Fulfilled

Neurath warned that engaging in semantic analyses of language would lead to a new era of metaphysics. And indeed many of the most influential metaphysical ideas of the twentieth century were articulated and defended through reasoning about semantics. What follows are three case studies in metaphysics: Carnap's linguistic frameworks, Armstrong's theory of truth-makers, and (early) Putnam's knowledge-transcendent correspondence relation. In each case, special attention is paid to the role semantics plays in establishing metaphysical conclusions. Each argument is then checked from the point of view of Neurath's physicalist syntacticism to show that semantic assumptions are necessary to drive the metaphysics.

Carnap (1950) argues that questions of ontology are trivial when raised inside a linguistic framework and cannot be coherently raised at all when no linguistic framework is assumed. But from the perspective of Neurathian physicalist syntacticism, Carnap has already taken the plunge into deep metaphysical waters by treating linguistic frameworks as semantically interpreted.

A linguistic framework is constructed by introducing a system of rules for a language. This procedure is not very different from constructing a logical syntax except that a linguistic framework includes a semantic interpretation. When one introduces a linguistic framework, the expressions of the framework are assigned designations: sentences designate propositions, predicates designate properties, and names designate things.

Carnap's discussion of linguistic frameworks provides an explicit characterization of the link from semantics to ontology. He walks us through an example. Start by accepting the linguistic framework of numbers. Within this framework, the statement "Five is a number" is analytic. From the fact that "Five is a number" is analytic, it follows that "There is some n such that n is a number" is also analytic.

Neurath would be perfectly happy to countenance rules of transformation of this kind if they were to be construed syntactically. We could include the string "There is some n such that n is a number" in the encyclopedia by inferring it from "Five is a number." Construed as parts of the program of unified science, these statements express merely their own physical structure. In syntax there is no ontology, no domain of facts beyond the language itself.

Carnap, however, has taken his semantic turn. Now, analytic sentences are *true*. So from the fact that "There is some n such that n is a number" is analytic, it follows that it is *true* that there are numbers. More generally, accepting a framework means regarding the analytic sentences of that framework as true. It is not surprising that questions of existence can be settled by adopting linguistic frameworks. One can adopt a framework that asserts the truth or falsity of the answers to those questions.

Carnap presents this picture as if it were deflationary. He says, "[n]obody who meant the question 'are there numbers?' in the internal sense would either assert or even seriously consider a negative answer." But that is because, in accepting the framework of numbers, they have accepted that "Numbers exist" is true and thus accepted that there are numbers. Carnap observes that once we accept the number framework, and thereby accept that there are numbers, the question of whether there are *really* numbers no longer makes sense. Carnap thinks that this conclusion indicates that questions of reality are pseudo-questions.

But this is not how the syntacticist sees the situation. From Neurath's point of view, once we have accepted Carnap's semantic framework and endorsed the idea that there are true statements of e.g. arithmetic, we have accepted something more than the existence of linguistic forms identified with physical properties. The person who accepts the framework of numbers thinks it is true that numbers exist in addition to the syntactic strings with which truths about numbers are expressed. The fact that we cannot identify a meaningful question of the reality of numbers that is separate from the question of whether to adopt a linguistic framework does not show that metaphysics has been avoided, only that adopting a framework fully determines our answers to metaphysical questions. This is because accepting linguistic frameworks requires accepting truths about non-linguistic reality. Despite his intentions to the contrary, Carnap demonstrates the bridge from semantics to metaphysics.

Let us consider a likely objection to this interpretation. A defender of Carnap in the spirit of Creath (1990) might maintain that analytic truth and its attendant ontology is not a significant commitment to metaphysics because *anything* can be regarded as a logical truth. Carnap's position in "Empiricism Semantics and Ontology" is the result of combining the principle of tolerance with a semantic theory of truth. The result is the view that any sentence can be treated as a truth. Doesn't this render the notion of logical truth harmless by making it trivial?

From Neurath's perspective, the question of harm is not to be assessed in terms of whether the notion of truth is trivial in this sense, but whether accepting this account of truth gives rise to metaphysical speculation, debate, and conviction. In these respects, Carnap's view constitutes harmful metaphysics. Consider some of the questions it raises. If you and I adopt different frameworks, then what is logically true for me is not logically true for you. Is logical truth therefore relative to a person at a time? I can adopt a framework according to which it is logically true that there are no electrons. Is the existence of electrons then to be decided by my choice of framework? Is the existence of things always relativized to people or is there an absolute existence in addition? Carnap mentions that we might refrain from speaking and thereby reject all linguistic frameworks. Does this mean the existence of numbers requires speech? Even metaphysical questions that Carnap explicitly aims to rule out are still open for discussion: Are numbers real, given their existence depends on the adoption of a linguistic framework? These are metaphysical questions that are legitimated by Carnap's endorsement of a semantic notion of truth, questions that would be rendered meaningless by Neurath's syntacticism.

Armstrong (1993, 1997) gives a semantical argument for a theory of truth makers that purports to be independent of linguistic frameworks. According to Armstrong a state of affairs exists if and only if a *particular* has a property, or if a relation holds between two or more particulars. States of affairs are constituted by the particulars, properties, and relations that are sufficient for their existence. Armstrong takes properties and relations to be universals. Neurath would see these as metaphysical statements par excellence.

The semantical argument for these posits follows closely on the tail of the assumption that natural language involves a semantic concept of truth. Armstrong (1993, 430) writes, "[w]hy is anything more needed than particulars and universals, monadic and polyadic? The answer to this comes from one of the fundamental assumptions that drive this ontology. It is the need for truths to have a truthmaker

[…] or an ontological ground […]." The idea is that the mere existence of e.g. a property F and a particular *a* is not sufficient to guarantee the truth of "*a* is F." If the sentence is true, then (according to Armstrong) some truthmaker is required, something that brings these elements together. "The state of affairs of *a's being F* is suggested as that truthmaker, as the ontological ground" (Armstrong 1993, 430). Hence the acceptance of a semantic notion of truth engenders discussion of states of affairs, a type of entity posited as the ground of truth, as well as speculation concerning the reality of properties, relations, and particulars as the things that constitute these states of affairs.

The notion of a state of affairs brings with it a host of metaphysical puzzles. What is the truthmaker for the sentence "Grass is not blue"? Does the negation sign designate a property of negation? Are there negative states of affairs? Are properties also particulars? If not then how can we name them? Armstrong's picture also gives rise to the challenge of Bradley's regress. He raises the question himself (1993, 432): "Is not bringing the constituents of a state of affairs, the particulars, the properties and the relations, together into states of affairs, a further relation in which all the constituents stand? But then the new relation is just a further element which requires to be integrated along with the other constituents." If being part of a state of affairs is itself a state of affairs, then each state of affairs requires infinitely many more states of affairs to hold it together.

Neurath's syntacticism undermines Armstrong's discussion of states of affairs in a most straightforward way. Since sentences do not have properties such as truth and falsehood, there is no need to explain the truth of "a is F" by positing some ontological structure in which the truth of the sentence is grounded. More generally, the metaphysics of truthmakers and truthmaking is rendered completely idle by the syntactic perspective.

Putnam (1975) initiates another major metaphysical program of the twentieth century through his much discussed Twin Earth thought experiment. The proper semantic insight to be drawn according to Putnam is that on Twin Earth the word "water" designates XYZ, not water. Yet the people on Twin Earth are physically and functionally the same as Earthlings. The conclusion is semantic content does not supervene on the internal physical nature or structure of an organism, but rather on relations between the organism and its environment. Concordantly, the truth of sentences is determined in part by such relations. Hence, Putnam conjures the idea of semantic relations of reference and truth that hold between organisms and their environments.

The metaphysical fallout of this semantic argument is immense. Our understanding of scientific truth must take into account not just the facts that we investigate but also the relationships between the organism and its environment that constitute the semantic content side of the correspondence between our theories and reality. These representation relationships can hold without our knowledge. Our "operational definitions" by which we control the use of our terms can turn out to be false given what our words actually refer to in virtue of this external reference-fixing relation. This provides for the possibility that even our best possible scientific theories may be false in ultimate reality. It also provides the possibility that some description available in our language or some more sophisticated language might be absolutely true, in the sense that descriptions made available by the ideal language would capture reality exactly. The notion that facts about reality beyond the grasp of even our best theories might determine the absolute truth or falsity of any given encyclopedia is perhaps the most extreme antithesis of Neurathian physicalism.

The syntacticist would reject Putnam's earliest premise. For the syntacticist there is no sense to be made of the claim that "water" refers to the substance water. The only significant relations that "water" enters into are relations with other syntactic objects. Nothing about the nature of reality or the nature of scientific inquiry could be concluded from analyses of these relations.

I have presented three examples, but there are many ways in which the semantic turn has given rise to contemporary metaphysics. This study could be expanded at great length. Further important examples of metaphysical conclusions being drawn from semantic assumptions include Davidson's (1967) project of giving a theory of meaning in terms of Tarski's theory of truth, Kripke's (1980) theory of metaphysical necessity, Burge's (1979) theory of social externalism for mental contents, the wide range of 'naturalistic' explanations of intentionality developed by Millikan (1984), Fodor (1987), Dretske (1988), and many others.

13.3 Conclusion

I conclude that Neurath was right about the danger semantics posed to logical empiricism. It was an empirical question whether Carnap was right to think that a semantic theory of truth could be added harmlessly to Neurath's scientific world-view. The evidence strongly suggests Carnap was wrong. A physical syntacticism of the kind advocated by Neurath would have provided a strong counter-position to the metaphysical discourse of the twentieth century. Perhaps that is why Neurath's opposition to a semantic theory of truth was so vehemently rejected. Whatever the historical reasons for its rejection, Neurath's syntacticism presents an intriguing and underdeveloped philosophical program with the potential to generate important critical resources for anti-metaphysical theorizing.

References

Armstrong, David M. 1993. A World of States of Affairs. *Philosophical Perspectives* 7 (3): 429–440.

———. 1997. *A World of States of Affairs*. Cambridge: Cambridge University Press.

Beth, Evert W. 1963. Carnap's Views on the Advantages of Constructed Systems Over Natural Languages in the Philosophy of Science. In *The Philosophy of Rudolf Carnap*, ed. Paul A. Schilpp, 469–502. La Salle: Open Court.

Burge, Tyler. 1979. Individualism and the Mental. *Midwest Studies in Philosophy* 4 (1): 73–122.

Carnap, Rudolf. 1934/1937. *The Logical Syntax of Language*. London: K. Paul, Trench, Trubner & Co., Ltd..

———. 1936. Testability and Meaning. *Philosophy of Science* 3 (4): 419–471.

———. 1950. Empiricism, Semantics, and Ontology. *Revue Internationale de Philosophie* 4 (2): 20–40.

———. 1963. Replies and systematic expositions. In *The Philosophy of Rudolf Carnap*, ed. Paul A. Schilpp, 859–1013. La Salle: Open Court.

Cartwright, Nancy, Jordi Cat, Lola Fleck, and Thomas Uebel. 1996. *Otto Neurath: Philosophy Between Science and Politics*. Cambridge: Cambridge University Press.

Cat, Jordi. 1995. The Popper-Neurath Debate and Neurath's Attack on Scientific Method. *Studies in History and Philosophy of Science Part A* 26 (2): 219–250.

Coffa, Alberto. 1987. Carnap, Tarski and the Search for Truth. *Noûs* 21 (4): 547–572.

Creath, Richard. 1990. The Unimportance of Semantics. *PSA: Proceedings of the Biennial Meeting of the Philosophy of Science Association* 1990: 405–416.

Davidson, Donald. 1967. Truth and Meaning. *Synthese* 17 (1): 304–323.

Dretske, Fred. 1988. *Explaining Behavior: Reasons in a World of Causes*. Cambridge, MA: MIT Press.

Fodor, Jerry A. 1987. *Psychosemantics: The Problem of Meaning in the Philosophy of Mind*. Cambridge, MA: MIT Press.

Friedman, Michael. 1999. *Reconsidering Logical Positivism*. Cambridge: Cambridge University Press.

Gödel, Kurt. 1931. Über formal unentscheidbare Sätze der Principia Mathematica und verwandter Systeme I. *Monatshefte für mathematik und physik* 38 (1): 173–198.

———. 1995. Is Mathematics the Syntax of Language? In *Kurt Gödel: Collected Works. Vol. 3. Unpublished Essays and Lectures*, ed. Solomon Feferman et al., 334–362. Oxford: Oxford University Press.

Kripke, Saul A. 1980. *Naming and Necessity*. Cambridge, MA: Harvard University Press.

Mancosu, Paolo. 2008. Tarski, Neurath, and Kokoszyńska on the Semantic Conception of Truth. In *New Essays on Tarski and Philosophy*, ed. Douglas Patterson, 192–224. Oxford: Oxford University Press.

Millikan, Ruth G. 1984. *Language, Thought and Other Biological Categories*. Cambridge, MA: MIT Press.

Mormann, Thomas. 1999. Neurath's opposition to Tarskian semantics. In *Alfred Tarski and the Vienna Circle*, ed. Jan Woleński and Eckehart Köhler, 165–178. Dordrecht: Springer.

Neurath, Otto. 1916/1983. On the Classification of Systems of Hypotheses (with Special Reference to Optics). In *Otto Neurath: Philosophical Papers 1913–1946*, ed. Robert S. Cohen and Marie Neurath, 13–31. Dordrecht: D. Reidel.

———. 1930/1983. Ways of the Scientific World Conception. In *Otto Neurath: Philosophical Papers 1913–1946*, ed. Robert S. Cohen and Marie Neurath, 32–47. Dordrecht: D. Reidel.

———. 1931a/1983. Physicalism: The Philosophy of the Viennese Circle. In *Otto Neurath: Philosophical Papers 1913–1946*, ed. Robert S. Cohen and Marie Neurath, 48–51. Dordrecht: D. Reidel.

———. 1931b/1983. Physicalism. In *Otto Neurath: Philosophical Papers 1913–1946*, ed. Robert S. Cohen and Marie Neurath, 52–57. Dordrecht: D. Reidel.

———. 1931c/1983. Sociology in the Framework of Physicalism. In *Otto Neurath: Philosophical Papers 1913–1946*, ed. Robert S. Cohen and Marie Neurath, 58–91. Dordrecht: D. Reidel.

———. 1934/1983. Radical Physicalism and the 'Real World'. In *Otto Neurath: Philosophical Papers 1913–1946*, ed. Robert S. Cohen and Marie Neurath, 100–114. Dordrecht: D. Reidel.

———. 1935/1983. The Unity of Science as a Task. In *Otto Neurath: Philosophical Papers 1913–1946*, ed. Robert S. Cohen and Marie Neurath, 115–120. Dordrecht: D. Reidel.

———. 1936a/1983. Encyclopedia as 'Model'. In *Otto Neurath: Philosophical Papers 1913–1946*, ed. Robert S. Cohen and Marie Neurath, 145–158. Dordrecht: D. Reidel.

———. 1936b/1983. Physicalism and the Investigation of Knowledge. In *Otto Neurath: Philosophical papers 1913–1946*, ed. Robert S. Cohen and Marie Neurath, 159–171. Dordrecht: D. Reidel.
———. 1937/1983. Unified Science and Its Encyclopedia. In *Otto Neurath: Philosophical Papers 1913–1946*, ed. Robert S. Cohen and Marie Neurath, 172–182. Dordrecht: D. Reidel.
Oberdan, Thomas. 1992. The Concept of Truth in Carnap's Logical Syntax of Language. *Synthese* 93 (1-2): 239–260.
Putnam, Hilary. 1975. The Meaning of 'Meaning.' In *Minnesota Studies in the Philosophy of Science*, vol. 7, 131–193.
Quine, Willard V.O. 1951. Two Dogmas of Empiricism. *Philosophical Review* 60 (1): 20–43.
Sarkar, Sahotra. 2013. Carnap and the Compulsions of Interpretation: Reining in the Liberalization of Empiricism. *European Journal for Philosophy of Science* 3 (3): 353–372.
Tarski, Alfred 1936/1956. The Concept of Truth in Formalized Languages. In *Logic, Semantics, Metamathematics: Papers from 1923 to 1938*. Trans. J. H. Woodger, 152–278. Oxford: Clarendon Press.
Tuboly, Adam Tamas. 2017. From 'Syntax' to 'Semantik' – Carnap's Inferentialism and Its Prospects. *Polish Journal of Philosophy* 11 (1): 57–78.
Uebel, Thomas. 1996. On Neurath's boat. In Cartwright, Cat, Fleck, and Uebel (1996): 89–166.
———. 2001. Carnap and Neurath in Exile: Can Their Disputes Be Resolved? *International Studies in the Philosophy of Science* 15 (2): 211–220.
———. 2007. *Empiricism at the Crossroads: The Vienna Circle's Protocol-Sentence Debate Revisited*. Chicago: Open Court.

Part IV
Context and Influence

Chapter 14
What a Difference a Decade Makes: The Planning Debates and the Fate of the Unity of Science Movement

George A. Reisch

Abstract This paper examines selected writings of the American science writer Waldemar Kaempffert, Science Editor for the *New York Times*, in public support of Otto Neurath, his Isotype projects, and his Unity of Science Movement. Attention is focused first on Kaempffert's writings in the 1930s, when some intellectuals, the American public, and their elected leaders were relatively sympathetic with Neurath's quest to unify the sciences in ways that would advance and direct scientific research toward practical goals. Attention then turns to the 1940s to examine the debate over the nature, scope, and limits of wartime research and Vannevar Bush's call for a national institution to support research. Against Bush, James Bryant Conant, and others, Kaempffert argued vigorously for a foundation that would adopt values and methods of the Unity of Science Movement, but he lost that argument as the National Science Foundation finally took shape. To suggest that this public debate influenced not only the decline of Neurath's Unity of Science Movement but the scholarly development of history and philosophy of science after the war, the paper considers early writings and events in the life of Conant's protégé Thomas Kuhn, whose *Structure of Scientific Revolutions* helped shape that development.

Research on the history of logical empiricism has shown the large extent to which it was originally rooted in cultural and social concerns and ambitions that fell out of favor in professional philosophy during and after the 1950s.[1] One largely unexplored marker of logical empiricism's vitality in the 1930s and early 40s is the coverage received by Otto Neurath in the popular press. Thanks to Waldemar Kaempffert, the Science Editor for the *New York Times* who happened to be

[1] See for example, Cartwright, Cat, Fleck, Uebel (1996), Stadler (2001 /2015), Howard (2003), Reisch (1994, 2005), Wuest (2015). On Neurath's early political concerns see Don Howard's and Günther Sandner's chapters in the present volume. On the rise of analytic philosophy in the United States and its relation to political currents, see Capps (2003) and McCumber (2001, 2016).

G. A. Reisch (✉)
Chicago, IL, USA
e-mail: reischg@georgereisch.com

J. Cat, A. T. Tuboly (eds.), *Neurath Reconsidered*,
Boston Studies in the Philosophy and History of Science 336,
https://doi.org/10.1007/978-3-030-02128-3_14

Neurath's cousin, goings-on within Neurath's museum and *Isotype* work, his Unity of Science Movement, and to some extent the Vienna Circle itself were effectively promoted to the American public and its elected leaders. To this extent, the dream Neurath once shared with Rudolf Carnap and Hans Hahn – to elevate and invigorate modern life with the insights, intellectual rigor, and enlightenment values of their *Wissenschaftliche Weltaufassung* – was not so dreamlike. Kaempffert's coverage of Neurath's various projects was frequent, enthusiastic, and well positioned (given the stature of *The New York Times*) to make good on the manifesto's claim that "endeavors toward a new organization of social and economic relations, toward the unification of mankind, toward a reform of school and education, all show an inner link with the scientific world-conception." Unlike those advocates who would "lead a withdrawn existence on the icy slopes of logic," Kaempffert was "a fighter" (Neurath et al. 1929/1973, 305, 317) who enlisted Neurath's projects in a crusade to teach Americans about science and to shape federal policy toward research and its financial support.

Here I will survey Kaempffert's coverage of Neurath and his projects against the backdrop of the larger international movement in the 1930s for understanding science as a social phenomenon whose research, discoveries, and applications, like other institutions – including schools, factories, economies, and popular cultures – can be consciously organized and planned for the collective benefit of modern society. By the early 1940s, we shall see, the debate between those who believed modern science could be profitably planned and their critics who believed that planning would harm modern science became fractured and complicated by the sensational wartime achievement of the atom bomb. On the one hand, the bomb's rapid and successful development seemed to demonstrate the powers and benefits of planning and organizing science and to validate the claims of figures like J.D. Bernal in England, Otto Neurath in Vienna and later England, and William Malisoff in the United States. On the other hand, critics such as James Bryant Conant, Warren Weaver, and Michael Polanyi believed that the war effort demonstrated at most the rapid and effective *application* of then-recent discoveries in "pure science." In the nature of scientific research, they insisted, pure research cannot be organized or planned, and attempts to do so will at least slow scientific progress, if not altogether prohibit the pursuit of novelties and serendipitous circumstances that sometimes lead to momentous discoveries.

Pressing political concerns cut across and complicated this debate. As the cold war took shape and Truman's United States and Stalin's Soviet Union – former allies in the fight against Germany and Japan – became mistrusting and anxious about each other's intentions, critics of planning positioned their objections within this geopolitical framework. Not only did planning not work (as demonstrated by Lysenkoism's harm to biology in Russia), they believed, efforts to plan science in the west would amount to a cultural concession in the cold war struggle, for under planning western science would become authoritarian and Stalinistic. For the most part, those who promoted planning in science were no fans of Stalin or the Kremlin. But, as the arc and evolution of Kaempffert's advocacy shows, these political anxieties deeply shaped national debate over science and the possible forms of national science policy that could be imagined – either free, creative, and progressive or, on the other hand, controlled, repressed, and unproductive. Kaempffert tried to articulate and defend a vision of nationally funded science that was neither "laissez-faire" and chaotic nor authoritarian and intellectually repressive. But his voice became marginalized as American political and intellectual leaders fell into line to vigorously oppose any form of planning and organization.

Finally, I will describe two aspects of a young Thomas Kuhn's relationship to the planning debates. One, his brief collaboration with Philipp Frank in the early 1950s when Frank struggled to keep Neurath's Unity of Science Movement strong and vital for science studies; and, two, Kuhn's Lowell Lectures of 1951, during which Kuhn evidently feared that he was being thrust into these contentious, public debates over planning and science's place in postwar society. Kuhn chose to avoid these debates, denied that his research had meaningful bearing on them, and crafted a theoretical approach to science's history – fully articulated in his *The Structure of Scientific Revolutions* published a decade later – that denied prospects for planning science. To the considerable extent that Kuhn's *Structure* itself became a paradigm for postwar science studies, this epilogue suggests one way in which the decline and fate of Neurath's unity of science project is linked to the programmatic foundations and assumptions of history and philosophy of science as they are practiced today.

14.1 Waldemar Kaempffert and the Unity of Science Movement

Waldemar Kaempffert, born 1877, was an American cousin of Otto Neurath, born 1882. They were related through Neurath's mother, Gertrud Kaempffert.[2] Waldemar was born in New York City, attended the City College of New York, earned a law degree, and began writing about science for popular audiences in magazines and books. In 1916, he became an editor of *Popular Science Monthly* and by 1922 had moved to the *New York Times*. After a brief, three-year relocation to Chicago where Kaempffert became the Director of the Museum of Science and Industry, he returned to New York and the *New York Times* in 1931, where he remained Science Editor until his retirement in 1956.[3]

Kaempffert shared Neurath's enthusiasm about science and his civic responsibility to help ordinary citizens understand as much as possible about science's roles – actual and potential – in modern life. While Kaempffert's efforts to promote scientific understanding were rooted neither in the fundamental epistemological concerns of the Vienna Circle nor the technical, scientific experience of British Marxists, he endorsed the view that advances in science were transforming the mod-

[2] According to Neurath's son, Paul, Waldemar and Otto were cousins, which suggests that Waldemar's father, Bernhard, in New York and Neurath's mother Gertrud in Austria were siblings or themselves cousins (Fleck et al. 2005, 285).

[3] This and other information about Kaempffert can be found in his obituary (*New York Times* 1956).

ern world for the better. In the *Book of Modern Marvels*, for example, assembled from articles written by others for *Popular Science Monthly*, Kaempffert (1917, 6) noted that developments in knowledge ranging from scientific research to criminology and warfare "make this the most dramatically interesting period in the whole history of the world."

The politics of this enthusiasm, however, presented a problem that would color much of Kaempffert's coverage of Neurath and his Unity of Science Movement in the coming decades. In broad terms, the movement to promote planning and organization in science can be understood to have swept westward in the wake of the Russian Revolution. In 1931, when Boris Hessen and Nikolai Bukharin brought Marxist sociology of science to the meeting in London of the International Congress of the History of Science and Technology, they converted a generation of British intellectuals, including Bernal, Lancelot Hogben, Joseph Needham, and J.G. Crowther. These and other like-minded intellectuals came to believe it was no coincidence that the new Soviet Union was growing by leaps and bounds (famously under Stalin's first and subsequent "five year plans") when western nations which had not learned to organize and plan scientific research remained mired in the great depression (Werskey 1978, 142–3, 2007). Neurath and his unity of science project, in turn, brought this sensibility to the United States manifest in the new scientific philosophy of logical empiricism, his visual language Isotype, designed to educate the masses to clearly understand social and economic information, and his *International Encyclopedia of Unified Science*, designed to facilitate interdisciplinary cooperation among different sciences (Reisch 1994).

Kaempffert supported his cousin's efforts faithfully and energetically. On at least 18 occasions in the 1930s, these projects appeared in notices, articles, features and reviews in the *New York Times*, usually written by Kaempffert but sometimes by other writers. In all cases, the coverage was sympathetic and championed the goals, methods, and democratic (if not socialist) ideals embedded in these projects.[4] (See Fig. 14.1.). Given the ravages of the depression, many of Kaempffert's readers in the United States were already sympathetic to socialism and scientific and economic planning as likely solutions to the urgent economic problems at hand. Kaempffert did not hide these socialistic sensibilities. In his piece "Staccato Speech for Silent Statistics" of 1933 (Kaempffert 1933), and most likely with knowledge of Neurath's own work for the Soviet government as an expert in visual statistics (*Survey Graphic* 1932), he pointed out that Isotype "is now definitely established by decree in the Soviet Union for the pictographic presentation of statistics in schools and in posters that are to instruct the masses" (the same point is made in Modley 1935). (See Fig. 14.2.). Again in 1937 he pointed out that Isotype overcomes one of the obstacles of "written and spoken language" which "varies with the class to which it is addressed." For "it is one purpose of Isotypes to bridge the intellectual gap between the college professor and the masses to whom he wishes to appeal" (Kaempffert 1937b).

[4] See e.g. Kaempffert (1933, 1935, 1936, 1937a, b, c, 1938), *New York Times* (1933, 1937a, b, 1939a, b, c, d), Barnard (1933), Modley (1935), Duffus (1939), and Thompson (1939).

2 THE NEW YORK TIMES BOOK REVIEW, *August 7, 1938.*

Toward Bridging the Gaps Between the Sciences

An Encyclopedia Designed to Correlate the Relationships Of Many Diverse Groups

INTERNATIONAL ENCYCLOPEDIA OF UNIFIED SCIENCE. Editor-in-Chief, Otto Neurath. Associate Editors, Rudolf Carnap, Charles W. Morris. Vol. 1, No. 1: Encyclopedia and Unified Science. By Otto Neurath, Niels Bohr, Bertrand Russell, Rudolf Carnap, Charles W. Morris. 75 pp. $1. Vol. 1, No. 2: Foundations of the Theory of Signs. By Charles W. Morris. 59 pp. $1. Chicago: University of Chicago Press.

By WALDEMAR KAEMPFFERT

MANY years ago Dr. Otto Neurath, a social scientist of international reputation, organized with Dr. Rudolf Carnap, now of the University of Chicago; the late Dr. Moritz Schlick; Dr. Philipp Frank of the University of Prague and others equally distinguished a group which soon made itself known all over the world as "the Viennese Circle." They were a hard-headed lot, with no more use for metaphysics than so many Irish foremen of railway section gangs. Strongly influenced by Ernst Mach and Bertrand Russell, they decided that only workaday scientific experience mattered. All else was so much verbal mocnshine. Out of the weekly discussions of the Viennese Circle came the modern school of logical empiricism. That the movement should win recruits in the United States was natural enough. At Harvard Peirce had been preaching what he called "commonsenseism" and William James "pragmatism" long before the Viennese Circle came into existence.

As practical thinkers these logical empiricists decided that they must act as well as talk and write. They agreed that the problem of language and logic should be attacked. Here, for example, were the physicists telling the world that within the atom an electron is "free" to choose certain orbits; and here poets, politicians and business men talking about "free will." Was the electron's "freedom" the same as that ascribed to the will? And then there were biologists discussing "stimulus" and "reaction." Did they mean "cause" and "effect"? If so, a special terminology of their own seemed unnecessary.

Even in the same science this vagueness was found to prevail. Psychiatrists described the case of a man who fell ill because he had to visit a relative whom he despised. The psychologists who believed in conditioned reflexes taught that a dog became ill when it was experimentally bewildered by a sound which had once meant that food was about tc be served but now meant nothing. The two situations were both the consequences of similar psychic disturbances, but the verbal descriptions were different. So with "Gestalt" psychologists and behaviorists. They could not or would not use the same language to describe the process whereby an animal seeks food. Sometimes more than language was involved. Problems were formulated inexactly, methods of investigation were defective.

Neurath and his Viennese associates saw the need of unifying the sciences. The need became more and more apparent as language was analyzed. It was far from the mind of the group to evolve a super-science to legislate for all the known disciplines. The object was integration, and integration was to be achieved by building bridges from one science to another. Thus the movement for the unification of science was inaugurated.

To achieve unification an international encyclopedia was proposed, discussed and endorsed at various international congresses. An advisory committee was appointed, including such celebrities as Niels Bohr, John Dewey, J. Lukasiewicz, Bertrand Russell, J. H. Woodger. And now come the first two volumes edited by Neurath and Professors Rudolf Carnap and Charles W. Morris of the University of Chicago.

If the innocent man in the street thinks that he has only to turn to the letter "R" and then to "Roosevelt" to discover the important facts about the great Theodore of that name, he is doomed to be disappointed. To be sure, the twenty monographs to be published will have a highly analytical index in which it is remotely possible that Theodore Roosevelt's name may be printed with a reference to particular volumes and pages. But it is improbable that the dates of Theodore Roosevelt's birth and death will be discovered or the part that he played in splitting the Republican party. On the other hand, there should be much about the place of Einstein in the history of science, though very little about his personal life.

This is a different encyclopedia. Its purpose, as we have already indicated, is not to supply information — though it must of necessity do that — but to act as a pilot to scientists. In the past they have been independent navigators who paid little attention to whistles, flags, semaphores, wireless or other means of intercommunication. The encyclopedia is intended to tell each navigator what other ships are doing and what he can learn from their signals, their movements, their errands. A heterogeneous collection of vessels on a common ocean, but each going its own way, is to be converted into a homogeneous fleet. There will be no admiral to give orders. Only this encyclopedia is to serve as chart and compass. To it all the navigators will contribute, and all will revise it from time to time to make it more useful to themselves and the educated public.

Dr. Otto Neurath.
(New York Times Studio.)

What we have, then, is an extension of that cooperation which has always been the glory of science. Physicists now make common cause with other physicists, regardless of race, nationality or religion. And so with biologists and chemists. The next step is to bring physicists, chemists and biologists together, a natural evolution.

With philosophy any such unification is impossible. From time immemorial philosophers have been engaged chiefly in pointing oul defects in the "systems" devised by their opponents and exploiting the beauties of their own. But science? Cooperation is its life blood. By broadening this cooperation unrealized possibilities will be revealed.

Neurath has little use for "system," as the philosophers use the word:

> Encyclopedism may be regarded as a special attitude; one may also speak of encyclopedism as a program. An encyclopedia and not a system is the genuine model of the scientific knowledge of science as a whole. An encyclopedic integration of scientific statements, with all the discrepancies and difficulties which appear, is the maximum of integration which we can achieve. It is useful to avoid dogmatism and bumptiousness in scientism and empirical panlogism. One can love exactness and nevertheless consciously tolerate a certain amount of vagueness.

Neurath thus describes the structure of the work which is to be the instrument of cooperation:

> The encyclopedia is to be constructed like an onion. The heart of this onion is formed by twenty pamphlets which constitute two introductory volumes. These volumes, entitled Foundation of the Unity of Science, are complete in themselves but also serve as the introduction to what will follow.

It is difficult to judge from the two introductory volumes how effectively this plan will be carried out. But they are promising. They whet the appetite for more.

Dr. Neurath opens the first volume with a historical résumé of past encyclopedias and shows the need for this one in a discussion of "Unified Science as Encyclopedic Integration." Niels Bohr follows with a mere scrap on "Analysis and Synthesis of Science" which would bear elabo-

(Continued on Page 18)

The Pragmatic Approach to Aesthetics

AESTHETIC QUALITY: A Contextualistic Theory of Beauty. By Stephen C. Pepper. 247 pp. New York: Charles Scribner's Sons. $2.

SINCE the days of Plato and Aristotle men's minds have sought to unravel the knotty problem of aethetic—"beauty." An enormous amount of writings has poured from the presses—some tackling the problem directly, that is, working from the particular object of art to an elaboration of aesthetic theory; others (like Kant and Hegel) reversed the process by seeking the specific value of a given work of art, indirectly, in terms of abstract metaphysical schemes of values; and still others, more recent theorists (like Croce, James and Bergson), have utilized both methods in dealing with this question.

Yet, notwithstanding the vast amount of thought and energy expended in tracking down this elusive quality of beauty within some sort of generally recognizable form, we are no nearer to a universal agreement as to what beauty, or art, is.

And that is all to the good: for beauty, like life—of which, indeed, it is a manifestation—obviously must undergo constant renewal of itself, to be, and to satisfy man's needs. In short, like life, beauty has its own raison d'être for being and becoming.

The aim of Professor Pepper in writing this book on the nature of aesthetics is an attempt, within the covers of a brief volume, "to clarify the implications for aesthetics of the pragmatic or contextualistic movement." In other words, it is an attempt to explain the nature of "beauty," manifested in life or expressed by or through the medium of "concrete work of art, in terms of pragmatic experience and values."

What, then, is the "nature of aesthetic experience," in terms of the "pragmatic or contextualistic" scheme of values or theory?

According to the author's findings, grounded as they are on personal aesthetic experience and checked against the average common sense notion of "beauty," the "nature of aesthetic experience" is, to put it in a nutshell, the "enhanced emotional quality" of a given "event"; so that, to complete the circle, "beauty," whether in a concrete work of art or in life generally, is this very "enhanced quality" of an "event" emotionally intensified and experienced. (Obviously, this implies a "personal-impersonal relationship" between cognating subject and incognative object.) This "enhanced quality," again in life generally as in a given work of art, is begotten and conditioned by the elements of novelty and "dramatic conflict" of a given event, organized into a balanced attention-begetting pattern and interest-sustaining design. So that the "beauty" we "realize" in experiencing a work of art, whether in the spatial or temporal arts, is, by its very nature, a harmonious fusing of these diverse, and often contradictory or paradoxical, elements into a single emotionally heightened whole.

So far, then, so good. But, as Professor Pepper himself would be the first one to admit, and as he indeed very circumspectly and disarmingly does admit, there are other "justified ways," besides the pragmatic or "contextualistic" one, "of approaching the field of aesthetics"—such, for instance, as the transcendental approach of St. Thomas Aquinas and Kant, or the ideological approach of Hegel, or the lyrical one of Croce and others.

However, within the limits Professor Pepper has set to his field of aesthetic observation, based as it is on the pragmatism of James and Dewey, and further elaborated in this book into a more consistent and formal frame of reference, he has acquitted himself admirably. "Aesthetic Quality" is the most clear-cut, concise and consistent —if somewhat hard and dry in thought and prose—presentation of the pragmatic or "contextualistic" point of view on aesthetics that the reviewer has read.

DINO FERRARI.

Fig. 14.1 One of Kaempffert's more prominent features about Neurath and the *International Encyclopedia of Unified Science*, in the New York Times Book Review, Aug. 7, 1938

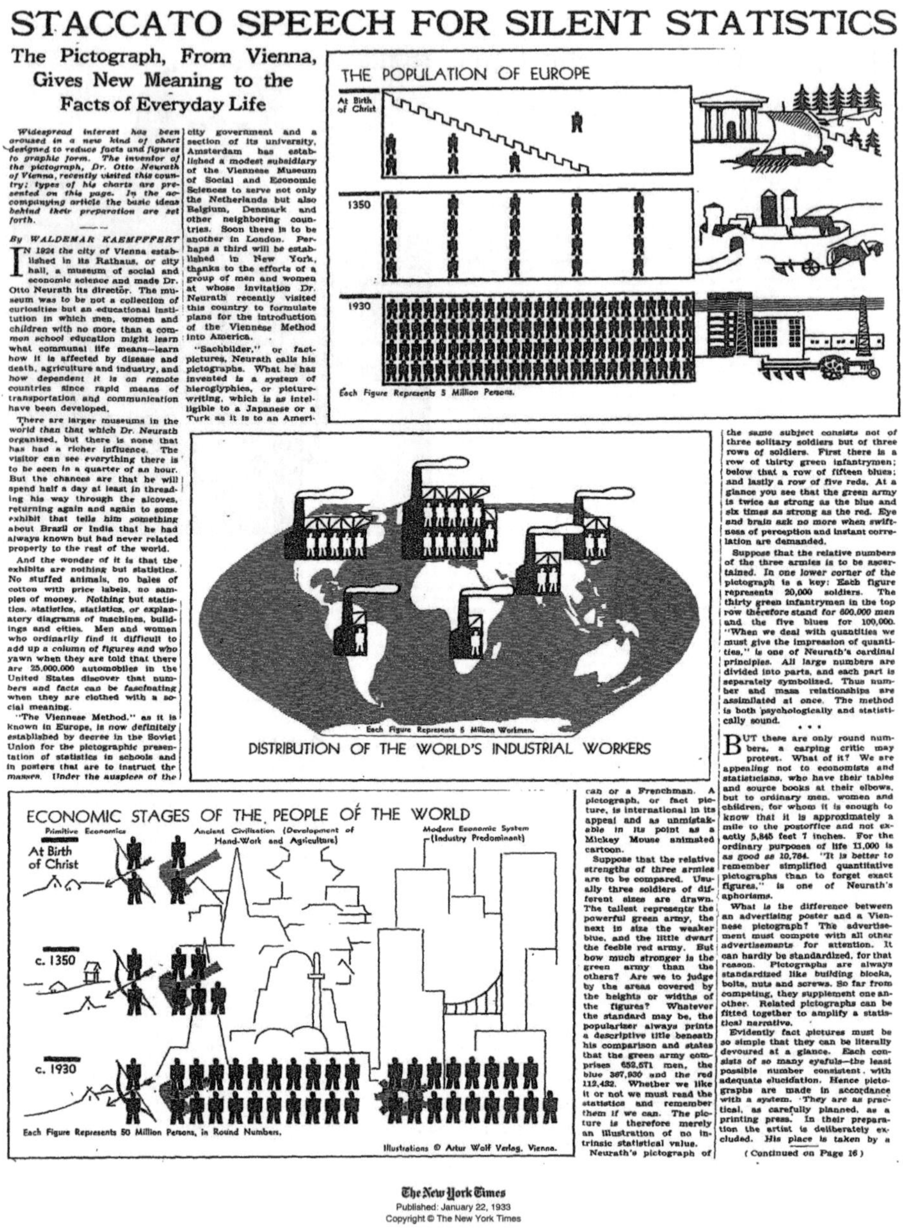

STACCATO SPEECH FOR SILENT STATISTICS

The Pictograph, From Vienna, Gives New Meaning to the Facts of Everyday Life

Widespread interest has been aroused in a new kind of chart designed to reduce facts and figures to graphic form. The inventor of the pictograph, Dr. Otto Neurath of Vienna, recently visited this country; types of his charts are presented on this page. In the accompanying article the basic ideas behind their preparation are set forth.

By WALDEMAR KAEMPFFERT

IN 1924 the city of Vienna established in its Rathaus, or city hall, a museum of social and economic science and made Dr. Otto Neurath its director. The museum was to be not a collection of curiosities but an educational institution in which men, women and children with no more than a common school education might learn what communal life means—learn how it is affected by disease and death, agriculture and industry, and how dependent it is on remote countries since rapid means of transportation and communication have been developed.

There are larger museums in the world than that which Dr. Neurath organized, but there is none that has had a richer influence. The visitor can see everything there is to be seen in a quarter of an hour. But the chances are that he will spend half a day at least in threading his way through the alcoves, returning again and again to some exhibit that tells him something about Brazil or India that he had always known but had never related properly to the rest of the world.

And the wonder of it is that the exhibits are nothing but statistics. No stuffed animals, no bales of cotton with price labels, no samples of money. Nothing but statistics, statistics, statistics, or explanatory diagrams of machines, buildings and cities. Men and women who ordinarily find it difficult to add up a column of figures and who yawn when they are told that there are 25,000,000 automobiles in the United States discover that numbers and facts can be fascinating when they are clothed with a social meaning.

"The Viennese Method," as it is known in Europe, is now definitely established by decree in the Soviet Union for the pictographic presentation of statistics in schools and in posters that are to instruct the masses. Under the auspices of the city government and a section of its university, Amsterdam has established a modest subsidiary of the Viennese Museum of Social and Economic Sciences to serve not only the Netherlands but also Belgium, Denmark and other neighboring countries. Soon there is to be another in London. Perhaps a third will be established in New York, thanks to the efforts of a group of men and women at whose invitation Dr. Neurath recently visited this country to formulate plans for the introduction of the Viennese Method into America.

"Sachbilder," or fact-pictures, Neurath calls his pictographs. What he has invented is a system of hieroglyphics, or picture-writing, which is as intelligible to a Japanese or a Turk as it is to an American or a Frenchman. A pictograph, or fact picture, is international in its appeal and as unmistakable in its point as a Mickey Mouse animated cartoon.

Suppose that the relative strengths of three armies are to be compared. Usually three soldiers of different sizes are drawn. The tallest represents the powerful green army, the next in size the weaker blue, and the little dwarf the feeble red army. But how much stronger is the green army than the others? Are we to judge by the areas covered by the heights or widths of the figures? Whatever the standard may be, the popularizer always prints a descriptive title beneath his comparison and states that the green army comprises 652,571 men, the blue 367,930 and the red 112,432. Whether we like it or not we must read the statistics and remember them if we can. The picture is therefore merely an illustration of no intrinsic statistical value.

Neurath's pictograph of the same subject consists not of three solitary soldiers but of three rows of soldiers. First there is a row of thirty green infantrymen; below that a row of fifteen blues; and lastly a row of five reds. At a glance you see that the green army is twice as strong as the blue and six times as strong as the red. Eye and brain ask no more when swiftness of perception and instant correlation are demanded.

Suppose that the relative numbers of the three armies is to be ascertained. In one lower corner of the pictograph is a key: Each figure represents 20,000 soldiers. The thirty green infantrymen in the top row therefore stand for 600,000 men and the five blues for 100,000. "When we deal with quantities we must give the impression of quantities," is one of Neurath's cardinal principles. All large numbers are divided into parts, and each part is separately symbolized. Thus number and mass relationships are assimilated at once. The method is both psychologically and statistically sound.

• • •

BUT these are only round numbers, a carping critic may protest. What of it? We are appealing not to economists and statisticians, who have their tables and source books at their elbows, but to ordinary men, women and children, for whom it is enough to know that it is approximately a mile to the postoffice and not exactly 5,845 feet 7 inches. For the ordinary purposes of life 11,000 is as good as 10,784. "It is better to remember simplified quantitative pictographs than to forget exact figures," is one of Neurath's aphorisms.

What is the difference between an advertising poster and a Viennese pictograph? The advertisement must compete with all other advertisements for attention. It can hardly be standardized, for that reason. Pictographs are always standardized like building blocks, bolts, nuts and screws. So far from competing, they supplement one another. Related pictographs can be fitted together to amplify a statistical narrative.

Evidently fact pictures must be so simple that they can be literally devoured at a glance. Each consists of so many eyefuls—the least possible number consistent with adequate elucidation. Hence pictographs are made in accordance with a system. They are as practical, as carefully planned, as a printing press. In their preparation the artist is deliberately excluded. His place is taken by a

(Continued on Page 16)

The New York Times
Published: January 22, 1933

Fig. 14.2 Kaempffert's coverage of Neurath's ISOTYPE project, Jan 22, 1933

While Isotype aimed to spread and democratize social and economic information, Neurath's new encyclopedia of science aimed to invigorate science itself, partly by calling for interdisciplinary cooperation and the cultivation and use of a universal language of science. It would allow researchers in different fields to collaborate and locate the gaps and inconsistencies in current knowledge. This too would help facilitate communication between scientists and the public. Envisioning the forthcoming encyclopedia as making great use of Isotypes (which is not accurate of the monographs that would appear beginning about a year later) Kaempffert mused that this will be "the strangest and most useful encyclopedia ever published."

> Give it to Jack Smith in the fo'c'stle of a ship or to Einstein and each will absorb more information from it in five minutes than he could from the millions of words in an ordinary encyclopedia. (Kaempffert 1937b)

While portraying Neurath's projects as part of a progressive, international vanguard reaching across the disciplines and across social classes, Kaempffert knew that he ran some risk of exciting ideological and xenophobic anxieties about Communism that took root in the United States in the wake of the Russian Revolution. These anxieties would have been well known to Kaempffert because they were on full display during the so-called first red scare of the early 1920s. But in the 1930s there was a strong case to make that history was on Neurath's side and that his projects belonged to America's future. This was at a time, for example, when not just Russia but also the Works Progress Administration under Roosevelt's New Deal adopted Isotype methods and styles into their literature and public posters (See Fig. 14.3.). Thanks in part to Rudolf Modley, who assisted Neurath at his *Gesellschafts- und Wirtschaftsmuseum* in Vienna and who later came to the United States to work with Kaempffert at the Museum of Science and Industry, the distinctive aesthetic and political sensibilities of the Vienna Method of pictorial statistics were not uncommon in 1930s America. In Chicago, Isotype exhibits were familiar to those visiting the Museum and Science and Industry, while the Roosevelt Administration's so-called alphabet agencies, including the WPA (Works Progress Administration) and FERA (Federal Emergency Relief Agency), employed Modley and others to create posters and charts about the nation's efforts to lift itself out of the depression (Charles and Giraud 2013). A review in the *New York Times* of Modley's book *How to Use Pictorial Statistics* noted his aim "to present picture of social and economic facts" in ways that are both engaging and memorable to the "man in the street" (*New York Times* 1937b).

While Neurath's projects within the Unity of Science Movement aimed mainly to educate adults about the ways of modern science, the Vienna Method was also embraced by progressive educators of children. In 1933, one Eunice Barnard reported on this trend in the *New York Times* and wondered, "can the children of tomorrow throw away many of their wordy textbooks and learn instead through the 'fact picture' – a kind of cross between the ordinary picture and the statistician's graph?" According to presentations given at Columbia University's Teachers College, Barnard reported, "America is just beginning to see the educational possibilities of the fact picture, which have heretofore been developed chiefly by Dr. Otto Neurath of the Social-Economic Museum in Vienna" (Barnard 1933).

Given this cultural landscape, marked on one side by the nation's conservative political and religious sensibilities and on the other by the growing belief that progress in science, government, and education offered the nation a way forward and out of the depression, Kaempffert steered a cautious, middle course in his coverage and commentary about Neurath's projects. On some occasions, he seems to have toned

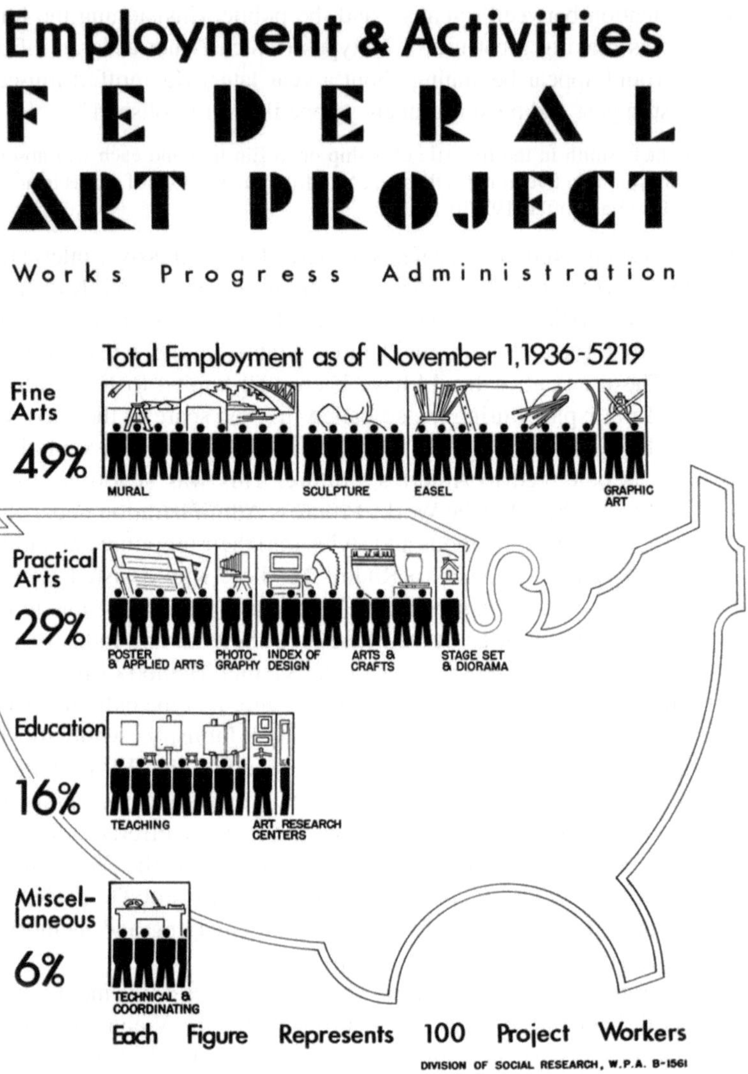

Fig. 14.3 U.S. Government posters utilizing and adapting Neurath's Vienna Method of pictorial statistics in the 1930s

down the Marxist content and overtones in what Neurath wrote and said about his projects. In his book "Personal Life and Class Struggle," of 1928, for example, Neurath clearly associated his vision of unified science with a Marxist sensibility that aimed to solve problems for the masses: "The Marxist always aims at informing himself about the main problem of the proletarian class struggle: how will the suffering of the capitalist order come to an end?" To Bourgeois scientists, on the other hand, "it is left to chance whether a man thinks about some linguistic formations in Chinese or about a medieval legal text, about African beetles or about wind

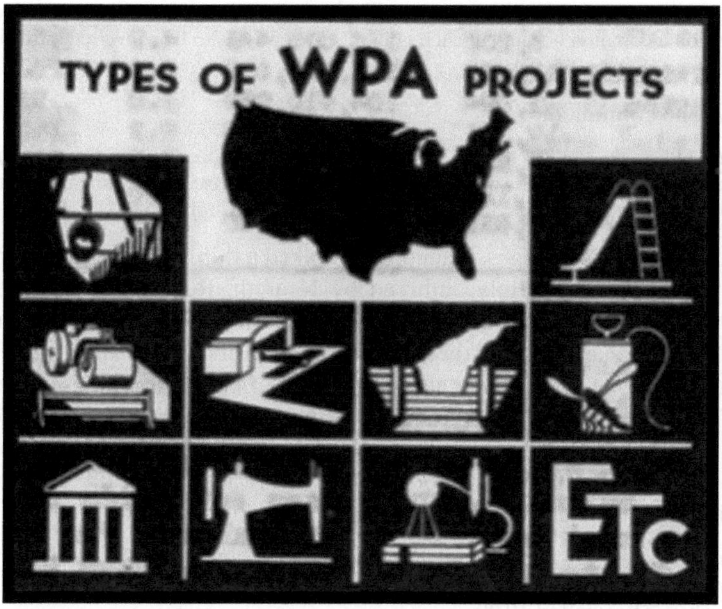

Fig. 14.3 (continued)

conditions at the North Pole" (Neurath 1928/1973, 294–5). Kaempffert discussed the unification of the sciences in similar ways, but without using this Marxist terminology. For example, he wrote,

> Naturally, it requires an organization to carry out this conception of unity. The different specialists cannot be left to their own devices – the old method. They must be brought together and made to understand one another. (Kaempffert 1937c.)

Thus Neurath's "bourgeois science" became Kaempffert's "old method" of science, one he later called a "*laissez-faire*" approach to research.

Kaempffert also emphasized that Neurath's Unity of Science Movement did not aim to legislate or impose unity on scientific research. Writing about the Vienna Circle's views of unified science in 1938, for example, he wrote, the need for unity

> became more and more apparent as language was analyzed. It was far from the mind of the group to evolve a super-science to legislate for all the known disciplines. The object was integration, and integration was to be achieved by building bridges form one science to another. (Kaempffert 1938)

To explain how the new encyclopedia would help achieve this goal, Kaempffert borrowed Neurath's (and later Quine's) well-known metaphor of science as the Ship of Theseus whose beams and planks are slowly replaced over time. Kaempffert's metaphor was also nautical, but better suited to this anti-authoritarian point: The encyclopedia's function, he explained, is

> to act as a pilot to scientists. In the past they have been independent navigators [...]. [But] [t]he encyclopedia is intended to tell each navigator what other ships are doing and what he

> can learn from their signals, their movements, their errands. A heterogeneous collection of vessels on a common ocean, but each going its own way, is to be converted into a homogenous fleet. (Kaempffert 1938)

Crucially, he continued,

> [t]here will be no admiral to give orders. Only this encyclopedia is to serve as chart and compass. To it all navigators will contribute, and all will revise it from time to time to make it more useful to themselves and the educated public. (Kaempffert 1938)

In these and similar passages, Kaempffert walked a thin, political line. In reviewing the first monograph multiply authored by Neurath, Rudolf Carnap, and Charles Morris (the *Encyclopedia* editors) as well as Bertrand Russell, and John Dewey, Kaempffert emphasized that these "sensible contributors to this work make no attempt to club one another into accepting ideas." In comparing these new encyclopedists to those around Diderot, and at a time when most American intellectuals were alarmed by threats to intellectual freedom in Germany, Spain, and the Soviet Union, Kaempffert attempted to portray the effort as intrinsically unpolitical, and yet politically important in its effects:

> The men who are making this encyclopedia have no desire to enter the political arena but every desire to influence intellectual leaders and to insure freedom of scientific inquiry everywhere. (Kaempffert 1938)

Several years later, we shall see, some of Kaempffert's fellow opinion leaders would emphatically disagree with this view. To them, any such effort to organize and plan research was itself a threat to intellectual freedom and the advancement of knowledge.

14.2 Kaempffert and Kilgore Versus Bush and Conant

World War Two made the United States' relationship to Marxism and the Soviet Union even more complicated and unstable than the political terrain Kaempffert navigated in the 1930s. Once the nation declared war on both Japan and Germany, Stalin and his Soviet Union became an ally in the nation's crusade to defeat fascism and totalitarianism. Not least because Hitler's campaign against the Soviets to his east weakened his grip on Western Europe and North Africa, and because it was the Soviets who finally captured Berlin and ended Hitler's regime, even soon-to-be cold warriors like James Bryant Conant recognized America's debt to the Soviets. As Conant (1970, 561) put it, "gratitude for the prowess of the armies which finally overwhelmed the Germans on the Eastern Front was mixed with fear of the ambitions of the dwellers in the Kremlin."

Beginning soon after the close of the war 1945, the fears Conant mentioned began to dominate and simplify U.S. and Soviet relations. By the McCarthy era of the early 1950s, these fears defined the national mood, the United States moved to the political right, and all things Soviet, Communistic, or totalitarian were seen as powerful, sometimes immanent threats to the nation's well being. Anticommunist politicians, administrators, and intellectuals thrived in this cold war climate, while American communists who did not reinvent themselves politically retreated into nervous silence or went "underground" to avoid capture and prosecution by J. Edgar Hoover's FBI.

This transition was rapid but not instantaneous. Beginning in the mid 1940s, we can see in some detail how the nation's move to the anticommunist right overwhelmed and discredited the philosophical and scientific ideals that Kaempffert championed. This story begins during the war with debate over the nation's ability to utilize its scientific and industrial resources for maximum military effectiveness. In Congress, West Virginia Senator Harley Kilgore called for a national, institutional effort to organize war research and was quickly and powerfully supported by Kaempffert. As he saw it, the senator's proposals would implement Neurath's ideas about unifying the sciences at a national scale. In 1943, in the pages of the magazine *The American Mercury*, for example, Kaempffert defended Kilgore's ideas in much the same terms that he used to present and defend Neurath's encyclopedia in the *New York Times*. At present, he wrote,

> [e]ach government agency, each industrial laboratory goes its own way. There is no overall direction and control. [...] All this Senator Harley Kilgore of West Virginia would correct by two bills of his [that ...] would marshal the country's scientists and engineers in a vast research organization to solve the pressing war problems of the armed forces and industry. (Kaempffert 1943, 441.)

Echoing Neurath's own complaints about "Bourgeois" science under which "it is left to chance" what different investigators study, Kaempffert criticized the popular *laissez-faire* approach to research. The problem is, he wrote, "[t]here is no systematic attack on fundamentals. Everything is left to chance – the chance that some able investigator will be stirred into action and that he will somehow manage to raise the money that he needs to make his inquiries" (Kaempffert 1943, 447).

In 1944, as the end of the war came into sight, President Roosevelt commissioned Vannevar Bush to formulate a proposal for scientific research in the upcoming postwar era. Bush was at that time in charge of the Office of Scientific Research and Development (OSRD) and working closely with Conant, J. Robert Oppenheimer, and others in the Manhattan Project to develop and test the atomic bomb. Roosevelt was not a philosopher of science and may not have known about Neurath's Unity of Science Movement or the ambitions of the Bernalists. But by acknowledging in his letter to Bush the successful and "tangible results" of the OSRD's "unique experiment of teamwork and cooperation in coordinating scientific research," he took for granted Kaempffert's and Neurath's view that research stood to gain by being coordinated, organized, and planned. He also shared their hope that science be applied to reform and improve modern life. In particular, Roosevelt insisted that these enormous strides in organized military research be somehow translated or adapted "for the improvement of national health, the creation of new enterprises bringing new jobs, and the betterment of the national standard of living." "There is," he wrote, "no reason why the lessons found in this experiment cannot be profitably employed in times of peace."[5]

Bush's reply came in the form of his report, published in 1945 as *Science, the Endless Frontier*. Among Bush's conclusions was that the United States "can no longer count on ravaged Europe as a source of fundamental knowledge" or basic, pure science. It must therefore begin to cultivate its own and it should do so through "the creation of a National Research Foundation" (Bush 1945, 22, 34). Yet Bush's vision of how such a foundation would operate was rather different from Kilgore's proposals. The main difference reflected the longstanding divide between science planners and their critics – a divide which had intensified in Europe during the war with the formation of the Society for Freedom in Science (SFS) by Michael Polanyi and colleagues. Directly opposing the views and clout of the Bernalists, the new society was formed in 1940 and premised on the convictions that "science can only flourish […] when research is conducted in an atmosphere of freedom" and "[s]cientific life should be autonomous and not subject to outside control in the appointment of personnel or in the allocation of funds assigned by Society to Science."[6] In the United States, a similar reaction had been formulated by Abraham Flexner, founder of the Institute for Advanced Study in Princeton, New Jersey. In his essay, "The Usefulness of Useless Knowledge," first published to a popular audience in *Harpers Magazine* the year before, Flexner (1939, 550) urged that even the thought of potential applications and uses in researchers' minds would hamper and impair the forward march of research – just as it was then being impaired in the great universities in Europe, where "the freedom of the human spirit" was tragically suppressed by politicians and administrators presuming to guide and control scientific research.[7]

Bush emphasized this categorical difference between applied and pure (or "fundamental") science and repeatedly spoke of "freedom," "complete freedom," and "independence" from political or practical constraints as "essential" for scientific progress. Like Flexner, he insisted that this simple point was demonstrated by the history of science itself.[8] He also echoed the SFS's view that funds granted to researchers by governments cannot be used to guide research. It is of "utmost importance," he wrote, that basic research funded in universities and research

[5] Roosevelt to Vannevar Bush, Nov. 17, 1944. The letter is reprinted in Bush (1945, 3–4).

[6] Quoted in Shils (1947, 80–2). See also Nye (2011, 204–6).

[7] Of Michael Faraday, Flexner (1939, 546) wrote, "[a]ny suspicion of utility would have restricted his restless creativity" – as if the mere *thought* of prototype electric motors or other applications would have harmed his research in magnetic forces and electric currents. For a historical survey of the pure-versus applied distinction at midcentury and the difficulty of maintaining it credibly, see Douglas (2014).

[8] Bush (1945, 56) wrote, "[t]he entire history of science bears testimony to the supreme importance of affording the prepared mind complete freedom for the exercise of initiative." Flexner (1939, 545) had earlier written, "[…] throughout the whole history of science most of the really great discoveries which had ultimately proved to be beneficial to mankind had been made by men and women who were driven not by the desire to be useful but merely the desire to satisfy their curiosity."

institutes "leave the internal control of policy, personnel, and the method and scope of research to the institutions themselves" (Bush 1945, 33). Given that some degree of control over research comes with the choice of which researchers and institutions get funded and which do not, Bush (1945, 33) added that these choices should be made not by the public or its representatives, but by an agency staffed by "persons of broad interest in and understanding of the peculiarities of scientific research and education."

Kaempffert was unimpressed and disappointed with Bush's report. It rubbed against his political sensibilities, for within Bush's vision of publicly funded research there was no mechanism by which "Jack Smith in the fo'c'stle of a ship," as Kaempffert had written eight years before, or other ordinary citizens would have much say in what research goals were pursued with public funds. Its insistent equation of scientific progress with freedom and independence denied his and Neurath's methodological conviction that attempts to plan and guide research could be, and often were, scientifically fruitful. In response, he reviewed *Science, The Endless Frontier* and attacked the fundamental assumption on which Bush's proposal depended: "The plain truth," he wrote, "is that there is no difference between 'pure' and 'applied' science. Science is science, whether it is engaged in solving the problem of television or the constitution of matter" (Kaempffert 1945a). Success in pure science, even Nobel prizes, he pointed out, had come from industrial laboratories in Europe and in America, and despite the tragedy of Lysenkoism impressive achievements in Soviet medical research had been overseen by the Kremlin.

Most "astonishing" to Kaempffert was Bush's failure to see his own leadership of the OSRD as plain evidence for the potential effectiveness of planned, organized research. Writing just weeks before the nation learned of the new atomic bomb after its use in Japan, Kaempffert pointed out that the OSRD was widely considered a spectacular success for its developments and improvements in medicine, logistics, and technologies like RADAR. The OSRD, Kaempffert reminded the nation, "mapped out the whole field of military and medical science" and "sought out the best research scientists and laboratories and engaged them by contract to develop projects; in a word, it organized, it planned, it directed." Yet instead of carrying forward the acquired organizational experience into the coming era of peacetime research, Bush's report reverted to "a kind of laissez-faire system, much like that adopted by philanthropic foundations." Instead of directing and commissioning research, the new foundation would merely "wait for ideas to be submitted" for potential funding (Kaempffert 1945a). In terms of the nautical metaphor he used to describe Neurath's project, Bush knew very well how to organize a fleet of researchers to pay attention to each other and build on each others' successes. But now he was suddenly and inexplicably an advocate of "the old method" and the *laissez-faire* approach to modern science.

Kaempffert recognized the geopolitical realities animating Bush's recommendations. "All through the report," he wrote, "runs the fear of Government dictation and interference." Of course this was a real fear, as demonstrated by the state of genetics in Russia. "I hold no brief for the Marxist ideology which has permeated Russian science," he wrote. "America wants no restriction of that kind of scientific thinking."

But the fact remained that in other areas of science, "Russia has made astonishing progress" in spite of "the ideological limitations imposed on it" – a fact overlooked in Bush's report. Despite its exaltation of the wondrous, "endless frontier" of science, Kaempffert argued, Bush's report ignored recent history and common sense to impose a methodological frontier on American science:

> This country showed that it could mobilize science and plan research for war. There is no reason why it should not similarly mobilize and plan research for peace, without departing from democratic principles; no reason why it should not take over the best features of the Russian system and adapt them to its own needs, no reason why it should not first map out the whole field of science to reveal gaps in our knowledge and see to it that every science is developed rationally in every field. (Kaempffert 1945a)

While pointing readers to Bernal's *The Social Function of Science*, Kaempffert (1945a) insisted that "[a]n all-embracing plan which embodies the best in Soviet and American systems need not be dictatorial."

Kaempffert was not alone in his dismay. The day before his review appeared, the *Times* featured an unsigned editorial, titled "Research for Defense," that made similar points with similar language. The *New York Times* itself, in other words, was taking Kaempffert's side and calling for OSRD-style direction of research within any national institution created to facilitate publicly funded research (*New York Times* 1945a).

At this point, Conant, writing as Chairman of the National Defence Research Committee within the OSRD, stepped in to dispute Kaempffert's and the paper's position. In a letter to the editor, he too recognized the geopolitical dimensions of this debate. He conceded that were the United States poised to go "down the road of socialism" like Britain and to nationalize some or all of its industries, Kaempffert's position would make sense. But the United States was not so poised, as Conant reminded Kaempffert by pointing to the recent elections of 1944 in which neither major party called for British-style reforms. In addition, Conant insisted that Kaempffert misunderstood the OSRD and how it operated. Yes, administrators set goals for researchers to meet; but if those goals were met it was precisely because researchers had "the maximum of autonomy" and "the minimum of centralized control" from OSRD. Following Bush, that is, Conant invoked a substantive distinction between pure and applied research in order to sequester and isolate science from threats of dictatorial control. "Almost by definition," he wrote, advances in pure science consist not in getting closer to specified goals but in being surprised and intellectually sparked by something unexpected and inexplicable. The story of penicillin's discovery, he remarked, should be studied by "all who are interested in this subject" for it underscores the serendipity and unpredictability of basic research and the intellectual freedom it requires:

> There is only one proved method of assisting the advancement of pure science – that of picking men of genius, backing them heavily and leaving them to direct themselves. There is only one proved method of getting results in applied science – picking men of genius, backing them heavily, and keeping their aim on the target chosen. (Conant 1945)

"A reply to Dr. Conant's interesting letter will be published in a later issue," the editors wrote after Conant signed off.

Kaempffert girded for battle and replied a week later, more or less insisting that Conant's letter did not make sense. His effort to portray OSRD-style research as somehow supporting, instead of problematizing, a robust distinction between pure and applied science "is lost on us," he wrote. As counterpoint to Conant's case study involving penicillin, he pointed to the discovery of radioactivity and its impact on atomic theory to illustrate how pure and applied research are usually joined together as they push knowledge forward: "Technology leads as naturally to fundamentals as fundamentals lead to technology" (Kaempffert 1945b). Whether or not they did so intentionally, Kaempffert and his fellow editors pushed back against Conant's tenuous arguments again two months later as they congratulated Alexander Fleming, Howard Florey, and Ernst Chain for the Nobel Prize the discovery had brought them. They editorialized,

> [w]e have here another example of what research can do when it is organized, planned, and competently directed. A group of bacteriologists, [and] physiologists, biochemists working as a team scored a triumph in months instead of the usual years. (*New York Times* 1945b)

14.3 The Kilgore Bill

By the fall of 1945, two bills in Congress competed to create the new national foundation for scientific research that Bush had proposed. Kilgore's proposals for wartime research had evolved into what was known as the Kilgore Bill, while the Magnuson bill, sponsored by the Washington State senator Warren Magnuson, offered a different vision of postwar science. Among the controversial differences were the status and ownership of patents coming from publicly funded research, the place of social sciences in the new foundation and, most importantly, how much power the officers would have to approve or commission projects and thus guide research in various directions. Also at issue were rules establishing whether officers commissioned would represent public or private, corporate interests. The Magnuson bill called for a board of nine leaders from science and industry, one of whom they would elect to be the director. The Kilgore bill, on the other hand, called for a single director who would actively direct research and remain more closely accountable to the public through its elected leaders. In Kaempffert's terms, the Magnuson Bill was largely committed to a *laissez-faire* research sensitive to the needs of private industry and corporations, while the Kilgore bill aimed to organize and plan research for the public good and to emphasize democratic control of publicly funded research.

President Truman himself tilted in Kaempffert's and Kilgore's direction. In a message to the nation in September, he agreed that the OSRD and the Manhattan Project had demonstrated the viability of planned research and he questioned the practicality of adhering to a strict distinction between pure and applied science. The future of science in America, that is, should not be hampered by a hands-off, wait-and-see approach that exalted the unpredictability and depended on "brilliant inspiration or sudden flights of genius" to move science and technology ahead. The nation instead required a powerful institution to "coordinate and control diverse scientific activities now conducted by the several departments and agencies of the Federal Government" (*New York Times* 1945c). A month later, one of Truman's officers reportedly sent a note to both Senators Kilgore and Magnuson expressing the president's support for a "straightline administration" (*New York Times* 1945d) of the new institution, consisting of a single administrator and advisory board, both appointed by the president.

With this, Bush, Conant, and about forty other leaders of foundations, businesses, and universities sprang into action. They formed an official public-relations committee, called The Committee Supporting the Bush Report, and wrote an open letter to President Truman on behalf of "the great majority of American scientists" (*New York Times* 1945e). While they agreed that the board of the new organization could be "composed of laymen and scientists," they insisted that the President appoint them "without reference to political affiliation" and that scientists have a say in those appointments. Most importantly, they argued,

> it would be most unwise to subordinate the board to a single director appointed by the President, as is done in the Kilgore bill. No single person, however eminent or competent, could, except in a great emergency, command the confidence and support of all branches of science and of the many organizations and agencies, private and public, whose cooperation will be required. (*New York Times* 1945e)

Perhaps because Truman seemed sympathetic with the view that methods of wartime science involving cooperation among "all branches of science" could be fruitful in times of peace, the new committee's argument quietly shifted gears. It conceded that "all branches of science" could be organized and directed to cooperate fruitfully – but that is only in times of "great emergency" such as the recently concluded world war. The task at hand was to conduct science freely and without any taint of Soviet, authoritarian methods and values. Postwar science must involve "a high degree of institutional and individual freedom and responsibility," the Committee insisted. Similarly, the board of the new foundation may appoint from within its ranks "a chief administrative officer," but that officer "should not be in a position to dictate or interfere" with the board's activities. "He should be the agent – not the master – of the board." Nor must the board itself "be empowered to control or coordinate other Government scientific agencies" (*New York Times* 1945e).

The committee's position was framed within the specter of dictatorial control that Kaempffert had been trying to dispel for almost a decade as misleading and irrelevant to the future prospects available. In his coverage of Neurath's new encyclopedia, his advocacy of Kilgore's earlier proposals for wartime research, and his grappling with Bush and Conant over the methods of postwar science, he tried to disconnect the idea of planned, organized research from political fears of dictators and thugs. But he did not succeed. Congress eventually passed in 1950 the National Science Foundation Act which embodied much more of the Magnuson-Bill proposals than those of Kilgore and Kaempffert. Today's the NSF remains mainly a grant-giving foundation that funds but does not commission, guide, or organize research as Kaempffert and others once hoped it would. Its long and controversial creation, only partially illustrated by these debates in 1945, allowed other government agencies, like the National Institute of Health, the Atomic Energy Commission, and NASA, to grow and lay claim to specific areas and kinds of research. The comprehensive, unified approach to research that Neurath and Kaempffert advocated would never be realized within the United States government.

14.4 The Planning Debates and the Postwar Unity of Science Movement

Kaempffert's smart and determined advocacy of Neurath's project reminds us that the original promise of logical empiricism was not only intellectual but also practical and inescapably political. Advocates for and against planning agreed that modern science, as well as studies of its history and philosophical foundations, was inevitably connected to the cultural, intellectual, and economic struggles at hand. Like Clausewitz's understanding of war, debates over the nature of science, its roles in society, and its modes of progress were politics by other means.

The member of Bush and Conant's committee that most aggressively blended anticommunist politics and philosophy of science was Warren Weaver, Director of Natural Sciences at the Rockefeller Foundation. Perhaps stung by Kaempffert's view of Bush's proposed foundation as merely "a glorified Rockefeller Foundation" (Kaempffert 1945a) whose effectiveness would pale next to OSRD-style management, Weaver personally sustained the committee's attack on Kaempffert and the *New York Times*. In a letter to the editor in September 1945, Weaver angrily denounced Kaempffert and the *New York Times* for recommending that,

> all science should be mapped out and the gaps discovered, and that an all-high, all-embracing central organization, presumably set up by the Federal Government, should plan and direct scientific activities. (Weaver 1945)

Comparing the fruits of science to the dividends a corporation might pay to its shareholders, Weaver (1945) wrote: "The earnings of science are not to be gained by organizing a super-control which holds guns at the heads of scientists and tells them what to do. The earnings of science are gained only by setting the scientists free." "For the nth time," Kaempffert (1945c) wrote in frustration as he responded a week later, "we insist that no scientist would be compelled to join the organization and no director would tell anyone how to proceed in solving a problem."

Weaver's participation in this debate documents another link to Neurath's Unity of Science Movement. Somewhat ironically, Weaver himself was something of a "super control" behind the Unity of Science Movement because his Rockefeller Foundation first supported Philipp Frank and the Institute for the Unity of Science that he and others created after the war in Boston. Initially, at least, Weaver and his colleagues happily supported Frank because he was an intellectual refugee (Frank was touring the United States in 1938 when Hitler annexed Czechoslovakia, where

Frank had lived).[9] But Frank's relationships to Weaver and his underling officers deteriorated and fulfilled Weaver's initial doubts about whether Frank, then in his sixties, was capable of building and growing the Unity of Science Movement.[10] This was one reason why Frank was informed in 1952 that his second three-year grant would be his last. A Foundation report on this grant states that "longer and larger support [...] would be amply justified" were Frank and others "surely passed their prime" to recruit younger intellectual talent into their movement.[11]

14.5 Philipp Frank, Thomas Kuhn, and the Sociology of Science

Whether or not it was because of these sentiments within the Rockefeller Foundation, at the end of 1952 Frank reached out to a young Thomas Kuhn, then about 30 years old. Though Frank and Kuhn would surely have crossed paths as instructors within Conant's General Education program at Harvard, Frank wrote a letter to Kuhn in which he invited him to join a new committee within his Institute. It would include Frank, Robert Merton, Ernest Nagel, and Kuhn himself and would promote research in the sociology of science by providing small grants to researchers. The theoretical interests Frank expressed in his correspondence concerned the complicated relationships among epistemological, methodological, and social aspects of science and, in particular, cases of underdetermination when social interests or values can enter into scientific reasoning. As he put it in the précis he sent Kuhn, this project would aim to identify and understand "the role which sociological factors have played in the acceptance of scientific laws and theories."[12] Kuhn eagerly responded to Frank's

[9] Frank and other logical empiricists, one internal report on the Unity of Science Movement points out, were refugees from "the stifling atmosphere of totalitarianism" ("Toward Integration of the Sciences," hand-dated March 1949, Rockefeller Archive Center, Tarrytown, NY, hereafter RAC.). For more on Frank and the foundation of the Institute, see Reisch (2005). For more on Frank's exile and his relationships to James Bryant Conant, see Reisch (2017) and the other essays in Tuboly (2017) prepared for the 50th anniversary of Frank's death.

[10] As early as 1946, before Frank's Institute received funding from the Rockefeller Foundation, Weaver noted private concern after talking to Frank that the movement "may be developing in to too much of a personal venture on the part of F[rank]" ("Interview: WW, Friday Dec. 13 1946," RAC). By 1951, Rockefeller documents frequently express reservations about "old-timers in the Unity of Science Program" who are "repeating themselves" (Letter to Weaver, Dec. 9, 1951, RAC), Frank's organizational skills (Memo from C[hadbourne] [G]ilpatric, Nov. 5, 1951, RAC), and Frank's conviction (as Weaver put it) that "it is politically important to support activities which emphasize empirical and pragmatic philosophies, since Fascism and Communism depend essentially on metaphysical doctrines" – an argument Weaver found "very tenuous and unconvincing" ("WW's Diary, Thursday Sept. 27, 1951," RAC).

[11] Trustee Report, Jan 18, 1952, RAC.

[12] A copy of this document and Kuhn's reply remains in Kuhn's archival papers at MIT (hereafter, TSK-MIT), box 25, folder 53.

offer and, in a letter remaining in his archival papers, jumped right into theoretical debate about what, exactly, the new committee should take "sociology of science" to mean.

In a subsequent report to Weaver's Foundation, Frank mentioned this new committee and this collaboration with Kuhn (as well as the sociologist Bernard Barber, whom Frank evidently also invited to join). But it does not appear that the committee bore much scholarly fruit. Research done under it was largely parochial to Frank's particular interests. In his final report, Frank noted that the sociology project would henceforth be supported by the new NSF, that the six papers "will be published together with other results of the research supported by the National Science Foundation."[13] Though the NSF did begin to support research in social science and related fields (including history and philosophy of science) in the mid-to late 1950s, the committee's work and the research it supported was not published. Frank's own career was nearing its end and there is little evidence that his interests in the mutual effects of science, philosophy, and politics gained traction in history, philosophy, or sociology of science.[14]

Kuhn's collaboration with Frank suggests that he did not share Frank's political interests in this project. His reply to Frank's offer reads "not sent" in the top margin, written in Kuhn's hand. The notation may simply mean that Kuhn discussed the issues with Frank in person, or that he sent a different letter in reply. But against the backdrop of the planning debates, there is programmatic significance in Kuhn's effort to convince Frank that the future of research in the sociology of science lay in a different direction. The sociological dynamics that matter, Kuhn said, operate strictly inside scientific communities and are largely screened off from culture and popular politics – from matters of "religion, social status, economic organization." For this reason, Kuhn wrote,

> [i]t would seem to me more fruitful to examine the ubiquitous role of the sociology of the professional group than to concentrate solely on those factors (like government, church, etc.) which at this time and place have relatively little impact upon decisions made by professional scientists about problems arising within their sciences. (Unsent letter from Kuhn to Frank, TSK-MIT, box 25, folder 53)

Though Kuhn (1962/2012) had yet to formulate his conceptions of scientific communities, normal scientists, and paradigms, he was inclined to acknowledge only the internal sociology of professional communities as properly within the scope of sociological study.

---

[13] "Report of the Institute for the Unity of Science to the Rockefeller Foundation for the Period of July 1952-1953," RAC. Frank's final report to the Rockefeller foundation lists seven "research assistants" who wrote six papers under the guidance or support of the committee (Anatol Rapoport, Jeremy Bernstein, Ernst Topitsch, Benoit Hepner, John Wilkinson and William H. Meyer, and Lewis Feuer). The majority of subjects of research listed represent Frank's theoretical interests, including criteria for acceptance and rejection of theories (Rapoport) and philosophical interpretations of general relativity (Bernstein, Topitsch, Hepner).

[14] There is positive evidence, on the other hand, that Frank's research project was viewed with sufficient disdain to be attacked as intellectually irresponsible in the pages of *Philosophy of Science* (Kegley 1959). On the relation of sociological studies of science and logical empiricism see Uebel (2000), with particular reference to Frank, see Reisch (2005, Chap. 11.).

Kuhn was aware of the political controversies surrounding planning as well as his mentor Conant's debate with Kaempffert and the formation of the NSF. If he had not followed this dispute in the *New York Times* in 1945, it was brought to his attention two years later when he first joined Conant's General Education program to teach history of science and read Conant's book *On Understanding Science* (Conant 1947). In this book Conant outlined a general education course that complemented Bush's report, specifically his call for research policy and the allocation of funds to be made by "persons of broad interest in and understanding of the peculiarities of scientific research." Graduates of Conant's new General Education program – future leaders in politics, business, and government administration – would gain just such a detailed, but nonspecialized understanding of science by working through historical case studies in science. Though he remained convinced that science could not fruitfully be planned, that accidents and acts of individual, unpredictable genius were essential to scientific progress, Conant acknowledged complex relationships between science and society, that "science is indeed a social process." Later in the book, he noted the "storm signals of present controversy" that surround the pure versus applied distinction and prospects for organizing and directing research for progressive social ends. In a detailed endnote, Conant directed interested readers to texts by Hessen, Crowther, Bernal and others (Conant 1947, 64, 107, n. 43).

The most vivid evidence that Kuhn was aware of and concerned about these "storm signals" comes from his preparations to deliver the prestigious Lowell Lectures at Boston's Public Library in early 1951, a year and a half before Kuhn articulated to Frank his conception of sociology of science. Upon seeing an advertisement for his upcoming lectures in the newspaper *The Boston Globe*, Kuhn immediately telephoned his contacts at the Institute and later followed up with letters that reveal his "acute distress," as he put it, over being presented to the public as an expert authority on contemporary research.

What he saw was an advertisement (Fig. 14.4.) which read:

What are the Problems of Scientific Research Today?

under which appeared the particulars about his upcoming lectures. The Institute evidently agreed to cease publishing the ad, but a day or two later Kuhn became incensed again upon discovering a paper flyer distributed around Boston. It read, in part,

> In a world in which science's quest for physical theory has already had results that promise to change the course of history, the fate of mankind may depend upon solving the problems of research [...].

The final, ominous ellipses were intended to intrigue readers, to bring more Bostonians than usual to the public library. But it appeared to Kuhn that the publicity was dangerous, irresponsible, and perilous for his future career. On the eve of his debut as a public intellectual, he evidently saw himself presented as an advocate of scientific planning. The tagline "What are the problems of scientific research today?" echoed Kaempffert's advocacy of Senator Kilgore's wartime plans for "scientists and engineers in a vast research organization *to solve the pressing war*

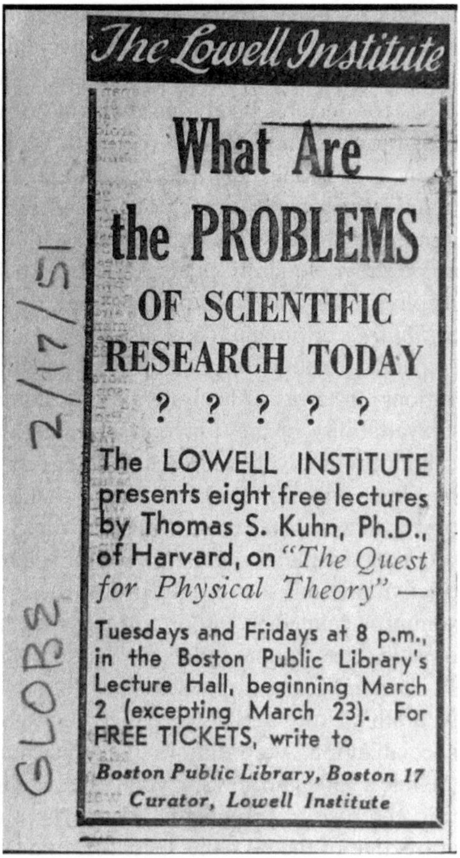

Fig. 14.4 Advertisement in the *Boston Globe* for Thomas Kuhn's Lowell Lectures (clipping from Kuhn's papers, TSK-MIT, box 3, folder 10)

problems of the armed forces and industry" (Kaempffert 1943; emphasis added). It invited the public to believe that Kuhn was a scientific planner who knew – or, worse, falsely presumed to know – how today's scientists could change the course of history and save mankind.

Kuhn had earlier provided Ralph Lowell with a title for his lectures, "The Quest for Physical Theory – Problems in the Methodology of Scientific Research." The problem with the Institute's tagline evidently was its failure to include "methodology." Without this word, Kuhn seemed to fear, his stated interest in *methodological* problems could be seen as interest in *scientific* problems thought to bear on contemporary issues in science and society. "It may help you to understand my dismay," he wrote, "if I explain that the fascinating topic your copy writer has so clearly stated is one to which I believe no serious and responsible student of science would address himself."[15] A month before, describing his current research to one of his Deans, Kuhn had said the same thing: methodological study can help us understand the

[15] Kuhn to Lowell, Feb. 20, 1951. TSK-MIT, box 3 folder 10.

nature of science – but, he emphasized, it "cannot ever venture to prescribe fruitful research procedures to the working scientist."[16]

Though *Structure* first appeared as a monograph within Neurath's encyclopedia, it has for decades been read as a kind of theoretical Trojan Horse designed to discredit logical empiricism along with Whig or presentistic bias in historiography of science. It should also be understood, however, in light of Kuhn's personal and practical reaction to these contentious cold war debates over scientific planning. In a number of ways, his theoretical arguments bleed into these political debates and come down squarely against the leftist, Marxist tradition in science studies that inspired Neurath, Kaempffert, Frank and the larger Unity of Science Movement. *Structure*'s critique of formalism, for example, marshals psychology, sociology, and semantics to deny the possibility of a neutral, universal language of scientific observations. In terms of Neurath's and Kaempffert's conception of unified science, the unavailability of any universal language of science implies that unified scientists cannot productively survey knowledge as a whole in order to identify the "gaps" to be filled, to build "bridges" from science to science, and thus create Kaempffert's powerful scientific fleet. Against Frank's interests in sociology of science, for a second example, *Structure*'s view of scientific communities dominated by their distinctive internal sociologies of knowledge denies that scientific knowledge or reasoning is related in important and specifiable ways to social or political realities outside those communities.

Structure also opposes the foundational methodological premise behind Neurath's Unity of Science Movement: its claim that science suffers from excessive specialization. According to flyers distributed by the University of Chicago Press in the late 1930s, the new encyclopedia would be primarily "concerned with the scientific enterprise as a whole." This synoptic view was offered as an "indispensable corrective of the extreme specialization of scientific research." "It is an urgent task of science," one flyer continued,

> to work out the synthesis of its results and methods. Otherwise, science will not have carried to the limit the fulfillment of its own task as science, nor will it perform adequately its educational role in the modern world. Science is gradually rousing itself for the performance of its total task. (University of Chicago Press Papers, University of Chicago Library, box 347, folder 2.)

As his distress over the Lowell Institute's publicity suggests, Kuhn was uncomfortable being cast in this or any similar role that told scientists what to do or how they should do it. But Kuhn did more than reject this role; he theorized why it was unserious and irresponsible to accept it. In the first of his lectures, he mentioned the Institute's publicity took care to set the record straight with his audience. "If any of you happens to have followed closely the advance notices of this series of lectures," his prepared lectures read,

[16] Kuhn to Owen, Jan. 6, 1951, TSK-MIT, box 3 folder 10.

> you may have remarked that the topic just described bears very little relation to the one announced in some of the flyers prepared by the Lowell Institute's copy writer. That topic was, I believe, described under a banner head reading: "What are the Problems of Scientific Research Today?

"I can scarcely imagine a more fascinating question; I should gladly attend a series of lectures devoted to it," Kuhn continued. "Except," he emphasized,

> that I doubt whether any serious student of science or scientific method would consider himself equipped to address such a subject. Therefore with apologies for any confusion that the misrepresentation may have created, I should like to announce that I do not intend to deal with any of the problems raised by that question at any point in this series of lectures. (Kuhn, "Introduction: Textbook Science and Creative Science." Surviving drafts of Kuhn's Lowell Lectures are at TSK-MIT, box 3, folders 10–11.)

This is because, Kuhn explained, the study of methodology can only bear fruit when directed at practices and beliefs superseded and rendered obsolete or antique. If we seek to understand where theories and "finished conceptual schemes" come from, he explained, this

> is peculiarly hard to achieve in dealing with the science in which we happen to believe. For the theory in which we believe is necessarily and uniquely characterized by the apparent inevitability of its relation to the facts from which it arose. Our belief itself represents a commitment to the double position that only this theory will account for the facts which we know and that this theory will account for all the relevant facts. We may admit that in the future there will be other facts and other theories, but for the moment we cannot conceive these, so the appearance of the inevitable connection remains. (Kuhn, "Introduction: Textbook Science and Creative Science." TSK-MIT, box 3, folders 10–11., pp. 7, 8–9.)

Science cannot be planned, that is, because the mechanisms of scientific progress (such as the dynamics of paradigms Kuhn would later articulate in *Structure*) are obscured by contemporary beliefs that scientific knowledge is true and complete ("that only this theory will account for the fact which we know and that this theory will account for all the relevant facts"). They are available for inspection and analysis only by future historians or methodologist who are not so blinded by their epistemological and methodological commitments.[17]

[17] Not only Kuhn's focus on science of the past, but a tacitly invoked pure-versus-applied distinction additionally isolated his Lowell Lectures from debates about science's proper or potential roles in modern life: "We shall be concerned with the sort of research that led to the Newtonian laws of motion, not with the manner in which these laws were applied in building new machines or instruments. We shall be concerned with the work of such men as Boyle and Dalton, in so far as this led to a new understanding and a new set of laws governing the formation of chemical compounds, but we shall not be concerned with the manner in which these laws, once arrived at and confirmed, were applied to the production of dyes, explosives, or plastics" (Kuhn, "Introduction: Textbook Science and Creative Science," op. cit., p. 7).

14.6 *The Structure of Scientific Revolutions* and the Postwar Future of Science Studies

By the time that *Structure* was published in 1962, Kuhn had reformulated his theoretical language and refined his theory of how science works. But his new language of paradigms, normal science, crisis, and revolution made clear (arguably clearer than his Lowell Lectures, which remain unpublished) his understanding of why the Unity of Science Movement's aspirations were not only utopian but fundamentally mistaken. The real engine of scientific progress, *Structure* argued, is not unification, synthesis, and the overcoming of specialization; it is specialization itself and the impulse to push any paradigm to its epistemological limits.

On this point, *Structure* stands with Bush's, Conant's, Flexner's and Weaver's shared contention that progress depends essentially on the freedom of scientists to conduct research independently, by their own lights, and without external constraint. Importantly, this is not because scientists themselves are the only effective and legitimate planners of science's future. As he implied in his Lowell Lectures, Kuhn believed that no living persons – not even scientists themselves – are equipped to plan the future of science. Instead, scientists are equipped and trained only to solve the problems their paradigmatic tradition bequeaths to them. But this modest function is vitally important for the advancement of science, for scientists in their communities are uniquely positioned to notice and recognize when these puzzles become recalcitrant and something may be fundamentally wrong with the current paradigm. As he put it in his essay "The Function of Dogma in Scientific Research," the dogmatic determination of scientists to solve all of the puzzles their paradigm offers renders the scientific community "an immensely sensitive detector of the trouble spots" that may spark future revolutions (Kuhn 1963, 349). As he put it in *Structure*, this mental "rigidity" of scientists "provides the community with a sensitive indicator that something has gone wrong" with the paradigm in question (Kuhn 1962/2012, 165). So it is, according to *Structure*, that while future scientific change is real and inevitable, no one can reliably predict in advance, much less plan or legislate, what future paradigms will be like. Modern science is essentially unplannable.

As this feature of Kuhn's *Structure*-era theorizing may suggest, Kuhn did not share all of the libertarian politics of science defended by Conant, Bush, Weaver and others during the planning debates. While they believed productive and creative scientists and their innovations were among the West's most convincing proof of the geopolitical superiority of liberal democracy over eastern totalitarianism, Kuhn saw professional scientists in a very different light as dogmatic, narrowly trained, and not in fact creative, open minded, and intellectually adventurous. As he explained in his chapter "The Invisibility of Revolutions," most scientists are not even aware of the revolutionary dynamics by which their beliefs and commitments became established and accepted. As some scholars would later notice, by recognizing only a single, commanding paradigm at the heart of any scientific community, *Structure* seemed to take leave of Bush, Weaver, and Conant and to accept and recommend centralized forms of epistemic organization.[18] But Kuhn agreed nonetheless with these cold warriors that the Unity of Science Movement and all attempts to plan and guide the future of science had little credibility and therefore deserved little programmatic support within the academic study of science.

Acknowledgement I thank Günther Sandner for pointing out Modley's importance to me, as well as spirited commentary from others at the conference 'The Socio-Ethical Dimension of Knowledge: The Mission of Logical Empiricism,' Budapest, December 2017.

References

Barnard, Eunice. 1933. In the Classroom and On the Campus. *New York Times*, Aug. 20.

Bush, Vannevar. 1945. *Science: The Endless Frontier*. Washington, DC: National Science Foundation.

Capps, John. 2003. Pragmatism and the McCarthy Era. *Transactions of the Charles S. Peirce Society* 39 (1): 61–76.

Cartwright, Nancy, Jordi Cat, Lola Fleck, and Thomas Uebel. 1996. *Otto Neurath: Philosophy Between Science and Politics*. Cambridge: Cambridge University Press.

Charles, Loïc, and Yann Giraud. 2013. Economics for the Masses: The Visual Display of Economic Knowledge in the United States (1910-1945). *History of Political Economy* 45 (4): 567–612.

Conant, James Bryant. 1945. Letter to the Editor ("National Research Argued"). *New York Times*, Aug. 13.

———. 1947. *On Understanding Science*. New Haven: Yale University Press.

———. 1970. *My Several Lives: Memoirs of a Social Inventor*. New York: Harper and Row.

Douglas, Heather. 2014. Pure Science and the Problem of Progress. *Studies in History and Philosophy of Science Part A* 46: 55–63.

Duffus, R.L. 1939. A Graphic View of Our World, review of Neurath (1939). *New York Times*, Oct. 1.

Fleck, Christian, Albert Mueller, and Nico Stehr. 2005. Afterword in Paul Neurath, *Society of Terror: Inside the Dachau and Buchenwald Concentration Camps*. New York: Routledge.

Flexner, Abraham. 1939. The Usefulness of Useless Knowledge. *Harpers* 179, June/November, 544–552. Reprinted by Princeton University Press, 2017.

Fuller, Steve. 2000. *Thomas Kuhn: A Philosophical History for Our Time*. Chicago: University of Chicago Press.

———. 2004. *Popper vs. Kuhn: The Struggle for the Soul of Science*. New York: Columbia.

Hollinger, David. 1996. *Science, Jews, and Secular Culture*. Princeton: Princeton University Press.

Howard, Don. 2003. Two Left Turns Make a Right: On the Curious Political Career of North American Philosophy of Science at Midcentury. In *Logical Empiricism in North America*, ed. Gary L. Hardcastle and Alan W. Richardson, 25–96. Minneapolis: University of Minnesota Press.

Kaempffert, Waldemar, ed. 1917. *Book of Modern Marvels*. New York: Leslie-Judge, co.

———. 1933. Staccato Speech for Silent Statistics. *New York Times*, Jan. 22.

[18] Conant himself, for example, initially objected to Kuhn's theory of paradigms on the grounds that it portrayed all members of a community bound by just one scientific point of view. Other objections came from Paul Feyerabend, Karl Popper and his circle, and others treated in detail in Reisch (2019). More recently, David Hollinger noted the ironically "totalitarian" character of *Structure*'s theory of science (Hollinger 1996, 169, 170). The Popperian/Conantian line of argument against *Structure* was later taken up by Steve Fuller (2000, 2004).

———. 1935. Words, Words, Words. A School of Thought in Vienna Makes a Test of Language. *New York Times*, Feb. 3.
———. 1936. Congress in Copenhagen. The New Concept of Causality Will Receive Emphasis. *New York Times*, April 12.
———. 1937a. Unifying the Sciences. A Language Must Be Developed Common to All Fields. *New York Times*, Jan. 10.
———. 1937b. Language of Isotypes. New, International Means of Communication Gains Use. *New York Times*, Jan. 17.
———. 1937c. Science Encyclopedia. Sciences to Be Unified Through Common Language. *New York Times*, Feb. 14.
———. 1938. Toward Bridging the Gaps Between the Sciences, review of first 2 volumes of the *International Encyclopedia of Unified Science*, *New York Times*, Aug. 7.
———. 1943. Horizons of Science: The Case for Planned Research. *American Mercury* 57: 442–447.
———. 1945a. Science in Review: Dr. Bush Outlines a Plan for the Creation of a National Research Foundation. *New York Times*, July 22.
———. 1945b. Science in Review: Further Arguments in Favor of Research Organized on a National Scale. *New York Times*, Aug. 19.
———. 1945c. Science in Review: Organization and Planning Held Necessary to Close Gaps in Our Knowledge. *New York Times*, Sept. 9.
Kegley, Charles. 1959. Reflections on Philipp Frank's Philosophy of Science. *Philosophy of Science* 26: 35–40.
Kuhn, Thomas. 1962/2012. *The Structure of Scientific Revolutions*. 4th edition. Chicago: University of Chicago Press.
———. 1963. The Function of Dogma in Scientific Research. In *Scientific Change*, ed. A.C. Crombie, 347–369. London: Heinemann.
McCumber, John. 2001. *Time in the Ditch*. Evanston, IL: Northwestern University Press.
———. 2016. *The Philosophy Scare*. Chicago: University of Chicago Press.
Modley, Rudolf. 1935. Facts Told Pictorially. *New York Times*, Sept. 15.
Neurath, Otto. 1928/1973. Personal Life and Class Struggle. In *Empiricism and Sociology*, ed. Marie Neurath and Robert S. Cohen, 249–298. Dordrecht: D. Reidel.
Neurath, Otto, Rudolf Carnap and Hans Hahn. 1929/1973. The Scientific Conception of the World: The Vienna Circle. In *Empiricism and Sociology*, ed. Marie Neurath and Robert S. Cohen, 299–319. Dordrecht: D. Reidel.
New York Times. 1933. Educator Describes Picture Esperanto, Jan. 10.
———. 1937a. Real Estate Expects Fair to Spur Market, Jan. 24.
———. 1937b. Review of Rudolf Modley, *How to Use Pictorial Statistics*. New York: Harper & Bros., Nov. 14.
———. 1939a. Magnetic Blackboard, May 28.
———. 1939b. Book Notes, announcement of Neurath, Otto, *Modern Man in the Making* (Neurath 1939), Aug. 17.
———. 1939c. Books Published Today, announcement of (Neurath 1939), Sept. 12.
———. 1939d. Books for Christmas, includes (Neurath 1939), Dec. 3.
———. 1945a. Research for Defence, July 21, 1945.
———. 1945b. The Nobel Awards, Oct. 27.
———. 1945c. Large Problems of Post-War America Are Laid Before Congress by President in Message, Sept. 7.
———. 1945d. Truman on Research Bill, Oct. 30.
———. 1945e. Truman Aid Asked for Magnuson Bill, Nov. 27.
———. 1956. Waldemar B. Kaempffert Dies, Science Editor of the Times, 79, *New York Times*, Nov. 28.
Nye, Mary Jo. 2011. *Michael Polanyi and His Generation*. Chicago: University of Chicago Press.

Reisch, George A. 1994. Planning Science: Otto Neurath and the International Encyclopedia of Unified Science. *British Journal for the History of Science* 27: 153–175.

———. 2005. *How the Cold War Transformed Philosophy of Science*. New York: Cambridge University Press.

———. 2017. Pragmatic Engagements: Philipp Frank and James Bryant Conant on Science, Education, and Democracy. *Studies in East European Thought* 69: 227–244.

———. 2019. *The Politics of Paradigms: Thomas Kuhn, James Bryant Conant, and the "Struggle for Men's Minds" in the Cold War*. New York: SUNY Press.

Shils, Edward. 1947. A Critique of Planning – The Society for Freedom in Science. *Bulletin of Atomic Scientists* 3.

Stadler, Friedrich. 2001/2015. *The Vienna Circle: Studies in the Origins, Development, and Influence of Logical Empiricism*. 2nd edition. Dordrecht: Springer.

Survey Graphic (Anonymous). 1932. Otto Neurath Visits Russia. *Survey Graphic* 21 (Nov. 1): 538–539.

Thompson, Ralph. 1939. Books of the Times, review of (Neurath 1939), *New York Times*, Sept. 13.

Tuboly, Adam Tamas (ed.). 2017. The Life and Works of Philipp Frank, Special Issue. *Studies in East European Thought* 69 (3): 199–276.

Uebel, Thomas. 2000. Logical empiricism and sociology of knowledge: The case of Neurath and Frank. *Philosophy of Science* 67: 138–150.

Weaver, Warren. 1945. Letter to the Editor, "Free Science Sought," *New York Times*, Sept. 2.

Werskey, Gary. 1978. *The Visible College*. London: Allen Lane.

———. 2007. The Marxist Critique of Capitalist Science: A History in Three Movements? *Science as Culture* 11: 397–461.

Wuest, Amy. 2015. *Philipp Frank: Philosophy of science, pragmatism, and social engagement*. Ph.D Dissertation. University of Western Ontario.

Chapter 15
Of Tennis Courts and Fireplaces: Neurath's Internment on the Isle of Man and his Politics of Design

Michelle Henning

Abstract Otto Neurath's version of functionalism is one that begins with people "as we find them," a proposition first set out in his 1917 essay "The Converse Taylor System." Any attempt to redesign the existing furnishings of everyday life must take into account "functions" that go beyond the obvious purpose of objects: functions that are to do with sociability, happiness, familiarity, the love of "coziness," and that address the diversity and contradictoriness of people. This essay considers how Neurath applied and made use of these ideas about design in 1940s Britain, during and after his internment on the Isle of Man between 1940–1941 and in talks, papers and correspondence from this period. It does not focus on the Isotype Institute, which would usually be considered his principal intervention in design, but on his commentary on everyday objects and practices. In particular it centres on four items – tennis courts, fireplaces, chairs and shoes – and through these elaborates some of the connections between Neurath's ideas about the design of everyday life, and the significance of everyday practices, and his logical empiricism.

15.1 Introduction[1]

During the early 1920s, as part of his work in the *Österreichischer Verband für Siedlungs- und Kleingartenwesen* (Settlement and Allotment Garden Association), and then through his work in the *Gesellschafts-und Wirtschaftsmuseum* (Social and

[1] This essay is partly based in AHRC-funded archival research in Otto Neurath's correspondence and papers during 2007–2009, and later research at the Manx Museum in Douglas, a visit to the site of the Onchan internment camp and discussion with friends and relatives of internees in nearby camps, as well as secondary reading. I am grateful to the AHRC, Eric Kindel at the Otto and Marie Neurath Isotype Collection (Department of Typography and Graphic Design, University of Reading), to Ádam Tuboly and Jordi Cat, and to Sabrina Rahman for all their help. This essay began life as a paper at *Politics, Democratic Education and Empowerment: The Case of Otto*

M. Henning (✉)
London School of Film, Media and Design, University of West London, London, UK
e-mail: Michelle.Henning@uwl.ac.uk

J. Cat, A. T. Tuboly (eds.), *Neurath Reconsidered*,
Boston Studies in the Philosophy and History of Science 336,
https://doi.org/10.1007/978-3-030-02128-3_15

Economic Museum) in Vienna, Otto Neurath had developed strong views about design as it pertained to everyday life. His views regarding functionalism in design, and regarding the social importance of design, were strongly influenced by his close friend, the architect and textile designer Josef Frank. With Frank, he oversaw the development of the innovative housing estate, the *Werkbundsiedlung*, which opened on the outskirts of Vienna in 1932 (Rahman 2014, 22; Sophie Hochhäusl in the present volume).

Frank's practice stood in stark contrast to the functionalist aesthetic of the *neue Sachlichkeit* and the Bauhaus. His houses built for the 1927 German *Werkbund* exhibition ("*Die Wohnung*") in Stuttgart, had been criticised on the grounds that his work was too decorative: critics described the interiors as "femininely appointed," filled with "frippery," like a "bordello" (cited in Long 2002, 108; Galison 1990, 723). Frank responded by arguing that an empty and affected "functionalism" did not address psychological needs for comfort, coziness, and liveability. He also argued that the fashion for bare furnishings was more oriented toward intellectuals than to the working classes: "The demand for bareness is made particularly by those who think continuously, or at least need to be able to do so, and who can obtain comfort and rest by other means" (cited in Blau 1999, 196; Frank 1927).

In his own work, Neurath recognized the importance of everyday household objects and architecture in making possible certain ways of living, allowing a tolerable and viable way of life. However, like his friend, he was a strong critic of the ideas of shaping a way of life that were held by some modernist designers (of the *neue Sachlichkeit* and the Bauhaus) and the concepts of function and causality these implied. Together with Frank, he took the view that functionalism in design was actually a specific aesthetic style, and that there was no such thing as a true or complete functionalism, since that would require knowing in advance the full range of uses to which a designed object might be put.

Neurath's version of functionalism is one that begins with people "as we find them," a proposition first set out in his essay "The Converse Taylor System" (1917/1973). Any attempt to redesign the existing furnishings of everyday life must take into account "functions" that go beyond the obvious purpose of objects: functions that are to do with sociability, happiness, familiarity, the love of "coziness," and that address the diversity and contradictoriness of people. This essay considers how Neurath applied and made use of these ideas about design in 1940s Britain, during and after his internment on the Isle of Man between 1940–1941 and in talks, papers and correspondence from this period. It does not focus on the Isotype Institute, which would usually be considered his principal intervention in design, but on his commentary on everyday objects and practices. In particular I shall focus on four items – tennis courts, fireplaces, chairs and shoes – and through these elaborate some of the connections between Neurath's ideas about the design of everyday life and the significance of everyday practices, and his logical empiricism.

Neurath (1992–1945) Universität Wien/Institute Vienna Circle Symposium, 28 May 2015, and I am grateful to Günther Sandner, Chris Burke, Elisabeth Nemeth, Friedrich Stadler as well as the other participants in the symposium.

15.2 Internment and the Tennis Court

Details of Otto Neurath and Marie Reidemeister's arrival in Britain and their internment on the Isle of Man are given in Ádám Tamás Tuboly's chapter in this volume, "United by Action: Neurath in England." Here, I will give additional background to contextualize my discussion of how Neurath's ideas about design and everyday life both shaped and were shaped by his experience of internment. This background will help to elucidate a lecture given by Neurath while he was interned.

The Isle of Man is not part of the United Kingdom but is classed as a "self-governing British Crown dependency." One of its principal trades in the 1930s was tourism. In 1939–40, the holiday trade on the Isle of Man was severely affected by the War, and the Manx Chamber of Trade suggested the island be a site for Internment camps as it was in World War One. These previous camps had held civilians who were mostly German nationals living in Britain. However, this time, the home secretary decided not to build camps in the countryside but to requisition the large, terraced Victorian boarding houses, which were central to the bed and breakfast trade on the island. The decision did not please the Manx landladies, who had to vacate their houses very quickly at the end of May 1940. But it was designed to placate the press and the public, coinciding with a growing media panic about spies and "fifth columnists." World War Two internment on the Isle of Man began in late 1939.

When Neurath and Marie arrived in England on 15th May 1940, the numbers of refugees arriving from Europe each day was increasing. The internment program expanded to take in men and women who had lived in Britain for years, together with the newly arrived refugees from Europe, most of whom were Jewish. Neurath, who at 57 was one of the oldest internees (the cut-off age was 60), was held in Onchan camp, in the north of Douglas bay on the Isle of Man, one of around 1200–1300 German-speaking men. The camp was made up of four streets in Onchan, surrounded by double fences of barbed wire, and consisted of around 56 to 60 large furnished houses, many with nine bedrooms or more. With two or three men to a bedroom, Onchan camp was still less overcrowded than other camps on the island.

The internment program was indiscriminate and conditions uncomfortable. In the beginning, Nazi sympathizers were sometimes housed with Jews. Some would have arrived without full identification papers, and could be using false names. This made it very difficult to know who to trust among the other prisoners. Neurath, to my knowledge, did not commit much description of the camp to writing, but from other refugee accounts, we know that the emotional impact of internment was very varied. For some prisoners, particularly those who had already experienced the Nazi concentration camps, it was traumatic. For Jews and known opponents of Nazism there was another danger: in 1940, no-one could know the outcome of the war, and the internees had no access to news and communications, but they realized that should Germany take Britain, they would have no escape. Some felt that they were effectively "sitting ducks." Moreover, winter was brutally cold on the Isle of Man, with only Victorian fireplaces for heating.

Nevertheless, in his history of the Isle of Man internment camps in World War Two, *Island of Barbed Wire*, Connery Chappell (1984, 40) suggests that the size of the houses, the beautiful sea views from the headland and the presence of football pitches and tennis courts meant that "Onchan Camp could reasonably have been regarded as the 'best' male internment camp on the island." Local residents generally had no contact with the interned men but would see them accompanied by soldiers, going down to the sea, to go swimming. The camp included recreation facilities because a social club was a part of the requisitioned area.[2] At first there was a ban on communications but radios were allowed after a while and the men produced their own newspaper – *The Onchan Pioneer.* A Popular University was founded and between May 1940 and February 1941, 496 lectures were held.

According to *The Onchan Pioneer* it was Neurath's lecture, given in January 1941, that held the record of the highest attendance for an indoor lecture. 250 men came to hear him give a talk cryptically titled (according to the *Pioneer*), "How do you make the tennis court so durable?". This title can be read very literally as meaning "Why is the tennis court hard?". That is, why have a tarmacked tennis court (such as the one in Onchan camp) rather than a grass court? However, Neurath's equally cryptic notes for this paper are in the Otto Neurath Nachlass, with the title "Wie Machen Sie's nur dass der Tennis-Rasen so Dauerhaft ist," perhaps better translated as "How Does the Tennis Court Endure?"[3] The notes are in German and it is likely that the lecture was given in German, but unfortunately they do not give a clear sense of the full content, since they are little more than a list of prompts. Neurath subtitled it "A social-critical reflection," and we can see in the list some hints of the ideas that would continue to preoccupy him over the next five years.

One prompt listed in the lecture notes is the phrase: "*Wo ist das Pferd?*" ("Where is the horse?"). This peculiar question makes more sense in the context of Neurath's notes for a later lecture, given after his release from internment, at Bedford College, Cambridge, on 9 November 1941. There, Neurath described how in the early nineteenth century, in some countries, members of the ruling classes wore tailcoats and high boots, clothes that were originally associated with horse riding, but now worn when not riding. His point was that the fashion signified a kind of modern status, because of the direct connection with horse riding, but he cautioned that not all customs could be read in this way: fashion loses its connection with function, but the function does not necessarily become a residual, subliminal or "subconscious" purpose or meaning (see Neurath 1942/1973).

This is part of a larger argument where he sets out his philosophy of logical empiricism and argues against thinking in terms of cause and effect or attempting to deduce too much, a tendency Neurath had long associated with what he termed "pseudo-rationalism" (Neurath 1913/1983). But it is also a critique of idealist

[2] It was called The Royal Avenue Social Club. See Onchan District Commissioners Flickr site: https://www.flickr.com/photos/88093414@N03/9520895627/. accessed 18 October 2018.

[3] "Tennis Rasen" 6. Jan. 1941. Otto Neurath Nachlass (ONN), Wiener Kreis Stichtung, Noord-Hollands Archief, 206/K. 82. Thanks to Sabrina Rahman for help with the translation of these notes.

theories of culture which tried to establish a connection between the will, spirit or mentality of an age or a generation, and its artistic or aesthetic manifestations, understood in terms of "style." Neurath's antipathy to such theories of national character and *Geist* was set out early in his 1921 essay "Anti-Spengler". This attack on Oswald Spengler's influential *Decline of the West* emphasized Spengler's pseudo-rationalism – "through method and proof he wants to compel our approval" – in what is, Neurath (1921/1973, 160) argued, essentially a work of speculative prophecy. In particular, Neurath saw Spengler's popularity as dangerous because of his insistence on "proofs" of decline, his confusion of history and biology, his treatment of culture as independent of social and environmental context or conditions.

Against the influence of such accounts Neurath attempted to set out other ways of describing and accounting for cultural difference. Indeed, the tennis court lecture notes reveal evidence of interest in questions of national difference, and in the possible role of German philosophical tendencies in the success of Nazism. There are mentions of traditions and taboos, of co-existence and tolerance of different aspirations and ways of life: all themes that will preoccupy his later writing in Britain. The notes end with mention of happiness and then, at the end, the words "*Tennis-Rasen*"– tennis court.

15.3 The English Fireplace

It is clear from his notes that Neurath meant "enduring" rather than durable, which has a subtly different meaning. If we ask the question "how do you make the tennis court endure?" we suddenly seem to be on more familiar Neurath territory and we can begin to imagine a possible lecture. Why, for instance, keep a tennis court in a prison? How to keep playing in such a situation? What is the importance of play, of pleasure?[4] What if the question about the tennis court is actually about how to preserve joy, against the odds?

Play mattered for Neurath. Above all, he valued human happiness and in his writings had frequently suggested that it ought to be the basis and the starting point for planning (town planning, and social and economic planning more broadly).[5] Later he would be misrepresented by Friedrich Hayek, author of the influential *The Road to Serfdom* (1944), as an apologist for planning as enforced social conformism and social engineering. Yet Neurath's understanding of planning and of design could not

[4] I don't know if the courts were used. I do know of one game – called Witness – played by Imre Goth and other internees in a different male camp several streets away. A good friend of his, the artist Marcia Farquhar reports: "The game involved an elected group staging an incident with all sorts of details to be recalled, or not, by the rest of the group watching. Even though the audience group were looking to remember there was a high instance of contradictory/fabricated memories. Imre only told me of this game in relation to the unreliability of witnesses." (Farquhar, Marcia Email to Michelle Henning, 6th May 2015).

[5] Neurath's Epicurean understanding of happiness is discussed in several texts including: Sandner (2007), O'Neill (2008), Cartwright, Cat, Fleck, and Uebel (1996) and Stuchlik (2011).

be further from this. His view was that planning done well would enable the individual freedom and non-conformism necessary for happiness. On Hayek he wrote,

> Professor von Hayek thinks of planning exclusively as something dictatorial. I do not know why… planning can be connected with a suppression of individuals hardly heard before, but also to enable us to be free to an extent hardly heard before, "free" i.e. a multiplicity of ways of life possible, non-conformism supported by planned institutions. (Neurath, "Physicalism, Planning and the Social Sciences: Bricks Prepared for a Discussion v. Hayek, 26 July 1945." 202/K.56. Otto Neurath Nachlass.)[6]

It seems likely that the tennis court played a similar role as the "English fireplace" did in Neurath's thought. The latter comes up several times in Neurath's notes and effects after internment. It is mentioned in his 1941 lecture at Bedford College Cambridge on logical empiricism. It comes up again in his 1942 essay "International Planning for Freedom." It also appears in notes for a talk, not by Neurath, but by Henry N. Winter titled "The Englishman Abroad" which is among Neurath's papers in the Otto and Marie Neurath Isotype Collection at the University of Reading.

In these writings the fireplace example serves a dual purpose. First, it seems to represent the importance of pleasure and human happiness as against the tendency of social planners and designers to emphasize efficiency and function. For example, in the Cambridge lecture, Neurath contrasted the German use of the fire with the British. In Germany, he suggested, a fire is "a tool for making warm," but in Britain it also has a social function, "centralising, grouping people, … giving an opportunity to be together."

Grouping and gathering around the fire had been a necessity in the Onchan camp during the bitter winter nights. In the Manx Museum in Douglas, one painting made by an internee shows a large group of internees around a fireplace. The museum also includes a Manx fireplace, which in the traditional Isle of Man cottages would typically take up an entire wall and incorporated seating and ovens. The fireplaces of the Onchan boarding houses are more conventionally Victorian in design, similar to suburban houses built throughout Britain in the late nineteenth-century. These large houses would have had fireplaces in bedrooms, as well as in the main living room, but availability of coal may well have limited fires to the main room, forcing the men to cluster together around the fire.

In his 1941 lecture, Neurath made the point that, if to German and Austrian eyes, the English fireplace is a "waste of calories" (since 80% of the heat goes up the chimney), we might say the same about skiing. He argued that what one person views in terms of efficiency, calories and waste, another calls pleasure. His discussion of the fireplace is a riposte to those functionalists in design who have a limited notion of function, and a means of showing that planning is more difficult than it might seem, since the impact of a fireplace is not quantifiable. He had made this argument in an earlier article, "Inventory of the Standard of Living", where he

[6]As John O'Neill (2006, 2) says, Neurath was "a central target" of Hayek's papers "The Counter-Revolution of Science" and "Scientism and the Study of Society" published between 1941 and 1944. Hayek misrepresents Neurath as more concerned with scientific measures and centralization than he actually was.

recognized that "the usual standard of living research, though very useful, does not tell the whole story of human happiness" (Neurath 1937/2004, 517). The things that go toward our happiness are of different orders, he wrote, We can speak about trebling the mortality rate, but perhaps not about trebling the beauty of an ocean view" (Neurath 1937/2004, 517).

Even as early as 1912, Neurath had argued against the attempts to measure pleasure made by the Austrian School of economists and the Utilitarians. In a lecture entitled "The Problem of the Pleasure Maximum," he set out to demonstrate the impossibility of such calculations since, as Jordi Cat (2014) explains "cardinal measures for comparative utility, or pleasure values, could not be determined for the same individual, much less for different individuals" (cf. Neurath 1912/1973, 118–119).[7] In "Inventory of the Standard of Living," he developed this critique further, describing the dominant approach in economics as an "atomistic, utilitarian approach" in which human feelings appear only as the pleasures and pains correlated with "commodities" and "discommodities" (Neurath 1937/2004, 513–526). Neurath argued that a sense of well-being is not the aggregate of various pleasures and pains, and therefore cannot be measured according to a pleasure calculus.[8]

The point about the function of the fireplace comes up again in this 1942 essay, where he makes the point that fireplaces are not "happiness neutral:"

> Let us take an uncontroversial example. Assume the scientists tell the English people that their fireplaces waste calories – of course they do so enormously. But the fireplaces as an element of our environment are not "happiness-neutral" as it were, as is e.g., the cable shaft below the surface of the street. The fireplaces are related to homely comfort and to many customs of our private life. (Neurath 1942/1973, 427.)

The second purpose of the fireplace example is as a means for Neurath to distance himself from any straightforward idea that design can produce or cause certain forms of sociability. It should be clear that Neurath was not claiming that the fireplace contributed to an English immunity to Nazi or Fascist government, although at times in his writing he seems to be implying this. Rather, he uses it as an example to warn against deducing too much from phenomena. Design carries a great responsibility but its consequences cannot be determined in advance: as he says in the Cambridge lecture, "changing the fireplace institution means changing many things; we cannot say what. It is very difficult for sociologist to find out what things are related with that" (Neurath 1941)[9].

To deduce too much, Neurath argued, is dangerous: "relations are always interesting, but all relations are so to speak without direction. If you give a relation a direction you are adding more. That is dangerous" (1941). He began his Cambridge

---

[7] See also Uebel (2004) and his chapter in the present volume.

[8] Neurath used the term "Lebensstimmung" which has been variously translated as "states of felicity" or "quality of life" but which could mean "sense of well-being" or, more clumsily "life-feeling."

[9] This is connected to his idea of unpredictability in principle, which he considered to be one of his most important contributions to the philosophy of science. See his *Encyclopedia* monograph, Neurath (1944, Sect. 12.).

lecture with a discussion of the development of logical empiricism out of the opposition between an over-systematizing "rationalistic attitude" and an empiricism that focused on scattered detail with no coherence. Logical empiricism, Neurath (1941) argued, is not an attempt to build a totalizing system, but a reflective approach, "an attempt to analyse more carefully than before the terms and expressions used in arguing." This means avoiding certain terminology and statements which assume more definite knowledge than can possibly be known, and accepting the provisional nature of decisions and solutions.[10] In everyday life we have to make decisions and find solutions all the time, and this understandably leads to "a tendency to overstate the possibility of unambiguous judgments." The uncertainty associated with the fireplace example demonstrated one of the main difficulties of planning: the impossibility of knowing the world in its entirety, and the importance of assembling diverse perspectives (and biographies of objects and people) rather than trying to impose an overarching view.

"International Planning for Freedom" is an explicit call for social planning – for the need to "consciously cultivate the future and the possible" (Neurath 1919/1973, 155). But the fireplace serves as a warning about how nuanced and complex this is, and a reminder of how pleasure as well as efficiency must take a central role. What makes people happy is very hard to anticipate since "all homely comfort relate to certain traditional customs and environments and that joy sometimes might depend solely on the fact that something should *not* be changed [...]. How much 'discomfort' is liked because it is 'ours.' And yet other people like changes and adventure" (Neurath 1942/1973, 423; original emphasis).

Neurath did not specify which of his German friends had been so damning about the wasteful English fireplace. But we can look at commentary from the period to see that the fireplace is under fire, so to speak. William Gaunt (1934, 605), writing in the *Journal of the Royal Society of the Arts* in 1934, argued "we no longer have any need for a huge black cave in the room in which a fire burns" and "round which a shivering family crouch" on the grounds that the wireless now provided an alternate focus (similar arguments have been made about the television). In 1942, in the same journal, R. Fitzmaurice (1942, 501), anticipating the postwar rebuilding of Britain emphasized that the key factors in the design of heating systems were the economy and efficiency of fuel but that this was blocked in Britain by "a violent prejudice" in favor of the "open domestic grate." In other words, Neurath's example of the fire was not simply plucked from the air or even from experience, but came

[10] It means avoiding, for example, the language of cause and effect, in which we deduce one thing from another. Neurath gave an entertaining list of the kinds of accounts this would disallow: such as Max Weber's account of Protestantism as facilitating capitalism; accounts of the "concealed intentions" revealed in styles of dress or customs; arguments of the origins of a torturer in his childhood experiences; or about war as a necessary outlet for destructive tendencies in humanity, and similar arguments regarding film. Whether film produces aggression or acts as an outlet is a debate on which Neurath has "not the slightest hypothesis" – instead his aim is to point to the problem of the kind of definite assertions such speculations lead to – "if you are reading as a boy such things then the results are..." (Neurath 1941). On Neurath and films, see Cat and Alford (forthcoming).

from a recognition that heating systems were a key part of the debates in Britain surrounding postwar planning.

15.4 (In)Tolerance and Diversity

Of the three mentions of the English fireplace I cited above, the third is in a talk given by a correspondent of Neurath's called Henry N. Winter, later the author of a book called *Fluency in German*. In early January 1944 Winter had sent a copy of his notes to Neurath, at Neurath's request. Winter referred to his talk as being on his "impressions of Germany" and "the riddle of the German character," although the paper he enclosed was titled "Notes for a Talk: The Englishman Abroad." The talk was divided into headings: "The Englishman;" "The Foreigner;" "Home Life" and "Position of Women." Under "The Foreigner" Winter's notes include the following:

> In Germany every provincial town has its municipal theatre with own company, opera house, orchestra, art gallery, academy of music. Puritan tradition in England – suspicion of art and social pleasures.
>
> Englishman at heart a countryman, brings cottage and garden tradition into his towns. Retires to the country, whereas the German retires to some idealised town. Healthy interflow between town and country […].
>
> Large blocks of flats in continental cities compared with the English ideal of "one family, one house" with garden attached. Social significance of this difference. Significance of the open fireplace. (Winter to Neurath, January 1944. The Otto and Marie Neurath Isotype Collection, Department of Typography and Graphic Design, University of Reading.)

The notes characterize both English and Germans with highly dubious stereotypes, particularly in relation to attitudes of men toward women.[11]

Neurath's letter in return (on 15 January 1944) thanked Winter very politely but also provided some gentle criticism: he emphasized his own ability to see Germans from the outside since he is Austrian, not German; he emphasized the problems with proceeding from anecdote, or observation of the "puzzling multiplicity of German behavior," adding "you have to proof your case. And that is, as you know, difficult." He also commented, "I am just looking through the literature of the nineteenth century to find out, how that goes together the freedom of criticism, sometimes free to an unexpected extent, and the obedience and acceptance of militarism etc."[12]

In other words, Neurath himself was trying to find out why the German cultural environment might lend itself to militarism, propaganda, and a culture of obedience.[13]

---

[11] In his letter, Winter also refers to another part of the talk, where "I wished to attempt some explanation of the riddle of the German character, based upon the idea of a 'collective neurosis' (Freud und Jung)."

[12] Neurath to Winter, 15 January 1944. The Otto and Marie Neurath Isotype Collection, Department of Typography and Graphic Design, University of Reading.

[13] See Ádám Tamás Tuboly's and Antonia Soulez's chapter in the present volume, and Sandner (2011). Neurath was working on a book project which was never finished, provisionally titled *Tolerance and Persecution*.

He also seems to lean toward broad-brush generalizations which, though not as crude as Winter's, are still premised on polar oppositions between German and "Anglo" (British) character, though in an attempt to avoid essentializing or personalizing it, Neurath talks in terms of "atmosphere" and "climate." As Günther Sandner has argued,

> [t]he "German climate" [...] was not identical to the national character. For Neurath, not every German was automatically a representative of the "German climate". What he wanted to address were specific relations between certain features of German philosophy and literature and the behaviour of people. A human climate represented an ensemble of certain modes of behaviour, statements and articulations. (Sandner 2011, 76.)

Anticipating the issues of the "denazification" of German youth, which he understood not in terms of collective guilt, but in terms of the ideological consequences of Nazi propaganda, Neurath compiled lists or "questionnaires" (though not intended to be used to question people) as a means to collect "descriptive material." These include binary oppositions such as an opposition between the German trust of great leaders, and the British distrust of leaders, or contrasting the tendency of the "German atmosphere" to treat "lack of enthusiasm" as a defect, with the attitude in Britain (where it might even constitute a virtue!).[14] Similarly, he suggested, that to the British "compromise appears humane," while in the German culture "compromise appears bad." The same themes emerge in his correspondence with Carnap:

> It is impressive to listen to plain people here, how they avoid boasting and overstatements in daily matters. I collect "expressions", e.g. fire guard leaders speaking seriously, used e.g. once the term "happiness" explaining how people should get a feeling to be sheltered by the neighbours etc and then explaining, what is needed to act "quickly", to be "calm" and to have the "usual commonsense". I like this type of habit much more than the continental one, with "highest duty", "national community", "selfsacrifice", "obedience", "subordination", etc "eternal ideals", wherever you give a chance to open the mouth. (Neurath to Carnap, 25 September 1943. RC 102-55-03. See letter 22 in this volume.)

For Neurath as for Winter, fireplaces are linked to differences in national tendencies. Despite his own cautions about correlations between behaviors, environments and ideologies, the horrors of Nazism led Neurath to what appear, from a contemporary perspective, as untenable generalizations about cultural difference. In at least one talk he gave, Neurath acknowledged his own feelings, as an Austrian, about German culture:

> I hesitate a little to speak on this subject, for you see, resentment is not a good for scientific deductions and scientific discussions, and as a citizen of one occupied country and a refugee from another occupied country, I have sufficient resentment, but I know from history that sometimes hate and love are not the worst teachers. (Neurath, "Contributing features in the emotional and intellectual isolation of the German." K.48. Otto Neurath Nachlass, no date.)

---

[14] Neurath, "Questionnaire (IV)." 202/K.58. ONN. He also published an essay on this topic in *The Journal of Education* (Neurath 1945).

He acknowledged that he came from a very different intellectual tradition, which had more in common with English and French philosophy than with the German tradition informed by Kant and Hegel. But in a private letter to Ina Carnap, he also acknowledged a more personal resentment, linking his falling-out with her husband to the latter's Germanic attitude of unyielding principle:

> As you say Carnap is inflicting pain in the name of 'science, impartiality and suchlike gods', that is just, what I try to fight, and what my German friends usually try to defend, whereas my English friends in most cases agree with my attitude, which is based on compromise, muddle, happiness and not on some unhuman 'principles' (Neurath to Ina Carnap, 24 September 1945. RC 102-55-13. See letter 35 in the volume.)

It seems ironic that though he valued what he saw as the British "compromise habit, the not believing in too many arguments,"[15] Neurath was unable to refrain from such arguments in his correspondence with Carnap, despite the fact both were writing in English. There is a stark juxtaposition between their affectionate exchanges of household and personal news, and their strikingly uncompromising, blunt criticisms of one another's work. Perhaps neither man had a great enough grasp of the English language to make use of its many means for "beating around the bush," or its tendency toward politeness and euphemism. In any case, it seems that compromise and avoidance of argument were qualities Neurath aspired to, not ones that he could enact in his relationship with Carnap.

If Neurath could not separate his attitudes to the "German atmosphere" from his personal resentment and hurt, and his own intellectual distance from the dominant German philosophical tradition, it is also the case that, to some extent at least, Neurath was attempting to analyze German and British "atmospheres" or "climates" in the specific context of a broader discussion of reconstruction and Nazi education in Germany. In common with numerous exiled German-speaking scholars at the time, Neurath wanted to make sense of the culture, physical environment and even "intellectual and emotional environment" in which Nazism had taken root, not simply to form a theory of national difference but to counter the impact of Nazi education on a generation of young Germans.[16]

His notes on the meeting of the Belgium committee on 15 June 1945 reveal that he had argued there that the Nazi view that the war was a historical test, the victor crowned as fit to rule the world, had been challenged by the defeat of Nazism. Indeed, "[t]he war taught the Nazis the lesson that just the Nations with muddle defeated the nation which praised always the over efficiency of army, navy, air force and everything under the sun."[17] The victory of the allies was, in this sense, not a lesson in who was fit to rule the world, but in the potential of co-operation and compromise.

[15] Neurath to Carnap, 25 September 1943. RC 102-55-03. See letter 22 in this volume.

[16] Neurath to Joyce, 27 November 1944. The Otto and Marie Neurath Isotype Collection. See also Sandner (2011) and Antonia Soulez's chapter in the present volume.

[17] Neurath, "Meeting, Belgium committee... chairman Lauwerys, 15th June 1945." K.79, Otto Neurath Nachlass.

Neurath's diagnosis of the British "atmosphere" of muddle and compromise predates his arrival in Britain. It seems remarkable, but likely, that he addressed these themes in the tennis court lecture while interned in Onchan. Held captive by the British as a result of a xenophobic moral panic, Neurath was still prepared to speak of the British tradition of tolerance and diversity. The notes for the lecture include the following list:

> "Tolerance and toleration
> Coexistence and interpenetration of different aspirations
> Not even unity of the majority"

Though Neurath was an avowed Anglophile, and given he would also have had to be careful what he said within the context of the internment camp, we should not assume that he was celebrating this tolerance and mutual coexistence as a certain or unassailable fact of wartime Britain. He may well have been holding Britain to its own standards, and perhaps, using the microcosm of the group around the fireplace in the internment camp to help his audience picture that ideal democratic atmosphere. Later, in "International Planning for Freedom" he would argue that not only would a democratic society tolerate the diversity of people but the diversity and contradictions within individuals themselves. He makes the point through a quotation from the Swiss author Conrad Ferdinand Meyer's epic poem *Huttens letzte Tage* (Hutten's Last Days): "I am not a wittily constructed work of fiction; I am a human being and full of contradiction" (Neurath 1942/1973, 429).

That Neurath was able to talk about tolerance and diversity in a lecture in Onchan Camp might have had something to do with the shift in policy that had happened during his period of internment. In May 1940, when the internment policy was first put in place, the numbers of refugees arriving from Europe each day was increasing. The social survey organization Mass Observation had carried out a survey in April 1940 that suggested that very few people felt that mass internment of refugees was necessary: Tom Harrisson (1940, 36) of Mass Observation wrote, "literally not a single person contacted during the investigation felt that aliens should be interned *en masse*." A policy of classifying "enemy aliens" according to the risk they posed was already in place, but both the *Daily Mail* and the *Daily Mirror* had been pressing for extending internment to all "enemy aliens." This pressure came from journalists and editors who only a few years before had expressed pro-Nazi views. Their campaign succeeded in increasing hostility toward Germans, Austrians and (later) to Italians. When Mass Observation repeated their survey in mid-May the press campaigns seem to have had an effect. Harrisson (1940, 36) reported in *The New Statesman* that "many people who a month before were inclined to be tolerant of aliens were now almost pogrom minded."

Yet opposition to the policy came right away, in parliament and in the press, and it was compounded by the torpedoing of the *Arandora Star* in July 1940, a ship which was deporting internees to Canada. While some in parliament and the press attempted to present the drowned victims as Nazi sympathizers, the sinking of the *Arandora Star* affected attitudes towards the internment policy. This change in attitudes may also have been influenced by a book, *The Internment of Aliens* by the

27 year old François Lafitte, which was published by Penguin books in November 1940. Copies were smuggled into the camps, and although the book did not in itself-change internment policy, it was widely read.

Lafitte had been in Vienna in 1934 when the Austrian fascists invaded and had contacts among the Viennese left. In Britain, he was active in communist circles. His book mentioned Neurath and Marie Reidemeister, describing Neurath as a "world-famous pioneer of pictorial statistics" who "fled from Vienna in 1934 because he was a Social Democrat" (Lafitte 1940, 80). The book was filled with statistics and surveyed how members of the press and parliament, "individuals who should have known better" had drummed up xenophobic feeling and the pressure to "intern the lot." He emphasized that the majority of those interned were Jews, and detailed some of the cruel separations caused by the deportation policy, and how non-Nazis were forced together with Nazis in the camps. By the time Lafitte's book was published, tribunals for the possible release of internees were already underway on the Isle of Man, and during 1941 most internees would be freed – Neurath and Reidemeister were released in early February.

15.5 Happiness and Muddle

Neurath's principal opportunity to put his ideas about planning for human happiness into action in England was in his involvement with the redevelopment of the town of Bilston near Wolverhampton, in the West Midlands.[18] This work was fraught with local political difficulties and that it proved extremely stressful for Neurath is evident from his letters. He felt the contradiction between what he was touted in the media as doing – "bringing happiness to Bilston" – and the limited room for maneuver or influence that he had been given. Neurath died before his work in Bilston was completed, but from his letters we can see that he was trying to put his idea of a nuanced approach to planning into practice – as he wrote "I am looking at all these items from a personal point of view, how a single person in your society may look at it, as a father, as a tired person, as a person who would like to read a book."[19]

In beginning with where people are, what they actually do and enjoy (instead of where they ideally "ought to be") Neurath was being remarkably consistent with one of his earliest writings, "The Converse Taylor System" of 1917, where he argues for an approach to social planning that does not impose structures from above but builds upwards, from the diversity of people "as we find them" (Neurath 1917/1973, 131). He was also distancing himself from a certain tradition in German and British thought, which associated planning with moral reform. In 1940s Britain, there was

---

[18] See Nikolow (2004), Henning (2007), Rahman (2014). Sabrina Rahman's text is a short article about an exhibition she co-curated in Bilston, based on her research on Neurath's impact on the redevelopment of Bilston.

[19] Neurath's letter to A.V. Williams, 5 November 1945. Isotype 1/12-13. In the Otto and Marie Neurath Isotype Collection.

an influential discourse around "problem families," which, as Gillian Swanson (2007, 56–57) has argued, was shaped by the eugenics movement and pathologized "domestic failure". Another factor was the longer tradition of "social hygiene" reform that made sexuality and personal life the subject of social planning.

Nevertheless, in Britain, Neurath also found an environment open to debates about empathy and fellow-feeling and about happiness. As Swanson (2013, 141, 135) suggests, there was a significant difference between early twentieth-century British psychological models which emphasized "the cultivation of social feeling [...] towards 'human sympathies', feelings of 'fellowship' and universal 'brotherhood'" and other European models which "held group behaviour (and mass culture) in lower regard." Additionally, in the 1930s and '40s, British commentators repeatedly invoked the United States Declaration of Independence in order to argue for the role of government in facilitating the pursuit of happiness. The Liberal MP and author of the 1942 Beveridge Report which paved the way for the post-war welfare state, William Beveridge (1946, 56), wrote that one of the "primary duties" of government was "making possible for all the pursuit of happiness."

For Neurath, this attention to happiness was closely tied to British or English muddle (it is unclear in Neurath's writings whether he conflates Englishness and Britishness). This notion of "muddle" is often mentioned in writings on Neurath, where it tends to be described in terms of the absence of strict regulations. However, the term is more nuanced, as Neurath was aware. "Muddle" can be defined as frustrating, disorganized confusion, and we also have the British English expressions "muddling along" or "muddling through" which means to get by, to make do. It is associated with "botching," with the fix that is just good enough, and with making it up as you go along.

The wartime meanings of "muddle" were distinctive, and differed from the meanings it had accrued in other variants of English (such as American English). In one nineteenth-century American publication, Richard Soule's 1871 *Dictionary of English Synonyms*, muddle is defined primarily in relation to drunkenness and wastefulness: to "stupefy, fuddle, inebriate" and to "muddle away" was to "waste, misuse, squander." But in 1930s and '40s Britain, one could, quite successfully and tolerably, muddle along through life; muddling along is the opposite to grand ambitions, dreams of ideal society, or organized planning.[20] In this period, it also had a specific meaning linked to British identity and politics, which was to do with the absence of ideology, of policy and of economic planning. It was understood as a positive national characteristic, almost a virtue, at times. The term was used in both British and American contexts to characterize wartime Britain.[21]

[20] Botching is at the more creative end of muddling along and it is of course not an exclusively British trait: so for example, when the Onchan families finally got their homes back they discovered that the men had knocked doors through to get from one house in a terrace to another, had filled attics with soil to grow mushrooms, and had blocked the drains with radio parts, from the home-made radios they had cobbled together.

[21] See for example: "We Americans, younger in form of self-government by many years than the English, can learn something from the manner in which the English 'muddle' through adversity" (Mundt 1941).

Neurath connected British "muddle" to its origins in an old country. Germany, by contrast was a new country and "when people have no long tradition in civilization and no established type of living, what do they do? They make rules."[22] He perhaps overstated his admiration for muddle to his British correspondents, sensitive to the political climate in wartime and also possibly conscious of the censor. Nevertheless, and whether or not it was actually true that the British were not rule-bound and "do not fear differences" (a position that overlooks the bureaucratic and racist nature of the British Empire), he wanted to make the point that "the muddle is related to democracy"(ibid.). In a letter to Ina Carnap he also directly related it to happiness: "when thinking of human happiness one has to bear muddle, which is also essential for any evolved democracy."[23]

Given his lifelong commitment to social planning, Neurath's feelings on this score were also, and understandably, quite mixed. On the one hand he recognized in "muddle" a quality that might be necessary for preventing any kind of cultural hospitality toward Nazism, on the other hand he wanted to see botching and making do as merely a rational response to imperfect design, and therefore something that can be designed – or planned – out. I gave an example of this in an essay about Neurath's visit to Bilston (Henning 2007, 11–12). The town clerk of Bilston, A.V. Williams, wrote that the town councilors were worried that slum-dwellers moved into new modern houses would simply turn those houses into slums by putting coal in the bath. Against this, Neurath "stressed most emphatically that people only put coals in the bathtub for some very good reason" such as inadequate fuel storage places, or expensive hot water systems (Williams 1973, 76). He went on to mention that he knew a man in Vienna who kept a pig in his bath. Putting coal (or pigs) in the bath is a way of muddling along that supports (rather than undermines) Neurath's faith in human ingenuity, rationality and creativity. However, with a proper heating system or fuel storage, Neurath imagined that the muddle, or botch, would no longer be necessary.

Even so, Neurath's sensitivity to human feeling and the tendency to love what is not necessarily efficient or functional, is much greater than that of most commentators of the period, and despite his largely positive representation of Englishness, he was critical of certain practices of social planning and intervention in England. One famous and influential example is the Peckham Experiment which was initiated in 1925, and then rolled out in full from 1935–1939. The Peckham Experiment had set out to explore the impact of the environment on children's health and development through direct social intervention, by establishing a social club with health and leisure facilities, called the Pioneer Health Centre, in Peckham, South London. As Ben Highmore explains:

> For a smallish membership fee families could belong to the health centre and make daily use of its facilities, which included: a swimming pool, a gymnasium, crèche facilities, snooker, darts and table tennis, a cafeteria, covered play areas, a room for dances, a theatre

[22] Neurath, "Contributing features in the emotional and intellectual isolation of the German." K.48, ONN.

[23] Neurath to Ina Carnap, 24 September 1945. RC 102-55-13. See letter 35 in this volume.

> (for acting rather than surgically operating), badminton court, and so on. As well as using it for all forms of socialising and play, families could undertake regular "health overhauls." (Highmore 2006, 74–75.)

Actually, these "health overhauls," central to the "experiment," were compulsory conditions of membership. They were not conceived as medical appointments (since the consultants were biologists and the participants had not identified themselves as ill), but as opportunities to test and record the health of the families and to intervene pedagogically in practices of parenting, family planning and self-help. As Highmore (2006, 74) describes it, the Pioneer Health Centre was a modernist project "dedicated to the study and production of health." Yet it was also informed by the eugenics debates mentioned earlier. According to Innes Pearse, one of its founders, the center was designed on principles of surveillance:

> It was [...] necessary that the observers should be able to note the effect of the new environment upon family action. Hence the building was planned for visibility and free circulation throughout [...]. Everything was visible. One object of this provision was to test the hypothesis we had laid down, namely that the sight of action was a natural stimulus to action [...]. (Pearse 1945, 48–55.)

Although Neurath may have approved of the use of "exuberant social practices and learning through play" (Highmore 2006) at the Health Centre, he was concerned by and wished to distance himself from the experimental emphasis, which made participants the objects of study. In a letter to R.C. Kirk of the department of Zoology at Birmingham University, Neurath wrote: "please do not speak of 'experiment,' it is the very life of people at stake. I dislike the speaking of the Peckham experiment, it is more than that, because the life of families is altered and one cannot repeat the action [...] and using some people as experimental material for other people is against my feelings [...]."[24]

15.6 Chairs, Shoes and Functionalism

Even so, Neurath's modernist attention to the mundane aspects of British life was in some ways consistent with the approach in Peckham, and with the statistical, survey-based work of British organizations such as Mass Observation and Le Play House (Neurath was familiar with both).[25] His concern with ordinary objects was also consistent with the 1920s *neue Sachlichkeit* interest in everyday things.[26] For

[24] Neurath's letter to R.C. Kirk, 7 November 1945. The Otto and Marie Neurath Isotype Collection.

[25] His and Marie Neurath's correspondence includes letters to and from Dorothea Farquharson of the Institute of Sociology at Le Play House, in The Otto and Marie Neurath Isotype Collection. A 1943 letter to Josef Frank states "I just looked through the Mass Observation book on housing and so many common sense remarks from plain people." Neurath to Frank, 28 September 1943, Osterreichische Nationalbibliotek 1230/43.

[26] Neurath's relationship to the *neue Sachlichkeit* is discussed in Dahms (2004) and Damböck (2017).

him, these mundane things play three roles: they are a means of exploring and exemplifying the task of sociology and social theory, and what logical empiricism is capable of; they enable him to debate the nature of functionalism in design, debates begun in the 1920s in his interaction with the Bauhaus and with modern architecture in Austria and Germany; and third, they are the material of Isotype charts – the stuff out of which data can be produced, ways of life described and analyzed.[27] This last use is also mentioned in the Carnap correspondence:

> I should highly appreciate it if you were kind enough to send interesting newspaper cuttings and reprints and such stuff. We have now a nice studio again, with many files full of interesting material, but it is not our old richness, which was evolved in years. We like very much statistical data, interesting pictures of single objects, e.g. certain characteristic animals, busses, chairs, teapots, coffeepots etc, lists of knifes and forks, cups and pots, plates etc refrigerators etc. We are buying LIFE, LOOK etc for catching such material. Today we found in this way the shape of an American telephone apparatus, but there are thousands of apparatus, you know. (Neurath to Carnap, 17 July 1942. RC 102-56-04. See letter 11 in this volume.)

This attention to objects for the purposes of Isotype is not at all trivial. In this period, Isotype is becoming increasingly international and so the Isotype Institute had to take into account the recognizability and meaning of the pictograms in different cultural contexts. Since the beginning of the Vienna Method in the 1920s, close attention had been paid by Neurath, and by the chief artist in Vienna, Gerd Arntz, to the specific choice and design of symbols that had to fulfill criteria of recognizability and repeatability, since, as Christopher Burke puts it, "it was necessary that these signs be suitable for repetition in sequence along a line, to indicate statistical quantity; this differentiates Isotype pictograms from their successors in public signing" (Burke et al. 2013, 501–502). At the *Gesellschafts-und Wirtschaftsmuseum* they put together a systematic card index or "picture dictionary" of Arntz's designs, but without Arntz in Britain, with limited access to their old files, and with the changing shape of designed technical objects, there was a continuous need to update old pictograms and develop new ones.

Through Isotype and in his writings and lectures, Neurath began to plot a correlation between the design and uses of everyday objects and forms of sociability, as part of his attempt to arrive at a more complex, pluralist functionalism – understood from the ground up, that is, from the empirical basis of everyday experience.

This is best demonstrated via the example of chairs. Tennis courts and fireplaces endure: the tennis courts because of the necessity for play even in the most constrained circumstances, and fireplaces because of British obstinacy, and the inadequacy of efficiency calculations (or calorie counting). Yet chairs are amongst the objects most easily and frequently reinvented in modernism. The centrality of the chair in modernist design is remarkable and linked to the fact that chairs are most evidently a means to shape people by positioning their bodies. Chairs are anthropomorphic: literally taking on the shape of people, but also becoming person-like. In

---

[27] Isotype is discussed by Angélique Groß and Sophie Hochhäusl in this volume. See also Twyman (1975), Burke, Kindel, and Walker (2013) and Henning (2010).

the 1935 essay "Art as Experience" the Bauhaus teacher Josef Albers (1935, 391–392) wrote: "We should try to see a chair as a living creature […] as an apparatus willing to hold us, to carry, to surround or embrace us." Adolf Loos (1998, 65), another Viennese observer of British or English muddle noted: "following the principle that every type of tiredness requires a different chair, an English room is never furnished with one type of seat alone." Loos neglected the fact that being tired is not the only precondition for sitting. In Britain at least, fireplaces and chairs were intimately connected – you pulled up a chair to the fireside.

Similarly in one of his draft questionnaires, mentioned earlier, Neurath wrote about the German attitude that the house and furniture are not "indifferent places of happy living, adapted to people of various inclinations and tastes," and he contrasted this to "the Anglo-Saxon atmosphere [which] supports the attitude that house and furniture should remain relatively indifferent, not presenting any 'expression' of a certain person (father, mother or some architect), but to be a centre for different persons, therefore mixing up various kinds of seats, tables, etc."[28]

Neurath decorated one letter to J.K. Hunt, a member of the Ministry of Production committee, with a little cartoon captioned "The higher the seat, the lower the salary."[29] This wry joke points to another aspect of British seating arrangements that Loos overlooked: their function in maintaining forms of social distinction. Changing the seating arrangements, moving the chairs around, may be easier and more predictable in its impact than changing something as durable as the fireplace but, as with fireplaces, chairs have functions that go beyond their obvious use and beyond the purely symbolic. They facilitate interaction, they allow for certain kinds of discussion and they shape social behavior. In Neurath's view, to understand how they do this would be a task for a careful and nuanced empirical sociology.

For Neurath, the cultural and symbolic role of designed objects was as important as their social function even if one could not be extrapolated from the other. Shoes, for instance, have the ability to shape ways of being in the world – how we stand, run, walk. On the 19th March 1944, Neurath wrote to the British photographer John Hinde who he had met through the book-packaging firm Adprint, with some thoughts about a possible Isotype chart on shoes. During the war Hinde was mostly employed doing highly-staged, well-crafted, wartime propaganda photographs in full color. He was also working on a book project with Neurath, based around a Mass Observation study of the village of Luccombe, on Exmoor (the project was eventually published as *An Exmoor Village* by George G. Harrap and co. ltd. 1947). However, Neurath may have known that Hinde had also produced shoe advertisements for the company Clarks. Indeed, he was grandson of the company founder, James Clark, and lived in a village near Clarks' Somerset factory.

In his letter, Neurath offered plenty of suggestions for ways of researching shoes. Although it is unclear whether this is with reference to the Exmoor village project or a different project, the letter demonstrates that the Isotype Institute was much

[28] Neurath, "Questionnaire (IV)." 202/K.58. Otto Neurath Nachlass.

[29] Neurath's letter to Mr. J.K. Hunt, 18 September 1943. The Otto and Marie Neurath Isotype Collection.

more than a design organization that produced statistical charts and diagrams, and that it was involved in initiating and commissioning social research. Neurath wanted to know what kind of shoes people wore in "this part of the country," and suggested some ways of classifying them according to use: did people go barefoot at all, did they wear shoes and stockings, did they wear specific shoes for specific occupations or for "festivals, dancing, church, etc. or everyday life?" He also wanted to pay attention to the differences between shoes, between boys and girls, men and women, fashionable and old-fashioned people. Additionally, he asked Hinde for some comparative data: "with the next town, with other countrysides, with London etc." He requested details of heels, color, the materials the shoes were made of and of the extent to which shoes were repaired, or damaged shoes still worn. He even wondered what "names and expressions" were used to describe shoes. Neurath acknowledged that nothing might come of these questions: "Perhaps the result will not be very stimulating and only used in the text, perhaps something comes out worthwhile for 'Isotypizing' it."[30]

This is possibly not the first time Neurath has thought about shoes in relation to everyday experience. At the *Gesellschafts-und Wirtschaftsmuseum*, the museum in Vienna that Neurath opened in 1925, the staff photographer took a number of photographs that seem to attend to feet and footwear. The explicit content of these photographs is work, specifically factory labor, but the images draw our attention to the male and female workers' shoes. This is to do with the fact that the machinery they were using was partly foot-operated, but also the photographer would have been aware that shoes were an indicator of wealth or deprivation, and of types of labor (the workman's steel boot, the woman worker's comfortable slipper and swollen ankle indicating long periods spent standing, the fashionable Mary-Janes of the younger women). Among the *Gesellschafts-und Wirtschaftsmuseum*'s collections of photographs of the new state kindergartens, part of the social housing built by the socialist municipal government, are images of children tying their shoes. Such images have an obvious significance, indicating the growing prosperity of child and city as well as the development of independence through the civic kindergarten education.

Neurath had used shoes as an example in his Cambridge lecture of 1941 and possibly in the Tennis Court lecture too. As I suggested near the beginning of this chapter, he used the example of a tailcoat and riding boots to show how style is not purely functional ("where is the horse?"), but nor is it an unconscious expression. His example is a dancer in tailcoat and low shoes:

> He has very low shoes, very nice low shoes and he has also perhaps tails. What is that for a strange combination? The low shoes are shoes of the Red Indians and of other people who are running on plains and the tails are horseman's clothes. So I imagine [...] somebody might write I see the comprehensive modern man in his feeling combine all types of human life: on horseback subconsciously in the tails and running on the plains subconsciously in his shoes. (Neurath 1941.)

Neurath argues there are "dozens and dozens of books, seriously written of such a type." He could be referencing any number of Spengler-influenced texts, but there

[30] Neurath's letter to John Hinde, 19 March 1944. The Otto and Marie Neurath Isotype Collection.

is also an evident connection with the art historian Heinrich Wölfflin. In his dissertation (published in 1886) Wölfflin had argued that the Gothic shoe expressed the very same "historical state" of the human will or mind as the Gothic cathedral did. The three-pointed style, he argued, developed in contradiction to the demands of function and materials in order to express the Gothic spirit. Wölfflin claimed that "we feel forms by analogy to our bodies and forms are created as the unconscious expression of the corporeal feeling of an age." The closeness of shoes to the body connects body and spirit, or *Geist*.[31]

Whether Neurath was aware of Wölfflin's discussion of the Gothic shoe, it is unclear, but he certainly uses the example of shoes to demonstrate how culture and custom cannot be explained solely with reference to function nor with reference to "unconscious expression." Shoes, like fireplaces, are meaningful objects – and it is from such everyday, basic objects that we can learn lessons both for planning and design and for sociological study. These lessons are to do with the difficulties of disentangling overt symbolism and actual use, efficiency and meaning, of establishing causality and of the risk in making uninformed, under-researched changes to the everyday environment. Through these objects, and the everyday "muddling along" that they represent, Neurath was refining his politics of design and of decentralized planning.

Even in the 1940s, Neurath still wrote of the need to avoid dangerous, imprecise terms, and he bemoaned the difficulty people have in letting go of such terms. Yet as a number of Neurath experts have explained (notably Cartwright Cat, Fleck and Uebel 1996) he also recognized that ordinary language was necessarily formed of *Ballungen* – imprecise clusters of concepts. Cartwright and her co-authors also link "muddle" (as a specifically British quality) to Neurath's opposition to over-centralized planning. The present essay has attempted to show how Neurath's interest in muddle in the 1940s was accompanied by an increasingly thoughtful attention to the everyday objects through which daily life was made bearable, comfortable and pleasurable. These would form the basis of an approach to planning in which human happiness, not moral improvement, was the core value.

References

Albers, Josef. 1935. Art as Experience. *Progressive Education Journal* 12 (October): 391–393.

Beveridge, William. 1946. Life, Liberty, and the Pursuit of Happiness (1950 Model). *The Review of Economics and Statistics* 28 (2): 53–59.

Blau, Eve. 1999. *The Architecture of Red Vienna 1919–1934*. Cambridge, MA: MIT Press.

Burke, Christopher, Eric Kindel, and Sue Walker, eds. 2013. *Isotype: Design and Contexts: 1925–1971*. London: Hyphen Press.

Cartwright, Nancy, Jordi Cat, Lola Fleck, and Thomas Uebel. 1996. *Otto Neurath: Philosophy Between Science and Politics*. Cambridge: Cambridge University Press.

[31] Wölfflin cited in Schwarz (2005, 5–6). See Schwarz (2005, 4) for a discussion of Wölfflin's writing on the Gothic shoe.

Cat, Jordi. 2014. Political Economy: Theory, Practice, and Philosophical Consequence (Supplement to Otto Neurath), *Stanford Encyclopaedia of Philosophy*. http://plato.stanford.edu/entries/neurath/political-economy.html.

Cat, Jordi, and Benjamin Alford. (forthcoming). *Drawing Neurath and Polanyi Together and Apart: the Social, Scientific and Cinematographic Projects and Cultures that Animated their Diagrams*. Forthcoming manuscript.

Chappell, Connery. 1984. *Island of Barbed Wire: The Remarkable Story of World War Two Internment on the Isle of Man*. London: Robert Hale Ltd.

Dahms, Hans-Joachim. 2004. Neue Sachlichkeit in the Architecture and Philosophy of the 1920s. In *Carnap Brought Home. The View from Jena*, ed. Steve Awodey and Carsten Klein, 357–375. Chicago: Open Court.

Damböck, Christian. 2017. *‹Deutscher Empirismus› Studien zur Philosophie im deutschsprachigen Raum 1830–1930*. Dordrecht: Springer.

Fitzmaurice, R. 1942. Lighting, Heating and Ventilation. *Journal of the Royal Society of Arts* 90 (4616): 494–506.

Frank, Josef. 1927. Der Gschnas fürs G'müt und der Gschnas als Problem. In *Bau und Wohnung*, exhibition catalogue, ed. Deutscher Werkbund, 48–57. Stuttgart: Akademischer Verlag Dr. Fr. Wedekind.

Galison, Peter. 1990. Aufbau/Bauhaus: Logical Positivism and Architectural Modernism. *Critical Inquiry* 16: 709–752.

Gaunt, William. 1934. How to Improve the Design of British Goods. *Journal of the Royal Society of the Arts* 82 (4248): 603–620.

Harrisson, Tom. 1940. Should Leaders Lead? *The New Statesman and Nation* 20 (July 13): 36–37.

Henning, Michelle. 2007. The Pig in the Bath: New Materialisms and Cultural Studies. *Radical Philosophy* 145: 11–19.

———. 2010. Living Life in Pictures: Isotype as Modernist Cultural Practice. *New Formations* 70: 41–59.

Highmore, Ben. 2006. Hopscotch Modernism: On Everyday Life and the Blurring of Art and Social Science. *Modernist Cultures* 2 (1): 70–79.

Lafitte, François. 1940. *The Internment of Aliens*. Harmondsworth: Penguin.

Long, Christopher. 2002. *Josef Frank: Life and Work*. Chicago: University of Chicago Press.

Loos, Adolf. 1998. *Ornament and Crime: Selected Essays*. ed. Adolf Open, Trans. Michael Mitchell. Riverside: Ariadne Press.

Mundt, Karl. 1941. A Strong Opposition Is Needed: The Proper Functions Of A Minority Party. Delivered in the House of Representatives, August 1. *Vital Speeches of the Day* VII: 696–698 Online at http://www.ibiblio.org/pha/policy/1941/1941-08-01b.html.

Neurath, Otto. 1912/1973. The Problem of the Pleasure Maximum. In *Empiricism and Sociology*, ed. Marie Neurath and Robert S. Cohen, 113–122. Dordrecht: D. Reidel.

———. 1913/1983. The Lost Wanderers of Descartes and the Auxiliary Motive. In *Otto Neurath: Philosophical Papers 1913–1946*, ed. Robert S. Cohen and Marie Neurath, 1–12. Dordrecht: D. Reidel.

———. 1917/1973. The Converse Taylor System: Reflections on the Selection of the Fittest. In *Empiricism and Sociology*, ed. Marie Neurath and Robert S. Cohen, 130–134. Dordrecht: D. Reidel.

———. 1919/1973. Utopia as a Social Engineer's Construction. In *Empiricism and Sociology*, ed. Marie Neurath and Robert S. Cohen, 150–155. Dordrecht: D. Reidel.

———. 1921/1973. Anti-Spengler. In *Empiricism and Sociology*, ed. Marie Neurath and Robert S. Cohen, 158–213. Dordrecht: D. Reidel.

———. 1937/2004. Inventory of the Standard of Living. In *Economic Writings. Selections 1904–1945*, ed. Thomas E. Uebel and Robert S. Cohen, 513–526. Dordrecht: Kluwer.

———. 1941. Logical Empiricism and everyday problems. Lecture given at Bedford College, Cambridge, 19 November 1941. K.50. ONN.

———. 1942/1973. International Planning for Freedom. In *Empiricism and Sociology*, ed. Marie Neurath and Robert S. Cohen, 422–440. Dordrecht: D. Reidel.
———. 1944. *Foundations of the Social Sciences*. Chicago: University of Chicago Press.
———. 1945. Germany's Education and Democracy. *The Journal of Education* 77 (912): 370–371.
Nikolow, Sybilla. 2004. Planning, Democratization and Popularization with Isotype ca.1945: A Study of Otto Neurath's Pictorial Statistics with the Example of Bilston, England. In *Induction and Deduction in the Sciences*, ed. Friedrich Stadler, 299–329. Dordrecht: Kluwer.
O'Neill, John. 2006. Knowledge, Planning, and Markets: A Missing Chapter in the Socialist Calculation Debates. *Economics and Philosophy* 22: 1–24.
———. 2008. Happiness and the Good Life. *Environmental Values* 17 (2): 125–144.
Pearse, Innes H. 1945. The Peckham Experiment. *The Eugenics Review* 37 (2): 48–55.
Rahman, Sabrina. 2014. Happiness and Housing in 1940s Bilston. *WAVE Art and Heritage*: 22–23.
Sandner, Günther. 2007. Economy, Ideology and Culture: Otto Neurath's Approach to a Precarious Relationship. In *Otto Neurath's Economics in Context*, ed. Elisabeth Nemeth, Stefan W. Schmitz, and Thomas Uebel, 141–155. Dordrecht: Springer.
———. 2011. The German Climate and Its Opposite: Otto Neurath in England, 1940-45. In *Political Exile and Exile Politics in Britain after 1933*, ed. A. Grenville and A. Reiter, 67–85. Amsterdam/New York: Rodopi.
Schwarz, Frederic J. 2005. *Blind Spots: Critical Theory and the History of Art in Twentieth-Century Germany*. New Haven/London: Yale University Press.
Soule, Richard. 1871. *Dictionary of English Synonyms and Synonymous or Parallel Expressions*. Boston: Little Brown and Company.
Stuchlik, Joshua. 2011. Felicitology: Neurath's Naturalization of Ethics. *HOPOS* 1 (2): 183–208.
Swanson, Gillian. 2007. Serenity, Self-Regard and the Genetic Sequence: Social Psychiatry and Preventive Eugenics in Britain, 1930s–1950s. *New Formations* 60: 50–66.
———. 2013. The Tender Instinct is the Hope of the World': Human Feeling and Social Change Before Empathy. *New Formations* 79: 126–150.
Twyman, Michael. 1975. The Significance of Isotype. In *Graphic Communication Through Isotype*, 7–17. Reading: Reading University.
Uebel, Thomas. 2004. Neurath's Economics in Critical Context. In *Otto Neurath: Economic Writings. Selections 1904–1945*, ed. Thomas Uebel and Robert S. Cohen, 1–108. Dordrecht: Kluwer.
Williams, A.V. 1973. Excerpt from the 'Memorandum of the Visit by Dr. Otto Neurath to the Borough of Bilston, July 24, 1945'. In *Empiricism and Sociology*, ed. Marie Neurath and Robert S. Cohen, 75–79. Dordrecht: D. Reidel.

Chapter 16
Does Understanding Mean Forgiveness? Otto Neurath and Plato's "Republic" in 1944–45

Antonia Soulez

En hommage to Robert S. Cohen, recently deceased

To Philippe Soulez for and before whom this text was to be presented

Abstract In this paper I consider Otto Neurath's late discussion of the political and social context of Plato's *Republic*, especially how Neurath conceived them in the 1940s. Neurath's argumentation is contrasted with the ideas of Karl Popper, both with regard to the latter's reading of Plato and to his general methodology. The distinction between Neurath's treatments of epistemology and politics is also discussed, by highlighting how these two were interwoven in the discussion, and how they differentiated Neurath's articles from Popper's considerations in the *Open Society*.

16.1 Introduction

As Karl Popper (1973, 55) pointed out in his testimonies presented to readers of Otto Neurath's "Memories" in the *Empiricism and Sociology* volume, both himself and Neurath planned to elaborate a theory of knowledge capable of dealing with problems of political philosophy. Popper, having left Marxism and wearied by what he calls its "prophecies" (which he had believed in during his youth) welcomed, after the failure in Munich, in Neurath the quality of a social philosopher who hoped to lead the future Vienna Circle towards a "philosophical reform of politics."[1]

[1] For the relation between politics and philosophy, see the chapters of Günther Sandner and Don Howard in the present volume.

A. Soulez (✉)
Université de Paris VIII, Paris, France
e-mail: antonia.soulez2@gmail.com

J. Cat, A. T. Tuboly (eds.), *Neurath Reconsidered*,
Boston Studies in the Philosophy and History of Science 336,
https://doi.org/10.1007/978-3-030-02128-3_16

We know that Popper afterward became the official critic of the Vienna Circle. His thesis of "falsificationism," presented within his *Logic of Scientific Discovery* in 1934, has remained notable. He did, however, appreciate Neurath's efforts to put into practice a "new philosophy of science" able to contribute to the development of political theories, even if this resulted in greater distance from Neurath's professed socialism. To state it in his own terms, Neurath and he differed profoundly in practically all the areas that interested both of them, except for one thing: the fact that a theory of knowledge was crucial for an understanding of history – the Vienna Circle had not given sufficient weight to this – and for an understanding of political issues.

This unique point of encounter between epistemology and politics is precisely what I should like to clarify, with the help of Plato's *Republic*. At the moment, Popper's battle with Plato as precursor to modern totalitarianism is well known. But when *The Open Society and its Enemies* (Popper 1945/1966) first appeared in England in 1945, with E. Gombrich's help, there were protests from specialists who felt that Popper mistreated Plato's text. It is interesting though not surprising to note that British writers such as Warner Fite, G. C. Field (Bristol) and C. E. M. Joad (London), as well as, to a lesser extent, Richard Crossman[2] – in short, many well-known commentators and translators of the *Republic*, are mentioned or discussed in Popper's work with a good deal of asperity. They are the same disputants who can also be found in Neurath's Plato-controversy around the same time, still a decade before Levinson's[3] counter-reply.

The interesting aspect of what must be called Popper's violent criticism is that it shows that a philosophical position that is usually considered an eminently refined one, is argued to be the source of a most brutal and dangerous political theory. Popper proposes a careful reading of the *Republic*, but I do not want to dwell on that reading, which would divert me from Neurath, the subject of this paper.

I would simply like to bring to your attention a meeting of minds between Neurath and Popper concerning Plato, so that the outside observer has the impression that Plato is being tried in the same way by both of them. And there are reasons for that: it is quite true that there is a family resemblance between these Austrians, like Gomperz, father and son, both with a weakness for Epicurean philosophy, and not disguising their preference for philosophies which might give rise to materialist and utilitarian theories. A study comparing early twentieth century Austrian and German approaches to ancient Greek thought would certainly be worthwhile. It should probably also be mentioned that young Marx, author of a thesis on Democritus and Epicurus, influenced Popper and, supposedly, probably also Neurath (see Sandner 2014a).

[2] Crossman, as the author of *Plato To-day*, he is the first to have contraposed Plato's *Republic* with contemporary political issues. He accuses the *Republic* of being a "polite form of fascism": "The more I read the *Republic*" Crossman (1937, 190) writes, "the more I hate it." Cornford (1941) refers to this in his *Republic of Plato*.

[3] Levinson (1953), in his *Defense of Plato*, argues against Popper's and Neurath's imputation of a Nazi-type thesis of eugenics to Plato.

My choice of presenting the opposition between two conceptions of "social engineering" is based, however, on a communion of ideas between the two Austrian critics of the budding totalitarianism of Plato's *Republic*. Instead of describing a contrast for its own sake, I wish to point out, by means of a number of divergences between their respective epistemologies, that Plato's *Republic* and its interpretative problems during the years 1944–45 reveal serious methodological differences which distinguish Popper and Neurath. I am thinking of Popper's criticism of the Vienna Circle's verificationism, and of Neurath's criticism of Popper's pseudo-rationalism. Yet in spite of their agreement on the two principle dangers – racism and the idea of a natural enemy who must be vanquished – there is one aspect of the Platonic *Republic* which Popper challenges implacably, but which apparently escapes Neurath's polemic: it is the utopian aspect and its prophetic pathos.

This final comment leads me to demarcate Neurath's "social engineering" from Popper's.[4] I want to demonstrate how it was possible that the ideas of Neurath – a planner at heart and supporter of social techniques for building utopias – might be subject to the attacks of Popper's criticism of Platonic utopias, denounced in chapter 9 of *Open Society*.

16.2 The Republic According to Neurath

16.2.1 The Republic in Political History

I shall simply follow Neurath's statements, reorganizing them slightly, and drawing on articles published with Joseph A. Lauwerys in the British *Journal of Education* of November, February, May and August of 1944 and 1945, presenting the attacks and counterattacks concerning the reading of the *Republic*.

First of all, Plato's *Republic* periodically gives rise to politically problematic interpretations when read outside the narrow circle of specialists. In the eighteenth century, there was such an episode, "Plato and the French Revolution," marked by the discussions of the figure of the legislator – less totalitarian in truth, in the *Republic* than in the *Laws*. In such discussions, the model of Sparta against Athen's one was prevailing, as shown in writings by Rousseau and Malby. In these respects, the reader is here invited to read Pierre Vidal-Naquet (1975) on the conflict between Sparta and Athens represented by the characters of Lycurgus and Solon, the law-makers. Then, the "Plato and Bolshevism," or Russell's examination of communist totalitarianism and the ideology of a total regeneration of the social fabric through a

[4] A term usually having a negative connotation in countries of German language or culture, as I was able to confirm: It is striking that Neurath and Popper used it with the positive tone of a verifiable project conceived by a social and political philosopher. Victor Goldschmidt (1970) refers to it in a note about "utopia," as being a key-term of a "piecemeal" rather than "holist" program of social technique in Popper's case – in reference to Popper's *Poverty of Historicism* (1957).

politicized ethics (cf. H. Marcuse as well). And finally, there is the episode that interests me here, "Plato and Nazism."

16.2.2 Neurath and the Greeks

Neurath's interest in antiquity began very early, but from an economic angle. In 1904 and then in 1909 he wrote "Interest on Money in Antiquity" ("Geldzins in Altertum," 1904/2004) and *Economic History of Antiquity* (*Antike Wirtschaftsgeschichte*, 1909/2004). His thesis was published in 1906 and entitled: *Zur Anschauung der Antike ueber Handel, Gewerbe und Landwirtschaft* (Neurath 1906–1907). His thesis director was Eduard Meyer, who, according to Marie Neurath,[5] was an "historian and good linguist, acquainted with all the central European languages." It was from 1932 that Nazism was an evident threat to Otto Neurath – becoming at the same time also a theoretical problem for him –, as Marie Neurath also told me in a private talk in London, shortly before her death. She was not yet his wife at the time, but his associate. Of German origin (the mathematician Kurt Reidemeister's sister), she made her choice in 1925: she would live in Vienna rather than return to Germany; the decision was made a year after meeting Otto Neurath.

To Neurath, the system closest to his program of "empirical sociology" was Marxism. He had in mind particularly the Marxism of the earliest writings: the *Manifesto, 18 Brumaire, German Ideology*. Given his social-democratic sympathies, Neurath should have felt close to the Austrian Marxist Otto Bauer, whom he greatly admired, even though Bauer could hardly believe in the attempt at socialism in Bavaria, in which Neurath had actively participated. So there was only a distant sympathy between the two partisans of socialism.[6]

The Neurath who reads the *Republic* in 1944–45 is not exactly the historian of ancient Greece. He is an empirical materialist who, as a victim of persecution, had to flee his country like so many other people, going first to Holland in 1934 and then to England in 1940, where he was interned on arrival, before being able to live a normal life in Oxford from 1941 to 1945, in pursuit of his work (he died on the 22nd of December 1945).[7] He is, as I said, the empirical materialist who, beginning in 1932, asked himself about how Nazism could have been possible in Germany (according to Marie Neurath).

[5] This is information I received personally from Marie Neurath, during private consultations that soon became friendly chats, when she was kind enough to receive me at her home in London on March 3 and 4, 1983. My thanks go to this great lady who has since died, and to Bob Cohen who introduced me to her in 1982. On Neurath's early years and his relation to Meyer see Sandner (2014b, 45–48).

[6] On Neurath's relation to Otto Bauer see Cartwright, Cat, Fleck and Uebel (1996, Part I).

[7] Neurath's English years is described by Sander (2011) and Michelle Henning's, Silke Körber's, and Adam Tamas Tuboly's chapters in the present volume.

His articles of 1944–1945 particularly communicate a desire to educate. During the Twenties in Vienna, Neurath had experienced the Austrian movement for educational reforms, led by the social democrat, Otto Glöckel. In order to understand how Nazism was possible in Germany, the education of German youth must be examined. Reflections on education are very timely, furthermore. Nazism is in full sway when Neurath thinks about what seems to him the only real solution: to re-educate the German literate population, a solution which associates an enlightened humanism – Neurath always considered himself the modern heir to that tradition – with, (on a broader scale), a strategy of de-Nazification made to activate the international conscience.

Some 20 years earlier, as I have said, certain critics had diagnosed the seeds of a Bolshevik sort of oriental dictatorship in Plato's great political dialogue. Russell, disenchanted by his voyage in Russia, reads the *Republic* differently since he sees communism differently.[8] Then, beginning in the Thirties and Forties, the *Republic* is seen to become a reference book for certain Nazis eager to base their theories of eugenics on great masters of the past.[9] To give an idea of what could be said in these times in reference namely to Plato, let us read for instance the following lines taken from Hans F. G. Günther's later preface to the third edition of his book, *Platon, eugéniste et vitaliste* (*Plato, Eugenicist and Vitalist*)[10]:

> From 1933 to 1945, Germany, in the wake of North America, had fully justified legislation concerning heredity (a law to control genetically transmitted illnesses), as well as all sorts of absurd misinterpretations of Galton's ideas[11] and consequently, of Plato's. Under the influence of an equally absurd type of reeducation, eugenics or racial hygiene – rules of health to conserve a good genetic stock maintaining and multiplying hereditary qualities – appeared or still appears to many Germans as an inhuman and even "animal" practice, while Plato's teachings already proposed eugenic practices as the only way to safeguard and increase human dignity [...]. (Quoted by Goldschmidt 1970, 135. Cf. Günther 1965.)[12]

[8] Cf. Ronald W. Clark's (1976) biography of Russell, and Jacques Bouveresse's (1978) review of it.

[9] On this subject see Goldschmidt (1970), particularly the section, "Quarrels over Platonism."

[10] The book was first published in German in 1928 as *Platon als Hüter des Lebens*. In the introduction to the 3rd edition (1965) there is a significant "state of research" on Plato, citing Windelband, Taylor, Jules Stenzel, P. Friedlander, C. Ritter, W. Jaeger, L. Robin, G. Kruger, E. Hoffmann, H. Herter, all from publications or reeditions dating between 1928 and 1950, as well as comments on the favorable reception of his book of 1928, especially in Hans Leisegang's "*La signification actuelle de Platon*."

[11] Francis Galton, nineteenth century physiologist who – with Gobineau and Mendel – established bases for eugenics or racial purification. He is known for his table of men of great talent, and is the author of *Hereditary Genius*. Neurath (1921/1973, 179) cites him in reference to an "archetypal" method used by Spengler to derive forms of culture, in his *Anti-Spengler*. Galton is rather well estimated, cited with Goethe by Neurath and even by Wittgenstein as a forerunner of ideas on family resemblance.

[12] And he adds several titles (by W. Braeucker, H. Siemans ...) useful in examining this domain.

That such lines could be written again in 1965 makes us realize the possibility of presenting these ideas again and have "public debates" about them in western Germany.[13]

The educational importance given to reading such texts and all the eugenic literature associated with them shows that Otto Neurath had correctly chosen his terrain. What troubled Neurath the educator was the racist and imperialist-German propaganda which evolved during the Thirties through works destined to be educational. He remembered especially the German atlas and manuals, the "Nazi textbooks" which served as title for an article in 1944 and gave example of the Putzger school atlas, comparing the editions from 1910 and 1937 (Leipzig) which demonstrate the progress of alarming ideas (Neurath and Lauwerys 1944).

The danger is "continental," Neurath says, and not at all a product of our accidental culture (he contrasts continental culture with the culture of "Western countries").[14] Specialists and British scholars calmly study Plato's *Republic*, innocent of the fact that they are supporting ideas which serve the ends of racist and warlike propaganda. Enlightened and liberally minded scholars do not realize that they are playing into the hands of dangerous teachers of German youth, which needs, says Neurath, to be re-educated. It is his warning cry: take care not to drag our youth along the same road!

Thus, the specialist in ancient economy is now clothed as interpreter of Plato's dialogue. His message is straightforward and momentous: in the *Republic*, he says, "everything is based on the idea of war" (cf. Neurath and Lauwerys 1945). Neurath gets right to the heart of the matter. Referring to the program of extreme specialization of duties in the fifth part of the dialogue, and the idea that the only worthwhile war is the one between Greeks and Barbarians, the pre-eminent "natural enemy" (Neurath and Lauwerys 1945, 470–471), Neurath puts his finger on Greek ethnocentrism as opposed to Epicurean cosmopolitism, on the reactionary character of the criterion of inherited qualities to be respected in the education of the guardians of the Platonic republic (415a-435b), and the elite and police-like character of the army formed by the guardians, the class especially trained to guard the city. All of these traits describe the training in the *Republic*, an anti-democratic, repressive and selective system that may be compared with Hitlerism. Add to this the censorship in the arts (music must be military), the training of the young (compared with the raising of animals [459]) and the development of a cast system. And there is still the cult of familial identity, the control of marriage and birth which, with the elimination of the unsuitable (406, 460), must contribute to preserving a *katharon genos*, a pure and renewed race. All these are markedly typical aspects, he says, of the German "*weltanschaulichen Grundgedanken*."

[13] Precisely since 1963, following work by the fashionable geneticists like Hans Nachtscheim, author of one of the statements on eugenics, in the newspaper *Die Welt* of May 1965.

[14] "On the continent, things are different, for the tradition of scholarship differs somewhat from the tradition over here. When a German philosopher characterized Hitler's advent as the victory of Platonism, he was expressing wide felt sentiments […]" (Neurath and Lauwerys 1944, 575).

16.2.3 What Have Platonism's Defenders to Say in 1944–45?

Neurath's critics refer to works by such authors as Garforth, G. C. Field, C. E. M. Joad who, when the *Republic* seems equivocal to them, reply that it evokes Soviet Russian communism far more than German Nazism, particularly considering Plato's violent indictment of tyranny.[15] Readers who interpret it as a salute to a Mussolini or a Hitler should rather consider the way Plato treated the Greek tyrants.

And where, asks Field, do we see a program of "racial purification" in Plato, "as we understand it today?" Not only is there a basic reasoning about the inevitability of war – which escaped Neurath – not only is Plato's rhetoric of high ideals – which aroused Neurath's contempt. Different from Hitler's, but accusing Plato of racism is an anachronism and proof of a total lack of "understanding." Neurath reads Plato in the first degree, with the prejudice of his century, without considering the Greek context.

Joad adds to these defending voices that Neurath's preference for Epicurean cosmopolitism instead of Plato's republicanism was another aspect of an alienation from a political position, marked by a cult for personal relations and far from public affairs. Last but not least, Plato does not proffer the mystique, but the ethics of a Führer. They keynote of Platonic politics is a vision of the Good. The vision prescribes – according to an idea of rationality – the principle of selection of the fittest, as the way to deal with a redefinition of the citizen.

Plato's defenders accuse Neurath of incomprehension. Neurath does not "understand," from several points of view:

1. He does not read the text in its context but judges it from the distance of his own century.
2. In short, he does not know how to read the *Republic*, he is not a specialist. From there stems his elementary – not to say primitive and unrefined – reading.
3. Furthermore, he is a poor philosopher, since he has missed the rational point of departure. Closed to ideas, to discussion, he has reacted to certain "details" without seeing the "totality" of their situation.

Neurath replies to this argument about his lack of comprehension, in an article of May 1945, and that date gives his answer its interest. Neurath does not want to "understand," because "to understand means that all is forgiven" (Neurath and Lauwerys 1945, 222). That gives crucial importance to a simple or seemingly naïve reading. To him, a refined and complex interpretation is the gateway to ideology: here, erudite and respectable appearances hide justifications of the worst ideals. One has only to think of the "Manifesto of Pan-Germanism," signed by so many high-ranking intellectuals and professors. The responsibility of the intellectual is in question. Yet Field's argument is that the semantic ambiguity of the expression such as "purification of race," as he put in the mouth of ancient Greeks allows more than one

[15] It should be noted, that Joad (1950) later accused logical positivism (especially A. J. Ayer's famous *Language, Logic, and Truth*) of ensuring a positive atmosphere for fascism in Oxford.

interpretation – at least theirs (the Greeks) and ours in our contemporary time. Two thousand years of distance and quotation marks used, seem to him to reinforce the criticized project.

16.3 Epistemological Background to Neurath's Replies to the Questions About Understanding

I do not know which formulation to choose: whether to read Neurath, the political man in the light of epistemology or to read Neurath the epistemologist in a political light. But under these circumstances, I deliberately choose the first mode, in opposition to Gideon Freudenthal's (1989) recommendation in his "Otto Neurath from Authoritarian Liberalism to Empiricism." If it is, in fact, legitimate to relate Neurath's theory of knowledge to his political formation where history and the evolution of his ideas are concerned, it is impossible to understand the full meaning of his response without transposing it into terms of his theory of knowledge: the theory he defended in his role as representative of the Vienna Circle, even when its defense might have meant distancing himself from other members of the Vienna Circle.

Three types of responses may be distinguished, shedding light on the exact nature of his attacks against Platonic conceptions or race and war which he feels make of the *Republic* a forerunner of German Nazism.

1. Practice and theory cannot be separated. Theory is just as much of a practical commitment. According to this principle, confirmed in an article of 1913 about Descartes' "lost wanderers in the forest" – the famous parable where Descartes illustrates the second maxim of provisional morals.

Neurath (1913/1983) rejects the idea of a difference in kind between thought and action. This rejection governs his conception of "social technique" (Soulez 1988) in the construction of utopias here and now; there is no need to wait for the ideal moment to apply the technique. This is why "to understand" the theory would be "to forgive" the unforgivable political action.

Let us apply Neurath's refusal to understand the "noble lie." We shall see that this means to refuse two levels of falsehood.

The myth of a pure *genos*[16] in Plato, a lie termed "Phoenician" (*Republic*, book 4, 546a-547a), is intended to convince citizens that they are all brothers. This lie, which supposes that they are not brothers (qualifications begin by being associated, like metals to be refined for specific uses)[17] is also a lie at the level of internal control, where the magistrates are supposed to convince themselves so they can convince others. This myth of belonging to a *genos*, *"gennaion pseudos,"* is presented

[16] A myth very often cited by German eugenists; Cf. Gunther (1965) in which German eugenics found fuel for their myth of "blood and earth" (Blut und Boden).

[17] Gold (authority) is distinguished from silver (auxiliaries to authorities) and from iron and copper (peasants and working classes).

as a myth, not as a truth. However – and this is the educational impact essential to maintaining the high standards of the city – the myth lies only at the level of words and not at the level of the soul: as Julia Annas (1981, 107) reminds us in the introduction to Plato's *Republic*, there is "the noble falsehood," which is in the soul, and the tolerable falsehood, which is only in the spoken word.

But, as Neurath sees it, we are not dealing with pure thought on one hand and action on the other; they are on a par. There is no falsehood that would be a truth in the soul. A lie can have only one meaning. This single meaning corresponds to the pragmatic maxim formulated by J. L. Austin and expressed in his initial lecture, *How to do things with words*: Rigorous truth and morality are both characteristic of someone who simply says, "[o]ur word is our bond."[18]

2. It is this difference of degree which justifies the coupling of theory and politics, epistemology and realpolitik, also defended by Popper, as I pointed out earlier. It is therefore completely coherent, and coherent in its strongest sense, that Neurath should apply to Plato himself that anti-Cartesian maxim corresponding to his necessarily provisory political morals. Plato's text is guilty of totalitarianism. The theory of the *Republic* and political totalitarianism belong to the same "system," are parts of the same integrating whole. Correspondingly, Platonists are virtual signers of the manifesto of Pan-Germanism.[19] Criticism of the *Republic* must not remain the monopoly of historians of Greek philosophy.

Putting Plato back into a Greek context dissimulates a bias. Taking context into account, it is evident that "Greekness" is not synonymous with totalitarianism. There, Neurath – as Popper did – invokes the example of Pericles' funeral oration according to Thucydides. There were, then, even in Plato's time, liberal and democratic ideas!

Specialization is dangerous from every standpoint, whether it concerns the pre-established qualifications ("fitness") for a particular activity in the city – "selection" – or the selective attention to detail proffered by professional interpreters of Plato.

3. These are "overall" observations, like Neurath's reply to critics: "If you don't like the details, revise the ensemble."

Concerning the issue of war, Neurath remains faithful to his own "war economics," declaring that, in times of peace, preparation for war should not be an aim, as Plato sees it. The advantages of wartime organization should be used to build a

[18] "Accuracy and morality alike are on the side of the plain saying that our word is our bond" (Austin 1962, 10).

[19] "Manifesto of Intellectuals" read June 20, 1915 at an assembly of German professors, diplomats and officials in Berlin's Künstlerhaus. Unpublished, it circulated as a "strictly confidential document" signed by 1341 supporters (352 university scholars, 158 schoolmasters and clergymen, 148 judges, 252 artists, writers and editors), not counting the popular support which would have meant free circulation of the *Manifesto* in the countryside, fief od the "Junkers" (Who formed a real class together with the landowners, manufacturers and big industrialists in the Rhine-Westphalian region). Cf. Bevan (1918).

liberalized and durably peaceful society. There is no natural enemy, there are only hostilities. Hopefully, the resolution of these hostilities may finally make friends of former enemies.

On the subject of economy, however, Neurath's explanations regarding his articles on Plato and there.[20]

16.4 The Spirit of Planning Aspirations vs. Historicism

Epicurianism, empiricism, a sort of Austro-Marxism (there is no time to go into all this)[21] all of which nourish Otto Neurath's convictions, do not explain everything about his reading of Plato, or of the present polemic, perhaps, in its epistemological – political duality. I should like to discuss an aspect of Neurath's ideas which differs completely from Popper's reading of Plato: the concept of planned economy.

In one of his *addenda* of 1961, where he replies to a criticism of his *Open Society* by Ronald Levinson (1953) in his important book, *In Defense of Plato*, Popper (1945/1966, 323. ff.) defends himself against what seems to him an unfortunate amalgam of his reading of Plato with the reading by the "two-headed monster" formed by the couple, Neurath and J. A. Lauwerys, in the same controversy. He adds incidentally, that in spite of what would seem a fraternal combat, there was no particular affinity between the philosophies of the two, as an attentive reader of their respective writings would discern.

Certainly, the desire for dissociation may be exaggerated, if we remember Popper's statement mentioned early in this talk. Detailed analyses follow, with recourse to the Greek text for point by point replies to Levinson. Here, an effort at interpretation by the historian of philosophy is implicit; Neurath refused to make such an effort, which he might have considered futile. He goes on to avow not having read Neurath's attacks and being aware of them only through Levinson's mention of them in his book of 1953. Philosophically, he says, nothing would seem to unite them because Neurath is really for Hegel and against Kant, while Popper not only sides with Kant against logical positivism, but carries over to Hegel (and then to Marx) his well-known criticism of Plato's historicism which introduces this very disastrous lineage into the history of ideas.

We might say that the essence of Popper's criticism is the subject of historic fatality, a philosophical attitude at the same time irrational and politically dangerous. "We don't need a sense of history," says Popper (1957), in *The Poverty of*

---

[20] On Neurath's economy see Uebel (2004) and his chapter in the present volume.

[21] Neurath's background is explained in details by Sandner (2014b) and in his chapter in the present volume.

Historicism, but just in that need consists the Platonic "poverty of historicism"[22] with its very noble "theory of ideas" leading straight to various forms of totalitarianism. Besides many other references, roughly the same passages of the *Republic* which Neurath points out are invoked to illustrate Popper's thesis. The same errors or weaknesses in the planning of the *Republic* are challenged: the ideas of war, eugenics, and even the noble lie which obliges the Guardians of the state to remain true to a false discourse on the status of the pure race and its comparison to gold, in the myth of metals which they must recite to others and believe themselves.[23] This is not to say that Neurath did not have an anti-historicist bent resembling Popper's. His *Anti-Spengler* (Neurath 1921/1973) suffices to convince us, I should say, that Neurath's Spengler makes a kind of echo to Popper's Plato; it gives us a modern German version of historical fatalism. But Neurath is above all a social engineer of reconstruction. He believes that the idea of planning will counteract the ideas of decline. Planning is not the counterpart of historical degeneration, as Popper views Platonic utopia to be: it is the way to avoid just that. We might notice that in 1921 – the same date as *Anti-Spengler* – Neurath (1921) using the pseudonym of Karl Wilhelm, writes his project for Jewish economy in Palestine for the "construction" (*"Aufbau"*) of Palestine. The exact title is, "Jewish planned economy in Palestine; application of social technique."

Does this mean that because of his idea of a planned economy, which would be directed by the governing administration, Neurath exposes himself to the same criticisms that Popper directs at supporters of "utopian social engineering" such as Plato (title of *Open Society*'s chapter 9), or Hippodamos de Milet, the first "city planner," whom Aristotle mentions in his *Politics* (1276b22)?

All of Neurath's explicit efforts went towards combining freedom and planning. But, as Gideon Freudenthal (1989, 207) writes, "the way to liberalism is not liberal." Neurath cannot be likened to the kinds of liberal Austrians including Popper; he actually maintains a form of "authoritarian liberalism." In an important article on the "orchestration of the sciences," where he defends his encyclopedic idea of science, opposed to the traditional pyramidal classifications,[24] Neurath (1946/1983, 237–238) wants to demonstrate that the task of reconstructing a community has a very different character from a Platonic utopia, contrary to what Horace Kallen proclaims.[25] And this is so, precisely because a liberal outcome both motivates and must ultimately crown an international planning program. Far from being "imperi-

[22] *The Poverty of Historicism* was first published as articles in the journal *Economica* in 1944-45.

[23] I discussed this myth in a lecture entitled "The Sophism of Belonging to the Race," at a symposium, "Is the word 'race' superfluous in the French constitution?" Senate and Sorbonne, 27–28 March, 1992. See Soulez (1992).

[24] The pyramidal classifications, inherited from Comte, Spencer, Wundt, are still to be found in Wilhelm Ostwald. On Ostwald's ideas in the context of logical empiricism see Dahms (2016).

[25] This is a reaction to a 1939 lecture by Kallen, "The Meaning of 'Unity' among the Sciences," see Kallen (1940). Neurath says that he forged the word "orchestration" after having listened to Kallen at a Harvard meeting. On Neurath and Kallen see Reisch (2005, 167–190).

alist," his plan for unity in science, which would be a linguistic internationalism, is democratic in spirit.[26]

A number of traits separate Neurath's encyclopedic approach from totalitarianism, which could menace even a planning program oriented towards freedom:

(A) Its anti-separatism. Neurath eliminates a clear distinction between the linguistic and the empirical, and renders impossible the sort of foundationalism which relies on strictly imposed protocol statements meaning all – even logical-linguistic – forms of *tabula rasa* á la Descartes, and consequently, all divisions and institutional separations of knowledge into specialized areas arranged in a hierarchical manner.

(B) Its anti-experimentalism. Neurath reproaches experimentalism for the "pseudo-rationality" of the idea that scientific progress is a "step by step" affair advancing "from one test to another." Neurath places his holistic, encyclopedic model in opposition to Popper's incomplete model tainted with the metaphysics of an idiom of pure experience and governed by "piecemeal social engineering." One cannot calculate the future; one can only make decisions. In contrast to Popper's probabilism concerning scientific ideas, supposedly rationally regulated by historical verification, Neurath (1913/1983) proposes throwing dice to draw lots, recalling the decisive paradigm case in his study of democratic institutions in ancient Athens.

(C) Its voluntarism. Neurath's idea of planning, an idea more arbitrary than natural, gives more credence to a voluntary decision than to a system of empirical proofs. This attitude introduces a social engineering called utopian science, which owes its spirit to J. Popper-Lynkeus, Austrian social reformer, demographer and statistician of an *Aufklärung* – understood as an utilitarian tendency, as well as a moral doctrine provisionally built upon a reevaluated Cartesianism – that is, cleared of the Cartesian duality between pure thought and action, therefore equally applied to thought.

The ensemble of these elements makes Neurath's planning the contrary of a deterministic project. In Neurath's thinking, this means that the *"Zukunftbild"* of the regulation of the future, if correctly understood, no longer owes anything to Marxist "historicism."[27]

---

[26]At this point, Neurath refers to his "International Planning for Freedom." See Neurath (1942/1973). His most detailed discussion of democracy (and its relation to pedagogy) is to be found in his posthumously published manuscript, "Visual Education: Humanisation versus Popularisation." See Neurath (1996).

[27]Lack of competence in a precise approach to Neurath's economy makes me prefer to leave this area to connoisseurs. Clearly, Neurath's idea of freedom, although strongly marked by the Epicurean idea of happiness, requires some technical precisions in economy. Controlled finance demands economic administration, with a distinction between state economy and the democratic social order which must be directed towards the happiness of the individual, and not exclusively the group to the detriment of the individual. Lastly, a socialist economy as he understands it ceases to depend on money as its operating force. It excludes profit and must control finance even if that function cannot, by definition, be socialized. Cf. for example Neurath (1919/1973).

16.5 Conclusion

If on a deliberate shocking tone of anachronism, Neurath answers "yes" to the question, "Was Plato a Nazi?," it is by virtue of an opposition to the hermeneutic paradigm of understanding, an opposition in principle which he states loudly and clearly and which is rooted in his conception of methodology and epistemology of science. A holistic vision forms the background of his reading of Plato, calling for a judgment of the *Republic* within the framework of a theory and a political attitude; this means that he criticizes a certain conception of "social engineering" which a refined and detailed reading might mask, even though it were acknowledged. Holism confirms a philosophy of social and political action which also implicated the ideas concerning this action which also implicates the ideas concerning this action, in the same way and to the same degree as the action itself. Neurath's holism endorses utopian social engineering. Does this utopia revert to the "Platonic utopia's social engineering" denounced by Popper? No, answers Neurath to Kallen's warning. Rather "utopia" means – instead of the dream of someone disappointed by history – a project for building the future, without awaiting the blessing of experience or ideal conditions which are by definition unattainable. The ship must be repaired at sea and this means working with doubtful premises which may always be revised.

This utilitarian and Epicurean principle calls for construction, as indicated in the "plan for a Jewish economy" of 1921 ("the plan of plans")[28]; it seems to be founded on the imaginative projection of a potential reality, always on the horizon. The tendency of this attitude is analyzed by Ernst Bloch in his *The Spirit of Utopia*, 1918: it seems to be seeking "a way of making forerunners of possibilities." All philosophy about reconstruction could be directed to reflecting on projecting into the future a program of social intervention which assumes that men are capable of acting to modify the future, contrary to a fatalist approach. This means that the human "agency" should not be determined by an overall causal system. The "science of utopia" is aiming precisely at the very "social technique" which an "image of the future" forces ahead into a "doctrine of happiness" which Otto Neurath called "social Epicureanism" (see Stuchlik 2011). This representation, imbued with voluntarism, is clearly an impressive instrument for combatting historicism's ravaged. But what guarantee do we have – in Ernst Bloch's (1918/2000) terms – that the "dream of advancing" in a regeneration of the human social community is not a shadowy apparition cast by a "backsliding dream"? We can wager that Neurath would have answered this as he replied to Kallen's criticism of unconsciously extending Plato's totalitarianism when he invoked his big, working, methodological model of the "encyclopedia." But the answer belongs more to epistemology than to politics: the latter might have its reasons that the former cannot know! Perhaps Neurath's prudent holism does not exclude a certain visionary character.

[28] This plan is translated for the first time into French as well as the article on Neurath's text "The lost wanderer of Descartes and the Auxiliary Motive" in our special issue on Neurath: *Otto Neurath un philosophe entre science et guerre*, *Cahiers de philosophie du langage,* n° 2, en hommage à Philippe Soulez, in coll. with Elisabeth Nemeth (University of Vienna), 1997, publ. L'Harmattan.

Acknowledgement My thanks go to Jacqueline Kiang for the original English translation, and also for Thomas Uebel's earlier revising. I am especially indebted to Elisabeth Nemeth. This is a slightly updated and extended version of an earlier paper that appeared in 1999.

References

Annas, Julia. 1981. *An Introduction to Plato's Republic*. Oxford: Clarendon Press.
Austin, John L. 1962. *How to Do Things with Words*. Oxford: Clarendon Press.
Bevan, Edwyn. 1918. *The Pan-German Program*. London: Allen & Unwin.
Bloch, Ernst. 1918/2000. *The Spirit of Utopia*. Stanford: Stanford University Press.
Bouveresse, Jacques. 1978. Bertrand Russell, La Sagesse, La Politique et le Reste. *Le Monde*, 21 February 1978.
Cartwright, Nancy, Jordi Cat, Lola Fleck, and Thomas Uebel. 1996. *Otto Neurath: Philosophy Between Science and Politics*. Cambridge: Cambridge University Press.
Clark, Ronald W. 1976. *The Life of Bertrand Russell*. New York: Alfred Knopf.
Cornford, Francis M. 1941. *The Republic of Plato*. Oxford: Clarendon Press.
Crossman, Richard. 1937. *Plato To-Day*. London: G. Allen and Unwin.
Dahms, Hans-Joachim. 2016. Carnap's Early Conception of a 'System of All Concepts': The Importance of Wilhelm Ostwald. In *Influences on the Aufbau*, ed. Christian Damböck, 163–185. Cham: Springer.
Freudenthal, Gideon. 1989. Otto Neurath from Authoritarian Liberalism to Empiricism. In *Knowledge and Politics*, ed. M. Dascal and O. Gruengard, 207–240. Westwood: Westwood Press.
Goldschmidt, Victor. 1970. *Platonisme et pensée contemporaine*. Paris: Aubier Montaigne.
Günther, Hans F. K. 1965. *Platon eugéniste et vitaliste*. Éditions Pardès.
Joad, C.E.M. 1950. *A Critique of Logical Positivism*. London: Victor Gollancz Ltd..
Kallen, Horace. 1940. The Meanings of 'Unity' Among the Sciences. *Educational Administration and Supervision* 26 (2): 81–97.
Levinson, Ronald B. 1953. *Defense of Plato*. Cambridge: Harvard University Press.
Neurath, Otto. 1904/2004. Interest on Money in Antiquity. In *Economic Writings. Selections 1904–1945*, ed. Thomas Uebel and Robert S. Cohen, 111–119. Dordrecht: Kluwer.
———. 1906–1907. Zur Anschauung der Antike über Handel, Gewerbe und Landwirtschaft. *Jahrbücher für Nationalökonomie und Statistik*, 3rd ser., 32 (1906): 577–606 and 34 (1907): 145–205.
———. 1909/2004. Economic History of Antiquity [Excerpts]. In *Economic Writings. Selections 1904–1945*, ed. Thomas Uebel and Robert S. Cohen, 120–152. Dordrecht: Kluwer.
———. 1913/1983. The Lost Wanderers of Descartes and the Auxiliary Motive. In *Otto Neurath: Philosophical Papers 1913–1946*, ed. Robert S. Cohen and Marie Neurath, 1–12. Dordrecht: D. Reidel.
———. 1919/1973. Utopia as a Social Engineer's Construction. In *Empiricism and Sociology*, ed. Marie Neurath and Robert S. Cohen, 150–155. Dordrecht: D. Reidel.
———. 1921. *Jüdische Planwirtschaft in Palestina*. (Signed as Karl Wilhelm.) Berlin.
———. 1921/1973. Anti-Spengler. In *Empiricism and Sociology*, ed. Marie Neurath and Robert S. Cohen, 158–213. Dordrecht: D. Reidel.
———. 1942/1973. International Planning for Freedom. In *Empiricism and Sociology*, ed. Marie Neurath and Robert S. Cohen, 422–440. Dordrecht: D. Reidel.
———. 1946/1983. The Orchestration of the Sciences by the Encyclopedism of Logical Empiricism. In *Otto Neurath: Philosophical Papers 1913-1946*, ed. Robert S. Cohen and Marie Neurath, 230–242. Dordrecht: D. Reidel.

———. 1996. Visual Education: Humanisation versus Popularisation. In *Encyclopedia and Utopia. The Life and Work of Otto Neurath (1882–1945)*, ed. Elisabeth Nemeth and Friedrich Stadler, 245–335. Dordrecht: Kluwer.

Neurath, Otto, and Joseph A. Lauwerys. 1944. Nazi-Textbooks and the Future. *The Journal of Education* 76 (904): 521–522; (95): 574–576.

———. 1945. Plato's Republic and German Education. *The Journal of Education* 77: 57–59, 222–224, 394.

Popper, Karl. 1945/1966. *The Open Society and Its Enemies. Volume 1: Plato*. London: Routledge.

———. 1957. *The Poverty of Historicism*. London: Routledge.

———. 1973. On Otto Neurath. In *Empiricism and Sociology*, ed. Marie Neurath and Robert S. Cohen, 51–56. Dordrecht: D. Reidel.

Reisch, George. 2005. *How the Cold War Transformed Philosophy of Science: To the Icy Slopes of Logic*. New York: Cambridge University Press.

Sandner, Günther. 2011. The German Climate and Its Opposite: Otto Neurath in England, 1940-45. In *Political Exile and Exile Politics in Britain After 1933*, ed. A. Grenville and A. Reiter, 67–85. Amsterdam/New York: Rodopi.

———. 2014a. Political Polyphony. Otto Neurath and Politics Reconsidered. In *European Philosophy of Science – Philosophy of Science in Europe and the Viennese Heritage*, ed. Maria Carla Galavotti, Elisabeth Nemeth, and Friedrich Stadler, 211–222. Dordrecht: Springer.

———. 2014b. *Otto Neurath. Eine politische Biographie*. Wien: Zsolnay.

Soulez, Antonia. 1988. La construction des utopies comme tâche de l'Ingénieur Social, selon O. Neurath en 1919. In *Les Philosophes et la guerre de 14*, ed. Philippe Soulez and Etienne Balibar, 237–250. Saint-Denis: Presses universitaires de Vincennes.

———. 1992. Dire non au sophisme de l'appartenance à la race. *Mots. Les langages du politique* 33: 23–33.

Stuchlik, Joshua. 2011. Felicitology: Neurath's Naturalization of Ethics. *HOPOS* 1 (2): 183–208.

Uebel, Thomas. 2004. Neurath's Economics in Critical Context. In *Economic Writings. Selections 1904–1945*, ed. Thomas Uebel and Robert S. Cohen, 1–108. Dordrecht: Kluwer.

Vidal-Naquet, Pierre. 1975. Le mirage grec á l'époque des Lumières. *Esprit* 12.

Chapter 17
Thinking About the "Common Reader:" Otto Neurath, L. Susan Stebbing and the (Modern) Picture-Text Style

Silke Körber

Abstract When Otto Neurath went into exile in 1934, first to Holland and then to England, he succeeded in establishing important new connections within the context of the international Unity of Science movement, for which he was largely responsible. A notable example was the British philosopher L. Susan Stebbing, who supported his pragmatic ideas on the "humanization" of knowledge. Both Neurath and Stebbing were looking for ways to apply modern logic and linguistic analysis, not only to the transfer of information in science and teaching, but above all in publication projects for the "common reader." In 1941, Stebbing became the first president of the Isotype Institute in Oxford, which Neurath directed until 1945. Soon after ISOTYPE was founded, long-term relations began between it and the book-packaging company Adprint managed by German-speaking emigrés in London, as well as its successors and British clients (publishers). A technically and organizationally sophisticated process for the production of illustrated non-fiction books was gradually established. The "picture-text style" developed by Neurath and epitomized in *Modern Man in the Making* (1939) was applied to non-fiction books and series with "integrated layouts," then professionalized and successively transformed into a production model for illustrated books which enabled scientific information to be prepared for the mass market and the "common reader" – in the service of a modern, democratic (post-war) society.

S. Körber (✉)
Institute of Philosophy, Doctoral Candidate, Humboldt-Universität zu Berlin, Berlin, Germany
e-mail: koerbers@hu-berlin.de

J. Cat, A. T. Tuboly (eds.), *Neurath Reconsidered*,
Boston Studies in the Philosophy and History of Science 336,
https://doi.org/10.1007/978-3-030-02128-3_17

17.1 L. Susan Stebbing and Otto Neurath: Clear Thinking and the Democratization of Knowledge

In exile in The Hague, Otto Neurath wrote a letter on 30 July 1934 to the British philosopher Professor L. Susan Stebbing (1885–1943) in London to invite her to the *First International Congress for the Unity of Science* in Paris in 1935 and to a preparatory meeting in Marienbad in August 1934, which was to take place shortly before the *International Congress of Philosophy* in Prague.[1] This formal invitation was the prelude to a regular exchange of letters which lasted until Neurath's flight to England and the end of his internment on the Isle of Man in early 1941. It was also the beginning of a friendship, which lasted until Stebbing's early death in 1943 and was marked by a continuous exchange of ideas and collaboration. The latter took place not only within the framework of the Unity of Science movement, of which Stebbing was an active member, but also in an attempt to democratize knowledge. Stebbing soon became a member of the committee that organized congresses, and was especially involved in the *Fourth International Congress* at Girton College in Cambridge (14–19 July 1938) on the topic of "Scientific Language." She also gradually became involved in the scientific activities and publication projects of the Unity of Science movement (see Sandner 2014, 257ff.). In addition, she became the first president of the Isotype Institute founded in Oxford in 1941, where she supported Neurath's efforts to make the principles of logical thinking and the critical use of language into a basis for the transfer of knowledge outside science. The fact that Neurath made this philosopher the president of the Isotype Institute makes sense, given that he found her not only an equal companion in implementing the scientific goals of the Unity of Science Movement, but also an individual who critically questioned the political and social situation during the war and whose own interests had focused increasingly on linking analytical logical thinking with the world of practical life and whose own texts were directed at the "common reader," i.e. the "citizen" who is politically aware and active. In order to better understand Stebbing's role and significance, some brief biographical information is needed.

L. Susan Stebbing was born in 1885 as the youngest of six children of a merchant. She studied history at Girton College in Cambridge from 1904 to 1908 – which was exceptional at that time.[2] Following her interests, she began studying philosophy, which she completed in 1912 with an MA in Moral Science at King's College London. In 1933 she was the first woman in Great Britain to be appointed a professor of philosophy, at Bedford College in London. Stebbing was co-founder of the journal *Analysis*, chairman of the Aristotelian Society (1933–1934) and presi-

[1] See correspondence between Otto Neurath and L. Susan Stebbing (No. 303, 304; 30.07.1934 to ca. Jan./Feb.1941) at Otto Neurath Nachlass (ONN), Wiener Kreis Archief, Noord-Hollands Archief, Haarlem (NL), partial estate of Otto Neurath. Stebbing replied on 14 July 1934 that unfortunately she could not attend the pre-conference and the philosophy congress in Prague, but intended to attend the conference in Paris planned for 1935, which she did.

[2] See Beaney and Chapman (2017), Beaney (2003) and Chapman (2013). The biographical information in this section is taken primarily from Beaney's and Chapman's publications.

dent of the Mind Association (1931–1932). During her studies she was influenced primarily by the philosophy of G. E. Moore, whom she first met in 1917 at one of her talks at the Aristotelian Society. She was also particularly interested in modern logic, philosophy of science and philosophy of language, as represented by the philosopher/mathematicians Bertrand Russell and Alfred N. Whitehead. Stebbing can therefore be regarded as a representative of what later became known as the Cambridge School within analytical philosophy, which was committed to the logical analysis and interpretation of common language to solve philosophical and individual scientific problems – even before the rise of ordinary language philosophy in Oxford after the end of the Second World War.

In her standard work *A Modern Introduction to Logic* (1930), which went through several editions and reprints, modern mathematical logic was already treated on an equal footing with Aristotelian logic, and the work is regarded as the "first textbook of analytic philosophy." As Beaney and Chapman found, she also played a decisive role in establishing the philosophy of the Vienna Circle and logical empiricism in Great Britain: Stebbing "was at the centre of the debate about the relationship between the Cambridge School and the Vienna Circle, which formed the two main traditions of analytic philosophy in the 1930s" (Beaney 2003, 339). When Neurath contacted Stebbing in 1934, she had already gained an international reputation[3] and been a visiting professor at Columbia University in New York in 1931–1932. Having expressed great interest in the Vienna Circle,[4] she had met Moritz Schlick at the *Seventh International Congress of Philosophy* in Oxford in 1930, and in 1934 invited Rudolf Carnap to attend a series of lectures at Bedford College in London where he was also introduced to Bertrand Russell and Alfred J. Ayer.

In the regular and increasingly personal correspondence housed at the Noord-Hollands Archief in Haarlem, Neurath also informed Stebbing about his plans in the field of visual education. In April 1939 he wrote, "I read your blue booklet [*Thinking to Some Purpose*[5]] through and was very pleased to find the remarks about Visual Miseducation. Excuse me, please, that I beg you to write a review on my booklet[s] *International Picture Language. The first rules of ISOTYPE* and *Basic by ISOTYPE*. I seek always a reviewer for these books, but nobody of our people who are interested in Logic are interested in Visual Education and *ISOTYPE* too. You are the first; I am very glad that you are full of educational ideas. It is nice that you are a member of the board of directors of our International Foundation of Visual Education."[6]

Neurath's statement is relevant in several ways. It indicates reflection on his own position within the Vienna Circle and the logical empiricism movement in exile,

[3] In a letter to Stebbing dated 12 Feb. 1938 Neurath describes her as "famous logician." See Correspondence Neurath, Stebbing (1934–1941).

[4] See Stebbing to Neurath 14 July 1934, Corr. No. 303. (ONN). Cf. Beaney (2003, 339).

[5] Neurath is probably referring to a passage in the chapter "On being misled by half, and other fractions" in which Stebbing (1939, 154f.) deals in particular with the misleading and erroneous graphical conversion of statistical data. Since the 1920s, Neurath and his team had been working on preventing such errors.

[6] Neurath to Stebbing, 8 April 1939; original emphases. Corr. No. 304. (ONN).

while also identifying Stebbing as like-minded in so far as she combines the method of logical analysis with critique schooled in everyday use and the effort to objectify ordinary language, especially in the public sphere. At the same time it reveals that Stebbing's interest and involvement in the Visual Education project did not begin in 1941 when she became president of the Isotype Institute, but that she had belonged since 1939 to the Foundation of Visual Education which had been founded in The Hague in 1934.[7] In *Thinking to Some Purpose* Stebbing (1939, 153ff.) had already discussed statistics and their pictorial representation and pointed out that these are fundamentally helpful and meaningful means of grasping contexts, but only if they are systematically and carefully executed. On a more fundamental level, however, Neurath's letter also suggests that he saw himself increasingly – as has been noted on numerous occasions[8] – as an outsider in the scientific milieu of logical empiricism. This seems also to have been the case to some extent with Stebbing, at least with regard to her interest in the concrete communication of scientific content and the methodology of critical-analytical thinking across social boundaries. As Beaney and Chapman (2017) observe, "[s]he combined her commitment to formal logic with a belief in the importance of practical analysis of everyday texts and as a result in the necessity of public engagement by philosophers, in a manner which was at the time rare in academia."

Stebbing's interest in Neurath's ideas for visualization thus becomes plausible, especially against the background of her own general approach to logical thinking, which she regarded as purposeful thinking and acting in the struggle against political ideologization and for democracy. She had initially treated language as a means of communication and conveyance of new scientific findings in an academic context. Her interest in modern mathematical logic, however, did not lead to the call for an ideal scientific language. Instead, and in contrast to representatives of Unified Science and logical empiricism, she doubted that an artificial symbolic language beyond mathematics could be developed for all sciences.[9] However, she also saw an increasing need to make progress in the sciences accessible to the general public, and sought to direct scientific publications not only to experts and philosophers but also to the "common reader" for educational purposes. Her main interest lay in a clarifying, critical analysis of ordinary language and the disclosure of implicit prejudices and cognitive errors.

[7] See Neurath to Stebbing, 4 May 1939, Corr. No. 304. (ONN).

[8] Neurath's difficult position within the logical empiricism movement and his increasingly strained relationship with Carnap during exile in the debate on the "semantic turn" have been noted on various occasions. Neurath and Carnap, who emigrated to the USA in 1936 and taught at the University of Chicago, belonged to the "left wing" of the Vienna Circle and increasingly disagreed in exile about the goals and methodological foundations of unified science and the jointly edited publication project, the *International Encyclopedia of Unified Science*. For a recent review, see also Derek Anderson's and André Carus's chapters in this volume and Tuboly (2017).

[9] See Stebbing (1937, 111). As she explains here, "[t]he only appropriate language is that of mathematics. To those who cannot use the symbolism of mathematics such scientific theories must remain largely incomprehensible." Stebbing's understanding of logical analysis, also in demarcation to logical empiricism, cannot be further addressed in this context. See above all Beaney (2003).

In 1939, the same year that Neurath's *Modern Man in the Making* was published in the US by Alfred A. Knopf, she published *Thinking to Some Purpose*, based on a series of lectures for the BBC, with Penguin in its Pelican book series for popular non-fiction. In this publication, mentioned above by Neurath, logical thinking is called "purposive thinking," which should be clear, systematic and reflect the author's own prejudices in order to form judgments and act accordingly (Stebbing 1939, 22ff., 27, 30). Stebbing also warned against popularizing complex scientific content in the service of entertainment without using sophisticated methodologies and appropriately precise and clear language. Neurath (1996, 257) explained his position in "Visual Education: Humanisation versus Popularisation," as follows: "Sometimes writers think that it suffices to translate a translation of well-selected terms into popular terms is sufficient, whereas it is common knowledge that the insuffiency of these terms was the main reason for the introduction of scientific terms. This kind of translating from the complicated to the simple, from top to bottom, as it were, we shall call, Popularisation of knowledge."

Stebbing explicitly opposed those authors and scientists who sought to win over the lay public by emotional evocation, personification and improper language and metaphor – which serve precisely to hinder logical thinking on the part of the reader. Since Stebbing's writings are little known, a longer passage on this topic from *Philosophy and the Physicists* (1937) is cited below. Here she criticized particularly the writings of Sir James Jeans and Sir Arthur Stanley Eddington, well-known physicists, astronomers and scientific popularizers:

> In these days of popular expositions, both written and broadcast, of Outlines, and of mammoth Guides to the Intelligent Man – guides through science, guides through economics [...] the common reader cannot be unaware that the sciences in general and the physical sciences in particular have been developing rapidly [...]. These developments in science have a twofold interest. First, their results have given us information, often surprising, about the world we live in. Secondly, the following out of scientific method is in itself exciting, affording us the purest of all satisfactions – intellectual satisfaction. There is among common readers a genuine interest in scientific research, a desire to follow as far as a layman can what is being found and to understand the implications of these findings [...] there are not a few scientists who have written books that to some extent satisfy our needs. Unfortunately, however, there are other famous scientists who do not seem to realize that their subject has an intrinsic interest for the common reader, and accordingly they seek to arouse his emotions, thereby inducing a frame of mind inimical to intellectual discernment. Popularizations of such a kind constitute a grave danger to thinking clearly. [...] Yet we common readers surely have a right to expect that a scientist setting out to discuss for our benefit philosophical problems arising from his special studies will do so in a scientific spirit. He would seem to be under a special obligation to avoid cheap emotionalism and specious appeals, and to write as clearly as the difficult nature of the subject-matter permits. (Stebbing 1937, 4ff.)

Similar to Neurath's approach, Stebbing's rejection is directed ultimately against the ignorance of experts that leads to inappropriate simplifications, imprecise generalizations or abstractions and distortions of facts. As she also stated in *Thinking to Some Purpose* (1939, 30, 238f.), this favors that "muddled thinking" which stands in the way of purposeful logical thinking and makes it susceptible to instrumentalization, for example by political authorities. Moreover, she called for a use of language that is appropriate to the subject and that reflects one's own type of communication, especially from those who influence a society's politics, social affairs and science. She also emphasized that there could be no absolute, "neutral" and objective knowledge, and that even attempts to enlighten should always be examined.

Her understanding of knowledge seems to resemble Neurath's approach as described in his well-known ship simile (Cartwright, Cat, Fleck and Uebel 1996, 89–95). Also for Stebbing, science is a collective undertaking based on intersubjectively achieved knowledge that remains contextually bound and revisable:

> Science is the work of scientists, who, profiting by each other's labours, come gradually to achieve an agreed body of knowledge, and in the course of this achievement continually develop new and more powerful technical methods. The natural scientist observes, formulates hypotheses, performs experiments, and verifies his theoretical constructions. A scientific law worthy of the name of a Principle [...] is achieved only by the labours of scientists working within a certain context of agreed theory, subject no doubt from time to time to considerable revision, but capable of being taken as the main basis for further advance. (Stebbing 1937, 69.)

Overall, Stebbing's approach is to combine theory and practice, i.e. to link the principles of logic with pedagogical goals in everyday contexts and make them usable for the public sphere and for social and political debate. This ultimately connects her approach with Neurath's.[10] He writes the following in his essay "Universal Jargon and Terminology" (1941/1983, 217): "It is not only by accident that L. Susan Stebbing wrote on the one hand a book criticizing highly metaphysical speculations of modern physicists and on the other hand her *Thinking to Some Purpose* and her *Ideals and Illusions*. It will be stimulating when we new critics of our language will be criticized by means of the procedures we proposed."

Regardless of her interest in principle, however, Stebbing was quite critical of some positions of logical empiricism represented by members of the Vienna Circle, including Otto Neurath, especially of their criticism of metaphysics as Beaney and Chapman (2017) have emphasized. Nevertheless, she can be seen as a mediator between positions in the Vienna Circle and the Cambridge School – although her achievements extend beyond that. Beaney and Chapman also note that

> she saw no discrepancy between the rigours of logical argument and the requirements of practical problem solving, and stressed the need for clarity and transparency in language use. In this, she championed the importance to philosophers of paying attention to ordinary language and the varieties of its everyday use somewhat in advance of the rise of 'ordinary language philosophy' in Oxford after the Second World War. (Beaney and Chapman 2017.)

Neurath himself was convinced that in Stebbing the logician he had found an interlocutor who shared his interest in visual communication, language criticism and socio-political thinking. On 12 February 1938, he wrote to her, "you are flesh of the flesh of logical empiricism."[11]

In his examination of her writings and in the correspondence between the two which has yet to be studied in detail, Neurath clearly received ideas for his own

[10] For the relation of (social and political) practice and theory, see Don Howard's chapter in the present volume.

[11] Neurath to Stebbing, 12 February 1938, Corr. No. 303. (ONN).

work. In exile he was also looking for other avenues to explore and other ways to further expand his readership, for example by working with the Adprint book packager and via publications for the Anglo-American book market. As president of ISOTYPE in Oxford, Stebbing was only able to guide the further development of the institute until 1943.[12] In a letter to Rudolf Carnap dated 25 September 1943, Neurath wrote that "Stebbing died after a cancer operation. [...] Our best friend. A very brave and sincere personality. Many people in England feel her death a heavy loss. She represented a kind of public conscience for some circles. We loved her. [...] She acted as chairman, as I lectured in Cambridge, in a very nice way, and in agreement with most of my statements."[13]

17.2 Clarity in Thought and Expression: The Picture-Text Style and the Integrated Book

In his preface to *Modern Man in the Making*, Otto Neurath described the visual style developed to present the book's content – a historical-sociological view of the global development of modern man, cultures, wars, business and scientific and medical progress – as follows:

> An attempt has been made to evolve for this purpose a special picture-text-style which should enable anybody to walk through the modern world that is beginning to appear about us and see it as he may see a landscape with its hills and plains, woods and meadows. [...] The principle of visualization applied in this book is based on the ISOTYPE method, developed by me together with my collaborators during the last fifteen years. It shows connexions between facts instead of discussing them. Impressive visual aids do not merely act as illustrations or as eye-bait in this book; they are parts of the explanations themselves. The reader may not understand the contents by reading the text only; he must 'read' the pictures as carefully as the text. An international picture language is combined with a world language. (Neurath 1939, 7f.)

The form of knowledge transfer described here was not simply a further development of the Viennese method of visualizing statistical data that arose in Austria in the 1920s and was used extensively at the Österreichisches Gesellschafts- und Wirtschaftsmuseum (Social and Economic Museum of Vienna).[14] It was a question

[12] The ISOTYPE collection at the archive of the University of Reading apparently contains no further documents or letters by L. Susan Stebbing.

[13] Neurath to Carnap, 25 September 1943 (RC 102–55-03); see letter 22 in the present volume.

[14] As part of Neurath's work at the Museum for Social and Economic Affairs which he directed from 1925 on, a major focus was on disseminating new scientific and socio-politically relevant information on behalf of the Vienna City Administration, the Chamber of Labour and the Social Insurance Institutes. For example, quantitative statistical data and facts that had been widely distributed in newspapers and magazines since the end of the nineteenth century were to be translated into graphical form for the purposes of popular enlightenment. The "Vienna Method," which had become known in Austria and abroad via exhibitions, publications and lectures, ended in 1934 with the expulsion of almost all permanent employees into exile. Also in exile, Neurath continued the

of finding a new style of communication for another medium,[15] which in view of the changes brought by war and exile could continue pursuing pedagogical aims in the service of the general public just as effectively as scientific work in the Unity of Science movement. For Neurath (1996, 259), the two aims were inseparable: "The educational background for Visual Education is that of Unified Science." On 8 April 1939 he wrote to Stebbing, "You know I plan to prepare a Visual Thesaurus in our institute, as a 'pendant' (companion pair) to our Encyclopedia. The one Unification by Visualization, the other Unification by Logicalisation."[16]

The aim was to promote the conditions for cooperation across social boundaries and national borders, because for Neurath the (re-)construction of a tolerant and democratic world society depended on this. Just as Stebbing called for critical thinking, precise analysis and the conscious use of language, i.e. non-manipulative, emotional and streamlined transfer of knowledge to facilitate logical thinking and the formation of one's own judgment in the direction of practical democratic action, Neurath (1996, 263) too strove for a consistent, internationally understandable "neutral" transfer of information made possible by the visualization of data – "a comprehensive Visual Education Scheme." This visual style should be objective and easily accessible without being boring or aesthetically unattractive, applicable equally to the humanities and natural sciences, addressing high and mass culture or even different target groups, adults and children alike. Neurath (1996, 262) did see some exceptions, such as content with a high degree of abstraction: "Subjects that depend on verbal expression only cannot be taught by pictures, e.g. theological or philosophical doctrines. It is however perfectly possible to make quite complicated matters of fact intelligible by pictures." Generally speaking, therefore, he was concerned with the search for "[…] possible ways of transferring simple scientific knowledge by means of a common visual language, as it were, in a common visual style" (Neurath 1996, 253).

Neurath had devoted great attention to historical models and developing the visualization of information, of both a scientific and non-scientific nature, in his other pictorial educational writings, especially in manuscripts that were only published posthumously (*From Hieroglyphics to Isotype. A visual autobiography,* and "Visual Education: Humanisation versus Popularisation"). As he explained in his publication planned for the "common reader" created as a supplement to the scientific text ("Visual Education"), he was not interested in the aesthetic quality of the images,

work with a small staff (especially Marie Reidemeister and Gerd Arntz); without the concrete framework of the museum, however, the content and type of work had to adapt to new requirements. See Burke, Kindle, Walker (2013), Nikolow (2006), (2007), and Angélique Groß's chapter in the present volume.

[15] Marie Neurath (1973, 63f.) recalled the situation as follows: "I remember when Otto and I were walking together one evening along the street to post several letters and we were talking about *Modern Man*. Otto remarked that we had only to put together some of the many things we had done in the past; and for once it was my turn to suggest: why not use this chance to make something different? It was all in our hands. And so the picture-text style idea was born that night. It certainly came more from his brain than from mine, but I did present one real gift to him at that time, the word 'Isotype.' He heard it and liked it, and asked Arntz the next day to design an Isotype trademark." See also the role of former employee Rudolf Modley in Ihara (2013, 325).

[16] Neurath to Stebbing, 8 April 1939, Corr. No. 304. (ONN).

but in the image as a sign, as a means of communication, as an "informative picture." However, given the long historical tradition of visualizing content such as the knowledge of prehistoric cultures, the magical and religious ideas of the ancient Egyptians, astronomical observations and geometric drawings, or works such as maps up to and including the *Orbis Pictus* of Amos Comenius, Humboldt's *Cosmos* and the French *Encyclopedia*, Neurath considered that the time had come to combine visualization with modern logical reasoning. In his *Visual Education*, for example, he explained that modern transfer of knowledge required closer links between verbal and visual education.

> Two streams come together in modern visual education. One is visualisation in general and the other is logical arguing. Arguing requires a tradition. Our tradition in arguing and visualisation developed in the Church of the Middle Ages. Both can become secularised; this process has been supported by printing. Perhaps there is something in the idea that the parallel – scholasticism and visual aids in the Church – may find its counterpart in modern scientific arguing, together with visual aids. (Neurath 1996, 280).

Two things are expressed here. On the one hand, the method of close argumentational connections between text and image that Neurath strove for and achieved in *Modern Man in the Making*, which had also been possible in technically non-complicated ways even in the colorfully illuminated pages of medieval codexes was now facilitated by current technical innovations. These innovations made it increasingly easy to produce books and print illustrations in high quantity and quality. They included typesetting and rotary machines for print and reproduction purposes, coated paper (not only for plate sections) to gradually eliminate separating text and plate into different printed sheets, the successive use of offset printing and not least of all the rapid spread of photography at the turn of the nineteenth to the twentieth century.[17] The combination of text and images in newspapers, magazines and advertising became more closely linked and complex, which affected the style of book illustrations.[18]

On the other hand, this type of knowledge transfer became particularly relevant for Neurath as a basis for socially responsible action against the background of social and political crises. "Sincere arguing is something of international importance because it creates a common basis for discussions and decisions. All tendencies towards what may be called logical empiricism are therefore particularly closely connected with consistent visual education. Both start from ordinary man's commonsense and are in full harmony with highly developed scientific activities" (Neurath 1996, 285). It is precisely here that Neurath's unbroken trust in education and science can be seen. As Cartwright, Cat, Fleck and Uebel (1996, 3) observe, "[a]lthough he was a pluralist about knowledge systems and took seriously their historical and cultural roots, he trusted firmly in the power of science."

During World War Two, Adprint developed two series for the British publisher G. G. Harrap with financial support from the Ministry of Information. *The Soviets*

---

[17] See Heiting (2012, 6–10); Lucius (2012, 315–320) and Peters (2012, 21–22).

[18] One example is the work of the medical doctor and non-fiction author Fritz Kahn, who collaborated with his friend Paul Steiner on several book projects in exile. A business partner of Wolfgang Foges, Steiner had opened a branch of Adprint in New York and later very successfully managed it himself. On Fritz Kahn see Debschitz, (2009) and (2013) and on Steiner see Körber (2014).

Fig. 17.1 *The New Democracy: Education. The New Horizon* Book cover of a title from 1947. The books were produced by Adprint, the sociologist Karl Mannheim was a member of the editorial committee, and Neurath's Isotype Institute was responsible for the visualization of the volume in coordination with the authors

and Ourselves and *America and Britain* were intended to make the British people more familiar with the everyday life and the political and economic systems of the other Allies. The *New Democracy* series for the Nicholson & Watson publishing house was intended to provide post-war British society with information and expert advice on important matters. The books were produced by Adprint and use Isotype graphics and photographs to illustrate the content. (Otto and Marie Neurath Isotype Collection, University of Reading, Figs. 17.1, 17.2, and 17.3)

As mentioned above, however, Neurath's visual picture-text style in books was associated with special system-based, organizational and technical requirements. Moreover, due to the limited resources available in exile, finding a suitable partner seemed the only way to secure the requisite technical know-how and the considerable funding needed for color reproduction and printing in order to produce further publications of this kind. Because Neurath's *Modern Man in the Making* had proven to be a success the American publisher Alfred A. Knopf offered him a follow-up project, but this eventually did not come to fruition on account of his having to flee again, this time from Holland. Even though Neurath was able to continue later his research on the planned topic ("Persecution and Brotherhood") in Oxford as well, he never got more than a few pages of notes.

A more sustainable form of collaboration was launched following Neurath's release from the internment camp to which he and Marie Reidemeister (later Neurath) had been sent after their flight. The Neuraths and their staff at the Isotype institute participated in these book projects with a small group of German-language publishers in exile[19] in Great Britain who fled Nazi persecution in the 1930s and

[19] The terms exile, emigration and forced migration are understood here synonymously as non-freely chosen flight from the country of origin to a country of refuge on account of political and/or racist persecution, although exile research has long differentiated between groups of exile and emigrants in terms of causes, return interests, language behaviour etc. See especially the publication series *Exilforschung – Ein internationales Jahrbuch*, e.g.: Krohn and Winckler (2012, VII-XIV).

Fig. 17.2 *The New Democracy: Woman and a New Society*
This book, published in 1946 in *The New Democracy* series, was written by Karl Mannheim's research assistant, the German sociologist Charlotte Lütkens

Fig. 17.3 *The Soviets and Ourselves: How do you do Tovarish?*
This volume, published in 1947 in the series *The Soviets and Ourselves*, was created for the publisher G.G. Harrap, subsidized by the Ministry of Information

1940s, including the publisher Wolfgang Foges (1910–1986). According to Marie Neurath, Foges was already familiar with Otto Neurath's ideas when he emigrated to Great Britain (Neurath and Kinross 2009, 65). Both came from the same intellectual Viennese milieu and shared an interest in modern science, art and a visual culture established through new media.[20] In an interview in 1964, Foges described Otto Neurath as "the most important influence on my thinking" (cited in Hamilton 1964). Foges lived in London since 1937, and had previously published an illustrated magazine.[21] His Adprint company produced primarily illustrated non-fiction

[20] See Burke (2010, XIII). On the influence of emigrated Jewish publishers, scientists and editors in the UK see James (1999, 371–377).

[21] Foges was editor-in-chief of the Moderne Welt Verlag and the illustrated magazine *Moderne Welt* for art, literature and fashion in Vienna. It was founded in 1918 by the international company

books in series, initially with the financial support of Lord Glenconner (Tennant & Son).[22] Foges worked with other German-speaking emigrants, including Eva Feuchtwang (1908–1999), Walter Neurath (1903–1967, no relation to Otto Neurath) and Paul Steiner (1913–1996)[23] who headed the US branch of Adprint in New York. The core group was supported by other creative emigrants from Germany and Austria, such as graphic artists, photographers and picture editors – a new professional occupation at the time.[24]

Adprint achieved its first successes with the production of the illustrated *King Penguin* series starting in 1939, which essentially copied the German *Insel-Bücherei* series launched in 1912 (see Lambert 2009, 114). Penguin's small-format introductions to history, nature and wildlife, preferably those of Great Britain, contained an average of 16 color illustrations plus black-and-white images and made ideal and relatively inexpensive gifts (see Edwards and Hall 1988). The Adprint programme consisted of popular general education works ranging from culture and history to nature and technology. Adprint was probably the first publisher to be known as a "book packager," and thanks to this organizational form was also able to engage in productive joint projects (Ridler 1976, 9). A book packager was a book producer that designed books on behalf of or for publishers and produced them up to a prepress stage, or also did commission printing (Schlitzer 2007, 13). Paul Steiner, Wolfgang Foges' friend and later business partner, wrote in his unpublished autobiography that this was a new business model that focussed on technically and organizationally complex publications with extensive picture material and text (see also Ridler 1976, 4).

> Foges had the brilliant idea to establish a publishing house based on the principle of creating books, that is a publishing house whose sole purpose was to conceive books and book series, to commission the texts from competent authors, but not to worry about distribution […], instead selling […] the edition to an established publisher. Since the books in question were invariably non-fiction books that were conceived by an in-house editorial team and because those non-fiction book series lend themselves to illustration, Foges' book creation house was not just responsible for the text of the individual volumes, but also for the artwork. (Steiner, undated, 421f.)

This new production process required a modified organizational structure with a more team-oriented approach than was customary at traditional publishing houses.

Wiener Weltmode (Chic Parisienne Bachwitz AG) and changed hands several times before Foges became involved. See ANNO, the database of the Austrian National Library http://anno.onb.ac.at/info/dmw_info.htm

[22] According to John Spiers, innovative book series can express publishers' self-image and creative visions. At the same time, such serial artefacts function as an independent system that can define public and personal fields of meaning and convey perspectives on the respective historical situations. See Spiers (2011, 1–61), also Lambert (2009, 113f).

[23] For this and other biographical information on the individuals mentioned, see Fischer (2011, 78f., 227f., 312), and for Paul Steiner see Körber (2014).

[24] For an in-depth analysis from an exile research perspective of the influence of German-language Jewish publishers on the British book market, especially art books, see Nyberg (2009, 51).

Fig. 17.4 Double page spread from the *The New Democracy series: Women and a New Society* (1946) *Women and a New Society*: "Comparison is an educational element," says Neurath in his posthumously published "Visual Education." A double-page spread from the Adprint publication shows how this principle can be implemented in layout design. Neurath also noted that this would encourage recipients to draw comparisons themselves – whether on the level of individual characteristics or topics. (Otto and Marie Neurath Isotype Collection, University of Reading)

In particular, in order to establish the successive text-image connections created with Isotype in what were known as "integrated layouts," close coordination was needed between internal employees such as text and image editors, and external authors, photographers and designers. This type of cooperation would also become important in further developing illustrated non-fiction books.[25] The increasingly sophisticated production of heavily illustrated non-fiction books, which became successively differentiated and professionalized in the second half of the twentieth century, was most efficiently managed by creative teams working in concert (Körber 2016) (Figs. 17.4 and 17.5).

The principle of cooperation among editors, authors and graphic designers in planning and producing publications can already be seen in the first edition of

[25] For reasons of space further attention cannot be devoted to the development of the non-fiction or illustrated non-fiction book. In brief, these are non-fiction, entertaining and comprehensible presentations of current and complex information, which can also contain illustrations, for readers without specific training in the respective field. Although such books have been produced at the latest since the Enlightenment, the idea of popularizing specialized knowledge gained a new quality following World War One. For more information see Voges (2012), Körber (2016).

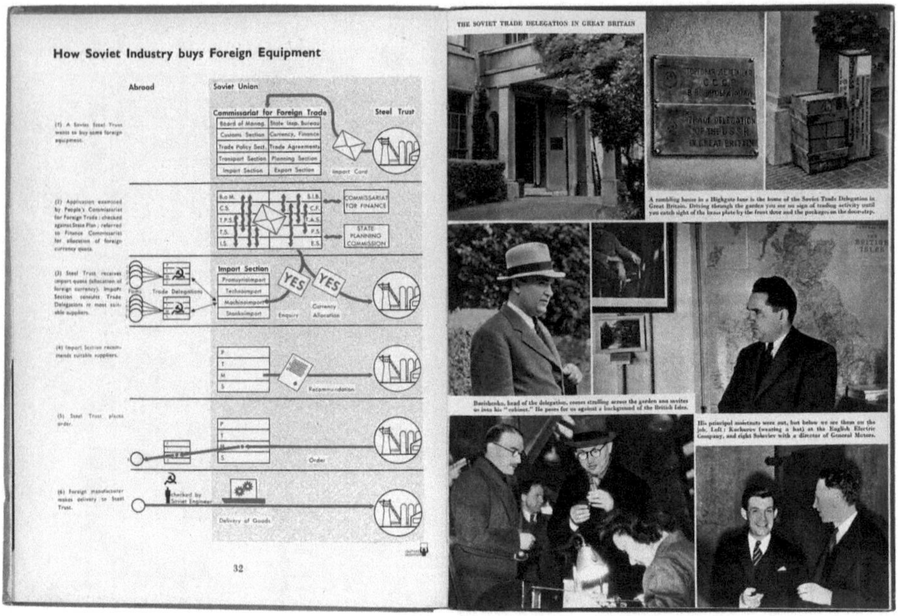

Fig. 17.5 Double page spread from the first volume of the *Future Book series* (1946)
The *Future Books* series was produced by Adprint from 1946 to 1952. The series concept is based on a closely linked structure of text, photographs, maps and ISOTYPE graphics as can be seen on this double page from the first volume entitled *Overture*. The modern design of the books and the magazine of the same name was developed in close creative coordination and cooperation among Adprint employees, well-known authors, photographers and the Isotype Institute. (Otto and Marie Neurath Isotype Collection, University of Reading)

Adprint's *Future Books: Overture* series which started appearing in 1946. Like the similarly named *Future Magazine*, the series addressed socially relevant topics from culture to technology. "So far as method is concerned teamwork is the idea behind FUTURE books. It governs the planning of the series […] it underlies the marriage of written and visual exposition which marks these books out from other miscellanies" (Milne 1946, 1). Especially with respect to technology, illustrated newspapers and magazines tended to make more professional and flexible use of pictorial material at this time than book publishers, as Paul Steiner (undated, 422) explains in his memoirs. Since the time at the museum in Vienna,[26] close cooperation among specialists and experimentation with the processes of conveying information had played a key role at the Isotype Institute, and now fit in well with the developing organizational structures at Adprint. In addition, Foges and his staff cooperated with external serial editors and editors-in-chief of the publishing houses involved, for example in the case of the *Britain in Pictures* series (Carney 1995, 14–17).

[26] Teams of scientists, statisticians and other specialists compiled information as part of the "Vienna Method," employing a modernist, clear design under the direction of the progressive graphic designer Gerd Arntz, as was also taught at the Bauhaus with which Neurath was connected. See Galison (1990, 709–752) and Sophie Hochhäusl's chapter in the present volume.

The concept behind Adprint's book publications on general educational topics was clearly based essentially on the interplay of image and text, or rather focused primarily on staging pictorial material for the construction of meaning. In general, the content was created in the course of its conceptualization in the interplay of text and visual design of double pages as smallest units up to longer sequences within the book. The types and formats of images for the respective publication were combined with units of text that could take the form of main, special or image texts. Comprehension of the content was therefore no longer bound to the text alone. Germano Facetti, as art director for Foges and later as the designer responsible for reworking the *Penguin Classics* series, described the works produced by Foges and integrated books in general as follows: "The flow of images, captions and diagrams was planned like a documentary film; but unlike a film the book could be opened at any page to provide leads onward and backward in the text [...] the images and the graphics follow a track where the essence of the text is underlined [...] in 'close-ups' enhancing the art works and in 'sequences' integrated in the text" (Facetti 1964, 53, 57).

The influence of this new and variegated media, which according to Neurath could be used most effectively for transferring knowledge, was now clearly evident when different media and presentation techniques were used. "It is manifest how fruitful a film presentation may become if it intentionally combines documentary and diagrammatic techniques, just as a book can combine photographs and diagrams with the text" (Neurath 2010, 120). While a close connection between text and image on double pages and over longer sequences was a distinctive feature of many of Adprint publications, the image material itself could vary – from photos and maps to pictograms and art reproductions.

Adprint was very successful in the UK book market and worked with publishers such as Collins, Hamlyn, Harrap, Muller, Pitman and the UK Ministry of Information (see Lambert 2009, 114). As Jim Aulich (2012, 343–366) explains, it was the contemporary combination of affordable but visually innovative and high-quality books that made Adprint an interesting business partner, particularly for the Ministry of Information – and especially with regard to positive propaganda on "mutual understanding" amongst the Allies against National Socialism. With the support of the Ministry of Information, the book series entitled *The Soviets and Ourselves* and *America and Britain* were published by G.G. Harrap and distinguished themselves with elaborate color prints of Isotype graphics and photographs. Their covers were designed by the German artist John Heartfield (né Helmut Herzfelde), a pioneer of photomontage, among others.

The special style of the books developed by Adprint was regarded by the British book trade as belonging to the German-language book culture, for example in the tradition of the Bauhaus Verlag (publishing house) (see Nyburg 2009, 226). However, the actual design of the individual series varied from case to case: the modern, diversified publications which were made in collaboration with the Isotype Institute and which could be produced particularly elaborately thanks to the support of the Ministry of Information, stood out in the field (Burke, Kindel and Walker 2013, 360). But continuous improvement of the quality of images under the direction of production manager Walter Neurath also affected the overall impression of other series. However, the scope for content and design was limited not least of all by factors such as paper shortages, high overall production costs for color prints and an inadequate production infrastructure in the UK (see Hogben 1949, 262; Kinross 2010, 120–145; James 1999, 371–377). After Otto Neurath's death, Foges continued to produce books with Marie Neurath and the Isotype Institute, especially for children and young people (Walker 2013). He continued his activity as a book producer and founded a new book packaging company called Aldus Books Ltd., which worked together with houses like Doubleday in New York (Lambert 2009, 116f.).

Walter Neurath, who played a major role in many of Adprint's successes, founded his own publishing house in 1949 with Eva Feuchtwang whom he would later marry. Over the decades to come, Thames & Hudson became an international leader in the art and culture sectors (Rosenthal 2009, 111–122; Fischer 2011, 227f.). Here, too, the principle of democratizing art books continued to yield mass-produced but inexpensive works of high quality. The same applied to art books from Phaidon Press, another house founded by German-speaking emigrants in England (Fischer 1999, 289–309). But unlike Phaidon and similar publishers, Thames & Hudson used as many color illustrations as possible, which were presented in the integrated layouts already developed and proven by Adprint. The rising demands on print quality and the greater number of color illustrations were financed by a system of international co-productions perfected by Walter Neurath, which spread the production and printing costs as well as the financial risk across various publishing houses (Craker 1985, 9f.). This enabled long print runs, which in turn lowered the production price per copy. As a consequence, book concepts and designs had to be developed to enable extensive international compatibility among texts and illustrations. Foges had already striven for this type of international distribution and cooperation by establishing a branch office in New York, but was not in a position to achieve these goals until after the war (Lambert 2009, 118). Thames & Hudson, by contrast, began in the 1950s under completely different conditions. It enjoyed increasing political and economic stability in a society with growing purchasing power and a consistent demand for illustrative book formats, including high-quality color prints.[27]

As the book trade acquired greater degrees of internationalization, the combination of proven design and production principles with a range of topics suitable for international distribution proved to be sustainable over the long term for popular illustrated works of non-fiction – as perfected for art books at Thames & Hudson and by Foges' partner Paul Steiner at Chanticleer Press in the USA, for which space

[27] The effects of these correlations on book production conditions can be seen in the business correspondence between Adprint (in particular Wolfgang Foges and Walter Neurath) and Otto Neurath/Isotype. It is housed in the Isotype collection at the archive of the Institute for Typography and Graphic Communication at the University of Reading, UK: Isotype Collection 1: Isotype Institute Correspondence (1941–1945). See also Eve (2009) for difficulties in the cooperation after Otto Neurath's death.

considerations here prevent further elaboration.[28] In Germany, some books by Foges and especially those by Thames & Hudson and Paul Steiners have been published very successfully by Droemer Knaur in collaboration with the publisher Willy Droemer as *Bildsachbücher* (illustrated non-fiction books),[29] and this production and distribution model has survived internationally to this day.

17.3 Conclusion

Even in exile, Neurath continued to study potential new methods of transferring knowledge in science and society. These efforts gained a new dynamic in his exchange with the philosopher L. Susan Stebbing, as can only be touched on in this article. In combination with the tried and tested transformation of visual data into a densely woven "picture-text style," the logical principles of the subsequent critical examination of common language became an additional means of making knowledge content available to a mass audience for Neurath. Against the background of his theoretical examination of the historical development of visual communication, this was intended to broaden the possibilities of purely linguistic transfer of knowledge toward a more structural-contextual understanding of content that could also keep pace with the diversification of media in books at the time. Conscious and careful handling of the selection of information and of the choice of language, such as the avoidance of technical terminology, was directed against the inappropriate and popularizing simplification that both Neurath and Stebbing condemned. Both demanded a respectful and serious approach to the transfer of knowledge in order to strengthen judgment and logical thinking on the part of the "common reader" as a prerequisite for social participation in modern democracies. Education and training were therefore not understood as simply passing on information, but rather were associated with the ability to think, analyze and argue in structured ways.

As of 1941 Neurath pursued this approach by further developing the "picture-text style" in publication projects with the book packager Adprint. The resulting so called integrated layouts were an overall attempt to develop book series in cooperation with image and text editors, graphic designers, authors and photographers, forming an argumentative syntax of information in terms of content and physical integration. The idea that this should not be a "naïve" objective but nevertheless appropriate reproduction of knowledge was already evident in the self-image and differentiated processes of selecting and transforming information which were also fundamental to the creation of ISOTYPE symbols. As Neurath (1939, 7) wrote, "How can facts be presented without causing confusion by their overwhelming

[28] See Fischer (2011) and Körber (2014) for further information on Paul Steiner.

[29] As stated in the publisher's autumn 1961 programme, the "new book type" was a "requirement of the times:" the illustrated book of the past was no longer sufficient; a new, closer connection between images and words was to be created, a new type of *Bild-Band* (picture book) which was also non-fiction. See *Droemer Gesamtkatalog* (Autumn 1961, 3f).

diversity? The visualization of selected primary material connected with simple statements is one solution. Even such 'selection' influences readers in a certain direction, but one can interpret the same facts in different senses and augment them."

This approach to book design differed markedly from that of non-fiction volumes that conveyed their contents primarily through text accompanied by a few secondary illustrations. Via collaboration between ISOTYPE and Adprint and their successors, these ideas and concepts found their way into the practices and methods of structuring and systematizing text and images in the illustrated non-fiction book genre – a development that can only be hinted at here. While individual elements of visual design and content structure were not in themselves innovations in book production, the systematic approach and the exploitation of new technical possibilities for more cost-effective and therefore mass reproduction and printing technology must be seen as influential for modern, highly visual book design.[30] As has been noted on numerous occasions, these possibilities were assessed by a generation of German-speaking emigrants in the twentieth century for a new culture of visualizing information in the Anglo-American book market.[31] The aim was to reconcile the strong socio-political awareness, pedagogical goals and influences of the new visual mass culture. Especially in the field of conveying information for general and further education, the "picture-text style" for books was to no small extent a material expression of changing cultural techniques and concepts of knowledge in response to political and social crises and upheavals of the time.

Acknowledgments This article is the result of an ongoing doctoral research project at the Department of Philosophy, Humboldt-University zu Berlin, which was supported by scholarships from the German Historical Institute in London and Washington, DC, and an Elsa-Neumann-Scholarship.

References

Abel, Richard, and Gordon Graham. 2009. *The Impact of Expatriate Publishers in Britain and America in the 20th Century*. New Brunswick/London: Immigrant Publishers/Transaction Publishers.

Aulich, Jim. 2012. Stealing the Thunder: The Soviet Union and Graphic Propaganda on the Home Front during the Second World War. *Visual Culture in Britain* 13 (2): 343–366.

Beaney, Michael. 2003. Susan Stebbing on Cambridge and Vienna Analysis. In *The Vienna Circle and Logical Empiricism*, ed. Friedrich Stadler, 339–350. Dordrecht: Springer.

Beaney, Michael, and Siobhan Chapman. 2017. Susan Stebbing. In *The Stanford Encyclopedia of Philosophy*, ed. Edward N. Zalta. https://plato.stanford.edu/archives/sum2017/entries/stebbing/ Accessed 15 Apr 2018.

Burke, Christopher. 2010. Introduction. In Neurath (2010): VII–XVIII.

[30] See James (1999); Fischer (2003); Kinross (2009), Lambert (2009); Nyburg (2009).

[31] See e.g. Kinross (2009) and Abel – Graham (2009).

Burke, Christopher, Eric Kindel, and Sue Walker, eds. 2013. *Isotype: Design and Contexts 1925–1971*. London: Hyphen Press.
Carney, Michael. 1995. *Britain in Pictures. A History and Bibliography*. London: Werner Shaw Ltd.
Cartwright, Nancy, Jordi Cat, Lola Fleck, and Thomas Uebel. 1996. *Otto Neurath: Philosophy Between Science and Politics*. Cambridge: Cambridge University Press.
Chapman, Siobhan. 2013. *Susan Stebbing and the Language of Common Sense*. Basingstoke: Palgrave Macmillan.
Craker, Trevor. 1985. *Opening Accounts and Closing Memories. Thirty Years with Thames and Hudson*. London: Thames & Hudson.
Debschitz, Uta von, and Thilo von Debschitz. 2009. *Fritz Kahn – Man Machine*. Vienna: Springer.
———. 2013. *Fritz Kahn*. Cologne: Taschen.
Droemer Gesamtkatalog. 1961 (Autumn). Munich: Droemersche Verlgsanstalt.
Edwards, Russell, and David J. Hall. 1988. *So Much Admired: Die Insel Bücherei and the King Penguin Series*. Edinburgh: Salvia Books.
Eve, Matthew. 2009. Isotype in trouble, 1946–1948. In *Modern Typography in Britain: Graphic Design, Politics, and Society*, Typography Papers 8, ed. Paul Stiff, 129–134. London: Hyphen Press.
Facetti, Germano. 1964. The integrated book. *The Penrose Annual* 57: 51–58.
Fischer, Ernst. 1999. The Phaidon Press in Vienna, 1923–1938. *Visual Resources* 15 (3): 289–309.
———. 2003. *Buchgestaltung im Exil 1933–1950. Eine Ausstellung des Deutschen Exilarchivs*. Wiesbaden: Harrassowitz.
———. 2011. *Verleger, Buchhändler und Antiquare aus Deutschland und Österreich in der Emigration nach 1933. Ein biographisches Handbuch*. Elbingen: Verband deutscher Antiquare e.V.
Galison, Peter. 1990. Aufbau/Bauhaus. Logical Positivism and Arcitectural Modernism. *Critical Inquiry* 16 (4): 709–752.
Hamilton, Ian. 1964. *Visionary bookman*. The Guardian, 10 November 1964.
Heiting, Manfred. 2012. Autopsie: Anschauung, Information und Analyse. Über die Sammlung zur Wissenschaft. In *Autopsie. Deutschsprachige Fotobücher 1918 bis 1945*, ed. R. Jäger, 6–10. Göttingen: Steidl.
Hogben, Lancelot. 1949. *From Cave Painting to Comic Strip*. New York: Chanticleer Press.
Ihara, Hisayasu. 2013. Isotype in America: Otto Neurath and Rudolf Modley. In Burke, Kindel, and Walker (2013): 298–353.
James, Clive. 1999. On the Library Coffee-Table. *Visual Resources* 15 (3): 371–377. Special Issue: The Early History of the Phaidon Press, 1926–1967.
Kinross, Robin. 2009. Design in Central-European London: Interactions between Émigrés and Natives in the 1940s. In *Modern Typography in Britain: Graphic Design, Politics, and Society*, Typography Papers 8, ed. Paul Stiff, 105–112. London: Hyphen Press.
———. 2010. *Modern Typography. An Essay in Critical History*. London: Hyphen Press.
Körber, Silke. 2014. Paul Steiner. *Transatlantic Perspectives*. http://www.transatlanticperspectives.org/entry.php?rec=162 Accessed on 15 Apr 2018.
———. 2016. The Democratization of Knowledge and a New Visual Culture: Popular Illustrated Non-Fiction Books by German-speaking Émigré Publishers in Britain during and after the Second World War. *Publishing History* 76: 57–76.
Krohn, Claus-Dieter, and Lutz Winckler (eds.). 2012. *Exilforschung. Ein internationales Jahrbuch*, ed. Gesellschaft für Exilforschung 30, 2012: Exilforschungen im historischen Prozess. Hamburg: edition text + kritik.
Lambert, David. 2009. Wolfgang Foges and the new illustrated book in Britain: Adprint, Rathbone Books, and Aldus Books. In *Modern Typography in Britain: Graphic Design, Politics, and Society, Typography Papers 8*, ed. Paul Stiff, 113–128. London: Hyphen Press.
Lucius, Wulf D. von. 2012. Buchgestaltung und Buchkunst. In *Geschichte des deutschen Buchhandels im 19. und 20. Jahrhundert. Die Weimarer Republik 1918–1933*. Bd. 2, Teil 1, ed. E. Fischer and S. Füssel, 315–340. Munich: De Gruyter.

Milne, Marjorie B., ed. 1946. *Future Books: Volume 1, Overture*. London: Leathley Publications.
Neurath, Otto. 1939. *Modern Man in the Making*. New York: Alfred A. Knopf.
———. 1941/1983. Universal Jargon and Terminology. In *Otto Neurath: Philosophical Papers 1913–1946*, ed. Marie Neurath and Robert S. Cohen, 213–229. Dordrecht: D. Reidel.
———. 1996. Visual Education: Humanisation versus Popularisation. In *Encyclopedia and Utopia. The Life and Work of Otto Neurath (1882–1945)*, ed. Elisabeth Nemeth and Friedrich Stadler, 245–335. Dordrecht: Kluwer.
———. 2010. *From hieroglyphics to Isotype. A visual autobiography.* Ed. Matthew Eve and Christopher Burke. London: Hyphen Press.
Neurath, Marie. 1973. Marie Neurath: 1940–1945. In *Empiricism and Sociology*, ed. Marie Neurath and Robert S. Cohen, 68–75. Dordrecht: D. Reidel.
Neurath, Marie, and Robin Kinross. 2009. *The Transformer: Principles of Making Isotype Charts*. London: Hyphen Press.
Nikolow, Sybilla. 2006. Imaginäre Gemeinschaften. Statistische Bilder der Bevölkerung. In *Konstruierte Sichtbarkeiten. Wissenschafts- und Technikbilder seit der Frühen Neuzeit*, ed. M. Heßler, 263–278. Munich: Wilhelm Fink Verlag.
———. 2007. Aufklärung durch und mit Beobachtungstatsachen. Otto Neuraths Bildstatistik als Vehikel zur Verbreitung der wissenschaftlichen Weltauffassung des Wiener Kreises. In *Wissenschaft und Öffentlichkeit als Ressourcen füreinander. Studien zur Wissenschaftsgeschichte im 20. Jahrhundert*, ed. S. Nikolow and A. Schirrmacher, 245–272. Frankfurt/New York: Campus.
Nyburg, Anna. 2009. *'Hardly a trace left of Danube or Spree?' A contribution to the study of art book publishing and illustrated book production in Britain by German-speaking exiles from National Socialism*. Unpublished dissertation, University of London.
Peters, Dorothea. 2012. Fotografie, Buch und grafisches Gewerbe: Zur Entwicklung der Druckverfahren von Fotobüchern. In *Autopsie. Deutschsprachige Fotobücher 1918 bis 1945*, ed. R. Jäger, 12–23. Göttingen: Steidl.
Ridler, Colin. 1976. *Out-House Publishing*. Oxford: Oxford Polytechnic Press.
Rosenthal, Tom. 2009. Walter and Eva Neurath: Their books married words with pictures. In *Immigrant Publishers*, ed. R. Abel and G. Graham, 111–122. New Brunswick/London: Transaction Publishers.
Sandner, Günther. 2014. *Otto Neurath: Eine politische Biographie*. Vienna: Zsolnay.
Schlitzer, Monika. 2007. *Langenscheidt Praxiswörterbuch Verlagswesen und Buchhandel Englisch*. Munich: Langenscheidt.
Spiers, John. 2011. *The Culture of the Publisher's Series, Vol. 1.: Authors, Publishers and the Shaping of Taste, ed. J. Spiers*. London: Palgrave Macmillan.
Stebbing, L. Susan. 1937. *Philosophy and the Physicists*. London: Methuen.
———. 1939. *Thinking to Some Purpose*. Harmondsworth: Penguin.
Steiner, Paul. *Unpublished autobiography,* undated. Typescript at the Leo Baeck Institute. New York (Acc. No. ME938).
Tuboly, Adam Tamas. 2017. From 'syntax' to 'Semantik' – Carnap's Inferentialism and its prospects. *Polish Journal of Philosophy* 12 (2): 57–78.
Voges, Britt. 2012. Sachbuch- und Ratgeberverlage. In *Geschichte des deutschen Buchhandels im 19. und 20. Jahrhundert. Die Weimarer Republik 1918–1933. Vol. 2, Teil 2*, ed. E. Fischer and S. Füssel, 241–270. Munich: De Gruyter.
Walker, Sue. 2013. Graphic explanation for children 1944–71. In Burke, Kindel, and Walker (2013): 390–419.

Chapter 18
Logical Empiricism and Art: The Correspondence Otto Neurath/Meyer Schapiro

Hans-Joachim Dahms

Abstract Logical Positivists had a very lively interest in the revolutionary science of their time, but also in modern art and especially in 'international style' architecture. Surprisingly they never published a representative volume or longer statement on art and architecture. But: it is not well known that Otto Neurath, their leading organizer and spokesman, invited the (later on) eminent art historian and critic Meyer Schapiro (professor at Columbia University NY) to contribute a volume on art to the *International Encyclopedia of Unified Science*. Schapiro failed to deliver the promised book. But from the extended correspondence and some material in the Schapiro papers one can describe the general direction the thing would have taken. The correspondence is also very interesting, because it also covers questions of the endangered peace and the approaching war, the academic scene in Europe and in the USA, and surprisingly: Martin Heidegger. Neurath died in 1945, but Schapiro came back to the Heidegger theme in 1968 when he wrote his famous harsh criticism of Heidegger's programmatic long paper "Das Kunstwerk" (the work of art) where he interprets one of Van Gogh's shoe-paintings. Schapiro's short article caused much controversy then (even Jacques Derrida intervened).

After the defeat of the national-socialist dictatorship, logical empiricism was either completely forgotten in middle Europe, where it had its roots, or had become a subject of polemics. In Austria, its criticism of religion and metaphysics were exposed as subversive and even its alleged communist tendencies laid bare[1]; in the Federal Republic of Germany, its focus on philosophy of science and logic to the detriment of ethics and aesthetics received devastating criticisms. In leftists circles such as the Frankfurt school of critical theory in Germany, the argumentative figure of a "double positivism" was created: on the one hand it made the

[1] See for an overview on the situation of philosophy in Vienna after 1945 Dahms and Stadler (2015, 115–123) and the literature mentioned there.

H.-J. Dahms (✉)
Institute Vienna Circle, Vienna, Austria
e-mail: hans-joachim.dahms@univie.ac.at

J. Cat, A. T. Tuboly (eds.), *Neurath Reconsidered*,
Boston Studies in the Philosophy and History of Science 336,
https://doi.org/10.1007/978-3-030-02128-3_18

(phenomenological) given as the starting point for all knowledge and, on the other, it "accepted the (political) given, that is, the social and political status quo, in an attitude of conservative quietism" (on this, see Dahms 1994, 303–307).

It took a younger generation of contemporary historians of philosophy to unmask these views as prejudices, using research in many archives and interviews with eyewitnesses. It started – around the late 70s – mainly with the deconstruction of the view that "positivists" were worshippers of the socially and politically "given" on issues of *politics, ethics and morals.* In the course of these investigations it emerged that exactly the opposite was the case: logical empiricists had belonged in the inter-war period in Austria and during the short-lived Weimar Republic in Germany to the small minority, which had favored and proclaimed far-reaching reforms of economy, society and culture and had promoted these goals through a whole number of activities. Nowadays it seems that this new view of the historical "positivism" has prevailed.

In the 90s, although with less intensity, a number of publications examined the relationship of the Vienna Circle to culture and the arts. They focused in the beginning on the encounters of its members with the Bauhaus in Dessau.[2] This line of research is in need of an accompanying deeper look into the Austrian scene in modern architecture and interior design. Josef Frank (brother of the Circle member Philipp Frank, the physicist) played a leading role in the vast housing program of "Red Vienna", together with Otto Neurath, who immediately after his return from the short-lived political experiments of the Bavarian revolution in Munich, had quickly assumed a leading role in the settlers movement in Vienna (see Thun-Hohenstein, Czech, and Hackenschmidt (2016), and Sophie Hochhäusl's chapter in the present volume). The cooperation between these two men proved successful first in the reconstruction of the Austrian Werkbund in November of 1928 and reached its zenith in the International Werkbund exhibition of modern housing in the summer of 1932 in Vienna (see Nierhaus et al. 2012).

Up to now, the relationship of the logical empiricists with art in a narrower sense – leaving architecture aside – has received less attention. That is a surprising fact, since Franz Roh, Rudolf Carnap's friend from common years as students in Jena (and also Otto Neurath's savior after the fall of the Soviet Republic of Bavaria in may 1919) became later on, in the middle of the 20s, the leading theoretician of the movement of "Neue Sachlichkeit" (New Objectivity) (Dahms 2004). Also Gerd Arntz, a painter from Cologne and there a member of the left-leaning group "Cologne Progressives" of painters and also photographers, was hired by Otto Neurath in 1929 for his Social and Economic Museum of Vienna. There he designed many pictograms for the Neurathian picture statistics according to the "Vienna method" (and later on: ISOTYPE).[3]

That it took so long before all these themes and issues became research material (or even now have to wait for investigations) is in part the logical empiricists' own fault: they were not able to publish anything comprehensive about them (apart from

[2] See, e.g., Galison (1990), Stadler (1995), Dahms (2004), (2019), Bernhard (2015), (2019a), (2019b).

[3] See for autobiographical information Arntz (1982) and (1988).

much cited programmatic trombone sounds in Carnap's *Logical Structure of the World* (1928/2003) and in the Vienna Circle's Manifesto; see Carnap, Hahn, Neurath (1929/1973).

That the situation stayed that way is, at least in part, the result of a historical misfortune, namely, the circumstance that a planned und agreed upon contribution about art for Neurath's *International Encyclopedia of Unified Science* (IEUS) was not finished and delivered. I have in mind the pamphlet *Interpretation and Judgment of Art*, which Meyer Schapiro (Columbia University) had agreed to write. My description of this episode is based mainly on the correspondence between Neurath and Schapiro (and the publications of the two men, of course). As far as I can see, up to now nobody cared to dig into those unpublished papers.[4] It would be worthwhile to take a closer look at the notes and preliminary papers for the IEUS in Meyer Schapiro's papers at Columbia University.[5] There are indeed a number of items that belong in the vicinity of the planned contribution, for instance, the piece titled "Physicalism in art." But the required sorting of the material is a difficult endeavor: there are preliminary notes and drafts etc. for different encyclopedias, among them one for the *Encyclopedia of the Social Sciences*, which are not marked among the papers for the later use. To disentangle this knot cannot be achieved in this article.

18.1 Themes in the Neurath-Schapiro Correspondence

18.1.1 Schapiro's Planned Contribution for the IEUS

Neurath and Schapiro knew each other since 1936, when Neurath visited the exiled Horkheimer Institute for Social Research, and when was later invited to Columbia University. Schapiro (and also Ernest Nagel) accompanied him (Dahms 1999). The correspondence between Neurath and Schapiro started in October 1937, when Schapiro sent reprints of articles[6] after a meeting in which also Schapiro's wife had participated. It continued – with interruptions during World War Two – till the very last days before Neurath's death on 22 December 1945 in Oxford.[7] All in all Neurath sent 10 letters of sometimes astonishing length,[8] and Schapiro sent 10 letters to Neurath (among them a telegram from London at the 29th of July 1939, shortly before the German aggression of Poland, which started the war).

[4] In the otherwise indispensable Neurath biography in Sandner (2014), Schapiro is mentioned nowhere.

[5] A few of them have been published already in Schapiro (1999a).

[6] Schapiro to Neurath, 17 October 1937.

[7] The last letters are: Schapiro to Neurath, 9 September, and Neurath to Schapiro 18 September 1945.

[8] Neurath's last letters have 9 pages (29 September 1942), 6 pages (4 September 1945) and 7 pages (18 September 1945). Two of Neurath's letters (29 March 1940 and the very last of 18 September 1945) carry his typical elephants at the end.

When, from where and how exactly Schapiro was invited to contribute to the IEUS cannot be determined from the correspondence. Aesthetics had not belonged to the themes in the initial planning of the IEUS. Still, one comes across ideas for such a volume in 1938, when the first contributions had already appeared. In March of that year Neurath wrote a letter about the necessity to have such a pamphlet in the IEUS to his co-editor Charles Morris and also included a proposal for its author: "[…] I am definitely in favour of inserting a booklet by Meyer Schapiro about arts. The title should express, that 'arts' is the object of scientific analysis. History of art should be discussed prominently."[9]

While introducing Neurath is not required in this volume, Schapiro on the other hand will only be known to very few interested in Neurath – or in general, interested in logical empiricism. Therefore some remarks here about his life and work are in order. Schapiro was born on the 23rd of April 1904 in Schaulen in the government district of Kowno (Russia). Already in his younger years he emigrated with his family to the USA. He stayed for the rest of his life in New York, where he also lived from 1932 till the end of his life in the same house. He studied art history and philosophy at Columbia University and stayed there during his whole academic career, from lecturer to assistant professor and to full professor. In contrast to his very settled life in New York, in Europe he travelled a lot on trips to museums and other sites of artistic significance.

His work shows an unusual span of interests, from antique Roman, medieval Romanesque architecture to modern and contemporary art and almost everything in between. His monographs about Vincent van Gogh and Paul Cezanne were published with many editions and reprints and translated into many languages. He also wrote extensively about overarching themes in art history like the problem of style.

That Neurath thought of Schapiro as an author for the IEUS already 80 years ago, when he was not known as "the most important art historian that America has produced" or – shortly before his death in march 1996 – even named "art history's only living hero," is not an accident (citations from Holly 1997, 6). He would have impressed Neurath with a combination of attributes. In art history he propagated and followed a decidedly social and historical approach; this can be seen, for instance, in his polemics against the Viennese school of art criticism published in 1936. Michael Holly has summed up its main points as follows:

1. a concentration on the way social structures impinge on the formal structure of works of art,
2. a focus on concrete, historical objects,
3. a refusal to admit any transhistorical, metaphysical forces into an analysis,
4. an adequate conception of historical process […]
5. a scientific rigor which can only result from an empirical study of historical factors and conditions. (Holly 1997, 7.)[10]

[9] Neurath to Charles Morris, 14 March 1938.

[10] See also the other articles of the symposium about Meyer Schapiro in the *Journal of Aesthetics and Art Criticism*, volume 55, issue 1 (1997) and the contributions on Schapiro to the Oxford *Art Journal* volume 17, no. 1 (1994).

All historians and theorists should investigate their own method and inquire into its relation to contemporary values in art and social life. One can easily assume that these methodological principles pleased Neurath equally, whereas they are now criticized by many as mere "factology." Furthermore, at that time Schapiro began to develop an interest in abstract art, which was in line with Neurath's avantgardist tendencies, whereas it prompted furious criticisms from some of Schapiro's more orthodox Marxist friends.

Schapiro was not only interested in his subject, but in addition belonged to those left-leaning "New York Intellectuals" such as John Dewey, Sidney Hook, Ernest Nagel etc. who entered many important actual debates in journals such as *Partisan Review* about the question of whether to go to war against Hitler's Germany or the famous "failure of nerve" debate against religious and metaphysical philosophers and writers. Furthermore, Schapiro had a small discussion-circle with some of his friends of an open-minded empiricist tendency about philosophical themes.

As already said above, Schapiro's contribution never reached the chief editor of the IEUS, Neurath. But it is possible to reconstruct the fate of the planned booklet from the correspondence. It seems to me that the "blame" over this failure has to be assigned to both sides, that is: not only to the prospective author Schapiro, but also to the editor Neurath.

Neurath was very much interested to know what Schapiro was about to write in his contribution, and so urged him time and again to present preliminary ideas and to observe the fundamentals of logical empiricism. Already in his very first letter from 18 February 1938, Neurath had asked him to "say something about physicalistic language and analysis of art or something similar" at the congress for unity of science from 14 to 19 July 1938 in Cambridge, UK. Whether that happened, the records do not say. In any case Neurath invited Schapiro, who at that time was again on a visit to Europe, to come to his Institute in den Haag, in the Netherlands, for one or two evenings in order to "tell a small circle about CONSIDERATIONS OF ART AND PHYSICALISM or what else you want."[11] This meeting must have taken place, it seems, since Schapiro announced his arrival in a telegram on 29 July 1939. Gerd Arntz, who had moved from Vienna to den Haag too, belonged surely to the "small circle" as well as Marie Reidemeister, Neurath's third wife. In the next letter from Neurath from 19th March 1940, the meeting is mentioned as well as discussions in New York. Neurath complained that Schapiro had not delivered – as promised – to send "a short article for our JOURNAL OF UNIFIED SCIENCE about the physicalistic approach in studies of art."[12]

In all these letters about the booklet for the IEUS the emphasis on physicalism plays the leading role. What was meant by "physicalism" and even the matter of who had invented the whole idea had been a subject of controversy in the times of the Vienna Circle around 1930 (see for an overview Haller 1993, 166–171).

[11] Neurath to Schapiro, 10 July 1939.

[12] Neurath to Schapiro, 29 March 1940.

Seemingly, the postulate that every term of science should be defined (or at least reduced) in a long chain of definitions to physical ones was not in the focus of its adherents and even less the idea that all general statements of all the sciences should be deduced in the last resort from the laws of physics. The main idea was, instead, that all sentences should be about entities existing in space and time. This postulate is of course fulfilled by the observation sentences of physics (and every other natural science), and, according to Neurath, also by the sentences and objects of every day life. His main goal was to establish a sort of common-sense materialism that was also in line with contemporary knowledge in the most advanced sciences. In this way physicalism should be able to deliver a sufficient basis for scientific inquiry and to exclude at the same time religious and metaphysical speculation and talk about dubious entities. As part of his project to reform the sciences, Neurath also tried to establish behaviorism as the physicalistic version of psychology and social behaviorism in sociology (including a portion of a behavioristics of scientists, which should substitute the sociology of knowledge).[13] As Karl Popper told in an interview, Neurath even envisaged to export this reformed materialism – or physicalism – to the Soviet Union. It would be of course interesting to know how he might have conceived a physicalistic methodology of art criticism and history.

In any case, a few days after his last letter to Schapiro about physicalism Neurath had to flee the Netherlands on a life-saving boat to England, where he was in turn interned as a possible (national-socialist!) "enemy alien" on the Isle of Man.[14] When he was released and went back to Oxford, the correspondence started again. Still, in a long letter from 29 September 1942 about Schapiro's booklet for the IEUS (as one item among many others) Neurath wrote:

> I am very glad – by the way – that you are prepared to write a monograph about analysis of arts. Please be kind enough to tell a little about your plan. I think it will be of importance that such a monograph will appear and written by you. (Neurath to Schapiro, 21 September 1941.)

Later on, in this nine-page-latter he underlined why it would be so important to have the monograph in the IEUS:

> I should like very much to hear a little more of the Encyclopedia monograph. Could you explain the main ideas and chapters? I should appreciate it very much. You see, I deplore very much, that Logical empiricism is mentioned mostly in relation to physics etc., and just in this field it is not so necessary to regenerate our arguing. Therefore it is important to show in what way we Logical Empiricists treat Arts, etc. I also think – more and more – that "logical" mistakes in the narrower sense, play not a great role and that the Language Making is much more important, and the theory of that, what I want to call TERMINOLOGY. (Neurath to Schapiro, 21 September 1941.)

Since it seemed that Neurath's emphasis on empiricist principles shifted slightly from the former physicalism to what he now called "terminology," this latter conception is in need of clarification as well. This is a program to establish a

[13] For a description and critique of these programs, see Dahms (1997, 100–110).

[14] On Neurath's period in England see Michelle Henning's, Silke Körber's and Adam Tamas Tuboly's chapters in the present volume.

(relatively short) list of both admitted terms for scientific use and an "index verborum prohibitorum." As already the Latin name for the latter list indicates, this is an allusion to the papal "index librorum prohibitorum." Neurath's idea of such an index for forbidden terms dates back to the times before World War One, and so it has no connection with logical empiricism, and among its adherents found no followers. Nevertheless, he pursued the idea in his talks and publications (and also in his reading of books). So he is reported to have said during the 1937 *International Philosophical Congress in Paris* that it was sufficient to read a book only diagonally and judge its relevance from the presence (or absence) of terms on his "index." In the correspondence with other authors of the IEUS, it played also a role when Neurath tried in vain to persuade John Dewey to substitute his title "Theory of Values" by "Empirical Axiology" (they reached a compromise nearer to Dewey's proposal, namely "Theory of Valuation"). Neurath also rejected a proposed booklet by Herbert Feigl about Scientific Explanation because the term "explanation" appeared to him a very dangerous one. In Neurath's own booklet or the IEUS, the program of terminology and the "index" are prominent: pages 9 till 19 (of his relatively short contribution of only 51 pages) are devoted to this program and titled "Accepted and Rejected Statements" and "The Richness of a Sociological Vocabulary." In its Appendix one finds an "Index of Terms" with allowed "phraseology used in this monograph" and prohibited concepts ("expressions avoided in this monograph"). "True (false)," "mind (matter)," "good (bad)" and "justice" all belong in the latter category (Neurath 1944/1970).

Now Neurath intended to urge Schapiro to change the already agreed upon title of his monograph "Interpretation and Judgment of Art," because both concepts were "very dangerous terms," as he wrote his co-editor Morris.[15] He instructed Morris to inform Schapiro accordingly. Such a request is not contained in the correspondence between Neurath and Schapiro, which became interrupted from the 29th September 1942 till the 4th of September 1945.

Now in a letter from 19 August 1942, in which Schapiro wrote also about his booklet for the IEUS, he had to admit not to have started with it. But it was "rarely out of my mind and almost everything I see and read provides some material for it."[16] Indeed Schapiro seems to have worked from time to time on it, but more often found other projects more important and, as a result, did not finish the task. It would be interesting to know, to what degree Neurath's insistence on "physicalism" and "terminology" contributed to the waning of Schapiro's intention to write the booklet. The prospect of the threat of a straightjacket or at least a corset was surely not an incentive for productive work.

After Neurath's death, his replacements as main editors of the IEUS, Charles Morris and Rudolf Carnap, put an end to the notorious "Meyer-Schapiro-problem." Carnap gave him a deadline and, when this was crossed, changed the whole subject

[15] Neurath to Morris, 26 February 1943.

[16] Schapiro to Neurath, 12 August 1942.

and he devoted the last booklet to economics and let his former pupil and colleague from Prague Gerhard Tintner write a booklet on *Methodology of Mathematical Economics and Econometrics*. At that time Schapiro was already so far removed from the development of mainstream logical empiricism that he only very seldom came back to their ideas and publications. In his long article "Philosophy and Worldview in Painting" Schapiro mentioned only Carnap's conception of art and its function, according to which works of art cannot claim to tell the truth, only to be an expression of emotions:

> Carnap speaks [...] simply to indicate that metaphysics, worldviews – broad approaches to a totality, to many a problem – are not so in a logical analytical sense but rather in the sense of expressing a life feeling or outlook which affects us emotionally in the art and metaphysics. But only the art gives us aesthetic pleasure; metaphysics gives us, at least for Carnap, perplexity or a headache or a wish to be done with it as a disease of language. (Schapiro 1999b, 17f.)

Schapiro sums up Carnap's ideas correctly. It is important to keep in mind that Carnap was not an enemy of art, music etc.; he was an enemy of a sort of idle word-music he believed to have found in the books of metaphysicians, because they proclaimed to have attained knowledge, whereas – according to him – they produced only bad music. Schapiro (1999b, 18) himself adds in his article that he too ranked (real) music above metaphysics.

18.1.2 National Socialism, Comments on the War, Hopes for Reconstruction After Its End

Since both correspondents were much interested to discuss the aggravating international situation, a considerable part is devoted to other themes besides art and the IEUS: the start of the war and its later stages, which also for Neurath himself led into dramatic developments. One only has to think of his narrow escape on a life saving boat from Holland to England and his internment on the Isle of Man. He mentioned also the fate of his son, who had remained in Austria and – as a leading member of the former socialist student organization – was deported to the German concentrations camps Dachau and Buchenwald after the "Anschluss" of Austria to the German Reich.[17] Later on, hopes of the defeat of the Nazi dictatorship and the measures needed for the denazification of Germany became dominating themes. They would be the worthwhile subjects of a separate article.

[17] Paul Neurath even wrote his dissertation in sociology (under the supervision of Paul Lazarsfeld) on "Social Structure in the German Concentration Camps Dachau and Buchenwald," which was published only 70 years later; see P. Neurath (2004).

18.1.3 Philosophy and Sociology: Martin Heidegger and Max Weber

But there are also themes that do not belong to contemporary history, but to the political circumstances of philosophy and sociology. These contexts are important, because they are apt to shed new light of the persons under discussion in the correspondence, but also on the biographies and descriptions both of Neurath and Schapiro.[18]

Among other things, we find a discussion about Plato as (an alleged) intellectual forerunner of fascism in ancient Greek times, about Martin Heidegger as a contemporary counterpart and – surprisingly – also about Max Weber. The correspondence partners soon reached an agreement about Plato: one should keep an eye during the denazification process on a possible prolongation or revival of platonistic tendencies in political philosophy in a new Germany, given Plato's support for a rigid dictatorship and of slavery.[19] The discussions of Heidegger and Weber took a more complicated course and seem more interesting.

18.1.3.1 Heidegger

Concerning Heidegger Neurath had remarked in his long letter of 29th September 1942, when he introduced his plans to cleanse language by impeccable terminology:

> [...] we shall demonstrate what a poor and empty shell this Heidegger is, it is OK that he became a Nazi [...]. Heidegger is a poor poet and his statements regarded as statements are monstrosities, there one should analyze his descriptive importance, as dealing with habits mostly, not discussed by our logicalized boys, unfortunately. (Neurath to Schapiro, 29 September 1942, p. 8.)

Schapiro entered this discussion when he wrote in the beginning in a more general way:

> Scientific criticism was never applied in a thorough way to the problems that the metaphysicians and neo-religionists thrive on, so that people haven't even the memory of a literature which did justice to them.

These remarks about "metaphysicians and neo-religionists" remind of the famous "failure of nerve-debate," in which the New York Intellectuals fought for enlightenment and a scientific worldview in times of a revival of antiscientific skepticism and the tendency to take consolation in super-natural entities and forces. Schapiro in his letter then continued more concretely:

---

[18] Since no biography of Schapiro was published up to now, one has to rely on obituaries and handbook-articles. On Neurath see Sandner (2014).

[19] On Neurath's critique of Plato, see Antonia Soulez's chapter in the present volume.

> In our circle no one read Heidegger seriously, one simply said "das Nichts nichtet" and laughed, so that when a newer generation is impressed by the lyric of anxiety in Heidegger, they have no means of dealing with it and at the same time cannot be satisfied with a criticism which dismisses Heidegger as a syntactical monstrosity. (Schapiro to Neurath, 12 August 1942.)[20]

Here Schapiro admonished his correspondence partner and his logical empiricist friends not to take the task of criticizing Heidegger too light-heartedly, as Carnap (1932/1959) had done it in his famous article about "Overcoming metaphysics…" from 1932, where utterances like "das nicht nichtet" were simply taken out of context and made the standard example of ridicule. Instead, Schapiro thought, one should try to decipher Heidegger's intentions behind his often extravagant formulations and then discuss them critically. This approach was taken by Michael Friedman (2000) and Peter E. Gordon (2010) more recently.

In his penultimate letter from the 22nd July 1945 Schapiro had found a recent occasion to come back to the Heidegger theme. He started again in a very general way:

> There is now a growing curiosity (more among literary people than philosophers) about existentialism, not for Kierkegaard, who had his day some years ago and is now uninteresting… but for Heidegger, whom very few have read (it is astounding with what faith he is received by poets who know only a few quotations, including "das Nichts nichtet"), but who has a considerable reactive value, because he is concerned with fatality, care, disgust, the trivial and nothingness, things which American and British philosophers completely underestimate, although they have an oppressive actuality. (Schapiro to Neurath, 22 July 1945.)

Schapiro informed Neurath of a special episode concerning existentialism:

> When the French writer, Sartre, lectured on Heidegger in New York last April – a poor piece of philosophizing, I thought – he made a strong impression. The optimism of "naturalism" is appallingly shallow, and sound too much like the league of democratic experts to satisfy anyone. (Ibid.)

Sartre indeed had come to New York in January 1945 – only half a year after the liberation of Paris from the German occupation in August 1944 – in order to serve as foreign correspondent for the French journal *Combat* (*Fight*) – edited by Albert Camus.[21] Sartre's lecture was given in the "Maison Francaise" of Columbia University on 117th Street (Guttenplan 2013).

Schapiro closed his remarks on Heidegger with the following words:

> Heidegger has to be criticized in his own terms, and not from outside, because he doesn't use the language properly: and the new interest in his philosophy has to be accepted as a challenge to deal with the same problems scientifically. And they are social and ethical problems on which empiricism has had little to say; and that little, often negative and worthless. (Ibid.)

---

[20] The "circle" was seemingly a philosophical discussion group, to which Sidney Hook, Ernest Nagel and Schapiro belonged and which tried to combine logical analysis with a prospect of political activism. The correspondence contains in various places information about the group and its decline during World War Two.

[21] For an account of Sartre's and Camus's relationship to the USA, see Martin (2012).

As we will see (Sect. 18.2.1), Schapiro himself dealt with Heidegger and one of his problems "scientifically." That problem was not a "social and ethical" one, though. It concerned a field, where Schapiro had a special competence, namely: art.

18.1.3.2 Max Weber

The most controversial part of the discussions between Neurath and Schapiro was about Max Weber. Neurath had witnessed him at a congress of the "Verein für Socialpolitik" 1909 in Vienna as a brilliant speaker and a passionate participant in discussions. Later on during Neurath's habilitation in Heidelberg 1916 he almost certainly would have met Weber, although he had resigned his professorship over a decade earlier because of poor health, but played an important role behind the scenes at the university. In any case, Neurath came again across Weber during the meetings at Lauenstein castle in southern Thuringia, where Weber fought a hard battle as proponent of a negotiated peace in World War One and a thorough democratization of Germany afterwards against the annexionist and chauvinist Max Maurenbrecher (see for the Lauenstein events Dahms and Neumann (1994) and Radkau 2011, 483–487). Weber's role in the trial for high treason against Neurath after the defeat of the Bavarian Soviet Republic in summer 1919 surely was less to Neurath's taste, because Weber had attempted to protect Neurath by depicting him as politically naïve (Dahms and Neumann 1994, 131f.). When one now adds to this the paragraphs in one of Weber's magna opera, *Wirtschaft und Gesellschaft*, where he criticizes Neurath's plans and activities for full socialization of the economy in revolutionary Bavaria (Dahms and Neumann 1994, 136–140), perhaps it becomes understandable that Neurath (1931/1973) wrote quite critically about Weber in his *Empirical Sociology* of 1931.

It comes all the more as a surprise that he defended Weber vehemently in the correspondence with Schapiro against the latter's disdainful invectives. Weber is mentioned in the correspondence for the first time in a letter of Neurath from the 29th March 1940. The context is the writing of Neurath's own booklet for the IEUS:

> I am writing my article about the Foundations of the Social Sciences for our encyclopedia. I want to speak about my own discipline in friendly words – you know the successes in knowledge are tremendous, but not in the clarification of expression. I think it is more important what Max Weber says about concrete fact than about the methods of his science "Wissenschaftslehre". Unfortunately it is the subject matter of "foundations": the structure of a discipline. I hope to tell as correct as possible what is done in the field without too sharp criticism. (Neurath to Schapiro, 29 March 1940.)

These remarks were written and sent only a few days before the invasion of German troops into the Netherlands on 10 May 1940 and the capitulation 5 days later. Neurath had to leave behind the manuscript for his *Foundations of the Social Sciences* on his flight in a sea rescue boat and then – after his release from internment, to start anew on his contribution.

Max Weber again popped up as topic after the war, when Schapiro (1945) had written a very critical short article about him and sent it to Neurath. He reacted as follows:

> [...] in general I agree with you. I read a lot of Weber, also his letters, biography etc. to understand better German "Liberalism" with its strong nationalist implications, lack of finger-feeling etc. see Weber's contacts with Ludendorf (sic). (Neurath to Schapiro, 4 September 1945.)

Later on in the same letter he sketched a more detailed picture of his character:

> WEBER was a very complicated person. Very sincere, very outspoken, charming, attractive, clever.. prepared to listen to other people's opinions. Very prepared to invite scientific opponents to discussions etc. Never avoiding a controversy.... Against tyrants, against William II. I met him more than once in difficult periods e.g. at the meeting on the Lauenstein etc. BUT full of studentic tradition, DISTANCE important to him, and a certain attitude. Always thinking in certain extremist terms. But his life nevertheless without a real energetic push towards a goal...
>
> He was a real fighter for the freedom of speech and his vision of a leader implies, that one is able to put him away, whenever one wants to do so. As Clemenceau put away Foch without hesitation and Churchill put away generals without hesitation and the British voters put away Churchill, without putting him to the gallows. (Neurath to Schapiro, 4 September 1945.)

Neurath criticized especially Schapiro's depiction of Weber as an opportunist: "If you call Weber an opportunist, you will find not many people who would not be allied with such a name...." Even hints to Weber's wild attacks against Karl Liebknecht were not accepted by Neurath as behavior that could undermine his integrity: the communists in Vienna had agitated against the social democrats in the same way, when they accused them of "treason" or accused the social democrats leader Otto Bauer to be "well paid." That sounds like playing down Weber's utterances, because he exclaimed during the election campaign for the left-liberal German Democratic Party (DDP) on 4 January 1919: "Liebknecht belongs in the madhouse and Rosa Luxemburg in the Zoological Garden" (see for this episode Radkau 2011, 507). A few days after this speech both Luxemburg and Liebknecht were murdered – within a short distance from the Berlin Zoo. It is no wonder that the correspondence partners reached no agreement.

18.2 Two Sequels

18.2.1 Schapiro's Criticism of Heidegger

Although nothing became out of his planned contribution to the IEUS, Schapiro nevertheless continued to reflect on the relation between scientific philosophy and art. Two remarkable results came out of this. The discussion with Neurath about Heidegger had been one reason (among others) that Schapiro reflected and ultimately published something on the latter – only decades after the publication of Heidegger's famous article on the origin of the work of art and an interpretation of a painting of Vincent van Gogh given there.

A friend of Schapiro, the psychiatrist Kurt Goldstein, an émigré from Germany, had called his attention to some ideas of Heidegger's about art contained in a number of lectures given in 1935 and finally published as "Zum Ursprung des Kunstwerks" ("On the Origin of the Work of Art") in a volume of articles under the title "Holzwege" ("Woodway," in German also meaning sometimes a wrong track) in 1950 (see Heidegger 1950). There he discussed not only the essence of works of art and useful artifacts, but also a painting by van Gogh of shoes taken as an example of useful objects. It so happened that exactly in the same year the first edition of a book by Schapiro (1950) on van Gogh appeared (later on it would be reprinted many times and translated into many languages). But it took almost two decades before Schapiro took up the subject again. Sadly the article about Heidegger and van Gogh appeared only in a volume in memory of the deceased Goldstein in 1968 (see Schapiro 1968).[22] This was Schapiro's attempt to achieve what he had missed in the writings of positivist critics of Heidegger in the correspondence with Neurath, namely: to analyze in detail his methods and results, without letting oneself be misled by his extravagant language and rhetoric.

That Heidegger had chosen the example of shoes had a somewhat complicate background: in order to arrive finally at finding out the essence of the work of art, he started with things (in general) and then took a middle step on useful artifacts (named by him with the – in German – unusual term "Zeug"), in order to arrive finally in the third stage at the category foremost on his mind in the article: the object of art. I concentrate on the middle step, because Heidegger tried to extract the essence of "Zeug" from the contemplation of works of art. Heidegger interpreted van Gogh's painting of shoes in the following way:

> From the dark opening of the worn insides of the shoes the toilsome tread of the worker stands forth. In the stiffly solid heaviness of the shoes there is the accumulated tenacity of her slow trudge through the far-spreading and ever-uniform furrows of the field, swept by a raw wind. On the leather there lies the dampness and saturation of the soil. Under the soles there slides the loneliness of the field-path as the evening declines. In the shoes there vibrates the silent call of the earth, its quiet gift of the ripening corn and its enigmatic self-refusal in the fallow desolation of the wintry field. This equipment is pervaded by uncomplaining anxiety about the certainty of bread, the wordless joy of having once more withstood want, the trembling before the advent of birth and shivering at the surround menace of death. This equipment belongs to the earth and it is protected in the world of the peasant woman. From out of this protected belonging the equipment itself rises to its resting-in-itself. (Heidegger 1950, 22f. Cited after Schapiro's translation in his 1994a, 135 f.)

Strangely, this interpretation hasn't been correlated till now to the time of its initial composition. As I said, a first version of Heidegger's article was given in the form of a lecture in 1935. Therefore it seems tempting to ask how it fits with the spiritual and political environment of the early Nazi regime. That Heidegger took a painting of van Gogh as a subject seems at first sight to indicate a self-conscious distance

[22] For Schapiro's further notes on Heidegger and van Gogh, see Schapiro (1994a) and (1994b).

from the official Nazi attitude to art. Their chief ideologue Alfred Rosenberg had made some detrimental comments on the artist already in his "Mythos des Zwanzigsten Jahrhunderts" ("Myth of the 20th Century"), published in 1930: "he painted apple trees, cabbage and street stones, till he got mad" (cited after Roh 1962, 72). Later on even some of van Gogh's paintings were confiscated in the preparations of the infamous exhibition "Entartete Kunst" ("Degenerate Art"). But they did not go to the Munich exhibition, but instead landed in the private collection of Hermann Göring. Apart from taking an ideologically dubious van Gogh painting as a starting point for his investigation into the essence of "Zeug," the interpretation of that artwork is perfectly in line with National Socialist ideology, namely the "Blut-und-Boden" ("blood and soil") mythology. That already applies to Heidegger's identification of the shoes as the shoes of farmers, but especially to the part, where a farmer woman experience at giving birth is evoked.

Now Schapiro aimed at the deconstruction of this interpretation. He knew that van Gogh had produced a number of shoe-themed paintings. So he wrote in 1965 a letter to Heidegger, asking him when and where he had seen that particular painting. Heidegger answered as follows:

> Very estimated Mr. Meyer-Schapiro (sic)!
>
> With pleasure I answer your question as far as I can. When I was on a lecture tour in Holland in March 1930, I visited in Amsterdam the great Van-Gogh-Exhibition which just was opened. There I saw "The Shoes". According to my memory it was one of the early versions of that theme. My "description" is related to that original seen in Amsterdam. I think, you will as an experienced expert find out easily with the means at your disposal, which painting is meant. (Heidegger to Schapiro, 6 May 1965.)

Heidegger went on in his letter to praise Schapiro's book on Cezanne and expressed hopes that one day an art historian would gather all the different views of Cezanne's beloved Mont Victoire and interpret them in their development. In any case his hints toward the Amsterdam exhibition of 1930 were sufficient to identify the painting. It was thus possible for Schapiro to offer a thorough refutation of Heidegger's identification of the depicted shoes with the ones of a peasant woman. His investigations demonstrated that they were van Gogh's own shoes. In other words, if van Gogh had wanted to make a statement at all about the owner of the shoes, it should have been an autobiographical remark. And that had to do with the fact that the painter was used to walk long ways on his own feet. For example, the young van Gogh traveled from his parents home in Holland to the coal region of Belgium in order to serve there as a lay priest and help workers in misery.

Later on, philosophers took up the challenge posed by Schapiro's interpretation. Jacques Derrida (1978/1987), for instance, tried to show the irrelevance of Schapiro's findings. The controversy between critics and adherents of Heidegger went on till recently.[23] In 2009 there was an exhibition in Cologne at the Museum Ludwig in

[23] I cannot comment on Derrida and all the others who cared to write about the Heidegger/Schapiro-argument.

which only one painting was shown: van Gogh's shoes (which Heidegger had seen in 1930), surrounded by a multitude of controversial interpretations.[24]

In this whole controversy not enough emphasis was put on the question whether and to what degree Heidegger's philosophy depends on the correctness of his (or the alternative) interpretation. He had declared in his article that philosophy in its quest to detect the essence of things, especially of "Zeug," should not look at those types of objects themselves, their fabrication and use etc., but to take instead the detour via the depiction of them through works of art:

> The essential being of equipment was found. But how? Not by description and explanation of a really present shoe equipment; not by a report about the making of shoes, also not by the observation of a here or there real occurring use of shoes. But – but only by us bringing us in front of the painting of van Gogh. It has spoken. In the nearness of the work we have suddenly been elsewhere than we use to be. The work of art let it know, what shoe equipment is in truth. (Heidegger 1950, 24.)

It seems that the essence of "Zeug" could better be detected by "positivistic" approaches excluded by Heidegger than by the interpretation of a work of art. Therefore the question is, which lesson to draw about a philosophical method that let slip through completely wrong interpretation of art works.

18.2.2 Schapiro Stimulating Nelson Goodman's Languages of Art

Before Schapiro published his article on Heidegger, he had pursued ideas in the correspondence with Neurath. He stimulated a close friend, Nelson Goodman, to write a philosophical treatise on art. Goodman was himself a former art-gallery manager in Boston and by then a Harvard professor famous for books such as *The structure of Appearance* and *Fact, Fiction, Forecast*. This plan was developed during joint visits to art exhibitions and galleries in New York.[25]

Goodman indeed followed this project. In the end he could publish his epoch-making book *Languages of Art*. This work appeared in 1968, two years before the final two-volume edition of the IEUS. It is the most remarkable writing on art in the tradition of analytic philosophy. The question of whether the book would have fitted in with the IEUS must be left open here. But let's suppose it had: how would Neurath's encyclopedia project, and with it the whole tradition of logical empiricism, be looked at and evaluated if Goodman's book – published immediately after Thomas Kuhn's book *The Structure of Scientific Revolutions* – had been the conclusion of the whole IEUS enterprise?

[24] See the booklet accompanying the exhibition: Batchen (2009).

[25] These "strolls" are mentioned variously in the correspondence Goodman/ Schapiro.

Unpublished Sources

Columbia University (New York, USA), Special collections department

Meyer Schapiro papers

Correspondence with

Nelson Goodman
Martin Heidegger
Otto Neurath

Noord Holland Archief, Harlem

Otto Neurath papers

Correspondence with

Meyer Schapiro
Charles Morris

References

Arntz, Gerd. 1982. Otto Neurath, ich, und die Bildstatistik. In *Arbeiterbildung in der Zwischenkriegszeit: Otto Neurath, Gerd Arntz*, ed. Friedrich Stadler, 31–34. Wien-München: Löcker.

———. 1988. *Zeit unterm Messer: Holz- und Linolschnitte 1920–1970*. Köln: Leske Verlag.

Batchen, Geoffrey. 2009. *What of Shoes? Van Gogh and Art History*. Leipzig: E.A. Seemann.

Bernhard, Peter. 2015. Zur "inneren Verwandtschaft" von Neopositivismus und Neuem Bauen. In *Architektur und Philosophie. Grundlagen, Standpunkte, Perspektiven*, ed. Jörg Gleiter and Ludger Schwarte, 162–175, 267–275. Bielefeld: Transcript.

———. 2019a. 'Sie diskutieren sehr gern, aber sehr dilettantisch.' Carnaps Vorträge am Bauhaus. In *Logical Empiricism, Life Reform, and the German Youth Movement*, ed. Christian Damböck, Günther Sandner, and Meike Werner. Cham: Springer (Forthcoming).

———. 2019b. Carnap und das Bauhaus. In *Young Carnap in an Historical Context: 1918–1935*, ed. Christian Damböck and Gereon Wolters. Veröffentlichungen des Instituts Wiener Kreis. Dordrecht: D. Reidel (Forthcoming).

Carnap, Rudolf. 1928/2003. 1928/2003. *The Logical Structure of the World*. Trans. R. George. Chicago, Ill.: Open Court.

———. 1932/1959. The elimination of metaphysics through logical analysis of language. In *Logical Positivism*, ed. A. J. Ayer, 60–81. New York: The Free Press.

Carnap, Rudolf, Hans Hahn, and Otto Neurath. 1929/1973. The scientific conception of the world: The Vienna circle. In *Empiricism and sociology*, ed. Marie Neurath and Robert S. Cohen, 299–318. Dordrecht: D. Reidel.

Dahms, Hans-Joachim. 1994. Positivismusstreit. In *Die Auseinandersetzungen der Frankfurter Schule mit dem logischen Positivismus, dem amerikanischen Pragmatismus und dem kritischen Rationalismus*. Frankfurt am Main: Suhrkamp.

———. 1997. Felix Kaufmann und der Physikalismus. In *Phänomenologie und Logischer Empirismus. Zentenarium Felix Kaufmann (1895–1949)*, ed. Friedrich Stadler, 97–114. Wien/ New York: Springer.

———. 1999. Otto Neuraths 'International Encyclopedia of Unified Science' als Torso. Bemerkungen über die geplanten, aber nicht erschienenen, Monographien der Enzyklopädie. In *Otto Neurath: Rationalität, Planung, Vielfalt*, ed. Elisabeth Nemeth and Richard Heinrich, 184–227. Wien-Berlin: Oldenbourg/Akademie.

———. 2004. Neue Sachlichkeit in the architecture and philosophy of the 1920s. In *Carnap Brought Home. The View from Jena*, ed. Steve Awodey and Carsten Klein, 357–375. Chicago: Open Court.

———. 2019. Rudolf Carnap: Philosopher of 'Neue Sachlichkeit.' In *Young Carnap in an Historical Context: 1918–1935*, ed. Christian Damböck and Gereon Wolters. Veröffentlichungen des Instituts Wiener Kreis. Dordrecht: D. Reidel. Forthcoming.

Dahms, Hans-Joachim, and Michael Neumann. 1994. Sozialwissenschaftler und Philosophen in der Münchener Räterepublik. In *Jahrbuch für Soziologiegeschichte 1992*, ed. Carsten Klingemann et al., 115–146. VS Verlag für Sozialwissenschaften.

Dahms, Hans-Joachim, and Friedrich Stadler. 2015. Die Philosophie an der Universität Wien von 1848 bis zur Gegenwart. In *Universität –Forschung – Lehre. Themen und Perspektiven im langen 20 Jahrhundert (= 650 Jahre Universität Wien – Aufbruch ins neue Jahrhundert. Band 1)*, ed. Katharina Kniefacz, Elisabeth Nemeth, Herbert Posch, Friedrich Stadler, 77–132. Wien: V&R unipress.

Derrida, Jacques. 1978/1987. *The Truth in Painting*. Chicago: University of Chicago Press.

Friedman, Michael. 2000. *A Parting of the Ways. Carnap, Cassirer, and Heidegger*. Chicago/La Salle: Open Court.

Galison, Galison. 1990. Aufbau/Bauhaus: Logical Positivism and Architectural Modernism. *Critical Inquiry* 16: 709–752.

Goodman, Nelson. 1968. *Languages of Art: An Approach to a Theory of Symbols*. Indianapolis: The Bobbs-Merrill Company.

Gordon, Peter. 2010. *Continental Divide. Heidegger, Cassirer, Davos*. Cambridge, MA: Harvard University Press.

Guttenplan, D. D. 2013. When Sartre and Camus met Hoover's G-men. *The New York Times*, 08 September 2013.

Haller, Rudolf. 1993. *Neopositivismus. Eine historische Einführung in die Philosophie des Wiener Kreises*. Darmstadt: Wissenschaftliche Buchgesellschaft.

Heidegger, Martin. 1950. Der Ursprung des Kunstwerks. In *Holzwege*, 7–68. Frankfurt am Main: Vittorio Klostermann.

Holly, Michael Ann. 1997. Schapiro Style. *The Journal of Aesthetics and Art Criticism* 55 (1): 6–10.

Martin, Andy. 2012. Sartre and Camus in New York. *The New York Times*, 14 July 2012.

Neurath, Otto. 1931/1973. Empirical Sociology. In *Empiricism and Sociology*, ed. Marie Neurath and Robert S. Cohen, 319–421. Dordrecht: D. Reidel.

———. 1944/1970. Foundations of the social sciences. In *Foundations of the Unity of science: Volume 1*, 1–51. Chicago: University of Chicago Press.

Neurath, Paul. 2004. *Die Gesellschaft des Terros. Innenansichten der Konzentrationslager Dachau und Buchenwald*, ed. Christian Fleck and Nico Stehr. Frankfurt am Main: Suhrkamp.

Nierhaus, Andreas, Eva-Maria Orosz, and Anita Aigner. 2012. *Werkbundsiedlung Wien 1932. Ein Manifest des neuen Wohnens*. Wien: Müry Salzmann.

Radkau, Joachim. 2011. *Max Weber. A Biography*. Cambridge and Malden: Polity.

Roh, Franz. 1962. *Entartete Kunst. Kunstbarbarei im Dritten Reich*. Hannover: Fackelträger.

Sandner, Günther. 2014. *Otto Neurath. Eine politische Biographie*. Wien: Zsolnay.

Schapiro, Meyer. 1945. A Note on Max Weber's Politics. *Politics* 2 (2): 44–48.

———. 1950. *Vincent van Gogh*. New York: Harry N. Abrams.

———. 1968. The still life as a personal object – A note on Heidegger and van Gogh. In *The reach of the mind: Essays in memory of Kurt Goldstein*, ed. M.L. Simmel, 203–209. New York: Springer.

———. 1994a. The still life as a personal object – A note on Heidegger and van Gogh. In *Selected papers IV: Theory and philosophy of art: Style, artist, and society*, 135–142. New York: George Braziller.

———. 1994b. Further notes on Heidegger and van Gogh. In *Selected papers IV: Theory and philosophy of art: Style, artist, and society*, 143–152. New York: George Braziller.

———. 1999a. *Selected papers V: Worldview in painting – Art and society*. New York: George Braziller.

———. 1999b. *Philosophy and Worldview in Painting*. In Schapiro (1999a): 11–73.

Stadler, Friedrich. 1995. Wissenschaftliche Weltauffassung und Kunst: Zur werttheoretischen Dimension im Wiener Kries. *Deutsche Zeitschrift für Philosophie* 43: 635–651.

Thun-Hohenstein, Christoph, Hermann Czech, and Sebastian Hackenschmidt, eds. 2016. *Josef Frank. Against design. Das anti-formalistische Werk des Architekten. Trans. Eva Ciabattoni et al*. Basel: Birkhäuser Verlag GmbH.

Appendices

Appendix 1: The Logic Articles of Otto Neurath and Olga Hahn

Translation and additional notes by Jordi Cat (Fig. A1)

1. Otto Neurath: Ernst Schröder's Proof of Theorem 12, that the Law of Commutativity Holds for Identical Operations (1909)

(Otto Neurath, 1909, "Ernst Schröders Beweis des 12. Theorems: Für die identischen Operationen gilt das 'Kommutationsgesetz.'" *Archiv für Philosophie*, 2. Abteilung: *Archiv für systematische Philosophie*, Neue Folge, Bd. 15: 104–106)[1]

(*Lessons on the Algebra of Logic*, p. 254 ff.)[2]

Schröder's proof is based on the following statements:

Principle I. $a \in a$.

Principle II. If $a \in b$ and simultaneously $b \in c$, then also $a \in c$.

Definition 1 of identical equality (identity) [*der identischen Gleicheit (Identität)*].

1' If $a \in b$ and simultaneously $b \in a$, it is said that $a = b$.

1" If $a = b$ holds, so must $a \in b$ and $b \in a$.

[1] Reprinted in Neurath (1981, 1–3). [Translator's note.]

[2] The reference is to Schröder (1890). [Translator's note.]

J. Cat, A. T. Tuboly (eds.), *Neurath Reconsidered*,
Boston Studies in the Philosophy and History of Science 336,
https://doi.org/10.1007/978-3-030-02128-3

Then we need

3_x’ If	$c \,€\, a, c \,€\, b,$		3_+’ If	$a \,€\, c, b \,€\, c,$
so holds	$c \,€\, ab.$		so holds	$a+b \,€\, c.$
3_x’’ If	$c \,€\, ab,$		3_+’’ If	$a+b \,€\, c,$
so holds	$c \,€\, a, c \,€\, b.$		so holds	$a \,€\, c, b \,€\, c.$

Moreover, Theorem 6:

$ab \,€\, a, ab \,€\, b$	$a \,€\, a+b, b \,€\, a+b.$

The proof goes as follows (p. 254)[3]:

‘12_x	$ab = ba$	12_+^{c}	$a+b=b+a$

Accordingly, both

factors of an identical product	members of an identical sum

are *exchanged* without any effect on the meaning of the value of each expression. The identical multiplication and addition – we could also say – are “*commutative*” operations; their result is “*symmetrical*” with respect to the elements in each operation.

Proof of the statements. According to the formula of Th.

6_x	$ab \,€\, b, ab \,€\, a$	6_+	$b \,€\, a+b, a \,€\, a+b,$

of which, according to the remark on Pr. I, p. 190, any form could be established first, it follows from Def. (3_x)’ and (3_+)’, respectively, that:

$ab \,€\, ba$	$b+a \,€\, a+b$

and in this generally proven formula, one may also exchange a and b and get:

$ba \,€\, ab$	$a+b \,€\, b+a,$

which merges with the previous result according to Def. (1) into

[3] Schröder expressed his principle of duality presenting results involving interchangeable operations with the double columns. The numbering is qualified with the corresponding operation signs, x and +, but the original publication and the reprint (Neurath 1981, 1–3) omit the operation signs distinguishing the respective versions of Theorem number 12. [Translator’s note.]

$$ab = ba \qquad \Big| \qquad a + b = b + a,$$

which was to be proven.

[The second result could also have been derived, analogously to the first, from the inclusion[4] relations yielded by Th. 6:

$$ba \text{ € } a, ba \text{ € } b \qquad \Big| \qquad a \text{ € } b + a, b \text{ € } b + a$$

according to Def. 3'; but this variant of the proof would have been in appearance somewhat less simple.]"

This proof provided by Schröder is *superfluous*. If in Definition 3 the statements c € a, c € b and a € c, b € c are combined, it is explicitly stated that both statements should be *valid* at the same time. The *order of the statements* is indifferent. So, it would be correct to write: If c € a, c € b, c € ab is valid, or, what is the same, c € ba.

A proof is superfluous, because *ab* and *ba are not only equal in terms of the result*, but each time both symbols denote the *same* relation between the *same* domains. Were the symbols for example for *a*: o, and for *b*: ×, the expression [*Schreibung*] of the product would be, without further ado: ⊗. The question of whether to write o× or ×o could only arise if *two relations* could be conceived with the same result. Two circumstances might have prompted Schröder to provide this superfluous proof, firstly, the fact that his symbolic notation, since it is one-dimensional, yields two signs of its own accord, and secondly, the fact that it is based on the commutation law of mathematics. It would be a case where the reliance on mathematical nomenclature in logic proved to be inadequate.

The present consideration suggests itself that, in those cases where the symbolism for the *same* relation between the *same* objects allows two or more expressions (*Schreibungen*), the equality of these symbols, which *does not hold only in the results*, should be indicated by its *own sign*, e.g. in our case:

$$ab \equiv ba$$

reading *ab* symbolically equal to *ba*. Otherwise it would be possible to remove the inconvenience by giving preference to any expression, e.g. the one in which the first letter in the alphabet stands, and the reversed expression only comes into consideration when a second relation between the same objects is found, which may eventually produce the same result. I.e., in our case we would only speak of e.g. "*ab*" and in this symbolic notation the expression "*ba*" would be completely meaningless.

[4] Translations, including Schröder's own use of English terms and disambiguations, and presentations such as Louis Couturat's *L'Algèbre de la Logique* (1905/1914) and recent discussions, follow this interpretive rule: using "inclusion" ["*Enthaltensein*"] when the terms are introduced explicitly to denote classes or extensions, "subsumption" ["*Subsumtionen*"] for the sign more generally, but also for the case where terms denote concepts, and "implication" ["*Einordnung*"], when the terms denote statements – and the algebra extends to a proper symbolic logic for rules of reasoning. [Translator's note]

Archiv für Philosophie.

II. Abteilung:

Archiv für systematische Philosophie.

Neue Folge. XV. Band, 2. Heft.

IX.

Zum Dualismus in der Logik.[1])

Von

Olga Hahn (Wien) und **Otto Neurath** (Wien).

Die Darstellung des Dualismus in Schröders grundlegenden Vorlesungen über die Algebra der Logik (exakte Logik) weist eine Reihe von Unvollkommenheiten auf, die teilweise bereits dem Autor selbst zum Bewußtsein gekommen sind. So sagt er u. a. (Vorl. ü. d. Alg. d. L. II, 1, 1891 S. XI, „Fernere Berichtigungen und Nachträge zum ersten Bande"): „Ich verhehle mir keineswegs, daß in der späten Stellung, welche wir dem Theorem 37) $(a \not\in b) = (b_1 \not\in a_1)$ ungeachtet seiner Einfachheit und seines hohen Grades von unmittelbarer Evidenz in dem System unserer Theorie anweisen mußten, sich möglicherweise noch eine Unvollkommenheit von deren, obzwar völlig korrektem, systematischem Aufbaue kundgibt." Unsere Dar-

[1]) Nach Ablieferung unseres Manuskripts erschien: E. Schröders „Abriß der Algebra der Logik" bearbeitet von E. Müller, in dem sich einige Ausführungen mit unserer Darstellung berühren. Es sei hier auf diese interessante kleine Schrift hingewiesen, die eine Reihe von Vorzügen gegenüber älteren Arbeiten Schröders aufweist. Die einführenden Paragraphen sind schärfer gefaßt als die analogen Schröderschen Ausführungen und nähern sich jener Auffassung der Logik, wie sie durch die Itelsonsche Definition, als der Lehre von den Gegenständen überhaupt, präzisiert wurde. Wir haben an unserer Darstellung auf Grund des Abrisses keine Änderung vorgenommen, da die Behandlung einiger Sätze des Aussagenkalkuls vor dem Gebietkalkul den Kern unserer Ausführungen nicht treffen würde und wir so weit als möglich den Kontakt mit Schröders Hauptwerk bewahren wollen.

Fig. A1 First page of Olga Hahn and Otto Neurath's "On Duality in Logic" (1909)

2. Olga Hahn and Otto Neurath: On Duality in Logic (1909)[5]

(Olga Hahn and Otto Neurath, 1909, "Zum Dualismus in der Logik." *Archiv für Philosophie*, 2. Abteilung: *Archiv für systematische Philosophie*, Neue Folge, Bd. 15, Heft 2: 149–162.)

The presentation of duality in *Schröder's* seminal Vorlesungen über die Algebra der Logik (exakte Logik) (Lessons on the Algebra of Logic (Exact Logic)) shows a number of imperfections of which the author is aware. Thus he says inter alia (Vorl. ü. d. Alg. d. L. vol. II (1), 1891, p. xi, "Additional Corrections and Appendices to the First Volume"): "I am not hiding the fact that in the late position we had to introduce Theorem 37, $(a \text{ € } b) = (b_{|} \text{ € } a_{|})$, regardless of its simplicity and high degree of direct evidence within the system of our theory, there *may* still be an imperfection in its completely correct systematic construction." Our presentation, which includes the first half of the aforementioned Theorem 37, namely,

$$(a \text{ € } b) \text{ € } (b_{|} \text{ € } a_{|})$$

incorporated among the first principles, was required by the effort to make clearer *the essence and origin of duality* to further develop the system and carry out a number of proofs more correctly. Although now Schröder wanted like us to see Theorem 37 shifted to the beginning from a totally different viewpoint, we believe we must nonetheless emphasize this circumstance as a positive agreement with our line of thought. We will come back some other expressions of unease in Schröder concerning duality. We would like to follow Schröder's presentation as closely as possible, without wanting to declare faultless everything we adopt unaltered. On the one hand, it is only a matter of strongly emphasizing certain alterations on principle, and on the other, we want to avoid hindering contact among researchers working in the same domain with innovations in terminology and presentation that are not absolutely necessary. In our presentation we will put Schröder's numbering for principles, theorems and definitions in brackets next to ours. Whenever we write theorems without any proof, it means that we adopt Schröder's own without modification.

---

[5] After delivery of our manuscript, appeared E. Schröder's "Abriß der Algebra der Logik" ("Outline of the Algebra of Logic"), edited by E. Müller, in which some of the presentation touches upon our account. It should be noted that this interesting small work presents a number of advantages over Schröder's older works. The introductory paragraphs are more precise than the analogous in Schröder and come closer to the conception of logic specified by Itelson's definition, that is *study of objects in general*. We haven't introduced any alteration in our presentation on account of the *Abriß*, since the treatment of several statements of the calculus of *statements* [*Aussagenkalkul*, which Schröder himself often translated by reference to Hugh MacColl's term for his own propositional logic, "calculus of equivalent statements"] from the calculus of *domains* [*Gebietkalkul*, more general than the calculus of classes, when the terms denote classes based on different selective judgments within a domain] would not affect the core of our presentation and we want to preserve as much as possible the connection with Schröder's main work.

For every conceptual demarcation of a domain *a* there exists one and only one domain *non-a*, the former is called the *positive* [*Posit*], the latter is called the *negative* [*Negat*].[6] We place these two fundamental concepts at the front of our presentation, while Schröder introduces the negation in a subsequent place by means of Definition 6. Our statement above corresponds approximately to Schröder's not very precisely formulated Postulate 3. We now begin, with Schröder, with

Principle I [I]: $a \in a$.

This statement holds for any *arbitrary* domain, therefore for *positives* as well as for *negatives*, and also contains the statement about it

$$a_| \in a_|.$$

We hereby reach the definition of identical equality,

Definition 1' [1']:
If $a \in b$ and at the same time $b \in a$, it is stated that $a = b$.

Definition 1" [1"]:
If, $a = b$ holds, so must $a \in b$ and $b \in a$.

Theorem 1 [1]: $a = a$.

Both the definition and the theorem are, of course, subject to what has just been said regarding their application to negatives.

Principle II [Theorem 37, first half]:
If $a \in b$, it holds that $b_| \in a_|$.

Written in the calculus of statements is that

$$(a \in b) \in (b_| \in a_|).$$

This statement declares that just as every positive determines a negative, every relation between two positives determines a relation between the two related negatives. We will see in the following that this principle is suitable to derive two series of statements in such a way that just as *multiplication* or *addition* corresponds, respectively, to relations between *positives*, *addition*, and respective *multiplication*, correspond to the associated relations between the assigned *negatives. The duality of these two operations seems to be traceable black to the fundamental connection between positive and negative.*

[6] Schröder introduced the term *Negat* in the first volume of the *Abriss* (1909, 24), as synonymous with *nicht* ("*a nicht*") and *non* ("*non* a") for the negative domain, or domain of negations. [Translator's note.]

Theorem 2 [32]: If $a = b$, $b_| = a_|$.

Proof: $a = b$ means according to Definition 1" that $a \in b$ and $b \in a$. From $a \in b$ follows according to Principle II that $b_| \in a_|$ and equally from $b \in a$ follows $a_| \in b_|$, from which according to Definition 1' results $b_| = a_|$.

Schröder is unable to derive his Theorem 32 in the way indicated, because he himself used it to derive his Theorem 37 – our Principle II. Schröder has not avoided the flaw that the equality of two domains is not derived from the definition of equality through double subsumption. About this he expresses himself as follows (Vol II (1), p. xi): "We feel compelled to turn to the proof [of Theorem 37] to invoke the theorems 30, 32 and 36 based on equations, whereas it seemed natural to establish the corresponding Contraposition Theorem 32 for equations, according to Definition 1, on the 37 for subsumptions."

Principle III [II]:

$$\begin{array}{rcl} \text{If} & a \in b & \text{and} \\ \text{at the same time} & \underline{b \in c,} & \\ \text{it holds that} & a \in c. & \end{array}$$

Applying Principle I, we obtain that

$$\begin{array}{rcl} \text{if} & a \in b & \text{and} \\ \text{at the same time} & \underline{b \in c} & \\ \text{it holds that} & c_| \in a_|. & \end{array}$$

We see that we can reach conclusions that contain only negatives from premises that contain only positives. However, we are not yet entitled to accept, in general, that the conclusiveness of the statements about relations between positives also means the conclusiveness of the corresponding relations between negatives, since our Principle II means *only* that, if $a \in b$, then $b_| \in a_|$, but not the other way around, that if $b_| \in a_|$, then also $a \in b$. In the present case, though, from the relations between negatives we can derive that

$$\begin{array}{ll} b_| \in a_| & \\ \underline{c_| \in b_|} & \text{draw the conclusion:} \\ c_| \in a_| & \end{array}$$

but only because Principle III seems again applicable, which holds generally for positives and negatives alike. Of all statements, we want to describe those reached through the application of Principle II as the first *duals*, although it should be noted that they are fundamentally different from the ones described as duals by Schröder. The principles of dual statements are obviously *only theorems*, which we describe with the corresponding Roman numerals and the negation stroke.

In an analogous way, for the dual theorems we use parentheses with Arabic numerals and the negation stroke. In Schröder's case, theorems 2 to 4 now follow – in our case, 3 to 5 –, which we adopt without modifications.[7]

Definition 2' [3_x']
If c € a and at the same time
c € b, we say
c € ab or ba^{2}

Definition 2'' [3_x'']
If we say c € ab,
It means that it holds that
c € a and at the same time c € b

Definition 3' [Th. 38_x]
If a € b, so we say that
$ab_| = 0$

Definition 3''
If we say that $ab_| = 0$, it
means that a € b

Theorem $2_|$' (in preliminary form)
If c € a and at the same time
c € a it holds that
$(ab)_|$ € $c_|$

Theorem $2_|$''
if c € ab,
it holds at the same time that
$a_|$ € $c_|$ and $b_|$ € $c_|$

Theorem $3_|$'
if a € b, then
$0_| = (ab_|)_|$

Theorem $3_|$''
if $ab_| = 0$ then it holds that
$b_|$ € $a_|$.

This definition of identical zero is more in keeping with the character of the other definitions, such as the dollows from Definitier and, as we show, accomplishes the same.

As always a € a, it follows from Definition 3'

Theorem 6 [30_x]
$aa_| = 0$

Theorem 7
If $ab = 0$ and at the same time
$a_|b_| = 0$ then
$a = b_|$ and
$b = a_|$

Theorem $6_|$
$0_| = (aa_|)_|$.

Theorem $7_|$
if $ab = 0$ and at the same time
$a_|b_| = 0$ it holds that
$a_| = (b_|)_|$ and
$b_| = (a_|)_|$.

Proof. From Theorem $3_|$" it follows that, if one replaces $b_|$ with b,

if $ab = 0$, then b € $a_|$.

It also follows from Definition 3" that, if one replaces a with $a_|$,

if $a_|b_|$ € 0, then $a_|$ € b.

According to Definition 1' also $b = a_|$.

[7] Cf. *Archiv für systematische Philosophie*, XV, vol 1, p. 104. [See article 1. in the Appendix; translator's note.]

From Theorem 3" it emerges that, if $ba_| = 0$, then $a_| € b_|$ and if one replaces $a_|$ with a, one obtains,

$$\text{if } ab = 0, \text{ then } a € b_|,$$

and when we replace b with $b_|$, we obtain from Definition 3" that

$$\text{if } a_| b_| = 0, \text{ then } b_| € a.$$

From Definition 1' also $b_| = a$

Theorem 8 [31]

$$(a_|)_| = a.$$

According to Theorem 6 it holds that $aa_| = 0$ and in the same way

$$a_|(a_|)_| = 0,$$

therefore, from Theorem 7, $a = (a_|)_|$.

We can see from this that the relation between a and $a_|$ is reversible. We can now draw from this an important conclusion.

Theorem 9 [37]: The statements $a € b$ and $b € a$ are equivalent, i.e., in the calculus of statements written as

$$(a € b) = (b_| € a_|)$$

which corresponds in Schröder to the complete Theorem 37.

Proof. If $b_| € a_|$, it holds from Principle II that

$$(a_|)_| € (b_|)_|,$$

therefore, from Theorem 8, $a € b$. Together with Principle II, this results in the theorem to be proven. *We see that the relation between positives and negatives is reversible, and we are therefore able to draw conclusions from the conclusiveness of the positive statements and vice versa.*

We note that Schröder's Theorems 2 and 3 should be actually described as dual statements, not in [logical] succession. If $a € b$ and $b = c$, then $a € c$, and, if $a = b$ and $b € c$, then also $a € c$, but side by side:

If $a = b$ and $b € c$, then $a € c$	If $b_| = a_|$ and $c_| € b_|$, then $c_| € a_|$.

In the new expression, the definition of multiplication is

If *c* € *a* and at the same time *c* € *b*, then we say *c* € *ab* and vice versa.	If $a_|$ € $c_|$ and at the same time $b_|$ € $c_|$, then we say $(ab)_|$ € $c_|$ and vice versa.

To the definition of multiplication we now add that of addition:

Definition 4' [3_+'] If *a* € *c* and at the same time *b* € *c* then we say that *a* + *b* € *c* *Definition* 4'' [3_+''] if we say that (*a*+*b*) € *c*, one says that *a* € *c* and at the same time *b* € *c*	*Theorem* $4_|$' if $c_|$ € $a_|$ and at the same time $c_|$ € $b_|$ it holds that $c_|$ € $(a+b)_|$. *Theorem* $4_|$'' if $c_|$ € $(a+b)_|$, it holds that $c_|$ € $a_|$ and at the same time $c_|$ € $b_|$.

From what we have stated it follows that each of the three statements

I. "*a* € *c* and $c_|$ € $b_|$ are simultaneously valid"
II. "*a* + *b* € *c*"
III. "$c_|$ € $a_|b_|$"

implies the other two.

If in II we replace *c* with *a*+*b*, then III follows in the form $(a+b)_|$ € $a_|b_|$.

If in III we replace $c_|$ with $a_|b_|$, the statement dual to II follows in the form $a_|b_|$ € $(a+b)_|$.

Through the application of Definition 1' we obtain

Theorem 10 [36] $(a+b)_| = a_|b_|$	*Theorem* $10_|$ $(a_|b_|)_| = ((a+b)_|)_| = a+b$.

From Theorem $10_|$ we obtain Schröder's 36, if we replace *a* with $a_|$ and *b* with $b_|$:

$$\left(\left(a_|\right)_|\left(b_|\right)_|\right)_| = a_| + b_|$$

and by applying Theorem 8 and Theorems $2_|$' and $2_|$'', $(ab)_| = a_| + b_|$.

We see that, if

ab € *c*, then also $c_|$ € $a_| + b_|$

and vice versa, if

c € $a_| + b_|$, then *ab* € *c*.

Likewise, if

$$c \mathrel{€} ab, \text{ then } a_{|} + b_{|} \mathrel{€} c_{|}$$

and vice versa, if

$$a_{|} + b_{|} \mathrel{€} c_{|}, \text{ then } c \mathrel{€} ab.$$

If a general statement is given, e.g.,

$$\begin{array}{rl} \text{If} & a \mathrel{€} c \\ \text{and at the same time} & b \mathrel{€} c, \end{array}$$

then we say $a + b$ € c, likewise, since a, b and c are completely general, also holds the statement:

$$\begin{array}{rl} \text{if} & a_{|} \mathrel{€} c_{|} \\ \text{and at the same time} & b_{|} \mathrel{€} c_{|}, \\ \hline \text{then} & a_{|} + b_{|} \mathrel{€} c_{|} \end{array}$$

where $a_{|}$, $b_{|}$, $c_{|}$, however, do not represent de negatives of the designated positives, and are again general symbols.

Likewise, in general statements negatives can be converted into positives.

From

$$\begin{array}{rc} & a \mathrel{€} c \\ & b \mathrel{€} c \\ \hline \text{I.} & a{+}b \mathrel{€} c \end{array}$$

we obtain the dual statement

$$\begin{array}{rc} & c_{|} \mathrel{€} a_{|} \\ & c_{|} \mathrel{€} b_{|} \\ \hline \text{II.} & c_{|} \mathrel{€} a_{|}b_{|} \end{array}$$

And from these, by converting negatives into positives, we obtain

$$\begin{array}{rc} & c \mathrel{€} a \\ & c \mathrel{€} b \\ \hline \text{III.} & c \mathrel{€} ab. \end{array}$$

While all three statements taken individually claim to be of the same generality, I and II are also distinguished by the fact that they apply at the same time to pairs of positives and negatives, whereas, e.g. statement III is not necessarily valid at the same time as II or I. The positives in I and III are in general not identical. This is one of the considerations that suggest themselves in Schröder in which he compares the calculus of statements and the calculus of domains (Vol. II, 1, p. 46). According to Schröder's terminology, statement III represents the dual of statement I. While in our formulation there is a logical connection between I and II, in Schröder statements I and II are connected only empirically. For us, statement II forms a bridge between I and III, and the connection between a statement and its dual counterpart seems to be due to the empirical element in the principles. In addition, Schröder's choices repeatedly introduce flaws in the course of the proof. From the fact that Schröder is eventually led to the peculiar Theorem 35, one may feel inclined to infer that this is in particular a fundamental deficiency in Schröder's otherwise clear presentation (Vol. I, p. 315): "In each statement and general formula of the identical calculus of domains, it is allowed to interchange at the same time the super- and subordination signs, the *0* and *1* as well as the times and plus signs, and this means that one must always obtain a valid sentence and a correct formula." The unfounded "must" is particularly disruptive in a theorem that lacks actual proof. Schröder himself declares (p. 318) in no satisfactory way: "Incidentally, as in the examples above, we will not be *compelled* to make any fundamental use of Theorem 35, of duality, while being able to prove the required statements individually and in detail. As far as we like, we can let Theorem 35 play here only the role of an *empirical* principle, which subsequently determines in each statement that its dual counterpart holds, in other words, it sums up all these determinations in a general statement with a comprehensive induction." Since Schröder relies on induction, Theorem 35 acquires the peculiar characteristic of being always valid for every statements found at that point; whereas, according to the strict wording of Schröder's statement, for subsequent dual statements duality would have to be determined inductively. If one claims Theorem 35 only for each statement found, then it is certainly found by complete induction; but if one assumes the application of duality for the ensuing statements, then one is applying an incomplete induction, which cannot have a place here. Schröder hints on p. 318 at an attempt to explain duality in relation to the connection between content and extension relations, but it is not developed, only suggesting that the author feels dissatisfied. Our presentation of duality shows that he finds sufficient grounds only in the extension of the comparison of positives and negatives, and that he does not feel compelled to refer back to the opposition between extension and content.

In the following we will use Theorems 8 and 10 to write the dual statements and will write *1* always instead of $0_{|}$.

In the new notation, it reads as follows,

Theorem 6	*Theorem* $6_{|}$
$aa_{|} = 0$	$1 = a + a_{|}$.
Theorem 7	*Theorem* $7_{|}$
if $ab = 0$ and at the same time $a_{|}b_{|} = 0$, then $a = b_{|}$ and $b = a_{|}$	if $1 = a_{|} + b_{|}$ and at the same time $1 = a + b$, then $a = b_{|}$ and $b = a_{|}$.

We could express Theorem 7 also as follows:

if $ab = 0$ and at the same time $1 = a + b$, then $a = b_{|}$ and $b = a_{|}$,
which corresponds to Schröder's *definition* of negation.

If $ab = 0$ and $a + b = 1$, and on the other hand, $ac = 0$ and $a + c = 1$, then $b = a_{|}$ and $c = a_{|}$, and therefore $b = c_{|}$, which in Schröder functions as the auxiliary Theorem 29.

To continue our presentation,

Theorem 11 [6_x]

ab € a

Theorem 12 [13_x]

$$a\,(bc) = (ab)\,c$$

Theorem 13 [14_x]

$$aa = a$$

Theorem 14 [15_x]

if a € b, then
ac € bc

Theorem 15 [16_x]

If $a = b$, then
$ac = bc$

Theorem 16 [17_x]

if a € b and c' € b',
then aa' € bb'

Theorem $11_{|}$

$a_{|}$ € $a_{|} + b_{|}$.

Theorem $12_{|}$

$$(a_{|} + b_{|}) + c_{|} = a_{|} + (b_{|} + c_{|}).$$

Theorem $13_{|}$

$$a_{|} = a_{|} + a_{|}.$$

Theorem $14_{|}$

if $b_{|}$ € $a_{|}$, then
$b_{|} + c_{|}$ € $a + c_{|}$.

Theorem $15_{|}$

if $b_{|} = a_{|}$, then
$b_{|} + c_{|} = a_{|} + c_{|}$.

Theorem $16_{|}$

if $b_{|}$ € $a_{|}$ and b' € a',
then $b_{|} + b_{|}'$ € $a_{|} + a_{|}'$.

Analogous to Theorems 17 [18] and 18 [19],

Theorem 19

$(a$ € $b) = (a = ab)$

Theorem $19_{|}$

$(a_{|} + b_{|} = a_{|}) + (b_{|}$ € $a_{|})$,

and replacing on the right side a with $b_{|}$ and $b_{|}$ with a, with the left side together with Schröder's Theorem 20, we obtain $(a = ab) = (a$ € $b) = ((a + b) = b)$.

Theorem 20

0 € a

Theorem $20_{|}$

$a_{|}$ € 1.

Proof: According to Theorem 6, $aa_{|} = 0$. According to theorem 11, $aa_{|}$ € a, therefore 0 € a for any desired a. In Schröder we are told this is in Definition 2. So we see that our zero definition leads to the same results as Schröder's.

Theorem 21 [5_x] if $a \in 0$, then $a = 0$	*Theorem* $21_\|$ if $1 \in a_\|$, then $1 = a_\|$.
Theorem 22 [21_x] $a \,.\, 1 = a$	*Theorem* $22_\|$ $1 = a_\| + 1$.
Theorem 23 [22_x] $a \,.\, 0 = 0$	*Theorem* $23_\|$ $1 = a_\| + 1$.
Theorem 24 [23_x] $a\,(a + b) = a$	*Theorem* $24_\|$ $a_\| = a_\| + a_\| b_\|$.
Theorem 25 [24_x] if $1 = ab$, then $a = 1$ and $b = 1$	*Theorem* $25_\|$ if $a_\| + b_\| = 0$, then $a_\| = 0$ and $b_\| = 0$.
Theorem 26 if $ac \in bc$ and at the same time $b_\| c_\| \in a_\| c_\|$, then $a \in b$	*Theorem* $26_\|$ if $b_\| + c_\| \in a_\| + c_\|$ and at the same time $a + c \in b + c$, then $b_\| \in a_\|$,

which, if one carries out the required substitutions, corresponds to Theorem 40 in Schröder.

Proof: If ac € bc, then from Theorem 11, ac € b, therefore from Definition 3' $acb_| = 0$, then from Definition 3", $ab_|$ € $c_|$, likewise $b_| c_|$ € $a_|$ and $b_| c_| a = 0$, thus $ab_|$ € c. From $ab_|$ € $c_|$ and $ab_|$ € c and from Definition 2', it follows that $ab_|$ € $cc_|$, from Theorem 6, then $ab_|$ € 0, *then* $ab_| = 0$ after Theorem 21, and as a result, after Definition 3", a € b.

Theorem 27 [41] if $ab \in c$, then this is equivalent to $a \in b_\| + c$	*Theorem* $27_\|$ if $c_\| \in a_\| + b_\|$, then this is equivalent to $bc_\| \in a_\|$.

Proof: From Definition 3', $abc_| = 0$, therefore, from Definition 3" a € $b_| + c$.

Theorem 28 $a \in b$ is equivalent to $ab_\| \in b + a_\|$	*Theorem* $28_\|$ $b_\| \in a_\|$ is equivalent to $ab_\| \in b + a_\|$.

Proof: If a € b, then from Definition 3, $ab_|$ € 0. Its dual statement is 1 € $b + a_|$,

Therefore, $ab_| \in 0 \in 1 \in b + a_|$,

on the other hand, if $ab_|$ € $b + a_|$, it holds that $ab_|\,(b + a) = 0$, i.e. $ab_| ab_|$ € 0, i.e. a € b. This theorem may be found in Schröder's Vol II, 1, p. 314.

Theorem 29 (corresponds to the Theorem in Vol. II, 1, p. 316) if $a \in a_\|$, then $a \in 0$	*Theorem* $29_\|$ if $a \in a_\|$, then $1 \in a_\|$.

Proof: $a\,(a_|)_| = 0$, therefore $aa = 0$, i.e., $a = 0$.

Theorem 30

if a € b and b € $a_|$,
then $a = 0$

Theorem $30_|$

if $b_|$ € $a_|$ and a € $b_|$,
then $1 = a_|$.

Proof: If a € b and b € $a_|$, then a € $a_|$, i.e. from Theorem 29, $a = 0$.

Theorem 31

if a € b and $b|$ € a,
then $b = 1$

Theorem $31_|$

if $b_|$ € $a_|$ and $a_|$ € b,
then $0 = b_|$.

Proof: $b_|$ € $a_|$ since a € b then $b_|$ € b, i.e. $b_|$ € $(b_|)_|$, and as a result of Theorem 29, $b = 1$.

Theorem 31

if a € b and $b_|$ € a,
then, $b = 1$

Theorem $31_|$

if $b_|$ € $a_|$ and $a_|$ € b,
then $0 = b_|$.

Proof: $b_|$ € $a_|$ as a € b, then $b_|$ € b, i.e. $b_|$ € $(b_|)_|$, therefore, from Theorem $29_|$, $b = 1$.

Theorem 32

a € b is equivalent to
$ab_|$ € b

Theorem $32_|$

$b_|$ € $a_|$ is equivalent to
$b_|$ € $a_| + b$.

Proof: From Definition 3, $ab_| = 0$ and 0 € b, thus $ab_|$ € b and, on the other hand, if $ab_|$ € b, then $ab_|b_| = 0$, thus a € b. If in the dual statement we replace $b_|$ with a and $a_|$ with b, we obtain a € b, equivalent to a € $b + a_|$, which can be found in Schröder's Vol II, 1, p. 314.

Theorem 33

$a\,(a_| + b)$ € b

Theorem $33_|$

$b_|$ € $a + a_|b$.

Proof: From Theorem 6 it holds that $ab_|\boldsymbol{.}\ (b + a) = 0$, from this follows from Definition 3

$a(b + a_|)$ € b.

Theorem 34

if bc € 0, then
a € $ab_| + ac_|$

Theorem $34_|$

if $1 = b_| + c_|$, then
$(a_| + b)(a_| + c)$ € $a_|$.

Proof: from Theorem 33 it holds that $a(a_| + b)$ € b and $a(a_| + c)$ € c, therefore $a(a_| + c)\,(a_| + b)$ € $bc = 0$; it follows from Definition 3 that a € $ab_| + ac_|$. This theorem corresponds in some respects to Schröder's Principle III: if $bc + 0$, then $a(b + c)$ € $ab + ac$, considering that, if $bc = 0$, then $b_| + c_| = 1$ holds, thus $a(b_| + c_|) = a$.

Theorem 35 [Theorem 34]	*Theorem* $35_|$
$0 = (a + b)(a + b_|)(a_| + b)(a_| + b_|)$	$a_|b_| + a_|b + ab_| + ab = 1.$

Proof: From Definition 3 it must hold that, if the statement above is correct, $(a + b)$ € $ab + a_|b + ab_|$; now we know from Theorem 34 that, since always $bb_| = 0$, a € $ab_| + ab$ as well as

$$b \in ba_| + ba,$$

hence $a + b$ € $ab + ab_| + a_|b$, in Schröder as his own Theorem 33, which proves the theorem.

Theorem 36 [Theorem 39]	*Theorem* $36_|$
$ab_| + a_|b = 0$ is equivalent	$b_| = a_|$ is equivalent to
$a = b$	$(a + b_|)(a_| + b) = 1.$

Proof: From Definition 3 it holds that, if a € b, then $ab_| = 0$, if b € a, then $ba_| = 0$ and therefore if $a = b$, $ab_| = 0$, $a_|b = 0$ and $ab_| + a_|b = 0$, and vice versa, we know from Theorem 25 that $ab = 0$ and $ba_| = 0$, thus it holds at the same time that a € b and b € a, i.e., $a = b$. In order to derive the second equivalence in Schröder from our dual sentence, we replace $a_|$ with b and $b_|$ with a, and, after applying Theorem 34, we obtain $ab + a_|b_| = 1$.

Theorem 37 [Theorem 27]	*Theorem* $37_|$
$a(b + c) = ab + ac$	$(a_| + b_|)(a_| + c_|) = a_| + b_|c_|.$

Proof: The first half, from Schröder's Theorem 25, $ab + ac$ € $a(b + c)$; the second half, from Definition 3, the theorem is

equivalent to: $a(b + c)(a_| + c_|) = 0$, it also holds that
$a(a_| + b_|) \in b_|$, from Theorem 33
$a(a_| + c_|) \in c_|$, likewise from Theorem 33
$\underline{b + c \in b + c}$

therefore, the product on both sides

$$a(a_| + b_|)(a_| + c_|)(b + c) \in b_|c_|(b + c) \in 0$$

from Theorem 6,

which proves that the law of distributivity applies to the whole system. While Schröder should acknowledge at this point that duality escapes him, and a certain amount of uncertainty on p. 294 and p. 310 is unmistakable in his expressions, our presentation seems more appropriate for a consistent structure, which Schröder himself has always sought. This is so not only by making duality clearer and more intelligible, but also by making it possible to lay out the proof without systematic difficulties.

3. Otto Neurath: Univocality and Commutativity of the Logical Product ab (1909)

(Otto Neurath, 1909, "Eindeutigkeit und Kommutativität des logischen Produktes ab." *Archiv für Philosophie*, 2. Abteilung: *Archiv für systematische Philosophie*, Neue Folge, Bd. 15, 1909, 342–344) [8]

In the recently published "Abriß der Algebra der Logik", by Ernst *Schröder* and edited by Eugen *Müller*,[9] the commutation law is derived again, which is why I want to present some additional remarks to my article.[10] Although in §2 of the *Abriss* it states that two *signs* or two *elements*, *a* and *b*, may be *identical*, the identity sign ($a \equiv b$), which corresponds to the symbolic equality ($a \equiv b$) I have proposed, is not always used where necessary. In "Presuppositions of the theory of statements" in §18 it states the following about the commutativity of the product: "If the product $\alpha\beta$ means the simultaneous validity of both statements, α and β, whichever is mentioned in the first position, so that $\beta\alpha$ has the same meaning as, or is identical to, $\alpha\beta$. The factors of a product are interchangeable." The author then continues: "Since identical statements are ordered and equal to each other, one has the "commutation statement"

$$\beta\alpha \in \alpha\beta, \qquad \beta\alpha = \alpha\beta."$$

It is in no way convenient to call this symbolic equality commutativity, since in arithmetic, from whose terminology the term is taken, there is a difference between multiplier and multiplicand. One obtains *two products*, with the names *ab* and *ba*, and then finds that both products have the same value, whereas here the same product has two names, because the two statements are *valid* at the same time and cannot

[8] Reprinted in Neurath (1981, 17-18). [Translator's note.]

[9] See Schröder (1909). [Translator's note.]

[10] Cf. *Archiv für systematische Philosophie* XV, vol. 1, 1909, pp. 104ff. [See article 1. in the Appendix; translator's note.]

be distinguished from each other as multiplier and multiplicand, and *only* the one-dimensional *descriptive ordering* automatically generates, as it were, the two names.

In "Presuppositions of domain theory" the "law of commutativity" is not applied to the definition of product. It means actually only

$$(x \in a)\,(x \in b) = (x \in ab),$$

whereas the following could be added:

$$(x \in a)(x \in b) \equiv (x \in ab).$$

Based on §18, which has introduced $\alpha\beta = \beta\alpha$, it now holds true that

$$(x \in a)(x \in b) \equiv (x \in b)(x \in a) \text{ and therefore also}$$

$$(x \in b)(x \in a) \equiv (x \in ab).$$

The symbol "*ab*" should now be discussed. It is manifestly independent of the order of the statements. What does *ba* mean, then? If the stipulation we put forward in the cited treatise holds, that only the alphabetical order should be used, *ba* is a meaningless sign. Otherwise, however, the following holds by definition:

$$(x \in ab) \equiv (x \in a)(x \in b) \equiv (x \in b)(x \in a).$$

It would be meaningful, therefore, besides the symbolic equality of statements:

$$\alpha\beta \equiv \beta\alpha, \quad \text{here the equality of domain statements}$$

$$ab \equiv ba \quad \text{also to introduce immediately.}$$

In the abovementioned *Abriß*, however, the concept of "commutativity" leads to further ambiguities.

In §42 there is a completely correct proof that a product of two domains *a* and *b* is *univocal*,

$$(ab)' = ab.$$

But then in §43 it continues: "As such a second product value (*ab*)' may be considered somewhat like the product *ba*, with interchanged factors, which corresponds to a swap in the premises in VI_x'."[11] Commutativity will then be proven in the *same*

[11] Axiom VI_x' is $(x \in a)(x \in b) \in (x \in ab)$. See Schröder (1909, 23). [Translator's note.]

way as univocality, a proof also that in essence it is only proof of univocality. Here the symbol "*ba*" is assigned only to the product (*x* € *b*)(*x* € *a*), which, as we have shown above, is not permitted. The proof of univocality shows that *two domains have only one product*, and the proof of commutativity should show that the *two products have the same value*. However, the *two products* do not appear defined *anywhere*; so, of the two proofs, only the proof of *univocality* has a place in the system.

4. Olga Hahn: On the Axiomatics of the Logical Calculus of Domains (1909)

(O. Hahn, 1909, "Zur Axiomatik des logischen Gebietkalkuls." *Archiv für Philosophie*, 2. Abteilung: *Archiv für systematische Philosophie*, Neue Folge, Bd. 15, 1909, 345–347)

In his Vorlesungen über die Algebra der Logik Ernst Schröder indicates that it might be expedient to place the statement $a = (a_{|})$ as an axiom at the top of the calculus of domains. Since it may be of some interest to discuss the various possibilities for an axiom system for this calculus, the following attempt is made to carry out Schröder's suggestion.

Each conceptual delimitation of a domain *a* determines at the same time one and only one domain $a_{|}$, which we call the negative belonging to the positive.

Like Schröder, we begin with the principle of identity:

Principle I. *a* € *a*.
Principle II. If *a* € *b* and at the same time *b* € *c*, then *a* € *c*.
Definition 1. If (*a* € *b*) and (*b* € *a*), then $a = b$, and vice versa.

The theorems resulting from these statements will be adopted from Schröder without modification. Just as all those statements of Schröder's are used implicitly, no new proof is necessary for this presentation.

Principle III. $(a_{|})_{|} = a$,

i.e., the negative of the negative is equal to the positive to which, according to our postulate, it appears assigned. The relation between positive and negative is univocal in an invertible manner, as the univocal operation of negation assigns to each negative one and only one positive. We could formulate our statement as: *the positive of a positive is equal to its negative*.

Definition 2. If *a* € $a_{|}$, then we say that $a = 0$ and $a_{|} = 1$, and vice versa. We will see in the following that this definition corresponds to that of *0* and *1* in Schröder.
Theorem 1. From Definition 2 it follows that $1 = 0_{|}$ and thereby, from Pr. III, $1_{|} = 0$.

Definition 3. If at the same time $a \in c$ and $c_| \in b_|$, then we say that $a + b \in c$ and $c_| \in a_| b_|$, and vice versa, whereby the definition of addition and of multiplication is given.[12]

Hereby follows

Theorem 2. The statements $a + b \in c$ and $c_| \in a_| b_|$ are equivalent. Our double definition allows us to contrast each statement in the form of Theorem 2 about addition (multiplication) with one about multiplication (addition).

Theorem 3. If $a_| \in c_|$ and at the same time $c \in b$, then it holds that $a_| + b_| \in c_|$ and $c \in ab$.

Proof: from Pr. III and Def. 2, therefore $a_| + b_| \in c_|$ is equivalent to $c \in ab$.

Theorem 4. $a + a \in a$, $a_| \in a_| a_|$.

Proof: from Pr. I and Def. 3.

Theorem 5. $a \in b$ and $b_| \in a_|$ are equivalent.

Proof: if $a \in b$, then from Th. 4, also $a + a \in b$; therefore, from Def. 3, $b_| \in a_| a_|$, and, from Th. 4, $b_| \in a_|$; therefore

$$(a \in b) \in (b_| \in a_|).$$

If $b_| \in a_|$, then $b_| \in a_| a_|$, therefore $a + a \in b$ and $a \in b$, then:

$$(b_| \in a_|) \in (a \in b), \text{ whereby}$$
$$(b_| \in a_|) = (a \in b) \text{ is proven.}$$

This theorem, whose significance for logical duality has been proven elsewhere,[13] makes it possible to write the other dual theorems.

Theorem 6. $a + b = b + a$.

Proof: From Th. 5 it follows that instead of $a \in c$ and $c_| \in b_|$ in Def. 3, one can write $c_| \in a_|$ and $b \in c$, from where, after Pr. I, $b + a \in a + b$.

Similarly, one can prove that $a + b \in b + a$, whereby the law of commutation for addition and multiplication is proven. While in Schröder's presentation there is no reason for the introduction of the commutation law,[14] here it becomes necessary due to the asymmetric expression of the definitions of addition and multiplication.

[12] The univocality proof is in accordance with Schröder and Müller's *Abriß* (Teubner, 1909).

[13] *Archiv f. system. Phil.* XV, 2, 1909. "On Duality in Logic" (O. Hahn and O. Neurath). [See article 2. in this Appendix; translator's note.]

[14] *Arch. f. system. Phil.* XV. 1. 1909. P. 104f. [See article 2. in this Appendix; translator's note.]

Theorem 7. $(ab)_{|} = a_{|} + b_{|}$.
Proof with Def. 3, Pr. III and Th. 5.

Theorem 8. $aa_{|} = 0$, $a + a_{|} = 1$
Proof: $aa_{|}$ € $0a + a_{|}$ – see Schröder – and since $a + a_{|} = (aa_{|})_{|}$, the statement follows from Def. 2.

From this, 0 € a, $a_{|}$ € 1, for any a, which correspond to Schröder's definitions of 0 and 1.

Principle IV. If $ab_{|}$ € $b + a_{|}$, then a € b.
Theorem 9. (a € b) = ($ab_{|}$ € 0).
Proof: if a € b, then $ab_{|}$ € $bb_{|}$ € 0. If $ab_{|}$ € 0, then from Def. 2, $ab_{|}$ € $(ab_{|})_{|}$ € $b + a_{|}$, therefore, from Pr. IV, a € b, which was to be proven.

Since the law of distribution can be derived from this theorem,[15] the principles and definitions presented here are sufficient to derive the entirety of Schröder's calculus of domains.

5. Otto Neurath: Definition Equality and Symbolic Equality (1910)

(Otto Neurath, 1910, "Definitionsgleichheit und symbolische Gleichheit." *Archiv für Philosophie*, 2. Abteilung: *Archiv für systematische Philosophie*, Neue Folge, Band 16, 1910, 142–144) [16]

Ernst Schröder and *Eugen Müller* speak of a commutation law in the logical calculus. Elsewhere[17] I have shown that the two symbols *ab* and *ba* appear to be defined at the same time by denoting the same relation between two objects, not two relations defined separately or two different instances of the type of relation with the same result as the mathematical products *ab* and *ba*. It turns out that it is appropriate to distinguish the case of equality of two symbols from the second case. I pressed for expressing the fact this way

$$ab \equiv ba,$$

to be read *ab* symbolically equal to *ba*. Taking this difference into account, it is possible to avoid oversights like Schröder and Müller's.

---

[15] *Arch. f. system. Phil.* XV. 2. 1909. "On Duality in Logic." [See article. 2. in this Appendix; translator's note.]

[16] Reprinted in Neurath (1981, 19-21). [Translator's note.]

[17] Cf. *Archiv f. system. Philosophie* 1909, Bd. XV, "Ernst Schröders Beweis des 12. Theorems: Für identische Operationen gilt das 'Kommutationsgesetz,'" p. 104ff, and "Eindeutigkeit und Kommutativität des logischen Produktes ab," p. 302ff. [See article 1. and article 3 in this Appendix; translator's note.]

I have recently noticed that J. Petzoldt and, in a sense, also W. Wundt have already proposed drawing a *similar* distinction. It is easy to find in the literature expressions along the same lines, especially by mathematicians and physicists. A collection of such points would likely help draw attention to the question of the essence of *definition*.

Petzoldt adopts Wundt's distinction between *definition* equality, e.g., $c = s/t$, and *causal* equality, e.g., $v = gt$.[18] Petzoldt now describes definition equalities as *identities* and hence suggests[19] writing

$$c \equiv s / t.$$

The question now is whether there is still a difference between this identity and symbolic equality, that is, whether there is yet a third case besides equality and symbolic equality.

The case $ab \equiv ba$ in algebra of logic is completely clear; we want to maintain the expression 'symbolic equality' for such types, because only *one* product appears defined in Schröder's formulation.[20] On the other hand, in the example given by Petzoldt,

$$(a+b)(a-b) = a^2 - b^2$$

[18] See Petzold (1895). Univocality is the property of determination at the basis of determinism associated with causal laws and, with Mach, their empirical significance. Petzold referred to an article by the empirical psychology pioneer Wilhelm Wundt (1894). The question arose for Wundt as a problem in the empirical interpretation of mathematical relations in physics and psychology. As a formal issue in logic – on psychological foundations and the basis of empirical knowledge –, Wundt had examined equality relations in the first volume of his *Logik* (revised edition, 1893), and as matter of natural knowledge in the two parts of the second volume, published separately in a revised edition (see vol. 2, section 1, 1894, 327-332, and vol. 2, section 2, 1895, 146-7,155, 194, 208 and 256). Neurath had been reading *Logik* for a review published the same year in the context of the debate over the unity of the natural and human sciences. See Neurath (1910/2004). [Translator's note].

[19] Cf. *Vierteljhschrft f. wiss. Phil.* 1895, XIX, "Das Gesetz der Eindeutigkeit," p. 150.

[20] It differs from O. Hahn's axiomatization (*Archiv f. system. Phil.* XV, 1909, p. 347): "Similarly, one can prove that whereby the law of commutation for addition and multiplication is proven. While in Schröder's presentation there is no reason for the introduction of the commutation law, here it becomes necessary due to the asymmetric expression of the definitions of addition and multiplication." [See article 4 in this Appendix; translator's note.]

The interesting fact that the commutation law in logic depends on the axiom system suggests the question of the degree to which commutation laws may be eliminated by variations of the axiom system. Today a systematic approach to axiomatics is becoming more and more a requirement.

is no doubt an instance of the usual "equality."[21]

What about the case $a + a + a + a + \ldots + a$ (b times) $= ab$, where Petzoldt rightly wants to see the identity sign used? We can establish a difference immediately. The left-hand side of the definition itself makes sense, not on the basis of *a* definition, but by combining multiple definitions. The right-hand side cannot be defined at the same time as the left-hand side; it is defined by the left-hand side. The situation is different for symbolic equality; the left-hand side has been determined first by definition; *the right-hand side could have been defined equally*, and then the left-hand side can be obtained by means of a symbolic equation. Both sides occupy the *same* place in relation to the definitions.

It might be worth considering, therefore, three different cases, which we can illustrate with examples from mathematics:

I. *Equality*: $(a + b)(a - b) = a^2 - b^2$.
II. *Definition equality*: $a + a + a + a + \ldots + a$ (b times) $\equiv ab$.
III. *Symbolic equality*: $\sqrt[b]{a} \equiv a^{1/b}$.

That the discrimination between III and I is effective has been shown by logical example, where the mixture of II and III led to confusion; whether the symbolic independence of II from III is in place would first have to be examined more closely, but I have shown that the three cases can be set apart and in the discussion of the essence of definition the distinction may even be necessary.

Translator's References

Couturat, Louis. 1905. *L'Algèbre de la Logique*. Paris: Gauthier-Villars.

Mach, Ernst. 1883/1919. *The Science of Mechanics. A Critical and Historical Account of Its Development*. 4th ed. Chicago/London: Open Court.

Neurath, Otto. 1910/2004. On the Theory of Social Science. In *Otto Neurath: Economic Writings. Selections 1904–1945*, ed. Thomas Uebel and Robert S. Cohen, 265–291. Dordrecht: Kluwer.

———. 1981. *Gesammelte philosophische und methodologische Schriften*. Ed. Rudolf Haller and Heiner Rutte. Wien: Hölder-Pichler-Tempsky.

Petzoldt, Joseph. 1895. Das Gesetz der Eindeutigkeit. *Vierteljhschrft für wissenschaftliche Philosophie* 19: 146–203.

Schröder, Ernst. 1890. *Vorlesungen über die Algebra der Logik (Exakte Logik)*. Vol. 1. Leipzig: Teubner.

———. 1909. *Abriss der Algebra der Logik*. Bearbeitet von E. Müller. In drei Teilen. I. Teil: Elementarlehre. Berlin/Leipzig: Teubner.

Wundt, Wilhelm. 1894. Über psychische Causalität und das Princip des psychophysichen Parallelismus. *Philosophischen Studien* 10 (1): 1–124.

[21] Although Petzoldt's article is on univocality, Neurath had neglected it in the process of writing his own earlier article in the context of algebraic logic, also Mach's emphasis on univocal determinations of position and motion in *Die Mechanik in ihrer Entwickelung* (1883/1919). [Translator's note.]

Appendix 2: The 1940–1945 Neurath-Carnap Correspondence

Edited with additional notes by Adam Tamas Tuboly and Jordi Cat (Fig. A2)

Editorial Introduction

Robert S. Cohen and Marie Neurath edited the first (*Empiricism and Sociology*) and the second (*Otto Neurath: Philosophical Papers 1913–1946*) selected writings of Neurath appearing in English. The volumes were published in the important "Vienna Circle Collection," under the general editorship of Henk L. Mulder, Robert S. Cohen, and Brian McGuinness (with, among others, A. J. Ayer, Herbert Feigl, Viktor Kraft, and Karl Menger on the Editorial Advisory Board). The second volume's updated bibliography of Neurath's writings ended with the following item: "The

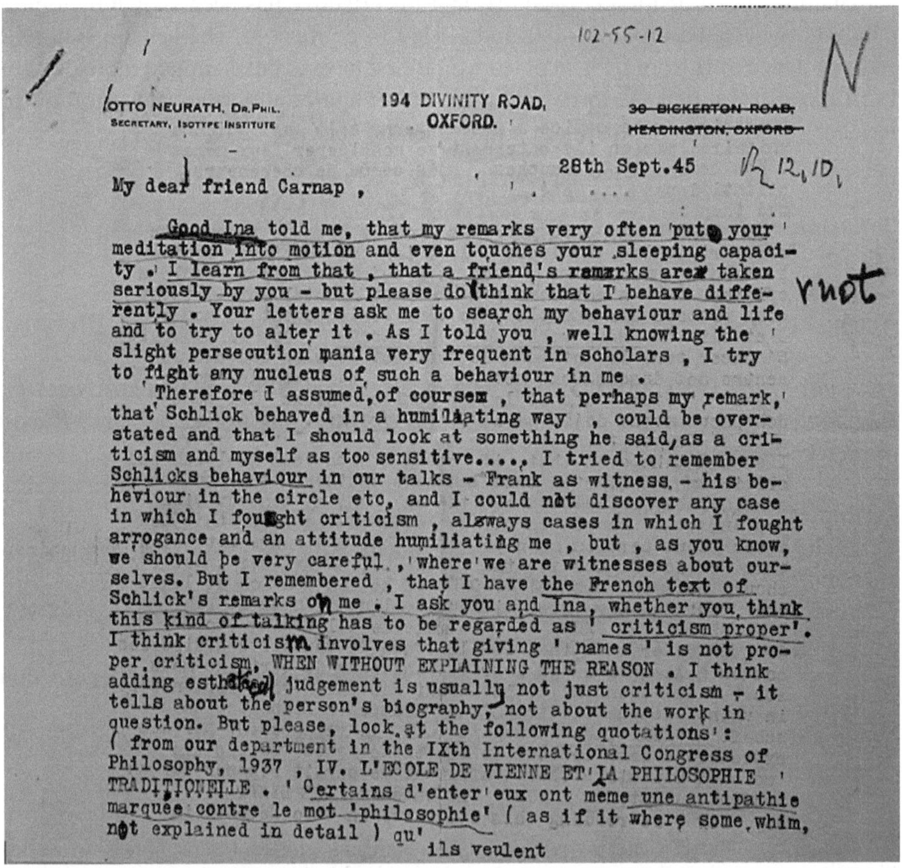
102-55-12 N

OTTO NEURATH, DR.PHIL.
SECRETARY, ISOTYPE INSTITUTE

194 DIVINITY ROAD,
OXFORD.

~~30 BICKERTON ROAD,~~
~~HEADINGTON, OXFORD~~

28th Sept.45

My dear friend Carnap ,

Good Ina told me, that my remarks very often puts your meditation into motion and even touches your sleeping capacity . I learn from that , that a friend's remarks are taken seriously by you - but please do not think that I behave differently . Your letters ask me to search my behaviour and life and to try to alter it . As I told you , well knowing the slight persecution mania very frequent in scholars , I try to fight any nucleus of such a behaviour in me .

Therefore I assumed,of coursem , that perhaps my remark, that Schlick behaved in a humiliating way , could be overstated and that I should look at something he said,as a criticism and myself as too sensitive...., I tried to remember Schlicks behaviour in our talks - Frank as witness - his beheviour in the circle etc, and I could not discover any case in which I fought criticism , always cases in which I fought arrogance and an attitude humiliating me , but , as you know, we should be very careful , where we are witnesses about ourselves. But I remembered , that I have the French text of Schlick's remarks on me . I ask you and Ina, whether you think this kind of talking has to be regarded as ' criticism proper'. I think criticism involves that giving ' names ' is not proper criticism, WHEN WITHOUT EXPLAINING THE REASON . I think adding esthetical judgement is usually not just criticism - it tells about the person's biography, not about the work in question. But please, look at the following quotations : (from our department in the IXth International Congress of Philosophy, 1937 , IV. L'ECOLE DE VIENNE ET LA PHILOSOPHIE TRADITIONELLE . ' Certains d'enter eux ont meme une antipathie marquee contre le mot 'philosophie' (as if it where some whim, not explained in detail) qu' ils veulent

Fig. A2 First page of Neurath's letter to Carnap

Correspondence of Otto Neurath and Rudolf Carnap," edited by Marie Neurath and Henk K. Mulder, forthcoming in the "Vienna Circle Collection." Nothing came out for years.

Two years later still, Rainer Hegselmann (1985, 276, n. 1.) indicated in his paper about the philosophical significance of the Neurath-Carnap correspondence that there would be a German-English edition of the 1940–45 correspondence under the editorship of Marie Neurath, Henk K. Mulder, and himself. As it turned out, there was an internal debate among the editors since Neurath's third wife, Marie, wanted to publish just a selection of letters in which all quarrels between Neurath and Carnap should be eliminated. After a few years of discussion, however, Rainer Hegselmann convinced Marie Neurath that everything should be published, the conflicts included, and she agreed. Nevertheless, due to the changing priorities and others reasons, the correspondence was never published.[22]

Although these letters have not appeared in print before, they have not been unknown to those working on the re-evaluation of logical empiricism, especially of Carnap's and Neurath's philosophies. They have been used regularly in philosophical and historical arguments.[23]

Content and Context of the Letters

The 1940–1945 correspondence between Neurath and Carnap is as intellectual as it is personal. The reader is able to get a quite comprehensive view not just of how these philosophers established their commitments, arguments, and researches in the 1940s, but also of their personal development and attitudes, especially in changing circumstances, and, no less significant, how the personal and the philosophical changes expressed each other.

Philosophical correspondences may be consulted not just for reasons of curiosity, to peek into the daily life of philosophers, they also open up a view into a special forum for discussions prior to published arguments. W. V. O. Quine wrote to Carnap, for example, that after their years-length debates about modality in their letters, they "did accomplish a good deal toward ironing out misunderstandings in the preliminary correspondence, and that it is all right to let the public in."[24] Also Carnap and Neurath mentioned the benefit of having a private, partly informal and freely developing debate about problematic issues before any publication on the given topic (see especially letters 17. and 23.). In fact, that is the situation in which Neurath's early death left matters.

[22] We are indebted for this information to Rainer Hegselmann (email correspondence with Ádám Tamás Tuboly, August 9, 2017).

[23] The most important and throughout going use of these letters are in Hegselmann 1985, 1987; Uebel 1992, 2007; and Reisch 2003a, 2003b and 2005.

[24] Quine to Carnap, 4 December 1946; in Creath (1990, 403).

The 1940–45 correspondence presented here can be consulted also with a historiographical interest in mind. In order to write *a* history of logical empiricism (it is quite dubious whether "the" history of logical empiricism could or should be written), the story-telling historian finds a whole gamut of possible perspectives and sources, for instance:

(1) The history of the received orthodox view (mainly by W. V. O. Quine, Nelson Goodman, A. J. Ayer, and in textbooks from the twentieth century).
(2) The recent work challenging the received view about logical empiricism (Coffa 1991; Uebel 1992, 2007; Richardson 1998; Friedman 1999; Reisch 2005; Carus 2007; Richardson – Uebel 2007; Damböck 2016).
(3) Original self-presentations of the movement in published memoirs, biographies, and historical introductions by logical empiricists themselves (Blumberg – Feigl 1931; Reichenbach 1936, 1951; Neurath 1936/1981; Morris 1937; Kaila 1939/2014; Frank 1949; Jörgensen 1951/1970; Kraft 1950/1953; von Mises 1956; Carnap 1963; Menger 1979a, 1979b, 1982, 1994; Rutte – Acham – Götschl 1973; Popper 1976/2002; Haller – Rutte 1977; Feigl 1969/1981a, 1969/1981b; Bergmann 1993; Næss 1993).[25]
(4) Histories documented in private letters and other archive files.
(5) Alternative accounts by contemporaneous actors, including critics (Popper, again, too).
(6) Contemporaneous public notices and narratives featured in the news media and other non-professional venues.
(7) The history of intellectual collectives such as schools and other self-organized and self-labeled groups.

These stories usually do not match and sometimes even contradict one another (especially in the case of (1) and (2)). It is also true that works in (2) built upon the narratives and themes emerging in the files of (4). Nevertheless, some unpublished drafts, notes, and conference presentations could undercut even those stories that already diverged from the usual received.

Correspondences might be important for another reason as well: they could tease out or re-evaluate ignored or forgotten strands of the movement. Take the case of Hans Hahn: he is not the most cited author among members of the Vienna Circle; his 'philosophical writings' is formally quite a thin booklet, and he was entirely dropped from the stories of (1) and even from many of (2). Still, according e.g. to Philipp Frank, Hahn was the 'real founder' of the Circle, teacher and mentor of many important members of the movement, a regular and persistent member of the evening discussions whose critical notices carried great weight. But that could be reconstructed only from certain works in (3), and mainly from (4).

A similar case could be presented regarding Philipp Frank too. After the 1950s, he seems to have been dropped from the canon in (1); he was usually mentioned as

[25] Some of these works are treated by Stadler 2001/2015, Sect. 4.1. and Hofer – Stöltzner 2014.

only occasionally attending the Circle's meetings in (3); he is more often mentioned as a peculiar case by (2). But according to (4), he seems to be an equal partner of Schlick (and not just in age), a leading figure of opinion-formation for some members, a sometimes sloppy, but still active organizer like Neurath, and a forerunner of important trends in history, philosophy, and sociology of science.[26]

The stories presented in the members' memoirs are sometimes similar to those in (1), other times to those in (2). The reasons behind this phenomenon are quite complex, but the time and place of writing, the language of the work, the milieu and context of philosophy could be relevant in all of the given cases. Thus, the histories documented in (3) might be counterbalanced also by the materials in (5), (6) and even (4): letters often contain valuable insights on the production and aims of these memoirs, stories and works of logical empiricists.

All of these considerations are especially relevant to the Neurath-Carnap correspondence. Though there are many (side) issues recurring frequently in the letters, seven themes provided the core issues of their correspondence. These are, in (sometimes overlapping) temporal order, (a) *emigration*, (b) *semantics*, (c) *Popper's philosophy*, (d) *Russell's philosophy*, (e) *Neurath's* Encyclopedia *monograph*, (f) *Neurath's own place in the movement*, (g) *the relation between Platonism in the history of philosophy and the oppressive German climate the led to Nazism.* However, the main issue is partly political, partly cultural, partly scientific, and partly methodological: namely the difference between absolutism and pluralism. We shall not go into the details here; all of these topics are treated in the general introduction to the present volume, and are discussed in detail by the authors of the volume in their chapters.

Form, Structure, Editorial Issues

The correspondence between Neurath and Carnap contains many side-references to various persons. We have provided biographical and bibliographical information (dates, fields of research, relevant works) for only those people (scientists, philosophers, politicians or just friends) who do not have a biography in Friedrich Stadler's *The Vienna Circle* (2001/2015, 397–592). Since Neurath and Carnap were always embedded in various scientific-cultural-social circles, these figures, occasionally mentioned in the letters, also provide further insight into the context of logical empiricism.

Bibliographical information for those works that are either mentioned or discussed in details by Neurath and Carnap is presented also in the footnotes. The English translations that became available after their correspondence are noted as well; in other cases, details of the original editions are given without any specific

[26] On Frank's rehabilitation see the special issue of *Studies in East European Thought* (Tuboly 2017), edited on occasion of the 50th anniversary of Frank's death.

references. Since Neurath often quoted from original German texts (without providing translations), we have provided the English translations from later editions in the footnotes. Since the *International Encyclopedia of Unified Science* was a frequent theme, and the correspondents referred to the various monographs as well, they are noted in the footnotes with the volume- and monograph number as "II-3," meaning Volume II, Monograph 3. As the most easily accessible edition is the one released in 1970,[27] we have provided page numbers referring to it.

Some words should be said about the selection-principles. Until 1940, and through the first months of that year, Neurath and Carnap corresponded in their native language. The fact that they started to exchange letters in English is significant for various reasons. It is known from the Carnap-Quine correspondence that Carnap started to write letters in English earlier to prepare for his journey to America (in Creath 1990, 126). Writing letters in English thus meant a means to practice the language of the future adoptive country in an informal setting. In Neurath's case, the first reason might have been that he was required to write in English by the rules enforced at the British internment camps: when he was interned during the summer of 1940 in England, all the letters sent out and received by internees were opened, read through, and often censored by the suitable officers (see e.g. Kochan 1983, 136). Accustomed to writing letters in the new language, staying in England for the rest of his life during wartime as a foreign national habit and caution would have prompted corresponding in English also with individuals in other countries engaged in war.

It was our editorial decision not to include those letters that were written in German during the relevant period: though there are a few of them between January and May 1940 (the last was sent out by Neurath on May 6, a few days before the Nazis entered Holland and he fled to England), all of them are about editorial issues regarding the *Encyclopedia* and its structure. They will be published under the general editorship of Christian Damböck in the official and critical edition of the Carnap correspondence.

The Neurath-Carnap correspondence practically stopped after May 1941 when Neurath became interned with Marie for eight months: they were released only in February 1941; the correspondence started then with renewed force as soon as they settled down in Oxford that year. Our selection of the 1940–1945 correspondence, despite the absent parts, still make up a coherent and continuous exchange of letters: their running thread is provided by the abovementioned philosophical and social questions.

Besides the selection process, publishing the English letters meant another editorial challenge. Though Neurath and Carnap wrote letters in their second language for years, in most of these letters are still often the use of English, especially by Neurath, can be idiosyncratic when not incorrect. We have not modified any of the stylistic features: all the texts are preserved as they were originally written and

[27] *Foundations of the Unity Science: Toward an International Encyclopedia of Unified Science*, Volumes 1–2, Chicago and London: University of Chicago Press.

received by the correspondents. The fact they both turned to English in their personal communication is a moving document of being in English-speaking countries that are both foreign and at war against Germany. Preserving their respective use of English also documents the degree of alienation and compromise involved in the experience of emigration.

On the other hand, writing in English might have another incentive. In their debate about auxiliary and international languages, Neurath wrote to Carnap the followings: "You see in the English speaking world, English is the auxiliary language also for foreigners, the various aliens here, all the foreign governments use English, it is astonishing how the whole business runs in this way, English is manifestly now the Lingua franca for very many people. Who will learn Esperanto in the USA? For what purpose?"[28] English was thus an instrument, imposed on Neurath by the requirement to communicate effectively in the international community. But in the usage of English he found other reliefs as well: "You see, how difficult it is and what monst[er]s of sentences there appear in my German. How to translate, as I told you: 'the pattern of the brotherhood of mankind.' I am glad you know now, that English is, for me, more adequate than German, which is full of Metaphysics and lacks CONCRETE Brotherhood phraseology. Funny, but an experience."[29]

Besides the stylistic issues, we should warn the readers of the many ungrammatical formulations and typos. Where obvious mistaken phrases or simple errors occurred, we have indicated them in [brackets], by replacing the mistaken letters, or providing the missing one. To see some examples, "I should like to see Brunswiks paper" became "I should like to see Brunswik[']s paper," or "much more then any other publication of our movement" became "much more th[a]n any other publication of our movement," "it would be allright" became "it would be a[l]right." In many other cases, when obviously idiosyncratic or mistaken words appeared, we tried to provide again in [brackets] what would have been the appropriate expression. For example, "I prefer the differences of the decisions, because then we get more people with civil courage and less people who always look to the 'formal' correctness of central decisions" became "I prefer the differences of [in] the decisions, because then we get more people with civil courage and less [fewer] people who always look to the 'formal' correctness of central decisions." In other cases, when entire words should have been deleted or entirely rewritten, we chose to indicate them with "[sic]" noting that the mistake or incorrect formulation was in the original text.

Thus, we haven't indicated and corrected every ungrammatical expression, but tried to do so with missing characters, mistaken prepositions, or misspellings. Our aim was not, after all, to produce a critical edition of the correspondence, but to make such a text that might be useful for a wide range of scholars, interested in the history of logical empiricism, early analytic philosophy, and twentieth century thought in general.

---

[28] Neurath to Carnap, 22 December 1942 (RC 115-07-61); letter 16. in the present appendix.

[29] Neurath to Bernhard Reichenbach, 22 April 1944 (Otto Neurath Nachlass (ONN, Wiener Kreis Archiv, Rijksarchief in Noord-Holland, Haarlem, The Netherlands).

Since the letters were written in an informal tone, Neurath and Carnap often referred to people, but notably not to each other, by their first name. In such cases we provided the surname as well in brackets to help the reader in identifying whom they talked about. In other cases, when it seemed to be necessary or helpful for the reader, we inserted the first name in brackets.

Finally, marginal handwritten notes are indicated in brackets and italicized: "[...] so important as my neighbour's happiness [marginal addition: *but would prefer suffering*] by making the man unhappy [...]." The same *italics* technique is used to indicate handwritten signatures, and we provided signatures in brackets where there were none.

Location of the Letters

Most of these letters are preserved in both of the Carnap and Neurath archives:

- Rudolf Carnap Papers, 1905–1970, ASP.1974.01, Archives of Scientific Philosophy, Archives & Special Collections, University of Pittsburgh Library System.
- Otto Neurath Nachlass (Wiener Kreis Archiv), Rijksarchief in Noord-Holland, Haarlem, The Netherlands.

There are a few exceptions though that could be located only in one of the Archives. In order to track down the original sources and letters, we have provided their Archive numbers underneath the name of each letter's correspondents and date.

Acknowledgements We are indebted to Gerard Alberts, Aaron Alvarez, Brigitta Arden, Michael Beaney, Hans-Joachim Dahms, Christian Damböck, Johannes Friedl, Rainer Hegselmann, Robert Kaller, A. J. Kox, Ferdinand Mertens, George Reisch, Günther Sandner, Jan Woleński, and Deodáth Zuh. During the preparation and edition of this volume, Adam Tamas Tuboly was supported by the MTA BTK Lendület Morals and Science Research Group; by the MTA Premium Postdoctoral and the János Bolyai Research Scholarship; and by the "Empiricism and atomism in the twentieth-century Anglo-Saxon philosophy" NKFI project (124970).

References

Bergmann, Gustav. 1993. Memories of the Vienna Circle. Letter to Otto Neurath. In *Scientific Philosophy: Origins and Developments*, ed. Friedrich Stadler, 193–208. Dordrecht: Kluwer.

Blumberg, Albert E., and Herbert Feigl. 1931. Logical Positivism. A new movement in European philosophy. *The Journal of Philosophy* 28 (11): 281–296.

Carnap, Rudolf. 1963. Intellectual Autobiography. In *The Philosophy of Rudolf Carnap*, ed. Paul A. Schilpp, 3–84. La Salle: Open Court.

Carus, André W. 2007. *Carnap in Twentieth-Century Thought: Explication as Enlightenment*. Cambridge: Cambridge University Press.
Coffa, Alberto. 1991. *The Semantic Tradition from Kant to Carnap: To the Vienna Station*. Cambridge: Cambridge University Press.
Creath, Richard, ed. 1990. *Dear Carnap, Dear Van: The Quine-Carnap Correspondence and Related Work*. Berkeley: University of California Press.
Damböck, Christian, ed. 2016. *Influences on the Aufbau*. Cham: Springer.
Feigl, Herbert. 1969/1981a. The Origin and Spirit of Logical Positivism. In *Herbert Feigl: Inquiries and Provocations. Selected Writings*, ed. Robert S. Cohen, 21–37. Dordrecht: Reidel.
———. 1969/1981b. The Wiener Kreis in America. In *Herbert Feigl: Inquiries and Provocations. Selected Writings*, ed. Robert S. Cohen, 57–94. Dordrecht: Reidel.
Frank, Philipp. 1949. Introduction – Historical Background. In *Modern Science and Its Philosophy*, 1–52. Cambridge, MA: Harvard University Press.
Friedman, Michael. 1999. *Reconsidering Logical Positivism*. Cambridge: Cambridge University Press.
Haller, Rudolf, and Heiner Rutte. 1977: Gespräch mit Heinrich Neider: Persönliche Erinnerungen an den Wiener Kreis. *Conceptus* 44: 21–42.
Hegselmann, Rainer. 1985. Die Korrespondenz zwischen Otto Neurath and Rudolf Carnap aus den Jahren 1934 bis 1945 – Ein vorläufiger Bericht. In *Philosophie, Wissenschaft, Aufklärung: Beiträge zur Geschichte und Wirkung des Wiener Kreises*, ed. Hans-Joachim Dahms, 276–290. Berlin/New York: Walter de Gruyter.
———. 1987. Introduction. In *Unified Science*, ed. Brian McGuinness, ix–xxi. Dordrecht: D. Reidel.
Hofer, Veronika, and Michael Stöltzner. 2014. Vienna Circle Historiographies. In *European Philosophy of Science – Philosophy of Science in Europe and the Viennese Heritage*, ed. Maria Carla Galavotti, Elisabeth Nemeth, and Friedrich Stadler, 295–318. Dordrecht: Springer.
Jörgensen, Jörgen. 1951/1970. The Development of Logical Empiricism. In *Foundations of the Unity Science: Toward an International Encyclopedia of Unified Science*, 845–936. Volume 2. Chicago/London: University of Chicago Press.
Kaila, Eino. 1939/2014. *Human Knowledge. A Classic Statement of Logical Empiricism*. Edited by Juha Manninen, Ilkka Niiniluoto, and George Reisch. Chicago: Open Court.
Kraft, Viktor. 1950/1953. *The Vienna Circle the Origin of Neo-Positivism: A Chapter in the History of Recent Philosophy*. New York: Philosophical Library.
Menger, Karl. 1979a. Introduction. In *Selected Papers in Logic and Foundations, Didactics, Economics*, 1–7. Dordrecht: D. Reidel.
———. 1979b. Logical Tolerance in the Vienna Circle. In *Selected Papers in Logic and Foundations, Didactics, Economics*, 11–16. Dordrecht: D. Reidel.
———. 1982. Memories of Moritz Schlick. In *Rationality and Science*, ed. Eugene T. Gadol, 83–103. Wien/New York: Springer.

———. 1994. *Reminiscences of the Vienna Circle and the Mathematical Colloquium*. Dordrecht: Kluwer.

Mises, Richard von. 1956. *Positivism. A Study in Human Understanding*. New York: George Brazillier, Inc.

Morris, Charles. 1937. *Logical Positivism, Pragmatism, and Scientific Empiricism.* Paris: Hermann & Édituers.

Naess, Arne. 1993. Logical Empiricism and the Uniqueness of the Schlick Seminar: A Personal Experience with Consequences. In *Scientific Philosophy: Origins and Developments*, ed. Friedrich Stadler, 11–26. Dordrecht: Kluwer.

Neurath, Otto. 1936/1981. Die Entwicklung des Wiener Kreises und die Zukunft des Logischen Empirismus. In *Otto Neurath: Gesammelte philosophische und methodologische Schriften. Band 2*, ed. Rudolf Haller and Heiner Rutte, 673–702. Wien: Verlag Hölder-Pichler-Tempsky.

Popper, Karl. 1976/2002. *Unended Quest: An Intellectual Autobiography*. London/New York: Routledge.

Reichenbach, Hans. 1936. Logistic Empiricism in Germany and the present State of its Problems. *The Journal of Philosophy* 33 (6): 141–160.

———. 1951. *The Rise of Scientific Philosophy*. Berkeley/Los Angeles: University of California Press.

Reisch, George. 2003a. Disunity in the International Encyclopedia of Unified Science. In *Logical Empiricism in North America*, ed. Gary L. Hardcastle and Alan W. Richardson, 197–215. Minneapolis: University of Minnesota Press.

———. 2003b. On the International Encyclopedia, the Neurath-Carnap Disputes, and the Second World War. In *Logical Empiricism. Historical and Contemporary Perspectives*, ed. Paolo Parrini, Wesley C. Salmon, and Merrilee H. Salmon, 94–108. Pittsburgh: University of Pittsburgh Press.

———. 2005. *How the Cold War Transformed Philosophy of Science. To the Icy Slopes of Logic*. New York: Cambridge University Press.

Richardson, Alan W. 1998. *Carnap's Construction of the World – The* Aufbau *and the Emergence of Logical Empiricism.* Cambridge: Cambridge University Press.

Richardson, Alan W., and Thomas Uebel, eds. 2007. *The Cambridge Companion to Logical Empiricism.* Cambridge: Cambridge University Press.

Rutte, Heinrich, Karl Acham, and Johann Götschl. 1973: Gespräch mit Viktor Kraft. *Conceptus* 7: 9–25.

Stadler, Friedrich. 2001/2015. *The Vienna Circle. Studies in the Origins, Development, and Influence of Logical Empiricism.* 2nd ed. Dordrecht: Springer.

Tuboly, Ádám Tamás, ed. 2017. The Life and Works Philipp Frank. Special Issue of *Studies in East European Thought* 69 (3): 199–276.

Uebel, Thomas. 1992. *Overcoming Logical Positivism from Within – The Emergence of Neurath's Naturalism in the Vienna Circle's Protocol Sentence Debate.* Amsterdam: Rodopi.

———. 2007. *Empiricism at the Crossroads. The Vienna Circle's Protocol-Sentence Debate*. Chicago/La Salle: Open Court.

Table of Contents

1.	Carnap to Neurath, January 22, 1940.	522
2.	Carnap to Neurath, January 31, 1940.	524
3.	Carnap to Morris, Copy to Neurath, January 31, 1940.	525
4.	Carnap to Neurath, November 1, 1940.	526
5.	Neurath to Carnap, April 4, 1941.	529
6.	Carnap to Neurath, June 5, 1941.	531
7.	Neurath to Carnap, September 21, 1941	535
8.	Carnap to Neurath, January 27, 1942.	538
9.	Neurath to Carnap, March 2, 1942.	540
10.	Carnap to Neurath, June 24, 1942.	542
11.	Neurath to Carnap, July 17, 1942.	544
12.	Neurath to Carnap, July 20, 1942.	552
13.	Neurath to Carnap, July 29, 1942.	553
14.	Neurath to Carnap, August 27, 1942.	559
15.	Carnap to Neurath, November 7, 1942.	560
16.	Neurath to Carnap, December 22, 1942.	563
17.	Neurath to Carnap, January 15, 1943.	570
18.	Carnap to Neurath, January 29, 1943.	573
19.	Carnap to Neurath, March 15, 1943.	575
20.	Carnap to Neurath, May 11, 1943.	579
21.	Neurath to Carnap, July 15, 1943.	582
22.	Neurath to Carnap, September 25, 1943.	584
23.	Neurath to Carnap, October 28, 1943	606
24.	Carnap to Neurath, February 4, 1944.	607
25.	Neurath to Carnap and Morris, Copy to Hogben, March 23, 1944.	614
26.	Neurath to Carnap, April 1, 1944.	614
27.	Carnap to Neurath, October 7, 1944.	624
28.	Neurath to Carnap, November 18, 1944.	626
29.	Carnap to Neurath, May 6, 1945.	632
30.	Carnap to Morris, Copy to Neurath, June 19, 1944.	635
31.	Neurath to Carnap, June 16, 1945.	637
32.	Carnap to Neurath, Aug. 23, 1945.	646
33.	Ina Carnap to Neurath, Aug. 24, 1945.	649
34.	Neurath to Carnap, September 22, 1945. Unsent.	653
35.	Neurath to Ina Carnap, September 24, 1945.	669
36.	Neurath to Carnap, September 24, 1945.	674
37.	Neurath to Carnap, September 28, 1945.	681
38.	Ina and Carnap to Marie Neurath, January 2, 1946.	683
39.	Marie Neurath to Ina and Carnap, January 14, 1946.	684
References		687
Index		695

1. Carnap to Neurath, January 22, 1940.

(ASP RC 115-07-36)
RUDOLF CARNAP
DEPARTMENT OF PHILOSOPHY
UNIVERSITY OF CHICAGO
CHICAGO, ILLINOIS
22. Jan. 1940

Dear Neurath:

I received your two cables and inferred from the second that my letter of Nov. 29th arrived.

Kaila sent me his book "Den Mänskliga Kunskapen" (Söderström, Helsingfors).[30] It seems to me a good introduction into logical empiricism written for a wider public; but taking into consideration also some more technical questions. It will appeal to many readers because it considers the historical background from Euclid and Aristotle on, much more th[a]n any other publication of our movement.

Content: I. Construction of Theories.

1. The searching for invariants.
2. How the searching for invariants created Greek science.
3. The Aristotelian concept of knowledge.
4. The Galilean concept of knowledge.
5. Induction.

II. Formal Truth of Theories.

1. Logical truth.
2. Mathematical truth.
3. The two first principal theses of logical empiricism.

III. Empirical truth of theories.

1. Principle of Verifiability. The third principal thesis of logical empiricism.
2. The logic of physical theory.
3. Logical behaviorism. The fourth principal thesis of logical empiricism.

Kaila asked about the possibility of an English translation. In this case he would change the book somewhat. But his letter was written in the beginning of September, and I do not know about his present conditions. It seems to me, an English translation would be very desirable. But I cannot make a final judgment about the book because my Swedish reading is rather slow, and, therefore, I read only certain parts.

[30] "Den Mänskliga Kunskapen," was the Swedish translation of Eino Kaila's Finnish "Inhilmillinen tieto: Mitä se on ja mitä se ei ole," published originally in 1939. An English translation was published recently, see Kaila (1939/2014).

I guess that you ask Jörgensen and Ness for judgment.[31] I suppose that they have the book. Please let me know when you got their opinion, and then we may decide whether or not an English translation of the book would be suitable for our Library.[32]

Do you have the material of the Congress discussions?[33] Did we not plan to send to everybody a typescript of his remarks? I did not get anything so far. Or did you give up the intention of publishing the discussion?

Best thanks for your New Years greetings. I am very pleased to see that the elephant waves the flower gaily. I infer from it that he is as well as it is possible under the present circumstances.

Cordially yours,
Carnap
January 22, 1940

Supplementary remarks concerning "Studies in Semantics"

The ms. of Part I is now finished. I will be typed during the next weeks, then I shall send it to some friends, and then perhaps make some changes. Thus, I suppose, it will not be ready for print before March or April. I have added two appendixes to be printed in small type. One indicating further problems which will perhaps be dealt with in later Parts; the second explaining the changes which are to be made with respect to my book "Logical Science"[34] from the new point of view of semantics.

[31] "Ness" was the usual spelling of Arne Næss (1912–2009) who attended Vienna Circle meetings in 1934–35. For his recollections of these events, see Næss (1993). Jørgen Jørgensen (1894–1969), sometimes mentioned as Jörgen Jörgensen, or Joergen Joergensen, was a Danish philosopher, co-editor of *Einheitswissenschaft* and member of the "Committee of Organization" for the *IEUS*. In the English-speaking world he is mainly known for his *A Treatise of Formal Logic* (1931), and for his monograph *The Development of Logical Empiricism* (1951). Neurath (1938a) wrote a paper about Jørgensen's work in Denmark in 1938.

[32] "Library" was the "Library of Unified Science," a continuation of Neurath's *Einheitswissenschaft* series (see letter 28). It consisted in a "monograph series" (Heinrich Gomperz's *Interpretation. Logical Analysis of a Method of Historical Research* (1939) were volumes 8–9), and a "book series," including Richard von Mises' *Kleines Lehrbuch des Positivismus. Einführung in die empiristische Wissenschaftsauffassung* (1939) and Hans Kelsen's *Vergeltung und Kausalität* (1941). Von Mises's book was translated into English in 1951. For more details see Hegselmann (1987).

[33] Presumably the Fifth International Congress for the Unity of Science Congress, Harvard University, Cambridge, Mass. (USA), September 3–9, 1939. Despite their intentions, the papers read at the Congress did not appear together.

[34] Carnap presumably refers to his *Logical Syntax of Language* (1937), originally published as *Logische Syntax der Sprache* in 1934.

According to my calculation, the present ms. of Part I would have about 130 printed pages of the size of the Encyclopedia or about 120 pages of the size of the "Logical Science." What will be the size of the monographs in the "Library"? How many English words on a page?

R. C.

2. Carnap to Neurath, January 31, 1940.

Rudolf Carnap
Faculty of Exchange
University of Chicago
CHICAGO, ILL.
January 31, 1940.

Dear Neurath:

Morris asks to write to you my opinion about an author for the biology pamphlet. Since I do not know anybody whom I could propose, I am willing to accept Morris' suggestion of Gerard,[35] although I do not know him sufficiently myself. (See enclosed copy of my letter to Morris.)

For the sake of safety, I repeat some item of my letter of Jan. 22nd.

1. We might consider an English translation for our Library of Kaila's new book "Den Mänskliga Kunskapen", (Söderstöm, Helsingfros) which is an introduction to logical empiricism. Please ask Jörgensen and Ness for their opinion.
2. Part I of my "Studies in Semantics"[36] will become longer than previously calculated, about 120 or 130 printed pages. It will not be ready for print before March or April.

Cordially,
Carnap

[35] Presumably Ralph Gerard (1900–1974) the American psychologist, specialized on neurophysiology. During the 1930s and 1940s, Gerard worked at the Department of Physiology, University of Chicago, where Carnap thought as well since 1936.

[36] Part I of Studies in Semantics was Carnap's *Introduction to Semantics* (1942).

3. Carnap to Morris, Copy to Neurath, January 31, 1940.

Rudolf Carnap
Faculty of Exchange
University of Chicago
CHICAGO, ILL.
January 31, 1940
Prof. Charles W. Morris
Emerson Hall
Harvard University
Cambridge, Mass.

Dear Charles:

In the meantime you will received my letter of Jan. 23rd and Ina's of 26th.

I am returning Neurath's letter. I am not sufficiently acquainted with Gerard's work and views in order to pass judgment about him and his suitability as an author for the Encyclopedia. But I am willing to accept your suggestion that we ask him. I am simultaneously writing this to Neurath by air mail.

I am surprised to learn that Santillana's ms. is already finished.[37] I was not quite satisfied with certain features in his previous abstract concerning the development of mathematics, and especially the nature of geometry from the point of view of science. I believe that at the partition of the subject matter had not been cleared up, and, therefore, I said (I believe wrote so to Neurath) that I wanted to postpone my remarks S. would write a new abstract. But I have not got anything from since that time. I hope, I may now make these remarks when I shall have read the ms., although they might possibly involve greater changes in the ms.

I agree with you that the next unit of the Enc. should not be announced until the present one is much nearer to completion.

Cordially,
C.

[37] Giorgio de Santillana (1902–1974) was an Italian-American historian of science. He wrote a monograph (to be more precise: the first part of it) for the *IEUS* with Edgar Zilsel (1941).

4. Carnap to Neurath, November 1, 1940.

(ASP RC 115-07-53)
RUDOLF CARNAP
~~DEPARTMENT OF PHILOSOPHY~~
~~UNIVERSITY OF CHICAGO~~
~~CHICAGO, ILLINOIS~~
Cambridge, Mass., Nov.1, 1940.
15 Everett SA

Dear Neurath:

We all were immensely glad to learn of your and Mieze's escape from Holland.[38] And now we are looking forward to the day when it will be possible for you both to come to this country. I have written testimonials for you and Mieze and sent them to Kaempffert [marginal addition: *already on August 13*].[39]

It was most comforting to see that the elephant still bears flowers and grinned quite happily. Some difficulties seem to have developed in the plan to get you in on a non-quota visa via Johnson's committee for refugee scholars.[40] I do not see clearly in the matter, but Kaufmann's letter do[es] not sound helpful.[41] I have also warmly recommended Waismann's case to the same committee, no action has been taken on his case either. Kaufmann is devoting himself greatly to the task of pressing the cases of our friends, but he does not seem to be very influential. Since Kaempffert planned a different mode of action, we hope very much that he will be [a] success in your case.

[38] Mieze was Marie Reidemeister, first a collaborator on Neurath's ISOTYPE project, and later (in 1941, see letter 6 below) his wife. Marie (often mentioned as "Mary") edited later the English translation of Neurath's selected works; see Neurath and Cohen (1973) and (1983). Regarding Otto Neurath's papers, where possible, we will refer to the English translations as they are in the mentioned selected works.

[39] Waldemar Kaempffert (1877–1956) was Neurath's cousin and the science editor of *New York Times*, advertising Neurath's ideas frequently (also in the American socio-political journal, the *Survey Graphic*).

[40] Alvin Saunders Johnson (1874–1971), an American economist who was a co-founder and director of The New School for Social Research. Johnson helped many European refugees, among others the Vienna Circle member Felix Kaufmann, who emigrated to the U.S. from Nazism.

[41] Presumably Felix Kaufmann, member of the Vienna Circle.

So much has happened since we last heard from each other that it is hard to select a beginning. Perhaps you want to hear about friends. Kotarbinski,[42] Rougier[43] and Peppi Frank[44] have been invited by Johnson to come here – Frank finds it hard to get a visa though, since he does not fulfill the requirement of previous two years of teaching. Jørgensen has been taken into consideration by the committee, likewise Grelling,[45] who is somewhere in a camp in the Pyrenees; nobody has heard from Jørgensen since the invasion. Ness has not been molested so far.

I have finished Semantics II[46]; though the Press would like to bring both volumes, they say they cannot do so without considerable subsidies (something like 1500 dollars per volume). The same applies to the publication of a translation of my Logistics.[47] Meiner has asked me what he is to do with the 60 copies of the Sonderausgabe des Kongressberichtes and has sent me a copy of the bill for 336

[42] Tadeusz Kotarbinski (1886–1981) was a Polish logician, one of the leading members of the Lwów-Warsaw School. He is known for his special form of realism, called reism.

[43] Louis Rougier (1889–1982) was a French philosopher of science who was responsible for the French advertisement and propagation of the Unity of Science Movement. He was also the organizer of the International Congresses for the Unity of science, and was a member of the Committee of Organization of the *IEUS*. His 1919 book on the philosophy of physics was translated into English in 1921.

[44] 'Pepi Frank' was the famous architect Jozef Frank, one of Philipp Frank's younger brothers, who held lectures in the *Ernst Mach Society* in Vienna, and worked with Neurath in his *Museum of Society and Economy in Vienna* (*Gesellschafts- und Wirtschaftsmuseum in Wien*). On Jozef Frank, see Thurm-Nemeth (1998).

[45] Kurt Grelling was a member of the so-called Berlin Group of logical empiricism under the guidance of Hans Reichenbach. Grelling was murdered in Auschwitz, on September 18, 1942.

[46] See Carnap's *Formalization of Logic* (1943).

[47] In 1938, Kurt Edward Rosinger (a philosopher from Harvard, later Princeton) translated Carnap's *Abriss der Logistik* into English. According to Richard Creath (1990, 255), Rosinger translated the book for Alfred N. Whitehead's son. Later Carnap wrote to Quine (October 25, 1938; in Creath 1990, 255) that University of Chicago Press would like to publish the English translation. Carnap wrote also to John Cooley (September 22, 1938; ibid. p. 256) that he aims at buying his mimeographed 'Outline of Symbolic Logic' because he is "writing a new and entirely changed edition of [his] 'Abriss der symbolischen Logik.'" (Cooley's manuscript was published in 1942 (and became a classic text) as *A Primer of Formal Logic*.) Both the rewritten German version (*Einführung in die symbolische Logik*, 1954) and the new translation (*Introduction to Symbolic Logic and its Applications*, 1958) appeared much later (see further letter 8. below). The book was translated, after all, by William H. Meyer and John Wilkinson. A new translation of the original *Abriss* will be published in the *Collected Works of Carnap*.

marks. He wrote that Hazebroek does not reply to him.[48] I do not suppose that you will have heard from Hazebroek, but I thought you might like to know about the matter. Meiner writes that he is willing to ship the copies to the individuals for whom they are destined if he gets an order to do so. It becomes clearer from day to day how vital your presence is for the Unity of Science Movement; without you, it simply does not seem to be a movement, but just individuals.

Cambridge is a place full of activities, learned and otherwise. We have two kinds of discussion groups, one on logic, which is necessarily small, and a large one where scientists are participating and which has the purpose of spreading the gospel.[49] Russell is here for this term, also Feigl, Tarski, Wundheiler[50]; Franks live in the same house as we.

Russell told me that he wrote to Morrison, Minister of Labor in London, in your behalf.[51] We hope that at the time this letter arrives, you are already released and that you and Mieze are married. Our very best gratulations and good wishes!

Tarski told me that Struik[52] (Prof. of Math. here, an old acquaintance of Frank, perhaps you saw him at the Congress) has some connections with President Cardenas of Mexico; he has succeeded in getting a visa to Mexico for a friend. If you have plans for going to Mexico and wish Struik to intervene, please let me know.

---

[48] Neurath presumably is referring to the Dutch Piet Hazebroek, (1907–1971), a mathematical physicist by profession, who was a friend of Neurath's secretary, Annie van Middelburg (later Van Ginkel), and thus belonged to Neurath's circle in The Hague. He was also a member of the socialist student group, called 'Politeia,' and worked with Shell from 1937. Though there isn't any recorded scientific publication of Hazebroek, it is not unusual for scientists working at Shell to have no official publications of their work even if they did research. See Merten (2007).

[49] By the "logic group" Carnap's presumably refers to those discussions that he had at Harvard with Alfred Tarski, Carl Hempel, and Nelson Goodman (occasionally with Bertrand Russell as well) in 1940 and 1941. For more details, see Frost-Arnold (2013). The other group was the so-called "Science of Science" discussion group; see Hardcastle (2003).

[50] Alexander Wundheiler (1902–1957), a Polish philosopher and logician, student of Tadeusz Kotarbiński at the University of Warsaw and belonged to the Lvov-Warsaw School. Wundheiler also attended the Fifth International Congress for the Unity of Science at Harvard in 1939. For more biographical and philosophical background see Artur Koterski and Thomas Uebel (2017). For the English translation of Wundheiler and Edward Poznański's important paper on truth in physics, see Wundheiler and Poznański's (1934/2017).

[51] Herbert Morrison (1888–1965) was a British Labour Politician, Minister of Supply (12 May 1940 – 4 October 1940), later Deputy Prime Minister of the United Kingdom (1945–1951).

[52] Dirk Jan Struik (1894–2000) was a Dutch mathematician working mainly in the United States (retired from MIT). He is known for his *A Concise History of Mathematics* (1948).

I am sending to Miss Stebbing[53] for you: 1. Zilsel's MS for Encycl., 2. a reprint of my "Testability." You will get my Encycl.-monograph together with all the other ones from the publisher. There is no other recent publication of mine. I shall send you the "Log. Syntax" as soon as you are settled down somewhere on this side of the ocean.[54]

To you and Mieze (we hope: Mrs. Neurath) our
love and best wishes,

yours,
Carnap

5. Neurath to Carnap, April 4, 1941.

(ASP RC 102-55-20)
64 Park Town, Oxford
4th April 1941

Dear Carnap,

What a pleasure to get my notes for the monograph in the Encyclopedia. That helps me very much in my attempt to reconstruct my own ideas on this matter. In the meantime I have learned a lot from modern literature I was in a position to consult here in the Bodleian Library.

We are very busy with our studies in different fields; first of all we are collecting [material] for the book TOLERATION AND PERSECUTION.[55] Many other things have to be done, too. We have now two technical collaborators who are in a position to use a printing press and we go on with full speed; we are buying books, collecting newspaper cuttings creating new files, so to speak the nucleus of a fine institute. We learned from friends that many people in London and elsewhere are interested in our work; I shall be invited to go to London to read a paper about actual problems in this field etc.

[53] Susan Stebbing (1885–1943) was a British philosopher, co-founder of the journal *Analysis*, and the first woman professor in England. She wrote many important logic textbooks during the 1930s, attended the *Unity of Science Congresses*, and helped Neurath during his English years. Though she also organized lectures for Carnap and Schlick in London, she was quite critical with logical empiricism. On Stebbing, see Beaney (2016) and Chapman (2013).

[54] See Carnap (1936–37); cf. further Carnap (1937) and (1939).

[55] "Tolerance and Persecution" in the Neurath Nachlass, 207/K.88. Wiener Kreis Archiv (Rijksarchief in Noord-Holland, Haarlem, The Netherlands). For a time, Alfred A. Knopf wanted to publish the manuscript as *Persecution and Brotherhood*. See Marie Neurath's memoirs in Cohen and Neurath (1973, 68).

The almond trees are blossoming, the birds are singing and the whole atmosphere is very nice, in spite of war and danger. I am buying my own books – I see Carnap's works in German and English [o]n the shelves, but I anticipate he will send me his books and articles very soon. Could you be kind enough to send me my EMPIRISCHE SOZIOLOGIE as soon as possible.[56] No copy is in the library here and I need it very much for my discussion class. Could you be kind enough to send me one copy of the yellow booklet and some of my articles of which you have reprints?

What do you think about Russell's last book?[57] It looks somewhat strange? Neo-neo-Platonic Mysticism – I think it is full of misunderstandings. He seems not to grasp our purpose. His position – an interesting coincidence – seems to be very similar to Schlick's conception in his latest days. How to safe the absolute remainder of relativism. Probably I shall write something about this theme.

I hope that Frank will be elaborate his monograph for the Encyclopedia and send the additional chapter on cosmology to Freundlich.[58] Please, press this lazy fellow to do something. Many thanks my dear Carnap for the money you provided for me. I hope we shall be very soon in a position to repay the sum.[59]

Please ask our friends to send us books, newspaper cuttings, reprints, letters etc. that we have our usual intellectual environment. We are completely recreated by our camp isolation and are prepared to work more than before.[60] It is some time necessary to create a good environment. Until now the Britons are very kind and helpful to us, especially the university people.

You did not answer my question why you do not intend to publish semantics in our LIBRARY OF UNIFIED SCIENCE. I asked you to cable me, whether you are prepared to do it or not. You see I prefer to make an agreement with a publisher who pays for the JOURNAL[61] and gives us free hand in publishing books in the library, as van Stockum did. I hope I shall settle the whole matter within a relatively short time.

---

[56] See Neurath (1931), translated as Neurath (1931/1973).

[57] See Russell (1941).

[58] See Frank (1946) and Finlay-Freundlich (1951).

[59] Many friends and colleagues have sent money and other materials to the Neuraths when they arrived to England. See Cohen and Neurath (1973, 72).

[60] When Neurath and Marie had to flee from The Hague, they used an overcrowded small boat, *The Seaman's Hope*. They were picked up by a British destroyer and since they had only German passports, they were imprisoned first in London and later one the Isle of Man; they were released on February 7 and 8, 1941. For further information see Cohen and Neurath (1973, 70–72) and Sandner (2011).

[61] Neurath is referring to *The Journal of Unified Science (Erkenntnis)*.

Tell me something about Harvard, about men and opinions, about Frank and Tarski etc., about the various scientists etc. We shall visit the US as soon as possible.

Our ideal is to live quietly but we have to meet many people and to attend lectures etc., sometimes I have to read a paper etc.

What do you know about Grelling? etc. We learned from friends that Prof. Joseph[62] (the Biologist) committed suicide at Vienna together with his wife, his wife's mother etc.

Paul stays as Goeteborg and tries to go to the US.[63] He asked for an affidavit but did not receive it until now. I hope he will succeed in this effort.

Please remember me and Mary to Ina. We should appreciate it if she would tell us something about the American life. Our only source of permanent information are [is] the New York Times.

We anticipate a long letter from you
with kind regards from both of us
ever yours

NB. Please have you no opportunity to help M. STRAUSS,[64] his grant will expire this month and he has no job. It would help him if he were in a position to tell the Society for the protection of Science and learning, that he has an opportunity to get something in the future in the US.

6. Carnap to Neurath, June 5, 1941.

(ASP RC 102-55-19)
R. Carnap
Faculty Exchange
University of Chicago
Chicago, Ill.
Cambridge, Mass., June 5, 1941.

[62] Neurath is presumably referring to the zoologist Heinrich Joseph (1875–1941).

[63] Paul Neurath (1911–2001) was Otto Neurath's son. He went to Sweden after he was released from the concentration camp of Buchenwald in 1939. See further letter 16. note 146.

[64] Martin Strauss (1907–1978) was a German physicist and philosopher, who tried to combine logical empiricism and the foundational searches of quantum mechanics. He was a persistent leftist (thus imprisoned twice during the 1930s in Germany), worked with Niels Bohr in Copenhagen and Philipp Frank in Prague, and he was associated with Hans Reichenbach and the Berlin Group during the late 1920s. In the major part of his works, he tried to extend Carnap's logical syntax project. His selected papers were published in 1972.

Dear Neuraths!

Congratulations to having merrily married Mary! We were very glad to hear about your new life, and about your optimism and courage.

Our year here at Harvard is nearly over (you had better send letters from now on to Chicago for forwarding). It has been a pleasant experience, the students being rather more sophisticated than the ones in Chicago, and an interested 'Science of Science' group consisting of members of various departments who were interested in our problems.[65] About the philosophers there is not much to say, they were very friendly toward me personally but not much in the way of discussions happened, except with Quine with whom I discussed frequently (together with Tarski). Frank has had a halt-time job at Harvard for the past year and expects that this appointment will be renewed for the coming year (in addition he has invitation from New York City College to lecture there the other half year).[66] Tarski has received a Guggenheim Fellowship for the coming year – his wife and children are still in Poland. Zilsel has been living on a grant for the past two years – his wife has had repeated nervous breakdowns and probably would be better off in an institution – or at least he probably would be better off with her in an institution. I have heard that he is rather depressed about it all, and he also seems to have difficulties in adapting himself.[67] It is getting harder and harder to find academic positions in this country. Not even Kelsen has found an appointment. One of the many reasons for this increasing difficulty is the cutting down of teaching personnel because of the reduced student enrolment due to conscription. The Harvard budget e.g. has been cut 10% and since they intend not to cut salaries, it results in not filling vacancies, appointing less tutors and instructors. This is also one of the many reasons why it seems impossible to find something for Strauss particularly as long as he is not in this country. If he were here, probably an organization could be induced to assist him financially, but overburdened as they are, they will in no case assume the responsibility to do so for someone who is not already here. You will have heard of the Johnson Action under which – with Rockefeller money – a number of refugee scientists have been called over. Kaufmann and I (together with Morris and friends)

---

[65] See Carnap to Neurath, letter 4.

[66] Frank become a 'Lecturer in Physics and Mathematics' in May, 1939; 2 years later, in the fall of 1941, he received his tenured part-time position: he had to teach somewhere else (usually in New York) every second semester. He became a member of the American Academy of Arts and Sciences in 1943 as the very first among the logical empiricist refugees. On Frank's reception and role in the United States see Reisch (2005, Chap. 11 and 15) and Tuboly (2017).

[67] Zilsel arrived to the United States in 1939 with his wife and son. Though he had some grants (e.g. a Rockefeller fellowship), he always lived in poor and insecure conditions; also his wife was treated occasionally in a sanatorium (with manic depression). After being isolated in poor conditions, Zilsel committed suicide on March 11, 1944 (in Mills College, San Francisco). See Raven (2003).

have done all we could to get you, Waismann, Jörgensen and a few Poles here under this scheme but no result whatsoever. My recommendation for Jörgensen has been refused because he still has a post, Waismann's case – I believe – is still pending; my recommendation also has been refused at first with the reasoning that he has a post; I have tried to argue that his position is not really a "post" but only a stipend etc. (though I actually do not know the particulars of his appointment) and have never heard about the final outcome. Will you, when you meet him, tell him about this stand of the matter? Kotarbinski actually got an appointment but refused to come – I believe because he has an old father in Poland whom he did not wish to leave. For the younger Poles – Lutman, Hosiasson – nothing could be done.[68] I have not understood the principle of selection of the Rockefeller+Johnson people but it seems fairly clear that on the one hand they want very well known names and on the other hand pull has an influence and not necessarily does scholarly merit speak the final word (at least that is the explanation I am inclined to give to the appointment of Rougier who is already here). But unfortunately it is not the sort of pull people like Morris, Nagel[69] and I are able to bring to bear. Pepi Frank has received an appointment – independent of the Johnson Action to the New School, Hanja[70] managed it in some way and he will come if he can secure passage. What happened with your own case I am not sure – I only know that the Rockefeller people (or somewhere else in the machinery) said no. Probably you have already better information from someone better informed. I am very sorry about it, because I think you should come and settle here. We all need you, and so does the Library, the Encyclopedia and the Journal. By the way: it is not that I did not want to have my "Semantics" [marginal addition: *& the Journal*] published in our Library. I offered it to the University of Chicago Press in the hope that the Library would be continued there and that my book could appear as one of its volumes. As I wrote you, the Press demanded a subsidy about equalling [matching] their out of pocket costs. In the meantime a new development has occurred: the Dept. of Philosophy here has allowed [awarded] me a grant toward the publication provided the book will appear at the Harvard University Press (the Press here also had demanded such a subsidy), and I am now

[68] Maria Kokoszyńska-Lutman (1905–1981) and Janina Hosiasson-Lindenbaum (1899–1942) were Polish logicians and philosophers, members of the Lvov-Warsaw School. Hosiasson (1940) is mainly known for her work on the so-called "raven paradox of confirmation." On Kokoszyńska-Lutman see Brożek (2017).

[69] Ernest Nagel (1901–1985) was an American philosopher, working mainly at Columbia University. He participated at the Vienna Circle's meetings at the turn of 1934–1935 and attended the *International Congresses for the Unity of Science*. Nagel wrote and edited many influential (text)books, among others *An Introduction to Logic and Scientific Method* (1934) with Morris R. Cohen and *The Structure of Science: Problems in the Logic of Scientific Explanation* (1961). He also wrote a monograph for the *IEUS*, *The Principles of the Theory of Probability* (1939).

[70] Hanja, or Hania Frank (1894–1967) was a former student, then the wife of Philipp Frank. On her life see Holton (2015).

negotiating with the Harvard Press about the publication. I do not think your idea about having it published in England is quite realistic. To send a manuscript there in these uncertain times seems to me inappropriate. And though I do not doubt your talent for finding a publisher there, I do not see any advantage in having it printed there. When we considered publication in Holland, one of the good reasons was that the printing costs would have been cheaper there than here. But printing in England would not be cheaper, and the fate of the manuscript and of the shipping of the books would be too uncertain. I also do not think that the English public at the time being would be interested in such highly technical material and that nearly all of the sales would be here. -- I have urged Frank again and again to write his encyclopedia contribution and each time he promises it for "in two months". -- I am sending you the yellow pamphlet and a number of reprints which I happen to have here – most of my things are in Chicago in storage and I can not get them before we are back in Chicago. Your "Empirische Soziologie" unfortunately has disappeared from my shelves and I do not know who has it -- I shall ask all suspects.[71] --- Grelling has a possibility to come provided he can find passage. Rockefellers refused his case also, but now were willing to invite him nominally i.e. Oppenheim[72] having promised them to finance their invitation (the invitation from them for the purpose of a non-quota visa); Grelling is still in a concentration camp -- but probably Hempels have written you about this already. -- A reprint of my "Testability" I have sent you on November 11th via Stebbing. I hope it has arrived safely, since it was one of my last copies. Please forward the enclosed letter to Hollitscher[73] after reading it.

All our love to Mieze and you,

Yours ina + C.

Yes, Russell's book is somewhat disappointing. But I think one can still call it empiricism though – as you say – it is the Schlick-denomination of empiricism.

Zilsel has asked to have his name mentioned on our letterheads [marginal addition: *as a member of committees*] of Congresses, Encyclopedia, etc. I believe that we can do that. I have asked Morris about it. What do you think? Z. wishes it in order to give some publicity to his name.

---

[71] See Neurath (1931/1973).

[72] Paul Oppenheim (1885–1977) was a German scientist and philosopher. He was a member of the Berlin Group with Reichenbach, but emigrated to the United States in 1939. He is mainly known for his various co-operations with Hempel (on types and explanations), Grelling (on Gestalts), Olaf Helmer (confirmation), Nicholas Rescher (Gestalt), John G. Kemeny (methodology) and Hilary Putnam (unity of science). See See Ziche and Müller (2013).

[73] Walter Hollitscher (1911–1986) studied philosophy, biology and medicine at the University of Vienna, and obtained a PhD under Moritz Schlick in 1933. During the 1930s he was a regular member of the Vienna Circle Meetings, but had to flee from Vienna upon the rise of National Socialism. After the war he returned to Vienna and became a full professor and published many articles and books e.g. on Marxism and the modern scientific world-conception.

7. Neurath to Carnap, September 21, 1941

(ASP RC 102-55-18)
21st September, 1941
24. Old Road. HEADINGTON. Oxford

My dear Carnap,

Many thanks for your congratulations. Now we are going on very well. No doubt about it. The fine loans we got from our friends enabled us to overcome the first weeks and to rebuild our office – now we have already finished our first ISOTYPE-film, a new type of films, on salvage for the Ministry of Information.[74] Another one on Blood Transfusion is in the making – a very interesting job and fairly paid. Mary likes this job too.

Our office looks already very impressive, I bought many books and get them partly rather cheap. I am reading and reading and writing. My discussion class at the University I had last term on LOGICAL EMPIRICISM AND THE SOCIAL SCIENCES[75] stimulated me to write something about the general principles of Language Making. I am sending you my article published by the ARISTOTELIAN SOCIETY.[76] I am very interested in your reaction and hope you will write me about it.

I am writing my Encyclopedia Monograph. I think it will be an interesting chapter. I am collecting material for my book on Toleration and Persecution.[77] Mankind is a sad thing, believe me that, if you should not be already convinced of this fact. We are reading many historical books. Killing and tormenting one another, that is man's business through the centuries.

In October we shall have a Unity-of-Science conference here.[78] Various people are interested in it, in spite of the war. Prof. Russell[79] will come from Birmingham.

---

[74] Otto and Marie Neurath was approached by the documentary film producer Paul Rotha shortly after their release from the internment camp. For several years, the Neuraths made animated pictures on various issues for documentary films, such as "A few Ounces a Day" (dir. by Paul Rotha, 1941) on salvage, and "Blood Transfusion" (dir. by H. M. Nieter, 1941); see Cohen and Neurath (1973, 72–74), and Neurath (1946). Cf. Sandner (2014) and Boon (2016, 172–180).

[75] See "Logical Empiricism and the Social Sciences" in the Neurath Nachlass, 206/K.84. Wiener Kreis Archiv (Rijksarchief in Noord-Holland, Haarlem, The Netherlands). See further letters 13. and 26.

[76] See Neurath (1941/1983).

[77] See Neurath to Carnap, letter 5. note 55.

[78] Because of the war many European scholars were unable to attend the *International Conference for the Unity of Science* in Chicago, so Neurath organized a small conference, "Terminology," with Susan Stebbing and Joseph A. Lauwerys in Oxford (October 2–5, 1941). See Stadler (2001/2015, 192–193).

[79] Presumably Leonard J. Russell (1884–1971), a British philosopher, and President of the Aristotelian Society and of the Mind Association (1932–33). He succeeded C. D. Broad at the

He discussed our ideas in the Aristotelian Society years ago relatively well.[80] We met him in Paris. Whether Strauss, Waismann etc. will read papers, or not I do not know, because the invitation are just sent out. There is always some discussion atmosphere, Pear[81] the Psychologist came twice and we discussed many problems for hours. There are also younger people interested in unified science or sociology. You see, we have many reason to be content. Now the problem of housing is solved too, in the beginning we had only one room, then two and now we were lucky enough to get a very small house as subtenants for some months. Sky and trees as environment, only a few houses. The bus goes from there directly to our office in the Institute of Social History (Cole,[82] he is interested in planning and writes together [with] his wife detective stories, you find one of them in the Pengui[n] series).

We have a great many acquaintances and friends in England, much more than in Holland. I read papers to students and scholars in Nottingham (London University, Educational Department), Exeter and shall read a paper in October in Cambridge.[83] The interest in our visual education work is apparently increasing. The intellectual and emotional atmosphere is OK. If all things remain as they are it would be a[l] right – more one cannot expect.

I am glad that you had such a good time at Harvard. Such is life, successively we, the opposition, become classics. Now we form a branch like others, as I learned from LAIRD's Introduction into modern philosophy.[84]

Please, give me Tarski's address. I would appreciate it very much if you were kind enough to give me addresses of other people, too. We have to reconstruct our files. Until now we have about 200 addresses, but that is only a fraction of our

---

University of Bristol, but later he went to the University of Birmingham. Leonard Russell corresponded with Neurath between 1941 and 1945.

[80] Neurath presumably refers to a Symposium on "Communication and Verification" with L. Susan Stebbing, Leonard J. Russell and A. E. Heath (1934) published in the *Supplementary Volumes* of the *Proceedings of the Aristotelian Society*.

[81] Tom Hatherley Pear (1886–1972) was a British psychologist (at Manchester) and the President of the British Psychological Society.

[82] George D. H. Cole (1889–1959), a political theorist, economist, historian, and Margaret Cole (1893–1980). After Neurath was released from the internment camp, Cole was helping him settling down in Oxford. See also letter 13.

[83] See "Logical empiricism (and everyday problems). Lecture at Bedford College, Cambridge, 19.11.1941." In the Neurath Nachlass, 201/K.50. Wiener Kreis Archiv (Rijksarchief in Noord-Holland, Haarlem, The Netherlands).

[84] Presumably John Laird (1936)'s *Recent Philosophy*.

original files we had at The Hague. I should be glad to have addresses of HELMER, WEINBERG,[85] ROUGIER, etc.

All our love to both of you,
with kind regards
Neurath

Poor Zilsel. He had always difficulties in adapting himself to other people, I think that is connected with his stomach difficulties. I always thought of that when I was angry with him. But I know there are always problems which cannot be solved by understanding their sources. I was therefore very glad to give him a monograph in the encyclopedia.[86] I hope it will help him a little to get new connections.

I hear[d] Kelsen got a job in the meantime – or not? Here the number of students is very reduced, as you may imagine. We are glad not to depend upon on [sic] universities in these times. Strauss has now an appointment, but rather disappointing from scientific point of view.

I did not expect much for me from Johnson or the Rockefeller people, when only a general action is in question. I did not press anything in connection with the US, because we are really happy here and would be glad to remain here and to help, as far as we can, in fighting the Nazis. Here is, so to speak, the front. And even in the future. The nicest would be, to stay here and to visit the States every year for 4 or 5 months. Our experience is, that we always find a realm of our activities when we are able to execute some unusual things which are our specialities. I think, if universities should be interested in me, it would be in connection with particular problems of social sciences, unified science or visual education. But we are not essentially interested in teaching, much more in science making and preparing educational aids. Rougier is much more the man who tries to get university posts and who is adapted to this business and what is around of [sic] it, therefore it is in accordance with the harmony of the world that he will be protected and supported by official committees etc. Such is life. You cannot want to be free and independent and simultaneously to be classified as a part of the globe which has to get certain prepared jobs. Johnson and other people are very conventional. I am not even conventional in my studies and my attitude but not opposed to collaborate with very different people. Johnson is at very good terms with me and I was glad that helps us very much in our visual education activities (he is member of our American Committee) but I did not assume that he would like to have me with him in his school, if I did not c[o]me as a representative of a strange opinion – then he would be interested in me, but not in me as a refugee only. Many thanks that you and Morris and other friends were busy to help me in this respect. Fortunately it was sufficient that I got the loans and that I am very

[85] Julius Weinberg (1908–1971) was an American philosopher, who obtained his PhD in 1935 at Cornell with a thesis on logical positivism, published as *An Examination of Logical Positivism* (1936). Though this was the very first book-length treatment of logical positivism in the United States, he became widely known for his *A Short History of Medieval Philosophy* (1964).

[86] See Carnap to Morris, copy to Neurath, letter 3. note 29.

pleased by your friendship, that is nectar for me, sweet drink – or soft drink, if you prefer this term, grape fruit juice or something like that – but in great quantities, too. I like things more in great quantities, as Ina knows.

Waismann does now act as a substitute for a metaphysical Professorship, but he is free to read what he wants. He tries to go to the US, as far as I know. I think he is here not really happy – as we are, e.g.

I am busy with reconstructing JOURNAL and LIBRARY here, because I want to see ENGLAND (as Europe's representative now) in the game. Why should all business of unified science be done in the US? The locality of publication is of importance. We have to rebuild Europe, do not forget it. The yellow pamphlets and other things are of great value for me. I now got: Psychologie und Einheitswiss. Emp. Soziologie, but I need: Was bedeutet rationale Wirtschaftsbetrachtung, the history of the Viennese circle (HERMANN) in French, reprints of my articles, then books and articles by a certain Carnap.[87]

We shall now form – as already discussed – and additional editing committee (advisory committee, or something like that) for our JOURNAL. Then shall invite ZILSEL, of course.

8. Carnap to Neurath, January 27, 1942.

(ASP RC 115-07-54)
RUDOLF CARNAP
DEPARTMENT OF PHILOSOPHY
UNIVERSITY OF CHICAGO
CHICAGO, ILLINOIS
January 27, 1942.

Dear Neurath:

We were very glad to get good news from you in your letter of last September. I see that you are as active and as optimistic as always and I do not doubt that you will have good success in your activities. I hoped that you would come to this country at least for a visit of several months. Therefore I delayed sending your reprints. Since, however, it seems uncertain when you will come, I am now sending to you in five small packages four reprints of myself and 31 reprints of and about yourself, inclined "Cercle De Vienne", and "Rationale Wirtschaftsbetrachtung", for which you asked.[88] I suppose that you have your and m[y] publications in Psyche

[87] See Neurath (1933/1987), (1931/1973), (1935/1987) and (1936/1981).

[88] See Neurath to Carnap, letter 7. note 87.

Miniatures,[89] Encyclopedia and Erkenntnis; if this is not the case please let me know, then I shall send you whatever I have. I address the packages to the Institute of Social History.

I read with great interest your article in the Aristotelian Society.[90] I am, of course, entirely in agreement with your general attitude, but there is also a number of particular points where I have doubts or objections which I should like very much to discuss with you, whenever we meet again. For instance, I am not quite satisfied with your reply to Russell. Although Russell, in his book, adds to his empiricism certain modifications, with which I can not agree, his criticism of your formulations is made on the basis of empiricism. And although he is not right with respect to what you actually mean, he is not quite wrong with respect to your formulations, which as I predicted would provoke a criticism of this kind. Russell does not at all think of "absolute truth in itself" (p. 147), but thinks of the empirical procedure of taking experiences as a basis for assertions. I should like you to read again my small paper on truth and verification (Congress 1935).[91] I am sorry I have no reprints of it. Unfortunately, it seems entirely forgotten; if for instance, Ayer had read it, he would have avoided his terribly misunderstandings of my views in his latest book.[92]

Here some addresses; Tarski (Dept. of Math.Harvard), Helmer (496 Hudson St., New York City), Dr. Julius Weinberg (359 Thurman St., Zanesville, Ohio), Rougier (St. Johns Coll., Annapolis, Md.). Kelsen has no job, not a stipend from Rockefeller Foundation.

We hope soon to get more detailed news about your plan with the Journal and the Library. If the Library is published in England, what would you think about an English translation of my "Introduction to Symbolic Logic" (Second Edition, entirely rewritten and much larger) as a volume in it?[93] Rosinger began the translation; but since the University of Chicago Press would not commit themselves definitely and did not make a contract with him, he stopped working at it after having translated about one third. If you agree, I shall ask the other editors of the Library for their opinion.

---

[89] In the *Psyche Miniatures General Series* (edited by Charles Ogden) were published Neurath's *International Picture Language* (No. 83, 1936) and his *BASIC by ISOTYPE* (No. 86, 1937). Furthermore Carnap's *The Unity of Science* (No. 63, 1934) and his *Philosophy and Logical Syntax* (No. 70, 1935): all of them in London by Kegan Paul, Trench, Trubner.

[90] See Neurath (1941/1983) That paper contains Neurath's response to Russell's *Inquiry into Meaning and Truth*, mentioned by Carnap above, one sentence later.

[91] See Carnap's "Wahrheit und Bewährung" (1936). Modified English translation as "Truth and Confirmation" (1949).

[92] Presumably Ayer's (1940) *The Foundations of Empirical Knowledge*.

[93] See letter 4, note 47.

My "Introduction to Semantics" will appear in February I hope.[94] I shall send you a copy as soon as it comes out. The second volume (Formalization of Logic) will soon go to print.[95] I am now working on a system of modalities, and also on probability and degree of confirmation.

When shall we see both of you here? We should like very much to have some good and detailed talks about you and friends and the whole world and the future. Our love to both of you.

Yours, Carnap

9. Neurath to Carnap, March 2, 1942.

(ASP RC 115-07-55)
21 old Road, Headington, Oxford.
2nd March, 1942

My dear Carnap,

Your two parcels with reprints were the nicest gifts I can imagine. Some of the reprints I could get only in the British Museum, others not at all. Most of them are of some use for me in writing my monograph for the Encyclopedia. All things are going on very well for us – we are as happy as one can be in times as sad as these are. We have now a library again, files, notes, newspaper clippings, etc. lettertypes, blocks of our ISOTYPEsigns etc. Our well furnished house with nice view on meadows and cows, garden, etc. And then, we like the democratic Anglo-saxon atmosphere very much. I shall send you my article INTERNATIONAL PLANNING FOR FREEDOM,[96] in which I explain, that a certain muddle is connected with all democracy and we should not complain of muddle and then want to have a democracy, too. On the long run, I think democracy is much more stable, you have not to fight suspicion etc. I just finished again the story of Aseff,[97] what a result of absolutism .... terrible. In the US you have the Lincoln tradition and others. I am much interested in the history of all these phenomena – the centre Holland and England, Merchantmen the bearer[s] of tolerance. In Prussia the 'Junker', the Prussian 'Teutonic Knights' etc., an old tradition of suppression. Think of the language in which not a few Germans were accustomed to speak of Poles, as of slaves. Problems after problems, when the war will be over.

---

[94] See Carnap (1942).

[95] See Carnap (1943).

[96] See Neurath (1942/1973).

[97] Perhaps Yevno Azef (1869–1918), who was a social revolutionist and a double agent. On this see what might have been available to Neurath as well, Nikolajewsky (1934).

I hope you get my Aristotelian Society paper.[98] I should like to hear from you what you think about. There are some difficult points.

I hope the JOURNAL will be settled very soon. Morris' cable arrived so late, that in the meantime, we got a particular license regulation, now the publisher is going on with this thing.

We are depressed by Susan [Stebbing]'s illness. You wrote me that Hanja did not feel well (or wrote Morris of that?) What about her? I tried to get from Frank an answer dealing with his own monograph on physics. I hope you press him, too, and with Freundlich's monograph.[99] After Freundlich's death we have to publish it with an introductory chapter by Frank and a last chapter, fitting the whole into our pattern. I hope that you will be able to force Frank to finish his job. Poor Freundlich. I met him last time in Amsterdam, he asked me about my opinion on his decisions whether he should go to Scotland or to US. Poor Fellow.

My son seems to successful at Columbia University.[100] Most people, we know are safe – Grelling and Walter Fraenkel[101] in France, do you know something of Grelling?

Now I think of substitutes for monographs which cannot be finished, now, Joergensen, Tinbergen, I hope Morris will write me about the Biologists.[102] Now I am thinking of the details of the following 60 monographs. I should like to see Brunswik[']s paper as soon, as possible.[103] I thought it is in the press. I only got Zilsel and Santillana.

We shall make an editorial board for our Journal and then invite, Zilsel, Nagel, Hempel, Feigl etc.

On the one hand – It needed about one year to be in the old position again, to continue all work, to have collaborators etc., on the other hand, many people say,

[98] See Neurath (1941/1983).

[99] See further Neurath to Carnap, letter 14.

[100] Paul Neurath was studying sociology and statistics at Columbia University and thought statistics at City College, New York, between 1943 and 1946.

[101] Presumably Walter Fraenkel (1879–1943) who was a Jewish painter and the husband of Hans Hahn's younger sister, Louise Hahn. Fraenkel was brought to the Drancy internment camp near to Paris – where Grelling was interned as well – and later to the concentration camps of Sobibor and Majdanek in Poland, where he was killed in 1943.

[102] In the 1937 plans of the *IEUS*, there were two monographs on biology: one by Felix Mainx and one on formal biology by Joseph H. Woodger. Mainx's monograph appeared in 1955 and Woodger wrote about *The Technique of Theory Construction* in 1939.

[103] See Brunswik (1952). For many years the psychology monograph was to be written jointly by Brunswik and Arne Naess under the title "The Theory of Behavior." See Morris (1937).

that we in a relatively short time rebuilt our 'fabric' – you can look at all things from various angels. We are happy, that is also a point of view, but the world science is gloomy and it will need much time before we shall be victorious. And afterwards, will there be a more stable world order? I am very interested in studies on this subject and accepted therefore the invitation to be on the editorial board of the NEW COMMONWEALTH QUARTERLY. I analyze the scientific arguments and think one should speak of 'international relations', 'world organization' etc., and drop the terms 'international law', 'world economy' etc. etc. If you read papers and books on this subjects, they are a mixture of history, proposals, and I do not know what, mostly including OUGHT-TO-BEs. These you can avoid by transformation, but not the 'natural right' etc.

What is going on with your books, articles etc., what happens with our friends? How the war affects your and their life?

Dear friend, we think thankfully of all friends, who did what they could to support our release. It was boring and wasting time, this internment business. We were not so bad off, because we were happy [marginal addition: *not*] to be with the Nazis – and after life-boat experience you look differently to many things. Robinson Crusoe ------- We collected experience by reading in our youth, now we were able to apply this experience to our personal life. Firstly separated, then Crusoe and the man Friday met, then release, as it were, departed from the island (in our case Isle of Man).[104]

We are making films and do so something for the war effort. We should like to do more for this purpose. I shall be glad to hear from you in the near future. With kind regards from both of us to both of you,

every yours

[Otto Neurath]

By the way: Can you or somebody else, send me a copy of my HARVARD paper, 1939. I need it urgently.[105] Each member of the congress got a copy and I think some remained undistributed.

10. Carnap to Neurath, June 24, 1942.

(ASP RC 115-07-56)

RUDOLF CARNAP
Faculty Exchange
University of Chicago
CHICAGO, ILL.
Chicago, June 24, 1942.

[104] The fictional characters of Crusoe and Friday featured prominently in Neurath's private language argument in "On Protocol Statements," see Neurath (1932a/1983, 96–97).

[105] See Neurath (1939/1983).

Dear Neurath:

Thanks for your letter of March 2. You got two of my parcels with reprints, and I hoped the other three, which I sent to exactly the same address, would also arrive soon. But short time ago I got back two of them, with remark: "Not known. Refused at University Registry, Oxford". Instead of inquiring a little further, they send them all the way back across the ocean! You said once that a little bit of easy-going inefficiency ("Muddling through") is quite nice. I am not so sure whether I like it much. On June 5, I sent you once more the two packages, this time to your private address, not to the Inst.f.Soc.Hist. I wonder what happened to the fifth package. And then I sent my book "Semantics" to the Inst. address on May 1; perhaps you could inquire there and give them your address.

I shall be free from teaching duties for one year from now on, with the help of the Rockef. Found., for continuing my work in semantics. Is it not remarkable that even in times like these, purely theoretical research is encouraged and supported? A second small book has gone to print and will, I hope, appear in the fall ("Formalization of Logic"). The work in the coming year will be devoted to new problems, among them probability and degree of confirmation. I do not yet know, where we shall be. Letters will always be forwarded.

I do not know whether you got my letter of Jan.27.[106] Will the Library be continued in spite of paper restrictions? And if so, what would you think of publishing in it my "Introduction to Symbolic Logic" (it has been entirely rewritten; so far one third of it has been translated).[107]

Please write me your son's address, and tell me if I can do anything for him.

Last year efforts were made to bring Grelling to this country. They seemed to develop well but had then to be interrupted because of the war.

I read with great interest Kaplan's Dr.-Thesis.[108] It seems the first systematic investigation in pragmatics from the present point of view, i.e. with explicit distinction between pragmatics, semantics and syntax. I talked it over with him in detail, especially the need of making more clear the explanations of his new concepts. He intends to work it over. I expect it to become a very good and interesting book. I think we might consider it for the Library. He could perhaps send you an abstract of it.

I talked with Morris about all the questions concerning the monographs in preparation or to be planned for Encycl. I and II. I suppose he will write you about it. I am

[106] See Carnap to Neurath, letter 8.

[107] See Carnap to Neurath, letter 4, note 47.

[108] Presumably Abraham Kaplan (1918–1993), an American philosopher who worked especially on ethics and the behavioral sciences; he was a student of Carnap in Chicago, and defended his PhD ("The Plurality of Language Structures") written under the direction of Hans Reichenbach at UCLA in 1942. Later he published an important paper in Paul A. Schilpp's *The Philosophy of Rudolf Carnap* volume on "Logical Empiricism and Value Judgements" (Kaplan 1963). See also Kaplan's (1991) memoires on Carnap.

very much in favor of getting the collaboration of Reichenbach and Feigl. Quite a number of monographs and supplements will have to be written, and R. and F. seem to fit in especially well.

I am looking forward to your article "Int. Planning f.Freedom". In your letter, you put the alternative as: muddling vs. democracy; and then of course we all prefer the second. The question is, whether democracy is actually incompatible with efficient planning and regulation.

We are glad to hear that you have a nice house, both of you living and working as happily as the world events allow, that you are again working at six or a dozen of projects at once, as in the old times. Our best wishes for you both, personally and for your work and for the big events upon the outcome of which all our lives depend.

Yours,

Carnap

11. Neurath to Carnap, July 17, 1942.

(ASP RC 102-56-04)

17th July, 1942

21 Old Road, Headington, Oxford

My dear Carnap, (and Carnapesse),

We always appreciate very much to get some news from you. We are living here happily in a small house (4 rooms, we are alone), with garden, flowers, fruit, vegetables, a cat visits us from time to time and a dog participates through the railings with remainders of our meals. We have many friends and acquaintances, I am lecturing in various places on Visual Education, Logical Empiricism and on, what is called 'social sciences' – they are more 'social' than 'sciences'. Nevermind .... The film making business is continuing, we are very successful in that, our first film was purely diagrammatic, now we are making diagrammatic scenes for other films, on food, tuberculosis, etc. The Ministry of Information is presenting them to the public. Some go to the US. I am writing my monograph for the encyclopedia and I think I could improve it very much – how I can fit together my arguing on 'true', 'statements on statements' etc. with your formulations, I do not know, perhaps you will find a way to put together these various strains. I think in Tarski's and Popper's arguings, which much influenced you is an essential metaphysical element of absolutism. But I shall wait for your books.

Most of the announced booklet and papers arrived, not all of them, Rationale Wirtschaftsbetrachtung fortunately and also one of my little Chinese stories, the only I have now, but NOT the Cercle de Vienne.[109] If you find a second copy of that,

[109] On these see booklets and papers, see Neurath to Carnap, letter 7. note 87. Neurath has published six papers about Chinese teachings and stories under the pseudonym "La-Se-Fe" in the 1920s. They were translated as "Six Lessons" (in Cohen and Neurath 1973, 84–100).

please send it me. In general we are very successful in rebuilding our library, we have some books by Mach, Boltzmann, etc. of course: Voltaire, Diderot, Condillac etc., and new books, such as Carnap, Russell, Weinberg, Stuart Chase[110], etc. The department [of] literature is a good one, history, statistics etc., I got most of my own books and even reprints in bookshops, our big atlas e.g. we are working as once in May with our three British collaborators. We came just back from a Visual Education meeting in Exeter with people from various parts of the country, including Lancelot Hogben, who referred on [to] his new AUXILIARY language INTERGLOSSA.[111] This and Basic are the only two auxiliary languages which seem worth to account for. The principle is similar to Peano's Latina sine flexione, but consistent. It is Chinese based on western roots. Question, imperative, etc. formed by means of particular words 'que' means question, then the statement as usual. Very simple. Lancelot Hogben analyzed the language making very carefully, it was for me and also for a philologist who was with us a great pleasure to listen to him. He selected the roots from internationally known worlds and found out that there are about 2000 such roots. If somebody does not know these roots, it is useful for him to learn them, even for his own language. Let me give some examples MICRO-PHON, PHONO-GRAPH, GRAPHO-LOGY, etc. Hogben says, therefore MICRO means small, GRAPH means writing etc., the roots are mainly [G]reek or [L]atin. A very fine idea. And no IDIOMATIC RULES.

BASIC is fine for reading and listening but not for writing, because you have to learn the idiomatic allowed combinations, that is sometimes more difficult than to learn separate words. This is the weak point in BASIC. Therefore BASIC remains more for reading and listening. INTERGLOSSA is an ideal language for writing scientific things and other things, the examples are very illuminating. The manuscript Hogben's is devoted to Ogden, whose work he much admires.

Please ask MORRIS he and you should be allow me to ask Hogben (perhaps together with Ogden) to write for our first two volumes of the encyclopedia a monograph FOUNDATIONS OF LANGUAGE MAKING. Hogben is prepared to do it and we shall learn a lot from him about comparative studies of language tools AS TOOLS. I think you are interested in such things very much. I should like to write a short introduction (perhaps you, too?) telling a little about our own problems, and how they are related to that. PLEASE ANSWER TOGETHER WITH MORRIS THIS PROPOSAL IN THE AFFIRMATIVE. If you had sufficient money send a

---

[110] Stuart Chase (1888–1985) was an American economist and polymath. Quite unfortunately, he wrote in his *The Tyranny of Words* (1938, 160) that "Dr. Neurath is a kind of pioneer in semantics. He believes in going to things wherever possible, rather than to words."

[111] Lancelot Hogben (1895–1975) was a British experimental zoologist and medical statistician, who also worked on popularizing science and international languages. In the late 1940s and 1950s Hogben acted as "honorary editor" of many ISOTYPE volumes of Otto and Marie Neurath. Their typical topics were *The First Great Inventions*, *How the First Men Lived*, *Visual Science*, *Living in Early Times*, *Living in Villages and Towns*, *Living in the World*. See further Hogben (1943).

cable with YES. I am sure this would be a GREAT SHOW. The manuscript of Hogben's book is full of stimulating remarks and you learn more about language in this way than usually. I think that brings LIFE into our business. Perhaps instead of the monograph by Tinbergen or another one. I am looking around, what can be done more for our encyclopedia, Journal, library etc. THERE ARE VERY GOOD PROSPECTS, but I shall tell better about that if they results of my efforts there realized. I think we shall perhaps get a kind of centre for our scientific enterprise, as we wish it to get, without any obligation to anybody. I hope so. Wait and see, please. We should not be impatient – there is a war on and a very serious and difficult one. I am astonished how well scientific work goes on in spite of all that and how many people are supporting our efforts.

I have MINIATURES, ERKENNTNIS (only partly) ENCYCLOPEDIA I should like to get reprints from other people, too. Please, be a nice fellow and tell that other people. Describe my situation, Hitler's gang has all my reprints, manuscripts etc., therefore I am thankful for all gifts, also for books of course. It is difficult to get American books and periodicals.

Please read again RUSSELL and my Aristotelian paper.[112] I shall tell you something in detail and I hope you will answer in detail.

RUSSELL, MEANING AND TRUTH,

p 14 he speaks of 'error' and 'knowledge' who is in the chair? Russell personally?

p 15 who is the 'hypothetically inerrant recorder of what actually happens'? We know this type of arguing, Laplace's demon etc.

p 15 'naïve realism' is a very late stage. A naive man does not know the split and therefore not the elimination of that split, the making of the split again etc.

p 15 split into person and things, effects made by things etc. – the old 'duplication' sufficiently criticized by Avenarius[113] etc.

'observation happening in him' the same.

p 16 'avoid error' implies somebody knows what the truth is.

p 22 if that is not TRUTH in the worst style, then I do not know, to what extent you agree with me: 'a proposition may be true although we can see no way of obtaining evidence ----'

p 50, can you transform into a tolerable statement: 'we do now know our present experience'. I think that is Schlick redivivus.

p 64 what about the assertion which has no antitheses. Can you explain that?

p 70 I think in Vienna already we eliminated the expression 'as many words as facts' as if words were no facts.

[112] See Neurath (1941/1983) and Russell (1941).

[113] Richard Avenarius (1843–1896) was a German philosopher, influencing Neurath and Philipp Frank. Besides Frank and Ernst Mach, Avenarius was attacked in Lenin's *Materialism and Empirio-criticism* too because of his positivist line of thought.

I avoid the term fact altogether, as you know. How you translate this absolute truth expression?

p 77 I see no difficulty to speak of 'desiring man', 'believing man' etc., why Russell speaks of 'desire'?

p 92 what is your translation of 'fact in the world'?

p 104 how you are translating 'no chronometer is exactly right?' This statement belongs to the language of absolute truth, there is some judge in the chair knowing the RIGHT time and then he declares no chronometer is right. I do not understand why not a chronometer should be 'right', as it were "by chance" IF I ACCEPTED THE JUDGE IN THE CHAIR. Without a judge in the chair I can only say that we select certain chronometers and perhaps we prefer a group of chronometers and form a MIDDLE or something like that, but why should not by chance one just have the hands at the figures of the middle? etc.

p 106 wavelengths have nothing to do with colours, you may see colours e.g. as result of a blow without any w[a]ve lengths in Russell[']s terminology. W[a]velengths are relateable to all sense fields, a blind man may treat optics (we agree in that)

p 108 characteristic that GOD and his impartiality is introduced ... ABSOLUTE TRUTH

p 0P0 do you know what is a 'non-mental' world? All these expressions are possible in a transcendent realm of truth, where one may distinguish between mental and non-mental etc. but we have only protocol statements in which nothing of this dualism is be found, as far as I can see. Only from the 'outside' such a distinction may be introduced.

p, 111 what a strange 'pure physical world' without 'words? Only in the TRUE WORLD such distinction has any place, not within the protocol realm.

p. 32 'minimum' ideology is also related to a certain absolutism of Truth – as if the whole body of statements were known are knowable.

p 133 never completely certain .... true. How to discuss such a statement? Behind the multiplicity of statements and the fact that were are selecting some 'ACCEPTING' them Russell imagines one particularly TRUE statement, we can never completely identify as what it is ....

p, 140 what I say is, that <u>discussion</u>, arguing, thinking etc. is SPEAKING, therefore I have fact-terms, statement-terms, statement of statement terms, etc. BUT NOT FACTS, etc. AS ELEMENTS OF DISCUSSION. A certain Carnap was sometimes partly of this opinion, wavering, but nevertheless more of this opinion. Now he seems more to be impressed by the other side of wavering.

p, 141 Neurath (and a certain Carnap with a certain time index, or always, I hope so?) did not speak of reality he denied not a comparison of reality and propositions, but in his language he did not ACKNOWLEDGE the term reality.... That is all.

p, 143 Neurath says (I hope Carnap, too) instead of saying there is no definite world with definite properties, we shall use the expression no definite aggregate of propositions.

p 144 I object to "continuity of ego', there is EGO, in time 1, Ego, index time 2 etc., that there is used the term 'genidentity' is another point, a proposal made by clever boys, Carnap included.[114]

p, 146 why should Crusoe not distinguish between image in a river and a nonimage on the bank of a river?

p, 148 I would say if a certain police is in action after a certain time people not only speaking to others but even to themselves are using different empiricist statements. They ACCEPT other statements, why not? We do not speak of TRUE and FALSEHOOD, but only of various groups of ACCEPTED STATEMENTS. And our acceptance is related to environment, certainly, there is no point outside the 'world' from where we may judge on TRUE and FALSE.

p, 148 what is the translation of 'I mean something' into a language without the term meaning? A statement I think is either accepted or rejected or it is not decided, therefore also a statement such as 'here is a table', it 'means' JUST 'here is a table' and I may deny or accept the statement. But Russell thinks it may be of importance whether 'really' there is a table, who is in the chair saying that or objecting to it? The bearer of absolute truth, who says statements he accepts or denies, not we. I say again: in the beginning of all discussion are statements, why not?

p, 154 again the 'immediate' past of Schlick's.

p, 156 who decides that some memory is 'true'? 'erroneus' etc., always from a speaker's point of view accepted or rejected, the judgment based on more or less accepted statements etc.

p 160 IT MUST BE TRUE, Who makes such a strange statement?

p 161 who finds out the 'CERTAIN DEGREE OF CREDENCE'.

p 164 who is the judge, that some propositions are 'psychological' ones?

p 168 I really do not understand how somebody may speak of a 'TRUE PROPOSITION' without an absolute standpoint, otherwise he has only ACCEPTED propositions with a person index.

171 Who tells us the story that an assertion has a 'subjective' and an 'objective' side? Manifestly a being only which is OUTSIDE our discussion between man and man, all our statements are in competition, some we are accepting others rejecting, but what is the translation of 'subjective' and 'objective' in the Protocol language and its family?

p 186 I should just say, that we only discuss 'social terms' and if TRUE is no social term, we cannot discuss it. How should we speak of a language precisely identical etc. WITHOUT ASSUMING THAT SOMEBODY COMPARES THE VARIOUS LANGUAGES 'from without' at is were.

[114] Kurt Lewin (1922) introduced "genidentity" in his *Der Begriff der Genese in Physik, Biologie und Entwicklungsgeschichte*: it is a relation to identify an object from one moment to the next. On Genidentity, see Padovani (2013). Cf. Carnap (1928/1967, §159).

p, 187 WE MUST SUPPOSE, who 'must'? Who says this statement? Russell? It is an hypothesis 'this and that language are the same' ....

p, 189 'psychological' comes into the story, how?

p-189, 191 etc. etc. always 'meaning' in itself discussed, i.e. absolute truth, not as a 'social' relation between men.

p, 218 the story of brain in which observable phenomena happen in the brain is often told and criticized by Avenarius and others. Duplication. You can discuss all these problems without speaking of brain. And then again the 'real' question, only 'meaningful' within a theory of 'absolute Truth'.

p, 221 what you think of the QUESTION 'how do I think ..." I would say: how to I use the term table in statements in which I do not use the terms I see, I touch etc.?

p, 221 Have we not sufficiently discussed the strange theory of 'correspondence'?

p, 245 as if 'facts' and 'sentences' were in different worlds, 'there are in a room e.g. tables: V 'In this room there is a table and a label on which is written ""there is a table in this room""" why not? that is all.

Russell seems to have sentence as a substitute for the old mind

p, 258 Who is in the chair who says 'if it is in fact true'?

p 268 As far as I can see, we are discussing whether we are accepting certain sentences or not, that is perhaps the statement most similar to Russell's 'we are speaking about what it means'.

p, 277 do you think that the 'real' word is different from the absolute truth? That is all of the same breeding.

p, 281 Russell objects to your opinion – why not you to his as far as ABSOLUTE TRUTH is in question.

p, 282 I did not speak of solipsism

p. 286 I do not speak of 'percepts',

p. 288 truth and knowledge – strange statement. Or not?

p, 289 I do not think we said something similar to Hegel.

p. 304 Russell always overlooks that my yesterday (remembering) is similar to another persons statements. Russell introduces a permanent EGO and a permanent REALITY, I think so.

p, 305 always the same.

p. 313 degree of certainty, as above said only in relation to an absolute certainty as test object,

p, 315 on one observation ... I do not speak of simple observations, and you, I think, only a short time.

p, 317 the absolutism of 'the moment of the occurrence' again Schlick.

p, 321 precision only within the mathematical language, not within the aggregational language, therefore no TRUTH problem in the discussion of empiricism.

p, 321 who tests the testing success?

329 to obviously sad

p, 340 inside my head ...

p, 347 there are universals ==== there is an absolute truth

p, 347 structure of the world, ===== absolute truth, something of it.

I sacrificed some time to put this together, because I want to know more about your position. What is told above is the story of absolute truth in itself, I really cannot see in the statements above any attempt to describe or analyze the procedures of taking [having] experiences.

I should like to read some explanation of yours on this point, perhaps in a letter to me. You see, just a statement like that on the ONE PERMANENT LANGUAGE, which is a real empiricist one, of no senseless at all, and the ONE PERMANENT INDIVIDUAL shows you how much Russell starts from unempiricist assumptions, because I can imagine experimental statements from which somebody infers statements such as: I have changed my language in relation to other people's language, which I accept as unchanged etc. etc. .... as far as some test statement is accepted etc. etc. NO DEFINITE PERMANENCE OF SOME EXCELLENT PERSON.

Many thanks for addresses.

Please, tell Morris and yourself, that I am trying to arrange Journal and Library. You see, the long to and fro reached just the period where the paper regulation dropped in and now I have to find out what to do. As you know, I do not cease to manage things and I shall think that we have a lot to do for Europe after this war.

Therefore, at the moment, I can say nothing on Kaplan's book (in all cases I would like to read manuscripts before, in spite of the fact that I appreciate very much your's and Morris' judgement, of course) and other books, which may be useful for the library of unified science. You see, Blackwell and others always are prepared to print without any payment by us, on the contrary paying to us, therefore I prefer an agreement, which gives us a good position and enables us to go on, without an interfering publisher. Therefore I appreciate very much our ENCYCLOPEDIA agreement which enables us to print, without particular permit of the publisher.

I would appreciate it very much, if you were prepared to publish the symbolic logic in our library, ask the co-editors, too. But, please, at the moment I cannot make a definite promise, because I have to get the general contract.

I am very interested in your semantics, I hope you found a way to give me an opportunity to translate your semantic statements into the 'accepting' language, it is not simple to transform the TRUE statements into ACCEPTING statements, the grammar is sometimes similar, but sometimes not, that depends upon your decision, how you want to introduce – or not to introduce – this damned 'absolute truth'. It is not only the TERM, but the GRAMMAR which is dangerous, as I pointed out again and again. I hope I shall be able to explain that more in detail in the future. I am also interested in the confirmation degree business. I have always great fear that something 'absolute' slides into an explanation. The same is with the probability business. I think the probability language is ONLY a mathematical one, but not an aggregational one, and one has to find out how to relate them to one another. I think, you not always make sufficient distinction between 'aggregational' terms and formula terms. And after analyzing many articles and books (covering the last century) I think that the main bulk of differences between various authors is related to this problem.[115] I found very remarkable older explanations of probability statements

[115] Cf. Neurath (1941/1983, 218–221). See further Neurath (1944, Sect. 8.).

and statements on the permanence of certain averages (quite different statements, very often confused). But that needs a longer letter or an article. And I have at the moment (fortunately) not much time, I had to write my, already published article (NEW COMMONWEALTH QUARTERLY) on INTERNATIONAL PLANNING FOR FREEDOM,[116] to prepare some lectures, one on TOLERATION, MUDDLE AND VICTORY with an stimulating discussion, have to make together with Mary films and films and illustrations for books, to continue my monograph (see above) and to organize a new institute for our visual education purposes – that all wants time, time. At the moment is [S]unday, three collaborator[s] are with us and are just discussing with Mary and me, Climate and Weather charts, Urban and Rural population, etc. all stimulating. Of course. I met many people and have a very interesting correspondence on classics with a philologist. Sometimes I get letters from Strauss and others. I think Waismann is less and less interested in our points of view. Our British acquaintances and friends are very kind and helpful, especially Susan Stebbing in spite of her illness. We just met her in Cambridge where we had a business meeting with her and other people. I should like to get from you a letter on friends and world, on Thomism in Chicago etc. I should highly appreciate it if you were kind enough to send interesting newspaper cuttings and reprints and such stuff. We have now a nice studio again, with many files full of interesting material, but it is not our old richness, which was evolved in years. We like very much statistical data, interesting pictures of single objects, e.g. certain characteristic animals, bu[s] es, chairs, teapots, coffeepots etc., lists of kni[v]es and forks, cups and pots, plates etc. refrigerators etc. We are buying LIFE, LOOK etc. for catching such material. Today we found in this way the shape of an American telephon[e] apparatus, but there are thousands of apparatus, you know.

In England is plenty of food, if you are not so interested in particular types of food, bread is free, as you know, there are many vegetables and a real glut in milk, we enjoy it very much, dried eggs now from the US come etc. we enjoy it, too. Fine fruit etc. fish free (if not canned). Not the slightest similarity with the Central European situation in the last war. Not much meat, but you can get Vitamin[s] in various forms, e.g. as pills if you are very interested in these things. We are healthy and happy, and only sorry that such a terrible war is necessary to fight these gangsters. Did you hear from Paul? He seems to be successful[l] in his studies and will get now a fellowship. What is with your health?

Morris wrote me about the separation from Trude – what a pity.[117] What about the daughter? Please give me Trude's address. What about other people, e.g. Moholy Nagy[118], the separated wife of whom I sometimes met here. What about Senior, and all other fellows? The youth of our Unity of Science Movement?

---

[116] See Neurath (1942/1973).

[117] Trude is Gertrude E. Thompson, the first wife of Charles Morris. In 1951, Morris married his second wife, Ellen Ruth Allen.

[118] László Moholy-Nagy (1895–1946) was a Hungarian painter and photographer, mainly associated with the Bauhaus. Moholy-Nagy was a friend of Carnap and Neurath during the 1920s. In 1937, he became the director of the New Bauhaus in Chicago, where he collaborated with Carnap and Charles Morris. On Carnap, Moholy-Nagy, and Bauhaus see Damböck (2017).

I hope and Mary, too, we shall meet again in peaceful valleys and talk, and talk ... and drink grapefruit from cans and orange juice .... as once in May. And then will be more planning in the world – perhaps – no unemployment, only the 'normal' difficulties of life based on love and hate, envy and prestige etc. It remains a sufficiently interesting lot of problems.

I am reading many biographies, besides my special books and articles. Sometimes I look in mathematical and physical books and enjoy the clearness and exactness of this field. I understand very well, why so many thinkers try to anticipate such a clearness in all sciences. BUT "es bleibt ein Erdenrest zu tragen peinlich und waer er von Asbest er ist nicht reinlich",[119] but aggregational.

With kind regards from both of us to you, the Carnapess and all our friends. Ever yours

Neurath

12. Neurath to Carnap, July 20, 1942.

(ASP RC 115-07-57)

20th July, 1942

My dear Carnap, one should not write to[o] soon (months means too soon) because something may happen. This morning one [of] your wonder packages appeared, I think it is the one I mentioned, it contained the French pamphlet on our circle, the Hypothesis System (I needed it very much and could not get it in any library) and many articles and your white pamphlet, white and wise.[120] Many thanks, my boy. I appreciate very much, that you are sacrificing so many reprints etc., that is friendship. I am really happy and pleased. On an average I am astonished how easily I bear the fact that my whole library, all my notes, all my manuscripts, all my reprints, all my belongings are in Hitler's dirty hands, but this lu[c]ky quality not to be depressed by a lack of things is not reducing my pleasure to get something, I think there are few things which please me so much (if we do not speak of persons and social events) as books and printed matter, files of material, living life, to thumb over things and to be stimulated by old and new ideas. It is pleasant, too, to rebuild my 'memory' my bulk of knowledge. I am reading books, I read first as a boy and I am pleased again. In Vienna or at The Hague, I had them in my library giving me the feeling that they are my own .... now I bought them and I am re-reading them, e.g. Lodge's nice arguments on mechanical models of electric currents, Mengenlehre by Fraenkel, etc. I always hope I shall find a Schroeder, I have now a VENN with

[119] See Goethe (1832/2014, 301); the recent English translation reads as: "This remainder of earth,/ it's distasteful to bear it;/even cremated,/it would still be impure."

[120] Neurath presumably refers to his 1916 paper, "Zur Klassifikation von Hypothesensystemen," translated as "On the Classification of Systems of Hypotheses," see Neurath (1916/1983).

his nice ellipses (I had them only in my Schroeder).[121] Perhaps you can get for me by chance (very [im]probale) JOSEPH POPPER LYNKEUS, Allgemeine Naehrpflicht.[122] I got Ballod-Atlanticus Zukunfstaats.[123] I want to have all these planning accounts. Mary is sitting in the same room, drawing a map of the USA for showing a cross section with the various plants – a nice business. Making charts is a great pleasure, really. And we support the war effort in this way (all this work is used by the Ministry of Information) I feel like a doctor who is fighting the plague. And our delicate tools are useful, too. I hope you will see some of our animated diagrams in USA.

The whole morning I was looking through the prints you have sent me – many thanks. Many thanks.

Now the reports on Russia and Libya are better – but better or worse, in the long run, we shall win the war, I do not doubt, and I did not doubt, as you remember. That does not change the pressure under which we are living, like in formers times, when the Black Death was waiting behind the quarantine walls. I prefer a situation in which no Black Death is waiting behind the walls, even if the walls are strong enough. What is with our friends, are they joining the USA army? Who?

What about the Franks? Males and females. What about other friends, please tell a little more. And do not forget to post copies of your letters, too. My proposal is to write on another even if no answer arrives after some months let me say, three month. That means 4 letters a year – there is a war on.

With kindest regards and many thanks
ever yours
[Otto Neurath]

13. Neurath to Carnap, July 29, 1942.

(ASP RC 115-07-58)
21 Old Road, Headington, OXFORD
29th July, 1942

Dear friend Carnap. It is really nice of you to send again the two parcels, I enjoy very much. Please send me semantics again, should I get the second copy (very improbable, as you will see from my following explanation) I shall deliver it to a person you will nominate [name]. I need your semantics very much.

You complain seriously of the package muddle and I shall answer seriously. Your ADDRESS was UNIVERSITY. You see, there is not such a centralized office as you

[121] Presumably Neurath refers to the following works: Lodge (1892), Fraenkel (1919), Schröder (1890–1905), and Venn (1881).

[122] See Popper-Lynkeus (1912).

[123] Ballod-Atlanticus (1898). 'Ballod-Atlanticus' was the pseudonym of Kārlis Balodis (1864–1931), a Latvian economist, financier, statistician, and also a demographist.

have in Chicago, because there is no such university. The main body is formed of the dozens of colleges (plus quasi colleges) each with its own rights and regulations, with its own 'Dons' who are not always related to the University. Lecturing is a funny thing here, sometimes you are invited by the University as such, sometimes by the college sometimes by a chair, you are reading for a professor (such an honour was mine, invited by Radcliffe Brown anthropologist)[124] – in all these cases you are on the list of the lectures. But there are cases in which discussion classes or even lectures are organized besides that etc. And now there are a great many institutes related not to the university but to colleges and then there are institutes, which are very loosely 'connected' and others very indirectly, e.g. INSTITUTE OF SOCIAL HISTORY, an institute made around a big part of Cole's private library, which can now be used by students in a particular reading room, with office etc. [on] BANBURY ROAD, the house has the No 19 as I wrote you. But the position of such institutes is changing and nobody is informed about it – a privacy unknown in Europe and perhaps in the US. Sometimes this Institute was with the International Institute of Social History in the same building (a DUTCH institute in exile), now this international institute changed its place and the building is now a part of a newly founded college, but Cole's library and his institute remained in the building nevertheless. I do not know how the 'rights' are divided, because Cole is head of this college. And so all things are going on smoothly without great decisions and to and fro, sometimes with to and fro.

What is the advantage of this pattern (which does not make simple to find an institute) with all this vagueness? You see if a man e.g. in Vienna could not become a '[P]rivatdo[z]ent' (let us assume he was a Jew and for decades there was an animosity) what could he do? fight on? or what? Here there many, many ways to teach and to make [do] research, you can be persona ingrata [non grata] in the one college but the other college is fond of you, here you cannot get a room for lecturing, there you can get a fellowship. Nothing is centralized, nothing is organized in the sense as you and many people think planning SHOULD MADE IT.

You speak of inefficiency of democracy ---- my dear boy, EFFICIENT IN [FOR] WHAT? [T]hat is the point. I disagree totally with all [the] people who think of sacrifice of personal freedom because they want planning. Planning avoids destruction of raw material, avoids unemployment, provides a society with raw materials, horse power etc., and now it depends of our organization, how we go on with the distribution. The most terrible thing would be if people got power to bully other people, therefore the centralization of decisions is dangerous, therefore the rights of Local Governments important, and the rights of smaller groups and individuals, too.

---

[124] Alfred Reginald Radcliffe-Brown (1881–1955) was an English social anthropologist mainly known for his theory of structural functionalism. Earlier in June 1941, Neurath invited him to give a talk at the Sixth International Congress for the Unity of Science (University of Chicago, Sept. 2–6, 1941), and Radcliff-Brown promised to write a paper about "Social Anthropology as a Natural Science of Human Society." After all, however, he did not participate at the Congress, or wrote the paper for Neurath. See their correspondence in the Neurath Nachlass (Wiener Kreis Archiv, Rijksarchief in Noord-Holland, Haarlem, The Netherlands).

MAJORITY PRINCIPLE, as the continentals think of 'democracy' is bad, if not reduced to the minimum of a minimum. Look: what would you say of a democracy in which the majority (60) would decide, that these 60 have 5 hours working day, the 40 but 12 hours working day[s], I remember the problem was in Vienna whether department houses or small buildings.[125] I propagated [argued] to avoid a majority decision, but asked for a distribution of building-power, as it were, in proportion to the wants of the people, if 60 want department houses and 40 small ones, give each of them, what he want, if the bricks, horse power etc. are more or less the same. You cannot make various railway gauges in one country without destroying the whole apparatus, but you can make hundreds of life pattern[s] in one country, why not? Then there is less [of a] possibility of bullying other people and there are not a few who like to do it, not only 'fanatics'. Better is education and propaganda than a law as e.g. the prohibition law in the US.

You see, when a judge tries to make decisions in accordance with his own conscience – and I think that is useful from my point of view, than you cannot give the courts of appeal to[o] many chances, otherwise they overrun the judge[']s decisions and he looks at the higher courts as on the continent and the civil courages vanishes, to be in accordance with the highest authorities becomes important. Here you have – that is acknowledged in the whole world – the best judges, with a minimum of corruption etc. Here you have to SELECT the judges in a different way when you want changes in practice or INFLUENCING THE PUBLIC OPINION OF THE GROUPS FROM WHICH THE JUDGES COME. But it is difficult to get equalisation of decisions, without weakening the intensity of the judge's activity. I prefer the differences of [in] the decisions, because then we get more people with civil courage and less [fewer] people who always look to the 'formal' correctness of central decisions. You know to what extent even Hitler is formally legal. I DO NOT MUCH BELIEVE IN CENTRAL DECISIONS BUT MUCH MORE IN THE TOTAL BEHAVIOUR OF A FREE NATION based on tradition.

These differences in decisions create necessarily a certain kind of difficulties usually called 'muddle'. You see, in free countries the police – in peace time – is [are] not interested in somebody's travels, no forms have to be filled in hotels etc.

FIRST PRINCIPLE: one of the many 'goods' produced are: love, freedom, multiplicity, etc. besides cars, houses, grape fruit juice, etc.

SECOND PRINCIPLE: we are prepared to pay for freedom with cars, grape fruit juices etc. ....

RESULT is – as far as I can see, not always less efficiency even in the production of cars, making war etc. because this nice environment of freedom reduces the TENSIONS between men. You know better what everybody wants to do. I think that this is one of the reasons why you have here a minimum of quislings. The German propaganda does not even try to induce sabotage here, but the British propaganda

[125] On Neurath's role in Vienna's city planning and the housing movement see Hochhäusl (2011).

tries to do that in Germany, and is right in that, because there are ten[s of] thousands of people in Germany who conceal their opinions totally.

There are people in high positions in Germany with the swastika around the arm who hate the system. I can hardly imagine that people who think Churchill is a damned fool would openly declare he is the finest boy. As far as I can see, people who dislike the government with, let me say, 80% will perhaps tell of 60% only or less, but not of 5%. The public discussion and the private remarks, the private arguings are similar. I would not say that some changes can appear, not expected, but not in the continental way. Perhaps I am wrong. This opportunity e.g. to be a conscientious objector (I would not say they are very well treated, but nevertheless you are not 'annihilated' or 'liquidate[d]' if you try to be one of them) is for me and some others of great importance. I think the reduced tension (partly depending of certain common features in education, creating a common attitude etc.) helps in a [sic] bearing hard shocks, and I think that just this situation (with what the continental call 'muddle' with a bad taste in the mouth) helps to win this war, really. Perhaps I am wrong. You see, one cannot bear muddle without liking it, that I am right in my thinking on the British behaviour I deduce from a book, in which a British author writes: 'There is always some "history" some "incident" – to be frank, some incipient muddle, that makes the social life worth living.'[126] I think you could not write that.

But only in comparison is power. Therefore I collected material from other countries, particularly on the deeply founded [i]nefficiency of the traditional Prussian and German organization. Read what Rathenau tells on the German civil service and the selection of collaborators, why he did not expect victory from the beginning, read the description of his interview with the war minister, what a story – the war has started, the minister full of pride tells him that his table is empty.[127] No recriminations, no questions (that means every soldier got his paper, with the hour of his train, carriage, no. of his seat and a remark that the lavatory on the rear is for the soldiers [and] on the front for the Herrn Offiziere etc. ABSOLUTELY EFFICIENT). Now Rathenau asks about the stores of raw materials .... you know the classic answer and that IMMEDIATELY it was necessary to make a new office and to use Rathenau's help (they did not like to use a Jew etc. you should read that and then think on muddle and democracy, that was not 'muddle' in the sense of multiplicity but well based stubbornness, one[-]sided and stupid).

Read on the Marne battle. Perhaps I shall write a satyric story of that. It is har[d] to believe. FROM GERMAN SOURCES you learn, that Moltke the boss of all bosses was selected stupidly, he had a nervous breakdown (I collected more nervous break downs, which perhaps are more dangerous than muddle) the generals commending armies did not report sufficiently, because they wanted to have all iron

[126] The author is Sir Bernard Pares (1867–1949), a historian, known mainly for his works on Russia. See Pares (1941, 193).

[127] Neurath presumably refers to Walther Rathenau (1867–1922), who was Foreign Minister of the Weimar Republic. See, for example, Rathenau (1919/1921).

crosses with diamonds and something else and did not want that the neighbour general knows too much of their intentions.

Result: in the period of telegraph and telephone the Boss of all Bosses put a simple colonel on a real horseback and gave him power to organize the retreat where necessary, and now this poor fellow riding on horseback from army to army put the whole thing in a terrible mess (please read the GERMAN version of these events). I PREFER OUR BRITISH MUDDLE. The same is with Red Tape. One should make comparative studies on that. That is that. I wait for your answer, my dear anti-muddleian boy. Perhaps my Viennese tradition helps me in liking the Anglo-Saxon way of living. I think the Americans overestimate TESTING and CUTTER EFFICIENCY more than the British. I know how many things here are defect[s], but it [there] is less bullying than at any other places in the world. The easy[-]going life is a great advantage.

I am glad that you will have time to study. I hope you will write me, how your semantics and its TRUTH is [are] related to my 'accepting' etc. I do not look without fear at your actions.

Library, Journal etc. will be attempted after a short time. As I already wrote to you, there is a new fine possibility to get money and apparatus for our unity-of-science work. That would be fine. I do not want to spoil the whole thing by pressing too much. WAIT AND SEE is always my principle. BUT DO NOT CEASE TO WAIT AND SEE. I am always prepared to work for our research and our publications, of course. Be sure that I always think of that.

PAUL's address is changing, please write c/o WALDEMAR KAEMPFFERT New York Times. He will get some fellowships and is successful. Please write him, he admires you and will be glad to get from you a letter and good suggestions or something else. He likes good books and grape fruit juice, too. I think so. Paul sometimes intended to go to Chicago (Ogburn).[128]

On the ENCYCLOPEDIA. I always think of the encyclopedia and I now think – and you and Morris will agree with me – we should put as much [many] monographs as possible into our pattern which deal with particular sciences or particular activities (e.g. LANGUAGE MAKING) BECAUSE WE NEED THE HELP OF SPECIALISTS IN SINGLE FIELDS. I just try to find such specialists. Therefore Schapiro[129] is a big show for us. Perhaps FEIGL and HEMPEL could make together the monograph joining the subject of 6 and 9, WE NEED A REVIEW ON THE REAL SITUATION IN OUR MOVEMENT, what logical empiricism tries to make and what kinds of systematization are reached. There are manifestly various groups of problems, the 'formula' and the 'aggregational expressions' e.g. (using my own

[128] Neurath is referring to the sociologist William F. Ogburn (1886–1959), who had taught at Columbia University and in 1927 became chair of the Department of Sociology at the University of Chicago.

[129] Meyer Shapiro (1904–1996) was an American art historian; Neurath arranged a monograph for him in the *IEUS* during the late 1930s. Rudolf Carnap and Charles Morris were still negotiating with Meyer during the 1950s, but the monograph was never completed.

terminology) are of different type and there are some of us which stresses this difference, others not. Some tendencies are towards systematization, others are cautious, the probability, induction, etc. by various authors discussed differently. I think such an analysis could be very useful. At the moment as have only books such as that by Weinberg etc. discussing a former stage.

If FEIGL would be prepared to make the general review and HEMPEL the half monograph on systematization it would be a fine show.

Collecting material on the second series of monographs I am preparing a survey on people who are related to us, but have some 'whims' – as it were – they should have an opportunity to express themselves with intensity (the first series of monographs stresses, what we have in common. I have in mind the psychologist Pear, then TARSKI with his TRUTH, then REICHENBACH. I re-read some of his enunciations and found out that he is really interested in stressing the differences. I want to give him a real chance to stress his theory of induction, his strange things on probability of hypotheses etc. (strange from my point of view, sometimes I do not grasp his intention). That is the right place for him after his very decided statements on our whole enterprise and our point of view. That is that.

The situation is this: FRANK writes, (he wrote me that he will do it immediately) introduction and finish chapter to FINLAY showing the importance of his explanations for the whole business. BIOLOGY is open, but has to be made. THEORY OF BEHAVIOUR will be written by BRUNSWIK and if Ness' manuscript is already there together with Ness, otherwise we shall get Brunswik alone.

We need a substitute for Rougier (he does not even answer Morris' letters) instead of Tinbergen we have Schapiro on arts. Wirth is writing, Hempel and Feigl together will make 6, I hope so and instead of 9 we make LANGUAGE MAKING that is a good tuba song. Bibliography and Index we shall make together. I collected a lot of articles etc. I think we should not only bring poor and pure titles but some short remarks and particularly mention books and articles in which are some paragraphs dealing with Logical Empiricism, PRO and CON. I asked Morris to ask Kaplan, how it is with his collection.

I should like to read Kaplan[']s ideas. In principle it would be nice to find for him a place in the LIBRARY, but until now we have not got the library again.

We have to leave our house, because the landlady comes back. BUT we just – we always have good luck even when we have bad luck – found a new one in our neighbourhood in spite of the fact that houses are very rare. Just around the corner. And the chance came from outside not from the corner.

Please ask Hempel whether he is prepared to do this job together with Feigl in the senses mentioned above. I had the impression that Hempel is much [quite] busy and then therefore a shortening of his task would be not so bad. Perhaps we find a common title and have th[e]n two nice half monographs.

I am writing my monograph and have such an amount of material together, that I now have to re-shorten the whole thing again. I hope it will be useful.

So is our life mixed up with war and science, love and thinking of victory. That a terrible thing such a war, but now can one eliminate these Nazis without war? In the US there are many tensions as I learn from English and American sources. Let us hope that all things will go on very well. I am confident, because we have more producing power together.

With kind regards to all friends,
every yours
[Otto Neurath]

14. Neurath to Carnap, August 27, 1942.

(ASP RC 115-07-59)

NEW ADDRESS 30 Bickerton Road, HEADINGTON, OXFORD
27th August, 1942

Dear Carnap,

I ask you a favour; my young friend Candida Kranold just wrote me, that Herman Kranold and his wife died one after another by heart attacks (fortunately without any pain) in a few days['] distance.[130] She is a physicist and has a brother and sister, both children. She just passed examinations, and wants to study, perhaps philosophy (Feigl, I shall write to him). As far as I can see, no money help is needed – she has a job, but some suggestions would be very helpful. Please be kind enough to write her and to ask her, whether you can help her with suggestions or introductions. She is very clever and full of energy and I have no doubt she will go on very well.

[The] Kranolds were not too happy at Taladega College, separated from the 'world', as you may imagine. As I told you, I visited them and lectured once there. A nice campus, a nice house, but really a kind of island, and sometimes very hot. The relations to [between] negroes and whites, of course, not without difficulties. Life is hard and difficult on an [sic] average.

He was a very brave man, always and I estimated his uprightness – she was clever and helpful. They loved one another very much. Less and less people remain, who can talk with me on common past .... How many died just in the last years, some in concentration camps or in Nazi prison, some committed suicide. We are glad that the most of our friends are somewhere in the Anglo-Saxon world, even very distant, but reachable.

We are just moving. Our landlord needs the house. We found another one well furnished with garden round the corner. Smaller rooms, but one more, that is an advantage, because sometimes friends stay with us and we have our own studio with library.

---

[130] Hermann Kranold (1888–1942) was a German political writer and member of the Social Democratic Party of Germany; he emigrated to the United States in 1936. Kranold developed a full-socialization program with Neurath and Wolfgang Schumann (known as the "Kranold-Neurath-Schumann Program"), that formed the basis for Neurath's plans in the Bavarian Revolution. See Cartwright, Cat, Fleck and Uebel (1996, 43–49). See further Schumann's memoirs in Cohen and Neurath (1973, 15–17).

We are making one animated diagram sequence after another, charts for books, writing, reading, etc. We are as happy as possible in so sad a time. We hear good news of Paul, he got a fellowship and succeeds in his studies. Joseph Frank is now in New York, too.

I wrote Philipp Frank to finish his Freundlich job, introduction and closing words – I think a few words on Freundlich should be said. I suggested to change the title, into, about FOUNDATIONS OF ASTRONOMICAL MEASUREMENT or something like that. Please talk it over with Morris, who will get a letter from me very soon. I am writing my monograph with intensity. I have a multiple [bundle] of pages together and I am now selecting the main points carefully, looking through the newest literature. I am very happy in doing so, buying books and discussing with friends many problems of Logical Empiricism, Visual Education etc.

I am a little depressed, that, as far as we can judge, even Vienna, not a few people, of which we did not expect it, have certain sympathies with the Nazis. Not that they are Nazis, but they think, something is very well done, e.g. the Jews are persecuted etc. [T]errible. And other things, too. In all cases after the war it will be there a sad situation. Whom you may trust? But life is always a complex and sad thing, only certain islands of happiness and now we compare these islands with the rest .... Sometimes we feel rather ashamed of our happy life (in spite of all the sorrow and all sadness in our life, too, but that is human fate). Look, we are not idle now, we can help to fight the Nazis in doing some work for the Ministry-of-Information films etc. The primitive life as far as eating, sleeping etc. is concerned is OK, the food situation is really very good. You see we have no food ideals, therefore we adapt ourselves to all the changes and therefore we enjoy this life. Through months we got sufficient milk and could drink it as sour milk (here unknown and despised).

I hope to get your semantics soon, did you register it?

I should like to get issues from American philosophical and scientific periodicals. Have you any opportunity to provide such stuff for me? Waldemar Kaempffert is very nice in sending me interesting odd numbers of periodicals, pamphlets etc. and I feel already comfortable surrounded by reprints, periodicals, books etc.

With kindest regards from both of us, to both of you, every yours [Otto Neurath]

Please, do not forget to give me Trude's address.
ADDRESS CANDIDA KRANOLD,
SPROUT OBSERVATORY, SWARTHMORE COLLEGE, SWARTHMORE, PA.

15. Carnap to Neurath, November 7, 1942.

(ASP RC 115-07-60)
Rudolf Carnap
University of Chicago
Chicago, Illinois.
Santa Fe, N.M., November 7, 1942.
(That is in New Mexico)

Dear Neuraths:

Thank you very much for your letters of July 17, 20, 29, and August 27.[131] We are very glad to see from all of them that you both are happy in your activities, as happy as is possible in a war. And that you are happy together and found a new nice place for living.

We are away from Chicago since July and until next June, on a leave of absence as I wrote you.[132] Unfortunately my back trouble has started again on the first day of our trip and has become worse than before. In consequence I have been in bed all the time since the beginning of July. But we are here in a nice landscape and nice climate; I am lying on a porch with two giant windows so that I can enjoy the landscape. And sometimes I can take short walks. Mostly I have no pain and can work. Therefore I am in general quite happy. Disturbing is only the idea that the future is so uncertain because so far no doctor has been able to find the real explanation and a proper cure beyond merely temporal relief. Strangely enough, Ernest Nagel seems to have similar back trouble, and he even went through an operation – all in vain.

I have asked the Oxford University Press, Oxford (they distribute the Semantics book in England) to send you a copy. If you do not get it soon, please inquire there. If the other copy should still arrive, please let me know. Likewise, if you have already or will find some of the reprints of yours which I sent you, please put aside my copies because I should like very much to get them back from you later. The second volume will soon appear; it is more of a technical nature, less concerned with general problems of the nature of semantics. (About its contents see preface to vol.I., p.ix.) Therefore I doubt whether this volume will be of interest to you. If, however, for some reason or another you should like to have it, please let me know.

I am very much interested in what you wrote about Hogben's ideas on language-making and about his own auxiliary language. I remember our previous discussions where I maintained the superiority of artificial languages in comparison to Basic English, while you were rather skeptical about their practical chances.[133] You say that Hogben's Interglossa seems better than the other artificial languages; do you know enough of the other ones to make a critical comparison or do you merely infer it from Hogben's good ideas about the method of language-making? As Morris wrote you already, we think it better to see first Hogben's book before we decide on a monograph of his for the Encyclopedia. Who will be the American publisher of Hogben's book and when is it to appear? I am looking forward to it with great interest.

[131] See Neurath to Carnap, letters 11., 12., 13., and 14.

[132] See Carnap to Neurath, letter 10.

[133] See Neurath to Carnap, letter 11.

I am very glad to have your detailed comments on Russell's book. However, in the moment I cannot study them in detail because I do not have the book with me. As soon as I have an opportunity I shall read them together with the book and answer you. I have myself many objections against this book, especially against his attempt of a revival of old epistemological questions. Please read Nagel's critical review in Journal of Philosophy 38, 1941, pp. 253–270.[134] I agree with most of Nagel's objections. On the other hand, I think that Nagel should also have indicated the positive values of the book; e.g. that it carries out certain logical analyses instead of metaphysical speculations. I am afraid that in certain points here as with respect to Tarski and Popper you are too critical, or rather too suspicious. You reject certain statements because they might perhaps be meant in a metaphysical way, although another interpretation is possible which makes them scientific, i.e. acceptable to an empiricist. At least this was the case with respect to many statements in Tarski's book which you criticized. Whether it is also the case with Russell I shall examine later. With people who stand in general on the same empiricist basis as we but who might deviate perhaps in some particular points (as e.g. Schlick, Tarski, Popper, and even Russell) I am more inclined to take the attitude of a cautious judge, that is to say, to give them in each particular point the benefit of the doubt. I have the impression that – in doubtful or ambiguous cases – you are inclined to condemn the accused of the crime of metaphysics.

You ask whether my semantical statements can be translated into your language of 'accepting'. This is not possible and it should not be required. The semantical concept of truth is fundamentally different from the pragmatical concepts of accepting or confirming. I suggest urgently that you read again my old paper "Wahrheit und Bewährung" (Congress Paris, 1935).[135] There I have tried to make the distinction clear; further, I made a short remark about it in "Semantics" p. 28. We should only require that every concept should be translatable in some way into an empiricist, scientific language; but we should not require that it be translatable into certain special terms into which we perhaps like to translate it. I hope that you will see from my book that the semantical term 'true' is definable in a scientific language (e.g. in yours), see e.g. p.26.

I should like to get a clear explanation from you of your distinction between aggregational terms and formula terms. I have never been able to understand what you said about them. Does 'formula term' mean 'uninterpreted' (i.e. belonging to a calculus or syntactical system) or does it mean what I call 'logical' in distinction to 'descriptive' (see "Semantics" §13)?

In addition to 'Life' and 'Look' you might perhaps find material of interest to you (statistics, report about cultural trends, etc.) in 'Survey Graphic', 'Time', and 'Fortune'.

I think, in the question of planning, our views are not very different. We both emphasize the advantages of planning, and I agree also with your description of the

[134] See Nagel (1941).

[135] See Carnap (1936a) and (1949).

dangers of overcentralization. However, I am still not convinced of the advantages of muddling through. I think that decentralization and democracy can well be combined with a procedure according to a plan instead of according to momentary whims or tradition. (I see, I made a slip in my last letter towards the end; instead of 'muddling' vs. 'democracy' I meant to say 'planning vs. democracy').

I wrote to Candid Kranold but did not get an answer.

Trude's address is: 5428 Ridgewood Court, Chicago. However, Ina thinks that Trude does not appreciate approaches by Charles' friends (including me) at the present time. Trude feels that a person is either Charles' friend or hers, and that both together does not work; also, that Charles' friends have no idea as to the actual situation and how it came about and that therefore they do not see her position properly.

You write that some people in Vienna have unexpectedly Nazi-sympathies. Are among them some whom I know?

I don't know anything about the Franks -- we are quite out of touch. Tarski teaches mathematics at Berkeley. Hempel has written something about confirmation; it seems near publication.[136] He is very busy teaching and Eva is secretary to a psychoanalyst. Among our common friends nobody is in the army: some are too old, some have dependents, some are not citizens and therefore cannot volunteer (though they can be drafted for service if they have taken out first papers; however, it so happens that they all have dependents and therefore the draft has not yet reached them). All the efforts in Washington – especially by Johnson, Oppenheim, Hempel – to get visas for the Grellings have failed; and now Hempel writes the terrible news that Grelling and his wife have been deported from France by the Gestapo; but we don't know any details.

I have not heard from Waismann since [for] years; do you know anything? Does he teach or how does he earn his living? Is he writing something? What became of the manuscript of his book? Please send me his present address.

With best regards and wished from us both to the two of you,

Cordially yours,

Carnap

16. Neurath to Carnap, December 22, 1942.

(ASP RC 115-07-61)

ISOTYPE REPORT 30 Bickerton Road, Headington, Oxford
REGISTERED 22nd December, 1942

[136] Presumably Hempel (1943).

Dear Carnap,

Now we have to send you seasonal greetings – the time goes on, and we with it. Very sorry, that you, Carnap are not so well off. I heard of Nagel's pain, but I did not imagine that your pain is also intense. I had the impression of a more superficial but nevertheless irritating thing. Poor boy, the whole time in bed. I hope the doctors will find out and fight your trouble, please, tell us, what will happen with you.

We are going on very well. We had to move, but got a new furnished house round the corner on the same hill, more adapted to our wants than the first one, which has been very nice indeed. We have now a nice studio in our house, where our collaborators work, a very nice team of British people. Now we are co-operating for 1½ year, no quarrel, no tension – a different atmosphere from the atmosphere on the continent. We like the British temper and habit very much. We are always healthy and cheerful, no bullying boss, no financial difficulties after our initial ones, which have been bridged by our kind friends – what a bargain, to have such good friends. It is so important not to be hampered just at the start.

Now we have our ISOTYPE INSTITUTE. Mary and I are directors of studies and secretaries, (Home Secretary permitted that) and now we are going on, as once in May, we have a permanent scientific collaborator in London, who collects material for us in the Libraries. Here we have, as I told you, first class libraries. Our own library is very fine, too. I got a lot of useful books, e.g. Ueberweg Heinze,[137] Encyclopedia Britannica etc. atlases etc. many book I wanted – I was furious that this Hitler gangsters have my library now, but now I am rebuilding my own, in spite of these beasts. All what happens in Europe is like a dark cloud --- what a world in which we are pleased by looking at the American fortresses in the sky, returning from Germany and France.

Oxford University Press, told me, she has no copies, Harvard would send me a copy. Please be kind enough and ask Harvard, whether they did send me a REGISTERED copy. I should like to answer some of your remarks after reading your book. Sure I want to get volume 2, too. Sooner or later I am reading everything in my library. "Auf die Postille gebueckt in der Sache des wärmenden Ofens".[138] Sure, I shall collect the copies I have twice and give you back yours. It was a great help for me to get them. I feel really at home, surrounded by books, my own articles and pamphlets. I bought reprints on the market.... What a time.

Oh sure I know a lot about artificial languages. You see, I first met VOLAPUEK, a language, like other languages, too, only "neutral". That was my impression, and I did not devote any further interest to it – then I met PEANO, Latino sine flexion, which interested me, because I was interested in Peano[']s axiomatization of math-

[137] Neurath presumably refers to Friedrich Überweg's *Grundriss der Geschichte der Philosophie*, which was republished at the beginnings of the twentieth century by Max Heinze.

[138] That quotation is from Johann Heinrich Voß's "Der siebzigste Geburtstag" (1802/1972, 69). It says exactly: "Auf die Postille gebückt, zur Seite des wärmenden Ofens."

ematics and in his attempt to write a mathematical deduction in symbols only. The use of Latin has something in it, but the Latin tradition is vanishing more and more and [---][139] this way? I think Peano's ideas did not long bother me. Again and again, I met ESPERANTO, and later on IDO, as a language it is a relatively complicated product with all the whimsicalities of relations between adjectives and noun, etc. I do not speak of the nonsense with the letters and accents not found in our composers drawers, Ido did abolish this nonsense, of which – people told me – business and monopoly have been responsible. But Interlingua and Ido are more or less like other languages, you have many "idiomatic" elements in, which are superfluous from a language point of view, as a tool of communication. It is difficult to get a sufficient number of books, written and printed in this language. ESPERANTO is successful – but very narrow in real usage. Poor people. We discussed with some of those people, who were related to the publishing centre the publication of ISOTYPE charts or books with Esperanto text, they confessed that their means are far away from such possibility – what a situation. That wants to be an international body. We found out, that it looks more like a hobby of many people, but without important effect. IF it were an international tool – why not, but no scientist is prepared to publish his books in Esperanto, he prefers English or French. Now the Esperanto centre in Germany is closed and in other countries Esperanto is weak, in spite of the fact that even chambers of commerce etc. are supporting it.

You see in the English speaking world, English is the auxiliary language also for foreigners, the various aliens here, all the foreign governments use English, it is astonishing how the whole business runs in this way, English is manifestly now the Lingua franca for very many people. Who will learn Esperanto in the USA? For what purpose?

Therefore BASIC ENGLISH has a great advantage, for READING and LISTENING; you see, it is very difficult to WRITE and SPEAK Basic, more difficult than trivial English. Why? You see, the goal is to speak CORRECT English by means of about 900 words, that means, to find out the IDIOMATIC WHIMSICALITIES which just allow to use these words only. Instead of a simply trivial English phrase you have to find out a complicated Basic phrase. The Advantage is the VOCABULARY simplicity, most important for reader and listener. Therefore the best thing for wireless. If I know only one of the 900 words will appear, I can guess what word was just said, otherwise the choice between dozens of words remains. The vocabulary is important for East Asia, for newspapers there, and periodicals. Important BASIC is a bridge to normal English. It is used to a wide extent.

Therefore I think that BASIC for reading and listening as I often explained has a particular chance. I see no serious objection to its usage for printing books and giving wireless talks, from tomorrow, if necessary – no particular organization is needed, no time devoted to that is wasted.

[139] Unreadable word.

The other international languages are "isolated" and not attractive as far as their vocabulary is in question. Why should one learn all these queer and odd terms, for nothing and nothing, when you have no hope to meet another Esperanto hobbyist, the chance to meet some man who understands Pidgin English is much greater. Therefore, I like some Pidgin English, BASIC or another one, it should be simpler in grammar and without idiomatic stuff....

INTERGLOSSA, you see, as I told you, all roots are useful and normal ones. Whatever you learn in Interglossa it will be part of the normal international terminology in science in any case. Some terms are rare, but why should I not learn rare scientific terms, which may be useful another day.

Writing INTERGLOSSA is the simplest thing you can imagine, international roots. USED IN SCIENCE, and Chinese Grammar, or Chinese lack of grammar, no relations between adjective and noun and all this stuff. For scientific writing really a bargain. Perhaps for other writing, too. Learning Interglossa, implies learning something on language making and language technique, whereas learning Esperanto is learning a new language, which is a little simpler ....

That is that ---------

What do you think of that?

I see no much positive value in Russell's book, because it does not lead anybody to scientific thinking, but, on the contrary, to unscientific epistemology of oldest style, behinds [sic] Avenarius. You see Duhem, remains the type of books, useful for scientific arguing, not Russell. He has his great merits, but not just here.

I again read Popper.[140] I hope that after so many years you will see, how empty all that stuff is. Just the points in which we are interested from point of view of scientific arguing, are NOT ACCEPTED BY HIM, just there he is full of statements on experimentum crucis, negation of all sentences etc. what a decrease after Duhem, Mach, etc. No feeling for scientific research. I am sorry that you have no time to give me NOW a short analysis of Popper's book, telling me, where the merits are. This queer idea that one can refute a hypothesis [so that it] cannot even be used in a weaker way. You know, that I think even your formulation on degrees of testing pro and contra are transgressing empiricism. Perhaps that is the reason, why you fell a little for Popper.

Of Tarski's metaphysics I do not [sic] longer say anything. It is trivial sad Aristotle redivivus, nothing more. THAT DOES NOT EVEN TOUCH HIS FINE ACCOUNTING BUSINESS. But you see, what does help Tarski, when somebody proposes, to use statements in principle as statements OF SOMEBODY, and therefore always combined with some "accepted". You may say, there are other possibilities, too; then I would like to see, how Tarski would be able to tell a story of throughout relativistic skepticism, without any ABSOLUTE relation "true", but

[140] Presumably Karl Popper's, *Logik der Forschung* (1935). For the English translation see Popper (1935/2002). The first English translation appeared in 1959.

only relations between changeable statements. I suggest not to use a statement THE SNOW IS WHITE, but only as abbreviation or a part of the sentence: WE ACCEPT THE STATEMENT THE SNOW IS WHITE. I accept the statement there are NAME PLATES which fit [correspond] to SNOW, other which fit [correspond] to ICE or WATER, why not, but even that is accepted by somebody, there are no STATEMENTS IN ITSELF …. I cannot see your point, perhaps in your Semantics you will tell of that.

That does not touch the calculus, the question is only where and when and how we may relate this calculus to scientific analysis. That I want to see. TARSKI tells about snow and white, I should like to see, how he analyzes meteorological analysis or biological analysis, or mechanics or something like that fruitful. Popper and Tarski should not be mentioned in the same breath, because Tarski is making things INDEPENDENT of the application, whereas Popper only speaks of applications of something trivial, which seems to me [i]napplicable, therefore when we do not accept the whims of refutation of universal statements, nothing remains of importance. It was some idea, combined with anti-Vienna-Circle resentment. Why not? It makes him breathing [sic] more cheerfully. [And] [s]o on.

You see, I test all these ideas by looking into the sciences, I am now reading carefully Maxwell's letters, speeches etc. and I found fine things. I am looking, how Lord Kelvin argued, Faraday, Marx, Max Weber, and then I try to find out, where we could sharpen our doubts. And I see more and more, that the important point is to find out, where we expect stable relations or instabilities, "chance" etc. I am really sorry that Zilsel, Hempel etc. talk of social sciences sometimes, but do not analyse these points. Usually you find the remark that group events are better predictable than individual events – THAT IS WRONG. It depends, what happens, there are stabilities in groups sometimes, but also instabilities ….

In my Empirische Soziologie, page 130. And, it is puzzling, many people wrote on my book, there appeared reviews, nobody neither one of our friends, nor one of our critics, did even mention this very, very important point, the NON-PREDICTABILITY of some social phenomena, related to INTERNAL SPEECH, as it were, is usually not even touched. Yes people, who have nothing to do with Logical Empiricism and with antiscientific attitude, are against all prediction etc., but "our people" do concentrate [o]n prediction and do not even mention the problem of non-predictability. I shall put it forward in my monograph with some energy. I do not speak of the trivial Marxists, who know all things, and speak only of ERROR afterwards …. what a lot of lawyers ….

About your "true" I shall discuss with you after reading your book.

Oh yes, I read SURVEY GRAPHIC regularly, I get it as a friend of this periodical, FORTUNE and TIMES I bought, whenever I see a copy, I have a collection of it, relatively seldom I see LOOK, Life is very popular here.

You see, analyzing various countries, creeds, etc. I think people who have the conviction they are right, even in the statement they are in error, or they are sinners, are very dangerous. Only planning together with skepticism and multiplicity will be nice. This implies, what the traditional German calls inefficiency, because he does not ask efficient in what [?] Planning for what? Output of autos? or what? I speak of

PLANNING FOR HAPPINESS, FOR FREEDOM, perhaps the "output" in machinery is smaller, but in freedom and happiness greater, what then? THIS PROBLEM IS NOT ANALYSED USUALLY BY PEOPLE INTERESTED IN PLANNIG, mostly propagated by people who are against planning. That is a pity. A fine muddling through is a pleasant thing, and you may learn a lot of that here. I like the British for that. There is a nice booklet with pictures MUDDLING THROUGH. By BENSON, ASKWITH and BENTLEY.[141] Much fun in it. THE AIM OF ENGLISH INEFFICIENCY IS THAT THINGS SHOULD GET DONE WITHOUT ANYONE SEEMING TO CARE. You see, we do not know, how traditions are related to one another and to our happiness, therefore going away from some tradition perhaps reduces our happiness – there are certain points, where the matter is simple, and the things done far away from our personal life, for instance, organizing coffee production and distribution, without burning a third of the crop is a planning matter, not dealing with the worker's tradition in Brazil and, I hope so, not with my coffee on the table. Why should no planning coal production be combined with old coal mining tradition and fire places wasting coal – the point is, that we do not destroy coal, not even used for our pleasure in wasting something. To produce less coal, because we like some singing and sprawling, means perhaps increase of happiness, but the rotting of coal in the docks does not help anybody, not even having more idleness or wasted coal in his fireplace.

Please, try to find out, what is with Candida Kranold, I did not get any answer, too. I hope nothing happened to her. Try to send a registered letter or a telegram, that is the way to find out officially what may be wrong.

I do not know Trude's opinion officially, therefore I wait how she will react. I have not the slightest idea how Charles and Trude have been together or not together, for me they are two separate individuals, one a more dreamlike phenomenon for me, related to some vague images along the road or in a room around a table, and some voice in the air, sometimes with a timbre I like and some irony, I like, and some broken German I like, and some habits I like – that is not much, but sufficient to want to remain in touch with somebody, called Trude in this case. I always try to continue nice contacts – that is that. Charles is a separate personality, I do not much know of his tricks and tracks, I know him as faithful and correct person as far as I am concerned. I highly estimate his correctness in our collaboration and I like his serious side of his life, as far as I know it – I have not the slightest ideas of his "private life" or "non-life" whatever it may be. Since his Paths of Life[142] I should not be astonished if he did found a new religion or something like that or become a painter, or an explorer – why not? The Human SOUL (= as Schnitzler says)[143] is a wide realm, and many various mansions may be there. Why should I not like him, and

[141] See Benson and Betty (1936).

[142] See Morris (1942).

[143] Neurath presumably refers to Arthur Schnitzler (1862–1931), who was an Austrian author and dramatist, oriented towards psychology. Schnitzler's books were burning by the Nazis in 1933.

like Carnaps, and Franks, each separately and sometimes together, Hempels, why not Eva being in Capstadt as a lonely nun in a MAITREYAN camp,[144] and Hempel as a lonely Bachelor in Haity, devoted to DIONYSIAN worship of some kind, MAITREYANLIFE, DIONYSIAN LIFE .... And all these MAITREYAN and DIONYSIANS may be regarded as members of a great family connected with Charles, why not? .... There are funny things in the world, I met somebody who told me of ghosts, he met just the day before, and others told me of a devil – why not? Some know something of THE truth, others of ERRORS, some make ISOTYPES and others a BAUHAUS, a strange world, ... why not? That is life. I must confess that I do not see any persons position properly, nevertheless I like some people and others not, some like much others like, other I dislike. I like e.g. Susan Stebbing very much, I do not think I judge her positions properly, I infer that from the experience, that I sometimes afterwards learn something about her I did not imagine before, but I liked her before and afterwards. Why should I not like Trude without knowing her position? You see even if the position were awkward, why not? We have friends, and we know their defects – just that is perhaps friendship, to like somebody in spite of this defects we know. That is that. If Trude has strange whimsicalities dealing with other people, why not, I knew people who could not transgress open places and others got skin trouble after eating strawberries, Andersen tells of a man who could not look at a parson. And Trude cannot correspond with Charles' acquaintances. Such is life – that is all. But, now wait and see.

I do not know whether you know people with Nazi sympathies, where we did not expect it – I had in mind some younger people, you know hardly. You did see perhaps sometimes Neubacher,[145] but one knew of this fellow that he had whimsicalities of national flavor. We despised the German Nationalism from the start, but some of these fellows without sensitiveness in their fingers spoke in grand style of national unity etc. not believing, that this kind of arguing in Germany was mostly related to sad things, including coarse antisemitism, plunder and robbery ....

Why you are out of touch with the Franks? Terrible Grelling's fate. He waited too long. He explained me he would go to USA, if there were some post for him etc. Many people thought so – unfortunately. I know cases in which people – antinazi Aryans – returned to Germany, disliking the situation abroad as an alien without background. It is right, you have to start from scratch, but why not? I myself take a [sic] pride, perhaps a foolish pride as other things, not to go away, if not immediately forced to do it by some gangsters etc. I did not want to leave Holland and I did not want to leave England as some friends suggested. Perhaps you feel a little like a soldier – against Hitler and this plague. I like to be with the British in these hard days.

---

[144] In Buddhism, a Maitreya is considered to be a future Buddha. On the Maitreyan path of life, see Morris (1942, Chap. 7).

[145] Possibly Hermann Neubacher (1893–1960), who was connected (just like Neurath) to the Social Democratic Party of Austria and to the housing project of Vienna. In the mid-1920s, Neubacher get closer at first to Pan-Germanism, then to Engelbert Dollfuß, the leader of "austrofascism." Later he entered the Nazi Party and became a diplomat in the Third Reich, working mainly in the foreign ministry for the Balkans.

Being in a new environment is rather thrilling and stimulating – I feel like a second youth here. On our life boat I thought of the future activity here. Perhaps others say we should go away sooner. Many did not like my return from the USA to Holland Autumn 1939, in war time .... Therefore I shall look at Grelling from this point of view. Sad, very sad. What was the reason the Washington did not give visa in time?

Waismann is teaching here in some college. I have seen him another day, but there is no real contact, his address: 104 Abingdon Road, Oxford.

Paul got a Columbia fellowship and is now assistant to the professor of statistics – that is Ok. Fine. After all his trouble now some real success – he is not behind his years now. He is preparing his doctor thesis[146] and hopes to finish this job, before he enters in some service, USA army business.

With kind regards from both of us both of you

Cordially yours

[Otto Neurath]

17. Neurath to Carnap, January 15, 1943.

(ASP RC 102-55-02)

15th January, 1943.

Dear Carnap,

Your Semantics copy did not arrive, I therefore tried to get one for a few days. I am just looking through the main chapters, particularly the chapters, you mentioned in your letter. I am really depressed to see here all the Aristotelian metaphysics in full glint and glamour, bewitching my dear fried Carnap through and through. As often, a formalist drapery and hangings seduce logically minded people, as you are very much. I anticipated that, as I anticipated the coming of a religion founder --- such is a certain behaviour of movements, which are based on empiricism. The analogy with Comte's positivism is not so far away. But why not – we are mortals, and therefore we have to be like mortals.

But, let us speak seriously, I mean business. It is for me obvious, that we should have a discussion between friends, as we have organized it about Protocol Statements. I do not like to criticize you, without admiring you and pulling your leg, I want to present my leg pulled by you. The tricks and tracks of such a discussion should be presented in the most kind and friendly way. I should like to do it in an American periodical (the JOURNAL matter is not definitely solved, and we shall repeat our discussion partly there but from another point of view).

[146] Paul Neurath defended his PhD thesis in 1943; his doctoral dissertation was on social life in the Dachau and Buchenwald Concentration Camps of Germany, where he was deported between March 30, 1938 and May 27, 1939, when he was released to Sweden. His book appeared only later as *The Society of Terror: Inside the Dachau and Buchenwald Concentration Camps* (Neurath 2005).

I think the nicest way, would be the following. I write a letter to you, you write an answer, and I make my finishing remarks, and you tell the public, now it is for the first sufficient given to other people's entertainment and studies.

I should suggest that we eliminate "misunderstandings" from our letters, that means it remains our unsolved problematic. And the whole affaire is useful and a kind of prototype, how to make a good discussion.

I do not like your way, to speak of me in your book of some empiricist who makes objections, without giving the reader any opportunity, to read the "objector's" (what an unenglish term) own position, as I published it again and again.

Should you have money in your pocket, cable me, if not write me – the world is able to wait for our 'famous' correspondence on Semantics. I shall discuss the "designatum" and "denotatum" terminology, too.

It is really stimulating to see, how the Roman Catholic Scholasticism find his way into our logical studies, which have been devoted to empiricism.

The Scholasticism created Brentanotism, Brentano begot Twardowski, Twardowski begot Kotarbinski, Lukasiewicz (you know is direct relations to the Neo-Scholasticism in Poland), both together begot now TARSKI etc. and now they are God fathers of OUR Carnap too, in this way THOMAS AQUINAS enters from another door Chicago, where he entered already via ADLER.[147] What an interesting story – that means TRADITION. You remember, I always have been full of mistrust, as far as Russell's Existence symbol was concerned, and Russell the man with the DUPLICATION (Avenarius called it INTERJECTION) is just extending this start, which is closely related to your and Tarski's and Aristoteles' start: THERE EXIST SOMETHING IN ITSELF, this statement I thought is in a language not acknowledged by us (?) or by me (sure).

But these historical remarks do not form a part of our future correspondence. I shall write you, how I am just preparing my monograph by the encyclopedia, and have to deal with the chronicler's language, and how I try to explain, what TERMINOLOGY implies, then your book arrives, and now I see, that the fine strings which have been Wittgensteinean before, or Schlickean, or Russellian, now become more and more Tarskian.... i.e. ARISTOTELEAN. And then I shall tell you my story, and you will answer ... OK.

Your remarks that Semantics may be misused by metaphysicians do not help you, because they are only continuing your actions.

I think the whole thing is very fundamental. I have now to think, what position your Semantics gets after the removal of the metaphysical elements. I think it remains something translatable into empiricism. We shall see – it is not so simple to find out the implications of such a fine structure. You know how to make a building – even if the fundaments were not so good. p. 53 "thoughts" come even into the picture, as in Wittgenstein, Russell etc.

[147] Mortimer J. Adler (1902–2001) was an Aristotelian and Thomist philosopher, and a colleague of Carnap at the University of Chicago.

I am prepared to think of an "Anwendungsbereich" of your system in any case, but first I want to find out, where the elements of non-empiricism may be found. You see I do not see, how all these Semantics problems fit into the discussions of scientific practice with which I am highly concerned at the moment.

I know not exactly what Strauss' objections are, and I asked him about his attitude towards your and my statements.

You see I am interested in statistics and in predictions in this field and therefore in the old problem how to give the whole story an empiricist flavor (the metaphysical flavor of all attempts is well known, also to you) but I doubt whether your way out, continuing the Wittgensteiniade will help us. I doubt it. But I am not very competent in analyzing your story. But sometimes I found something by starting in my own way and then looking at links with the statements made by others, e.g. by my admired friend Carnap, the logical syntax[148] of whom I prefer in many respects to Semantics. I shall be with CARNAP ONE against CARNAP TWO. You know that I did not agree with all your sayings. I am looking through your terminological appendix and I am highly interested in it. A hard chapter – of course. What I think of the Truth terms, and of the degrees etc., you know sufficiently.

We are busy with exhibition making, film making, charts making, reading, writing, designing, discussing social sciences, education etc. I have to write many letters on scientific matters, on educational projects, publishing projects etc. My monograph grows up like plants after a tropical rain. As I started I did not expect so rich a crops [sic] – perhaps it has been useful to go away from The Hague to let my manuscript with Hitler. Now it becomes much more vivid – I think so. I am just rewriting it. The material is ever-sufficient.

Now the war is going on very well; Paul became assistant to the professor of statistics,[149] we heard of friends that they are well in Holland and two of them got babies.... what a world, killing and births. An old story, a very old one, but a sad one. And then great discussions about birth rate and mortality rate. Human beings are rather mad. But as a scientist we have to look at all things just as we look at triangles and circles, but I think we may say sometimes: this damned triangle or this.... circle. Oh my dear Carnap, when shall we give names to bad things together and speaking [sic] of old days in Vienna?

Many greetings and good wishes, from both of us to both of you

every yours,

Otto Neurath

[148] See Carnap (1937).

[149] See Neurath to Carnap, letter 9. note 100 and letter 16. note 146.

18. Carnap to Neurath, January 29, 1943.

(ASP RC 115-07-62)
Rudolf Carnap
University of Chicago
Chicago, Illinois.
El Paso, Texas, January 29, 1943.

Dear Neurath,

Thank you for your detailed letter of December 22nd. I am hurrying to answer it because we shall leave here (Texas) in a few days in order to go to the Mayo Clinic (in Minnesota) -- this is the most famous hospital in the U.S. -- to get their opinion about my back since it does not improve at all. Normally I have no pain, but now much of the Rockefeller-year is over and I have to think of being able to teach again[150]; as I am today -- and have been for the past 7 months -- I could not teach, and therefore I have decided to make this long pilgrimage to Mayo's. I don't know yet that they will be able to do something for me -- some similar cases have been helped by an operation -- but it always depends what they think of the individual case. Well, we shall see.

I had a letter from the Harvard Press in October in which they said that they had no confirmation from the Oxford Press that the shipment containing copies of my book had arrived but that they felt certain that it had arrived because otherwise they would have been notified by the Oxford Press. Therefore I am surprised to learn from your letter that the Oxford Press has no copies. Anyway, I wrote yesterday to the Harvard Press, asking them to send you a copy, registered, if they have not already done so.

I regret very much that you are so intolerant with respect to some ideas of people who share with us the fundamental empiricist attitude. In the case of Popper, I believe your reaction is chiefly caused by the fact that he criticized the Vienna Circle quite unnecessarily. He was overcritical and so you are now. Even when he wrote the book he was in agreement with us on most fundamental points. When later he came into personal contact with us the agreement became even more strong and conscious to him. Some of his views which you criticize, e.g. the refutability of hypotheses, have the same defects -- and, I think, the same merits -- as many of our earlier views: they might be taken as first approximations but closer inspection shows that they are not entirely adequate but must be replaced by better approximations. I suppose that the same holds for many of our present views, including mine, where we do not see today how they should and can be improved.

---

[150] Carnap was on a leave of absence granted by the University of Chicago and financed by a Research Grant from Rockefeller Foundation since 1942. He has spent his time Santa Fe, New Mexico, working on a manuscript that became later *Meaning and Necessity* (1947). See further letter 24, note 239.

In the case of some of your views which are not shared by most of the people in our movement I often defend them in just the same way: I admit to those who criticize them that the formulations are not quite adequate but I point to the fact that they may be taken as first approximations. I like to defend your views but sometimes you make it, by Jove, hard for me to do so when you stubbornly stuck to your old formulations years after they have been shown to you to be inadequate (e.g. "No facts, only statements", your form of protocol sentences, "the semantical concept of truth is only applicable to calculi not to the language of science"). The views of all of us within the movement of empiricism do of course differ more or less from each other. I think for the sake of the movement it would be much better if we were more tolerant towards each other. If your intolerance would become the general custom, then I am afraid you would be among the first to be declared a heretic and excommunicated. By tolerance, of course, I do not mean acceptance of each others views. The differences of opinion should and will be discussed. But this discussion is not helped by labelling the views of the others as nonempiricist and metaphysical.

When you read my "Semantics" and especially when you write to me about it, please keep in mind that the semantical concepts used there are meant for application also to the language of science especially the concepts 'true', 'L-true', and similar ones. In this book I did not apply them to science and I made only brief remarks concerning their applicability. I regret now that I did not emphasize this point more, especially after reading Nagel's review in the Journ.Phil.[151] who in this and some other respects misunderstood the intention of the book, and after reading your remark about Tarski making things independent of the application. I do not know exactly what Tarski thinks of the question of applicability. In any case, the concept of truth as I deal with it is meant as a systematization of the inexact term 'true' as used by scientists and in everyday life.

What became of Waismann's MS?[152] Did you not have it once for examination with respect to the question of publication? I should have liked to see it.

We are very glad to hear that you and Mary are again successfully working in your Isotype Institute, that you are rebuilding your library, have good collaborators and friends. Now that the military situation looks more hopeful our thoughts turn

[151] See Nagel (1941).

[152] Waismann started to write a book (*Logik, Sprache, Philosopie*) on Wittgenstein's philosophy during the late 1920s: it was planned as the very first volume of *Schriften zur wissenschaftliche Weltauffassung*, edited by Moritz Schlick and Philipp Frank, to be published around 1929. Though for some time even Wittgenstein worked on the manuscript with Waismann, he rejected the publication of the material at various points. During the war a near-to-final version was lost, though Waismann arranged an English translation as well, under various titles. Charles Ogden at Routledge and Kegan Paul would have published that translation (in the series "International Library of Psychology, Philosophy and Scientific Method" which included the translation of Wittgenstein's *Tractatus* as well), but it never appeared in Waismann's life. The book in a new form was published in 1965 as *The Principles of Linguistic Philosophy*.

more and more to the question of the kind of peace that will emerge. And there are sometimes reasons to worry. Or do you think we may be hopeful here too?

Reichenbach wrote that Gomperz died in December.

When returning to Chicago, I shall probably teach elementary mathematics instead of philosophy! The conversion of our colleges for war needs has not left many students in philosophy, whereas many more (in Chicago we have a special instruction center for Navy-communication-men) will need mathematics. Therefore I have volunteered to switch over.

With warmest greetings from both of us to the two of you,

Yours,

Carnap

19. Carnap to Neurath, March 15, 1943.

(ASP RC 102-56-03)

Rudolf Carnap

El Paso, March 15, 1943.

Reply to Neurath's Comment on Russell, Meaning and Truth. (Letter of July 17, 1942)[153]

= =

After some unsuccessful attempts to locate your references in Russell's book I came to the conclusion that the English edition must have a different pagination, because of larger pages. I found a formula for the approximative calculation of the page numbers in the American edition, but still in some cases I could not find the places referred to. (The Amer.ed. has 445 pages.)

Engl. edit. | Amer. edit. | Abbreviations:

H: This formulation of R. is harmless because translatable into unobjectionable, empiricist language. (Some of these formulations I myself should prefer to avoid.)

T: This formulation about truth is correct if meant in the sense of the semantical concept of truth. (On these points we shall not easily

[153] See Neurath to Carnap, letter 11.

come to an agreement because you fail to see the difference between
the semantical concept of truth, which is scientific and not absolutistic, and the realistic concept of truth, which is absolutistic
and metaphysical and which is sometimes used by R., see R.)

R: Here is a point where R.'s realism comes in, which I, of course,
reject as metaphysical, like all of us.

p.15	p.14	I can explain your objection only by assuming that you did not read the whole paragraph. It seems to me that you attribute to R. a conception which he himself does not have but only reports and even criticizes: "The behaviorist is thinking of himself as …. he gives a false air of objectivity… As soon we remember the possible fallibility of the observer…"
15	15	(Naïve realism.) I agree.
15	15	(The split.) I agree.
22	23	In a benevolent interpretation, we might still take this as compatible with empiricism because a sentence may be confirmable without being testable. (Comp. "Testability" p.420ff.).[154] However, I am myself doubtful whether R. would agree with this interpretation.
50	59	"Like Schlick". Yes. Therefore not so bad. Even if not right, still compatible with empiricism.
64	78	I do not agree with R., and generally not with his interpretation of not-and or-sentences. (See below). But it is not contrary to empiricism.
70	86	"As if words were not facts". No! R. says himself immediately after it: "Words are some among facts". I do not see on this page any trace of absolutism.
77	94	H. By "desire", R. means nothing else but an occurrence in a desiring man.
92	113	H. Translatable into "something was yellow".
104	128	H. Translatable into "No chronometer is such that, for every time point t it shows t". You say "Why should not by chance…?" Yes, that is possible. Therefore R. has no sufficient reason for his universal assertion; but it is nevertheless scientific, not metaphysical.
106	132	(Wave length.) I agree with you.

[154] See Carnap (1936–37).

108	135	"God". H.
OPO(!)	135	H. Translatable into "world without organisms".
111	138	H. The discussion is entirely behavioristic; R. does not refer to "mind" but instead to a machine which uses words.
132	164	Entirely correct and scientific, no absolutism. You probably misunderstood him. He speaks about a minimum set of premises for a given set of propositions! The whole formulation is nothing else but the customary explanation of the axiomatic method.
133	166	T.
140 141 134	176ff. 178	As far as your formulations are concerned (which unfortunately are not in agreement with your opinions) I have told you since many years that I cannot accept them and hence I agree with the criticism of these formulations by R., Schlick, and many others. In distinction to R., I know your actual conception from conversations; and I am in agreement with it. It seems to me that in your reply in "Univ.Jargon"[155] to R.'s criticism you missed his point.
144	181	I agree with R. in his criticism of your triple-involved form of protocol sentences, especially as condensed in his paragraph "This is to say…" (p.147? – 183). (I do not like his formulation with "continuity of the ego", but it is translatable).
146	182	(Crusoe). I agree with you; but this has not much bearing on the question under discussion: the form of protocol sentences.
148	185	--
148	186	"I mean". T.
154	192	H.
156	194	T.
160	200	"Who…?" G.E.Moore, not R.
161	201	"Who…?" A psychologist. (By "degree of credence" R. means "degree of belief", not "degree of confirmation"!)
164	205	H.
168	210	T.
171	214	H.
186	234	T.
187	234	--
189	237	-- (H.)
189–191	237–9	You say that "meaning" in itself is discussed, not as a social relation. I do not see that; to me the discussion seems behavioristic, not absolutistic. (I should criticize R.'s discussion from the opposite side: To me it is too psychologistic.)

[155] See Neurath (1941/1983).

218	?	
221	?	
221	291?	(Correspondence theory of truth.) T.
245	307	"As if facts and sentences were in different worlds". Not at all. R. says explicitly and repeatedly that he takes sentences as utterances hence as a special kind of facts; thus they belong for him to the same world.
258	323	T.
268	?	
277	347	(Real world.) Perhaps R.
281	352?	R. misunderstands our rejection of realism.
282	?	
286	?	
288	361	T.
289	362	--
304	?	
305	?	
313	392?	No. The concept of degree of certainty (or confirmation) can be defined in a scientific language without any absolutism. That will be shown in a later volume of my Semantics.[156]
315	392?	--
317	398	--
321	?	
321	?	
329	?	
340	427	I too should reject this.
347	436	(Universals.) R.
347	438	(Structure of the world.) R.

My general opinion of R.'s book. As I wrote you once before, I agree with most of Nagel's objections in his review[157]; on the other hand, I think that Nagel failed to mention the merits of the book, especially in comparison with the general run of books in epistemology.

My chief points of disagreement with R. are the following two.

1. I object to what I should call R.'s psychologism. By this I mean the lack of distinction between logical problems and psychological problems.[158] He often

[156] That volume – though not in the "Studies in Semantics" series – became the *Logical Foundations of Probability* (1950).

[157] See Nagel (1941).

[158] On that distinction see Carnap (1936b).

begins discussing a logical problem and then inadvertently turns it into a psychological one. Herein, however, he is not guilty of metaphysics because he admits and often applies himself the behavioristic interpretation of psychology, although perhaps not quite consistently. More precisely: problems of pure semantics are misinterpreted as pragmatical problems. I don't think that this objection of mine is in any way related to your objections to the book. On the contrary I have the impression that you (together with others of our friends, e.g. Ness, Jørgensen, Strauss) often make the same mistake.

2. A more important criticism, in which I agree with you on the whole. Up to a certain point, R. proceeds in an empiricist and behavioristic way. Then, however, his realism appears on the scene and his discussion becomes metaphysical. He discusses pseudo-problems like those of the reality of non-perceived objects, of the physical world, of universals. – This general feature holds in particular for R.'s discussion of the concept of truth. Most of what he says on this concept is within the realm of an empiricist, scientific language, if we interpret it from the point of view of the semantical concept of truth. However, at certain places his realism comes through, and then his concept of truth takes on an absolutistic and metaphysical flavor. Thus I agree with your criticism of R.'s concept of truth only with respect to the places of the second kind (marked above by 'R.'), not those of the first kind (marked by 'T.').

I should like to emphasize the fact that the acceptance of the semantical concept of truth does by no means necessarily lead to the acceptance of realism. R.'s introduction of realism into the discussion of truth is an entirely unnecessary impurification. The semantical concept of truth (as used by Tarski and myself) is entirely scientific and has nothing to do with pseudo-problems like that of realism. It seems to me that in this point you make an error based on a misunderstanding of the semantical concept of truth. I am rather disappointed that all our conversations on this point throughout the years, beginning with our talk in the train to Paris in 1935, have not succeeded in clearing up of what seems to me just a misunderstanding. In my book I have again tried to make clear the nature of the semantical concept of truth (e.g. p. 26).[159] I refuse to give up the hope.

Carnap

20. Carnap to Neurath, May 11, 1943.

(ASP RC 102-55-01)

Santa Fe, N.M., May 11, 1943.

[159] See Carnap (1942, §7).

Dear Neurath,

I suppose that in the mean time you have received my letters of January 29th and March 15th (with reply to your comments on Russell).[160] About a month ago we have come back to Santa Fe; we bought a small and modest cottage high up on a hill with a wonderful view over Santa Fe and the distant mountains. It's all very simple (only an amateur built job which manifests itself each time when it rains -- it will need a few repairs to make it water-tight), but it has for us a great charm and the weather here is very pleasant, dry and sunny most of the year. Unfortunately my back does not show any "spontaneous improvement" which the Mayo people held before my eyes as a possibility. Therefore it seems probable that an operation will have to be done after all if the doctors can agree among themselves that it promises relief. Since they suggested a postponement of at least half a year probably nothing will be done before the end of August, however. I hope the operation will be successful so that I shall be able to teach in the Winter.

I hope you have received in the mean time vol. I and perhaps already vol. II of my Semantics. Vol. II does not contain much of interest to you, I suppose, but the Preface has some bearing on the question of the usefulness of Semantics by showing the role it could and should play in the development of contemporary technical logic.

When I wrote about fellow empiricists making objections against Semantics (in the Preface of vol. I) I was, of course, thinking in the first place of you and Nagel. I thought whether there was anything published to which I could refer. I could only think of Nagel's remarks in his paper at the Harvard Congress 1939.[161] But since these remarks were too short and that volume of the Journal has not appeared I thought it better not to refer to them.[162] As to you, I did not remember any place where you discuss Semantics in print in any detail so that it would be possible to reply to it. I know that you often criticized the use of the concept of truth or "TRUTH" by metaphysicians or (in your Aristotelian paper) by Russell.[163] But I do not remember any critical discussion of Semantics, that is to say the concept of truth

[160] See Carnap to Neurath, letters 18 and 19.

[161] Ernest Nagel delivered a talk ("Charles S. Peirce, Pioneer of Modern Empiricism") at the *Fifth International Congress for the Unity of Science*, held at Harvard University, Cambridge, Mass. (USA), September 3–9, 1939. See Nagel (1940).

[162] Felix Meiner suspended the publication of *Erkenntnis* in September 1937: since Reichenbach had Jewish ties, Meiner was not able to risk publishing the journal with him as editor-in-chief. Volume 7 was taken by Van Stockum & Son in the Netherlands, and from volume 8, the journal was renamed as *The Journal of Unified Science (Erkenntnis)*. Since the Third Reich has invaded the Netherlands in May 1940, volume 9 of the renamed journal never appeared. On the history of the Journal, see Hegselmann (1987).

[163] See Neurath (1941/1983, 226–229).

and related concepts as defined by Tarski or of the semantical concepts which I later applied to empirical science in my "Foundations."[164] Please tell me where you discussed these things so that I may refer to it at the next occasion.

I am now working on a book about probability and degree of confirmation.[165] I think, there I shall have to explain again the distinction between "true" and "confirmed" or "accepted". I explained it long ago in "Wahrheit und Bewährung";[166] the discussion there was meant especially against you. I don't remember that you ever tried to answer my arguments; but if you did, please tell me where. In your last letter you mention that I know sufficiently your view on degree of confirmation; but I don't remember that we ever talked sufficiently about it. I know only the fact that you are against this concept, but I do not know your reasons. If you can explain them in a letter I shall be very much interested. For me it is clear that the concept is entirely scientific and does not have the least metaphysical component in it. Nevertheless, I too regard it as problematic but only in the sense that it is not yet clear whether we can define a quantitative or at least a topological concept of degree of confirmation which will turn out to be fruitful for scientific work. Nagel has expressed serious doubts in his Encycl. monograph.[167]

As you can imagine, I am very sorry about the bad impression you got of my book, and that you even think it is a revival of Aristotelian metaphysics. I try to remember the many and sometimes long conversations we had in the past on Semantics. The first was in the train to Paris 1935. Then there was the public discussion at the Pre-Conference at Paris, with you and Ness on the one side, and Tarski and me on the other side.[168] After these two discussions, I remember I had the definite impression that there were no rational arguments left on your side. When Tarski and I showed that your arguments were based on misconceptions concerning the semantical concept of truth you had nothing to reply. What was left, as far as we saw it, were merely your emotional reactions, namely your dislike of the term "truth" and your vague fear that this way would finally lead us back to old metaphysics. Later we sometimes had discussions on the same topic in America; but I did not have the impression that we came any step forward towards a mutual understanding, still less to an agreement. Perhaps my book will now show you more clearly what we mean by Semantics and what the nature of its concepts is, and perhaps thereby make it possible to you to be more clear and more specific in your objections. In any case, in spite of the disappointing experiences in the past, I am very willing to continue the discussions with you. I promise to you that, when we meet again, I shall

---

[164] See Carnap's *Foundations of Logic and Mathematics* (1938).

[165] Carnap presumably refers to his *Logical Foundations of Probability* (1950).

[166] See Carnap, "Truth and Confirmation," (1936a) and (1949).

[167] See Nagel (1939, §8).

[168] See the remarks in Carnap's "Intellectual Autobiography," (1963, 61).

very patiently listen to all objections you have and, if I find again the old misunderstandings of the semantical concept of truth, I shall try my best to explain the definition to you. I do not know whether it would be possible to make a fruitful discussion by correspondence, where the formulations are necessarily so short, while the mess of things which must be cleared up is so big. But at least we could briefly indicate to each other our views even if we cannot discuss them in detail. Certainly I should be very much interested if you could indicate to me where you believe to find the traces of Aristotelian metaphysics. One point here seems to me very important; in criticizing a concept used by an author one should not criticize the term he uses or the bad things other people have said with the same term, but the meaning which the author gives to the term, that is to say, the way he uses it and the assertions which he makes with its help. Whether out of our correspondence or later out of conversations there will grow something that we shall regard as good basis for a public discussion, we shall see later.

What are your objections against Russell's sign of existence? It is in general use in modern logic and I do not see how a scientific language could do without something of this kind.

The attitude of the Press concerning the Encyclopedia is very regrettable. I am enclosing a copy of my letter to them.[169] I think the decisive word will have to come from you because the contract is between you and them.

One day, we hope in the not too distant future, we all shall sit here on our hill, look at the mountains, speak about Europe, about problems -- looking with tolerance at each other, even if sometimes we have to shake our heads.

With love to you both,

C.

21. Neurath to Carnap, July 15, 1943.

(ASP RC 115-07-63)

30 Bickerton Road, Headington, Oxford

15th July, 1943

My dear Carnap,

I am so busy with many things, that I shall write only a few lines today. Many thanks for both volumes on semantics, and your letters. We are very sorry that you have to take care of your health. We hope you will tell us, how you are going on now. Your letter to the Press is excellent. I wrote a long letter, have sent two copies to Morris to send one to you.[170] I hope you agree with me. IN NO CASE

---

[169] This letter is seemingly lost.

[170] Carnap has sent a letter about the *International Encyclopedia of Unified Science* to its publisher, the University of Chicago Press (see the former letter), thought it is seemingly lost, just like the one Neurath mentions in his letter.

INTERRUPTION, I should prefer to find another publisher, should the press really want to get rid of the enterprise, but first we shall try to make a nice loyal compromise. My monograph is waiting for the answer, then I shall send it to Morris for printing.[171]

I shall write you on Semantics in detail. I tried to find out how you introduce the "existential" statements, I could not find any information. The fact that Russell already introduced this "ontological" element into his logic does not help you and me. All his [bad][172] statements in this last book are more or less connected with his "ontological" start. I think so. The point is: We are using the sentences "there is an elephant here", when people accept a certain group of statements, but we cannot compare any statement with "an elephant is here" as long as we are discussing the usage of expressions, sentences etc. That is that.

War is going on very well, but it will need about two years to get rid of this plague in Europe. And the future does not look nice. What a big weight will be removed after the defeat of the Nazis, but imagine Germany, Austria, Holland .... most of our friends either away or killed, or suicide .... and the atmosphere full of distrust, envy, hate etc... I assume they will kill many of the Nazis but I fear there will be Nazis who will transform themselves into Antihitlerites and terrorize our friends as before only using another flag .... The Darlan story is sad and bitter.[173]

We personally are happy and lucky. The American Ambassador becoming [~~xyxy~~][174] DR in Birmingham spoke mainly on ISOTYPE as a future aid of international education .... fine. Many people interested in our work, our charts appreciated in the press etc. everything going on very well. Son is now Dr, Columbia, very proud – and he has a right to be proud after so short a time. Now he has two Dr degrees, law and social sciences, with fellowship etc. post etc. ....[175]

We have a nice team of nice collaborators, mostly British, we have a nice furnished house with a nice garden – yesterday an Owl on a pole, then flying away with heavy wings ... apples, gooseberries etc. flowers .... good friends, radio apparatus,

[171] Presumably Neurath's *Foundations of the Social Sciences* (1944).

[172] Unreadable word, presumably "bad" was in the original letter.

[173] Neurath probably refers to the case of François Darlan (1881–1942), who was the commander-in-chief of the French Navy in 1939 and after 1940 the deputy leader of the Vichy regime for a time. In 1942, the Allies took some parts of North Africa, and made a deal with Darlan: they gave him control of North African French forces in exchange for joining their side; nonetheless, a few weeks later Darlan was assassinated by a Frenchman recently escaped from a German prisoner-of-war camp.

[174] Unreadable, but John Winant (1889–1947) was the American ambassador to Britain between 1941 and 1946. He even wrote a preface to an ISOTYPE book, comparing the two countries; see Florence (1943).

[175] Paul Neurath studied first law, and obtained his doctoral degree in 1937, Vienna.

many books, and Kaempffert sending American periodicals and cuttings, reprints etc. Our files are growing up again and in a few years we shall feel like years ago, before Hitler grasped all our belongings.

What about our friends? Nagel, Feigl etc. ....

With kindest regards from both of us to both of you

Yours sincerely

[Otto Neurath]

22. Neurath to Carnap, September 25, 1943.

(ASP RC 102-55-03)

30 Bickerton Road, Headington, OXFORD

25th September, 1943

My dear Carnap,

Very bad news – you probably heard of it already – Stebbing died after a cancer operation. The first symptoms appeared some months ago, but the doctors thought, that perhaps radium or something may help. She continued her activity, and looked relatively well – always overworked. Operation suddenly appeared unavoidable. Unsuccessful and then waiting for the end. First very painful, later on better. We did not see her. Our best friend. A very brave and sincere personality. Many people in England feel her death [as] a heavy loss. She represented a kind of public conscience for some circles. We loved her. She has been at particular good terms with Mary, but of course with me, too. She acted as chairman, as I lectured in Cambridge, in a very nice way, and in agreement with most of my statements. And now life is going on. That is our generation, more or less. It is seldom to get new friends. Any loss is hard in itself and hard if you look at the garden of friendship, as I often do.

I have to thank you very much for your kind letters and all your efforts in dealing with my remarks. I have to apologize, that I did not answer you sooner, but I had a lot to do. I had to finish my monograph[176] and then to reduce its size, because the Press thinks now of the production cost more as before and we have now to bear in mind this viewpoint too. I think of the financial things, because we shall show, that the encyclopedia is well f[u]nded [by] in the public. We have to avoid losses. Even losses of our publisher. You agree with me, that we should go on properly. The war is now in the beginning of the end and it would be against my whole attitude to do anything else but go on, as usual. Only this attitude enabled us here after reaching the shores in our shoes to build up an institute with a nice team of collaborators and to reconstruct more or less all departments of our work, we have a full time collaborator in London for research work only. We are publishing charts after charts and seem to become a kind of British household furniture. That implies the making of arrangements, meeting many people, mostly nice ones. We feel here very at home

[176] See Neurath, *Foundations of the Social Sciences* (1944).

and feel with our British friends to be victorious on the battle front. The British way of living is nice, the compromise habit, the not believing in too many arguments, usual commonsense, instead of skyhigh principles from which one tries to deduce concrete details – in vain of course. This adaptation to a new aggregation of items forces us to be active every day, even when reading, listening to the radio, gardening, etc. besides our work. I have to lecture at various places, have to write articles, to attend meetings etc., yesterday evening fire guard meeting of my sector. And then I have to read a lot, Kaempffert sends me heaps of periodicals, reprints, surveys etc., I get regularly the Survey Graphic, etc. and try to imagine the American life, too. Now, before starting my new book, a big one, on visual education[177] (very probable) I shall write my long letter to my friend Carnap. I sometimes hear from Hempel, about you and your health. We are highly worried about you. It is not the question of danger, but of being in a sad state day for day. Unfortunately I have no idea, how I could please you by something you enjoy particularly. We always admire how you and Ina bear the burden.

What a world, millions killing and wounding one another intentionally and then sometimes diseases visit us, too, one reason more, that we should be as nice to one another as possible and enjoy our friends and neighbours as far as they are enjoyable. There are many nice people in this world of sorrow and pleasure. I think just people who are not in the foreground are very often nicer than the people acting and working. The same may be with learning. Usually people learn from mightful people, who may afterwards protect their pupils against the disasters of life: I learned very much from Itelson,[178] who never could do anything for me, except telling me wise arguments. Here in England we met many "small" people, who are good friends, without being in power. People in power are mostly too busy for nourishing friendship. And all our teaching is much more interested in efficiency, as something "making things" than in efficiency for friendship, and enjoying one another. The English education, from this point of view, is more human. I know very well, that caning pupils plays here its part etc., but nevertheless it remains here a kind of sense for personal happiness as something accepted – in the American constitution you have your "pursuit of happiness". It is impressive to listen to plain people here, how they avoid boasting and overstatements in daily matters. I collect "expressions", e.g. fire guard leaders speaking seriously, used e.g. once the term "happiness" explaining how people should get a feeling to be sheltered by the neighbours etc. and then explaining, what is needed to act "quickly", to be "calm" an to have the "usual commonsense". I like this type of habit much more than the continental one, with "high-

[177] Neurath probably refers to his manuscript that was written originally for Karl Mannheim's *The International Library of Sociology and Social Reconstruction* series. It was published recently as "Visual Education: Humanisation versus Popularisation," see Neurath (1996). Some parts were already published in Cohen and Neurath (1973, 227–248).

[178] Gregorius Itelson (1852–1926) was a Russian philosopher, who later worked in Berlin, where Neurath studied with him. Itelson died from the consequences of an anti-semiotic attack in Berlin. On Itelson (and Neurath) see Freudenthal and Karachentsev (2011).

est duty", "national community", "self[-]sacrifice", "obedience", "subordination", etc. "eternal ideals", wherever you give a chance to open the mouth.

I am just reading a lot of older books of German and Anglo-Saxon literature. It is astonishing, how e.g. our German scientists of high rank, such as Weber etc. unconsciously give a descriptive story of Franklin e.g. I should not give after reading his life in detail. I always thought (I only knew his autobiography and some scattered details of his life, of course meeting with Voltaire, action in England, his printing activities, electricity etc.) him a kind of enlighten[ne]d Puritan, devoted to a very exemplar daily routine (in this way as Weber and the Weberians describe it producing the environment of modern capitalism) and now, reading in detail his life, what appeared? A fine chap, acting like wise men do, with all human multiplicity life brings to us – we may agree with that or not, that is not the point, but that the common man's life may be found in the fine fellow, as which I regard Franklin, too. Always interested in humanity and human brotherhood but not in theological doctrines. Making jokes like Lichtenberg[179] – who is one of the Westerners in Germany – on the expressions dealing with drinking "boozy, tipsy, fuddled etc.". Toleration in Philadelphia – what a pleasant thing. A doubtful type of relation to his fiancée, who remains in America, when he is in England, hardly writing to her. Afterwards he marries her, who in the meantime married another man who run away. The story of Franklin's illegitimate child etc., many heartly friendships with women. Always in some way human and interested in human happiness, science, never gossiping but writing sharp and ironical articles, when important things at hand. He was not in the least ascetic. He did not desire to isolate himself from the common life. Nevertheless in addition to that ambitious. That is very characteristic, to like success but not to do something for success only, but for human happiness. And the Weberian strain of calculating prudence and morals not even in the writings, which contain the often quoted calculations (which are sometimes puzzling to me). But no "abstract" wealth as ideal. "Wealth is not his that has it, but his that enjoys is".[180] "Where there's marriage without love there will be love without marriage".[181] "An old young man will be a young old man".[182] (That is all about 1735, the time of the usually quoted calculating statements). And his "An egg today is better than a hen tomorrow"[183] is very

[179] Neurath presumably refers to Georg Christoph Lichtenberg (1742–1799) who was a German scientist – actually the very first professor of experimental physics in Germany – and a satirist who visited England twice. Many of his writings are translated into English; the latest one is Lichtenberg (2012).

[180] See Benjamin Franklin's (1914, 57), *Poor Richard's Almanack*, that was published by Franklin between 1732 and 1758.

[181] Franklin (1914, 60).

[182] Franklin (1914, 14).

[183] Franklin (1914, 13).

illuminating a certain kind of habit. He lived in a comfortable marriage. 1735 he writes to his friend Catherine "I can say, thanks to God, that I do not remember I was ever better. I still relish all the pleasures of life that a temperate man can in reason desire, and through favor, I have them all in my power... I must confess (but don't be jealous) that many more people love me now than ever did before...."[184] Franklin liked the Quakers (the group of Christians I like, too, because they are tolerant and human and active simultaneously). Franklin did not like the tendency of German writers to introduce new words always, but nevertheless his words "colonize" and "unshakable" remained in the English language. (I think not "introduced" by him, but used by him against the daily usage). He was against slavery, of course, in America but also in Britain (Scottish colliers etc.). A politician like modern American Ambassadors who do not come from diplomacy, as e.g. Winant[185] here and others in the Soviet Union. Paris. 74 he writes "I do not find that I grow any older".[186] His Dialogue between Franklin and the Gout tells something about his life. "You eat an aboundant breakfast, not less than four cups of tea with cream, and one or two slices of buttered toast covered with strips of smoked beef...sit down to write...without any kind of bodily exercise... chessboard... for two or three hours... amuse yourself with books, pamphlets, and newspapers most of which are not worth the trouble..."[187] Somebody said disapprovingly of Franklin that "at the age of seventy-odd had neither lost his love for beauty nor his taste for it".[188] Franklin writes to his step[-]ni[e]ce: "...civilest nation (the french) ... if 'tis understood that you like mutton, dine where you will you find mutton. Somebody, it seems, gave it out, that I loved ladies; and then everybody presented me their ladies (or the ladies presented themselves) to be embraced; that is, have their necks kissed. For as to the kissing of lips or checks it is not the mode here; the first is reckoned rude, and other may rub off the paint".[189] Madame Brillon writes him "People have the audacity to criticize my pleasant habit of sitting on your knee, and yours of always asking me for what I always refuse."[190] The correspondence on the life in Paradise together with all the women is a curios reading. His biographer says Statesman and scientist, profoundly masculine, he took women into account as well as any other force of

[184] See Franklin "to Miss Catherine Ray, September 11, 1755," in Franklin (1840, 89–90).

[185] Neurath presumably refers to John Winant, see letter 21, note 174.

[186] See Franklin "to Thomas Bond, March 16, 1780," in Franklin (1817, 59–60).

[187] See Franklin (1906, 19).

[188] See John Adams (1851, 134).

[189] See Franklin "to Mrs. Elizabeth Partridge. October 11, 1779," in Smyth (1907, 393–394).

[190] Quoted by Philipps Russell (1926, 263).

nature ... he treated every women as if she were a person too, and made her feel more truly one than ever. A full life, and very human.

I tell you that, that you imagine my state of arguing at the moment. I always appreciated the Bentham, Mill, strain in Anglo-Saxon arguing, but I always felt uneasy when thinking of Franklin, why? I quote from WEBER: "provisorische Veranschaulichung ... was mit dem 'Geist' des Kapitalismus gemeint ist ... wir halten uns zu diesem Behufe an ein Dokument, jenes 'Geistes' welches in nahezu klassischer Reinheit enthealt .... von aller direkten Beziehung zum Religoesen losgeloest...: Bedenke dass die Zeit Geld ist ... Die unbedeutendsten Handlungen, die den Kredit eines Mannes beeinflussen, muessen von ihm beachtet werden.... es laesst Dich als einen ebenso sorgfaeltigen wie ehrlichen Mann erscheinen und das vermehrt Deinen Kredit.[191] (A rather cynic remark, Franklin tells, how he tried to be decent on Sundays, avoiding actions in the public, therefore READING at home, that does not reduce the public credit etc. many remarks are rather ironical.) "That are remarks from "Necessary hints to those that would be rich 1736 and Advice to a young tradesman 1748." Es ist Benjamin Franklin, der in diesen Saetzen zu uns predigt. Dass es "Geist des Kapitalismus" ist, der aus ihm in charakteristischer Weise redet, wird niemand bezeifeln ----" Weber thinks that these quotations are used properly by Kuernberger in "Der Amerikamuede"[192] and continues "Dokument der (heute laengst verblassten) Gegensaetze deutschen und amerikanischen Empfindens, man kann auch sagen, jenes Innenlebens, wie es seit der deutschen Mystik des Mittelalters den deutschen Katholiken und Protestanten trotz alledem gemeinsam geblieben ist, gegen puritanisch-kapitalistische Tatkraft schlechthin unuebertroffen"[193] etc.

Just that I cannot see. I see that happiness habit in Franklin prevalent (of course I know sufficiently Puritan books, but I also know sufficiently ascetic books, written by Protestant and Catholic writers in Germany) and I think that just the acknowledging HAPPINESS and PLEASURE as "moral items" is the difference, whereas

---

[191] See Weber's *The Protestant Ethic and the Spirit of Capitalism* (2005, 15–16): "provisional description of what is here meant by the spirit of capitalism.... For this purpose we turn to a document of that spirit which contains what we are looking for in almost classical purity ... being free from all direct relationship to religion, being thus, for our purposes, free of preconceptions: Remember, that time is money... The most trifling actions that affect a man's credit are to be regarded ... it makes you appear a careful as well as an honest man, and that still increases your credit."

[192] See Kürnberger (1855).

[193] Weber (2005, 16): "It is Benjamin Franklin who preaches to us in these sentences [...]. That it is the spirit of capitalism which here speaks in characteristic fashion, no one will doubt [...]." And Weber's footnote, p. 138, n3: "[...] document of the (now long since blurred-over) differences between the German and the American outlook, one may even say of the type of spiritual life which, in spite of everything, has remained common to all Germans, Catholic and Protestant alike, since the German mysticism of the Middle Ages, as against the Puritan capitalistic valuation of action."

in Germany Kant stressed the point, that happiness and pleasure do not play any role within the moral sphere. Etc. You see, it is a long story, but I learn more and more, how strange German sociology tought us. I knew it before, but the extent becomes more and more clear.

I see, how well known philosophers in Germany always quote Plato, when speking of the ideal state, and I imagine how many young people accepting that, became weakened against Fascism. Plato is the only author in antiquity and in history, with some fame as moralist, who thought pure and simple cruelties pure and simple oppression as ideal. Children should look from horseback, when the parents disembowel enemies in battles, that they get, as he says the proper "taste of blood like young hounds". He supports censorship, allows only millitary [sic] music, doctors have not [do not have] to help ill people who are responsible for their illness, better for them and for the community when they die, all people of Hellenic blood should be united and then start the conquest of the barbarians, the "enemies by nature" that is the way to be irresistible and to do, what is the highest ideal of the leading groups: the purity of blood.

How can we expect that an enthusiastic youth full of preparedness to selfsacrifice, prepared to think of happiness as something dirty, English utilitarian, can reject the voice of the highly admired Plato-Hitler? Plato this Fifth columnist in Athens who supported Sparta. Prussia? (I remember how Marianne Weber and Gertrud Baeumer[194] after my lecture on planning for happiness said to me "Sehr interessant, wie immer, wenn Sie sprechen, aber warum mussten Sie das Niveau der Debatte senken? Gloeckkk is gemaiain." Of Professor Cohen,[195] Marburg Heinemann[196] told me the statement: "Die Schweine wollen gluecklich sein", Nohl[197]: "Dabei ist allem hoeheren Menschentum immer bewusst gewesen, dass Geniessen gemein macht und die Unabhengigkeit von der Lust die Basis alles hoeheren Menschentums ist. Jede Ethik muss darum zunaechst den Hedonismus erschlagen." I see the difference in habit and attitude much more in the happiness and common sense arguments of

[194] Presumably Marianne Weber (1870–1954) and Gertrud Bäumer (1873–1954). Marianna Weber was a sociologist, a public speaker, and an activist for women's rights, acting as the chairwoman of the Bund Deutscher Frauenvereine (League of German Women's Associations) for a few years. She was also the wife of Max Weber. Gertrud Bäumer was also a political and social activist in the feminist movement. She was a member of the German Democratic Party, working in the Reichstag between 1919 and 1932.

[195] Presumably Hermann Cohen (1842–1918), founder of the Marburg School of Neo-Kantianism. Carnap was a student of Cohen for a while; on their relation see Damböck (2017).

[196] Neurath presumably refers to Fritz Heinemann (1889–1970), who gave a presentation (entitled "Instauration Scientiarium") at the *First International Congress for the Unity of Science* (Paris, 1935). See Stadler (2001/2015, 178).

[197] Herman Nohl (1879–1960) was a German educator and philosopher, working in Jena (where Carnap was one of his students before the Great War) and Göttingen. On Nohl and Carnap see Damböck (2017) and Dahms (2004).

the Western group and in the overpersonal, transcendent, anti-happiness arguments of the German leaders in philosophy and moral discussions. And I think that the metaphysical strain in Germany is in closely connected with the "abstract" ideals, transcendent in eternity, Volkstum etc. AGAINST the human habit towards the own and other people's happiness. The difference between Franklin and Kant.

Perhaps you will think not too strong of my "intolerance" towards metaphysics, when you think of this possible correlation. The empiricist people, happiness people, like Epicure, Montaigne, Voltaire, Lichtenberg, Franklin, Mill, Bentham are less prepared to persecution [persecute] than others ON AN AVERAGE, but more prepared to help other people, because they feel UNHAPPY, PERSECUTED, BEING SLAVES, etc. Whereas in Plato, in spite of all transcendentalism, bodily lust plays a great role – in [the] Symposium the story of Alcibiades embracing Socrates a[ll] night long without success, or in the Republic the RIGHT of the brave soldier to embrace boys and girls, he likes. The price of victory. In Epicure, Franklin etc. the bodily lust [is] in a very temperate way accepted, Epicure telling people, that they have not much to expect from that, better to avoid it, but if, [sic] then thinking of the law, the customs, and other people's happiness. But the tradition is, that Plato is the rock of idealism and Epicure at least a flat and dull pleasure-teaching man. Think how Voltaire usually is treated, who was one of the few famous people in the 18th century who devoted years of his life to help immediately many persecuted people, suppressed people, not only by writing. If a man like Wieland[198] fought Plato's dirty and cruel habit, etc. the historians of literature did, what they could to destroy his fame, to label him as sensuous, etc. Crossman and your Fite[199] and a few others such as Kelsen protested against Plato's unrestricted fascism and mercilessness.

I am now re-reading our Austrian philosophers, too. I read them before and disliked them, but now you see how they formed an environment out of which grew Nazism. You see, lecturing on ZUR PSYCHOLOGIE DES ENTSCHLUSSES 1913 in Vienna,[200] I explained, that our arguments cannot be sufficient, therefore the importance of decision and lot – perhaps. Who protested violently? saying that this

---

[198] Christoph Martin Wieland (1733–1813) was a German poet and writer, mainly known for his romantic epic of "Oberon" (1780). Wieland, on the other hand, wrote a philosophical novel, "Geschichte des Agathon" (1766), about Greek life and culture, in which Plato is described quite sarcastically with critical overtones.

[199] Richard H. S. Crossman (1907–1974) was a British Labour Party Member of Parliament, and taught philosophy for a while at Oxford before 1935. He wrote in 1937 the *Plato Today* (originally broadcast by the BBC). Warner Fite (1867–1955) was an American Philosopher (teaching at the universities of Chicago, Texas, Indiana and Harvard), mainly concerned with ethics and moral philosophy. He wrote in 1934 the *Platonic Legend*. In Neurath's wording, they "unmasked teachings in [Plato's] Republic," paving the way for the comparison of Nazism and Plato's ideas in the *Republic*. See Neurath (1945/2004, 546).

[200] See Neurath (1913/1983).

kind of teaching is against the high ideals etc.? KOLBENHEYER[201] – and I am not astonished that he is now a leading NAZI writer after abandoning our Philosophische Gesellschaft an der Universitaet zu Wien, where a Neurath may teach such terrible things. Another famous Vienna philosopher, wrote during the first world war: "Connected with the growing up of one nation, through more power, better qualities, better environment, castles, temples, poets, science etc. other nations have to go down, less lucky, and nobody w[i]ll shed tears … into the eternal night … who ever fights for his own country always fights in this sense, for THE GOOD CA[U] SE. The hard law of nature (sic) victorious SHOULD be, who CAN be victorious … Hard and merciless history is a judge like nature, BENEFICIAL has been, was happened in reality, what not did happen, SHOULD not happen. … Tragedy of History … The success makes the decision GOOD, regarded from [the] viewpoint of future generations [it] is the case, which [that it] permanently and for ever has been victorious. … There are Poles, who feel they are Poles and want to rebuild tripartited Poland. But assum[ing] these Poles have been exterminated, nobody [any] longer wants a non-dissected Poland – who could judge differently but: a nation not suitable for life, dies, as it DESERVED TO DIE, and by this death space for other nations has to be gained with a better future.[202] (I despised this kind of writing [in] 1915 as [i]nhuman and full of senseless metaphysics. I doubt that 1915 one could write such [i]nhuman arguments without some metaphysics on "law of history", on "rights of a nation", etc. but today it sounds particularly [i]nhuman, when we read of the extermination of the Poles. It is not somewhat deplorable that just a Jewish professor of Philosophy wrote such terrible pogrom book? Would you mind, when, I say this professor with his metaphysics has been a characteristic item in an environment, out of which Nazidom grew up ?) Who succeeds in helping to make his country victorious for ever, he may be sure, that in the eyes of future generations this case will be regarded, too, as the really GOOD case, and appear as useful for the benefit of mankind (it sounds like a blasphemy from viewpoint of humanity to me, not to mention the metaphysical elements of this habit, with his un-pluralist outlook, there is only ONE possibility, not many.) It is not a proper thing for a nation in war to deliberate whether the own case is the good case, it is the nation's duty (sic) the nation's case to transform in the good case, BY BEING VICTORIOUS (I call that the pure and simple pirate philosophy full of metaphysical tricks and tracks). And [it is by] taking care in this way (sic) that the own case will by regarded by future generations as the good case. WE ARE CONVINCED THAT THIS OUR FATHERLAND AND THE GERMAN NATION WILL SUCCEED IN THIS". How much I prefer the attitude to find reasons to fight for something, one calls good, independent of victory and independent of the successful extermination of a nation. Sometimes "cant" and "hypocrisy" – of course, I can-

[201] Erwin Guido Kolbenheyer (1878–1962) was an Austrian novelist and poet, who became associated with Nazism.

[202] It is unclear whose work Neurath translated and quoted here.

not describe to you, how much I like an aggregation of human beings, where one has to apply hypocrisy and cant, when making cruel things, which are despised as such, and one cannot say, victory declares according to the hard law of history that your case is the good one… ugly and nauseati[ng] for me, today as 1915. I understand better people, who tell me, that they do their duty as long as they can in accordance with their nation, but that they would not go on with that, should the nation try to do something against their conscience, etc., of course, that is a temperate way of acting, but just this commonsense way enables us to think of people's happiness and does not overcome the primary feelings of mercy and humanity, so far as they are given in some individuals.

I think that this merciless habit in history very often is connected with absolutism in metaphysics and faith. If one thinks there are many possibilities in arguing then one cannot be very hard with argumentative conviction, only indirectly or by heart, but not in the argument: Of all the possible world systems one is the best in coherence, or the "relative" best (that does not alter the habit) one moral way of living is "the best" given by the "categoric[al] imperative", but "the decision of an authority" etc. may be combined with merciless destruction of other people. Whereas a sceptic habit as such does not give a "reason" for aggression and merciless action as deserved [determined] by THE ONLY BEST SOLUTION.

Therefore, I think, that PLURALIST arguing, which seems to be closely connected with empiricism, leads to a certain toleration, to a preparedness to look at the argument: you are intolerant, as a very serious one. The Pluralist cannot answer, of course, I am, how could I be different. You remember the story I told you of a friend of mine, who divorced his wife (with many children) letting her in a strange situation, and marrying another lady talked over his case with me. I knew all people in question and was simply sorry. I know, some things happen in life, not always clear-cut made, and we poor human beings try to go through all that, trying to avoid the own and other people's pain. Etc, you know how poorly all arguing is at such moments. That reduces our preparedness to judge other people's actions, as long as they are not clearly cruel and producing pain etc. There are no clear-cut situations … for me at least. But my friend did not like my sentimental habit, being sorrow[ful] and doubtful, taking the situation as it was, partly really lucky for the friend and his new wife, partly unlucky for the former wife, but the living together of both not lucky etc. etc. – you know how complicated such imaginations are, he started with explanations and finally presented me the CATEGORIC[AL] IMPERATIVE who induced him to act as he acted. I answered: "My dear, I did not criticize you, I listened, what you told me, and was in a rather doubtful mood, but what about the "categoric", other friends of mine in similar situations did similar things WITHOUT THE CATEGORIC."

You see, the pluralism of empiricism is closely connected with my speaking on a way of living and arguing. I would not say, that I should promote the pluralist arguing for reaching tolerance, but reaching it independently of that I found out that empiricists on an [sic] average are less prepared to become merciless persecutors, and not so frequently the enthusiast followers (for the higher glory of THE transcendent nation ideal etc. or something else) because they are not prepared to sacrifice the[ir] own and other people's happiness to something "idealist" and antihuman. The commonsense leads back to looking at human happiness. That is question of historical analysis. You know I am collecting material for my book o[n] persecution and human brotherhood.[203] We are all poor sinners and hardly able to judge everything --- that seems a mood, which does not lead to persecution and suppression. There are many mansions in the house of freedom and many different animals in zoo of a free democracy. I do not speak of a "law" or something like that, I only stress the point, that the multiplicity of outlooks supported simultaneously is connected with brotherhood of human beings as I see it evolved in the last centuries in Holland, England and America (I know, what one may say against that, and I say it with many reserves, but the descriptive silhouettes speak a certain language, which everybody may understand). What a difference between totalitarian Sparta and democratic Athens. In Sparta a custom to kill Helots as a kind of exercise for the Home Guard youth. Torquemada – what a terrible period.[204] I am analysing the writings of saints and philosophers. Particularly dangerous are poets like Dante, full of resentment and revenge. Like Plato. Always dangerous the aesthetical view of world, the pain of the damned in hell a kind of "bass in the harmony of the music of the deity", that goes through from St. Bernardino (15th century, continuing St. Augustine and others, where the sins form a part of the well managed picture, the beauty increased by shadows etc.) to Nohl, and other modern philosophers with their aestheticism of the word. Karl Moritz (Goethe period) as Mary told me, how impressed she was by his writings on that subject and later on feeling how doubtful the whole stuff is.

That is one of the reasons, why I am very careful in discussing pluralist and antipluralist arguing. You see, Popper's writing is essentially based on a non-pluralist view and I think that all of us, who have been continuing DUHEM, POINCARE, MACH, PEIRCE, etc. did maintain the pluralist empiricism. MORE than one theory possible, when we see one acceptable. MORE than one "history of the world" possible within any given empiricist frame, and – very important – each detailed concrete statement "here is a table" is pluralist from the start, only an attempt to go on tentatively with "here is a table", said by somebody. Another man, perhaps said "here is no table at all". Whatever we do, we start from possible DIVERGENT primary statements and try to put them together, dropping some today, perhaps not dropping them tomorrow accepting a statement, when making theory one for the area three, and accepting a contradictory statement, when making theory five for area two. Of course, we try to eliminate such contradictions, but the way of the sciences is just based on the tricks and tracks, which help us to avoid, as it were, the

[203] See Neurath, "Tolerance and Persecution" in the Neurath Nachlass, 207/K.88. Wiener Kreis Archiv (Rijksarchief in Noord- Holland, Haarlem, The Netherlands).

[204] Tomás de Torquemada (1420–1498) was a Castilian Dominican friar, and the first Grand Inquisitor of Spain. St. Bernardidno, or Bernardino of Siena (1380–1444) was an Italian priest and Franciscan missionary, and economic writer. Karl Philipp Moritz (1756–1793) was a German author and essayist during the late Enlightenment period, associated with *Sturm und Drang*.

INFECTION OF OUR COMPREHENSIVE SCIENCE THROUGH SINGLE CONTRADICTIONS.

Of course, I learned one of the first things from Itelson that any contradiction destroys all formulae in a system. BUT THAT IS THE POINT I DO NOT ASSUME THAT ALL SCIENTIFIC STATEMENTS TOGETHER FORM ONE COMPREHENSIVE SYSTEM, there are only SYSTEMATIZATIONS within empiricism, but not THE SYSTEMATIZATION of empiricism. I expected that Hempel would write about this problem in his monograph, giving the problematic situation, whatever he personally may prefer.[205] He – and you, too – are less objecting to Popper's anti-pluralist view, than I.[206] I did not get any real explanation of your attitude only objections to mine. I think it useful, to start with this point, because it runs through all our letters, yours, Hempel's and mine. It does not help to call me "intolerant" as long, as you do not explain [to] me, why the viewpoint of Popper which is rather connected with INTOLERANCE and TOLERANCE, as I explained above, has something in it from scientific point of view.

You see, Popper starts with the ONE system of the world, as the best possible at the moment. He speaks always of THE system, therefore ONE contradictory situation destroys any general statement. That leads to an asymmetry in TESTING and INDUCTION, whereas both are full of vagueness (PEIRCE), and rather symmetric from the viewpoint. First of all, why making absolutely general laws? Mach tries to restrict the area of validity, and I think we should try as empiricists to make [design] a procedure in such a way, that we may speak of limited uniformities. Further, I look at the procedures of scientists as follow[s]: (I re-read in the last two years and particularly during the last months many authors, Maxwell, Darwin, Newton, Kepler, Malthus, Marx, Smith, etc. and many single papers on various subjects) we start from certain observation statements, which may be dropped sometimes, and try to catch as many of them as possible by means of theoretical tools. Should we find holes for our pegs, we are very happy as research workers, and do not bother too much about the pegs without holes and the holes without pegs, feeling it a progress compared with a situation, without pegs wh[ich] fit into holes.

When we have a pattern, as I described it above, fitting into some area relatively well, and another pattern (contradicting the first in some statements) fitting into another area, we try to patternize [arrange] our material in such a way, that we get only one comprehensive pattern and all pegs fitting into all holes ---- a man who thinks as a research worker of such an ideal even as a "leading ideal" looks rather as a comic or jester. He teaches consistency and the infection power of contradictions, whereas the research worker needs some hints how to avoid the infection from contradictions within his encyclopedia, as it were. Of course nobody speaks of contradictions within a single argument. But it is astonishing how many primitive contradictions can be found in well[-]founded theories within single arguments, and

[205] Hempel's monograph appeared only in 1952 as *Concept Formation in Empirical Science*.

[206] See the reviews of Popper's book: Carnap (1935) and Carl G. Hempel (1937).

the whole machinery acts useful. My father[207] explained [to] me that by saying, that the great thinkers, are not very pedantic, stepping down a staircase made by them they use the railing of empiricism, and when single steps are badly made or lack at all [are missing] the great thinkers go on supported by the empiricist railings. THAT I THINK PICTURES MORE OR LESS THE SITUATION.

[With the r]esult, that I and other research worker, look particularly at – sometimes unexpected – POSITIVE INSTANCES and not so much at the NEGATIVE ones. Popper says, that the good research worker particularly tries to find the negative spots for testing his arguments. Unfortunately he does not tell [give] examples. I collected many examples of the contrary. When you have to start with tentative assumptions, more or less vague, you cannot even expect any sharp negative reaction in the corpore vili of our research field. Sometimes negative instances teach us something, but of course, the positive ones are much more convincing to me and other research workers.

There are sometimes even mathematical disciplines based on weak assumptions and successful, think of the history of differential calculus up to the second half the 19th century. I remember STOLZ, who tried to be more consistent, I think in the same strain as Dedekind.[208] The traditional technique, as in Kiepert – Stegemann textbook on differential calculus, full of man traps, but then Pareto's Differential calculus, tried to improve the consistency, in the German translation we got the first Pareto symbolism, too.[209] And how successful has been the primary differential calculus with all its contradictions, which remained "isolated" to a certain extent. To have a cunning in isolating the difficulties, as it were, seems to be important for research work. If I were a man of cunning in this field I should present some procedures, no general ones, of course, how one avoids the infection. A kind of local resistance against local infection, that is the point, not to be disturbed by contradictions in our rear, when we have to go on somewhere. No cowardice in scientific research; of course, "be cautious", is a rule, but not a general Popper-rule.

I always ask people to give examples for the Popper consistency and negative-test (Popper is only one name f[rom] a dozen names). And now to your statements on Semantics. I shall prefer to speak of the whole and only in a few cases deal with

[207] Neurath's father, Wilhelm Neurath (1840–1901) was an Austrian political economist, teaching at the Hochschule für Bodenkultur in Vienna. On Wilhelm Neurath, see Uebel (1993) and (1995a); cf. Sandner (2014, 209–228).

[208] Otto Stolz (1842–1905) was an Austrian mathematician, known for his work on analysis and infinitesimals. Richard Dedekind (1831–1916) was a German mathematician, working on abstract algebra, number theory and the definition of real numbers.

[209] See Kiepert and Stegemann (1897).Vilfredo Pareto (1848–1923) was an Italian polymath (engineer, sociologist, economist, philosopher etc.), his name occurs in the "Pareto principle" as well, claiming that in many cases 80% of the effects come from 20% of the causes (80/20 rule).

Russell, because the "T" points are just the important ones.[210] One question in between: do you think that Russell's book helps a student in understanding better his own empiricist research work, suggesting an alteration in dangerous terminology etc., and if not that, what this book helps? [sic] Telling me, that it is less disturbing than philosophical books is a weak consolation.

You see, I start from the many observation-statements, (not sense data or something like that), as made in a laboratory: professor X says here I see a certain retort with a fluid behave in a certain way. Another person tells a different story. What to do. Sometimes we can connect both, where the one man speaks of "grey" the other speaks of "green", then we speak of a certain item of the type "A", that implies "grey" in a X-story and "green" in an "Y" story etc. We try to find a COMMON statement, sometimes we find it, sometimes only under certain assumptions. Sometimes we say in our encyclopedia "here is an elephant" implying, that this statement is in concordance with certain observation statements, ...

That is the reason, why I do not know, how to transform your arguments in semantics, with my start, which seems to be a very humble thing. Modest and tolerant.

You say page 22, 23, Suppose Pierre says "... true, IF A CERTAIN OBJECT, PIERRE'S PENCIL HAS A CERTAIN COLOUR BLACK. ... TO FIND OUT WHETHER ... WHAT WE MUST DO IN THIS CASE IS TO OBSERVE THE COLOUR OF PIERRE'S PENCIL. You may think me hopeless, but please, try to talk with we in my way.

I should say X says, something is black, Y says something blue, I AM NOT USING A PHRASE LIKE, "we have to observe the colour of Pierre's pencil", but: how we may use both sentences. Perhaps in this way, X is saying he is in the state of a black-observing person, Y is saying he is in the state of a blue-observing person, but now the same story starts with the being there of a person Y or X. Y says here is a person Y and a person X, X says here is a person X only, no Person Y at all etc. The question is to find a pattern of statements we are finally to "accept" tentatively in our arguing.

And now I ask, how we may speak of accepting something as a "lie", you know I answer, when I accept X says this is brown, when I accept X says internally this is black and not brown, we may accept tentatively X is a liar. If we accept X is saying I see [a] brown table and X is internally saying I see [a] brown table, and we do not accept the statement here is a brown table (as a combined statement, as it were) then we call the X-statement a dream statement or an illusion statement.

Now somebody would say, we call this statement "true" only if X has an illusion, but this "has an illusion" implies in our language only, that it fits into a scheme of acceptance.

I see now various possibilities, e.g. (I asked sometimes Morris about that) that you may show me, that my explanations INCLUDE subconsciously the primary

[210] On the "T" notation see Carnap's letter to Neurath, letter 19.

habits of semantics, OK, please show me that, I confess that would be the nicest solution. As it were, you transform my statements in such a way, that we get a one-one correlation between your and my speaking, then everything is prepared for further discussion, or you may show me, where my empiricism reaches difficulties. Or where there are questions I cannot answer. It is for me startling that Tarski in Paris tried to give examples, always from certain mathematical or logical generalizations, with "all" or something like that, and Hempel trying to explain [to] me semantics in relation to empiricism mentions the Goedel business. You see, I do not want to deny that in speaking of calculi PERHAPS something appears, which forces – let me say a two-level discussion or an infinite number of levels [of] discussion, but then I should ask, whether this level-business touches directly [on] empiricist discussions, or only in so far as we need these higher calculation statements within an empiricist argument, but not as long as we speak of cows and calves, of brown and black only. Please allow me to be a little puzzled by this type of answers, when I as a humble empiricist want to understand the semantics secrets.

Page 29 you speak of "judicial proceedings" – that is my field. And just there I think, we should improve our empiricist view, by introducing my proposal, which does present only empiricist statements, but let undecided which is "true". Not even using this term or a substitute for it. Or can you show me, how "accepted" is only a way to be propositionally equivalent with you and Tarski but using other terms. I doubt that very much, because you are quoting Aristotle. His arguments, however we may interpret them, seem to me clear-cut metaphysics. And Kotarbinski is an Aristotelian, of course. You see I had a long discussion with him on the ONE world-system of Popper and asked him, [whether] that implies you agree with him, OF COURSE he answered. He agreed [o]n many points with me, but not [o]n the [sic] pluralism, and just this, this idea of the TRUE SYSTEM, THE TRUE WORLD, THE TRUE WORLD and OUR SENTENCES DIVERGENT FROM A TRUE WORLD etc. is the difference between pluralist empiricism, which is "monist" as it were in the language, but pluralist in the possibly acceptable statements. Therefore this Aristotelian metaphysics on the "true" is connected with the anti-pluralist viewpoint, that is a reason, why I am interested in this matter from a more general viewpoint in addition. Of course, that is not an ARGUMENT.

You see, as long as semantics appears as pure calculus I have nothing to say, assumed that your calculus is consistent, Martin Strauss is doubtful even on this point. But in the moment to bring any empiricist elements into the discussion, I ask how it affects empiricism.

You see, I am impressed by the behaviour of Hempel in social problems, Zilsel behaves similar[ly]. The SYSTEM assumption, excludes multiplicity of arguments, and does not put into account the UNPREDICTABILITY as something given IN PRINCIPLE. I treated this point in my monograph seriously.[211] Different kinds of unpredictability within empiricism. I do not say, that both things belong together,

[211] See Neurath's *Foundations of the Social Sciences*, (1944, Sect. 12). On Neurath's concept of "unpredictability in principle," see Reisch (2001)

who says the one, says the other, too, but there is a certain affinity, of course. Starting with pluralism, unpredictability does not shock you at all. Within a clear cut system unpredictability hardly may be discussed. Perhaps only when speaking of various levels, and we get an infinite number of sayings on saying – but I doubt whether that may be connected with the empiricist unpredictability in a proper way.

You always tell me, you agree with Russell's and Schlick's remarks on my protocol statements, but my statements intended are different.[212] Please, tell me first what you think how I should express my statements properly and then please tell me, why even then they are not in harmony with your opinion. I think it is important to come to some clearness therefore I write such a long letter, transferring to you the whole mood in which I treat the question, not concealing my far[-]reaching guesses, in as far as anti-pluralism is concerned, which I relate to Aristotelian-Platonic anti-pluralism, intolerance etc. etc., but that is NOT THE POINT OF OUR DISCUSSION. Historically only it would be remarkable, should we find out, after accepting the pluralism, and removing the anti-pluralist elements of semantics (SHOULD THERE BE SUCH ELEMENTS) that the strain in Polish-philosophy is an old Scholastic one. Lukasiewicz's co-operation with New-Thomism would be then fully understandable, the connection of Kotarbinski's Aristotelianism and Lukasiewicz free-will tendency in multi-valued logic etc. with Twardowski's logical trends, who was a pupil of Brentano, who combined the logical trend of Scholasticism with the trend towards modern French and English psychology.

There is sceptic trend in Scholasticism, too, DUHEM is a representative of this trend, very often scepticism in science, tries to get absolutism in religion as a kind of substitute. Therefore both trends multiplicity in action, multiplicity in arguing may be connected with Scholasticism, but it is a difference, whether the absolute solution appears as an additional element as in DUHEM or as within the empiricist scheme, as in ANTI-DUHEM Popper, with his religious and Kantian strain.

Historically, we may say:

PLURALISM IN EMPIRICISM, ABSOLUTISM IN TRANSCENDENT WORLD.
PLURALISM IN EMPIRICISM, NO ABSOLUTE WORLD OF TRANSCENDENCE
ABSOLUTISM IN EMPIRICISM, NO ABSOLUTE WORLD OF TRANSCENDENCE
1, is perhaps connected with DUHEM,
2, is perhaps connected with POINCARÉ,
3, is perhaps connected with POPPER and other pseudo-rationalists. But these are no arguments at all, only historical remarks.

Since I think it rather dangerous to speak of the DESIGNATUM of an expression I suggest to speak of an ACCEPTED SENTENCE (or of an [sic] designating sentence), instead of a denotatum I suggest to speak of an "acknowledged" expression (or of an [sic] denotating sentence). Now, it may be, that this change of terminology

[212] See Schlick (1934/1979) and Schlick (1935/1979).

enables you to formulate semantics in my terms, too, please, do it, if possible. IT WOULD BE OF GREAT IMPORTANCE FOR ME, BELIEVE ME I WOULD LIKE TO BE IN HARMONY WITH YOU EVEN IN THIS POINT, or this terminology prohibits semantics, then I should like to understand the defects of this terminology.

page 18, I go on in this way, I ask myself can somebody say: I see sentences, OK, can I say I see propositions – NO, therefore I try to eliminate the term proposition. In my monograph I suggest to speak of sentences propositionally equal. Have you any objections to that? Remains everything unaltered? or not? I know it needs some time to answer that. But what does not a friend perform for a friend, ...You know within the way of life I described above somebody may sacrifice happiness for other people's happiness, but not for something over-personal... that is the point. Perhaps you feel a little pain thinking of my unhappiness connected with the semantics difference and you want to reduce your little pain by removing my pain ....

There are other points in your book, too, I shall discuss after your answer. Perhaps we should start with these primary remarks.

You answer, that all my remarks touch even Russell's existence symbol in his logic. Of course, it does. I always knew that and I always looked with some suspicion to that. But as long as such symbols remain within [the] calculus not much may happen, we should not be too pedantic, but if they did enter the empiricist sphere through semantics, we have to be careful. But that is another long story, connected with the problems of introducing symbols, an action not very properly done [peformed] by Russell, as you may see, where he introduces letter combinations. How to say that ab and ba are the same or vice versa, only ab may be used.[213] Russell's remark is a very weak one and only an additional one – Stebbing made the right remark on that in her book.[214] This whole symmetry business is a difficult problem in itself. I am not sure, what to do in detail with it. My remarks on monogrammatic writing touch only one point, and my remarks on symbolically equivalence etc.[215]

I have the impression that Russell has an anti-pluralist tendency, that may be the main point in his "realism". People sometimes cannot bear, that we start with many divergent statements, and remain with divergent statements FOREVER, as it were.

---

[213] Neurath had addressed this issue in two of his early papers on logic. Between 1909 and 1910, Neurath wrote several papers on algebraic logic while one paper with his future second wife, Olga Hahn-Neurath (who was the sister of Hans Hahn, the founder of the Vienna Circle) was writing her doctoral dissertation on the topic. They co-authored one of them "Zum Dualismus in der Logik" (Hahn and Neurath 1909). Neurath's other papers are Neurath (1909a), (1909b) and (1910). Neurath treated the question of symbolic equality and commutativity in the last two papers. Olga Hahn wrote also Hahn (1909) and (1910). On a brief, critical discussion of Neurath's logical ideas see Köhler (1982/1991).

[214] See Stebbing (1943, Chap. XXII, §4). Stebbing addresses the relation of definition and linguistic analysis criticizing Russell on the very question Neurath touched on in two of his logic papers. This is the reason Neurath might have brought them up in the exchanges about logic and Russell.

[215] See Neurath (1910).

There HAS TO BE SOMETHING O N E. The persons in our group who deal with empiricist research are, as far as I can see, pluralist by habit, DUHEM, POINCARE, FRANK, NEURATH etc., whereas the others like HEMPEL, I think you, too, to a certain extent, of course TARSKI, POPPER, present a non-pluralist tendency. I think ZILSEL is non-pluralist on an [sic] average. Therefore the importance of the NEGATIVE instances in Popper. Hempel wrote [to] me as if it were something without doubt that general procedures of science have to be based on something like that, I answered, that I do not think there are general procedures, neither in induction nor in testing, but there are things you may learn. I should like to build up a kind of correspondence on that for publication, it is more friendly and not so systematic. I think you are often more systematic, than empiricism allows us to be, e.g. in your Meaning article,[216] which I read and re-read, admiring your cunning and skill, pleased by common features in our arguing and a little sorry about the differences. Analyzing that, I see clearly some historical features, we disagreed, when you agreed with WITTGENSTEIN (I think him an antiscientific metaphysician, who helped us a lot), with POPPER (I think him anti-pluralist metaphysician, who helped us a lot) with TARSKI (I think him an antipluralist, too, who helped us a lot). I guess that the pleasant systems of logic and mathematics seduce people like you, Hempel, Tarski to desire systems of a comprehensive kind in empiricism, where just that is – I think so – a characteristic of empiricism, that there are only islands of systematization, as above mentioned, and, in principle, some unpredictability, as long as we are inventing people, as long as we etc. It would be nice if we could fit into that correspondence Hempel['s] letters, Frank['s] letter, too. Perhaps Nagel, Strauss, etc.

You speak of my intolerance – and I want to avoid any kind of intolerance. I have the feeling that I build up a kind of tolerance arguments within our LOGICAL EMPIRICISM. Of course, we shall not be pedantic and not always call other people's opinion metaphysical between us, of course. We often see, as Frank said the metaphysical dots in another person's features but not the big spots in our own.

Perhaps you will find analogies to POPPER's experimentum crucis, the negative instance, the one world system in my own arguing. I should like to learn from you that.

A few remarks on your kind RUSSELL['s] remarks.[217]

I first wrote some pages [with] answers to each remark, but that is rather silly. The "T" points we should discuss full and pure, please do it. Of course I do not agree with the [sic] most of your "H" characteristics, because we have to look at the whole context and not whether a good lawyer could perhaps find a translation for something. The question is, whether Russell and his readers would translate [it] in this way. You doubt self that sometimes. But let us not discuss trifles, the main points is: "T" and your criticism of my standpoint, perhaps in the improved shape you will present to me (assumed I shall accept it).

[216] See Carnap (1936–37)

[217] See Carnap to Neurath, letter 19.

15/14 of course R. discusses with the behaviourist, but he uses in principle the same argument, speaking of "FALLIBILITY" that implies, that he does not accept, that the Beh. accepts a certain statement naively as infallible, whereas RUSSELL wants to find out that there is fallibility, how [should it be] defined? Who takes the chair? Russell? The error-fallibility terminology seems to be acknowledged by Russell.

50/59 knowing = more or less making a statement; having an experience = more or less making an experimental statement, saying: we cannot repeat a statement we are telling of [sic], seems to be objected by you in your Syntax.

64/78 what implies "antithesis" without assertion within empiricism?

70/86 when we drop the term "fact" and use only the term "factual statement" then we drop the expression "facts left unexpressed". Have you no objections to this latter expression?

104/128 you think one can differentiate in empiricism between THE right time and the chronometer time? I think empiricism has only different chronometer times and by convention we may call some computation "right time", but that is not what Russell has in mind. What you think about that in detail? It is one of MY POINTS again and again.

108/135 "God" is here the hypothetical substitute for Laplace's hypothetical spirit. THERE IS N[O] SUCH ANALOGY IN EMPIRICISM. physics deals with CONVENTIONAL unity only, which may be altered, not with SOME DEFINITE UNITY, as assumed in the God analogy. Otherwise I do not know, what the analogy intends to perform.

132/164 I think even the "aesthetic" test does not lead always to ONE solution. We may toss the coin, when not having time to try various possibilities. Just this aesthetic argument here, indicates (for me of course only) the tendency to find any "rational" way to reach ONE solution, whereas I insist that we have to think always of more than one solution, whatever may happen, as possible.

160/200 Russell should ask "who makes the statement".

160/201 I think the Moore discussion is analogical to the Behaviourist discussion. I wanted to say, how one has to go on, by asking WHO says something, and not where we find the definite answer. Russell thinks one cannot speak of the absolute item as such but only given to a certain degree, what involves the assumption that we have some absolute item, which cannot reach completely but only to a certain degree. That implies e.g. the one dimensional arrangement of the items etc. etc. You know, that is my objection to your MEANING article,[218] that you speak of "degrees". Not even degrees – I think so.

For me Moore and Russell are here in the same boat, because [an] absolute item and [an] item to a certain degree different from an absolute item makes no great difference for my argument. It seems to make a great difference for you as I infer from many of your writings. The DEGREE business is a very serious one. The PREFERENCE business could be branching and could be like the p[e]cking order of hens, i.e. circular. Degrees of "true" page 133.

---

[218] See Carnap (1936–37).

p, 218 in my edition, a possibly strange solipsistic duplication theory [of the] brain and phenomena.

245/307 I think that Russell is not consistent in the treating of statements as facts and treating differently both, when you say about page 22, where he speaks of II, subjective and objective side of statements…

what [do] you think about page 320 where R. speaks of [a] relation between events and the propositions. page 330 whether anything corresponds in the non-verbal world. You think that OK?

you see all these and other remarks are around the statement on page 340 on the perceptual whole within our head, what your reject. I do not know exactly, how you separate the accepted Russell from the non-accepted one. It would be interesting, perhaps one of your pupils could do that, to present the REALISM structure in Russell. I think most of his arguments belong to that or can as parts of it be interpreted.

I hope you will speak in detail on my remarks on 133–144. We have to discuss that seriously. See above.

I shall see, how you define "degrees" without limes.

I think you are such a sorcerer, that something may be possible, I cannot see today.

Let us drop [the] labels unempiricist etc. when discussing with one another, but you allow labels like pluralist, non-pluralist for not always repeating the single remarks. What I needed for my Social Sciences I explained in my monograph.[219] I think you will not like many of my attitudes – unfortunately, but your tolerance will overcome that and you will explain to other people, that I am not so bad as I look. I think, I should explain more in detail my point later on, but the main points are clear: unified language without dualism, asymmetry etc. from the start, pluralism with encyclopedism, avoiding infecting contradictions, unpredictabilities within empiricism.

I think I may guess what you have in mind by speaking of the language of science and not of science, but you see, my way is trying to find the point where our rivers meet one another and in semantics I did not find the point, whereas in your Logical Syntax I found the point, even when we did not agree.[220] The same about [the] MEANING article. I do not always agree but see, where we meet, where not. Nagel did not understand your viewpoint. Did somebody write about it from an empiricist point of view? I shall expect that some Neo-Thomists will find a way, how to use your and Tarski's semantics. That does not imply, that your arguments are similar to Thomism.

Waismann re-wrote his book. [The] English translation made for Ogden did not enjoy [please] him, therefore he started with altering it.[221] Ogden another day asked

---

[219] See Neurath (1944).

[220] See Carnap (1937).

[221] See Carnap to Neurath, letter 18. note 152.

me about this business, I could not answer. I did [have] not see[n] Waismann for a long time. By chance I heard of his wife's suicide. We visited him, wife and son another day, and felt not very happy there.[222] I am always sorry when seeing people, who escaped from the hell and live then depressed in some way or another. It is not the question of guilt but of good and bad luck in all the constellations. I do not think that Waismann is very happy in all his intellectual relations. The break [up] with Wittgenstein – [a] disaster for him. The main ideas remain and therefore he seems to be relatively isolated. Sad experience, as he told me, with Braithwaite.[223] He is lecturing at the University here, that is a great advantage. But that implies certain adapt[at]ions perhaps, I do not know. The world is so full of sadness. I heard a lecture [marginal addition: *by W*] on geometry. I did not like it very much. A fine sequence of lectures on Mengenlehre in the camp. He did not like very much my tendency to ask about finitism in Mengenlehre as one reached it in differential calculus and occasionally in probability etc. But I heard that such tendencies are now relatively strong. Have you [any] knowledge of that? What could I read? Or better look into it. Poor boy, I am. No friend, who can tell me that as many years ago. Perhaps that is a reason, why I now deal more with the theoretical background of the Social Science language.

Poor Gomperz dead. He behaved very kind[ly] to us when we arrived in Holland, helpful in spite of the fact, that he rather disliked my arguing. I disliked his arguing very much, particularly his involved moral reflections on topical subjects. Rather disgusting to me. Unfortunately a frequent type of arguing in middle Europe. What is about Gomperz's wife? I like her very much, a brave, handsome person, who knows how to deal with life. I always appreciated it very much that he married her, a simple Viennese merchant woman, continuing her business at the "Tuchlauben". Happy in travelling.

I hope you are well again and will teach Mathematics in Chicago; tell more about you and your life. It is a great advantage that you both together can be in [a] nice environment.

[222] Waismann's life was dogged by misfortune and filled with misery: he did not submit a doctoral dissertation until 1936, so he worked as a librarian and taught at adult education institutes, sometimes in precarious conditions. His relation to Wittgenstein was quite conflicted: according to Gustav Bergmann (1993, 204), Waismann became a "self-denying disciple" of Wittgenstein. Later Waismann moved to Oxford from Cambridge partly because of Wittgenstein's reappearance as the successor to G. E. Moore's professorship. Even though Waismann became a faculty member at Oxford, and later a Fellow of the British Academy, he was never able to adjust to the conditions of life in Britain; he became increasingly isolated, especially after the tragic suicide of his wife (1943) and son (1952, at the age of 16).

[223] Richard Braithwaite (1900–1990) was an English philosopher of science who is known for his 1953 book, *Scientific Explanation: A Study of the Function of Theory, Probability and Law in Science*. Braithwaite attended also some of the *International Congresses for the Unity of Science*.

On average our friends reached Anglo-Saxon countries, a few others, South America. Some of our Polish friends I think the Soviet Union.[224] But some have been killed, committed suicide or have been deported, or in concentration camps. What a sad world. What you know about Maue?[225]

I often ask me, to what extent we are responsible, too, for all what happened, by doing something or by failing to do something… I know historical analysis etc., but I want to imagine a little, what kinds of streams lead to the Nazidom. Direct ones, as the quoted chapter on exterminated Poles and the moral qualities of Victory and the obedience to the nation, whatever may happen, but also indirect ones, e.g. the supporting of totalitarian habits as such, and so on. A difficult problem, really.

Thanks for the article on planning.[226] I go further than that, and think we can be more multifarious within a planned economy, should we try to be so, whereas competition has been unifying. It is not so much the question of decentralisation, that too, of course, but more a question of [a] planned multiplicity [of] possibilities. Not even majority decisions should be overestimated. When a group of 40 likes flats and a group of 60 small houses, why not 4:6 flats and small houses? etc. Railroad gauges have to be unified. I shall ask Kaempffert to send Morris my article, he will send it to you.[227]

We have been in the Wales mountains, very happy, walking every day many hours, sleeping, eating, reading – not much, enjoying nice landscape and nice people. A few hours discussion with Heinemanns – a wholly un-intellectual recreation, our own talks more in [like] reflections on world and life. We try to find out how people behave in various countries under certain circumstances. The aggregation [sic] in the Anglo-Saxon countries and in Holland seems to be preferable to us. And we think these aggregations allow the growing up of more happiness, even if more

---

[224] After Germany attacked the Soviet Union on June 22, 1941, Leon Chwistek (1884–1944), a prominent member of the Lvov-Warsaw school of logic and philosophy escaped from Lvov and went to Tbilisi (the largest city of Georgia under the direction of the Red Army), and in 1943 to Moscow, where he died. Regarding South America, Neurath presumably refers to Hans A. Lindemann (1882–?), who regularly attended the Vienna Circle meetings during the early 1930s while he studied philosophy, psychology and art history at Vienna. He also wrote a dissertation under the direction of Schlick on behaviorism in 1932 and fled later to Argentina. Cf. Stadler (2001/2015, 594). See further Cordero (2010, 370–371).

[225] Maue Gramm was the wife of Josef Gramm, an art historian who lectured in Munich. She and Carnap were lovers for some time and he fathered two children with her, Gerhard (1929–2013) and Birgit (1927–).

[226] There is no information on what this might be. Even Carnap seemed to be puzzled about this remark since on the margin of the letter there is a shorthand note in German asking "What was that?"

[227] The only place where Neurath discusses that example of flats, houses and railroad gauges is his manuscript "Visual Education: Humanisation versus Popularisation," which was published only posthumously in 1996. Another article that might be relevant is Neurath's "International Planning for Freedom" (1942/1973), which Neurath had already meant to send to Carnap.

temperate. Less [fewer] tensions as something [is] loved and admired. Holidays, weekend, hobbies important, even when very restful and without excitement. Of course, there [are] deplorable things, too. But the whole aggregation [sic] compared with others is fine. We enjoy together with our British friends the victories. You know from Modern Man in the Making[228] that I always guessed the Allies will be much stronger than the Axis, much stronger. I did not expect that the Axis could finish the whole show in a Blitz. Now the whole show is manifestly over. It may be up to 1945, hardly longer in Europe, perhaps shorter. The Nazis fight for their naked life. Even Non-Nazis are often full of obedience in abstracto and think it a duty to fight on. Of course there are Anti-nazis there. But the aggregation [sic] is not prepared for opposition and fight against government.

The peace, I do not think, it will be a fine peace, but that the Plague will be over will be in any case pleasant. We should talk over all these things some day. Perhaps visiting you by plane will be possible in a near future. Who does know anything? Unpredictability teaches us, to think of our own decisions – what is called moral personality – as of more importance and to remain hopeful in difficult situations. Historical laws teach some people cowardice other[s] fanaticism. I do not like either.

How speaks Heine: Beat your march on the drum, no fear, that is my whole philosophy, that is my whole science, I know, that is the result, because I am a good tambour. etc.[229] Go oh, boy, something will happen.

I can repeat only, that we go on very well with ISOTYPE and that more and more people are interested in it, that we have many friends and acquaintances here and feel us very well sheltered. Our health is in good state, our mood, too. Paul, as Lynd wrote me, got a fine degree from Columbia and will be successful in his career, he predicts.[230] Philipp and Pepi Frank safe. Some people of whom I thought they are in Germany, appeared to be in America or England. That is always a day of particular pleasure, when such an [sic] information reaches us. Many good friends from camp remained good friends afterwards, therefore the environment is OK.[231] And we know we have good friends in other countries, too. Morris writes in detail on his paths of life.[232] I like his attitude and his analyzing paths of life as patterns, but I cannot agree wholly with his technique in detail and some distinctions. But the brotherhood of mankind is more important than such details. It is some kindness in his attitude and much tolerance.

---

[228] See Neurath (1939).

[229] Neurath is quoting the first verse of Heine's poem 'Doktrin' in *Zeitgedichte*: 'Schlage die Trommel und fürchte dich nicht.'

[230] Presumably Robert S. Lynd (1892–1970), the American sociologist, who was a friend and colleague of Paul Lazarsfeld, who helped Paul Neurath study in New York. Lynd is mainly known for his systematic sociological work at Muncie, Indiana ("Middletown" studies).

[231] Neurath presumably refers to his time in the internment camp on the Isle of Man.

[232] See Morris (1942).

Now, it is evening and I want to finish this letter. I hope it helps to find a way to Semantics, too ---- that is that. Oh boy, my dear.

What kind regards from both of us to both of you

ever yours

Otto Neurath

[Dear Carnaps

I hope very much that you have no pain and that the doctors can help you to get well again soon. I was very glad when I met you both in a cinema queue ... it was a dream only, unfortunately.

Good luck! My to Ina!

Yours Mieze][233]

23. Neurath to Carnap, October 28, 1943

[October 28] 1943[234]

Dear Carnap,

Enclosed I am sending you Freundlich's syllabus please return it with your remarks immediately.[235] I suggested he should start writing, but we should like to make (1) and (6) more "logical". We speak in the Foundations of the theories mainly. I am writing in hurry.

I hope my manuscript will be in print already and I shall read the proofs soon. I just signed a contract with Kegan Paul and shall write a big book on Visual Education.[236]

We hope you feel better now. I expect a letter from you. We are very well. Paul Dr and instructor at University college [sic]. I think he is OK now.

With good wishes and greetings

Ever yours

[Neurath]

[233] Handwritten note by Marie Neurath at the end of the letter.

[234] This letter was originally without exact dates in the Archive; only "1943" was indicated as a handwritten marginalia. But Carnap thanked Neurath in his next letter (February 4, 1944; see letter 24) for a certain note on Freundlich's syllabus from October 28. This letter might be that note.

[235] Freundlich's syllabus is lost/or not located in the Neurath Archive.

[236] See Neurath (1996).

24. Carnap to Neurath, February 4, 1944.

(ASP RC 102-55-04)
Rudolf Carnap
P.O.B. 1214
Santa Fe, N.M.
February 4, 1944

Dear Neurath,

Thank you very much for your letters of July 15, the long one of September 25, and the note of October 28 concerning Freundlich.[237] Your detailed letter raises many interesting problems and I shall now reply to some of them. In addition I was glad to get a picture of your activities, your personal life and moods etc. I should have written you much earlier if Ina had not developed a backache too which makes it hard for her to type; I hope, you will appreciate this letter all the more.

It was indeed very sad news about Stebbing's death.[238] We both like her very much. And now I have a tragic story to tell you, too. In the middle of January Eva Hempel died a few days after giving birth to a son, in consequence of a second operation. Now Hempel is alone with the baby, and you can imagine how terrible a blow it is for him. You know how dependent he was on Eva. The present plan is to have the baby at home, with a housekeeper-nurse but we do not yet know how well that will work out.

My back is still not essentially better. I expected to go this winter again to the Mayo Clinic, perhaps for an operation; but in the mean time I consulted two good specialists who came to Santa Fe's new Army Hospital, and they both believe that it is not a ruptured intervertebral disk – as the Mayo people believed – therefore they do not consider an operation and they try instead a conservative treatment. It is still too early to say how much success it will have but there is some progress: I can walk now up to about 40 minutes as against twenty before, and I am slowly increasing it. Fortunately the Rockefeller grant has been renewed for another year[239] – so we can stay here until September and the doctors give me hope that I shall be able to teach again at that time.

---

[237] See Neurath to Carnap, letters 21., 22. and 23.

[238] See Neurath to Carnap, letter 22.

[239] Carnap had a Research Grant from Rockefeller Foundation since 1942, which was renewed for another academic year in the summer of 1943. During 1942 and 1943, Carnap suffered crippling back pains that confined him to bed. His wife Ina wrote to W. V. O. Quine about it (May 23, 1944): "Carnap's back is much better – he is now up an average of six hours a day (not consecutively) [...]. Isn't it maddening to think that if we had the right doctors in time there would not have been any need for his staying in bed so long? On the other hand, Rockefeller would not have come through with a second year if it had not been for his illness." In Creath (1990, 374).

And now let us plunge into the philosophical problems. I don't think that it is very useful for us to discuss Russell'sbook much more. (By the way, I heard that he got a fellowship at Cambridge and that he plans to return to England in the summer.) I am myself very critical with respect to the book; but your many critical remarks about it (with which I do sometimes not quite agree and some of which I even regards as absurdly exaggerated in their attacks against harmless formulations), you push me against my will into the role of a defender of the book. It is certainly not a formulation of my opinions. When you sometimes criticize books by Schlick, Tarski or Popper, that is quite useful because I agree in the basic points with these books. But that is not the case with Russell's last book. I have discussed it in our correspondence only on your request. I do not know why you always come back to this book with which we both do not agree. I have explained my chief differences with Russell in a previous letter.[240] I think it would not be very fruitful to go now again into all the details. But if you have a specific question concerning my view on some points I shall be glad to answer it.

On Semantics. If I see it correctly, you raise objections or doubts in two points:

1. Semantics may be a[l]right as a mere calculus, but if it is applied to the language of empirical science, it seems doubtful or at least an open question whether it is fruitful and useful.
2. Some of the chief concepts of semantics are metaphysical; therefore empiricists have to reject them as meaningless.

I am not quite clear how you intend to combine these two objections; they do not seem to me to fit well together.

On (1). I am not much interested in semantics as a mere calculus, but chiefly in semantics as analysis of the language of science. I believe that it will here by very useful and important when it will be further developed and applied. I admit that today we have in this direction not more than the very first beginnings. Therefore there is not much point in discussing the usefulness. Examples for the application of semantical see [I], p. 61.f[241]; every scientist uses these and similar concepts; remember e.g. that Frank says that certain statements of physicists are "tautological".[242] You ask whether I could perhaps show you that you yourself use semantical concepts without noticing it. Yes, it would indeed be easy to give hundreds of examples. Here some taken from a few pages of your letter of September 25th[243]: You speak of "contradictions" (p.9); you say that Weber gives "a descriptive story of" Franklin; you speak (p.10) about "general laws", "area of

[240] See Carnap to Neurath, letter 19.

[241] See Carnap (1942, §14).

[242] Frank claimed that the law of causality was a convention, thus tautological. See Frank (1907/1949).

[243] See Neurath to Carnap, letter 22.

validity", "observation statements"; you say that a certain statement "pictures the situation" (a synonym for "is true"); you say (p.11) that you will "deal with" Russell; you say (p.11) that one item "implies" another one. (In a previous letter I have referred to the examples on p.61f. of my book[244]; now you express again your doubts and you ask for examples (without saying a word about those examples.) Whether the applications will be really fruitful or not will be shown by further work in this field. A new logico-mathematical theory (e.g. set theory, group theory) or empirical theory (e.g. Marx's conception of history, psychoanalysis) usually meets first many doubts about its fruitfulness; these doubts are justified because the majority of newly proposed ideas are not worth much. However, I think if a new theory is free of metaphysics and is proposed by serious people, then those who have doubts about its fruitfulness should be tolerant and should wait without polemizing too much; the further development will pass judgment. I say "if free of metaphysics", hence, above all, we have to decide question (2).

On (2). That the semantical concept of truth is not metaphysical can very easily be shown by the following translation: "The sentence 'this tree is green' is true" means not more and not less than "This tree is green". (If the latter sentence does not occur in your strangely restricted language you may take instead any other sentence which you regard as meaningful.) (By the way, if the term 'true' were to occur only in connection with a direct quotation as in the example given, the term would indeed hardly be useful; but that is another question.) This translation shows that the concept of truth is not metaphysical but scientific. Furthermore, the translation makes it clear that the term 'true' is not at all meant in the sense of 'absolutely certain', 'indubitable' or anything like that as you sometimes seem to believe. And the translation also shows that 'true' has nothing to do with 'accepted'; you make time and again the mistake of demanding that I should translate my semantical sentences into sentences with the term 'accepted'.

To your letter p.11, the last two paragraphs.[245] I am in complete agreement with your description of the scientific procedure. I should classify this as belonging to the methodology of science. I do not see what it has to do with semantics. Which assertion of mine concerning semantics seems to you to be in contradiction to your description? – Likewise p.12 you see[m] to believe that semantics intends to propose a new conception concerning scientific procedure; that is certainly not the case.

To p.13, "judicial proceedings". The question is not whether you might be able by certain tricks to avoid the word 'true' in such proceedings. My point is rather that the word 'true' in its semantical sense is very frequently used in science, jud. proc., etc.

---

[244] Carnap mentioned the application of semantical terms (though without the page numbers) in letter 18.

[245] See letter 22, page 596 in the appendix.

To p.14, "Sentences propositionally equal". O.K., no objection. This is a good semantical concept; I use for it the term 'logically equivalent' or 'L-equivalent'. If you wish to avoid propositions as extra-linguistic entities, you may take instead classes of L-equivalent sentences (this is e.g. done by Russell, Inquiry, p. 209).[246] However, in this way you do not avoid semantics! -- You seem to demand that we should not speak about things which we cannot see. You will hardly expect that I take this seriously?

On finitism. The best reading material is: Weyl, "Die heutige Erkenntnislage in der Mathematik",[247] Felix Kaufmann's book,[248] Bridgman's pamphlet on set theory,[249] Wittgenstein's book,[250] Waismann's "Theses" (mimeographed in Vienna).[251] With the exception of Bridgman (whose pamphlet however is rather weak) these are not authors which you otherwise estimate very highly. My opinion on finitism: it contains a sound basic idea, but none of the attempts made so far is practicable for the construction of a logico-mathematical system which could be used as a basis for the language of science.[252] -- You have mentioned several times that you have objections against Russell's symbol of existence; but you do not say what the objections are. Could you tell me any logician or mathematician or scientist who has made a proposal how to do in the language of science without the concept of existence and the concept of 'every' (which are definable by each other)?

On pluralism. I believe that in this point I am in agreement with your attitude. I say "I believe" and "attitude", not "opinion" because your formulation of pluralism (p.9) is so vague that I am not able to see in it any clear thesis. As you formulate it now, nearly everybody would agree, including Schlick, Popper, Russell. Since pluralism seems to you very important, and in particular also the question whether I agree with you in this point, please give me a more precise formulation of what is

[246] See Russell (1941).

[247] Hermann Weyl (1885–1955) was a German mathematician and physicist, mainly known for his work on the foundations of mathematics (he was influenced by phenomenology and intuitionism) and space-time theories. See Weyl (1925/1998).

[248] See Kaufmann (1930/1978).

[249] Percy W. Bridgman (1882–1961) was an American physicist and won the 1946 Nobel Prize in physics. He is usually known also for his approach called "operationalism" which influenced, among others, Philipp Frank. Carnap is referring presumably to Bridgman (1934).

[250] See Wittgenstein (1922).

[251] Friedrich Waismann prepared a longer list of theses of Wittgenstein's philosophy during the early 1930s for the members of the Vienna Circle. For the English translation see Waismann (1979).

[252] On Carnap and finitism see Frost-Arnold (2013).

asserted by it. -- Perhaps the difference here is not a difference of opinion but of emphasis. We emphasize the importance of the task of systematization in science; you, on the other hand, emphasize the fact that the statements accepted by scientists at a certain time do not form a well[-]connected system and you point to the dangers involved in overlooking this fact. I think you are right in both points. I should be still more inclined to agree with you were it not for the impression that you exaggerate very much in laying more stress on the dangers of systematization than on its usefulness, importance and indispensability; I doubt whether your overstressed warning is fruitful. I am afraid that it may do more harm than good in the education of young scientists. Do you remember how all of us in Vienna, especially Menger, were angry about the influence of Wittgenstein-Waismann ideas concerning mathematics upon the young mathematicians? We might call those ideas "defeatism in mathematics". They were apt to discourage the students to work in mathematics. Later I (together with the Warsaw people and Gödel) criticized Wittgenstein's "defeatism in speaking about language"; it led the young people to the position: if it is not possible to speak about language in an exact way, then let's not try to do it. Your warning against the dangers of an oversystematization an oversimplification, and schematization is quite a[l]right to some extent. But I have the impression that by exaggerating your warning you turn it into a new kind of defeatism. The students may think: if systematization, logicalisation, etc. in science is not advisable and even dangerous, then why take pains to eliminate contradictions and try to fit together the accepted statements more and more into a system. I do not say that this is your position, but your polemic against systematization may have this effect. -- I believe, if we are careful to avoid exaggerations on both sides, we shall easily be able to come to an agreement on this point; "we" included Popper, Tarski, etc. For, here there is really no serious difference of opinion; it is more a difference in temperament and therefore in the direction of interests. I think, it would be best for the development of science if the people on the one side who see more the turbulent whirl of material in all its colorfulness and vagueness, and those on the other side who love nice structural schemata would not polemize against each other but rather realize that the work of both is necessary for science.

On Popper. I have explained to you my view on Popper's position in my letter of Jan. 29, 1943[253]: It seems to me good as a first approximation. I think in the question of asymmetry we have to distinguish two assertions of Poppers:

1. In testing a universal law, there is an asymmetry between favorable (confirming) and unfavorable (disconfirming) cases. If somebody proposes a law hypothetically then he usually has already a number of favorable cases, and he is looking around for further cases. The asymmetry is this: an unfavorable case has a much stronger influence upon our judgment concerning the law than an additional favorable case.

[253] Carnap to Neurath, letter 18.

2. One single unfavorable case refutes the law.

Some objections might perhaps be raised against (2). But still I would think that it is acceptable as a first approximation. On the other hand, your rejection of (1) in your last letter is surprising to me.[254] I think that all empiricists and all good scientists agree in this; Who, in your opinion, does not? (The remark on this point which you quote from a letter of Hempels seems to me quite right[255]; your reply to it is incomprehensible to me.) -- As to the importance of negative cases: Darwin said that he was especially interested in all phenomena which did not seem in accordance with his hypotheses of evolution. -- You remember how our opponents called us solipsistic because we rejected the thesis of realism as meaningless? We always emphasized that if a sentence is metaphysical and hence meaningless, then the same holds for its negation. Now Popper makes certain assertions concerning universal laws, negative instances etc. You deny these assertions. This is your good right. But then you go on to say that Popper's assertions are metaphysical. If so, however, your counter-assertions are likewise metaphysical and meaningless. Seriously, I think it would be better if you would at long last abandon your habit of calling people who are empiricists and antimetaphysicians metaphysicians if you do not share their opinions. First, it is absurd; and second it does not help a successful discussion.

Negativistic exaggerations. Your write that you are reading just now many books by scientists concerning their theories and procedures. I suppose that we agree that when we speak about method and language of science it is important for us to see what good scientists do. But in cases of this kind we ought to be especially cautious. I think that in Vienna we sometimes were too rash in condemning something accepted by all or most scientists. (e.g. universal sentences); today I am inclined to be more cautious and to think that if a procedure or concept is accepted by the majority of good scientists, then it may still be that it needs some modification; but it is highly improbable that it is entirely wrong. Therefore I am astonished to see how many procedures or concepts used by the overwhelming majority of good scientists are rejected by you without your offering arguments as weighty as this situation would demand. I regard the following items as examples for this:

the concept of existence (Russell),
asymmetry of negative and positive cases (see above, Popper, (1
the simple semantical concepts e.g. 'true', 'designation', 'analytic', 'consequence', 'contradiction'
'error' – 'fallibility' -- terminology, used from the point of view of a behavioristic observer (your letter p.16).
'fact'.
'unobserved facts', 'unexperienced facts' (p.16).
'more confirmed' (the topological concept of degree of confirmation; I admit that the quantitative concept is more problematical).

[254] Neurath to Carnap, letter 22.

[255] Neurath to Carnap, letter 22.

'infinite class' (and hence, because defined by this, the concepts of limit, differential coefficient, velocity, acceleration, etc.).
terms referring to the things which we cannot see (p.14).
'explanation' (Feigl says that you object to this).

On Freundlich's syllabus. I have not much to comment because first the syllabus is extremely short, and then you know my old doubts about the suitability of this topic for the first volumes.[256] But I think, it will become an interesting monograph. Since the first two vols. are chiefly methodological it might be well if he would stress this point of view as far as possible. You say that we speak about theories; I should rather say, we speak about methods.

Your new monograph. I hope to get the proofs soon. I did not see the MS because Morris, in accordance with your request, gave it to the Press in order to avoid delay. I did not object to this because you wanted it this way. But I should have preferred if you had given the coeditors an opportunity to read the MS and to write comments to you before you finished it for the Press, as we have done with all other monographs.

I agree with the choice of Hogben for the biology monograph as Morris will have written you. When and where will his "Interglossa" appear?[257] I am looking forward to it with great interest.

We did not know about Mrs. Waismann's suicide. What was the reason? Do you know more about him now?[258]

I have heard from Switzerland that my children are well; one daughter has just married and the other is engaged. Heaven knows what the husbands will be like! The boy is allright; no news from or about friends.

From the snapshots you see how we live; from Spring to Fall I was out there on the porch (and the manuscript on probability and degree of confirmation grew well in this wonderful climate)[259]; Feigl was with us in September.

To you and Mieze our best.

Yours,

[Carnap]

---

[256] Carnap had claimed in an earlier letter to Neurath that 'cosmology' should not occur in the title of the monographs, since it is in this form entirely incomprehensible. See Carnap to Neurath, July 24, 1939 (ASP RC 102-53-05).

[257] Seemingly Hogben planned to write a monograph on biology for the *IEUS*. On the biology-monograph, see further letters 1., 25., 26. and letter 9. note 102. On "Interglossa" see letters 11, 16, 27 and 28.

[258] On Waismann see letter 22, note 221 and letter 18, note 152.

[259] Presumably Carnap's *Logical Foundations of Probability* (1950).

25. Neurath to Carnap and Morris, Copy to Hogben, March 23, 1944.

(ASP RC 115-07-64)
23rd March, 44

Dear co-editors,

I have the great pleasure to inform you, that Lancelot Hogben will write for our Encyclopedia:

FOUNDATIONS OF BIOLOGY, VOLUME I, Monograph 9.

The monograph will be of about 9000 words. He promised to write it before December 44. Let us hope he will finish it in good time. He is at the moment in a responsible position in the war office and therefore very busy with his activities.

I got the galley proofs of my monograph and shall return them very quickly. Since Frank, Freundlich and others are at work, too, we shall finish the first two volumes within a reasonable time – I hope so.

With the best wishes
Ever yours
[Otto Neurath]

26. Neurath to Carnap, April 1, 1944.

(ASP RC 102-55-05)
1st April, 44

Dear Carnap,

Many thanks for your long letter, particular thanks to Ina and her efforts to perform the task properly to such an extent.

Eva's death is something very terrible.[260] You see we liked the Hempels as two extraordinary fine and decent people. There are only a few specimens of this type there. Poor Hempel with his baby. Both together lived on very well. He needs some personal care, and he got it from Eva. You see, we are depressed by many losses. A few weeks ago died a very fine Viennese friend of ours a young man, just over 30. Also an exceptional decent person .... and Susan [Stebbing]. And all the deaths in the concentration camps, and the suicides.... I heard from some in the last time. A colleague of mine in the Commercial Academy. The son of a friend came with the parents via Russia from Denmark to England (the Father, my friend, perhaps you remember him, Otto Simon interested in Esperanto, Mathematician died in USA) and told me of this suicide committed in Vienna, as the Nazis entered the city. Now these hordes will go down – but what will be then? We assume that not a few Nazis will hide themselves [sic] and prefer to think of the next war instead of adapting themselves [sic] to a new world. What a world.

[260] See Carnap to Neurath, letter 24.

I am now reading some older books, to discover the "filiation" of the Nazi arguments. The main line Lagarde,[261] Langbehn (Rembrandt als Erzieher)[262] Chamberlain,[263] Spengler, etc. then all the typological literature, with Nohl, who now discovered, that Hitler implies the victory of the Platonic educational ideal. (In accordance with the Republic, of course – I do not object to the comparison, only that Nohl accept this evolution as something nice, but he supported Hans Grimm, Volk ohne Raum, before).[264] The types are the forerunners of races, the physiognomy of all kinds of race rubbish. Graphology, characterology etc. very dangerous spots. By the way, do you know by chance, when Br[oder] Christiansen entered [embraced] the Nazi ideology?[265] It is manifest for me in his older writings, too, but later on he expresses himself very clearly. When has been the decisive step.

I knew something about him and his influence and I am always astonished, when people tell me of the isolated ascetic thinker in the mountains. My picture is different. His writing is aggressive and reflects his contacts with many people. He quotes, of course, Nohl etc. I should like to get a vivid picture of his life, thinking and behaving…

What a pity, that your backbone is always troubling you. And what a situation not to know, what creates all the trouble. Good Nagel also complains about the doctors and the low stage of medical knowledge. It is tremendous this knowledge, but our body is very complicated and complex – that is the point. Let us hope that something will come out in the near future and the doctors can make you pain[-]free at least. What a [sic] good luck, that "Rockefeller" is helping you.

I am really sorry that you agree to such an extent with Schlick, Tarski and Popper that you feel more with them than with me. I should like to know, what Philipp Frank thinks on that subject. I feel that what you like, is a tendency towards CALCULUS, connected with some, what I would call a mixture of crude realism and metaphysical absolutism. You are very seldom speak[ing] of EMPIRICIST problems as such. I feel very uneasy. I should like to see, how we together with Philipp Frank and others connected with research in an empiricist field could reach some common field, which leads us to Semantics, as you and Tarski teach it.

---

[261] Paul Anton de Lagarde (1827–1891) was a German orientalist.

[262] August Julius Langbehn (1851–1907) was a German art historian and philosopher. He published "Rembrandt als Erzieher" (1890) anonymously ("Von einem Deutschen").

[263] Houston Stewart Chamberlain (1855–1927) was a British-born German philosopher, known for his two-volume book, *The Foundations of the Nineteenth Century*, which became an influential work in German right-wing national circles.

[264] Hans Grimm (1875–1959) was a German writer, and a believer in Nazism; for the mentioned novel, see Grimm (1926).

[265] Neurath refers presumably to Broder Christiansen (1869–1958), a philosopher of language and art, and old friend of Carnap. In 1933, he published a book about the foundations of graphology with Carnap's first wife, Elisabeth Carnap; see Christiansen and Carnap (1933).

The two arguments fit together:

(1) Something may form a fine and consistent calculus – I am not an expert in testing calculus. Strauss has some doubts about the consistency, too. – that does not prohibit, that
(2) certain "names" used, lead to a metaphysical "application" as it were. Imagine Cantor presenting his Mengenlehre as something on the Trinity and the unknowable qualities of angles etc. One could object to this "names" without criticizing the calculus as such.

(1) "pictures a situation" would be expressed cautiously by me "is an accepted observation-statement" or is an "accepted statement, because in accepted harmony with observation-statements". "descriptive story" would be cautiously expressed: "using observation terms and accepting the statements" … to your page 61, instead of needs observation, I should say: need observation statements for comparison … My objections to "degree" of confirmation I have often explained. We do not have any reason to assume a one-dimensional ranking of theories.

I am not much polemizing, just I think it my "duty" at least to tell of my uneasiness. Otherwise people say, who is silent, accepts. I do not accept Semantics. I feel wholly uneasy myself and should try to remain within the "acceptance" phraseology, where I personally have to analyze science. More I did not in my note in my monograph. Very carefully. (2) I think the subject becomes so difficult, because you start from "this tree is green", whereas I start from "we are using the sentence "this tree is green"" and do not overstep this threshold.

I think, that wherever in judicial procedures or scientific books you are using the term "true" a careful analysis shows you that the use leads to difficulties from [an] empiricist viewpoint. In a court, it is used instead of "what the court accepts" or something like that. Otherwise you get the situation that A says something is true, B says something is true, and nobody can say what is "really" true. As far as I can see, every person speaking carefully, would finally say "the following possibilities seem acceptable". That is all. A "pluri" formulation from the start, a group of pluriformulations.[266]

I do not say we should "not speak about things we do not see" I suggest not to use this kind of phraseology when speaking cautiously we should not speak of a "thing seen by somebody", but only "somebody <u>says</u> I see something" etc.

I speak of finitism in the field of observation-statements and think it is a particular question how to relate these statements to a CALCULUS, with "infinite" terms.

I am very doubtful about the term "every" within the field of experient[i]al discussions, I think it belongs to the field of calculus and very often difficulties come out from the scientific discussions because, I think so, the calculus terminology and the experiential terminology appear mixed up together.

You see the "systematization" disease, as I may call it is very dangerous. You see, when we assault every contradiction we cannot go on in science – the history of

[266] On "pluri-terms" and "pluri-formulations," see Neurath (1941/1983).

Newton's theory is very significant. Of course I learned by heart, that any contradiction in a system may "infect" each part of it. That is the logical lesson for the youngest. But in scientific research a particular quality of a research worker is connected with his ability to give certain hypotheses not a comprehensive validity. Sometimes people apply contradictory hypotheses in different parts of their studies, avoiding their clash. OF COURSE WE TRY TO ELIMINATE THAT, BUT FINDING SUCH CONTRADICTION SHOULD NOT ALWAYS PROHIBIT US TO GO ON.

Therefore we need a teaching which stresses the two points:

(1) how to avoid contradictions,
(2) how to go on even when contradictions appear, if connected with "success".

As far as I can see, the Popperism is just fighting that. One contradiction – everything is lost. THAT IS OK WITHIN A CALCULUS.

More. In the Social Sciences it is epidemic to say, that one does not know at the moment every element needed for complete prediction, but "in principle" everything is predictable, even today, if… I maintained in my EMPIRISCHE SOZIOLOGIE that there is a field of unpredictability "in principle"… very important.[267] Our youth did never discuss this point but maintained always the complete and consistent prediction scheme. You see all this talk on "unchangeable historical laws", of "historical necessity" – some Marxians [sic] now are pride of that (Marx himself has been mainly pluralist) etc. It is important for decent decisions, to maintain the pluralist situation and the unpredictability.

I think what you say about [the] indispensability of systematization is not contradicting my attitude, as long as one does not think

(1) that in principle the Laplace's Demon is the picture of the scientist in action
(2) that one will reach the LIMIT of knowledge more and more.

But I do not see that you, Popper etc. stress this point. Just at the moment it is important to see, that decisions will be made based on some scientifically reached results, but that these decisions will be based on [i]nsufficient material, in principle.

One has to stress the importance of mathematics and its application to science, but to add that the systematization work is LIMITED TO THE CALCULUS PART and that the concrete predictions always have some elements of unpredictability in them, the predictability is some casual opportunity – that is all.

Why should we not speak about language, of course one should, but one should not give the impression, that the systematization of a construed language is in any extent transferable AS SUCH to the experiential scientific language, whereas the ITEMS OF SYSTEMATIZATION are of greatest importance. NO DEFAITISM [DEFEATISM], [o]n the contrary. I think the system fetishism of Popper is [a] defaitist [defeatist].[268]

---

[267] See Neurath (1931/1973, 369. ff.).

[268] See Carnap's observations on defeatism in letter 24.

I should like to read one day an application of Popper to scientific analysis as such. As far as I can see, he is "absolute" – as I have maintained – in the idea of the BEST WORLD[-]SYSTEM, compared with others and in his contradiction principle as THE LEADING PRINCIPLE, whereas we try to eliminate contradictions or to localize them. Just this skill, as mentioned above, characterizes the experiential scientist.

I do not agree with your BEHAVIOUR statement, that unfavorable cases have a much stronger influence than unfavorable. That depends upon situation, people etc. Did you check up your assumption? You see, to find some extraordinary positive case is so exciting, that one tries to invalidate the trivial negative ones. In the empiricist sciences you are in a position to give the cases different weight and to "hope" that something may be eliminated afterwards.

I know relatively well the Ehrenhaft discussion.[269] Where certain research workers got extraordinary points outside the expected curve, they assumed "the tram has been disturbing the electric apparatus". To what extent this assumption is "sound" or not is not so simply to say. Ehrenhaft fought such "tram" cases, of course, but the leading physicists did not. Further, Ehrenhaft presented dozens of count[er]-experiments against the electricity elementary unit, but the others remained unshaken the NUMBER of positive instances seemed too big. I had a correspondence with Thirring[270] about this point from Holland, just because I am really interested in how science goes on. In sociology I know that the Popper habit would kill most investigations and every accepted going on in some fields, and I wanted to know whether physicists behave differently – not at all. I should like to find a collaborator, [so] that we could analyze the Ehrenhaft discussion. I convinced some people that Ehrenhafts remarks are more sound than they thought and nevertheless said that their opposition remains sound from point of view of the positive instances. You never entered this argument. A pity.

POPPER is NOT a first approximation, on the contrary. He kills the bird in the egg.

The same with (2). Of course any wise scientist is interested in negative cases, but he is particularly interested in positive ones. Predicting correctly some earthquakes is something and the negative results may be regarded as based on mistakes in assumptions, mistakes in the calculus etc. etc. The thinking of the clear cut scheme as THE model is dangerous.

---

[269] Neurath possibly refers Felix Ehrenhaft (1879–1952), an Austrian physicist who worked in Vienna. He is mainly known for his contribution to the measurement of electrical charges. Neurath might refer to the debate between the "Viennese physicists" (including Ehrenhaft) and Ernest Rutherford who pursued physical experiments in the Cavendish Laboratory at the University of Cambridge. On the debate about the results and interpretation of the Cambridge-Vienna experiments, see Hughes (2016).

[270] Hans Thirring (1888–1976) was an Austrian theoretical physicist, who worked in Vienna as the head of the institute for theoretical physics. Thirring was the leading member of a committee investigating alleged cases of psychic powers; other members were Moritz Schlick and Hans Hahn.

You see, as often, you treat complexes of arguments as if they were carefully selected. In the case of [P]opper. SOME ELEMENTS ARE METAPHYSICAL – absolutism, OTHERS are not in accordance with the scientific practice. But both items are connected in his writings. You see the idea of THE system – optimum etc. seems to be metaphysically approached, but the story of the universal laws perhaps only an objected application of some calculus. Should he connect it with the WORLD system, then it would be metaphysical, too.

In your language: EITHER I give Popper the OPTIMUM-SYSTEM interpretation, where he speaks of the importance of negative instances – then I reject the whole approach.

OR – I speak only of the application of a calculus principle then I reject the application only.

We know from experience, that very often the metaphysical jump starts [sic], where a mere CALCULUS approach has been regarded as an experiential statement, a world-statement e.g.

I think Popper being metaphysician proper. You never answered my ERKENNTNIS article in your letters, only in general terms.[271]

POPPER page 6 e.g. speaks of "strengen Nachpruefungen", as if any negative instance would be per se sufficiently "streng".[272] You always can assume, that the negative instance will be weakened afterwards as you may weaken a positive instance. This assumption that the [i]nequality starts here is not based on anything experiential.

p. 9 he fights the intention to base all scientific statements on observation-statements as "induction logic"....[273]

p. 11. 'wir vermuten, dass wissenschaftliche Forschung psychologisch gesehen, ohne einen... wenn man will "metaphysischen" Glauben an manchmal hoechst unklare theoretische Ideen wohl gar nicht moeglich ist. WHY "GLAUBEN" C[274]

then

jenes System, das wir "empirische Wissenschaft" nennen, soll aber nur die eine "wirkliche Welt", die "Welt" unserer Erfahrungswirklichkeit darstellen.[275]

---

[271] Neurath presumably refers to his review of Popper's *Logik der Forschung*; see Neurath (1935/1983).

[272] Popper (1935/2002, 10): "severe tests." All the quotations in the following footnotes are to the English translation of Popper's book.

[273] Popper (1935/2002, 13).

[274] Popper (1935/2002, 16): "[...] looking at the matter from the psychological angle, I am inclined to think that scientific discovery is impossible without faith [...] which is completely unwarranted from the point of view of science, and which, to that extent, is 'metaphysical.'" Neurath asks, "WHY FAITH?"

[275] Popper (1935/2002, 16): "Yet the system called 'empirical science' is intended to represent only one world: the 'real world' or the 'world of our experience.'"

I AM JUST FIGHTING THAT AS A METAPHYSICAL SENTENCE. THE ONE REALITY.

ausgezeichnet "unsere Erfahrungswelt" I speak of pluri-statements and therefore I see no way how to reach THE ONE WORLD, or THE ONE PREFERRED SYSTEM OF STATEMENTS.

I should like to learn from you, what you say about that. I think Philipp Frank fought sufficiently the ONE REAL WORLD besides the MANY POSSIBLE WORLDS, I myself and he and Duhem and Poincare belong together, we are only more consistent than Duhem and Poincare – I think so.

p. 13 Ein empirisch-wissenschaf. System muss an der Erfahrung scheitern koennen.[276] (since no EXACTNESS exists in experiential complexes or statements, but only in CALCULUS, what does this sentence intend according to your opinion?)

p. 15 Beziehung zwischen Basissaetzen und den Wahrnehmungs erlebnisse[277] – that is Schlick's story with all its difficulties I criticized in a particular article.[278]

p. 17 der die Versuchsanordnung nach Vorschrift aufbaut ALWAYS place and time are different, therefore the "ceteris paribus" always questionable. p. 18 intersubjectiv nachpruefbar....[279] There are many statements, we may call "empiricist" because combinable with observation-statements, but not "assayable" in the way, Popper thinks. You agree with me, I think so, on this point.

p. 22 Methodenlehre nicht empirische Wissenschaft....[280] What you say about that?

p. 28. spezifische und numerische Allgemeinheit.[281] That belongs to the existence formulations and the every formulation by Russell. Why not in principle speak within the empiricist part of the science only of "numerische Allgemeinheit" and what is more put into the calculus.

---

[276] Popper (1935/2002, 18): "[...] *it must be possible for an empirical scientific system to be refuted by experience*." Original emphasis.

[277] Popper (1935/2002, 21): "This is especially true of the relation between perceptual experiences and basic statements."

[278] Presumably Neurath (1934/1983).

[279] Popper (1935/2002, 24): "[...] by anyone who carries out the appropriate experiment in the way prescribed."

[280] Popper (1935/2002, 31): "But what I call 'methodology' should not be taken for an empirical science."

[281] Popper (1935/2002, 40): "Strict and numerical universality."

Universelle Es-gibt-Saetze sind nicht falsifizierbar. Wir werden deshalb sie… als nichtempirisch ("metaphysisch") bezeichnen muessen.[282] JUST THAT IS THE POINT. In my notes, you find the remark "PLEASE".

I can bring forward heaps of examples, that one starts with such predictions, hypotheses upon "we shall find…" and then we find, what we tried to find… just that seems to be extraordinary empiricist… what you think about it? Usually, like a lawyer you will belittle all these sentences, but then I do not know, for what purpose Popper wrote the whole book, if not, for maintaining just these points. He is not the only ANTISCIENTIFIC person, who starts from EXACTNESS, fighting common sense empiricism as I suggest.

p. 60 beobachtbar nicht psychologisch[283] – I do not know how to make head or tail of that.

p. 63 fast immer die experimentelle FASIFIKATION einer als bewaehrt anerkannten Theorie die den Fortschritt erzwingt- also wieder die von der Theorie geleitete Nachpruefung.[284] I think Popper made this statement without looking into the history of the sciences. Usually hypotheses are so vague in some points, that one does not even know how to "disconfirm" them. Soe [sic] "aberrations" one just bears, others not…. usually NEW CORRELATIONS APPEAR, NOT COVERED UP TO NOW, not contradicted by the theories in action. Etc… it is a long story. Do you think this statement by Popper is in accordance with the history of the sciences?

Zufallsentdeckungen ….

p. 67 Pruefbarkeits-oder Falsifizierbarkeitsgrad.[285] You know my objections to that. What you think about it?

p. 199. Wir betrachten also im allgemeinen eine… intersubj-nachpruefbare Falsifikation als endgiltig. Darin eben drueckt sich die Assymetrie zwischen Verifi. und Fals. aus.

p. 206. experimentum crucis ….[286]

What do you say about all that. Do you think it is in accordance with observation-statements in which scientists appear going on with scientific research with "definitely" killed theories ---- oh my dear, what a vision….

---

[282] Popper (1935/2002, 48): "Strictly existential statements […] cannot be falsified. I shall therefore have to treat [them] as non-empirical or 'metaphysical.'" German original on p. 33.

[283] Popper (1935/2002, 85): 'observable not psychological.'

[284] Popper (1935/2002, 90): "What compels the theorist […] in these cases, is almost always the experimental falsification of a theory, so far accepted and corroborated: it is, again, the outcome of tests guided by theory."

[285] Popper (1935/2002, 90): "Accidental discoveries..." [German phrase on p. 64.], and p. 95: "various degrees of testability or falsifiability."

[286] Popper (1935/2002, 267): "In general we regard an inter-subjectively testable falsification as final […]: this is the way in which the asymmetry between verification and falsification of theories makes itself felt." For *experimentum crusis*, see Popper (1935/2002, 224).

Again, as long as we remain within the negative and positive instances, I REJECT THE STATEMENTS, AS FAR AS HISTORY OF THE SCIENCES IS CONCERNED, but as far as this assumptions are connected with the ONE WORLD, I drop the expression as such, "isolated" – "metaphysical".

I should appreciate it very much, if you asked Philipp Frank in a letter about the expressions mentioned page 5 by you, as challenged by me. Of course I should suggest to avoid in CAREFUL ANALYSIS "fact", "error", "explanation" ----- Of course I am against the expression "explanation" because we – that is the Mach school, if you would use this term, – Philipp Frank, etc. try to avoid "explanation" as something besides finding correlations. You know the discussion about "Erklaerung" and Kirchhoff[']s statement on "description".[287] Of course it would be pedantic always to avoid the term "explanation" but I personally should dislike to use the term as the heading of scientific analysis as such. Of course "description" may involve, what Stebbing calls "constructive description" or whatever,[288] but it has been just a success, as we think, that Kirchhoff lead us away from the "Erklaerung" ("explanation") to the correlation point of view. I should suggest that Frank discusses that with you, Morris and Feigl. I cannot think, that Feigl's point of view is chained to the dangerous term "explanation" which just indicated through decades, what we are fighting. I think so. Occasionally using the term "explanation" is not of importance. I think it is SERIOUS. Of Course, if Feigl insists I should not prevent him going on, we are in a free republic of scientists, but I should make it a condition that at least Philipp Frank together with you, Morris will talk over the matter with Feigl. Frank knows the discussion in all details. I know many statements in Pearson[289] and other people who mean something for us, just supporting our Kirchhoff.

I do not use the term "method" because it becomes often a very metaphysical implication, perhaps scientific technique would be better, it sounds very concrete.

Yes, the greatest pleasure would it be to have a long talk and correspondence about my manuscript. But, I have been a little worried by the attitude of the Press, and therefore I tried to avoid any delay. I did not thought [think] – I confess – of anything else, but the continuation of our Encyclopedia. Now the end of the war is coming and therefore the encyclopedia will be of educational importance now. You see, coming out of the internment, I had to take care of our Visual Education insti-

---

287 Gustav R. Kirchhoff (1824–1887), German physicist who also coined the term "black-body radiation," advocated his so-called doctrine of description, with a rejection of causal explanation as a goal of science, in 1875 in a series of inaugural lectures on mechanics at the University of Berlin; see the Preface to his *Vorlesungen über Mathematische Physik. Mechanik* (1875).

288 Since Susan Stebbing did not use the phrase "constructive description" (Neurath's uncertainty as "'constructive description' or whatever" was not accidental thus), the best candidate for what Neurath might have in mind is Stebbing (1933), where she discussed various forms and means of analysis and construction.

289 Karl Pearson (1857–1936) was an English mathematician and statistician, who wrote the influential *The Grammar of Science* (1892).

tute and to prepare everything for that – we succeeded completely in going on with our work. Then I had to lecture at Oxford University – of course helping me in making my manuscript but nevertheless I had to look through many books not immediately necessary, the Professor invited me and attended kindly the lectures is an anthropologist and therefore I spoke more of anthropology as I did usually.[290] I liked it very much, but it needs some time to see how "functionalism" goes on, and how the discussion stands just at the moment. Of course I know the main lines by heart. I had to look through the history of anthropology as you had to look another day through the history of philosophy. Then I had to start with reconstructing my manuscript, in Hitler[']s hands. That is sometimes less pleasant, than starting from scratch. You feel "I said it SOOOOOO fine", and then you want to remember your slogans... Boy, boy. That is not always joy only. But then I started writing and enjoyed it very much and I think I improved many parts. I could go on in one style, whereas in Holland I always "mended" and "altered" already written chapters.... I had preferred it of course to have your remarks before. Please, look into the last proofs I shall get and if you think I should alter some overstatement, please tell me so. I like to be in harmony with you. I have the feeling to continue your Logical Syntax period[291] before you became Tarskisized [sic] with some Aristotelian flavour, which I detest. I always fear, that you, a calculatory genius, supports a kind of possible scholasticism who leads away from scientific empiricism. Historically it seems a dangerous strain, as you may judge from Lukasiewicz['s] contacts with Thomism etc. I do not want to overstate this point. But, I repeat I feel uneasy.

Please, tell me, how Frank thinks of all that, he is such an empiricist commonsense boy.

You will get the INTERGLOSSA. I shall be interested in your remarks.[292]

Fine, that you and Morris agree with Hogben's collaboration. Let us hope it will be some fine piece of work. He is sometimes strange in his formulations – let us hope the best. He is clever and an empiricist interested in language, but in a different way as [than] we are.

I do not know anything about Waismann, but that he teaches Philosophy. [The] [s]uicide['s] reason? We only know, that certain mental difficulties appeared some time ago before she committed suicide.[293]

Thanks for the snapshot. I should like a snapshot which includes Ina. When we shall be together again, surrounded by fruit juice tins. Grape fruit juice etc... Now we have much pleasure in eating and drinking oranges and lemons. You remember the oranges arrived together with time bombs. Now they removed the time bombs

---

[290] The anthropologist is Alfred Reginald Radcliffe-Brown; see letter 13, note 124.

[291] See Carnap (1937).

[292] See Carnap's letter to Neurath, letter 27.

[293] See letter 22, note 222.

and we got the oranges. Food is excellent and the British are fine. We have more and more acquaintances and feel wholly at home. Less [Fewer] tensions than in Austria or in other countries.

I should like, if you did take care in your writings on probability, degree of confirmation etc., of the distinction between CALCULUS and EXPERIENTIAL statements. Probability seems to be a pure calculus term and the degree of confirmation seem perhaps only possible within the framework of a MODEL, not within a concrete theory. But, of course, the results may allow us some applications. Duhem tried to show what types of formulas cannot be applied to physics, and I think similar remarks could be made always.

We wait for the collection of photographs, you promised to send. We are in good health, happy and active. Everything succeeds up to now. We always wish that all our friends would be in a similar position. We do not like to know friends being in difficult situations, ill, sorry, or something like that. How limited is our power to do something for friends.

Now we, as Austrians, shall become sooner or later co-belligerents, we hope so. Kind regards from both of us to both of you

Yours ever,
Neurath

27. Carnap to Neurath, October 7, 1944.

(ASP RC 102-55-07)
Rudolf Carnap
University of Chicago
Chicago, Illinois
October 7, 1944

Dear Neurath:

We are back in Chicago after two years of absence. We are still in a Hotel, for weeks now, but we hope to come into an apartment next week at last, though not a satisfactory one. Everything is overcrowded here. My back is much improved, so that I have no difficulty in giving my lectures. Hempel was with us in Santa Fe for several weeks and we enjoyed it very much. Although the hard shock which he suffered from Eva's death is still noticeable, he seems to be over the hump by now.[294]

Now about <u>your monograph</u>.[295] Last December I did not see your ms. because you wished to be rushed to the printer and I agreed. I expected then to get the proofs soon and then to have an opportunity to write to you still if necessary. In March I asked Morris to send me either the ms. or if it were already in print, the proofs.

[294] See Carnap to Neurath, letter 24.

[295] Neurath, *Foundations of the Social Sciences* (1944).

However, by some mistake I did not get the proofs before June. And then I saw they were already in pages. I read the monograph with great interest, there are many interesting and stimulating discussions in it. However, I must tell you frankly, then I found the formulations in many places unclear; they were apparently formulated in great haste. I understand your reason for this; you wanted to finish it quickly because the Press was already impatient. But I think it would have been much more important to have the monograph in good shape than to publish it as quickly as possible. On my question, Morris answered that any changes were by that time impossible because they would cause too great expenses and delays. Thus I could not do anything more about it, although I regretted it very much. You will understand that I did not wish to be made responsible for something that I had not seen and that if I had seen it, I should not have approved. On the other hand, I did not wish to take any conspicuous step, as e.g. the removal of my name from the title page. Therefore I choose a quite non-conspicuous way which will not be noticed by most readers[296]; I asked Morris to have a note printed on the back of the title page above the copy right saying that because of special circumstances I do not share the editorial responsibility for this monograph. I think there is no point in going into the details which I disapprove because it is too late anyway. Let me only remark that these points concern language and careless formulations, lack of explanations of your new terms, etc... but not differences between our opinions. In most points I agree with the general attitude which you take in the monograph. There are a few points, some old, some new in which we differ. Those I hope to discuss with you sometimes in the future. But I must say that in many points I would not know whether I agree with you or not because your opinion is not formulated in a sufficiently clear and understandable way.

Many thanks for sending me Hogben's "Interglossa". I read it with very great interest, and I own and read likewise "The Loom of Language"[297] which I found likewise very interesting and which in some points gives more detailed explanations of the reasons for Hogben's decisions. In some points this new language has made important improvements in comparison with the earlier projects. I am not quite sure whether the choice of mostly Greek word roots is the best possible. I have always thought that any new project should make improvements in two important respects: (1) to utilize the improvements made in Basic English, especially the principle of word economy; (2) certain improvements in the logical structure of the language which might be learned from symbolic logic. Hogben has done very well in the first point; and his construction of a very simple syntax is a great achievement. However, with respect to logic his language seems to me to have some weak points; e.g. the

[296] Actually one review noticed the fact that Carnap had withdrawn his name. The author was "V. C. A.," presumably Virgil C. Aldrich (1945, 470).

[297] On "Interglossa" see letters 11, 16 and 28. Hogben edited Frederick Bodmer's *The Loom of Language: A Guide to Foreign Languages for the Home Student* (1944). The book was the third in a series that aimed at the education of masses, at interpretation of science to the layman.

line between observable things and properties is not always drawn in the right place, and the whole matter of this distinction between the word classes could probably be simplified still more; further his treatment of "all", "every", "any", "some" etc. is not satisfactory. The examples of translations which he gives at the end deviate in many points considerably from the original; thus a retranslation into English would lead to quite different texts. Tests of this kind made with earlier languages (e.g. Esperanto and Ido) had much better results.

I have worked much on the construction of a new system of inductive logic, i.e. a theory of the degree of confirmation. It will become a large book. Part of it is written, but it will take a long time until it is finished. Therefore, I have now written a paper in which I outline the chief results.[298] I hope it will appear about next winter, then I shall send you a reprint. I am very much interested in getting your comments and criticisms on it. You wrote recently that you assume that I know your reasons for the rejection of the concept of the degree of confirmation. However, I know only the fact of your rejection of it, not the reasons. I should be very glad if you could outline or at least briefly indicate them to me.

At present I do not have your letters with me. Therefore, I shall take an opportunity of answering questions in them at a later time.

Cordially yours,

C.

[Handwritten note:
P.S. Please excuse my long silence. The reason is that Ina could not type either, because of sciatica.]

28. Neurath to Carnap, November 18, 1944.

(ASP RC 102-55-06)
18th Nov. 44

Dear Carnap,

I am glad to hear that you are now relatively well and starting lecturing again. Poor Hempel. Very often I think of his fate and of nice Eva, too.[299]

On INTERGLOSSA.[300] I think its word economy is sufficient, but word economy in itself is a very dangerous principle. Since we wrote a book in BASIC ENGLISH,[301] we have some experience in handling it. You see the word economy

[298] The book is *The Logical Foundations of Probability* published in 1950 after all; and the paper is presumably "The Two Concepts of Probability," Carnap (1945a).

[299] See Carnap to Neurath, letter 24.

[300] Hogben (1943), see letters 11, 16 and 27.

[301] Neurath might have in mind his *International Picture Language* (1936).

implies, that you are using the same word in very different, sometimes strange ways, that you have to think, how can we find an IDIOMATIC English expression, composed of the words permitted and then something very metaphorical or allegorical comes out. To remember the idiomatically permitted strange combinations is much more depressing the beginner, than the learning of some clearcut expressions, which cover, what you want to say perfectly.

I did not see, that you and many other people who like [the] WORD ECONOMY did realize this point sufficiently. Just, where you want to be exact the word economy is preventing you from doing so very often. Of course, very often you learn to use a simple word combination instead of a strange term, but sometimes it is different.

What source of words, would you suggest? Hogben's argument runs as follow: when using the Greek roots in already USED SCIENTIFIC TERMS, you either know the term, or when not, you are learning something useful for your scientific reading and writing. INTERGLOSSA is particularly thought of as a congress and academy language.

The other points are more important. It would be useful if you could write something about that. I like The Loom of Language – a little too large, I think at least. I think again of the possibility to ask the author for a paper for the encyclopedia on artificial languages, perhaps a paper written by him and you together. At the moment is a boom in artificial language discussion and we should have a great public. I did not suggest that to Morris und you, before asking you about your opinion. You remember I thought of HOGBEN as author of such a paper but now I think this new combination is even better than Hogben, because he would be [the] judge in his own case. Please think it over carefully.

I am very doubtful about your system of inductive logic i.e. a theory of the degree of confirmation. You see the degrees of confirmation etc. seem to anticipate that you are sure of ONE LIMIT OF THE SERIES, as it were, how you can say so? I do not know any empiricist material, which permits us to apply all this stuff on confirmation. Assum[ing] that the calculus in itself is OK – which I think is not sure, – how should a group of protocol statements look like, that you can speak of "degree of confirmation"? Can you give me any example? Of course you can always make an utopian structure on which you can demonstrate that, but then – you will see – you always will speak of THE REALITY (in some or another way, more or less concealed) and not of possibilities (pluralism). I should like to know, how you now look at Reichenbach's attempts to create a kind of inductive logic.[302] I personally see not even a possible way to overcome the difficulty to substitute [replace] a "decision" by [with] a calculus.

You never answered these points, e.g. by presenting a simple example, how you are using the words. Reading your papers, Kaplan's papers,[303] Hempel's papers,

[302] See, for example, Reichenbach (1938) and (1940).

[303] Presumably Kaplan (1942) and (1943).

Morris' papers, I always feel that they are, what I would call anti-pluralist from start. I know very well to what extent we agree, when removing the usual metaphysical speculations, but I think the anti-pluralist attitude, which starts from ONE world[-]system as more plausible than all others – the substitute for THE ONE REALITY – comes very often in the foreground. I do not deny that the calculi so evolved may be of great use sometimes, even I want to stress, that some of the more-than-one-level discussions, which form the basis of semantics are very useful, AS LONG AS THEY ARE USED FOR THE REALM OF STATEMENTS – therefore with a pluralist possibility – and not in the realm of statement-things, because one thing-phraseology as you use it is an anti-pluralist one. I see the scientific enterprise in comparing protocol statements and their derivates and in discovering some ways how [sic] to fit them together, as far as possible, whereby we have to select already ARBITRARILY certain formulations. Any protocol-statement is a pluralist one and therefore any selection already brings with it the other possibilities, too.

I think this antagonism between us is an old one. And perhaps something in your unexpected behaviour may be connected with that. I should even prefer this hypothesis to the assumption that you like to insult me. You see, you write in your letter of October 7,[304] "BY SOME MISTAKE I DID NOT GET THE PROOFS BEFORE JUNE (no guilt from my side, therefore)… You will understand that I did not wish to be made responsible for something that I had not seen und [sic] that if I had seen it I should not have approved." That sounds as if we as editors had to approve papers. After we accepted an author (perhaps after reading his syllabus) he has only to send us the manuscript, we may make remarks and ask for alterations, should he decline that, we have to print the paper. WE ARE ONLY RESPONSIBLE FOR THE SELECTION OF THE AUTHOR. Remember, what difficulties we had with Bloomfield in inserting even that spoken and written languages may be regarded side by side. Assume he had declined to mention that, then we had to print it, of course.[305] We are not a censoring body only an editing one, which forces the authors to read our remarks, to GIVE HIM THE OPPORTUNITY TO KNOW WHAT HE IS DOING, THAT IS ALL.

Then you continue with quasi-kindness: "on the other hand I did not wish to take any conspicuous step, as e.g. the removal of my name from the title page. Therefore I chose the quite non-conspicuous way which will not be noticed by most readers; I asked Morris to have a note printed on the back of the title page… that because of special circumstances (A VERY BAD FORMULA WHICH CAN IMPLY EVERYTHING) I do not share the editorial responsibility for this monograph." You could imagine that you suggested alterations and then I declined to make them. Then the paper should appear without touching any editor's scruples [sic] - or did you ever think differently of our rights?

---

[304] See Neurath to Carnap, letter 27.

[305] Leonard Bloomfield (1887–1949) was an American linguist who wrote a monograph on the *Linguistic Aspects of Science* (1939) for the *IEUS* (I-4).

"Points concern language and careless formulations lack of explanation of your new terms etc." You see without giving any examples, that sounds rather aggressive. Imagine, that you say together with that, that MORRIS did not do his duty in not mentioning that to me. You stress the point, that not "differences between our opinions" did move you. I should like to know something about the defects, you mention.

But now about the personal side of this matter. I ask me, how can a friend grieve me so much and without any manifest reason. Listen, Carnap, if this were the first, I should look into my own behaviour, but you see, you grieved me again and again. It started particularly, as you wrote your article about physics as a universal language of science.[306] In spite of the fact that you used results of our discussions of my remarks made in the Vienna Circle, you did not even mention me, walking over me as a nonentitiy [sic]. You remember I protested intensely and you inserted then the note, which tries to state my part in the whole realm of discussions. This experience depressed me very much.

But that was not the first and not the last time you grieved me. One day you wrote me a letter, in a similar teacher[-]like tone, that you can hardly accept an article of mine for the ERKENNTNIS, since my style of presentation did not reach the level, wanted there etc., etc.[307] Carnap, Carnap, when I look at the articles the ERKENNTNIS published usually and then look at my articles with all their defects, I think they do not belong to the worst there and in the literature they are sufficiently quoted and discussed. You are usually not very critical towards other people, but particularly irritated by me. And now after some smaller incidents, not so important, you produce this newest piece of unkindness.

You see, we are living in difficult times, we have to try to be as kind to another as possible, there are difficulties in making anything. One has to [be] glad, that one overcomes all difficulties, etc. I can hardly imagine, that a kind of feeling of "duty" or something like that lead you to such a decision and action. But it may be that or that. You grieved me, did you anticipate that, then I [would] tell you, that you reached your goal, did you not imagine that, then I tell you that you did it.

When I think of our movement, any such behaviour seems to me unwise. In my letter to Morris I touch [on] this point. I can hardly assume that the reactions by [from] you are fully independent of your attitude towards the problems themselves. I see three periods of your evolution, which always lead to a possible irritation.

It started with the Wittgenstein boom. I remember how you and Feigl pressed for inserting a high praise of Wittgenstein into the WIENER KREIS pamphlet.[308] I

---

[306] See Carnap (1931/1934). The most comprehensive article on Neurath's and Carnap's debate over priority is Uebel (1995b).

[307] Neurath presumably refers to Carnap's letter from March 2, 1932 (RC 029-12-61) where he stated that Neurath's "Sociology in the Framework of Physicalism" paper (1932b/1983) was unclear and its main points were not understood in the Circle.

[308] See Carnap, Hahn and Neurath (1929/1973). On the history of the pamphlet (usually called as the "manifesto of the Circle"), see Uebel (2008).

looked at Wittgenstein – and said so and wrote so – as an antiscientific thinker full of metaphysics, etc. I never could realize, how Schlick, Waismann, you and Hahn have been so impressed by him. Of course he produced many very clever and stimulating ideas, but the "negative" items are very high. I had the feeling that my resolute behaviour against Wittgenstein in some way or another disturbed all of you. When I look at the empiricist Elements of our circle I should arrange them as follow:

FRANK, who in my eyes is the most balanced empiricist, we have. Perhaps less interested in some sharp formulations than sometimes may be useful, but far from any kind of scholasticism. Usually in Prague.

NEURATH	
HAHN	
CARNAP	Centre of the empiricists in Vienna.
SCHLICK	Centre of all together in Vienna.
WAISMANN	
KAUFMANN	Two anti-empi-
NEUMANN.[309]	ricists.

You see this visual joke helps me to present my case clearly. You always had a certain inclination towards PYRAMIDISM, how [as] I call it. You want – as I explained to you some day in Vienna – to have some top of a pyramid and then nice deductive chains leading to the bottom. You answered that you really like that and that you are rather astonished that I reject this ideal totally. You see, in LOGISCHER AUFBAU DER WELT you did even show some inclination towards the GEISTESWISSENSCHAFTEN etc.[310] And now I appeared as something irritating you, because much more "strong" than others in presenting my case, perhaps because I did not publish regularly my ideas, as you and others did and used the discussion as my medium.

The next period may be characterized by your inclination towards Popper, Logik der Forschung[311] – you preferred his attitude to mine. I think today, what I thought then, that you like the strong antithesis of YES and NO in his looking at positive and negative instances, and then his tendency to start from ONE world system, as the most complete one.

And then came Tarski, whose Aristotelian Metaphysics seem to you useful in building up semantics. What may be helpful in building up a calculus, can become very dangerous in preparing an instrument for empiricism. Should my guess be

---

[309] Presumably Robert Neumann, a high-school teacher, who was a regular attendant of the Vienna Circle meetings on Thursdays.

[310] See Carnap (1928/1967, §§23–24). On the *Aufbau* and the *Geisteswissenschaften*, see Dewulf (2017) and Tuboly (2019).

[311] See Popper (1935/2002).

acceptable, that you FEEL A LITTLE THAT YOU ARE IN SOME WAY UNEMPIRICIST, then I should understand that I myself as your bad conscience irritate you enormously. That would make me more prepared to take your tendency to grieve me easier.

Let me come to an end. All that is guess work, but I think I should write it to you since I would say it to you. Of course writing and speaking are different, scripta litera manet, but I tried to write as peacefully [calmly] as possible, whereas in a conversation I should call you names, as usual.

In spite of the fact that all such events touch me very much, I usually overcome them and discover a way how [sic] to go on better than before. You see, your treating me badly by not mentioning me, induced me to start with publishing my main ideas regularly – I like that now. Your letter on my articles hardly acceptable for the Erkenntnis immediately induced me to create the series EINHEITSWISSENSCHAFT, where I could publish my own stuff whenever I wanted to do it, without any consent from my strong teacher.[312] Of course I did not insult you but invited you to be with me and to publish there, you by me highly admired giant of Logical analysis and a man who unveils the secrets of so many metaphysicians .... And so life could go on in a rather smooth way, as far as we are concerned. Now I got my fill again – let us wait, what nice activity will come out from that, what new plan, which will enable us to co-operate better than before....... Let me assume that you will grieve me next time within one decade, then in the following decade – but I think there will hardly be more than three decades, therefore no more than three grievances – I shall prepare myself for that.

I am rather an APARESH – should you know of these people – who are well acquainted with living peacefully together, but not with fighting one another. They bear any insult very intensely and hardly know how to answer... The same is with me. I am sad and feel myself very clumsy in writing such a letter, but I should also feel myself rather clumsy writing no letter at all. And I manifestly am not sufficiently educated to write the letter, which "SHOULD" be written. ....

That is that. Let us hope that we shall have good time together when we shall meet again. I hope we shall have a good air communication after the European war will be over.

With our best wishes for your health, remember us to Ina,
Always yours
Otto Neurath

OTTO NEURATH, 30 Bickerton Road, Headington, OXFORD

---

[312]The pamphlets from the series *Einheitswissenschaft* were translated and republished later in English, see McGuinnes (1987).

29. Carnap to Neurath, May 6, 1945.

(ASP RC 102-55-21)
RUDOLF CARNAP
Faculty Exchange
University of Chicago
CHICAGO, ILL.
May 6, 1945

Dear Neurath,

A short time ago, we had the pleasure of seeing Paul,[313] who was in Chicago for a few hours. We had not seen him since the days in the The Hague in 1934. In these years he has developed to a very fine young man. The grave experiences he has gone through have not made him bitter against fate and humanity. He seems quite cheerful. And although he is somewhat gloomy, like all of us, about the prospect of the political development in the peace time ahead, he has, like you and me, not given up a certain optimism about the future development "in the long run". He has a well-matured and independent judgment on persons, situations, books etc. From what he reported about arrangements he makes to get his students into contact with institutions where they can practically apply their theoretical statistical methods, he seems to be an excellent teacher. Also he seems very industrious and is studying in spite of his heavy teaching load, various things outside of but in connection with his special field, so that his work will gain a broader basis. I am sure he has a fine career ahead of him.

Now to the sad affair of your monograph and your letter of Nov.18.[314] I am very sorry that refusal to share the editorial responsibility for your monograph, caused you so much grief, but I do not see how I could help it. I had not the least intention to insult you. You should not take the matter so tragically, and above all, you should not take it so personally and with so violent emotions. Everyone of us has sometimes the experience that a publication of his is regarded by some of his friends as rather weak. We have to bear it and not take it as a personal insult. (For instance, I was rather shocked to read Nagel's extremely critical review of my "Semantics", but I never thought of regarding it as a personal insult.)[315] If everybody took such things personally and always reacted with violent moral reproaches (as you do in your letter, and you did before several times towards me when I simply followed my own judgment as editor or author, and another time towards Schlick) how would good cooperation and friendship be possible? I did not let my personal relation with Nagel be disturbed by what he did to my book. And I wish very much that you

[313] Neurath's son, Paul Neurath, see letter 5, note 63; letter 9, note 100; letter 16, note 146.

[314] Neurath to Carnap, see letter 28.

[315] See Nagel (1942).

would take the same attitude towards me. In fact, I have waited all the time whether another letter might not come from you in a more conciliatory tone, so that I would be allowed to forget the bitter, ironical and sarcastic remarks in your last letter.

There seems to be one point which must be cleared up among the three of us;[316] that is the question of our editorial rights. I shall outline my conception of them; if I am mistaken, please let me know. (i) I agree with you that in questions of content the author has the final decision. If an editor has an opinion different from the author in some point, he can only make a suggestion and state his reasons; the author is free to accept the suggestion or to stick to his point. (Thus, e.g. Bloomfield would have had the right to maintain his point of view in the question of "written languages".) (ii) On the other hand, the editors are the final judges in points of the following kind: choice of the topic (in general lines, not in all details), arrangement of the subject-matter, way of representation, including comprehensibility for our readers and correctness of language, and the like. These and similar points are usually regarded as editorial matters in a collective work like the Encycl. Therefore, if our present arrangement – contrary to my assumption – does not include these things among our editorial rights, then I suggest strongly that we change it accordingly. For, if a reader criticizes points of this kind, he will always make the editors responsible too, not only the author.

You ask what are the points of my criticism, I think you will understand them best from my letter to Morris (June 19, 1944), a copy of which is enclosed.[317] (You will see from the letter also that I tried my best to find a solution which would avoid any offense to you.) You see that most, and perhaps all, of the points criticized belong to what I regard as editorial matters. The same holds for most of the points criticized independently of me in a letter from Nagel to Morris. The fact that Nagel addressed his critical remarks (which were lengthy and in detail, and much sharper than mine) not to the author but to an editor, shows that he likewise regards these things as matters of editorial responsibility. Let us suppose the case that the note on the back of the title page were not printed at present, but that the printing of the text of the monograph were finished, so that no alterations in the text were possible. Would you in this case actually demand that I share the editorial responsibility for the monograph? I must tell you that I should regard such a demand as very unfair. I certainly should never make a demand of this kind if I were in a similar situation. Please remember the fact that the monograph did not reach me before it was printed in pages, and that I was not guilty for this fact. (I am astonished at your ironical remark in this point; it seems to me clear that the responsibility for this fact rests not upon me, but upon you, Morris, the Press, and perhaps Wirth,[318] in a proportion

[316] The "three of us" refers to the editors of the *IEUS*, Carnap, Neurath and Charles Morris.

[317] See Carnap to Morris, copy to Neurath, letter 30.

[318] Louis Wirth (1897–1952) was an American sociologist who mainly worked on the sociology of city, urban life, and immigration. At one point, Morris suggested to Neurath that Wirth could write

unknown to me.) Why then should I be required to bear innocently the blame by readers like Nagel and many others? (Nagel writes that his empiricist friends in N.Y. agree with his judgment and are "simply dismayed".)[319]

I wrote you earlier that my criticism has nothing to do with our differences of opinion with respect to semantics, degree of confirmation, etc. Perhaps the fact of Nagel's criticism will show you more clearly this independence, because in the two points mentioned Nagel is rather sceptical about what I am doing, and hence much closer to your point of view than to mine.

I think more important for us than the present particular incident is the problem what we could do to safeguard the quality of future publications. In this connection, I suppose, Morris has already taken up with you the question of a more active function of the Advisory Committee as suggested by Nagel.

My two papers on probability and degree of confirmation will probably appear this summer.[320] I shall send you reprints as soon as they come out. My best thanks for the booklet on health with your charts in it, and the reprints of Plato; did your opponent reply?[321]

I just read Sheldon's "The Varieties of Temperament", which is very interesting.[322] Sh. is now here, Morris is working with him, and I hope to see him soon. I suggest strongly that you read this book. If you do, write me <u>in which of his types you would classify yourself (and in which me).</u>

I wish with you that some day in the future we shall meet again under a friendlier star. And even now, let us try as best we can to be tolerant, understanding and peaceful to each other personally, no matter whether one criticizes the other's writing (as I do yours) or ideas (as you do mine).

With best regards,
Yours,
C.

---

a monograph on the sociology of science for the *IEUS*. See Morris' letter to Neurath, copy to Carnap, Feb. 21, 1937 (ASP RC 102-51-63).

319 See Reisch (2003, 208).

320 Possibly Carnap (1945a, 1945b).

321 The booklet is possibly *Health Education by Isotype*, written with H. E. Kleinschmidt. See Neurath and Kleinschmidt (1939). The "Plato" reference is presumably to Neurath and Lauwerys (1944).

322 See Sheldon (1942). William Herbert Sheldon (1898–1977) was an American psychologist and worked on so-called "somatotype psychology:" he tried to find correlations between body types, behavior and social settings. At some point, Sheldon was working Charles Morris alienated thus the latter from Neurath and others. For further details see Reisch (2005, Chap. 16).

30. Carnap to Morris, Copy to Neurath, June 19, 1944.

R. Carnap
P.O.B. 1914
Santa Fe, N. M.
June 19, 1944.

Dear Charles,

A few days ago I received at long last, the proofs of Neurath's monograph from the Press.[323] Since the pagination and the running titles are printed, I suppose that these are already the final proofs. I never got the galley proofs; I do not understand why the Press didn't send them to me automatically as in earlier cases.

After reading the monograph, I must say frankly that I find it in a rather unsatisfactory state. First, from the linguistic-stylistic point of view, it seems to me in a rather bad shape. I see from your letters that you made some minor changes. But I think it is still in need of a thoroughgoing revision. Now it will, of course, cause additional expenses: but I think, if the Press is not willing to bear them, we should cover them from the Encyclopedia royalties. A publication in the present form will do no good to the reputation of either the Encyclopedia or the Press. As you will remember, at some earlier occasion some readers expressed sharp criticism of the poor English. I am astonished that the editorial staff of the Press did not raise objections. I suppose they would not have accepted the ms. in its present form if it were a separate book; but here they probably thought that the responsibility for editing is not theirs but ours.

The second point is still more important. It is the way of formulation and representation of the whole. Here I mean not linguistic questions, but certain features that would be the same if N. had written in German. It seems that the whole has been formulated in a very hasty and careless way; no care and time has been taken to work it over, make things clearer, and give it some coherence. It jumps from one idea to another, while the poor reader looks in vain for a connecting thread. N. uses many of his rather obscure pet terms without explaining what he means by them, e.g. 'Universal Jargon', 'Encyclopedia', 'absolutistic', 'aggregation', etc.; also quite new ones, e.g. 'Terminological Empiricism', where not even I have been able to find out what is meant by it. Especially in the first part, dealing with more general questions of empiricism, the representation is often not clear and sometimes quite confuse. Some sentences are incomprehensible to me, and how many more will be so for the average reader. The later part are not quite so bad, especially the discussion of more specific problems of the foundations of the social sciences. On the whole, the monograph seems to me to below that level which so far we have succeeded in maintaining in the Encyclopedia, and the weakest thing N. has written (with possibly one exception, the Aristotelian Society paper).[324] The motivation for

[323] See Neurath (1944).

[324] See Neurath (1941/1983).

the hasty work is clear: N. wanted to hurry because the Press considered to discontinue the whole. I understand very well that you, even if you felt likewise critical, hesitated to write to N., first because of the lack of time, and then also because of his sensitivity and violent emotional reactions to criticism and his obstinacy and unwillingness to accept suggestions for improvements from anybody. Nevertheless, if I had seen the ms., I should have written to N. and asked him to work it over.

Now I think, this second point is more important, but it is also more difficult, maybe impossible, to do anything about it. I wish I could talk over with you the whole matter. Please write me frankly your opinion on both points. Do you think I am overcritical? And then, I should like to get your advice what we or I could do now. I do not wish to make an issue of it if I know beforehand that no good will come of it. I do not wish to insist on principles and ideals standards when I see that thereby I endanger my friendship with Neurath the Volcanic and the continuation of our good cooperation for the Encyclopedia and other things. Therefore, if (as I assume) it is now too late to ask N. to revise the whole, without making him furious and the Press unwilling and impatient, then I shall not veto the publication. This is a compromise and a hard concession to make, because I am convinced that the monograph will do more harm than good for the Encyclopedia and for the movement of empiricism in general. But I realize that to antagonize and offence N. now would possible do still more harm.

On the other hand, I think I ought not appear to be responsible for something which I am in fact responsible because I did not see the ms. before it was printed and I would not have accepted it in its present form if I had seen it. I should like to have your advice as to a way to release me from the editorial responsibility without, however, an explicit and public expression of my disapproval. Would it perhaps be best simply to omit my name from the title page of this monograph? (This would be like Roosevelt letting a bill become law without either signing it or voting it). Or do you think that this would not be right because our names are on the left-hand title page refers to the whole Encyclopedia and not to the monograph? Would you think it better to put a small, inconspicuous note on the back of the right-hand title page, perhaps above the Copyright, to the effect that, due to special circumstances (not to be specified or perhaps wartime circumstances or something else?) I do not share the editorial responsibility (or, I have not participated in the editorial function) for this monograph? Or something similar?

I can report good news about my back. It has continuously improved and I am now up about six hours daily, intermittently, with rest periods in between. This I am confident that I shall be able to teach in the autumn.

My work on probability proceeds well. It is growing ever more, so that it is still very far from being finished. How is your Theory of Signs developing?[325]

Cordially,

C.

---

[325] Morris's book on the theory of signs appeared in 1946 as *Signs, Language and Behavior*.

31. Neurath to Carnap, June 16, 1945.

(ASP RC 102-55-11)
16th June, 45

Dear Carnap,

....

Thanks for letter and the enclosed letter to Morris dealing with the FOUNDATIONS case. Let me tell you frankly and freely, that I think this letter to Morris is even a more serious thing than your behaviour. And since I think I should inform Morris of what I think about this writing of yours, I think the best is I am sending a copy of this letter to Morris.

You are – as far as I can see – mixing up totally different things, criticism and offence. You see, we human beings very often look at ourselves not in the same way as other people do. Therefore, of course, I am always prepared to re-adjust my opinion about myself. I think that is in harmony with my pretending to promote Epicurean Stoicism, i.e. on the one hand to think of people's happiness, on the other hand to think how to behave in harmony with the place in life we want to take. Therefore as far as I decided to go on as a non-conformist I have to take it easy, what comes out of it and as far as I decided to be a good collaborator, I have to behave as such – and I try. And, as far as I can judge from many examples, not without success. At least I am collaborating with many people, even writing together with others articles etc.[326] That implies much adaptation and preparedness to give way ... Of course, I have certain defects. My ability to judge rightly, when other people feel bored by me, is [has] not very much evolved. That my attitude towards life irritates many people, particularly conformists, I cannot deny, but that is partly unavoidable, in spite of the fact, that I try to act politely, whenever possible. This I stress, that you see I am aware of the problems involved. I am not so sure of your preparedness to re-adjust your own judgement about your grieving other people at many occasions, partly connected with certain non-conformist habits in your way of life, which leads – as in my case – to some intellectual and emotional success, but partly perhaps with some lack in thinking of other people's feelings. Perhaps. To add this: I myself very often grieve other people, I presume, but I think I should be prepared to analyze very carefully their pains.

What did you do? First of all your editorial remarks are absolutely sound, if applied with discretion and one could inform future collaborators of these principles. But, please, do not think that I even for a moment, objected to your criticising me. The idea that I dislike changes in my style or presentation of subjects suggested by other people sounds rather comic to me, since I am rather glad, to get such suggestions. I really have re-written whole papers gladly under such circumstances. I

[326] Presumably Neurath and Lauwerys (1944) and (1945): the papers are about Plato, Germany's education and Nazism.

cannot remember any case in which I disliked to alter something for the sake of understand[a]bility, as it were etc.

But, how you are going on in the letter to Morris, you are writing as follows: "I understand very well, that you, even if you felt likewise critical, hesitated to write to N. first because of the lack of time, and then also because of his sensitivity and violent emotional reactions to criticism and his obstinacy and unwillingness to accept suggestions for improvements from anybody." ...... "I am convinced that the monograph will do more harm than good for the Encyclopedia and for the movement of empiricism in general. But I realize that to antagonize and offend N. now could possible do still more harm" ... and so you dropped your name, without contacting me before.[327]

Look, there are so many fine modern inventions, e.g. "CABLE". You could tell me of that – even at my expenses [sic] – be sure I would have preferred TO PAY [FOR] ALL CHANGES SUGGESTED. The dropping of the name has nothing to do with a fair criticism but is GRIEVING. You cannot compare that with a hard criticism of your semantics by Nagel, because it is different.[328] I SHOULD BE GLAD TO READ A CRITICISM OF MY POINT OF VIEW BY YOU....... Believe me [about] that, please.

But, what about your remarks to Morris. Please, think for a moment, that a third person should read it – does it not sound like a denouncement? Either Morris knows me as a man of a violent and obstinate character, why the description? Or, if not, how did you think it suitable to bring forward such a heap of accusations? Tell the words of the sentence and the number of hard names given to me?

I ask you, please give me verse and line for that. You see we are now friends for two decades, we know very well that we can trust one another, whenever we should be in a position to come to one another's help, but on the other hand throughout our connections appears a certain strain of tension, which does not become softer up to now. I guess we shall be good friends the following decades, and I hope that perhaps with increasing wisdom, we shall be able even to reduce the tension, but how? You see the whole correspondence is a rather "continental" one, but we cannot go on in the Anglo-saxon way as long as the tension exists – unfortunately.

You see, prepared to learn from my friends, whenever possible, I searched my past for examples, which might induce you to formulate so condemning a judgement. I searched in vain – but that is perhaps my blindness. Please inform me about myself.

Please, tell me of YOUR EXPERIENCE, not of the slander and gossip created by other people within and without our circle.

But to explain to you better, what I see, a few sentences: I am, of course, a man who puts some pep into his utterances, and like clarifying slogans and expressions. I am sometimes rather noisy and not always restricted in the good Anglo-saxon way

[327] Quotation from Carnap's letter to Morris, copy to Neurath; letter 30, page 636 in the appendix.
[328] See Nagel (1942).

during discussions, interrupting other people … and so on. But I do not remember any case, where I behaved, as you are describing my CHARACTER, as it were, not only occasionally.

Since I am a sociologist and historian, I ask me, how even other people might have told you "stories" about me, which should prove such a character. I tried hard to discover cases, were I really angry pressed my case. Let me remember one tension: Hans Hahn and I have been [were] friends for many years – since our Gymnasium time. He, the older, taught me a lot of things. We, Frank, and other read Spinoza together in the "Rahnhof" etc.[329] Becoming older he had a certain inclination towards, what we call "Bonzentum", but since I estimated highly his frank attitude within society, his empiricism, etc.[330] I took it easy, even if he kidded me sometimes more, than I did.

I remember that I reacted sourly and some little clashes started. First as he supported [the] professorship of Eibl,[331] whose knowledge in scholasticism was above doubt, but whose nationalist propaganda was a real danger. Hahn thought he should think of the scholar only and swore Eibl would never use his chair for propaganda, such a well educated, fine man etc. I thought that a real danger for our own existence. ---- How disappointed was Hahn, when Eibl became a real propagandist within the faculty. Next. I liked to speak contemptuously – I do not deny that – of Innitzer (you know the Heil-Hitler Innitzer),[332] giving his names (as he deserved as far as his records I knew) whereas Hahn opposed in a rather irritated way and telling me, what an impossible fellow I am, overstating every thing. And when I warned him, he told me, what a cultured fine man I. was, how tolerant, how prepared to see everything from more than one angle etc. whereas I being really obstinate in all such situations and a violent grumbler through and through, not prepared to accept any criticism of my attitude by anybody, even my best friends, in this way I should make difficult contacts etc. and what "Bonzen" say at such occasions. I did not give way, I knew my Sudentenboy sufficiently. I have not necessary to stress, that Hahn would be disappointed again by Innitzer's later deeds; but this tensions coloured partly our contacts in scientific fields and I can realize, that Hahn thought my whole attitude

[329] Neurath, Philipp Frank and Hans Hahn formed what was called (by Rudolf Haller) later the "First Vienna Circle," a regular discussion group on Thursday evenings in the Viennese coffee houses between 1907 and 1912. Hahn died on July 24, 1934. On the First Circle see Uebel (2003).

[330] "Bonzen" and "Bonzentum" were commonly used to mean "bosses," "big shots," or "bigwigs," and "boss rule," and were often directed against leaders of the socialist movement, for fear that they might lose contact with the members of the working class and would not fight for their cause.

[331] Hans Eibl (1882–1958) taught philosophy at the University of Vienna. He was interested in the history of philosophy, especially in patristic and scholasticism. During the 1930s, Eibl became associated with Catholicism and National Socialism. On Eibl see Stadler (2001/2015, 296–297).

[332] Theodor Innitzer (1875–1955) was Archbishop of Vienna. He signed the declaration endorsing the Anschluss as "Heil Hitler."

an obstinate one. That I should understand very well in connection with such antecedents (there have been many of this kind as far as I can judge, my sensitiveness was right, his kind of professor-neutrality wrong, as far as later observation-statements prove). And I got the impression, that he liked to tease me, e.g. with the permanent repetition of the stale joke "Einheiz-Wissenschaft" (which is well done once a year, or twice, but not as something permanent).

Schlick liked this kind of treating me from above and even he himself, the highly refined man, behaved rather coarsely towards me, as particularly Philipp Frank remembers well and as you can judge from the paper presented in French to the Paris congress, where the "unesthetic name" of the Neurath brand has been mentioned – but unfortunately in the French translation, which is not "unesthetic" at all. Do you think of stories about me from this source? I do not think that covers your denouncement.[333]

What more? I behaved sometimes noisy and intensely, when speaking of Wittgenstein. I regarded him from the start as a mystic and metaphysician of the refined type, as an antiscientific person through and through and I dared to say so as the admiration of Wittgenstein was the fashion in the Vienna Circle. I remember in the discussion meetings I made again and again the remark "metaphysics" [and] Hahn suggested I should reduce my remarks to "M" for shortening the interruptions, then finally he suggested I should only tell, when I thought something "Non-M" to waste less time. He – that was my feeling – liked very much to be in harmony, as much as possible, with Schlick, and Schlick enjoyed his attacks on me, on "Einheiz-Wissenschaft" and my violent remarks on this chap Wittgenstein. I cannot deny that I did not give way to suggestions, I should look at Wittgenstein from a different angle etc., BUT NEVERTHELESS AS YOU CAN TESTIFY, I behaved very collaboratively as we put together the WIENER KREIS pamphlet.[334] Feigl, [was] at the time fully occupied by [with] Wittgenstein's greatness and Waismann tried to put into that pamphlet an eulogy on Wittgenstein – and there it stands, supported by you and Hahn. Of course I am "responsible" for that as one of the three editors, but on the other hand, I got the permission to put the "EINHEITSWISSENSCHAFT" into the pamphlet, too. I remember, how Hahn disliked such a slogan-word.... Even you tried to find excuses for Schlick's and Waismann's worshipping Wittgenstein. At this time in my behaviour evolved [developed] a certain sharpness, when Wittgenstein and his case ha[ve] been discussed. I do not deny that. Do you think my sensitiveness for [to] empiricism lead me astray? I am sure, that the Wittgensteinians gossiped a lot about my obstinacy, my unwillingness to take part in the worship etc. .... but why you are [sic] continuing such stories about me?

[333] For the English translation of the mentioned paper see Schlick (1936/1979).

[334] See Carnap, Hahn, and Neurath (1929/1973).

In all these and similar cases, I never pressed MY OWN OPINIONS VIOLENTLY, but tried to push away things I thought dangerous. I remember, how I disliked the Popper admiration, whom I thought as an antiempiricist man with many empiricist arguments, full of clever ideas, but not reliable, when empiricism of the unified science is at stake, he with his ONE WORLD SCHEME AS THE BEST etc. Antipluralist through and through. The defects of his probability arguments I felt strongly, but not being able to reach a real judgement on that, I behaved with restrictions there. I cannot deny, that I did not give way in the case of Popper, in spite of the friendly judgement you and Hempel found appropriate.[335] But I do not remember, that may [my] own opinion on something has been at the stake, both of you did not criticize my anti-absolutist attitude (you stress in your letter to me, that you do not know sufficiently how this term should be used, you think the term "Universal Jargon" is not sufficiently explained, really?, and "Encyclopedia" – in spite of the fact, that Morris and I wrote about this in monograph one.[336] I do not grasp your doubt, that one could understand "aggregational" attitude – as far as I explained it, it seems to be clear: making a statement on something today implies making a statement on the cosmic aggregation today.[337] Is that something strange? But that is another chapter, which we shall discuss, I hope so, in public some day or another, if be [we] both think it suitable therefore I could not behave obstinate[ly] as far as I am concerned. But, perhaps you know better examples for your heap of names given to me.

One point to the subject in question, the FOUNDATIONS, two people read the manuscript very carefully, one for English style, and Morris has been free to suggest any alteration he liked. He suggested some alterations. I did not object in the least.

To a certain extent you are making Morris responsible, because he did not suggest more alterations, as he could do. You are trying to suggest, that he did so under the menace of the violent volcano .... oh boy, oh boy, what a story. Detective novel in the unity of science movement ....

You speak of the "hurry" under which I worked. I do not think that is the essential point, I wrote the whole manuscript more than once. And you make the remark that my Aristotelian paper[338] is even worse ---- that I certainly did not write under pressure.

I cannot deny that I have difficulties in conveying something I think is new, when applying my phraseology – of course that is a funny behaviour. Let it that [sic] be. But PERHAPS, PERHAPS it is unavoidable in the period of growing up [learner's] phraseologies and the expressions will be better later on, PERHAPS. I do not feel at ease with my writings [sic]. And I know one can use this argument given above for covering any nonsense. How to know, what is the case?

[335] See Carnap's (1935) and Hempel's (1937) reviews of Popper.

[336] See Neurath (1938b) and Morris (1938).

[337] See especially Neurath (1944, Sect. 8.).

[338] See Neurath (1941/1983).

You see looking back at former writings I discovered this: one day I wrote a paper together with a friend,[339] who thought my style could be improved (in German) and I thought it wonderful. I agreed with this booklet, but when now reading again the papers of mine and this combined paper, I have to confess, that I can accept more or less today my own papers, with all their defects, whereas this book, in spite of his better style, seems to me avoiding certain important arguments and forming even deviations from the main point. But that is no pro[of], of course. I only tell you of my experiential statements.

You see, you tell me that so many of our friends complain about my style, order of arguments etc. (I know that since my youth) but – unfortunately, they do not discuss my arguments, not even the new ones – should that not make me a little suspicious. My kind father, when people attacked serious scholars in such a way, was accustomed to say: “Der Herrgott schaut nicht auf die Orthographie”.[340]

I am seriously thinking over [about] that. Scholars have an inclination to think that other scholars intentionally try to reduce their activities etc.: I do not make such a complain[t] because I think that for a non-conformist I am treated well, even in the field of education, where I am able to earn a living in spite of my being an educational non-conformist. I look at a very startling phenomenon. Take Hempel, he is writing on laws of history etc., he quotes a through and through metaphysician as his choice in discussing problems.[341] Then he continues in explaining the old[-] fashioned approach. But he does not even mention the fact that there does exist another opinion, too, about unpredictability IN PRINCIPLE.

What should I think of that. I selected Hempel as an example, because I am so sure of his kindness, his goodwill and his sincerity. Why did he not even mention my unpredictability approach, not even objecting to it, in spite of the fact that I printed that in my EMPIRISCHE SOZIOLOGIE and repeated it again and again?[342] Should I believe the story that I am not understandable by an average reader, assumed he reads naively my books? The “Orthography”, the “style” should be so bad .... What I guess is, that Hempel READING my papers LISTENING to my arguments, did not REMEMBER afterwards these points, BECAUSE THEY ARE TO[O] FOREIGN TO HIM.

[339] Neurath wrote two papers in German with someone (not counting now his formal logical papers with Olga Hahn-Neurath), these were (1) “Der Kompensationsverkehr im zwischenstaatlichen Warenhandel,” with Wilhelm Heilpern in 1928; and (2) “Können wir heute sozialisieren? Eine Darstellung der sozialistischen Lebensordnung und ihres Werdens,” with Wolfgang Schumann in 1919. Given that Schumann was a close friend, Neurath presumably is referring to that paper.

[340] “The Lord doesn’t look at orthography.” On Neurath’s father Uebel (1993) and (1995a).

[341] In his “The Function of General Laws in History,” Hempel (1942, 37. note 1) referred to Maurice Mandelbaum. Cf. Reisch (2001, 208).

[342] See Neurath (1931/1973, 369. ff.).

In analyzing my past, after reading your letter of May and the letter to Morris of June 19, 44 I discovered a very interesting instance.[343] A friend of mine wrote a fine review of a book of mine.[344] Before printing it he asked me, whether I should be prepared to read it to avoid any misunderstandings. And now look, in this review he had forgotten not only what I wrote positively, he complained also that I did not criticize something etc. Then I have shown him point after point in my book, and he said: strange, everything is here. This I thought a good opportunity to learn something about my style and asked him, what is difficult to grasp in my paper. And he, without hesitation, answered: everything is expressed fully and clearly ....

Is that not interesting. Of course it does not prove my case, but it explains to you, why I am not so sure of the criticism of my "Orthography". I shall try to find readers who are relatively naïve and ask them to tell of the contents of FOUNDATIONS.

I further searched my past to find violent anger. Yes, I have sometimes expressed wild emotions, when people used my own ideas, without telling so and deforming them. In the field of Visual Education that is not unusual, but also in our field. But that does not belong to the field of obstinacy.

But perhaps, when I stressed certain names, as e.g. LOGICAL EMPIRICISM. I remember very well, how in a coffeehouse where we meet, Schlick wanted me to use the term RADICAL EMPIRICISM, thinking of James, and I stressed the point, that a James term is not advisable, because James whom we have to admire for so many things is full of BERGSONIANISM and to a certain extent a highly skilled obscurant, more or less. And I remember, that many of our circle did not like LOGICAL EMPIRICISM – I was the only one who tried to push forward it. The other argument of mine was, that we should not speak of Logical Positivism, because Comte was such a metaphysician, full of antiscientific approaches, later on full of religious phantasy and the Positivist church here is not a real bargain for positivism. But later on, that is fate, people do not remember, why I promoted LOGICAL EMPIRICISM, they are even accepting the term, but NEVERTHELESS REMEMBER PERHAPS THAT THERE IS SUCH A VOLCANIC FELLOW, VIOLENT, OBSTINATE etc., who always wants to promote funny names, nobody wants to accept etc. ..... rather comic, sometimes depressing, not as far as I am concerned, I am all living happily on our hill, surrounded by flowers, birds and many kind people, but as far as other people's happiness is concerned, who are putting their whole personalities in such details, and then have to face such fate.

You think my FOUNDATIONS will do not good for our empiricist movement. Oh my dear, how cautiously one should make such judgements .... Looking back into history, it is strange how things are going on. You see, I am thinking that the FORMALIST strain, combined with ARISTOTELIAN duplication which seems to me now prevalent in many members of our movement, are real dangers, but I am very reluctant to bring that forward, because tensions between us might reduce the

[343] See Carnap to Neurath, letter 28.; Carnap to Morris, copy to Neurath, letter 30.

[344] Neurath is presumably referring to Jan Tinbergen (1936) who reviewed his "Was bedeutet rationale Wirtschaftsbetrachtung?" in *Erkenntnis*.

chances of the movement even more. You see every movement is based on TOLERANCE and COMPROMISE, and the ORCHESTRATION of the multiplicity of activities is just the point.... I have the feeling, that you are now on a not so good way. A kind of "popery" speaks to me, when I am reading your letter, a certain tendency to reduce other people's status, etc. .... but I am not so sure of that I ask you only to think of that if you are analyzing yourself.

When I am trying to stress: Protocol-statements, Logical Empiricism, Terminological Empiricism, Unity of Science, Pluralism of approach, Unpredictability as a basic item, economics-theory as a kind of organized folklore, etc. I am not "fighting" calculus-absolutism, etc. but I try to find a way of "orchestration", otherwise we cannot build up something of importance. I get the impression, that PLURALISM, UNPREDICTABILITY etc. are something strange to you, in spite of the fact that you stress your harmony with my main arguments, when despising my style and expressions. You see your "degrees of confirmation" do not fit into the pluralistic and non-predictability scheme etc. Partial schematizations ARE NOT FORERUNNERS OF AN INCREASING TOTAL SCHEMATIZATION .... that is just my point of view, but that both might be expressed is important, not the concealing of the difference.

I assume that some day one of our members will find the time to discuss my FOUNDATIONS not only from point of view of higher orthography and well-order [sic] of arguments, but also from the point of view of EMPIRICISM.

You see I have to make pleas for myself because nobody is prepared to do it for me .... at the moment, it could be could be, perhaps, perhaps, that the order of arguments is not so strange to people with a similar start ...... there exist such examples in history.

I personally always fear that I do queer things and therefore I like people like Frank who try to reduce my overestimating protocol-statements etc. without grieving me. And you see, that is the point. My letters dealt with YOUR UNKINDNESS. That is the main point. One has no "right" to ask for kindness, but one can object to unkindness. Look at the letter to Morris. You only tell of my bad qualities and that you drop ONLY your name, – as a concession – because he is such a violent boy .... No kind word about me in this letter. You see, it could e.g. come into your wise brain, that even strong elephants[345] sometimes need much of their energies to overcome the obstacles of life. You see, to reach the shores in shoes only is not just the best start for a new life in a foreign country and an internment is also not the best introduction.

If you were kind, you could write e.g. "and then we should not grieve Neurath, he has a difficult life ahead" etc., no, only the fear, I could become violent.... oh boy, oh boy, what a world, a sad world. Full of unkindness.

I do not want to analyze you, as you suggest, because it is not always useful to do that with friends. I learned that principle from a good writer and behaviourist

[345] Neurath used to sign his letter with the drawing of an elephant.

……. But I shall read SHELDON'S THE VARIETIES OF TEMPERAMENT.[346] I am rather suspicious of this business. Of course there is scientific research possible, but we have not reached sufficient results for practical purposes. I like Morris' idea to treat Paths of Life,[347] but I am doubtful whether it is useful to connect them with characterological types. The Nazis like characterology, and all kinds of physiognomics etc. … of course that is no real counter-argument, I am only sensitive in this field. NOT TOO MUCH OF THAT STUFF.

And what can we do with the knowledge? I remember, tested before the first world war by a psychologist I got such a pure visual type, that it is usually only with idiotic children – that is perhaps the reason why I am so adapted to visual judgement, perhaps but how can one discover that? What implies having a certain mixture? Overcompensation seems to be important, the st[u]tterer has more chance[s] to become a great orator, than the non-st[u]tterer ---- but I do not want to criticize a very important branch of science, I only tell you of my doubts. Often such analysis is FIXING your activities, because you think you e.g. in agreement with your "type" or "fate" etc., whereas the Epicurean Stoicism tries to FORM EVERY LIFE WHATEVER THE BASIC FACTS MIGHT BE …. strange, is it not? In Vienna we learned scepticism and used Freudian analysis, which we estimated highly, for so many jokes, that we never became addicts of Freudianism, we – that is people of the Epicurean Stoicism ….

Nice, what you tell about the "Herr Sohn".[348] I also think he has a well prepared future before him. I should like to meet him, and you and all the others and to talk about cows and calves, tricks and tracks, to kid one another ….

The world situation is bad, but the Hitlerit[e]s are away. Now we got some good news from people who lived through the hell and bad news of people who died. What about Grelling?

I am lecturing sometimes, even brain[-]trusting and making jokes to [sic] local groups, I have sometimes to do with education, here and there, Mary is very industrious in Isotype work and as housewife. We shall now get a bigger house, everything goes on well. Please remember us to all friends, particularly to Ina. We should like to get a vivid letter from her about you and your health. How you are now?

Thanks for the newspaper clippings, I should like to get more of that stuff. Via Paris we get sometimes unpleasant news, about behaviour of people during the occupation, cannot discover what happened really. A long story.

With kind regards, good wishes, greetings etc.

Ever yours,

N.

---

[346] See Sheldon (1942).

[347] See Morris (1942).

[348] Neurath presumably refers to his son Paul, who was mentioned at the beginning of Carnap's previous letter. See letter 29.

32. Carnap to Neurath, Aug. 23, 1945.

(ASP RC 102-55-09)
Rudolf Carnap
University of Chicago
Chicago 37, Illinois
August 23, 1945.

Dear Neurath,

I was very much discouraged by your letter of June 16th.[349] Not only because I see that my attempt to explain to you the reasons for dropping my name were entirely without success; more important, because my assurance that no personal question was involved and that I had not the least intention of offending you does not seem to have much effect. I thought I had extended to you the hand of reconciliation; instead of accepting it you continue reproaching me that I "grieve" you. Do you realize how distressing and disturbing your attitude towards me in these two letters has been to me? If you knew how many sleepless hours at night they have caused me, and how much inability to work in the daytime!

The remarks about you in my letter to Morris[350] were not at all meant as a denunciation but merely an allusion in half-facetious terms ("volcano") to facts obvious to everybody. The reason I wrote these remarks was, that I felt that Morris had a share in the responsibility for the publication of an unsatisfactory part in our common work and [marginal addition: *I referred to this*] possible motive of his, [marginal addition: *as an excuse for him*].

I think it would be best if we dropped this whole affair now and tried to start anew as best we can, instead of dragging it on indefinitely. However, since you ask so insistently what I meant when I spoke your violent emotional reactions, I will mention the two occasions uppermost in my mind: your quarrel with Schlick about your ms., the second, your quarrel with me when I was in Prague and you sent the long wires from Moskow.[351] I do not say that you alone were wrong on these two occasions; I admit that much was wrong in Schlick's attitude in that affair, and I suppose on objective judge could show me wrong things in my behavior too. The question now is not who was right or who wrong; probably nobody was entirely right but everybody has some good things to say for his side. The question now is only in what way you handled those differences once they had occurred, and whether or not there occurred violent emotional explosions. For me your present blindness about yourself is quite amazing: you speak at great length about the relatively mild

[349] See Neurath to Carnap, letter 31.

[350] See Carnap to Morris, copy to Neurath, letter 30.

[351] Carnap presumably refers to those wires of Neurath that he has sent on January 28 and February 8, 1932 (RC 029-12-68 and RC 029-12-65). In these messages he was asking for more citation of his works in Carnap's papers regarding physicalism. On this issue see Uebel (1995b).

tensions with Hahn and seem to forget entirely those two occasions which belong to the most depressing experiences of my life. On those earlier occasions and on the present one where you reacted with a furious outburst to the withdrawal of my name, your violent reactions seemed to me entirely out of proportion to the causes by taking forms which were for me (and for Schlick too) indescribably depressing. You will remember how dear Olga afterwards, when we met in the "Auge Gottes", brought about a reconciliation between us.[352] I felt greatly relieved; the reconciliation was for me, as I am sure it was for you too, not merely a diplomatic patch-work, but genuine and sincere. But you may also remember the serious warning I expressed. I said, jokingly, that you had been an elephant in a china shop; but I added, seriously, that I thought I should not be able to stand another time your method of outbursts, threats, and ultimatums. Now please don't misunderstand me: I did not mind at all the demand you made in your cables for being given credit; you deserved credit and I was glad to give it to you. What I minded was only the violent emotional way with outbursts and moral pressure by which you induced me to give you what seemed to me an exaggerated amount of credit. I gave it for the sake of peace and preservation of friendship. But I resent to the present day that this one time in my life I was bullied by another man into saying something not in accord with my conviction.

Well, those were the two chief occasions. I am happy to say that no explosions quite of that magnitude have occurred since. But there were a number of occasions of minor tensions where not actual explosions occurred but where the threat, more or less implicit or explicit, of impending explosions compelled me or tried to compel me and others to give in against our convictions. You must understand that, although the outside peace is preserved, such things are resented by those involved.

Let us not go any longer into details of such things. I am quite sure that my way of behavior likewise often annoys or hurts people. (You are quite right in saying this. But in another point you are not right; I assure you that I do not take these things lightly. Whenever I notice something of this kind -- however, you may again be right in saying that often I fail to notice it -- I try hard to examine myself conscientiously. And if I find that I was wrong I do my best to right the wrong.) Let me only make this general remark about how I see your role in our movement. Your temper and way of acting is different from most of us; it is more energetic, active, driving, aggressive. Consequently, it has fallen to you to be the driving force in our movement and all its various activities. We all are grateful and appreciative for this;

[352] There is an entry in Carnap's diaries (March 24, 1932) claiming that he met Otto and Olga Neurath-Hahn at the Auge Gottes in the evening. Olga helped to clarify their debate and some misunderstandings: eventually she was successful by the end of the day (RC 025-75-10). "Auge Gottes" may refer to the building called Zum Auge Gottes, which included apartments, a sanatorium and a student dormitory in which Carnap may have been staying while visiting Vienna from Prague, where he was living and working at the time.

we all realize where our train would still be stuck if we hadn't had the big locomotive. But then there come some occasions, mostly of a minor nature, where we suffer somewhat from what seems to us an overexertion of emotion, temper, energy, willpower. We milder and more peaceful creatures complain in those moments about that over-exuberance and aggressiveness. We are apt to forget in those moments that whoever is aggressive as a lion in tackling difficult tasks for our cause, cannot be mild as a lamb in personal relations. (Here however, I am happy to testify in your favor that you are mostly a well-tamed lion; that you often check your stubbornness and yield to suggestions of others, or discipline your aggressiveness into politeness and friendliness; in short, that the lion often whispers when everybody present feels with trembling how we would like to roar). But even if we may forget it on those occasions, the next time we see the achievements of your activity, we all are glad and grateful for having the lion in our midst.

In a few weeks I hope to get, after long delay, the reprints of my two papers on probability. I wonder how all my friends, especially you, will react. I am aware that here I deviate from what has been so far the general view on probability in our movement. But I already notice some changes in the attitude of Hempel, Feigl, Nagel, and others, who begin to think that the frequency interpretation is not the only one. I am now, during the vacations, working again all the time on the probability book. It will become a large book, and it will take my time still for several years. But in writing it, I become more and more convinced by developing the consequences, that my general conception of probability is fundamentally sound and fruitful for scientific thinking, although the technical details of the system are of course tentative and in need of further improvement.

We are happy about the change of the regime in England. We try not to be overoptimistic, since the new government has to solve so enormously difficult tasks. But still there is good reason for a cautious hope for the future. I have read some things by Laski and find myself in agreement with most of his basic ideas.[353] I should like very much to get your opinion about his views and also about his personality if you know him. I am disappointed that he does not seem to get an influential position in the new government; or do you think he has much influence even without an official post?

I just had the first letters from Switzerland in a long time. Walter (Dr. Emil J. Walter, Frohburgstr.95, Zürich 6) is asking after you – he does not even know that you have safely escaped. Dürr has written, likewise Tschichold.[354] Tarski just wrote,

[353] Harold Laski (1893–1950) was a British political theorist and served as the chairman of the British Labour Party between 1945 and 1946.

[354] Emil J. Walter (1897–1984) was a Swiss sociologist, interested mainly in the intersection questions of politics, science, and sociology. He also published some logical works that were reviewed in *The Journal of Symbolic Logic* during the 1930s. Karl Dürr (1888–1979) was a Swiss philosopher, working mainly on the history of logic. Both Walter and Dürr participated at the *International Congresses for the Unity of Science*. Jan Tschichold (1902–1974) was teaching typography in Munich; he was also a close friend of Carnap and Neurath. On Tschichold, see de Jong (2008).

saying that he knows that his wife and children are still alive. Have you heard that Popper has been appointed to a Readership in Logic and Scientific Method at the London School of Economics?[355] I suppose, you have heard that Philipp Frank had an accident – this time, I believe, he was hit by a bus, and had to lie in bed for quite a while.[356] But he wrote several months ago that he felt quite well again. Ina just had a letter from an American soldier in Vienna saying that her older brother and his Jewish wife are O.K. – first word from Vienna, we or anyone else of our acquaintances has had. Hempel wrote that he has had news from the mathematician Freudenthal[357] a former assistant of Brouwer's who had been in prison and in a camp. The Nagels are expecting a baby very shortly. --- That's all the news which I can think of that may interest you.

With best greetings and wishes to you and Mary,
Yours,
[Carnap]

33. Ina Carnap to Neurath, Aug. 24, 1945.

(ASP RC 102-55-10)
August 24, 1945.

Dear Neurath,

I have been following the correspondence between Carnap and you with ever increasing head-shaking. Now it has reached the point of sore feelings on all sides when I – the "of course impartial" third – feel obliged to toss in a few remarks of mine. It seems to me sort of tragic-comic that you two good old friends fight with each other rather than against the hostile world. The strange part is that each of you

[355] Karl Popper moved to Britain from New Zealand in 1946 to teach at the London School of Economics, and became professor at the University of London in 1949; he retired from his London position in 1969.

[356] Frank had other accidents earlier as well. Nina Holton (2015, 71) tells the followings in her memoirs on Frank's wife, Hania: "[Frank] also had a slight limp from an injury from a brief encounter with an autobus. For some reason, Professor Frank rarely bothered to clean his eyeglasses. Maybe he felt that he saw enough as is. But on the other hand, every now and then, one does run the risk of walking in the way of a bus." Besides his accident in Vienna before 1930, Frank had another accident, as it was described by Carnap in his letter to Charles Ogden, July 17, 1934: "Prof. Frank had the misfortunate in April in Paris to be injured by a motor-car and to break thereby his leg. He was compelled to remain in Paris in a hospital. Therefore he could not come to England. His recovery has taken very much time. But he is hoping to return to Prague in the next time." (ASP RC 102-58-02)

[357] Hans Freudenthal (1905–1990) was a German-born Dutch mathematician, who worked on algebraic topology. In 1943 Freudenthal was sent to a labor camp in the village of Havelte in the Netherlands, but in 1944 he escaped and went into hiding. After the war he taught at Utrecht University.

feels hurt and that I am sure that none of you means to hurt the other one. Well, such is life! If I had you both together with me and you were to ask for my confirmation about your main cause of justified grievance: "Is Carnap ponderous and pedantic and a rubber-in-of-salt-into-sores"? I would absolutely agree with you because that's what I have been telling him for 15 years. But if you were to ask me whether he is purposely unkind and insensitive to other peoples hurt feelings -- I would not agree or only to that extent that I would admit that he is not considerate when he thinks there is no good cause for the other's feelings to be hurt. I can assure you that in a marriage of 15 years the matter has come up between us more than once and that's the conclusion to which I have come: he does not respect sore spots in the other fellow; if he thinks what he does or says is right he will do or say it and it's the other fellow's job to overcome his soreness. And furthermore, the nearer a person is to him, the less consideration he will take for his weaknesses and he is absolutely uncomfortable in talks with people with whom he has to take special precautions as to his [their] sincerity. I assure you, it does not always make for an easy life, and I have been trying for all these years to get him to treat my weaknesses more carefully but with not much success. On the other hand, I do realize that most of the time when I feel hurt by him it is because of sore spots of mine, inferiority feelings or what have you. And thus, if in our imagined conversation he should appeal to me for confirmation of his main point – that you are extremely touchy, have rather violent emotional reactions, do bully people into agreeing with you, etc. – I should agree with him. You write that there has been a strain of tension all along through your relations, and that's perfectly true; as far as I have observed his side of the matter I know that it comes from his feeling that he always has to be on the defensive, and that – though you may find it hard to believe – he always has to restrain himself in order not to provoke you to highly emotional reactions. I suppose you don't know exactly how much aggressiveness and coercion there is in your way of doing things, but it's a fact. You already physically overpower Carnap by your amazing ability to discuss for unlimited hours and by the sheer volume of your voice; time and again I have observed Carnap trying to outshout you and ending up with a sore throat. Now you may rightly ask how come that you are having trouble with Carnap and e.g. not with Morris or Frank. Well, for that too I have my explanation: Carnap is really emotionally more involved with you than the two others (he has much more gratitude towards you for certain things than you would give him credit for) and he is very sensitive to reproaches (justified and unjustified) for not being a good friend. His feelings for you personally are full of friendliness and kindness, but he feels (and I agree) that these feelings for a friend should not influence professional decisions (of reviewers, editors, etc.), and still more he feels that nobody should have to be afraid that he risks personal friendship by such decisions. Morris is quite different chap (and what I say is of course influenced by the fact that I do not like him): he is extremely polite, in my opinion bordering on insincerity; he has the famous American talent for backslapping easy relations, staying on the surface of matters, and being a good fellow. Philipp Frank again is none of all that: he does not have Carnap's earnest righteousness and also he has a certain mellowness, half wisdom, half "Wurstigkeit"; it would never have occurred to him to drop his name even if he had felt like Carnap did, for the simple reason that he would not have thought it important enough and that he would have been too lazy to write the attending letters. Ah, but's [sic] the difference between the Viennese temperament and the zealous Lutheran from Prussia! I do not try to excuse Carnap, I am just trying to point out the way he is made. Yes, he could be induced to treat you with greater circumspection but that would not improve the friendship. He does not have the saving grace of a light touch and of a felicitous formulation which might soften the blows which he is striking in the name of science, impartiality, and other suchlike gods. But then again, perhaps you cannot have the charm and the reliability in one person! – You are quite mistaken in thinking that he grieves you lightly, without troubling what your feelings will be. I saw him stewing for days, deeply unhappy, when suddenly the page proofs of your book came and Morris gave him to understand that major changes were out of the question (not only because of the problems of your willingness to change and expense involved, but also because of the somewhat tenuous relations between the Press and the encyclopedia project which might be strained still further by such changes -- they were very short on paper, etc.). And you should see how he sits and broods whenever one of your accusing letters arrives -- for days he is unable to settle down to his work! If you keep in mind his good personal feelings for you and the fact that no unkindness is meant, perhaps it will ease the tension; and if ever again you should be grieved by a professional act of his, you could take it – at worst – as a misjudgement and not as a personal offence.

Our life has not changed very much during the past years, particularly now that Carnap is out of bed again and able to do many things which make up normal life. The 2 years in bed, combined with the uncertainty about an improvement were rather trying. But since he managed last year's teaching without a breakdown – though not without considerable strain and effort – we feel confident about the future even if he should not recuperate completely. At present we are in our little house in the New Mexico Mountains. (I have the car fixed so that he can make the trip lying). For the past two years we have a big German Shepherd dog who has become an important member of the family – last winter we had her in Chicago, it's an ordeal for her and for us but there is no other choice. Carnap is spending all his time on his big manuscript on probability – which he thinks is going to be his most important book.[358] A few papers of his have appeared in the last weeks, and you will get reprints as soon as they are available. But the probability book will take several more years.

News begin to come slowly from Europe via American soldiers. Carnap had a letter from his oldest daughter whose husband has been killed in the war, his other daughter's husband is a prisoner, they have no news about his son.[359] No word as yet about Grelling. The sad news that Hosiasson and her husband Lindenbaum have

[358] Presumably Carnap's *Logical Foundations of Probability* (1950).

[359] On Carnap's children, see letter 22, note 225.

been killed by the Nazis,[360] but Kotarbinski (of whom it is said the he has behaved exceedingly well with regard to Jewish colleagues and the he had refused an offer to come to America because he felt that he had to stay with them -- again typically American to offer refuge to the non-Jew rather than the Jews) is well and is active again. Odd bits about surviving Jewish friends: Flitner[361] in Hamburg (whose wife is Jewish) wrote recently that they are allright, a Jewish Frenchman whose wife is a friend of mine has safely come back from a German prison camp though without teeth and emaciated -- just astray bits of news. Let us know when you hear about common acquaintances!

The last weeks have brought so many historic news that personal ones appear less significant; the atomic bomb, the election in England, the likely peace with Japan. Grateful though we are for the peace, we wish it had not come about through the atomic bomb. The existence of this terrible power in the hands of a few nations appears a great danger. Much of the research has been going on practically under our very noses: the super-hush research lab is only 18 miles away from Santa Fe and we have seen some of the explosions and clouds for years.

Exactly a month from today we shall be leaving again for Chicago, where classes begin on the first of October. We have a rather dreary apartment there (the only one we could find last year) and we envy you your hill. But on the other hand we have been unusually lucky with our long stay here in Santa Fe during the Rockefeller-grant years.

Give our love to Mary -- and my admiration for her doing much of the Isotype work in addition to being a housewife, since I am only a housewife!

With best regards and wishes,

Yours,

ina

---

360 Adolf Lindenbaum (1904–1941) was a Polish logician and philosopher, mainly known for the "Lindenbaum lemma" and for his collaboration with Alfred Tarski (Lindenbaum-Tarski lemma).

361 Wilhelm Flitner (1889–1990) was a German pedagogy professor and a close friend of Carnap from their student's years in Jena before World War One. On their relation see, for example, Damböck (2017) and Dahms (2004).

34. Neurath to Carnap, September 22, 1945. Unsent.

(ASP RC 115-07-66)
22nd Sept. 45

My dear friend Carnap,

May be you felt discouraged when reading my Jun[e] letter,[362] because I wrote it in a somewhat hopeless mood. And I confess that this hopeless mood is continuing. I really do not know what to do. I am sure, that both of us would help one another in days of danger, that we are pleased in some way when being together, that we, what is called: like one another, but that very often, very, very often some minor items disturb a soft and comfortable being together and that from time to time, perhaps by accumulation the items become more important and sometimes lead to a kind of tension. In your cases characterized, please let me say so, by some grieving action (I do not say, that you are trying to grieve me) in my cases, characterized by some noisy explosion. That is that.

Since I regard friendship, love, harmonious contacts the most important joys of my life, I am not soon tired when making efforts to support a situation, which gives any chance of possible kind atmosphere within the framework of human relations. Perhaps I do not know much about the right way, how to make twisted things again straight .... I do not know, whether it is better to go on with you and just to try to suppress, what I have to say and to assume it is hopeless to alter a rigid person like you or to speak with you in a more outspoken way. It is a pity, that the situation reached this point .... In principle I do not believe in the German habit "Aussprachen haben", but when with Germans I often see myself pressed into "Aussprache", because the more emotional little contacts, delicate spider webs and other means of human intercourse do not work.

I feel really helpless and somewhat hopeless. I assume, that perhaps similar feelings are working in you and I interpret so that fact, that you are writing so long letters and that even Ina comes down the silent goddess from Olympus to bring the thundering heroes together before they start sitting sulkingly in their respective tents. I do not think, that by partial silence one reaches much and I do not guess that speaking is good .... but "tossing the coin" I reach the decision that we perhaps come near to one another, when we tell one another a little more about us. You started with that and I appreciate very much your various informations [sic] about you and your reactions.

You see, my dear boy, it is not the question of "reconciliation". If somebody feels himself "humiliated" by somebody else, "reconciliation" is not the right word for making the situation better. On the other hand I know from experience – I have seen many people and nations and their respective difficulties in coming to terms with one another – that it is extremely difficult to an atmosphere in which the irritation created by such feelings is prevalent.

---

[362] See Neurath to Carnap, letter 31.

I am assuming, that I do and did something, which irritated you considerably, without knowing it myself and as far as I am knowing it [marginal addition: *I will*] redress my attitude .... Your pedantry and rigidity is to me very often something strange and even hostile. I asked me sometimes, how that is, because in itself I cannot say that it looks so bad. I think that is a very complicated thing and any hypothesis is rather vague and sophisticated. But let me tell you, how I want to think about me, why I have the attitude of sheltering me [sic] against something I do not regard as directly ugly. It is not nice, but not ugly.

Since my youth I appreciated friendship, love, smooth atmospheres, where one never comes to clashes, etc. that is one of the reasons, why I try to live "procul negotiis" wherever possible[363] – but well adapted to go on with masses and groups, when needed. I dislike the tensions in such situations, I dislike particularly any kind of string pulling – in spite of the fact, that I should know how to do it. I dislike any kind of "officialdom", because here is a minimum of friendship and love. I prefer collaborators who are my friends, too – only a hard education through life, thought me that a certain reserved attitude is needed in a society which through and through is full of competition and difficulties. I finally reached a stage, which enables me to create organizations with a sufficiently kind atmosphere and to avoid the spheres of tensions. Mary and I enjoy ourselves in Oxford, because we have the nicest team of collaborators, we ever had before. We have almost only nice contacts and are far away from the centres of string pulling in London. I try to make comprehensive agreements for many years, which are reducing the official contacts to a minimum. I have mainly work-contacts (preparing something in common) and meeting contacts (meetings for two days, a week, etc.) which here are of a wonderful peaceful character and lead, as experience teaches us, to remarkable friendly contacts, even with people who disagree with our opinions .... In addition we are surrounded by friendship and have now more of that than before. We now see, what we got in Holland, we got heaps of long letters, full of intimacy, friendship, and the wish to continue personal contact by letter. And again and again they tell us, what Mary and I helped them by our opinions, attitudes, etc. I tell you that, because you as many other people look at me too much as a kind of organizer and promoter of things, whereas I think in terms of possible friendships, etc. any organization I and Mary build up tries to enlarge the sphere of friendships and friendly contacts. We are hesitating to invite collaborators, of which we think they could disturb the happiness of our team .... sometimes I prepare an outside post just for creating the human distance which helps to avoid the destruction of our harmonious social life ....

I am not very happy, when people, who should know better, look at all that as a mere tendency towards activity etc. It is something of that in, NEVER RULING THE SHOW. I try to get publishers which whom I can speak in a kind way etc. and I try even to educate them, if they are behaving differently. I am explaining all that somewhat in detail, hoping, that perhaps a slight feeling for my attitude enters your judgement. You see I never felt that you appreciate this side of my existence.

363 From the first verse in Latin in Epode n. 2, by the Roman poet Horace, "Beatus ille qui procul negotii" ("Blessed he who away from business").

I always think that friends try to make life one another as pleasant as possible, not because one has an advantage from that, but because I as a friend enjoy to see my friend happy. That is the reason, why I try to help them, not

(1) I want tomorrow their help
(2) it is a "duty" to help friends.

Both attitudes are not "bad" from my viewpoint, but they are not, what I am longing for.

Further I like any kind of DIRECTNESS, from human being to human being, as far as kindness, friendliness is concerned. Unsophisticated spontaneity, that is wonderful .... I know how are [sic] this bird is. I know that I myself very often disturb kind atmospheres by being, as I am. On the other hand I do not think that my destructive habit is prevalent. Too often people tell me, without being asked for, that they expect me to be successful, when in difficulties, because I am not only able to do many things, but also "charming". It is difficult to know how one behaves. But now I am beginning to think, that the traditional story of me as a wild man with some friendly grunt is very incomplete and rather a kind of "rationalization", which enables people to humiliate me, instead of taking me into their circle of friendship and contacts. And here is the point: apparently I am far away from "Bonzentum".[364] And I should guess, that people, who have something of that, of "snobishness" etc. have some aversion, when seeing how I cannot go on with that. And this difference – I think so – is very, very decisive in human relations. I think Schlick's extraordinary uncorrect behaviour again and again is only understandable in this way. He felt himself endangered – AND CORRECTLY – by the very existence of a person who does not acknowledge this way of behaviour.

That seemed to me a rather "personal" problem. But gradually I evolved a kind of hypothesis, which runs as follows. I should not publish that before I did not check up it better:

There are two way of behaviour particularly important to me, and as I think also in mankind. Of course they are not exclusive, they are not a dichotomy, etc. there are many others etc. I do not give this description a higher dignity, just "as I see it". My studies on BROTHERHOOD AND PERSECUTION lead me to that.[365]

One attitude tries to find HIGHEST ideals, as it were, in justice, duties, etc. everything regarded as something within a systematical structure, which we know already or at least should try to know. These people try to create a kind of scheme of "correctness". They are judging themselves and other people, whether they are correct in this sense or not, whether they have the right duties or not, the right racial background or not, the right morale or not etc. .... One has to have certain highest principles, and within this realm strong "convictions" and something of that type. The human relations have to subordinated to such HIGHEST PEAKS of

[364] On "Bonzentum" see letter 31, note 330.

[365] See Neurath, "Tolerance and Persecution" in the Neurath Nachlass, 207/K.88. Wiener Kreis Archiv (Rijksarchief in Noord- Holland, Haarlem, The Netherlands).

something – deity, duty, nation, leader, ascetic attitude, religion, enthusiasm, etc. Within this realm grows up a certain tendency towards rigid adherence to something, which often leads hardship, mercilessness etc. Everything becomes relative weak and unimportant, when the own "conviction", "ideals", etc. are on the stage – and they are always on the stage, as long as these people take themselves seriously. And they do it, by Jove. Even if such people start with "kind" ideals, this attitude as such is dangerous and leads to anything, as history – that is my impression – shows us. Just this way of behaviour enables scoundrels to use "idealists", who are prepared to be hard and merciless, as soon as their "ideals", "convictions" are in question – and they are always in question. Sometimes mercy appears as a duty, sometimes not, it is "occasional" as it were. There are ENDS and MEANS and such terrible items in the picture, which rule the show. That is an attitude, which I personally dislike since my youth, but now regard socially as very dangerous, too.

The other attitude is more like the following: people living together are able to create a friendly and kind atmosphere, to think, how they may make one another as happy as possible. They often are not successful and very often unhappiness appears, but this result is NEVER ACCEPTED AS UNAVOIDABLE WHEN CERTAIN "ENDS" should be reached .... there are not such "ends" within this group. Friendship and brotherhood are the basic attitude and NOTHING ELSE COUNTS, no conviction, no faith, no enthusiasm. On the contrary, after some experience collected, such people of the brotherhood kind become suspicious, when people instead of liking hobbies and activities supporting welfare and happiness, start with promoting ENTHUSIASM as such; in the realm of kindness [there] is not much space for "ecsta[s]y" and "enthusiasm" as such, it is a possibility to enjoy anything either alone or in good company, wonderful music, plays, movies, festivals, fairs, landscapes, etc. but without stressing the point how wonderful "enthusiasm" in itself is, "aufgehen in irgend etwas", "volle Hingabe" etc. These people of the human brotherhood do not care much for snob[b]ism of any kind. Kind people are more wanted than clever people, ordinary folk who have many interests are more wanted than sophisticated adventurers, the big and small events count more or less on the same scale, a friendly afternoon with a friend may be of the same "order" as a symphony of highest musical impressiveness .... Not comparable, of course, but the preparedness to design [assign] time for these things, energy, may be the same .... A kind member of a meeting, which asks for the size of my shoes for getting slippers for me, after hearing that I had difficulty to get them, whereas she has a friend who makes slippers etc. pleases me at least as much as a speaker who tells of the most impressive new invention, arguments etc.

The first group of the serious people with convictions, rules, duties, justice, high ideals, enthusiasm for something high and lofty, with consistent habits, sure of predictability at least with "probability" seems to me represented by many famous men and by many people in the streets, but also the other group of the people, who are rather gay and sociable, and not too much interested in principles, convictions, etc. but in a kind atmosphere. Who would avoid to do something "fine", when grieving other people. There are no HIGHER ideals, for which human pain could be taken as not too important. Often may happiness fight with happiness, and grieving one may be connected with pleasing another – an often sad situation, but always the discussion starts on happiness and not on anything else. Everything seems uncertain, and therefore the love and friendship of the living time so important.

Look, my dear and take it as attempt to come in closer contact with one another, when I tell you, that the first group of people has certain signs, which we find in Plato's REPUBLIC, or in Rousseau or in some other people, which for me are the real danger, and that your habit is often to a certain degree, as I see it not so far away from the description I have given above. Perhaps I see you not so as I should according to your opinion, or it is not reasonable to think the group dangerous, etc. Please, believe me I say all these things hesitatingly, because should I not succeed in altering your own judgement and likings a little, such an explanation makes things even worse.

You see, I do not think that arguments are estranging people much, but attitudes do it [sic]. You see I have more and more the feeling that the Platonic attitude is more or less connected with Nazidom. And the [sic] German Christianity – as I see it, perhaps I shall alter my opinion some day after studying more in detail the matter – is full of this rigidity and also the philosophy. Even people who dropped religion, who dropped metaphysics, – this cruel type of attitude represented by Kant, who preached many nice things in addition – could not overcome their traditional rigidity and puritan attitude. One of my very good friends – dead already – said that one day to me. All attempts to be gay, to have a well arranged home with comfort etc., reading, etc. did not help, it remained that "duty" played the central role. Helping other people not based on the intention to make other people happy, but because it is our duty to help, etc. If some attitude or action irritated other people, the answer was: I did what I could to be mild, etc. but since my conviction… sorry, that they others were hurt etc. I must say, that this type of behaviour is very often to be found in Germany, much rarer in Austria, where terrible brutality is a traditional thing, but not consistent rigidity, neither in kindness nor in brutality …. Easy going in both ways, as I explained to an English acquaintance, who thought the Austrians are kind and charming, I told her they are more easy going and often cruel etc. …. think of Franz Joseph, who ordered the hanging of the Hungarian Generals deliberately, after protest made by many people from outside …. But not because he was a Platonist …. Whereas the killing of Jews in Germany how, was much more based on "Platonism", as it were. You have to sacrifice your own mildness ("den verdammten Schweinehung niederringen"). In this way I try to make an hypothesis how kind people could bear that and even help in doing that …… And now I think, that in spite of all your personal charm and kindness you have many serious signs of a Platonic attitude. The second way of life is rather strange to you.

I cannot exclude, that you perhaps will be able to SEE that, in a similar way, and PERHAPS you will try to support the brotherhood elements in your attitude and to suppress the Platonic elements. I cannot say, that I succeeded in my own field, and I cannot deny, that friends are not so bad off when saying me that some of [marginal addition: *my*] attitudes are somewhat "Boche-like".[366] I confess that they are and I try to alter that. I NEVER STAND TO ANYTHING LIKE THAT. I do not say, that I am kinder than you, perhaps I grieved more persons than you did, but I think I should hardly fight for a conviction or something like that knowing that it is grieving somebody .... Mary tells me often that I grieve people in discussion about "Platonism", e.g. when trying to show a Jewish refugee that his or her approach to arts and life is in principle a kind of Platonism and therefore connected indirectly with Nazidom .... The only answer I can make is: here I am not grieving people for a principle's sake, BUT BECAUSE I THINK THAT NAZIDOM, PLATONISM, PURITANISM, ETC. as principles make masses of people UNHAPPY. It is a defence of happiness and here are standing happiness against happiness. In former years I fought for my own happiness, but now I think of my son and friends, who suffered from concentration camps etc. PERHAPS I AM OVERSTATING THE CONNECTIONS – please tell me so, I am prepared to revise my views. I should prefer to be "tolerant" in these cases, but against people who spread unhappiness, how should one be "tolerant", without being prepared to suffer as a martyr and to tell others to do it. I heard a philosopher explain that --- that is a third way of life, not even for human happiness to fight, but looking at brotherhood as the central point of human life, nevertheless to BEAR PERSECUTION. As Christ said: I said you so, I did not promise happy society, but persecution and your pain .... But that is not my attitude. I should feel myself as a hero, when suffering pain for making other people happy, but I should not feel myself a hero if I could secure other people's happiness (and my own happiness – which I think I should at least regard as so important as my neighbour's happiness [marginal addition: *but would prefer suffering*] by making the man unhappy, who tries to create unhappiness .... that is happiness against happiness. A HARD DECISION, but NEVER CONVICTION AGAINST HAPPINESS, that is the point, why I think Calvin so terrible a person, or Knox.[367] [A] [s]hort time ago I met a British socialist, who was unusual rigid and "absolute" in his political aspects. I wanted to ask him, whether he is connected with Puritanism, etc., but you know it is against custom to ask a person something intimate in this country. Therefore I was silent, but put forward my point, explaining something about unpredictability, kindness of direct contact, danger of terror as such, etc. and stressing how the other attitude looks like, finally he said with some emphasis: my own Calvinist tradition ... And I answered: in Scotland ...and he said I am Scotch. And then I said, what I

[366] The expression "Boche" was introduced by French soldiers in World War One to refer to Germans, especially German soldiers; it derives from the expression "tête de boche," used after the mid 19th century to designate Germans and Belgians – also "alboche" combining "allemande" and "boche" –, and, in turn, from the expression "tête de bois" – "wooden head" – for thick or heard-headed persons.

[367] John Knox (1513–1572) was a Scottish minister, theologian, and is considered to be the founder of the Presbyterian Church of Scotland.

thought of Cromwell, whom he regards as a fine, great man "we need something like that in modern socialism here". "Ah," I said "why not Knox" – "that was a fine boy, too" he answered. There you have the things .... That has nothing to do with Nazidom as a social structure, as killing Jews etc., but as hard and merciless attitude which thinks in terms of ENDS, CONVICTIONS, EFFICIENCY etc. and not in terms of brotherhood, happiness, friendship etc. which is a more Christian-Epicurean attitude, if one need a name for that. It is a misleading name.

If you were able to tell me, that you a little a very little only, are prepared to look at your attitude from this point of view, then I think perhaps we shall find a way to become friends who have not to [sic] fear the TENSIONS, of which you speak. If not – nothing will happen – we shall try to be good friends, but I should say, I should not be astonished to see a tension reappear, because this ATTITUDE is connected with tensions. I told you the tensions appear so rarely, that I should not fear them too much. Let us hope that the strain would not too much for you. But I should prefer, believe me, to have with you a serious and kind correspondence about the Platonism in you, as should call it, the Puritanism in you, the Prussianism in you etc. I know this kind of description is almost a kind of "giving [calling] names", a very risk, like an operation, but I think you are to[o] grand a personality to be afraid of such a discussion. It only shows you that the tension is of a type, which has NOTHING TO DO WITH RECONCILIATION. I am not in a mood which needs reconciliation .... I am only in a mood which is longing for your full friendship, for your preparedness to readjust your attitude, be sure, that I myself am fully prepared to listen to your explanations and to readjust myself. Since I do not look at life as something rigid I can anticipate changes in my outlook .... I do not think that my desire for friendship and brotherhood will be dropped, I think I am too much a spoiled child and the kindness is too important for me. Humiliation is irritating, not because I cannot bear it – I have strong bones – but because it is something destroying kindness and I have a sensitive skin .... I shall try to harden it, when you think it is necessary for our friendship, but better would be to be together soft and mild .... relativist, and not with any strong and rigid rules, applied at life without thinking of other people's unhappiness.

You are again repeating that your remarks to Morris about me are "obvious to everybody" --- I speak not of the volcano, but about what you say, that this volcano becomes active, when criticised. I tried to analyse my behaviour in detail, I asked other people, and JUST [marginal addition: *THAT*] DOES NOT COME OUT. Would I see that in me, I should think a terrible quality and extirpate it with all my power. Astonishingly you are not GIVING ME ONE INSTANCE IN YOUR LONG LETTERS IN WHICH I BECAME A VOLCANO AFTER FACTUAL CRITICISM.... I ask you now, NOT FOR DISCUSSION'S SAKE, but for my own "salvation", please, tell me of cases in which I reacted volcano-like after criticism of my arguments, my style, or something of this kind. Seriously I could not remember one case. I try to recollect outbursts and I usually find some ATTITUDE OF OTHER PEOPLE WHO CREATED MY PROTEST .... not because they criticised me. I dislike to be humiliated and treated with disrespect of the normal human relations ....

What you say, that you made th[i]s remark about me to find an excuse for MORRIS, what is that? My dear, dear friend, think, what you say. You the man who wants to be so very correct, is here saying something about another person, very like a denounciation [sic], for "excuses sake" .... It is so, as you say, it was also a criticism of Morris, what you did, BUT YOU WERE MORE INTERESTED IN NOT GRIEVING HIM THAN IN NOT GRIEVING ME. If you would tell me of my behaviour being so --- I should be really thankful, as I said more than once. I can hardly imagine, that any of my outburst were connected with criticism against my arguments or opinions ... perhaps as an exception, but I do not know at the moment such an exception.

If you want to drop this talk, please drop it, but perhaps here we may find the way of understanding one another and perhaps adapting us to one another. Perhaps I do not realise, when I am volcanic, or you think the "reasons" for my being volcanic are others than I think they are etc. ....

I DID NOT PROTEST AGAINST YOUR SAYING THAT I HAVE VIOLENT REACTION – that I never denied, and I am sorry, that I have this temper, what I tried to bring near to you, is that the assumption I am volcanic WHEN MY ARGUMENTS, MY STYLE etc. are criticized .... AND YOU CAREFULLY – I DO NOT SAY INTENTIONALLY – ARE AVOIDING IN YOUR LETTER JUST THIS POINT, only speaking of the violent emotional reactions, never denied by me.

Let us speak of SCHLICK. I reacted in my way, because SCHLICK, as I saw it treated me as a schoolmaster and higher BONZE[368] treats a schoolboy and a person not equal of standard. And THAT I NEVER BEAR .... and I do not think I shall alter this attitude. Aggressive impudence I dislike outmost. I confess that without remorse, I am not one of the soft people who suffer patiently, BUT I DENY THAT I REACTED EMOTIONALLY BECAUSE SCHLICK WANTED SOMETHING DIFFERENT .... I have hundreds of example, that I am altering manuscripts without any complaint, when people show me mistakes, arguments, style etc. which are defect[ive] .... I have too many witnesses for that, too many even in my files. But the case Schlick is different. It is very important to analyse such a case not to speak like you: "you had a quarrel", and "The question now is not who was right or who wrong; probably nobody was entirely right but everybody had some good things to say his side." You are bringing this whole show into the field of "rights" etc. "some good thing to say" --- as if it were a discussion, whereas the PERSON'S HAPPINESS IS IN QUESTION. How can you know, what "good things" Schlick had to say, when humiliating me? Philipp Frank was witness of conversations. As far as I remember, he never tried to conceal his opinion that Schlick treated me PERSONALLY BADLY. And I reacted against that, NOT AGAINST ANY CRITICISM...... Please, say something about this point. Otherwise our both statements are not linked up with one another ......

Please, tell me how you see the Schlick-Neurath conflict, which you regard as one of the most depressing experiences of your life. I really have not the slightest

[368] On "Bonze" see letter 31, note 330.

idea, why it did depress you that a BONZE who behaved badly, as he was accustomed to do, got a rebuff – a very seldom case, because usually the victims of such bosses are depending upon him. Therefore they EITHER "kuschen" or "transform the humiliation otherwise hardly to bear into admiration etc." .... Since Schlick regarded you as co-boss you did not suffer much from him, just occasionally.

I have more than one case, in which Schlick behaved without ANY REASON aggressive[ly] and humiliating me. One day Waismann invited me to tell of my ideas in the Schlick circle. I said, why not, I came and Schlick STARTED, introducing me: I do not know, why Dr Neurath wants to talk to us, but he may start .... or something like that. There are sufficient witnesses of this scene, which all people of the circle regarded as at east [sic] VERY UNUSUAL AND STRANGE – NOT THE SLIGHTEST AGGRESSION FROM MY SIDE, also not in the months before, nothing.[369] From the clear sky. And you remember the wholly unnecessary remark on the ugly word "Einheitswissenschaft" in Paris (in the paper read to the congress in French translation, where SCIENCE UNITAIRE is nothing ugly). Etc.[370] Wherever Schlick could he made a suffisante[371] remark. If it were not printed, I could think that I invent such details, as many scientists do, who are inclined to the usual mild form of scientific persecution mania. Since I know, that scientists have this disease I try to control myself and others as far as the symptoms of this disease are concerned. I asked Neider in detail about that.[372] He, who liked Schlick, DID NOT DENY THE DESCRIPTION, but tried to explain the case, that Schlick in his finer style of life, feared always I could destroy the delicacy of a mood, e.g. when he was sentimental at Christmas tree occasions with music etc., whereas, as you know, I myself like a certain sentimentality and never take it ba[dl]y, when other nice chaps are sentimental. I am only critical when bullying bosses, like Schlick behave sometimes overdelicately.... that is rather comic and remembers me of some delicacies of famous persecutors .... I do not want to overstate this subject. But, please, tell me a little more in detail, why the case Schlick did depress you so much, as far as I can see you are pitying Schlick more than me, and you are not speaking of HUMILIATION, but of arguments, points of view and such paraphernalia, which interest me never so much as human relations.

---

[369] Heinrich Neider tells the same story in an interview; see Haller and Rutte (1977).

[370] Neurath later recalls this event, with a few actual quotations. See Neurath to Carnap, letter 36.

[371] Here "suffisante" may correspond – given the context and his intentions – to the German word "süffisant," which may be translated as "smug."

[372] Heinrich Neider (1907–1990) studied philosophy and philology at the University of Vienna between 1926 and 1930. He was a member of the Vienna Circle and wrote his dissertation under Moritz Schlick on the question of understanding (Verstehen) and the human sciences (Geisteswissenschaften).

Poor Schlick, depressed by me …. such an aggressive person, full of unfriendly habits, depressed, when a victim answers …… Perhaps you looked at Schlick differently ….

The second case, again HAS NOTHING TO DO WITH CRITICISM. You wrote to Morris, of course you did not try to influence Neurath's manuscript, because you know like others, who [how] violently [marginal addition: *he*] reacts against criticism. Now I should expect you would tell me, when I behaved so, instead of that you tell the Moscow story.[373] Please, my dear, dear friend, put yourself for one moment into my way of arguing: "I say, Carnap tells people I am furious when criticises, that is too bad, I never did so", what do you tell me? That I became furious, of course I do not deny that.

The other case, you dropped your name. I did not become furious because you criticised me, but because you made, what one calls an affront. You did not drop your name as Bloomfield published his paper,[374] which many people do not think of sufficient value etc. … Do you think my paper is worse? Dropping a name is a very serious action …. But let us analyse the Moscow case. As I said, NOTHING TO DO WITH CRITICISM. I remember your warning very well. I surely did not regard our "reconciliation" as a diplomatic affaire, but I felt, that you did not realize, what humiliation means to people, who want kindness and some acknowledged status within a circle of scientific friends in accordance with the work done and not with the positions one reached in the hierarchy of the world.

Please, tell me, what you even today think an "exaggerated amount of credit" ---- you see I do not think that I asked you for a certain amount of credit (perhaps my memory deceives me and my files are, as you know with Hitler and his gang, i.e. somewhere in the [sic] hell, should they not reappear) I only wanted me properly quoted, otherwise my own papers would appear as a kind of Plagiarism. Please, believe me I am not very vain – I think rather somewhat [u]nder average. In any case: I do not design much energy to presenting myself, but I am much more interested in satisfying myself, discussing questions with others, etc. I dislike only that people are using my arguments without quoting me a little.

But that is not the point of irritation, but the feeling, that people would behave differently, when I were a "Bonze" a "professor" etc. Perhaps I am wrong in that. And that is humiliating. Not because I need being quoted --- I try to assure you that is not the point, but the feeling that I am treated less kindly than people in official positions, it is something [i]nhumane in that. Perhaps you will show me, how wrongly I see that. I dislike very much, when people complain that one does not quote that or that [sic] and therefore usually I do not make even an occasional remark, because it looks as if one interested too much in such little things. But perhaps it is sometimes useful to speak of that. And perhaps I can learn something from you. Let me include this case, then I shall continue the main line of my letter. Some

[373] See Carnap to Neurath, letter 32. note 351.

[374] On Bloomfield see letter 28, note 305.

day I read in Russell how he wants to use the term "Protocol" and "Protocol statement" etc. I said to me [myself]: strange so many people are using now this word, and in so different ways. How is that. One should look, how this word appears in various authors, since many people use it like a Schiboleth for Logical Empiricism. I know hardly a term of our movement so often mentioned with or without irony. Maybe it is only my hobby that I press this idea forward (I think it is a kind of key position – but that is my private matter), but IT IS HISTORICALLY USED BY AUTHORS WHO DISCUSS OUR MOVEMENT, even in such a little tractatus on modern philosophy as Laird's booklet,[375] the Protocol statements appear, of course in B. Russell, L. Russell, Weinberg, Popper, even – in Carnap. OK. I open THE DICTIONARY OF PHILOSOPHY,[376] which I used sometimes, and seek PROTOCOL, PROTOCOL STATEMENT, expecting to find a list of various usages of the word – nothing. I try other articles about Logical Empiricism – nothing about that. Do you think the EXPRESSION would disappear from the stage, if it were used by SCHLICK?

You see, I feel almost ashamed, when mentioning such trifles. And I do not think, that it is the non-quoting which irritates me, but the sign, that people do not respect me as a member of their community. Of course you cannot force people to accept you, but if that is connected with a certain disrespect it is grieving and irritating. You see e.g. I do not think that Feigl is a very important thinker, and that I need to being [sic] quoted by him – really not. But I sometimes thought we could become friends, having scientific intercourse etc. I remember very well a long discussion, I think somewhere, perhaps Geneva, on INDUCTION.[377] I told him about cross-induction, induction supporting one another etc. about logical problems etc. HE EVEN LATER ON, I THINK IN THE USA told me that he remembers our discussion, and how much it supported his thinking, how much he thinks that some remarks by Reichenbach are going into that direction – but Feigl never mentioned in his papers with one word this influence. I do not ask for telling this story – but I published sufficient papers, in which I touch this case, also [i]n our congresses, the mildest way could be, to quote only such a statement saying, that his arguments are somewhat in contact with Neurath's which he cannot accept, because etc. .... but this way of neglecting primitive rules of decent behaviour in the scientific society – I do not

[375] Presumably Laird (1936).

[376] Neurath is referring to Dagobert D. Runes's (1942) *The Dictionary of Philosophy*. Carnap wrote many entries for Runes's volume: Anti-metaphysics; Basic Sentences, Protocol Sentences; Denotation; Formal; Intersubjective; Meaning, Kinds of; Physicalism; Science of Science; Scientific Empiricism, Unity of Science Movement; Semiotic, Theory of Signs; Verification, Confirmation.

[377] Later in a historical paper about the Vienna Circle, Feigl (1969/1981, 65) mentioned that he had "spent a few days with Neurath in Geneva, serving as his French interpreter at a conference" in 1929. Feigl also wrote to Schlick (July 21, 1929, Moritz Schlick Nachlass) that Neurath had suggested to him to attend a pedagogy conference in Geneva, where he could meet important people.

complain too much that, because it is usual – is grieving as long as you think of a person as a possible friend .... Or another point. Feigl published a paper on LOGICAL EMPIRICISM.[378] I think there is nothing important in the fact, that I pressed this expression against Schlick's tendency to support the James term "radical empiricism" etc. and that I did not succeed very much, but I think that I, to a certain extent, am one of the pillars of this movement, not only its "promoter" as people sometimes like to treat me.

If you are reading Feigl's paper, you will find (I discovered that, as I wanted to note down, where he agrees with me, where not) that he almost CAREFULLY avoids to mention any of my many papers on logical empiricism and its foundations (I may be presenting unacceptable theories – that is not the point). He mentions me just as a sociologist, and – strange – did not even mention that I am the editor-in-chief of our encyclopedia. Please, do not tell me, that all that is "just by chance" .... think one moment HOW A KINDLY PERSON MAY FEEL WHEN PEOPLE WHO ARE FRIENDS OR POTENTIAL FRIENDS apparently do not think of me, when writing something, where I play a part. That is grieving, ... Of course there exist other interpretation of this behaviour, which I try not to evolve. Not even for my own purpose. You see kindness and brotherhood are based on faithfulness and we should do, what we can, not to fe[e]d suspicion of any kind, even when based on well acknowledged material. Suspicion destroys brotherhood – you see how the communist party destroys brotherhood by suspicion and by other things of this type. Therefore, please do not think I want to say something against Feigl, and to stir up something. Not at all. I tell you only, how it looks, when people do not like to acknowledge somebody, as they would acknowledge him, if he were a big boss. I THINK IT VERY [IM]PLAUSIBLE THAT FEIGL WOULD TREAT ME IN THE SAME WAY IF I WERE A PROFESSOR OR SCHLICK IN PERSON writing things of similar importance.

Perhaps I am "ein eingebildeter Laffe", but I think that my scientific work, is not less than Schlick's .... But that does not interest me much. I do not speak of the treatment by other people, but ONLY WHAT A PERSON FEELS IF FRIENDS AND POTENTIAL FRIENDS DO NOT ACKNOWLEDGE HIS EXISTENCE AS A MEMBER OF THEIR COMMUNITY ......

You are happy in saying that no further outbursts of that order appeared, I am happy to say that no further neglect of my scientific personality appeared of that order, but of smaller order – normally I should say. That is my fate. Let it go. Only in the next edition of the DICTIONARY, please insert PROTOCOL STATEMENT ...... It is not of importance, just a trifle ....

You see, when some day Feigl will do something, which is an additional disregard of my person and I shall be more violent, than the moment asks for, please remember this my information, which is not a complaint, only a scientific analysis of my behaviour.

---

[378] Neurath presumably refers to Feigl's paper "Logical Empiricism," see Feigl (1943, 406).

I am not saying you are "wrong", but you are not thinking of human contacts as more important than other things. When you think that your conviction presses you to drop your name – I should really like to know, how [sic] such a conviction looks like – did you really think of the pain you inflict, if not, why not? And if, do you mind, when I think it unbrotherly to inflict such a pain not for making other people happy etc. but ONLY FOR SATISFYING YOUR CONVICTION? That is him Hekuba[379] ...... I should like to know, how you see such things, perhaps here is the clue "von das Janze".[380] Sometimes I do not find through [sic] your attitude and your remarks.

I believe you, of course, when you tell me, how all these conflicts depress you, and how you seriously analyse yourself, but I cannot understand, why you are not mentioning the point, I always stress:

Do you think it proper to hurt a friend only for the conviction's sake? What are your convictions, compared with brotherhood, friendship and happiness? Do you think it is strange, what I say? I try to find a way to you and to lead you to me ....

I do not speak of the cases in which you [u]nvoluntarily hurt other people – I do it very often, we are poor people, who do not sufficiently see other people's impatience and pain .... That we both have in common. But I should never say "may other people swallow their pain, if I had to inflict it in accordance with my principles". Never. Perhaps I shall "rationalize" unkind actions, by seeing people's happiness increasing by my action etc., even my own – I do not pretend to be unselfish – but I should hardly sacrifice happiness to principles. What unhappiness could arose from your name remaining on the title page – THAT AND ONLY THAT IS A QUESTION IN HARMONY WITH MY WAY OF LIFE ...... if you think this question is never yours, then we know where we stand. We shall be friends, but this basic danger of tension remains. I think you and I will try to avoid clashes, but it would be nicer if you were prepared to have together with me a common approach to brotherhood and friendship based on happiness.

I DO NOT THINK YOU ARE MILDER THAN I AM .... AND I DO NOT THINK YOU ARE MORE PEACEFUL – as you think. Perhaps I shall agree with you after certain talks, but at the moment I only think, that I make more noise, that I am prepared to present emotions carelessly, etc., but I think that you are sooner prepared to put the pain of a situation on other people's shoulders, IF YOU THINK YOU ARE IN YOUR RIGHT. I perhaps put heavy weights on people's shoulders, but hardly by pretending that "I am in my right". Perhaps occasionally, one never knows how one may behave, but in principle I do not have this way of arguing, but I guess you have it, and that is "unmild" in itself, "unpeaceful" in itself .... I think it very important for my happiness – I do not know, whether it is for you of the same

---

[379] Hecuba was a queen in Greek mythology that appeared in works by Homer, Euripides and Shakespeare as a symbol of vengeance and the moving power of grief.

[380] The expression "von das Janze" means in Berlin dialect "from the whole" (with "Janze" instead of "Ganze").

importance – to have an opportunity to look at human actions as far as kindness I concerned together with you IN A SIMILAR WAY, and therefore also at us; it is not the question that we should agree about our judgement, but that we may find some compromise in applying certain attitudes towards life, particularly towards friendship and community life ....

You are speaking to me in your letter a little to[o] patronizing[ly], if you permit me to make this criticism. I think you take all these emotional things too much as a kind of manner, and not as something connected with a comprehensive attitude towards life. You see most organizers – I have experience in that – try only to put something in motion, whereas I, I think some people call that sentimentality – try, sometimes not in vain, to create a kind atmosphere. Of course, where people react against that, I become unexpectedly for these people emotional. Reichenbach, to give an example, made me emotional, because I thought it not brotherly, how he treated you in the various cases, we had to discuss. And, I cannot deny, that his theoretical remarks looked perhaps more strange to me than they perhaps would do, if he were not so far away from brotherhood in a community, in principle. He is very often charming and I know very well how to go on with him in the nicest way .... but I fear him, not as a scientist or debater, but as a person, who is not interested in creating brotherhood in this world.

I often – perhaps not seeing clearly – think, that theoretical remarks are covering dangerous attitudes. I know that it is a very dangerous field, but on the other hand it is something in it. In our movement I sometimes have the feeling that some members avoid discussing problems of decision, action etc. and are using logical analysis as a kind of escape from life. THAT IS NOT MY APPROACH. I am using Logical Empiricism as something that helps us going on more energetical [sic] than before, more tolerant than before etc. ....

I should like some time to learn from you how you look at my activities and my work, when not looking at me only as an engine or a lion .... sometimes I thought I had some wise arguments and important approaches to serve .... But from your letter to Morris I know how low you regard my Aristotelian paper. Perhaps I shall agree with you when I have learned more about your arguments in detail. Not to go into detail people like me feel also as unkindness, when friends are involved, in other people's behaviour I am not much interested. You see many people see themselves as member of the scientific republic, and in addition they have friends, I look first at friends and people who are potential friends, and then at the scientific republic ......

I shall read, what you have to say about probability, with great interest. I only complains that you and other hardly go into the criticism we have to make .... the same is with induction, semantics etc. Perhaps it is fruitful, to go on and not to discuss the problems on the border in detail .... I should prefer the latter, also from the movement's point of view. Frank's paper is fine, but I dislike how careless[ly] he speaks of "facts" "reality" etc.[381] Of course avoiding bad consequences, but there are weak brothers ....

---

[381] Philipp Frank spoke about "facts" and "reality" often in "The Philosophical Meaning of the Copernican Revolution," see Frank (1944/1949).

I am now waiting, what the various members of our movement will publish on the UNPREDICTABILITY problem. Some people who reacted vehemently against, as they started to show me, where my mistake is, did not succeed – but let us wait and see.

Of course Labour gives much hope and if they are able to have some success – e.g. less unemployment here than in the USA and get the next election again ... then it is something in our future. But the international situation is terrible, extraordinary terrible. We do not know what the policy of the Russians is, but if they REALLY WANT, WHAT THEY DO – it is a sad thing – but if what they do, is not intended, it is also unpleasant, that they do now know better, how to act in accordance with promises, etc. Of course all governments are at the moment somewhat unpleasant as far as foreign policy is concerned – but the links between all the problems are difficult and often very dirty .... such is life.

Laski has many good ideas.[382] I think that he does not always think how people take his remarks. Many difficulties come from that – unnecessary ones, as I think, and others. I heard him lecturing. Clever, at intense [sic], but I am not so sure that he always "means business". He is too much connected with policy that he can say, he wants to say something only theoretically. Of course I could go into details, where I agree or disagree. But I do not know, what questions you have particularly in mind.

Glad that Tarski's family is safe. I am very depressed by Hosiassons and Lindenbaum's death.[383] From Tschicholds we had some lines, Fraenzchen[384] and wife are safe. That is all, we know. I did not hear of Popper's Readership. That would be Hayek's work, who is connected with him. He hopes apparently to get from him the supports he needs for his philosophical position. He is not satisfied with his antisocialism he wants to explain also his antilogicalempiricism [sic]. With Popper he agrees, as he tells me whenever he has an opportunity to do so. In my mild way I try to have an exchange of articles with him. I treated his book as mildly as possible – it is rather a scandal, that a scholar publishes such a biased paper full of hardly provable statements.[385] But like his colleague Mises[386] – the economist – he is against socialist planning in any case ...... a sad figure on the social firma-

[382] See Carnap to Neurath, letter 32, note 353.

[383] See Ina Carnap to Neurath, letter 33, note 360.

[384] Neurath refers perhaps to Franz Roh (1890–1965), the German art historian and photographer. Roh was one of the founders of the so-called *Neue Sachlichkeit* movement, and he introduced Carnap to Neurath. On the relation of Roh and Carnap, see Damböck (2017) and Dahms (2004).

[385] See Neurath (1945/2004).

[386] Ludwig von Mises (1881–1973) was an Austrian economist and a member of the so-called Austrian School. On von Mises and Neurath, see for example Uebel (2007). Ludwig was the brother of Richard von Mises, who participated at the meetings of the First Vienna Circle, and was a close associate of the Berlin Group of Reichenbach.

ment; Popper, I guess, will support him in some way or another, or at least I shall not astonished to see him acting in this way ....

What about Grelling's family? It is terrible how one is hampered in all attempts to come in contact with people in Europe. France, Switzerland, Holland is [sic] OK, the British-American zones permit some information, the Russian zone is sealed off completely.

We are going well, as usual, enjoying life. We have to enlarge our institute – too much work to do. And now I am thinking of our congresses again. Ness wrote me a nice letter. He escaped the Gestapo, a friend died, tortured but [marginal addition: *not*] telling his hiding place. What a world.

I hope you will write me a kind and brotherly letter,

Yours ever

[Otto Neurath]

PS

Always fearing that my DEFENSIVE ACTION (as far as I can see I am in the defensive, first arguing SCHLICK, when ill[-]treating me, than again you [sic] – see Moscow case, you only are pleading, that I am overstepping self-defence) may again overstep according to your opinion self-defence I looked through the RUNES DICTIONARY. Under BASIC SENTENCES I discovered some remarks on various formulations, only one lacked – the NEURATH formulation, because in all the formulations, you mention the verba prohibita of my language appear, I do not speak of observation etc. as you know. MY CHARACTERISTIC NOTE IS NOT EVEN MENTIONED. If you were in a kindly mood and interested in my arguing, you would mention my version. And more, you would enable the reader to look into the now classic article on protocol statements,[387] which is quoted so often in the literature. IT IS [marginal addition: *a separate*] QUESTION WHETHER YOU AGREE OR NOT. Further, how do you think a reader of the dictionary will find anything of my publications by means of the articles you are providing for the public?

All this trifle is not worthwhile to mention. I only want to show you what I think I at least not a friendly treatment. I am not even sure whether it is professionally OK, perhaps not even that. Since I do not assume a primarily unkind attitude, I rather guess that you are not much interested in me and my work ....

What a good luck, that I am not depending upon my real and my potential friends in our movement .... but walk on as a free and independent citizen. But, I tell you, it is not nice, to feel so unkindly treated just within our movement, more unkindly than outside in the literature. Is it not rather strange? But, let us not talk over those things too much. I add such things because they are illuminating.

By the way, do you think a reader, who wants to find, as I wanted to do, PROTOCOL, he will look under BASIC? But all that were not worthwhile of mentioning if it did not fit into the whole pattern of treatment ......

---

[387] Neurath presumably refers to his famous "Protokollsätze" article; see Neurath (1932a/1983).

I only think that life is short and that therefore one should prepare as much happiness one another, I hoped that the tolerant pluralism of our movement would create more friendship than other movements, spreading further and further. There are some kind contacts, some ---- but the whole cooperative mood is lacking somewhat.

I am thinking of some book we should publish all together for telling the world of the situation in the field of Logical Empiricism etc. Many people are hungry to know, what is what. Perhaps I shall write you about such a plan. Of course only short informative articles .... The whole field indicated, as far as the scientific research activities are concerned, directly and indirectly ....

----

35. Neurath to Ina Carnap, September 24, 1945.

(ASP RC 102-55-13)
24th September, 45

My dear Ina,

What an extraordinary show – Ina writing a letter to the noisy man. RES AD TRIARIOS VENIT[388] – as we learned as boys in the Grammar School. ES WIRD ERNST. I am very thankful for your kind intervention. I think it is very wise, that you try to bring us together. Impartiality is not needed, only sympathy and interest in our friendship.

I really felt somewhat helpless, when reading Carnap's kind long letters in which he wanted to explain his and my behaviour, but hardly touching the points, which are important for me. I answered him today and I hope I succeeded in going on in a way more conform with his tendencies. You are particularly helpful, because your Austrian traditions give you a better start.

You are personally prepared to bear Carnap's attitude, that when he thinks he is in the right that he expects the victim should bear his pain bravely. I have to tell you, that not only as a victim I am against this attitude but in any case. Since I do not think that we can speak of 'right' properly, I decided all problems to discuss as happiness problems. If I have to isolate a person, because he or she is dangerous, then I should put into account the unhappiness of the person's isolation together with the happiness increase of the others reached by the isolation. That the isolation is a 'right' – as some people style that – should not led people to neglecting the isolated person's unhappiness. You may bear that, I shall not. REASON: not because I want to avoid unhappiness, I should even take that, but because I think this attitude as

[388] "It comes down to the triarii," Roman saying that referred to the legion's elite forces as measures of last resort.

such is dangerous for my human brethren and Carnap should try to see that. I tried to analyse human relations and I think that this habit is really dangerous.

Of course I am often unkind, grieving people, but I think I never let other people being unhappy, because I think I am 'right' (index word in my language). Discussing prison life etc. I often touch [on] this point. And I met many people here who are prepared to look at prison life and everything from this happiness point of view.

I must confess I often wanted to mention to Carnap this basic difference in our way-of-life attitude, but I hesitated, because it implies the assumption that Carnap's point of view is partly Platonic, Prussian, Puritanic etc. – i.e. preparing the soil for totalitarian persecution, which often leads to terrible things e.g. Nazidom. Please, do not misunderstand me, I am not saying that Carnap is politically unsound, not at all, but that this attitude has elements, which very often lead kind people to become gradually merciless even cruel. In the moment, in which you are leaving the field of brotherhood and happiness, looking at 'higher' things, you speak of godlike instances, science, impartiality, justice etc. (describing Carnap's attitude) you are entering the land of danger. Then people learn to look at these higher principles and disregard the happiness account.

You mention, that I bully people, when arguing – I am not sure of that, I think I am rather noisy, and mostly interested in creating doubts, weakening the strong positions of some absolutism. SHOULD I BULLYING [sic], I AM PREPARED TO ALTER THAT. That is what I ask from Carnap, that he should be prepared to alter his attitude (in general, not only as far as I am concerned).

You see I am a little doubtful, about 'coercion' exerted by me. Reason: when writing in a very conciliant [conciliatory] way (no noise, no bullying), e.g. in the Plato article,[389] the people who answer, answer exactly in the tone of irritation, I know, in the cases in which people tell me they feel themselves bullied – interesting, is it not? Secondly there are people, who after a certain 'education' say to me I should be less noisy etc., but never felt them coerced or bullied....

How do I explain that? Hypothesis of mine: most differences in life are based on ATTITUDES, not on opinions and not on actions themselves, but on the PATTERN OF ACTIONS. Many people feel that I do not agree with their habits and attitudes. I think Carnap feels that, too, even when we are peacefully together. Such a fundamental difference creates an atmosphere, which leads sometimes to explosions. I discovered that people of different religion, even atheists and people with faith can go on together as long as they look at love and friendship as something DECISIVE.

You see there are many ways of life, let me select two of them. The one the 'PLATONIC' one thinks in terms of a pyramid of principles, rules, rights, order, everything has it proper place, fixed by some structure, you can predict what will happen, rational arguing is decisive in life etc. This habit is dangerous in itself, the persecution of the Albigenses, Huguenots, Jews etc. is possible within such a framework.

[389] Presumably Neurath and Lauwerys (1945).

There is another way of life, where happiness and friendship are in the foreground. Why order? Love and happiness, – sometimes supported by order, justice and suchlike things, but they are not always applicable, when one wants to be kind. Of course not only to the next neighbour or relative but also to mankind. There are many loyalties you have to bear in mind, not only one. Happiness here, happiness there, pain here, pain there. Result: much muddle. And when thinking of human happiness one has to bear muddle, which is also essential for any evolved democracy……

I think I am entitled to discuss this matter, since you indicated that Carnap is a zealous Prussian Lutheran (I should not think of Lutheran, but of Sect[a]rian, but I know, what you want to indicate) – that implies not of the second type but rather of the former type.

I shall tell a story which illustrates the matter, I think. Arriving with Carnap from Bruxelles, the ticketcollector in Paris, tells me my ticket is the Belgian one. I tell her, I did get this back from the Belgian official and I do not have the other part. She asked the man with the red cap to decide, what should be done, he needed a few seconds for his decision 'OK' he said.

Now we left the station and tried to get a taxi. In this moment I feel in my pocket the French ticket; polite as I am I return and give the ticket to the collector. She is very impressed by that and expressed her thanks vividly.

CARNAP: 'Now I do not understand anything. She looks at the ticket, she asks the official in charge, he permits you to pass, and now the ticket was wrong. What was the intention in asking the official at all?' I: 'You see, he looked at me, and thought: that is not the attitude of a swindler and instead of bringing a form, calling witnesses, making a fuss, as in Germany it would be the rule, he made his decision'. Carnap moved calmly his wise head and did not understand the Western world. I did. Perhaps the story was a little different, but Carnap will tell it to you. In principle, that was the central point.

Dr. Eisenmenger the physician Franz Ferdinand's tells a story: Egypt, Cook Office. A German complains intensely that he put his luggage into [sic] the ship or carriage and did not get a receipt (you know that is English custom) and now it disappeared. Long statements etc. The official thinks one moment, then asks: 'how much value?' The German wholly abashed answered: "200 po[u]nds". "You will get the money, should the luggage not be here within an hour" said the official calmly and asked the next passenger, what he wants. Eisenmenger says, he was much impressed, thinking how such a problem would be solved in Germany, with oath, declarations, dozens of letters and counter-letters, etc. perhaps even court of law. Whereas this Englishman less interested in order and justice, just wanted to go on in some friendly way with a minimum of time designed [assigned] to that business.

Or. The commander of one of the internment camps saw how much depressed some of the people were by being not permitted to write frankly and freely to their wives etc. He said to them, 'give me your letters, I shall put them into my covers and post them as mine.' He did it openly, all other officers and the soldiers knew it. Imagine a German officer, separating 'professional' and 'personal' life, he would either not do it, or if, do it in a concealed way, fearing, he could reduce obedience, authority etc. ashamed that he was not able to 'bekaempfen den inneren Schweinehund' which wants mercy and such nonsense, when 'professional' duty is in question etc.

I think Carnap belongs more to the people, who sympathize with the Platonic-Prussian attitude. I like to be in a country where the difference personal and professional attitude is very small. One thinks of other people's happiness and tries not to forget of course public happiness, but also to think of the neighbours['] happiness, who is just in contact with the respective official.

What do you think about all that. I am now promoting muddle, democracy and brotherhood and I try to weaken all tendencies which go into the Platonic direction. You can realize how much it touches me, when I see how friends are supporting the other group with order, justice, pyramid of rules etc. (I know very well, that arbitrary administration is bad and that the poor are often interested in written, well[-] defined law, but nevertheless it is dangerous this rigidity) with ecsta[s]y, enthusiasm, etc.

I have no idea what kind of 'rights' and 'duties' of an editor pressed Carnap to drop his name from my monograph. I asked him, that he the 'correct' man should at least tell me that, not only in vague generalities. But I cannot see that it is increasing human happiness when [marginal addition: *someone is*] not thinking of the pain inflicted by executing some imagined 'duty'. Of course I understand an argument, which runs so: when I, Carnap have my name there, unhappy things will come out. Readers believing in me will take this monograph too seriously will make bad research etc. or will become bad parents or bad teachers; this unhappiness is so serious that I have to inflict pain on my friend Neurath …. or something like that. WHAT HAPPINESS WAS ENDANGERED BY HAVING HIS NAME ON THE MONOGRAPH. I am really inquisitive; how he will explain that?

You say, I think you regard it as a kind of exculpation: he is a zealous Lutheran etc. But you apparently overlook, that that would imply to object to that attitude in principle, it is not only some manner like another, it is some danger in that. Luther preached one should persecute the Jews, one should burn their synagogues, etc. he suggested one should massacre the peasants, some more some less does not matter, etc. You see it does not reduce my resistance against Carnap's attitude and behaviour when you tell me, that it belongs to the Luther group. 'Eh scho wissen'.

Please, realize my difficult situation. I have talks with German friends, who almost without exception do not realize to what extent the 'good' German attitude prepared Nazidom together with other items. I try to induce them to drop certain peculiarities and to accept more the Anglo-Saxon peculiarities. I want they should [sic] drop the interest in order as such, in stressing 'rights' more than peace and happiness, in stressing 'ideals' (ecsta[s]y, enthusiasm, arts, etc. science etc. impartiality, etc.) more than the daily happiness of the neighbours, the own happiness, the happiness of people more far away. As you say Carnap is inflicting pain in the name of 'science, impartiality and suchlike gods', that is just, what I try to fight, and what my German friends usually try to defend, whereas my English friends in most cases agree with my attitude, which is based on compromise, muddle, happiness and not on some [i]nhuman 'principles'. I should like to know from where Carnap got his statute book about the duties of an editor ...... You see, when this kind of Platonic ideology comes into the picture I become rather embittered. My friends and son and others in concentration camps etc. .... for so many years, some died. And the KIND GERMAN BOYS who performed cruel things, inflicted cruel things, inflicted pain, came from an environment, in which Platonic ideals count higher than human brotherhood, where performing some 'duty' implies inflicting pain as something normal. THAT IS, I THINK ONE OF THE DEEPER SOURCES OF OUR DIFFERENCE .... The Moraltheologists [sic] and Talmudists stressed the point of the law, and then they sometimes reduced its hardship. Carnap first creates a law according to which he is forced to withdraw his name and then he reduces [sic] this action by making a slight remark somewhere, not too obvious. You see I should think, when he thought that it is creating unhappiness, when people believe in the authority of Carnap, when reading my book, then he should do more, not less, than he did ...... It sounds now, as if some legal performance should be in principle executed but with some restrictions ...... Please, realize, that without being angry I look at that with not much joy and find it in its best rather comic ....

I hope you will not become impatient, when reading my long description of my attitude and why it is not a question of good will, [marginal addition: *as long as*] Carnap IN PRINCIPLE tries to perform, what I think is dangerous. Of course I shall try to be with him at [in] good terms, even when he does not alter his attitude, but it is hard for me, to think of a friend, who does not think of brotherhood, happiness, humanity as more important then some godlike phantoms .... Sorry. But perhaps he will understand a little the danger .... Let us hope. You see this inflicting pain remains, what it is, even when Carnap does it under tears .... The Kurfuerst killing his son under tears .... Prinz von Homburg.[390] Prinz von Homburg and Goethe's Iphigenie are my examples of real German lack of brotherhood and of the mercilessness which appears to be heroic or something like that. For me dangerous through and through ....

Joergensen and Ness are well. Poor friends who died. What will Lukasiewicz do with his nationalism? And Scholz with his preparedness to collaborate?[391] The half collaborator will persecute the full corroborator being persecuted by the quarters etc. What a world. Refugees will say that the others did not go away – what a shame – and these will call the refugees escapists. What a world.

Thanks for newspaper cuttings. I am very thankful for all such stuff. Our files are lively now, we enjoy our big library. The files are increasing, the collections of maps, pictures etc. People here are kind to us. Yesterday a neighbour whom we

---

[390] Neurath presumably refers to Heinrich von Kleist's famous play, *Der Prinz von Homburg oder die Schlacht bei Fehrbellin*, written in 1809–10.

[391] Heinrich Scholz (1884–1956) was a German logician, working especially on the history of logic. He had his own school of logic in Münster, whose members (like Friedrich Bachmann and Albrecht Becker) participated at the *International Congresses for the Unity of Science*.

never met, brought a telegram eronneously [sic] put under his door, adding – as a consolation – a basket full of wonderful apples.

Everything is full of some muddle, everything goes on smoothly and in a rather humane way. There are bad things here too – but people acknowledge that and want to alter it.

I am waiting for your and Carnap's letter. Perhaps it is helpful that you lead the discussion into the serious field of the way of life. Perhaps it is better, one says all that, instead of being delicate. Suppression of all these protests against Carnap's habits creates perhaps the temper, which leads to outbursts, because – not without reason – I regard his treating me badly as an result of his general outlook. A point in which you agree with me.

I hope you will bring us more closely together than we have been before, and without Carnapian sleepless nights. I am very prepared to create a kind and heartly [sic] Carnap-Neurath brotherhood. With love from Mary to both of you

Ever yours

[Otto Neurath]

36. Neurath to Carnap, September 24, 1945.

(ASP RC 102-55-14)

194 Divinity Rd.
OXFORD

My dear friend Carnap,

Perhaps you felt discouraged, when reading my letter, because I wrote it in a rather helpless mood. But let us go on and try – first of all, I have not the slightest doubt, that you and I would help one another when in danger, further, that we are pleased by our personal contact, etc. nevertheless there are manifestly some serious differences in outlook, which we as kind people should analyze for finding some compromise. Of course, should you be against compromise as such, then it becomes a difficult matter.

It is so, that I complain about you violating me by grieving me and you complain about me violating you by overstepping self-defence. Would you think I am not in harmony with your own stories, when saying, that you did not tell of any case in which I first violated you? Please think about this point, please.

And now I shall try to be analytic and to be restrained, not giving names from the beginning. I think, that you should tell me

(1) which are the points in my paper, which forced you to withdraw your name. You did tell me about that in the most vague terms only and looking through my paper I cannot discover things sufficiently bad.
(2) withdrawing a name is so serious an action, that I assume, that you can tell me what detrimental effects for our movement, readers, mankind as a whole you expected from your name being on the paper.

(3) if this is so serious, and your name on the paper would give too much authority to my bad paper that it becomes more dangerous, then why did you make the concession, to indicate the withdrawal to be – as you think – hardly conceivable, as you wrote me [?]
(4) do you really think that my paper is worse than all the others, e.g. Bloomfield's about which many people complaint [sic] to me?[392]

Please, be kind enough to answer these four questions. I do not conceal, that your behaviour looks less like an action made for the common good of readers, mankind, movement but rather like an action, which has to satisfy your own conscience only, without being made for the common good and the common happiness. I confess, that up to now I looked at this ambiguous behaviour as I look at certain actions suggested by Moral Theologists or Talmudists, who first give certain strong rules for satisfying the conscience and then tell how to make the strongness [sic] not too strong, again satisfying the conscience, but in both cases not telling us, how the happiness of the human brethren is increased by the suggestions and actions they are producing.

You see, I myself do not like the continental "Aussprachen haben", but then I see, that the continental atmosphere is forcing us to have them and therefore I try to make the best of it. The fact that you are writing long letters shows me, that you take the matter seriously, as I do it [sic], and that even Ina comes down from the high Olympus, the silent goddess to bring the thundering heroes together before they start sitting sulkingly in their respective tents shows me, that also she [marginal addition: *thinks*] things [sic] becoming serious and that she thinks we should take care of our friendship and come to some kind solution.

After answering the above questions please try to think a little about me as a human brother, who has his difficulties and whom to help would be a friend's task. You speak of "reconciliation", that is not the right word for making the situation better, when one of the partners does not feel being in a feud, but being "humiliated". I have seen so many human careers that I am entitled to say it is extremely difficult to alter the atmosphere of humiliation.

You see, when you tell me and others, that I am terrible in my outbursts, that is not humiliating – EVEN IF IT WERE NOT AN ACCEPTABLE DESCRIPTION. But I feel, as if you put away the basic elements of my life, when you are spreading the legend that I become furious, when criticized in my arguments, in my style, etc., as you did write to Morris.

Look, all my life I have two important pillars of my behaviour; to reserve a considerable part of my life to purely humane relations, kind relations, friendship, love, enjoying reading, looking at landscapes, etc., making people happy and becoming happy by being treated in a kind way. The other pillar is to take nothing for granted (without exception) and being prepared to alter anything which I do not regard as "folklore" (I should not be interested in altering trousers or ties etc.), i.e. my

[392] On Bloomfield, see letter 28, note 305.

arguments, my style, etc. that is the reason, why so many people are astonished with what extraordinary preparedness I am accepting suggestions dealing with my style, dealing with my arguments. It is normal, – there are many witnesses of that – that I say to people less educated than I am, alter the wording, if you think it better understandable. My publishers tell me how nice I am, never being angry when a "reader" of the office makes suggestions. I think it just characteristic that I never have an outburst, when people suggest alterations. Imagine that such a mythos, propagated by you (perhaps other people started with that and you accepted it without taking care of proving it – I have no idea) inflicts pain.

I think – you know it is difficult to get a picture of the own person – that I collected sufficiently data, which tell me, that I designed much energy and efforts to these two parts of my fabric, and succeeded. The number of friends is increasing – I should say rapidly, when looking at the long letters from so many people, who talk over private problems etc. Apparently people, who are not afraid of me and my "outbursts". But – perhaps that indicates something – they are in the vast majority non-Germans, they are Dutch, English, etc. Further I have many examples showing me that people, who first felt themselves irritated by my – very often badly noisy behaviour dropped one day this being irritated and go on nicely with me --- AFTER THEY DISCOVERED WHERE THE DIFFICULTY IS. And it is mainly in the feeling of humiliation, which people create in one another. Outbursts are not so bad, but outbursts which are originated from the humiliation feeling and outbursts which create humiliation feeling.

WHEN YOU ARE WRITING TO MORRIS THAT I BECOME VIOLENT WHEN CRITICISED I BECOME REALLY FURIOUS, because I feel it as denounciation [sic], based on NOTHING, as far as I can judge.

In your last letter you are writing that you wrote that to Morris, mainly for reducing his possible sorrow that he did not protest.[393] Imagine, my dear Carnap, how that humiliates me. You are thinking of his pain, not in the same way of mine, (1) inflicted by taking away the responsibility from Morris, – I assumed he did, what was necessary, I was giving him FREE HAND, as letters show – and in (2) inflicting another by telling him about my outbursts, when criticised. YOU DID NOT ANSWER THIS SERIOUS, VERY SERIOUS REMARK OF MINE, but started telling me, what I never denied, that I am violent in self-defence.

The way, in which you are criticizing my overstepping self-defence, is for me again humiliating, why? You are not much concerned with what Schlick did, but you speak in general of that fact, that both parties have something to say etc. THAT IS JUST THE WAY TO AVOID THE FRIENDLY ANALYSIS OF THE VICTIM'S FATE. What comes out: YOU are the victim, because you had one of the most depressing experiences.

When analysing Schlick's behaviour towards me, you would discover a continual tendency to humiliate and attack me. Whereas I, as you remember, did much for preparing the booklet for praising his coming back to Vienna. From my side not the slightest tendency to attack him.

---

[393] Carnap to Neurath, letter 32.

About his extraordinary rude and humiliating behaviour I have Frank as a witness, as we discussed my "SOZIOLOGIE".[394] If he did suggest far reaching changes as an equal to an equal I should never reacted irritatedly, but he behaved like a big boss who has a right to bully me and to treat me like a school boy [sic]. Perhaps I am an "eingebildeter Laffe", but I do not regard my work so much less, than his and EVEN IF IT WERE SO, I ASK FOR THE NORMAL RESPECT BETWEEN SCHOLARS.

I never got such a kind of cold humiliation in my life, than from him. Scene in the circle: Waismann suggested to me, I should some day tell about my points of view, he had spoken with Schlick and a certain day agreed. Schlick introduced me – WITHOUT ANY CONFLICT IN THE AIR – with words of this kind: I do not know, what Dr. Neurath has to tell us and what he really wants, but he likes to speak to us, please start, etc. COLLEAGUES OF THE CIRCLE WERE REALLY STARTLED BY THAT AND TOLD ME SO AFTERWARDS.[395]

How Schlick behaved at other occasions, you know perfectly well. His last paper read in French at our congress in Paris, spoke of EINHEITSWISSENSCHAFT – without any provocation – as an ugly word. Fortunately the rather stupid translator translated "Einheitswissenschaft" into SCIENCE UNITAIRE – and now nobody knows, why that is ugly. ETC.[396]

Neider, who liked Schlick, did never deny all that, but only tried to explain it by telling me, that the overdelicate and sensitive man always feared that I could touch his sentiments etc. I told Neider how much I personally like to be sentimental and never disturb other people's sentimentality, if not connected with brutality, as it sometimes is.

I do not deny that I exploded when Schlick behaved badly, I never denied that, but you should be a good friend and tell me, what the other side can say for HUMILIATING me.

The same is in your own case. The non-quoting me belongs to the humiliating things in many cases. The not regarding a person as sufficiently important, that his remarks have to be taken seriously. One may use them, but why quote them. You say, both sides have to tell their story. You never told me, why you thought it right not to quote me. Your story always starts AFTER THE ACTION, whereas I speak of the beginning, too.

Look, I always fear that I, like other scholars, could fall into the mean and poor scientific persecution mania. Therefore I suppress my anger in such cases, very often, saying to me – it is by chance etc. But if such cases increase in number I become angry. Look at the following example. Feigl wrote a paper on Logical

---

[394] See Neurath (1931/1973).

[395] See further Neurath to Carnap, letter 34. note 369.

[396] See Schlick (1936/1979).

Empiricism.[397] Do you think anybody reading this paper, will discover that I had a certain position in its history? I do not speak of the acceptability of my hobbies, protocol statements, index verborum, etc. but of the fact, that in MANY BOOKS, even in Laird's[398] little introduction to modern philosophy I am acknowledged as somebody of a certain validity within this movement. It is particularly interesting to note, that I fought for LOGICAL EMPIRICISM against Schlick, who wanted RADICAL EMPIRICISM – the term used by James. And I was against that, because James this half-Bergsonian boy with so much obscurantism should be treated kindly by us, because he taught us a lot, but never as a kind of authority, whose name [sic] proposals should be accepted. Please, look to the article by Feigl. He just put me into SOCIOLOGY – that is all. He does not even mention me as the initiator and editor-in-chief of the encyclopedia. DO YOU THINK HE WOULD TREAT ME SO IF I WERE [A] PROFESSOR, not to ask the question, what he did [sic] with all that if I were Schlick. Such things are normal. I analyzed some day – I WAS NOT INVOLVED – the big book of an author, who indicated he would not quote others (in the English way) but tell his story. He was PRIVATDOZENT, and sometimes a name appears, and ------ believe it or not, almost all names were the names of the professors of his university, where he hoped to get a professorship. Nice, as I am, I never published that in a review or so. I never complaint about Feigl. I am not regarding him as very important or influential, but for a long time I regarded him as a potential friend. I FEEL IRRITATED AS A FRIEND. You see, many years ago I had with Feigl a very fundamental talk about INDUCTION, particularly about cross-inductions etc. inductions supporting one another etc. Feigl himself told me in USA about that, how much he learned from me that day and how certain remarks by Reichenbach, he likes, go in this direction etc. But FEIGL never quoted me in his induction papers. Please, believe me "Es liegt mir stagelgrien auf"[399] whether he quotes me or not (that is not the case with CARNAP, RUSSELL, etc.), but a potential friend should not humiliate me, ... He could e.g. quote some of my many remarks on induction, as it were, to be kindly and perhaps ----- CORRECT, too. I have no idea, how you look at Feigl's paper on Logical Empiricism as far as he describes my activities in this movement. Perhaps you think it is just OK. Please, tell me, what you think about that.

This kind of behaviour I think somewhat irritating. If some day Feigl will get an "outburst", then he and others think that the [sic] Neurath is quickly irritated, whereas it is the outburst which covers many, many subjects ...... You see I am fighting even that in me, and I am fighting becoming suspicious, because that disturbs friendship and love. But it is not always simple to overcome this feeling.

---

[397] See Feigl (1943).

[398] Cf. Laird (1936).

[399] Adaptation of a Viennese expression popularized by the satirist Karl Krauss that may be translated as "it upsets me."

Imagine, some day I read in Russell, how he uses the word Protocol etc. and I thought strange, how many different definitions exist … And I looked into RUNES dictionary.[400] I could not find the term PROTOCOL at all. Strange. Then later on I discovered that it is under BASIC. But there I found various variations, BUT NOT MY OWN. I never speak of observation, sensation etc., but I start from sentences with certain types of words in, which belong to certain classes etc. Further. Do you think the various articles together in RUNES enable a person to know, what I have to say? You are e.g. not quoting my protocol-article nor any other, except on physicalism. You see I should look at me as over-suspicious, if I did not have the Moscow affaire,[401] the Feigl article etc. I should say so: my friends and potential friends do not regard me as much important within their circle – except as far as I act as a kind of manager …. OK. Let them. But it us somewhat humiliating, when on the other hand, OTHER PEOPLE WHO ARE NOT SO CLOSELY RELATED TO ME mention my work – as I think – properly. All that is not worth wile [sic], if our atmosphere were kind and homely, which it is not.

That the BONZEN ideology[402] is very normal, I know, but why also within our movement, which I thought should try to be more free and less hard than others, the pluralism with its tolerant features, etc. humane and kind relations should be possible here, contacts etc. more than in other places of our planet ….

I do not think you are "milder" than I am or more "peaceful" – as you think – I only make more noise than you. But I am, in principle, prepared to think of other people's happiness, mine included (I am not unselfish), but you, as far as I can read your arguments, you are thinking in terms of "right", "correct", "duty", "etc." which is far away from being kind and friendly, thinking of human happiness as a central item. You think – as far as I can see – that if you are in the right, other people's unhappiness is not counting much. If you are "just" the victim may bear his pain with courage etc., etc. Perhaps You see yourself differently. I am really waiting, [sic] how you will answer question (2). What human happiness you improved by making me unhappy, when dropping your name. THAT IS MY QUESTION. Can you indicate that you improved the happiness of mankind, some people etc. by inflicting unhappiness, then I shall think over the whole matter in a different way. Up to now I rather feel, that – SHOULD YOU BE ABLE TO SHOW THAT, I AM NOT SURE OF THAT – you were in the position as a "correct" man to inflict pain. Which I think a very sad attitude, indeed …… not only as far as friends are concerned.

This is a "Platonic" attitude, I should say. I am against this attitude, even if I did not appear as a victim. And differences in ATTITUDES are really dangerous, because they create often tensions. Please, believe me, that I am so fully prepared to find a living [sic] compromise, that I shall not be to[o] hard in my attitude towards

---

[400] See Runes (1942); cf. Neurath to Carnap, letter 34. note 376.

[401] See Carnap to Neurath, letter 32. note 351.

[402] See Neurath to Carnap, letter 31. note 330.

this Platonism, which I – to be frankly – dislike enormously and which I think is something unpleasant in your otherwise charming habits.

Of course, I shall do what I can to avoid clashes, but please, when you writing [sic] me a kind letter, try also to tell me, that you are not untouched by these my remarks on your Platonic or Puritan or how you may call it attitude. Please. But even if you cannot drop that – I shall try to be as nice as possible to you .... It is only simpler, when you could have a similar attitude with all its muddling through....

I hope we shall now discuss within our movement also such problems of attitude closely connected with pluralism, unpredictability, which – I hope so – will be discussed now, after I published my opinions as bluntly, as possible, and since they are of some importance – even if not acceptable to some of you.

Of course, Labour gives me some hope, e.g. in housing. But the USA attitude makes everything difficult, and the Russian attitude, too. I do not know, what the Russians are driving at, but if they want to do, what they are doing ---- what a terrible future before mankind. Perhaps much of that harsh behaviour is a kind of higher muddle, let us hope so.

Laski has many good ideas. But as a chairman of a party, he should [be] a little more distinguished [discriminating] between topical politics and general attitude.... I heard him lecture. I am not so sure, that he always "means business". I should like to go into details, if you did ask me concrete questions, dealing with him.

Glad that Tarski's family is sa[f]e. How terrible Hosiasonns and Lindenbaum's death.[403] I remember how they met one another and gradually became accustomed to one another, e.g. in Paris. I liked them and enjoyed their happiness.

No, I do not know of Popper's Readership. Probably Hayek's influence. He likes him very much indeed and I think Popper will support Hayek in his attacking Logical Empiricism. I, for peace and happiness, suggested to Hayek we should arrange a published symposium and not attack one another brutally. I wrote my review as restricted [restrained] as possible, but I think he is a fanatic with much bias....[404] Some of his remarks are rather a kind of scandal – I think so. But peace is something of importance.

We now get letters from Holland, France etc. .... Nothing from people in the Russian zone, only from people outside the Russian zone, which is "sealed off".

We are enjoying life, having many nice contacts a very nice correspondence with many people, enlarging our institute etc. Ness wrote me nicely, he escaped, but his friend died, tortured, but not telling of Ness' hiding place. He hopes we shall have our next congress at OSLO. Let us hope so....

Now I am expecting a particularly nice and kind letter from you, answering the famous four questions and telling me about your attitude and whether it is really so Platonic, as I see it.

[403] See Ina Carnap to Neurath, letter 33. note 360.

[404] See Neurath (1945/2004).

Be sure of my intention to calm down ill feeling and to discover a pleasant way of a pleasant compromise, if we cannot reach a more common basis of happiness and humanity.

Ever yours
[Otto Neurath]
24/Sept/45

37. Neurath to Carnap, September 28, 1945.

(ASP RC 102-55-12)

OTTO NEURATH, DR. PHIL.	194 DIVINITY ROAD,	~~30 DICKERTON ROAD,~~
SECRETARY, ISOTYPE INSTITUTE	OXFORD	~~HEADINGTON, OXFORD~~

28th Sept.45

My dear friend Carnap,

Good Ina told me, that my remarks very often put your meditation into motion and even touches your sleeping capacity. I learn from that, that a friend's remarks are taken seriously by you – but please do [marginal addition: *not*] think that I behave differently. Your letters ask me to search my behaviour and life and to try to alter it. As I told you, well knowing the slight persecution mania very frequent in scholars, I try to fight any nucleus of such a behaviour in me.

Therefore I assumed, of course, that perhaps my remark, that Schlick behaved in a humiliating way, could be overstated and that I should look at something he said, as a criticism and myself as too sensitive .... I tried to remember Schlicks behaviour in our talks – Frank as witness – his behaviour in the circle etc., and I could not discover any case in which I fought criticism, always cases in which I fought arrogance and an attitude humiliating me, but, as you know, we should be very careful, where we are witnesses about ourselves. But I remembered, that I have the French text of Schlick's remarks on me. I ask you and Ina, whether you think this kind of talking has to be regarded as 'criticism proper'. I think criticism involves that giving 'names' is not proper criticism. WHEN WITHOUT EXPLAINING THE REASON. I think adding esthetical judgement is usually not just criticism – it tells about the person's biography, not about the work in question. But please, look at the following quotations: (from our department in the IXth International Congress of Philosophy, 1937, IV. L'ECOLE DE VIENNE ET LA PHILOSOPHIE TRADITIONELLE. 'Certains d'enter eux ont même une antipathie marquée contre le mot 'philosophie' (as if it where some whim, not explained in detail) qu'ils veulent proscrire au bénéfice d'autres terms tels que 'science unitaire' par où ils entendent se considérer eux-mêmes, non comme des philosophes, mais comme de chercheurs scientifiques .... etc.'

and then:

'Ce que l'homme fait par peur produit presque tojours in effet ridicule. (please, look into my writings, whether I am using such terms, even when objecting other people's opinions) Ce n'est donc pas etonnant, si tel anti-metaphycien propose par example, le plus sérieusement du monde, d'établir un Index verborum prohibitorum, où il faudrait denouncer tous le mots qui se trouvent le plus souvent au centre des discussions métaphysiques, par example 'monde', 'âme', 'être', etc. et le mot 'philosophie' lui-même, dont nous avons déjà parlé. C'est vraiment une drôle d'idée (Do you think this is a word of scholarly criticism? I should take even that, if combined with real analysis and criticism. I started this language business as a student, considerably influenced by Itelson,[405] and my contact with you and other people is partly based on that fact, that we agree about certain main points. What a strange situation, that Schlick fights for the 'soul' ...... oh my dear. But I should not mind his OBJECTIONS, if he did mention at least one, but he only tries, without telling [mentioning] my name (NICHT-GENANNT SOLL ER WERDEN –, to make me a laughing-stock. Some crazy crank – that is all) que de vouloir conduire les hommes à la vérité en leur faisant peur de certains mots. etc.' and then the [a]estheticist has the word, [sic] fortunately in vain, because the translator dropped 'Einheitswissenschaft' 'Einheizwissenschaft' ... le replacement par des expressions sans couleur et peu esthétiques telles que 'science unitaire'. Cette attitude me parait reposer sur un profond malentendu 'JUST' malentendu' – HE DIXIT ......[406] Boy, boy, re-reading this I believe in Otto Neurath in the witness box, when telling that SCHLICK liked to humiliate Otto Neurath, not even thinking him as sufficiently equal that one has to explain the objections; the grand spirit gives names ---- how simple. And I cannot see, that you or Feigl or any other of Schlick's friends ever tried to give me a kind of satisfaction, by mentioning, that Schlick's way of criticism does not give sufficient credit to Neurath's hard work, – acceptable or unacceptable, that is another question – who tried for about 40 and more years to alter his own language carefully, thinking seriously of this matter and explaining again and again, why dropping 'soul' etc. Not because it is his whim, explaining in detail, why what we are striving for is an EINHEITSWISSENSCHAFT, – perhaps with some unacceptable details, about which we may start discussion. But how can you discuss with libel-like remarks without reasons?

---

[405] On Itelson see Freudenthal and Karachentsev (2011); cf. letter 22, note 178.

[406] Schlick (1936/1979, 492, 495): "Certain members even have marked antipathy to the word 'philosophy', which they wish to proscribe in favour of other terms, such as 'unitary science', whereby they mean to consider themselves not as philosophers, but as exponents of scientific research. [...] What a man does through fear nearly always produces a ridiculous effect. So it is not surprising if such an anti-metaphysician proposes, for example, in all seriousness, to establish an *Index verborum prohibitorum*, to which must be consigned all the words most often found at the centre of metaphysical discussions, such as 'world', 'soul', 'being', etc., and the word 'philosophy' itself, of we have already spoken. It is truly a quaint idea to wish to lead men to truth by making them fearful of certain words. [...] Replacing it by colourless and unlovely expressions such as 'unitary science'. This attitude seems to me to rest on a profound misunderstandings."

You see, this very often presses me in a position to tell of my own work, because others sometimes go too far – as far as I can see – in deteriorating my whole work and sometimes humiliating me. I am for peaceful contacts, BUT I AM NOT PREPARED TO SWALLOW HUMILIATION. Really not. But I am wholly prepared to learn, that what I think an unproper behaviour of other people is just the usual thing. Question: DO YOU THINK THAT THIS WAY OF SCHLICK TO TELL INTERNATIONAL READERS ABOUT MY POSITION WITHIN OUR MOVEMENT SHOULD BE CALLED PROPER?

I know DE MORTIS NIL NISI BENE[407] – but that seems to be only valid, when writing necrologs [obituaries] .... And another quotation (from Goethe) Man sagt Eigenlob stinkt, was aber ungerechter Tadel fuer einen Geruch hat, darnach fragen die Leute nicht (I quote from memory).[408]

That is that. I am looking forward to a very kind, very nice, very illuminating couple of letters from

both of you, every yours

[Otto Neurath]

38. Ina and Carnap to Marie Neurath, January 2, 1946.

(ASP RC 102-56-02)

January 2, 1946.

Very dear Mieze, *Neurath*[409]

We are very much distressed that this, our first letter to you, is caused by the sad news of Neurath's death. It came so unexpectedly that we could hardly believe it when we saw it by chance in a newspaper. Felix Kaufmann wrote that he had a very cheerful letter from Neurath only a short while ago. Thus we assume that you too have been unprepared for it. We are sad with you. The last years have not been easy for the two of you, and we always admired how courageously you took it. Not a single time has Neurath mentioned hardship in the life in England. He only wrote about finding new friends, reconstructing his work, enjoying life with you. It is good that he was still able to see the downfall of the Nazis and their cause. His death is a

[407] From the Latin expression 'De mortuis nil nisi bene dicendum' ('Of the dead, say nothing but good').

[408] Neurath refers to a passage from Goethe's *Sprüche in Prosa* (1908a, 16). The exact quotation would be: "Eitel Eigenlob stinkt: wörtlich: «Man sagt:. Das mag sein; was aber fremder und ungerechter Tadel für einen Geruch habe, dafür hat das Publikum keine Nase.»" It has been translated into English as: "It is said that vain self-praise stinks in the nostrils. That may be so; but for the kind of smell which comes from unjust blame by others the public has no nose at all." See Goethe (1908b, 81).

[409] A corner of the paper is missing, thus "[---]" stands for the missing words.

tragic loss not only to you and to his friends, and to the Unity of Science Movement, but also to the cause of democracy, to the fight against fascism. I think, he had at times a hard life, but it must have been a rich life; and though his tender soul caused him greater anguishes th[a]n the tougher ones experience, it must also have brought him deeper satisfaction.

I don't know how much you have followed the correspondence between Neurath and Carnap. If you have, you can imagine how deeply sorry Carnap now is when all the great arguments – advanced on both sides as safeguards for future occasions – have proven futile because there will be no future controversies. We only hope that both of you always knew that in spite of heated arguments Carnap always felt that he was a close friend of Neurath's. Ah, but there could have been more of the kindness which Neurath always considered as the most important consideration!

We wonder whether you intend to carry on the Institute, and whether you have enough experienced help for the work. We hope that you have good friends in England so that you will not be to[o] lonely; but we realize that friendships made later in [---] and in a new country hardly ever becomes as close as those [---] youth – we at least have found it that way in America. Is [---] anything we can do for you that might help? Will you let [---] if we can do something for you, financial or otherwise? Ka[---][410] thought you might want to come to America; if so, perhaps we could assist you?

Please remember that we both are your friends and that we grieve with you.

Yours,

I. and C.

39. Marie Neurath to Ina and Carnap, January 14, 1946.

(ASP RC 102-56-01)

14.1.46.

Dear Ina and dear Carnap

Very many thanks for your kind words. Yes it is difficult to believe. It was absolutely sudden, even for me who was with him to the last second. We had a day which was full of the usual activities, in fact our first holiday after very busy weeks, and we were looking forward to two quiet days quite for ourselves. We returned from a supper and vivid conversations after all, and were sitting down for a little chat at Otto's desk, and while I was reading something to him, and after he had just spoken a few words to me, his head sank down on the desk before him. He often looked very tired, and needed days of reading and rest in between, but he never complained about any pain – on the contrary, he stressed on his very last morning how lucky we were never to be ill. He rushed about so much that it was too natural that he was tired, and he took on more and more activities. You have a very wrong impression if

[410] The original letter is torn at this place, but given the content of Marie Neurath's response, "Ka" must be the first two character of Felix or Edgar Kaufmann's name.

you think that just the last years were difficult for him. They were in fact the most successful of his life, and he stressed again and again, how much he felt at home in this country. He had very good friends here, and an extraordinary response.

I shall stay in this country. I am carrying on the Isotype Institute and have very much to do. Kaufmann wrote me too – I shall try to answer, but there is an enormous number of letters to write, and I should be grateful if you would thank in my name for the time being. All of you are most kind and helpful. Yes indeed I need your help to find a way to carry on the work for the Unity of Science movement. Could not Hempel become the organizational centre over there? And perhaps Hazebroek over here? Our Dutch secretary could help in Holland too. And perhaps Martin Strauss in this country?

Please let know soon. Fortunately I don't need financial help, thank you so much. But if there are any interests to be looked after, as royalties from the Press, will you please see to it? I should be very grateful.

Yours ever,
Mieze

References

Adams, John. 1851. Diary. In *The Works of John Adams, second President of the United States*, vol. 3. Boston: Charles C. Little and James Brown.

Aldrich, Virgil C. 1945. Review of Neurath's Foundations of the Social Sciences. *The Journal of Philosophy* 42 (17): 470–472.

Ayer, Alfred J. 1940. *The Foundations of Empirical Knowledge*. London: Macmillan.

Ballod-Atlanticus. 1898. *Der Zukunftsstaat. Produktion und Konsum im Sozialstaat*. Stuttgart: J. H. W. Dietz Nachf.

Beaney, Michael. 2016. Susan Stebbing and the Early Reception of Logical Empiricism in Britain. In *Influences on the Aufbau*, ed. Christian Damböck, 233–256. Cham: Springer.

Benson, Theodora, and Askwith Betty. 1936. *Muddling Through or Britain in a Nutshell. Illustrated by Nicolas Bentley*. London: Victor Gollancz.

Bergmann, Gustav. 1993. Memories of the Vienna Circle: Letter to Otto Neurath (1938). In *Scientific Philosophy: Origins and Developments*, ed. Friedrich Stadler, 193–208. Dordrecht: Kluwer.

Bloomfield, Leonard. 1939. *Linguistic Aspects of Science*. Chicago: University of Chicago Press.

Bodmer, Frederick. 1944. *The Loom of Language: A Guide to Foreign Languages for the Home Student* (edited by Lancelot Hogben). New York: W. W. Norton.

Boon, Timothy. 2016. 'To formulate a plan for better living': Visual communication and scientific planning in Paul Rotha's documentary films, 1935–45. In *Pursuing the Unity of Science: Ideology and Practice from the Great War to the Cold War*, ed. Harmke Kamminga and Geert Somsen, 156–181. London: Routledge.

Braithwaite, Richard. 1953. *Scientific Explanation: A Study of the Function of Theory, Probability and Law in Science*. Cambridge: Cambridge University Press.

Bridgman, Percy W. 1934. A Physicist's Second Reaction to *Mengenlehre*. *Scripta Mathematica* 2: 101–117, 224–234.

Brożek, Anna. 2017. Maria Kokoszyńska: Between the Lvov-Warsaw School and the Vienna Circle. *Journal for the History of Analytical Philosophy* 5 (2): 18–36.

Brunswik, Egon. 1952. *The Conceptual Framework of Psychology*. Chicago: University of Chicago Press.

Carnap, Rudolf. 1928/1967. *The Logical Structure of the World*. Trans. R. George. Berkeley: University of California Press.

———. 1931/1934. *The Unity of Science*. Trans. Max Black. London: K. Paul, Trench, Trubner & Co., Ltd.

———. 1935. Karl Popper, *Logik der Forschung*. *Erkenntnis* 5: 290–294.

———. 1936–37. Testability and Meaning. *Philosophy of Science* 3 (4): 419–471 and 4 (4): 1–40.

J. Cat, A. T. Tuboly (eds.), *Neurath Reconsidered*,
Boston Studies in the Philosophy and History of Science 336,
https://doi.org/10.1007/978-3-030-02128-3

———. 1936a. Wahrheit und Bewährung. *Actes du Congrès international de philosophie scientifique et empirisme logique, fasc. 2: Induction et Probabilité*, 18–23. Paris: Hermann.

———. 1936b. Von der Erkenntnistheorie zur Wissenschaftslogik. *Actes du Congrès international de philosophie scientifique, fasc. 1: Philosophie scientifique et empirisme logique*, 36–41. Paris: Hermann.

———. 1937. *Logical Syntax of Language*. London: K. Paul, Trench, Trubner & Co (Originally published as *Logische Syntax der Sprache*. Wien: Springer, 1934).

———. 1939. *Foundations of Logic and Mathematics*. Chicago: University of Chicago Press.

———. 1942. *Introduction to Semantics*. Cambridge, MA: Harvard University Press.

———. 1943. *Formalization of Logic*. Cambridge, MA: Harvard University Press.

———. 1945a. The Two Concepts of Probability. *Philosophy and Phenomenological Research* 5 (4): 513–532.

———. 1945b. On Inductive Logic. *Philosophy of Science* 12 (2): 72–97.

———. 1947. *Meaning and Necessity*. Chicago: University of Chicago Press.

———. 1949. Truth and Confirmation. In *Readings in Philosophical Analysis*, ed. Herbert Feigl and Wilfrid Sellars, 119–127. New York: Appleton-Century-Crofts, Inc.

———. 1950. *Logical Foundations of Probability*. Chicago: University of Chicago Press.

———. 1963. Intellectual Autobiography. In *The Philosophy of Rudolf Carnap*, ed. Paul A. Schilpp, 3–84. La Salle: Open Court.

Carnap, Rudolf, Hans Hahn, and Otto Neurath. 1929/1973. The Scientific Conception of the World: The Vienna Circle. In Neurath and Cohen (1973): 299–318.

Cartwright, Nancy, Jordi Cat, Lola Fleck, and Thomas Uebel. 1996. *Otto Neurath: Philosophy Between Science and Politics*. Cambridge: Cambridge University Press.

Chapman, Siobhan. 2013. *Susan Stebbing and the Language of Common Sense*. Basingstoke: Palgrave Macmillan.

Chase, Stuart. 1938. *The Tyranny of Words*. London: Methuen & Co. Ltd.

Christiansen, Broder, and Eli Carnap. 1933. *Neue Grundlegung der Graphologie*. München: Felsen-Verlag.

Cordero, Alberto. 2010. Philosophy of Science. In *A Companion to Latin American Philosophy*, ed. Susana Nuccetelli, Ofelia Schutte, and Otávio Bueno, 370–382. Oxford: Wiley-Blackwell.

Creath, Richard, ed. 1990. *Dear Van, Dear Carnap: The Quine-Carnap Correspondence and Related Work*. Berkeley/Los Angeles/ London: University of California Press.

Crossman, Richard H.S. 1937. *Plato Today*. New York: Oxford University Press.

Dahms, Hans-Joachim. 2004. Neue Sachlichkeit in the Architecture and Philosophy. In *Carnap Brought Home – The View from Jena*, ed. Stewe Awodey and Carsten Klein, 357–375. Chicago: Open Court.

Damböck, Christian. 2017. *‹Deutscher Empirismus›. Studien zur Philosophie im deutschsprachigen Raum 1830–1930*. Dordrecht: Springer.

de Jong, Cees W., ed. 2008. *Jan Tschichold: Master Typographer. His Life, Work, and Legacy*. New York: Thames and Hudson.

de Santillana, Giorgio, and Edgar Zilsel. 1941. *The Development of Rationalism and Empiricism*. Chicago: University of Chicago Press.

Dewulf, Fons. 2017. Carnap's Incorporation of the Geisteswissenschaften in the *Aufbau*. *HOPOS* 7 (2): 199–225.

Feigl, Herbert. 1943. Logical Empiricism. In *Twentieth Century Philosophy*, ed. Dagobert D. Runes, 371–416. New York: Philosophical Library.

———. 1969/1981. The *Wiener Kreis* in America. In *Herbert Feigl: Inquiries and Provocations. Selected Writings 1919–1974*, ed. Robert S. Cohen, 57–94. Dordrecht: D. Reidel.

Finlay-Freundlich, Erwin. 1951. *Cosmology*. Chicago: University of Chicago Press.

Fite, Warner. 1934. *Platonic Legend*. New York: C. Scribner's Sons.

Florence, Lella Secor. 1943. *America and Britain. Only An Ocean Between*. Diagrams designed by the Isotype Institute. Foreword by John G. Winant. London: G.G. Harrap & Company Ltd.

Fraenkel, Abraham. 1919. *Einleitung in die Mengenlehre*. Berlin: Springer.

Frank, Philipp. 1907/1949. Experience and the Law of Causality. In *Modern Science and its Philosophy*, 53–60. Cambridge, MA.: Harvard University Press.
———. 1944/1949. The Philosophical Meaning of the Copernican Revolution. In *Modern Science and its Philosophy*, 216–227. Cambridge, MA.: Harvard University Press.
———. 1946. *Foundations of Physics*. Chicago: University of Chicago Press.
Franklin, Benjamin. 1840. *The Works of Benjamin Franklin*. Vol. VII. Boston: Hilliard, Gray and Company, 1840.
———. 1906. Dialogue Between Franklin and the Gout. In *The World's Wit and Humor*, vol. I, 18–25. New York: The Review of Reviews Company.
———. 1914. *Poor Richard's Almanack*. Waterloo: The U.S.C. Publishing Co.
Franklin, William Temple, ed. 1817. *The Private Correspondence of Benjamin Franklin*. 2nd ed. 2 vols. London: Printed for Henry Colburn.
Freudenthal, Gideon, and Tatiana Karachentsev. 2011. G. Itelson – A Socratic Philosopher. In *Otto Neurath and the Unity of Science*, ed. John Symons, Olga Pombo, and Juan Manuel Torres, 109–126. Dordrecht: Springer.
Frost-Arnold, Greg. 2013. *Carnap, Tarski, and Quine at Harvard: Conversations on Logic, Mathematics, and Science*. Chicago: Open Court.
Goethe, Johann Wolfgang von. 1832/2014. *Faust I & II*. Edited and Translated by Stuart Atkins, with a new introduction by David E. Wellbery. Princeton/Oxford: Princeton University Press.
———. 1908a. *Sprüche in Prosa: Maximen und Reflexionen*. Herausgeben und eingeleitet von Herman Krüger-Westend. Leipzig: Insel Verlag.
———. 1908b. *The Maxims and Reflections of Goethe*. Trans. Thomas Bailey Saunders. 2nd and Rev. ed. London: Macmillan and Co., Ltd.
Grimm, Hans. 1926. *Volk ohne Raum*. München: Verlag Albert Langen.
Hahn, Olga. 1909. Zur Axiomatik des logischen Gebietskalküls. *Archiv für Philosophie, 2 (1909). Abteilung: Archiv für systematische Philosophie* 15: 345–347 Translated in the present volume.
———. 1910. Über die Koeffizienten einer logischen Gleichung und ihre Beziehung zur Lehre von den Schlüssen. *Archiv für Philosophie, 2 (1910). Abteilung: Archiv für systematische Philosophie* 16: 149–176.
Haller, Rudolf, and Heiner Rutte. 1977. Gespräch mit Heinrich Neider: Persönliche Erinnerungen an den Wiener Kreis. *Conceptus* 44: 21–42.
Hardcastle, Gary L. 2003. Debabelizing Science: The Harvard Science of Science Discussion Group, 1940–41. In *Logical Empiricism in North America*, ed. Gary Hardcastle and Alan W. Richardson, 170–196. Minneapolis: University of Minnesota Press.
Hegselmann, Rainer. 1987. Introduction. In McGuinness (1987): ix–xxi.
Hempel, Carl G. 1937. Karl Popper, *Logik der Forschung*. *Deutsche Literaturzeitung* 58: 309–314.
———. 1942. The Function of General Laws in History. *The Journal of Philosophy* 39 (2): 35–48.
———. 1943. A Purely Syntactical Definition of Confirmation. *The Journal of Symbolic Logic* 8: 122–143.
———. 1952. *Concept Formation in Empirical Science*. Chicago: University of Chicago Press.
Hochhäusl, Sophie. 2011. *Otto Neurath – City Planning. Proposing a socio-political Map for Modern Urbanism*. Innsbruck: Innsbruck University Press.
Hogben, Lancelot. 1943. *Interglossa. A Draft of an Auxiliary for a Democratic World Order, Being an Attempt to Apply Semantic Principles to Language Design*. Harmondsworth/ New York: Penguin Books.
Holton, Nina. 2015. On Hania Frank. In *The Vienna Circle in Czechoslovakia: Pre-proceeding*, ed. Radek Schuster, Stefanie Dach, and Friedrich Stadler, 69–75. Pilsen: University of West Bohemia.
Hosiasson-Lindenbaum, Janina. 1940. On Confirmation. *The Journal of Symbolic Logic* 5 (4): 133–148.
Hughes, Jeff. 2016. Unity through experiment? Reductionism, rhetoric and the politics of nuclear physics, 1918–40. In *Pursuing the Unity of Science: Ideology and Practice from the Great War to the Cold War*, ed. Harmke Kamminga and Geert Somsen, 50–81. London: Routledge.

Jörgensen, Jörgen. 1931. *A Treatise of Formal Logic*. Vol. 1–3. Copenhagen/London: Levin & Munksgaard/Oxford University Press.

———. 1951. *The Development of Logical Empiricism*. Chicago: University of Chicago Press.

Kaila, Eino. 1939/2014. *Human Knowledge. A Classic Statement of Logical Empiricism*, ed. Juha Manninen, Ilkka Niiniluoto, and George A. Reisch. Chicago: Open Court.

Kaplan, Abraham. 1942. Are Moral Judgments Assertions? *The Philosophical Review* 51 (3): 280–303.

———. 1943. Content Analysis and the Theory of Signs. *Philosophy of Science* 10 (4): 230–247.

———. 1963. Logical Empiricism and Value Judgments. In *The Philosophy of Rudolf Carnap*, ed. Paul A. Schilpp, 827–856. La Salle: Open Court.

———. 1991. Rudolf Carnap. In *Remembering the University of Chicago: Teachers, Scientists, and Scholars*, ed. Edward Shils, 32–41. Chicago/London: University of Chicago Press.

Kaufmann, Felix. 1930/1978. The Infinite in Mathematics and its Elimination. In *The Infinite in Mathematics: Logico-mathematical Writings*, ed. Brian McGuiness, Trans. Paul Foulkes, 1–164. Dordrecht: D. Reidel.

Kiepert, Ludwig, and Max Stegemann. 1897. *Grundriss der Differential- und Integral-Rechnung*. Hannover: Helwing'sche Verlagsbuchhandlung.

Kirchhoff, Gustav. 1876. *Vorlesungen über Mathematische Physik. Mechanik*. Leipzig: Druck und Verlag von B. G. Teubner.

Köhler, Eckehart. 1982/1991. On Neurath's Writings on Logic, Ethics and Physics. In *Rediscovering the Forgotten Vienna Circle: Austrian Studies on Otto Neurath and the Vienna Circle*, ed. Thomas Uebel, 109–115. Dordrecht: Kluwer.

Koterski, Artur, and Thomas Uebel. 2017. Poznański and Wundheiler's 'The Concept of Truth in Physics': The Lvov-Warsaw School Contribution to Encyclopedism. In *The Significance of the Lvov-Warsaw School in the European Culture*, ed. Anna Brożek, Friedrich Stadler, and Jan Woleński, 291–307. Cham: Springer.

Kürnberger, Ferdinand. 1855. *Der Amerika-Müde, amerikanisches Kulturbild*. Frankfurt a.M.: Verlag von Meidinger Sohn & Cie.

Laird, John. 1936. *Recent Philosophy*. London: Home University Library.

Langbehn, August Julius. 1890. *Rembrandt als Erzieher*. Leipzig: Verlag von C. L. Hirschfeld.

Lewin, Kurt. 1922. *Der Begriff der Genese in Physik, Biologie und Entwicklungsgeschichte*. Berlin: Springer.

Lichtenberg, Georg C. 2012. *Philosophical Writings*. Trans. and ed. Steven Tester. Albany: State University of New York Press.

Lodge, Oliver J. 1892. *Modern Views of Electricity*. London/New York: Macmillan and Co.

Mainx, Felix. 1955. *Foundations of Biology*. Chicago: University of Chicago Press.

McGuiness, Brian, ed. 1987. *Unified Science*. Dordrecht: D. Reidel.

Merten, Ferdinand. 2007. *Otto Neurath in Den Haag: 1934–1940*. https://www.stroom.nl/media/MertensNeurathLezing.pdf. Accessed 6 Aug 2017.

Morris, Charles. 1937. Symposium of Unified Science. *Philosophy of Science* 4 (4): 496–498.

———. 1938. Scientific Empiricism. In *Encyclopedia and Unified Science*, 63–75. Chicago: University of Chicago Press.

———. 1942. *Paths of Life: Preface to a World Religion*. New York: Harper and Brothers.

———. 1946. *Signs, Language and Behavior*. New York: Prentice-Hall.

Naess, Arne. 1993. Logical Empiricism and the Uniqueness of the Schlick Seminar: A Personal Experience with Consequences. In *Scientific Philosophy: Origins and Developments*, ed. Friedrich Stadler, 11–25. Dordrecht: Kluwer.

Nagel, Ernest. 1939. *The Principles of the Theory of Probability*. Chicago: University of Chicago Press.

———. 1940. Charles S. Peirce, Pioneer of Modern Empiricism. *Philosophy of Science* 7 (1): 69–80.

———. 1941. Mr. Russell on Meaning and Truth. *The Journal of Philosophy* 38 (10): 253–270.

———. 1942. Review of Carnap's *Introduction to Semantics*. *The Journal of Philosophy* 39 (17): 468–473.

———. 1961. *The Structure of Science: Problems in the Logic of Scientific Explanation.* New York/Burlingame: Harcourt, Brace & World, Inc.
Nagel, Ernest, and Morris R. Cohen. 1934. *An Introduction to Logic and Scientific Method.* New York/Burlingame: Harcourt, Brace & World, Inc.
Neurath, Marie. 1946. Isotype in films. *Scannan: Journal of the Portlaoghise Film Society* 1 (3).
Neurath, Marie, and Robert S. Cohen, eds. 1973. *Empiricism and Sociology*. Dordrecht: D. Reidel.
———, eds. 1983. *Otto Neurath: Philosophical Papers 1913–1946.* Dordrecht: D. Reidel.
Neurath, Otto. 1909a. Ernst Schröders Beweis des 12. Theorems: Für die identischen Operationen gilt das 'Kommutationsgesetz.' *Archiv für Philosophie*, 2. Abteilung: *Archiv für systematische Philosophie*, Neue Folge, Bd. 15: 104–106. Translated in the present volume.
———. 1909b. Eindeutigkeit und Kommutativität des logischen Produktes ab. *Archiv für Philosophie*, 2. Abteilung: *Archiv für systematische Philosophie*, Neue Folge, Bd. 15: 342–344. Translated in the present volume.
———. 1910. Definitionsgleichheit und symbolische Gleichheit. *Archiv für Philosophie*, 2. Abteilung: *Archiv für systematische Philosophie*, Neue Folge, Band 16: 142–144. Translated in the present volume.
———. 1913/1983. The Lost Wanderers of Descartes and the Auxiliary Motive (On the Psychology of Decision). In Neurath and Cohen (1983): 1–12.
———. 1916/1983. On the Classification of Systems of Hypotheses. In Neurath and Cohen (1983): 13–31.
———. 1931. *Empirische Soziologie*. Berlin: Springer.
———. 1931/1973. Empirical Sociology. In Neurath and Cohen (1973): 319–421.
———. 1932a/1983. Protocol statements. In Neurath and Cohen (1983): 91–99.
———. 1932b/1983. Sociology in the Framework of Physicalism. In Neurath and Cohen (1983): 58–90.
———. 1933/1987. Unified Science and Psychology. In McGuinnes (1987): 1–23.
———. 1934/1983. Radical Physicalism and the 'Real Word.' In Neurath and Cohen (1983): 100–114.
———. 1935/1983. Pseudorationalism of Falsification. In Neurath and Cohen (1983): 121–131.
———. 1935/1987. What Is Meant by a Rational Economic Theory? In McGuinnes (1987): 67–109.
———. 1936. *International Picture Language*. London: Kegan Paul/Trench/Trubner & Co.
———. 1936/1981. Die Entwicklung des Wiener Kreises und die Zukunft des Logischen Empirismus. In *Otto Neurath: Gesammelte philosophische und methodologische Schriften.* Band 2, ed. Rudolf Haller and Heiner Rutte, 673–702. Wien: Verlag Hölder-Pichler-Tempsky.
———. 1938a. Encyclopaedism as a Pedagogical Aim: A Danish Approach. *Philosophy of Science* 5 (4): 484–492.
———. 1938b. Unified Science as Encyclopedic Integration. In *Encyclopedia and Unified Science*, 1–27. Chicago: University of Chicago Press.
———. 1939. *Modern Man in the Making*. New York: Alfred A. Knopf.
———. 1939/1983. The Social Sciences and Unified Science. In Neurath and Cohen (1983): 209–212.
———. 1941/1983. Universal Jargon and Terminology. In Neurath and Cohen (1983): 213–229.
———. 1942/1973. International Planning for Freedom. In Neurath and Cohen (1973): 422–440.
———. 1944. *Foundations of the Social Sciences*. Chicago: Chicago University Press.
———. 1945/2004. Alternatives to Market Competition [Review of Hayek's *The Road to Serfdom*]. In *Otto Neurath: Economic Writings. Selections 1904–1945*, ed. Thomas Uebel and Robert S. Cohen, 546–548. Dordrecht: Kluwer.
———. 1996. Visual Education: Humanisation versus Popularisation. In *Encyclopedia and Utopia: The Life and Work of Otto Neurath (1882–1945)*, ed. Elisabeth Nemeth and Friedrich Stadler, 245–335. Dordrecht: Kluwer.

Neurath, Otto, and Olga Hahn. 1909a. Zum Dualismus in der Logik. *Archiv für Philosophie, 2. Abteilung (1909). Archiv für systematische Philosophie* 15: 149–162 Translated in the present volume.

———. 1909b. Zum Dualismus in der Logik. *Archiv für Philosophie, 2. Abteilung (1909). Archiv für systematische Philosophie* 15: 149–162 Translated in the present volume.

Neurath, Otto, and H.E. Kleinschmidt. 1939. *Health Education by Isotype*. New York: American Public Health Association.

Neurath, Otto, and Joseph A. Lauwerys. 1944. Nazi Text-books and the Future. *The Journal of Education* 76 (904): 521–522, and No. 905: 574–576.

———. 1945. Plato's Republic and German Education. *The Journal of Education* 77 (907): 57–59; (910): 222, 224; (913): 394.

Neurath, Paul. 2005. *The Society of Terror: Inside the Dachau and Buchenwald Concentration Camps*, ed Nico Stehr and Christian Fleck. Paradigm Publishers.

Nikolajewsky, B. 1934. *Aseff the Spy: Russian Terrorist and Police Stool*. New York: Garden City.

Padovani, Flavia. 2013. Genidentity and Topology of Time: Kurt Lewin and Hans Reichenbach. In *The Berlin Group and the Philosophy of Logical Empiricism*, ed. Nikolay Milkov and Volker Peckhaus, 97–122. Dordrecht: Springer.

Pares, Bernard. 1941. *Russia*. London: A. Lane, Penguin Books.

Pearson, Karl. 1892. *The Grammar of Science*. London: Walter Scott.

Popper-Lynkeus, Josef. 1912. *Die allgemeine Nährpflicht als Lösung der sozialen Frage, eingehend bearbeitet und statistisch durchgerechnet. Mit einem Nachweis der theoretischen und praktischen Wertlosigkeit der Wirtschaftslehre*. Dresden: Reissner.

Popper, Karl. 1935. *Logik der Forschung*. Wien: Springer.

———. 1935/2002. The Logic of Scientific Discovery. London/New York: Routledge.

Rathenau, Walter. 1919/1921. *The New Society*. New York: Harcourt, Brace and Co..

Raven, Diederick. 2003. Edgar Zilsel in America. In *Logical Empiricism in North America*, ed. Gary Hardcastle and Alan W. Richardson, 129–148. Minneapolis: University of Minnesota Press.

Reichenbach, Hans. 1938. *Experience and Prediction. An Analysis of the Foundations and the Structure of Knowledge*. Chicago: University of Chicago Press.

———. 1940. On the Justification of Induction. *Journal of Philosophy* 37: 97–103.

Reisch, George. 2001. Against a third dogma of logical empiricism: Otto Neurath and 'unpredictability in principle'. *International Studies in the Philosophy of Science* 15 (2): 199–209.

———. 2003. Disunity in the *International Encyclopedia of Unified Science*. In *Logical Empiricism in North America*, ed. Gary Hardcastle and Alan W. Richardson, 197–215. Minneapolis: University of Minnesota Press.

———. 2005. *How the Cold War Transformed Philosophy of Science: To the Icy Slopes of Logic*. New York: Cambridge University Press.

Rougier, Louis. 1921. *Philosophy and the new physics; an essay on the relativity theory and the theory of quanta*. London: Routledge.

Runes, Dagobert D. 1942. *The Dictionary of Philosophy*. New York: Philosophical Library.

Russell, Bertrand. 1941. *An Inquiry into Meaning and Truth*. New York: W. W. Nortonpany.

Russell, Philipps. 1926. *Benjamin Franklin: The First Civilized American*. New York: Brentano's Publisher.

Sandner, Günther. 2011. The German Climate and Its Opposite: Otto Neurath in England, 1940–45. In *Political Exile and Exile Politics in Britain after 1933*, ed. A. Grenville and A. Reiter, 67–85. Amsterdam & New York: Rodopi.

———. 2014. *Otto Neurath: Eine politische Biographie*. Wien: Zsolnay.

Schlick, Moritz. 1934/1979. On the Foundations of Knowledge. In *Moritz Schlick: Philosophical Papers. Volume II [1924–1936]*, ed. Henk Mulder and Barbara F.B. van de Velde-Schlick, 370–387. Dordrecht: D. Reidel.

———. 1935/1979. Facts and Propositions. In *Moritz Schlick: Philosophical Papers. Volume II [1924–1936]*, ed. Henk Mulder and Barbara F.B. van de Velde-Schlick, 400–404. Dordrecht: D. Reidel.

———. 1936/1979. The Vienna School and Traditional Philosophy. In *Moritz Schlick: Philosophical Papers. Volume II [1924–1936]*, ed. Henk Mulder and Barbara F.B. van de Velde-Schlick, 491–498. Dordrecht: D. Reidel.

Schröder, Ernst. 1890–1905. *Vorlesungen über die Algebra der Logik*. Leipzig: B.G. Teubner.

Sheldon, William H. 1942. *The varieties of temperament: A psychology of constitutional differences*. New York/London: Harper & Brothers Publishers.

Smyth, Albert H., ed. 1907. *The Writings of Benjamin Franklin. Vol. VII. 1777–1779*. New York: The Macmillan Company.

Stadler, Friedrich. 2001/2015. *The Vienna Circle: Studies in the Origins, Development, and Influence of Logical Empiricism*. 2nd ed. Dordrecht: Springer.

Stebbing, Susan. 1933. Logical Positivism and Analysis. *Proceedings of the British Academy*: 53–87.

———. 1943. *A Modern Elementary Logic*. London: Methuen.

Stebbing, Susan, Leonard J. Russell, and A. E. Heath. 1934. Symposium: Communication and Verification. *Proceedings of the Aristotelian Society, Supplementary Volumes* 13: 159–202.

Strauss, Martin. 1972. *Modern Physics and its Philosophy: Selected Papers in the Logic, History, and Philosophy of Science*. Dordrecht: D. Reidel.

Struik, Dirk. Jan. 1948. *A Concise History of Mathematics*. New York: Dover.

Thurm-Nemeth, Volker, ed. 1998. *Konstruktion zwischen Werkbund und Bauhaus. Wissenschaft – Architektur – Wiener Kreis*. Wien: Hölder, Pichler, Tempsky.

Tinbergen. Jan. 1936. Besprechung, Otto Neurath: Was bedeutet rationale Wirtschaftsbetrachtung. *Erkenntnis* 6: 70–71.

Tuboly, Adam Tamas. 2017. Philipp Frank's decline and the crisis of logical empiricism. *Studies in East European Thought* 69 (3): 257–276.

———. 2019. The constitution of geistige Gegenstände in Carnap's Aufbau and the importance of Hans Freyer. In *Logical Empiricism, Life Reform, and the German Youth Movement*, ed. Christian Damböck, Günther Sandner, and Meike Werner. Cham: Springer (Forthcoming).

Uebel, Thomas. 1993. Wilhelm Neurath's Opposition to "Materialist" Darwinism. In *Scientific Philosophy: Origins and Development*, ed. Friedrich Stadler, 209–228. Dordrecht: Kluwer.

———. 1995a. Otto Neurath's Idealist Inheritance: 'The Social and Economic Thought of Wilhelm Neurath'. *Synthese* 103 (1): 87–121.

———. 1995b. Physicalism in Wittgenstein and the Vienna Circle. In *Physics, Philosophy, and the Scientific Community: Essays in the Philosophy of Science and Mathematics in Honor of Robert S. Cohen*, ed. Gavroglu Kostas, J.J. Stachel, and Marx W. Wartofsky, 327–356. Dordrecht: Kluwer.

———. 2003. On the Austrian Roots of Logical Empiricism: The First Vienna Circle. In *Logical Empiricism. Historical and Contemporary Perspectives*, ed. Paolo Parrini, Wesley C. Salmon, and Merrilee H. Salmon, 67–93. Pittsburgh: University of Pittsburgh Press.

———. 2007. Otto Neurath as an Austrian Economist: Behind the Scenes of the Early Socialist Calculation Debate. In *Otto Neurath's Economics in Context*, ed. Elisabeth Nemeth, Stefan W. Schmitz, and Thomas Uebel, 37–59. Dordrecht: Springer.

———. 2008. Writing a Revolution: On the Production and Early Reception of the Vienna Circle's Manifesto. *Perspectives on Science* 1: 71–87.

Venn, John. *Symbolic Logic*, 1881. London: Macmillan and Company.

Voß, Johann Heinrich. 1802/1972. Der siebzigste Geburtstag. In *Voß. Werke in einem Band*. Berlin: Aufbau-Verlag.

von Mises, Richard. 1951. *Positivism: A Study in Human Understanding*. Cambridge, MA: Harvard University Press.

Waismann, Friedrich. 1965. *The Principles of Linguistic Philosophy*, ed. Rom Harré. London/Melbourne/Toronto: Macmillan.

———. 1979. Theses. In *Wittgenstein and the Vienna Circle. Conversations Recorded by Friedrich Waismann*, ed. Brian McGuinness, 233–261. Oxford: Blackwell.

Weber, Max. 2005. *The Protestant Ethic and the Spirit of Capitalism*. Trans. Talcott Parsons. London/New York: Routledge.

Weinberg, Julius. 1936. *An Examination of Logical Positivism*. London: Routledge and Kegan Paul.

———. 1964. *A Short History of Medieval Philosophy*. Princeton: Princeton University Press.

Weyl, Hermann. 1925/1998. On the Current Epistemological Situation in Mathematics. In *From Brouwer to Hilbert: The Debate on the Foundations of Mathematics in the 1920s*, ed. Paolo Mancosu, 123–142. Oxford: Clarendon Press.

Wittgenstein, Ludwig. 1922. *Tractatus Logico-Philosophicus*. London: Kegan Paul, Trench, Trubner & Co..

Woodger, Joseph H. 1939. *The Technique of Theory Construction*. Chicago: University of Chicago Press.

Wundheiler, Alexander, and Edward Poznański. 1934/2017. The Concept of Truth in Physics. In *The Significance of the Lvov-Warsaw School in the European Culture*, ed. Anna Brożek, Friedrich Stadler, and Jan Woleński, 309–346. Cham: Springer.

Ziche, Paul, and Thomas Müller. 2013. Paul Oppenheim on Order – The Career of a Logico-Philosophical Concept. In *The Berlin Group and the Philosophy of Logical Empiricism*, ed. Nikolay Milkov and Volker Peckhaus, 265–290. Dordrecht: Springer.

Index

A
Absolutism, 549, 576–578, 592, 615
in Carnap, 544
merciless and metaphysics, 592
Accepting statements, 547–549
pragmatical concept of, 562
translation of true-statements, 550, 562
Adler, F., 50, 55, 57
Adler, M., 571
Adler, V., 50, 56
Adorno, T., 61, 198
Aesthetical view of world, 593
Aggregation, 547, 549, 550, 552, 557, 562, 604, 641
Akehurst, T.L., 109, 110
Albers, J., 430
Alcibiades, 590
Aquinas, T., 571
Aristotelian metaphysics, 582, 630
anti-pluralism, 597, 598
Carnap's semantics as, 570, 581
Aristotle, 118, 122, 244, 269, 272, 445, 566, 571, 597
Armstrong, D., 377–379
Arntz, G., 26, 134, 429, 457, 464, 472, 475
A.S.C., 4
Asymmetry
in testing and induction, 594, 611
Atomic bomb, 652
Aufbau and Geisteswissenschaften, 630
Augustine, St., 593
Austrian Association for Settlement and Allotment [*Österreichischer Verband für Siedlungs- und Kleingartenwesen*], 47, 76
Austrian Institute for Pictorial Statistics, 26
Avenarius, F., 69, 72, 75
Avenarius, R., 56, 69, 546, 549, 566, 571
Ayer, A.J., 90, 111, 441, 453, 539
Azef, Y. [Aseff], 540

B
Babbage, C., 245, 263
Bacon, F., 49
Bain, A., 263
Ballod-Atlanticus, 553
Barber, B., 403
Barnard, E., 388, 391
Barnes, B., 49
Basic English, 545, 561, 565, 566, 625
Bauer, H., 121, 202
Bauer, O., 25, 57, 75, 438, 482
Bauermeister, F., 158
Bauhaus, 27, 109, 147, 345, 414, 429, 430, 464, 465, 472, 569
Bäumer, G., 589
Baynes, T., 245
Behmann, H., 328
Benson, T., 568
Bentham, G., 245
Bentham, J., 97, 588, 590
Bentley, N., 568
Bernal, J.D., 386, 388, 398, 404
Bernardino, St., 593
Bernays, P., 328
Bernstein, J., 403
Besteiro, J., 46
Betty, A., 568
Beveridge, W., 426

J. Cat, A. T. Tuboly (eds.), *Neurath Reconsidered*,
Boston Studies in the Philosophy and History of Science 336,
https://doi.org/10.1007/978-3-030-02128-3

Biddle, J., 219, 220
Biot, J.-B., 230, 231
Black, M., 90
Blau, E., 118, 143, 152, 159, 162, 169, 170, 172, 414
Bloch, E., 447
Bloomfield, L., 628, 633, 662
Bloor, D., 49
Böckh, A., 275
Boltzmann, L., 270–272, 325, 545
Bolzano, B., 269, 271
Boole, G., 243, 245–248, 251, 253, 255–257, 260–263, 265, 266, 268, 271, 272, 281, 282, 290, 312, 317
Bradley, P., 94, 379
Braithwaite, R., 90, 603
Brandom, R., 339–341, 355–358
Brentano, F., 267–269, 271, 272, 329, 571, 598
Brewster, D., 230, 231, 233
Bridgman, P.W., 610
Brouwer, L.E.J., 312, 649
Brunswik, E., 94, 541, 558
Büchner, L., 55
Bühler, K., 99, 321
Bukharin, N., 388
Bunsen, R., 251
Burke, C., 9, 10, 32, 33, 118, 130, 197, 429, 458, 461, 465
Bush, V., 14, 394–401, 404, 408

C
Callwey, G.D.W., 72
Calvin, J., 658
Camus, A., 480
Cantor, G., 243, 616
Cardenas, L., 528
Carnap, I., 525, 531, 538, 563, 585, 607, 614, 623, 631, 645, 649, 653, 675, 681
  Austrian traditions of, 669
Carnap, R., 1, 28, 49, 67, 89, 123, 179, 198, 223, 244, 339, 363, 386, 422, 453, 472
  on auxiliary languages, 561
  on Cambridge discussion groups, 528
  not discussing Russell, 608
  on finitism, 610
  general opinion on Russell's book, 578
  on negativistic exaggerations, 612
  on Neurath grieving him, 646
  on Neurath's monograph, 613, 624, 625, 632, 633, 635, 636
  Neurath's Plato-paper, 634
  on Neurath on Russell, 539, 562
  on Neurath's role in the movement, 647, 648
  on Neurath the Volcanic, 636
  on pluralism, 610
  on Popper, 573, 611, 612
  quoting Aristotle, 597
  on Russell's empiricism, 539
  on semantics, 574, 579, 581, 608, 609
  reply to Neurath on Russell, 575
  suggesting Sheldon to Neurath, 634
  teaching elementary math., 575
  truth and verification, 539
Cartwright, E.S., 96
Cartwright, N., 7, 10, 17, 29, 38, 48, 62, 68, 91, 117, 119, 166, 206, 219, 223, 242, 297, 299, 323, 345, 364, 375, 385, 417, 432, 438, 456, 459
Castillon, F. de, 245
Categorical imperative, 592
Cayle, A., 243, 256
Central decisions, 555
Cezanne, P., 474, 484
Chain, E., 399
Chamberlain, H.S., 615
Chappell, C., 416
Chase, S., 545
Chasles, M., 257
Christiansen, B., 615
Church, A., 243, 288
Churchill, W., 556
CIAM, 27
Cicero, M.T., 120, 275
Clark, J., 430
Cohen, H., 589
Cohen, M.R., 39
Cohen, R.S., 2, 5, 8, 55, 68, 127, 218, 273, 274, 276–278
Cole, G.D.H., 76, 77, 94, 96, 536, 554
Colmar, A., 263
Comenius, A., 27, 31, 33, 459
Comte, A., 49, 60, 445, 570, 643
Conant, J.B., 14, 386, 394–398, 400–402, 404, 408, 409
Condillac, É.B. de, 545
Cornford, F.M., 436
Correspondence theory, 549
Couturat, L., 244, 247–249, 257, 258, 276, 288, 312, 313
Creath, R., 5, 99, 342, 363, 378
Crossman, R.H.S., 436, 590
Crowther, J.G., 388, 404
Crusoe, R., 542

D
Dante, A., 593
Darlan, F., 583
Darwin, C., 261, 272, 594, 612
Davidson, D., 3, 380
Dedekind, R., 243, 595
Degree of confirmation, 540, 550, 577, 578, 581, 616, 624, 626, 627, 634
Dekker, E., 108
Democracy, 540, 555, 556, 593, 672, 684
  compatibility with planning, 544, 563
  happiness and muddle, 671
  inefficiency of, 554
  in the long run, 540
  majority vote, 555
  vs. planning, 563
Democritus, 436
De Morgan, A., 243, 245–248, 262, 263, 312
Derrida, J., 484
Descartes, R., 52, 54, 218, 221, 222, 231, 244, 250, 261, 262, 297, 307, 356, 373, 442
Dewey, J., 94, 105, 307, 394, 475, 477
Diderot, D., 545
Dietzgen, J., 59
Differential calculus, 595, 603
  Kiepert-Stegemann textbook, 595
Dilthey, W., 104, 105, 286
Dollfuss, E., 48
Douglas, H., 62, 396
Dualism
  of mental and non-mental, 547
  unified language without, 602
Dubislav, W., 286
Duhem, P., 52, 54, 57, 224, 226, 234, 295, 298, 300, 316, 325, 368, 371, 566, 593, 598, 600, 620, 624
  absolutism, 598
  Scholasticism, 598
Duplication, 546, 549, 571, 643
Dupré, J., 7
Dürr, K., 648
Dvorak, J., 6

E
Eddington, A.S., 455
Ehrenfels, C. von, 268
Ehrenhaft, F., 618
Eibl, H., 639
Einstein, A., 9, 25, 51, 56, 94, 277, 320, 325, 391
Eisenmenger, V., 671
Elliott, K., 62
Ellis, J., 256
Empiricism, 550, 615, 630
  attitude, 573
  less merciless, 592
  movement of, 574, 636, 638
  pluralism of, 592, 593, 597
  and truth, 549
Encyclopedia
  Bibliography and Index, 558
  editorial rights, 633
  first series of monographs, 558
  Foundations of Language Making, 545, 558
  second series of monographs, 558
Encyclopedism, 602
Engels, F., 49, 50, 55, 59
English as auxiliary language, 565
Epicure, 590
  Stoicism, 637, 645
Epicurus, 80, 436
Erdmann, B., 286
Ermers, M., 158, 159, 163, 164
Ernst Mach Society [Verein Ernst Mach], 28, 37, 48, 99, 182, 310, 315
Esperanto, 565, 566, 614, 626
Eulenburg, F., 70
Euler, L., 229, 231–233, 235
Eve, M., 9, 465
Experimentum crucis, 566, 600, 621
Extra-linguistic entities, 610

F
Facetti, G., 465
Faraday, M., 396, 567
Fechner, G., 119, 265, 293, 297, 326
Feigl, H., 97, 98, 104, 107–109, 347, 477, 528, 541, 559, 584, 613, 622, 629, 640, 648, 678, 682
  for IEUS, 544, 557, 558
  paper on logical empiricism, 664, 677–679
  relation to Neurath, 663, 664, 678
Ferdinand, F., 671
Feuer, L., 403
Feuerbach, L., 55
Fichte, J.G., 35
Field, C.G., 309, 436, 441
Finitism, 603, 610, 616
Finlay-Freundlich, E., 530, 541, 558, 560, 606, 607, 613, 614
Fischer, A., 26
Fite, W., 436, 590
Fitzmaurice, R., 420

Fleck, L., 6, 7, 10, 17, 29, 38, 48, 62, 68, 91, 117, 119, 166, 206, 219, 223, 297, 299, 323, 345, 364, 375, 385, 387, 417, 432, 438, 456, 459
Fleming, A., 399
Flexner, A., 396, 408
Flitner, W., 652
Florey, H., 399
Foges, W., 459, 461, 462, 464–467
Formalization of Logic, 540, 543, 580
Foschungsinstitut für Gemeinwirtschaft, 76
Fraenkel, A., 552
Fraenkel, W., 541
Frank, H., 533, 541
Frank, J., 26, 151, 153–155, 157, 163, 172, 414, 428, 472
Frank, J. [Pepi], 527, 533, 560, 605
Frank, P., 2, 7, 25, 50, 54, 56, 62, 67, 91, 98, 101, 102, 147, 274, 295, 348, 387, 401, 402, 404, 405, 407, 472, 528, 531, 532, 541, 553, 563, 569, 600, 605, 608, 614, 615, 620, 622, 623, 630, 639, 640, 644, 650, 660, 666, 677, 681
  accident, 649
  chapter for Finlay-Freundlich, 541, 558, 560
  IEUS monograph, 530, 534, 541
Franklin, B., 586–588, 590, 608
Frazer, J., 96
Frege, G., 243, 248–250, 253, 259–261, 264, 271, 280, 282, 283, 285, 307, 311–315, 317, 322, 323
Fresnel, A.-J., 230, 231, 233
Freud, S., 645
Freudenthal, G., 276, 277, 442, 445
Freundlich, E., 147
Friedlander, P., 439
Fuller, S., 409
Functionalism in anthropology, 623

G
Galilei, G, 272
Galton, F., 278, 439
Gandhi, Mahatma, 80
Garforth, F.W., 441
Gätschenberger, R., 313, 321
Gaunt, W., 420
Gemeinwirtschaftliche Siedlungs- und Baustoffanstalt (GESIBA), 150, 155, 156, 161–164, 172
Genidentity, 548
George, H., 291
Gerard, R., 524, 525
Gergonne, J.D., 257
German Nationalism, 569
Glöckel, O., 26, 79, 439
Gödel, K., 320, 365, 374, 597, 611
Goethe, J.W. von, 225, 236, 439, 593, 673, 683
Goldschmidt, V., 437, 439
Goldstein, K., 483
Gomperz, H., 575, 603
Goodman, N., 485, 486
Gramm, M., 604
Grassmann, R., 243, 247, 252–254
Grattan-Guinness, I., 249, 250, 252, 256, 257, 259
Gregory, D., 245
Grelling, K., 102, 527, 531, 534, 543, 569, 570, 645, 651, 668
  deported by Gestapo, 563
  has no visa, 563
  in concentration camp, 534
Grimm, H., 615
Grossmann, H.G., 252, 257
Günther, H.F.G., 439, 442

H
Hack, H., 92
Hahn, H., 16, 25, 28, 55, 98, 99, 101, 179, 184, 198, 223, 274, 280, 295, 314–317, 348, 386, 473, 630, 639, 640, 647
  Einheiz-Wissenschaft, 640
  supporting Eibl, 639
  supporting Innitzer, 639
Hahn-Neurath, O., 99, 272–278, 288, 294, 295, 647
Haller, R., 5, 62, 68, 118, 178, 179, 226, 288, 321, 475
Hamilton, W., 245
Harrap, G.G., 430, 459, 465
Harrisson, T., 424
Hartley, D., 263
Hayek, F.A., 8, 35–40, 199, 204–206, 211, 680
  attacking logical empiricism, 667, 680
  helping Popper, 667
Hazebroek, P., 528, 685
Heartfield, J., 465
Hegel, G.W.F., 35, 104, 267, 423, 444, 549
Heidegger, M., 16, 54, 322, 479–486
Heijenoort, J. van, 248, 249, 260, 340
Heine, H., 605
Heinemann, F., 589, 604
Helmer, O., 537, 539

Helmholtz, H.L. von, 267, 286, 295, 316
Hempel, C.G., 2, 18, 90, 97, 102, 107, 109, 307, 308, 329, 343, 347, 534, 541, 563, 569, 585, 594, 597, 600, 607, 612, 614, 624, 626, 627, 641, 642, 648, 649, 685
  on confirmation, 563
  explaining semantics to Neurath, 597
  for IEUS, 557, 558
  on laws of history, 642
  on social science, 567
Hempel, E., 563, 569
  death of, 607, 614
Hepner, B., 403
Herschel, J., 245, 265, 300
Herter, H., 439
Hertz, H., 224
Hesse, O., 251
Hessen, B., 388, 404
Heyting, A., 312
Highmore, B., 427, 428
Hilbert, D., 243, 250, 257, 283, 312, 315, 320
Hill, A.V., 96
Hinde, J., 430, 431
Hintikka, J., 3, 323, 340, 342
Hitler, A., 11, 41, 394, 401, 440, 441, 475, 546, 552, 564, 569, 572, 584, 589, 615, 623, 645, 662
  is formally legal, 555
Hoffmann, E., 439
Höfler, A., 268–272
Hogben, L., 388, 466, 545, 546, 561, 623, 625, 627
  for biology monograph, 613, 614
  for IEUS, 545, 561
Hollitscher, W., 534
Hook, S., 475, 480
Hoover, J.E., 395
Horkheimer, M., 30, 61, 198
Hosiasson-Lindenbaum, Janina, 533, 651, 667, 680
Hume, D., 58, 96
Hunt, J.K., 430
Husserl, E., 249, 259, 262, 276
Huxley, J., 94, 96
Huygens, C., 232, 233

I

Ideal state, 589
Ido, 565, 626
Inductive logic, 626, 627
Innitzer, T., 639
Institute of Social History, 554
Institute for Social Research, 473
Institute for the Unity of Science, 2, 5, 401, 403
Institute for the Unity of Science in The Hague, 28
Institute for visual education, 551, 622–623
Interglossa, 545, 561, 566
International Encyclopedia of Unified Science (IEUS), 3, 5, 7, 16, 17, 28, 37, 49, 60, 90, 91, 96, 104, 388, 389, 454, 473–478, 481, 482, 485
International Institute of Social History, 94
International Planning for Freedom, 540, 544, 551
Introduction to Symbolic Logic
  to be published in Library, 539, 543, 550
  translation of Abriss, 539
Introduction to Semantics, 540, 543, 553, 560–562, 570, 572, 574, 580
  to be published by Harvard, 533
  to be published in Library, 530, 533
Isle of Man, 542
Isotype, 540, 569, 605, 645, 652
  animated diagrams, 553, 560
  Blood Transfusion, 535
  collaborator in London, 584
  with Esperanto, 565
  film, 535, 542, 544, 551, 560, 572
  Institute, 564, 574, 685
  international education, 583
Isotype Institute, 10, 15, 28, 48, 95, 106, 118, 185, 414, 429, 430, 452, 454, 460, 464–466
Itelson, G., 120, 276, 277, 288, 295, 312, 585, 594, 682

J

Jäckh, E., 94
Jaeger, W., 439
James, W., 247, 643, 664
  Bergsonianism, 643, 678
Jeans, J., 455
Jevons, W.S., 14, 243, 245, 246, 248, 251, 255, 260–266, 268, 271, 289, 291–296, 299–304, 306, 307, 312, 313
Joad, C.E.M., 96, 111, 436, 441
Jodl, F., 56, 269, 270, 272, 274, 295
Johnson, A., 537, 563
  committe for refugee scholars, 526
  Johnson Action, 532, 533, 537
Johnston, W., 24, 269, 273
Jörgensen, J., 101, 102, 327, 524, 527, 533, 541, 579, 673

Joseph, F., 657
Joseph, H., 531
The Journal of Unified Science, 530, 538, 539, 541, 546, 550, 557, 570, 580

K
Kaempffert, G., 24
Kaempffert, W., 14, 94, 386–388, 390, 391, 394–398, 400, 401, 404–406, 526, 557, 560, 584, 585, 604
Kahn, F., 459
Kaila, E., 524
Kallen, H., 60, 111, 445, 447
Kampffmeyer, H., 150, 152, 155, 157, 162, 173
Kant, I., 33, 37, 423, 444, 589, 590, 657
Kantorovich, V., 205
Kaplan, A., 543, 550, 558, 627
 for Library, 558
Kapp, K.W., 209
Kaufmann, E.J., 94
Kaufmann, F., 94, 98, 109, 315, 526, 532, 610, 630, 683–685
Kautsky, K., 119, 199
Kelsen, H., 532, 537, 539, 590
Kepler, J., 594
Kilgore, H., 394–398, 400, 404
Kindel, E., 10, 118, 127, 197, 413, 429, 465
Kirchhoff, G.R., 251, 346, 622
Kirk, R.C., 428
Kitcher, P., 49, 61, 224
Knopf, A.A., 94, 455, 460
Knox, J., 658
Köhler, E., 5, 242, 281, 282, 284, 288
Kokoszyńska-Lutman, M., 533
Kolbenheyer, E.G., 591
Kotarbiński, T., 329, 527, 533, 571, 597, 598, 652
Kraeutler, H., 10, 118, 144, 185, 197
Kranold, C., 559, 560, 563, 568
Kranold, H., 71, 74, 559
Kruger, G., 71, 439
Kuhn, T., 15, 49, 355, 387, 402, 404, 405, 407, 408, 485
Külpe, O., 321
Kürnbergerin, F., 588

L
Ladd-Franklin, C., 248, 280
Lafitte, F., 425
Lagarde, P.A. de, 615
Laing, G.J., 94
Laird, J., 536, 663, 678
Lalande, A., 312
Lambert, J.H., 245, 252, 261, 280, 317, 319, 322
Langbehn, A.J., 615
Lange, O., 205, 210
Laplace's Demon, 601, 617
Laski, H., 648, 667, 680
Lauwerys, J.A., 16, 47, 96, 110, 328, 423, 437, 440, 441, 444
Leibniz, G.W., 13, 31, 243–245, 247–249, 251, 252, 256, 259–263, 266, 269, 272, 277, 280, 286, 293, 300, 305, 307, 309, 311, 312, 314, 315, 317, 319, 321, 326
Leichter, K., 202
Leichter, O., 203
Lejeune-Dirichlet, P.G., 252
Lenin, V.I., 49, 50, 55
L-equivalent, 610
Levinson, R., 436, 444
Lévy-Bruhl, L., 96
Lewis, C.I., 242, 244–249, 257
Library of Unified Science, 524, 530, 533, 538, 539, 543, 546, 550, 557, 558
Lichtenberg, Georg Christoph, 586, 590
Liebknecht, K., 482
Lihotzky, M., 150, 152, 157, 160, 161, 163
Lindenbaum, A., 328, 651, 667, 680
Littrow, J., 221
Llull, R., 244, 245
Locke, J., 319, 345
Lodge, O.J., 552
Logical *vs.* psychological problems, 578, 579
Logical Syntax of Language, 572, 602, 623
Loos, A., 151, 152, 157, 163, 430
Lord Glenconner, 462
Lord Kelvin [William Thomson], 567
Lowell Institute, 406
Lowell, R., 405
Löwenheim, L., 249
Lukács, G., 47, 103
Łukasiewicz, J., 571, 673
 Thomism, 598, 623
Lundberg, G.A., 109
Luxemburg, R., 482
Lynd, R.S., 605

M
Mach, E., 24, 56, 57, 80, 136, 221, 227, 230, 236, 267–271, 274, 295, 300, 316, 318, 325, 345, 545, 566, 593, 594
 Mach school, 622
Mach, L., 80

Madame Brillon, 587
Magnuson, W., 399
Malisoff, W., 386
Malthus, T., 594
Malus, É.L., 224, 230, 231
Mannheim, K., 12, 39, 90, 96, 98–100, 102–104, 106–108, 251
Manninen, J., 107
Marcuse, H., 438
Marlow, F., 24, 28
Marsden, L., 93, 94
Marx, K., 35, 36, 73, 199, 202, 236, 436, 444, 567, 594, 617
  conception of history, 609
Masaryk, T., 46
Maurenbrecher, M., 481
Maxwell, J.C., 257, 370, 567, 594
Meiner, F., 527, 528
Meinong, A., 268, 270, 312
Menger, C., 38, 98, 290
Menger, K., 198, 283, 611
Meredith, P., 96
Merton, R., 103, 402
Metaphysics, 562, 566, 579, 581, 590–592, 597, 609, 612, 630, 657
Mettrie, J.O. de la, 55
Meyer, A., 163
Meyer, E., 24, 119, 274, 275, 438
Meyer, H., 109
Meyer, W.H., 403
Michael, L., 161
Migerka, F., 273
Migge, L., 157
Mill, J.S., 73, 263, 265, 268, 300, 307, 588, 590
Ministry of Information, 544, 553, 560
Mises, L. von, 8, 13, 38, 197–199, 201–203, 205–210, 212
Modal logic, modalities, 540
Modern Man in the Making, 605
Modley, R., 388, 391, 458
Moellendorff, W. von, 71
Moholy-Nagy, L., 551
Moleschott, J., 55
Moltke, H. von, 556
Montaigne, M. de, 590
Moore, G.E., 97, 453, 577, 601
Moritz, K., 593
Mormann, T., 219, 220, 222, 311, 323, 324, 340–343, 345, 346, 349, 353, 354, 375
Morris, C., 2, 5, 7, 28, 49, 60, 90, 94, 104, 327, 329, 347, 348, 394, 474, 477, 486, 524, 532–534, 537, 541, 543, 545, 550, 551, 557, 558, 560, 561, 563, 568, 569, 582, 583, 596, 604, 613, 622–625, 627–629, 633, 634, 637, 638, 641, 643, 644, 646, 650, 651, 659, 660, 662, 666, 675, 676
  Paths of Life, 605, 645
Morrison, H., 528
Muddle, 540, 544, 551, 555–557, 563, 671, 672
Mulder, H., 34, 80
Müller, A., 142, 148, 150, 154, 155, 158, 160, 162, 173
Müller, E., 251
Müller, K., 6
Müller, L., 272
Müller, M., 261
Müller, O., 225
Multiplicity of statements, 547
Mundaneum, 26, 48
Museum for Settlement and City Planning, 25
Museum on War Economy, 25, 47, 71, 74, 143, 146
Musil, R., 29, 354

N

Naess, A., 524, 527, 579, 581, 668, 673, 680
  IEUS monograph, 558
Nagel, E., 2, 94, 99, 100, 103, 402, 473, 475, 480, 533, 541, 561, 562, 564, 581, 584, 600, 602, 615, 634, 638, 648, 649
  against semantics, 580
  on Neurath's monograph, 633, 634
  review of Carnap, 632, 634
  review of Popper, 574
  review of Russell, 562, 578
Naïve realism, 576
Natorp, P., 286
Nazi arguments, 615
Needham, J., 388
Neider, H., 661, 677
Nemeth, E., 5, 7, 8, 12, 27, 49, 62, 70, 118, 144, 148, 179, 242, 305, 414
Neo-Scholasticism, 571
Neubacher, H., 569
Neue Sachlichkeit, 414, 428, 472
Neumann, J. von, 312
Neumann, L., 158
Neumann, R., 630
Neurath, Marie, 2, 26, 91, 127, 167, 189, 274, 413, 438, 458, 475
Neurath, Marie [Mary], 593

Neurath, Marie [Mieze], 526, 528, 531, 534, 535, 551–553, 564, 574, 645, 652, 654, 658, 674
escape from Holland, 526
marriage with O. Neurath, 528, 532
Neurath, O.
on accepted *vs.* true, 597
on Austrian philosophy, 590
on auxiliary languages, 564
avoiding the term 'fact, 547
calling everyone Platonic, 658
on Carnap's anti-pluralism, 628
Carnap's existential statements, 583
on Carnap's friendship, 653, 665
on Carnap and Hogben for IEUS, 627
on Carnap's letter to Morris, 637, 638, 644, 676
on Carnap and monograph, 628, 629
on Carnap on monograph, 637, 638, 641, 643, 662, 672, 674, 675, 679
on Carnap's Platonism, 659, 670, 672
on Carnap's semantics, 570, 571, 583, 595–597, 602, 616
Carnap is transgressing empiricism, 566
contract for Visual Education, 606
on contradictions, 594, 602, 616, 617
creating Einheitswissenschaft, 631
death of, 683, 684
eliminating proposition, 599
Erkenntnis article, 629
on Feigl, 663, 664, 678
on finitism, 616
on the friendly attitude, 656, 671
German attitude preparing Nazidom, 672
on Hahn, 639
on Hempel on laws, 642
on himself, 637
hypothesis about the way to Nazism, 604
on inductive logic, 627
influence on youth, 611
on Interglossa, 626, 627
invitation by Radcliff-Brown, 554
on James, 643
lecturing in England, 536, 544
on logical positivism, 643
on logical symbols, 599
life-boat experience, 542
M and non-M while reading Tractatus, 640
method as metaphysical, 622
moving to a new house, 558, 559, 564
Oxford lectures, 535, 623
on Popper, 566, 593, 594, 617, 619–621
Plato-article, 670
reading Spinoza with Frank and Hahn, 639
on Reichenbach, 666
review of Hayek, 667, 680
on Russell, 566, 599–602
on Russell, meaning, and truth, 546–549, 600–602
role in the movement, 679
semantics as a calculus, 597
semantics in his terminology, 598, 599
on 'the snow is white, 567
on Sparta *vs.* Athens, 593
suggesting logical empiricism, 643, 678
terminology, 571
on Testability and Meaning, 600–602
Tolerance and Persecution, 529, 535, 593, 655
on truth, 616
unpredictability in principle, 567, 597, 598, 602, 617, 642, 644, 667
using semantic concepts, 609
visiting a lecture on geometry, 603
after the war concerns, 583
on Wiener Kreis pamphlet, 629, 640
on Wittgenstein, 600
writing Visual Education, 585
Neurath, P., 6, 24, 34, 274, 478, 551, 557, 560, 572, 605, 606, 632
at Columbia University, 541
fellowship at Columbia, 570
in Göteborg, 531
having two Dr. degrees, 583
Neurath, Walter., 24, 120, 275, 462, 466
Neurath, Wilhelm, 24, 33, 120, 273, 274
New Commonwealth Quarterly, 551
Neurath as editor, 542
Newton, I., 222, 224–226, 231–233, 594, 617
Nohl, H., 589, 593, 615
Nussbaum, M., 121, 122

O

Oberdan, T., 6, 363
Observation statements, 596, 616
Ogden, C., 97, 545, 602
Oppenheim, P., 109, 307, 308, 534, 563
Oppenheimer, J.R., 395
Ortega y Gasset, J., 46

P

Pacioli, L., 303
Pareto, V., 291, 294, 595
Pascal, B., 263

Peacock, G., 245
Peano, G., 243, 248–250, 257, 261, 271, 307, 312, 313, 316, 317
  axiomatization of mathematics, 564
  Latino sine flexione, 545, 564, 565
Pear, T.H., 536
  for IEUS, 558
Pearson, K., 622
Peirce, C.S., 243, 246–248, 250, 253, 256–258, 261, 267, 268, 280, 294, 307, 312, 313, 316, 319, 321, 593, 594
Pesch, H., 80
Petzoldt, J., 267–269, 277, 284, 285
Philosophical Society of the University of Vienna, 218, 269, 295
Physicalism, 679
Pick, K., 150, 158, 159
Pidgin English, 566
Pinsent, A., 96
Planck, M., 271
Planning, 568
Plato, 11, 35, 45, 110, 309, 436, 479, 589, 590, 593, 657
  anti-pluralism, 598
  educational ideal, 615
  fascism, 589, 590
  Germans quoting, 589
  Republic, 590, 615
  the rock of idealism, 590
  Symposium, 590
Plato-Hitler, 589
Platonic attitude, 657, 679, 680
  and Nazism, 657, 670
  way of life, 670
Ploucquet, G., 245
Pluralist arguing, 592
Pluriformulations, 616
Poincaré, H., 57, 224, 226, 286, 295, 300, 316, 325, 368, 593, 598, 600, 620
Polanyi, M., 386, 396
Polish-philosophy
  Scholastic, 598
Pombo, O., 7
Poncelet, J.-V., 257
Popper, K., 35, 47, 100, 242, 345, 409, 435, 476, 544, 562, 566, 567, 573, 594, 595, 597, 600, 608, 610–612, 615, 617–621, 630, 641, 663, 667, 668
  anti-pluralist, 593, 594
  anti-Vienna Circle resentment, 567
  appointment at London, 649, 680
  Kantianism, 598
  one system of the world, 594
  supporting Hayek, 680
Popper-Lynkeus, J., 24, 37, 56, 120, 446, 553
Price, H.H., 96
Propaganda
  and education, 555
  German and British, 555
Protocol statements, 547, 570, 577, 628, 679
  Schibboleth for logical empiricism, 663
Pseudo-rationalism, 598
Psychoanalysis, 609
Psychology
  behavioristic interpretation, 579
Putnam, H., 211, 377, 379, 380
Pyramidism, 630

Q

Quine, W.V.O., 3, 29, 340, 341, 368, 393, 532

R

Radcliffe-Brown, A.R., 95, 96, 554
Radhakrishnan, S., 46
Ramus, P., 303
Rand, R., 37
Rapoport, A., 403
Rathenau, W., 71, 556
Reichenbach, B., 326
Reichenbach, H., 90, 98–100, 227, 286, 326, 347, 575, 627, 663, 678
  for IEUS, 544, 558
  relation to Neurath, 666
  theory of induction, 558
  theory of probability, 558
Reidemeister, K., 9, 438
Reidemeister, M., *see* Neurath, M.
Reininger, R., 270, 295–297, 305
Renner, K., 75
Reuman, J., 161
Revising statements, 593
Ricardo, D., 229
Richards, I.A., 97
Ritter, C., 439
Ritter, J., 257
Robin, L., 439
Rockefeller, 532–534, 537, 539, 543, 652
  grant for Carnap, 607
Roh, F., 472, 484
Roman Catholic Scholasticism, 571
Romilly, G., 111
Roosevelt, F.D., 391, 395, 396
Rorty, R., 339–341, 358
Rosenberg, A., 484
Rosinger, K., 539
Rotha, P., 9, 37

Rougier, L., 527, 533, 537, 539, 558
Rousseau, J.-J., 437, 657
Runes, D.R., 108, 668, 679
Russell, B., 17, 41, 51, 97, 243, 249, 250, 259, 283, 307, 312, 314–317, 320, 325, 326, 328–330, 351, 352, 394, 437, 439, 453, 528, 539, 545–549, 562, 566, 571, 575–578, 580, 582, 583, 596, 598, 599, 601, 602, 608–610, 612, 620, 663, 678, 679
  empiricist in Schlick's sense, 534
  existence symbol, 599
  psychologism, 578
  realism and anti-pluralism, 599
  realism and metaphysics, 579
  similar to the late Schlick, 530
  on truth, 579
  truth and realism, 579
  unempiricist assumptions, 550
Russell, L., 96, 535, 663
Rutte, H., 5, 68, 178, 288
Ryan, A., 4

S

Santillana, G. de, 525, 541
Sartre, J.-P., 480
Schäffle, A., 199
Schapire[-Neurath], A., 128, 291
Schapire, R., 24
Schapiro, M., 471–485, 557, 558
Schelling, F.W.J., 257
Schleiermacher, F., 257
Schlick, M., 5, 28, 48, 80, 89, 227, 283, 344, 369, 453, 530, 546, 548, 549, 562, 571, 576, 577, 608, 610, 615, 620, 630, 632, 640, 646, 647, 660–664, 668, 676–678, 681, 682
  denomination of empiricism, 534
  humiliating Neurath, 661, 676, 677, 681, 682
  on Neurath's manuscript, 660
  on protocol statements, 598
  suggesting radical empiricism, 643, 664, 678
  treating Neurath as a schoolboy, 660
  on unified science [Paris], 640, 661, 677, 681–683
Schmitz, S.W., 8, 118
Schmoller, G., 119, 274, 275, 290
Schnitzler, A., 568
Scholasticism, 623
Scholz, H., 328, 673
Schröder, E., 14, 218, 242, 552, 553
Schumann, W., 70–75, 77, 81, 200, 202, 204
Science of Science group, 532
Seidel, M., 102–104
Sekera, F., 153
Sellars, W., 339–341
Semantics
  as calculus, 608, 616
  as language of science, 608
  translatable into empiricism, 571
  usefulness of, 580, 608, 609
Sen, A., 121–123, 211
Sheldon, W., 634, 645
Sigwart, C. von, 267
Simon, O., 614
Simony, O., 276
Sismondi, J.C.L. de, 229
Smee, A., 263
Smith, A., 122, 229, 266, 594
Snow, C.P., 31
Social and Economic Museum [*Gesellschafts- und Wirtschaftsmuseum*], 25, 143, 182, 188, 391, 413, 429, 431, 457
Social sciences, 544, 567, 617, 635
Society for the protection of Science and Learning, 531
Socrates, 590
Solipsism, 549
Soule, R., 426
Soviet Union, 587, 604
Speak about language, 611
Spencer, H., 445
Spengler, O., 41, 417, 439, 615
Spinoza, B., 639
Stadler, F., 5, 7, 10, 11, 28, 38, 48, 49, 62, 69, 98, 117, 126, 134, 169, 179, 242, 268–270, 310, 312, 315, 385, 414, 471
Stanford, K., 227
Stebbing, S., 15, 37, 89, 90, 92–97, 244, 326, 328, 330, 331, 452–468, 529, 534, 569, 599, 614, 622
  death of, 584, 607
  illness, 541, 551
Stefan, J., 271
Stein, L., 286
Steiner, P., 459, 462, 464, 466, 467
Stenzel, J., 439
Stevens, S.S., 296
Stöhr, A., 267, 269–272, 279, 280
Stöltzner, M., 219–222, 228, 298
Stolz, O., 595
Strauss, M., 96, 532, 536, 537, 551, 579, 600, 685
  on Carnap's semantics, 572, 597, 616
  looking for a job, 531

Struik, D.J., 528
Studies in Semantics, 524
Stumpf, D., 7, 268
Swanson, G., 425
Symons, J., 7
Systematization, 557, 558, 594, 611, 617
  disease, 616
  islands of, 600
Székely, D.L., 5

T
Tarski, A., 107, 321, 328, 329, 340, 342, 344, 346, 347, 352, 376, 528, 531, 532, 536, 539, 544, 562, 566, 567, 571, 574, 579, 581, 597, 600, 608, 611, 615, 648, 667, 680
  as Aristotelian, 571
  Aristotelian metaphysics of, 566, 630
  at Berkeley, 563
  for IEUS, 558
  semantics and Thomism, 602
Taylor, A.E., 439
Testability, 576
Theory of probability, 550
  mathematics *vs.* aggregation, 550
Thirring, H., 618
Thomism in Chicago, 551
Thompson, G. [Trude], 551, 560, 563, 568, 569
Thomson, W., 245
Thünen, J.H. von, 277
Tinbergen, J., 541, 546, 558
Tönnies, F., 24, 119, 120, 274–276
Topitsch, E., 403
Torquemada, T. de, 41, 104, 593
Torres, J.M., 7
Trendelenburg, Adolf, 260, 261
True world, 547, 597
Truth
  absolute, 547–550
  correspondence theory of, 578
  for the language of science, 574
  Neurath's emotional reactions, 581
  not metaphysical, 609
  Paris discussion in 1935, 579, 581, 597
  Pre-Conference at Paris, 581
  semantic concept of, 562, 574, 579, 581, 609
  semantical concept of, 562
  as systematization, 574
Truth *vs.* confirmation, 581
Tschichold, J., 648, 667
Tufte, E.R., 127
Turner, F., 145
Twardowski, K., 269, 270, 329, 571, 598
Twyman, M., 34, 118, 127, 429

U
Uebel, T., 3, 24, 48, 67, 91, 117, 148, 178, 198, 219, 242, 364, 385, 417, 448, 456, 459
Unity of Science, 644
  Movement, 528, 551, 641, 684, 685
Unity-of-Science conference, Oxford, 535
Universals, 549, 578
Unwin, R., 157
Utilitarianism, 589

V
Vajda, S., 96
Van Gogh, V., 16, 474, 483–485
Venn, J., 243, 248, 249, 268, 271, 317, 552
Visual Education, 544
  meeting in Exeter, 545
Vogelsang, K. von, 73, 80
Vogt, K., 55
Vogt, P., 157
Volapuek [Volapük], 564
Voltaire, 545, 586, 590
von Mises, L., 667
Vossoughian, N., 10, 48, 62, 118, 143–147, 169, 193, 242

W
Waddington, C.H., 96
Waismann, F., 37, 98, 99, 283, 315, 347, 526, 533, 536, 538, 551, 563, 574, 623, 630, 640
  influence on youth, 611
  invites Neurath to talk in the Circle, 661, 677
  isolated and unhappy, 603
  in Oxford, 570
  suicide of wife, 603, 613
  Theses, 610
  translating book for Ogden, 602
Walker, S., 10, 118, 197, 429, 458, 465, 466
Walsh, V., 211
Walter, E.J., 648
Watkins, J.W.N., 4
Weaver, W., 386, 401–403, 408
Weber, Max, 13, 41, 70, 197–199, 201–203, 205–210, 212, 420, 479–482, 567, 586, 588, 608

Weber, Marianne, 589
Weinberg, J., 537, 539, 545, 558, 663
Weyl, H., 610
Whewell, W., 221, 232, 263, 265, 300
Whitehead, A.N., 243, 248–250, 259, 271, 311, 312, 315–317, 323, 453
Whiteley, C.H., 96
Wieland, C.M., 590
Wiener, N., 250
Wilkinson, J., 403
Williams, A.V., 425, 427
Wilson, F., 163
Wilson, M., 355
Winant, J.G., 587
Windelband, W., 276, 439
Winter, H.N., 418, 421, 422
Wirth, L., 104, 105, 558, 633
Wissell, R., 71
Wittgenstein, L., 31, 35, 99, 101, 249, 283, 315, 316, 320, 329, 330, 342, 344, 345, 347, 439, 571, 572, 600, 603, 610, 611, 629, 630, 640
Wölfflin, H., 432
Wolfram, L.H., 24
Wundheiler, A., 528
Wundt, W., 267, 268, 271, 272, 277, 279, 284, 285, 290, 292–297, 308, 445
Wyneken, G., 79

Y

Young, T., 230, 232, 233, 235

Z

Zeller, E., 286
Zilsel, E., 67, 105, 109, 529, 532, 534, 537, 538, 541, 597, 600
  on social science, 567
Zimmermann, R., 269, 270
Zipfinger, A., 157
Zolo, D., 5, 6, 226
Zweig, S., 11, 40, 41

MIX
Papier aus verantwortungsvollen Quellen
Paper from responsible sources
FSC® C105338

FSC
www.fsc.org

If you have any concerns about our products,
you can contact us on
ProductSafety@springernature.com

In case Publisher is established outside the EU,
the EU authorized representative is:
Springer Nature Customer Service Center GmbH
Europaplatz 3, 69115 Heidelberg, Germany

Printed by Libri Plureos GmbH
in Hamburg, Germany